SUPERCONDUCTIVITY

(IN TWO VOLUMES)

Volume 1

SUPERCONDUCTIVITY

(IN TWO VOLUMES)

Edited by R. D. PARKS

DEPARTMENT OF PHYSICS AND ASTRONOMY
UNIVERSITY OF ROCHESTER
ROCHESTER, NEW YORK

VOLUME 1

CRC Press
Taylor & Francis Group
Boca Raton London New York

CRC Press is an imprint of the
Taylor & Francis Group, an **informa** business

SUPERCONDUCTIVITY

Volume I

First published 1969 by MARCEL DEKKER, INC.

Published 2019 by CRC Press
Taylor & Francis Group
6000 Broken Sound Parkway NW, Suite 300
Boca Raton, FL 33487-2742

First issued in paperback 2019

ISBN 13: 978-0-367-45214-8 (pbk)
ISBN 13: 978-0-8247-1520-5 (hbk)

LIBRARY OF CONGRESS CATALOG CARD NUMBER 68-23775

PREFACE

During the preparation of this treatise one of the authors commented that it would be "the last nail in the coffin [of superconductivity]." While we may hope this is unduly pessimistic, it is clear that the field is well advanced in maturity and sophistication. That is what justifies the present undertaking. It would make far less sense to prepare so comprehensive a review of a field that was changing so rapidly as to render the work obsolete in a few years. Adequate reviews of superconductivity prepared in the mid-fifties, just prior to the Bardeen–Cooper–Schrieffer (BCS) theory, were considerably smaller than the present one, being roughly the size of Chapter 1. The explosion in superconductivity was ignited by the BCS theory, but a good share of the credit must go to the new breed of experimenters who were incubating in the fifties. They converged on the scene with their lock-in amplifiers, stacks of computer cards, and repertory of Feynman diagrams. They were schooled in the new tradition and could speak the theorists' language. Consequently, there occurred a cross-fertilization between the two groups, and an unforetold escalation of scientific discovery. This, coupled with other factors such as the important theoretical contributions from the Russian camp, abundant federal funding for research, and the profit incentive of industry (sharpened by the discovery of high-field superconductors), created the need for the 20 chapters which follow Chapter 2.

From the start it was my intention to make *Superconductivity* an integrated work instead of merely a collection of independent articles. This decision was costly in terms of time since it required that every manuscript be received before the type was set for Volume 1. The staggered preparation of the chapters had one good effect—it allowed the tardy authors to see the completed chapters, which in turn reduced the redundancy and increased the cohesiveness among the chapters.

Volume 1 covers what might be called *classical superconductivity*. It contains a full exposition of the BCS theory and its experimental verification, the Ginzburg–Landau theory, and the Gor'kov treatment of superconductivity. The fundamental experiments on macroscopic quantum phenomena and the Josephson effect are discussed in the same volume. The strong-coupling theory, together with the beautiful tunneling experiments on strong-coupling superconductors, presented in Chapters 10 and 11, should discourage all but the most fanatical opponents of the BCS theory. Volume 2, while also containing well understood topics, treats many unresolved problems. For example, it is certain that the last word has not yet been spoken on superconductivity in the transition metals (Chapter 13), dynamic effects in the vortex state (Chapter 19), or the intermediate state in type I superconductors (Chapter 21). One topic that flowered too late to obtain more than a mere mention is that of thermodynamic fluctuations and critical-point effects in superconductors, and is, at the time of this writing, an area of feverish activity.

In a work of this nature the bulk of the credit must go, of course, to the chapter authors. A particular debt of gratitude is due the authors of those chapters which deal more with historical and background material than with current research topics. These chapters are essential in making the treatise a useful pedagogical and reference work, and their preparation represents long and selfless work.

R. D. P.

Rochester, New York
October 1968

CONTRIBUTORS TO THIS TREATISE

Vinay Ambegaokar, *Department of Physics, Cornell University, Ithaca, New York*

P. W. Anderson, *Bell Telephone Laboratories, Inc., Murray Hill, New Jersey*

J. P. Burger, *Service de Physique des Solides (Laboratoire associé au Centre National de la Recherche Scientifique), Faculté des Sciences d'Orsay, Orsay, France*

B. S. Chandrasekhar, *Department of Physics and Condensed State Center, Case Western Reserve University, Cleveland, Ohio*

Marvin L. Cohen, *Department of Physics, University of California, Berkeley, California*

P. G. de Gennes, *Service de Physique des Solides (Laboratoire associé au Centre National de la Recherche Scientifique), Faculté des Sciences d'Orsay, Orsay, France*

W. DeSorbo, *Research and Development Center, General Electric Company, Schenectady, New York*

G. Deutscher, *Service de Physique des Solides (Laboratoire associé au Centre National de la Recherche Scientifique), Faculté des Sciences d'Orsay, Orsay, France*

Alexander L. Fetter, *Institute of Theoretical Physics, Department of Physics, Stanford University, Stanford, California*

D. M. Ginsberg, *Department of Physics and Materials Research Laboratory, University of Illinois, Urbana, Illinois*

G. Gladstone, *Department of Physics and Laboratory for Research on the Structure of Matter, University of Pennsylvania, Philadelphia, Pennsylvania*

L. C. Hebel, *Bell Telephone Laboratories, Inc., Whippany, New Jersey*

Pierre C. Hohenberg, *Bell Telephone Laboratories, Inc., Murray Hill, New Jersey*

M. A. Jensen, *Department of Physics and Laboratory for Research on the Structure of Matter, University of Pennsylvania, Philadelphia, Pennsylvania*

B. D. Josephson, *Royal Society Mond Laboratory, University of Cambridge, Cambridge, England*

Y. B. Kim, *Bell Telephone Laboratories, Inc., Murray Hill, New Jersey*

J. D. Livingston, *Research and Development Center, General Electric Company, Schenectady, New York*

W. L. McMillan, *Bell Telephone Laboratories, Inc., Murray Hill, New Jersey*

Kazumi Maki, *Department of Physics, University of California, San Diego, La Jolla, California [Present address: Department of Physics, Tohoku University, Sendai, Japan]*

Paul C. Martin, *Lyman Laboratory of Physics, Harvard University, Cambridge, Massachusetts*

James E. Mercereau, *Ford Scientific Laboratory, Newport Beach, California, and California Institute of Technology, Pasadena, California*

R. Meservey, *Francis Bitter National Magnet Laboratory, Massachusetts Institute of Technology, Cambridge, Massachusetts*

V. L. Newhouse, *Research and Development Center, General Electric Company, Schenectady, New York [Present address: School of Electrical Engineering, Purdue University, Lafayette, Indiana]*

G. Rickayzen, *University of Kent at Canterbury, Canterbury, England*

J. M. Rowell, *Bell Telephone Laboratories, Inc., Murray Hill, New Jersey*

D. Saint-James, *Service de Physique du Solide et de Résonance Magnétique, Centre d'Etudes Nucleaires de Saclay, Saclay, France*

Douglas J. Scalapino, *Department of Physics, University of Pennsylvania, Philadelphia, Pennsylvania*

J. R. Schrieffer, *Department of Physics and Laboratory for Research on the Structure of Matter, University of Pennsylvania, Philadelphia, Pennsylvania*

B. B. Schwartz, *Francis Bitter Magnet Laboratory, Massachusetts Institute of Technology, Cambridge, Massachusetts*

Bernard Serin, *Department of Physics, Rutgers—The State University, New Brunswick, New Jersey*

M. J. Stephen, *Bell Telephone Laboratories, Inc., Murray Hill, New Jersey [Present address: Department of Physics, Rutgers—The State University, New Brunswick, New Jersey]*

W. F. Vinen, *Department of Physics, University of Birmingham, Birmingham, England*

N. R. Werthamer, *Bell Telephone Laboratories, Inc., Murray Hill, New Jersey*

CONTENTS

VOLUME 2

SUPERCONDUCTIVITY

Volume 1

1

EARLY EXPERIMENTS AND PHENOMENOLOGICAL THEORIES

B. S. Chandrasekhar

DEPARTMENT OF PHYSICS AND CONDENSED STATE CENTER
CASE WESTERN RESERVE UNIVERSITY
CLEVELAND, OHIO

I. INTRODUCTION

A. Plan of the Chapter

This chapter reviews the state of our knowledge of superconductivity, both experimental and theoretical, as it was in 1957 when the microscopic theory was formulated by Bardeen, Cooper, and Schrieffer (*1*), which is now known universally as the *BCS theory*. The treatment is not historical; rather, an attempt is made to provide the background for the many topics which the succeeding chapters will discuss in detail. In the limited space available, it will not be possible to provide exhaustive compilations of experimental results or references. Representative recent references are quoted, however, and the reader is referred to them, and to the general references at the end of the chapter, for details. The first half of the chapter concerns itself with an account of the fundamental properties which distinguish a superconductor from other solids and of its behavior in mechanical, thermal, and electromagnetic fields. The latter half of the chapter discusses the thermodynamics of superconductors and the several phenomenological theories which have been proposed to explain these properties.

B. Discovery, and Fundamental Properties

Superconductivity was first discovered and so named by Kamerlingh Onnes (*2*) in 1911. In the course of an investigation of the electrical resistance of various metals at liquid helium temperatures, he observed that the resistance of a sample of mercury dropped from 0.08 Ω at above 4°K to less than 3×10^{-6} Ω at about 3°K, and that this drop occurred over a temperature interval of 0.01°K. There were subsequent attempts by him as well as by others to set limits to the temperature breadth of the transition and the resistance in the superconducting state. Neighbor et al. (*3*) have set an upper limit of 3×10^{-4} °K for the transition width in lead based on their specific heat and magnetic susceptibility measurements. Quinn and Ittner (*4*) observed the magnetic moment of a thin-film loop of lead carrying a persistent current as a function of time, and concluded that the resistivity in the superconducting state was less than 3.6×10^{-23} Ω–cm. This is to be compared with a (probable) resistivity of $\sim 10^{-7}$ Ω–cm in the normal state. While the breadth of the transition may increase if the sample is metallurgically imperfect, the extraordinary smallness of the resistance in the superconducting state appears to hold for all superconductors. Thus we arrive at the first characteristic property of a superconductor: Its electrical resistance, for all practical purposes, is zero, below a well-defined temperature T_c, called the *critical*, or *transition*, *temperature.*

At any temperature T below T_c, the application of a minimum magnetic field $H_c(T)$ destroys the superconductivity and restores the normal resistance appropriate to that field. In terms of reduced coordinates $t = T/T_c$ and $h_c = H_c(T)/H_c(0)$, it is found that for all superconductors the following relation approximately holds:

$$h_c \simeq 1 - t^2 \tag{1}$$

We shall see later that this critical field H_c for the destruction of superconductivity can be experimentally determined directly only for a few superconductors, the so-called type I superconductors. The great majority of superconductors are of type II, and their transition between the normal and superconducting states is spread out over a finite range of magnetic fields. From the thermodynamics of the transition, to be developed later, it is nevertheless possible to define an H_c for all superconductors, which is therefore called the *thermodynamic critical field.*

The critical field may be produced either from external sources or by passing a suitable electrical transport current through an appropriately shaped sample of superconductor. This quenching of superconductivity by the passage of a critical transport current is called the *Silsbee effect.*

In the rest of this chapter we shall confine ourselves almost entirely to type I superconductors; the clear recognition of the existence of type II superconductors happened at about the same time as the formulation of the BCS theory,

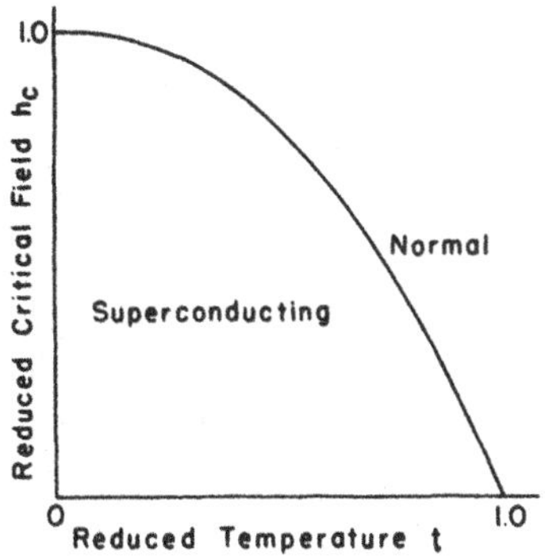

Fig. 1. Reduced critical field $h_c = H_c/H_0$, where $H_0 \equiv H_c(0)$, vs. reduced temperature $t = T/T_c$ (schematic).

and we are concerned here principally with the pre-BCS period. During this period it was generally believed that there was an "ideal" superconductor (typically a needle-shaped well-annealed single crystal), characterized by a sharp normal–superconducting transition in both temperature and magnetic field, and that nonideal behavior was due to "flaws" in the material. There will be ample discussion in the rest of this book of the fact that such "nonideal" superconductors are often also ideal in their own right, and form the type II class.

The quadratic variation of h_c with t can be represented on the phase diagram of Fig. 1. It turns out that this is indeed a true equilibrium phase diagram; i.e., the state of the system corresponding to any point on this diagram is independent of how the point is approached. This observation, combined with the zero resistance in the superconducting phase, leads to the result that the magnetic induction B inside the superconductor vanishes:

$$B = 0 \tag{2}$$

This is a statement of the Meissner–Ochsenfeld, or more popularly, if inequitably, the *Meissner effect* (*5*). We note that Eq. (2) holds inside a superconductor, irrespective of its connectivity, as long as $H < H_c$. For a long thin sample parallel to the applied field H_a, we have $H = H_a$, and so the intensity of magnetization I is given by

$$I = -H_a/4\pi \tag{3}$$

This leads to another formulation of the Meissner effect—that a superconductor shows perfect diamagnetism. The perfect conductivity is a necessary but not sufficient condition for the perfect diamagnetism, and so these two properties are to be regarded as distinct fundamental characteristics of a superconductor.

For a sample of arbitrary shape, the magnetic behavior is more complicated, since the field H in the sample may reach the value H_c locally while $H_a < H_c$. This leads to a breaking up of the sample into normal and superconducting regions, called the intermediate state, to be discussed further below.

C. Occurrence

The work of the last 50 years has shown that superconductivity is a very widespread phenomenon among conducting solids—elements, alloys, and compounds. An excellent compilation of data [T_c, $H_c(0)$, and crystal structure] on hundreds of superconductors has been published by Roberts (*6*). Transition temperatures range from 0.016°K for a 70% Mo–30% Nb alloy to 18.5°K for off-stoichiometric Nb_3Sn. Values of $H_c(0)$ are from 2.7 G for a 70% Mo–30% Zr alloy to about 2000 G for niobium. Superconductivity has been found in both transition and nontransition metals, in compounds where none, one, or more of the constituents may be superconducting, in metallic conductors as well as degenerate semiconductors. Bizarre compounds such as C_8K and $Ag_7O_8 \cdot NO_3$ are reported to be superconducting. Indeed, superconductivity appears to be by far the most commonly occurring ordering phenomenon in the condensed state of matter.

In the midst of such an embarrassment of riches, it was natural to ask for the criteria, if any, which determine whether a given material is likely to be superconducting or not. The alkali metals, the alkaline earths, and the noble metals have so far not been found superconducting. Ferromagnetic and ferroelectric ordering are found to inhibit superconductivity. On the other hand, several factors have been determined empirically to favor the occurrence of this phenomenon. Among these are Matthias's (*7*) observation that valence electron concentrations of 3, 5, or 7 per atom favor superconductivity, and the observation, documented by Matthias et al. (*8*) that certain crystal structures, notably the β-tungsten, the Laves phases, and the α-manganese structure, are conducive to the phenomenon. These empirical rules have played an important role in the discovery of new superconductors. The BCS theory, of course, gives its own fundamental criterion for the occurrence of superconductivity in terms of the interactions of the elementary excitations in a solid, as will be discussed later in this book. The application of this criterion to the prediction of new superconductors is not easy, and the search will probably continue at least for some time on an empirical basis.

II. EXPERIMENTAL PROPERTIES

A. Mechanical Properties

A superconductor undergoes no change in dimensions when the transition occurs in zero magnetic field. There is, however, a small magnetostriction in the superconducting state at lower temperatures, as well as an abrupt change in dimensions when the sample goes normal. This strain at $H_c(T)$ shows crystalline anisotropy, is of the order of a few times 10^{-8}, and in almost all cases corresponds to a contraction when the sample goes normal (*9*). This and the other reversible mechanical effects described below are all interrelated, as will become clear when the thermodynamics of the transition is discussed later.

The elastic constants in the superconducting state below T_c are smaller, by a few parts per million, than in the normal state (*10*). Furthermore, the velocity of a compressional wave in a cubic crystal shows a discontinuity at T_c, while that of a shear wave does not—a result that would be expected from symmetry considerations.

The above variations of the equilibrium mechanical properties during the superconducting–normal transition are thermodynamically related to the stress dependence of the free energy difference between the two phases. The extensive work in this area, which has been reviewed by Olsen and Rohrer (*11*) and by Seraphim and Marcus (*12*), has in summary led to the following results: (1) The thermodynamic interrelationships among the different effects have been adequately confirmed, and this provides additional evidence supporting the reversibility of the Meissner effect; (2) the stress dependence of the microscopic parameters (electronic density of states, electron–electron interaction parameter, and so on) which are involved in the BCS theory may be computed; and (3) several elegant experimental techniques which permit measuring changes in elastic constants of 1 part in 10^7, and strains of 1 part in 10^{10}, have been developed and successfully applied to other problems.

In contrast to the relative simplicity of the changes in these equilibrium mechanical properties is the behavior of ultrasonic attenuation in a superconductor. A review of the results up to about 1958 has been given by Morse (*13*). The attenuation is due to the interaction of the sound wave with the conduction electrons, the phonons, and lattice imperfections. The effects due to superconductivity will obviously be predominantly observed in the electronic attenuation, and this is dominant when $ql > 1$, where q is the phonon wave number and l is the electronic mean free path. This requirement implies pure superconductors, and relatively high sound frequencies (MHz range). Even with this restriction, one observes that complications due to the polarization and crystalline anisotropy are possible.

The essential experimental observations have been, when $ql > 1$, the following: (1) The electronic attenuation of both longitudinal (α_l) and transverse (α_t) waves goes to zero at very low temperatures exponentially with temperature; (2) α_l drops continuously, starting at T_c, as the temperature is lowered, while α_t shows a discontinuity at T_c; and (3) the attenuation shows a crystalline anisotropy. These features could be qualitatively understood in terms of a two-fluid model, by associating the attenuation with the normal electron fluid, but there were serious quantitative discrepancies. The detailed interpretation of these results was one of the early successes of the BCS theory.

B. Thermal Properties

In this section we shall concentrate on the work done on specific heats and thermal conductivities, both of which have received a great deal of attention from experimenters for many years. They provided a wealth of data for comparison with the earlier theories of superconductivity, and finally provided crucial evidence for the existence of an energy gap in the electronic excitation spectrum.

The specific heat C_n of a normal metal is usually regarded as composed additively of contributions from the lattice and from the conduction electrons, and at very low temperatures it may be expressed as

$$C_n = \gamma T + \beta T^3 \tag{4}$$

where the first term on the right is due to the electrons and the second is due to the lattice. We shall assume that the properties of the lattice are unchanged between the normal and superconducting states. This assumption, while it has never been rigorously proved experimentally, appears to be an extremely close approximation on the basis of a variety of observations such as the nuclear scattering of neutrons in lead and niobium (*14*), and most recently a determination of the Debye–Waller factor in normal and superconducting tin by means of the Mössbauer effect (*15*). We shall henceforth consider only the electronic contribution to the specific heat, and denote it by C_{es} and C_{en}, respectively, in the superconducting and normal states.

When the transition occurs in zero magnetic field, the electronic specific heat shows two striking features: (1) there is a discontinuous jump in the specific heat at T_c with $C_{es}(T_c) \simeq 3C_{en}(T_c)$, and (2) C_{es} drops rapidly and nonlinearly for $T < T_c$, going to zero at $T = 0$ as required by the third law of thermodynamics. The behavior is shown schematically in Fig. 2. This behavior was first clearly

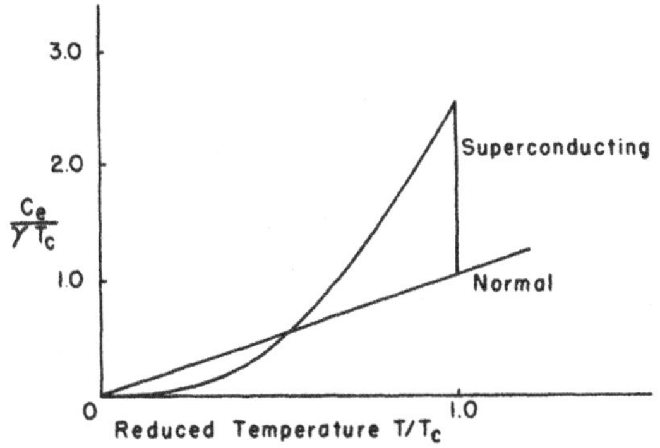

Fig. 2. Electronic specific heat, normalized to its value in the normal state at $T = T_c$, as a function of reduced temperature in the normal and superconducting states (schematic).

demonstrated by Keesom and Kok (*16*) in tin and has since been observed in other superconductors. There is no latent heat associated with the transition at T_c. If the transition occurs at $T < T_c$ in the presence of a magnetic field, there is a latent heat associated with the transition, corresponding to an absorption of heat when the sample goes normal; this transition is a first-order phase change. The transition in zero field at T_c is a second-order phase change, of the particular kind first formulated by Ehrenfest (*17*) and characterized by a jump in the specific heat and absence of a latent heat at the transition. A second-order phase change is continuous, in the sense that the state of the system changes continuously; an example is the normal↔superfluid transition in liquid helium at 2.17°K. What is important is that even such a continuous change in the system produces an abrupt change in the symmetry of the system. Landau (*18*) gave in 1937 a phenomenological theory of second-order transitions, introducing the important concept of an order parameter. The subsequent application of this theory by Landau and Ginzburg to superconductivity formed one of the great landmarks in the subject, and will be extensively discussed in this book.

One other point to be made about the jump in the specific heat is that its magnitude is thermodynamically related to the temperature dependence of the critical field, and this has been experimentally verified. It may be remarked here that this, and similar, verifications of the thermodynamic interrelationships among the various equilibrium properties relating to the normal–superconducting transition are to be viewed as confirmation of the Meissner effect basically, i.e., the reversibility of the transition.

The other aspect of the specific heat in the superconducting state C_{es} is its temperature dependence. It was the practice till the mid-1950s to fit C_{es} to a t^3-dependence, where $t = T/T_c$, even though it was recognized that this was only very approximate. This t^3-dependence of C_{es} seemed particularly attractive since it could be combined with the thermodynamic relation (see Section III)

$$\frac{1}{8\pi}\frac{d}{dT}H_c^2 = \frac{S_{es} - S_{en}}{V}$$

(which assumes that the lattice entropy is unchanged during the transition), to obtain both the parabolic variation of H_c with temperature and an expression for γ, the coefficient of the normal electronic specific heat ($C_{en} = \gamma T$):

$$\gamma = \frac{1}{2\pi}\frac{H_0^2}{T_c^2}$$

entirely in terms of superconducting parameters. Both of the above results are reasonable approximations to reality, at least as good as the t^3-dependence of C_{es}.

The temperature dependence of the specific heat contains important, albeit "averaged," information on the spectrum of elementary excitations in a solid, as evidenced for example by the T^3 specific heat due to phonons or the Schottky type of specific heat due to spin splitting. The discovery, in 1954, that the specific heats of superconducting vanadium and tin (*19*) could be fitted much better to an expression of the form $a\exp(-b/t)$ than to t^3 was a significant contribution to our understanding of superconductivity. This suggested very convincingly the existence of an energy gap in the excitation spectrum, a gap which in fact had earlier been proposed on the basis of thermoelectric and thermal conductivity measurements. The origin of this gap was of course explained soon thereafter by the BCS theory. It is remarkable that the two basic features of the superconducting specific heat, the jump at T_c and the temperature dependence, are intimately related to two significant contributions to our theoretical understanding of the phenomenon.

We shall now discuss the main features of the thermal conductivity of superconductors. The thermal conductivity of a metal is a vastly more complicated phenomenon than the specific heat, owing on the one hand to its being a nonequilibrium process, and on the other to the multiplicity of factors which contribute to it. The thermal energy in a metal is transported by both electrons and phonons, and each type of carrier suffers collisions with its own type, with the other type, and with lattice imperfections and boundaries. These different mechanisms have varying dependences on temperature, impurities, sample size, and so on. The added wealth of behavior one gets by going to the superconducting state, and perhaps switching on a magnetic field so as to set up the intermediate state, should be obvious. It will clearly not be possible to do justice to all the phenomena which have been reported in this area. The reader who wishes to go into the details is referred for normal metals to the exhaustive treatise by Ziman (*20*), and for superconductors to review articles by Olsen and Rosenberg (*21*) and by Mendelssohn (*22*).

In spite of the aforementioned complications, we may still single out certain simple cases, where the behavior is relatively easy to analyze. The experimental results for pure superconductors are typified in Figs. 3 and 4, based on the work

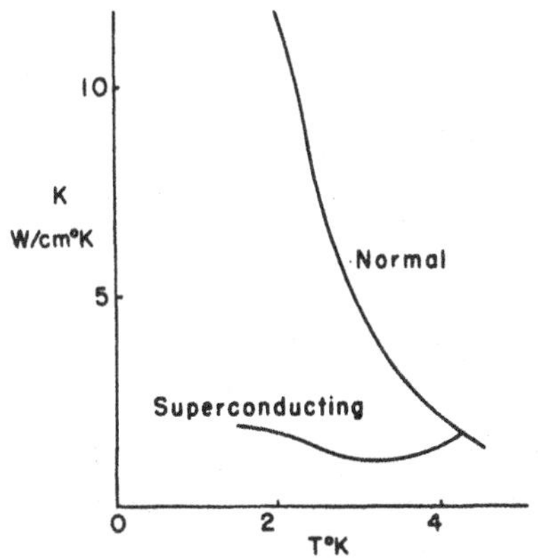

Fig. 3. Thermal conductivity K of mercury in the normal and superconducting states. [After Hulm (*23*).]

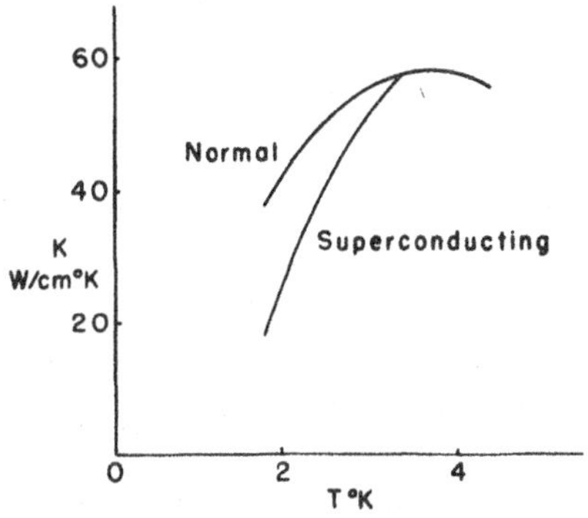

Fig. 4. Thermal conductivity K of tin in the normal and superconducting states. [After Hulm (*23*).]

of Hulm (*23*). The conductivity in the superconducting state is lower than that in the normal state, with no indication of a discontinuity at T_c. In concentrated alloys such as Pb + 10% Bi (*24*), on the other hand, this order can be reversed, as shown in Fig. 5. Finally, in a pure superconductor such as lead at 0.1°K, the ratio K_s/K_n can become as small as 10^{-5}.

The behavior of the pure superconductors was ingeniously analyzed by Hulm (*23*) as follows. The conduction may be assumed to be predominantly electronic, limited by scattering by impurities and phonons. Then we can write

$$\frac{1}{K_{en}} = \frac{1}{K_{eni}} + \frac{1}{K_{eng}}$$

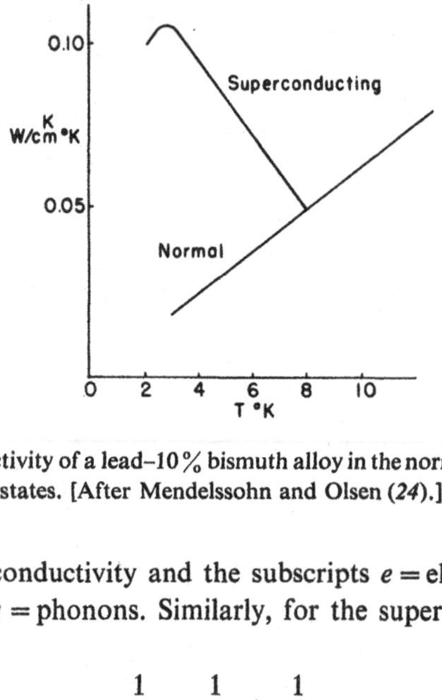

Fig. 5. Thermal conductivity of a lead–10% bismuth alloy in the normal and superconducting states. [After Mendelssohn and Olsen (*24*).]

where K = thermal conductivity and the subscripts e = electrons, n = normal, i = impurities, and g = phonons. Similarly, for the superconducting state, we have

$$\frac{1}{K_{es}} = \frac{1}{K_{esi}} + \frac{1}{K_{eng}}$$

Now the curve $K_{en}(T)$ shows a maximum at some temperature T_m, and the theory of K_{en} tells us that

$$\text{if } T > T_m, \quad \text{then } 1/K_{eng} \gg 1/K_{eni}$$
$$\text{if } T < T_m, \quad \text{then } 1/K_{eng} \ll 1/K_{eni}$$

We see that Figs. 3 and 4 show that scattering by phonons is dominant for Hg, and by impurities for Sn. By neglecting the weaker scattering mechanism in each case, we can use the measured conductivities to define functions $f(t)$ and

$g(t)$ of the reduced temperature t, where

$$f(t) = \frac{K_{esi}}{K_{eni}} \simeq \frac{K_s}{K_n} \qquad \text{for } T_c < T_m$$

$$g(t) = \frac{K_{esg}}{K_{eng}} \simeq \frac{K_s}{K_n} \qquad \text{for } T_c > T_m$$

It turns out that the variation of K_s/K_n for In, Sn, and Al ($T_c < T_m$) with reduced temperature t follows very closely the same curve, while for Hg and Pb ($T_c > T_m$) one gets a different curve; these two characteristic temperature dependences have been fairly successfully explained by the microscopic theory.

Qualitatively, the drop in K_{es} with decreasing temperature can be seen in terms of a two-fluid model: The concentration of the electron superfluid increases as T decreases. The superfluid does not carry an energy current, and so K_{es} drops. In the case of the alloys where $K_s > K_n$, one assumes that $K_g > K_e$ and the phonons are not scattered by the electron superfluid, so that K_s increases as T decreases. In a pure metal at very low temperatures, K_{es} finally becomes very small compared to K_{gs}, while $K_{en} \gg K_{gs}$, so that it is possible to achieve very large values of the ratio K_n/K_s. This last property has been applied to the construction of superconducting thermal switches.

Finally, we note the observation of Goodman (*25*) that $f(t)$ for tin below 1°K showed an exponential behavior of the form $a \exp(-b/t)$. He suggested that this indicated the existence of a gap in the excitation spectrum of the electrons in the superconductor, and provided perhaps the first clear-cut evidence for such a gap.

C. Electromagnetic Properties

1. Meissner Effect

The reversible exclusion of flux inside a superconductor was experimentally shown by Meissner and Ochsenfeld, as mentioned in Section I. In practice it turns out, however, that a precise *direct* verification of the effect for an actual superconducting sample is not a trivial matter, for a variety of reasons: the influence of sample shape upon the magnetic properties, the phenomenon of magnetic supercooling, and the occurrence of trapped flux when an applied magnetic field is removed. Perhaps the most convincing verification of the Meissner effect has been an indirect one, carried out by Mapother (*26*). He has shown that the measurements of the critical field H_c and the superconducting and normal specific heats C_s and C_n on tin and indium are consistent, within experimental error, with the thermodynamic relations connecting them which can be derived on the basis of the Meissner effect. These relations are discussed in Section III.

A needle-shaped sample in a parallel magnetic field would presumably show a sharp transition at H_c. But for any finite sample of arbitrary shape, the variation of the magnetization with applied field is more complex, and Chapter 21 will be devoted to this, and related, shape-dependent properties of superconductors. We shall make a few remarks here only on certain general features of the problem. The first is to point out that a two-phase domain structure, composed of superconducting and normal phases, is a reasonable expectation for what happens when the field over a portion of the surface reaches H_c, even though the applied field $H_a < H_c$. This is because a layer of material near this portion would go normal, thus causing the surface field to drop below H_c and reestablishing superconductivity there. A further increase in the applied field repeats the above process, producing a second layer of normal region and moving the first layer inward. The resulting magnetic state of the sample is called the *intermediate state*.

The need for assuming such a microstructure in some experiments was first recognized by Gorter and Casimir (*27*). The domain structure was directly observed by Meshkovsky and Shalnikov (*28*), and Meshkovsky (*29*), using the magnetoresistance of a fine bismuth wire to probe the field near the surface of a sample in the intermediate state. Subsequent investigators have used Faraday rotation and the Bitter powder technique for more detailed studies of the field distribution, as will be described in Chapter 21.

F. London (*30*) and Peierls applied thermodynamics to the intermediate state, and one can understand the magnetic behavior of certain simple shapes, such as ellipsoids, on the basis of their treatment. This topic is discussed further in Section III. But we can remark here that the thermodynamics of the pure superconducting to normal transition cannot give a complete quantitative picture of the intermediate state, which involves the contiguous coexistence of both phases and must therefore be influenced by the interphase surface energy. The importance of the interphase energy in determining the magnetic properties of superconductors was recognized by H. London (*31*), and was taken account of by Landau (*32*) in his theories of the intermediate state.

2. Penetration Depth

It was emphasized by F. London that the pure superconducting state in a magnetic field has a persistent shielding current associated with it. Since the Meissner effect ($B = 0$) is very well established for samples of macroscopic dimensions, this current must be confined to a region very close to the surface. Expressed in terms of the field, one may conclude that B drops to a vanishingly small value over a characteristic penetration depth λ, whose value has been determined, as explained later, to be about 10^{-5} cm.

One may ask why a careful study of the penetration depth is of interest when its magnitude is so small as to influence the Meissner effect only for samples comparable to λ in at least one dimension. The point here is that the penetration depth is the region where the magnetic contribution to the free energy density changes from its "normal" or "vacuum" value to that appropriate for the superconductor, and its dependence on various physical parameters can provide a sensitive test of the quantitative aspects of any theory of superconductivity.

The following methods have been used to get information about the penetration depth: the dependence of the magnetic susceptibility of a superconducting sample upon its size (*33*); the dependence upon appropriate parameters of the mutual inductance of a pair of coaxial coils containing the superconducting sample as a tightly fitting core, as the applied field or temperature is modulated at a low frequency (*34,35*); the high-frequency impedance of a superconducting sample (*36*); and the direct measurement of the penetration of an applied field through a thin cylindrical sample (*37*). We shall review here the important experimental results on the penetration depth, and emphasize those that played a significant role in advancing the theory, as will be further elaborated in Section III.

a. ***Magnitude of λ***

The magnitude of λ is temperature-dependent, as will be discussed later, but it is possible to obtain by extrapolation its value $\lambda(0)$ at $T = 0°K$. It is different for different superconducting elements, ranging from 3.9×10^{-6} cm for lead (*38*) to 13×10^{-6} cm for cadmium (*39*). The point to note is that it is typically a few hundred angstroms for all superconductors that have been measured so far. The penetration depth can also show crystalline anisotropy, as clearly demonstrated by Pippard (*40*) for tin. Figure 6 shows the variation of λ with the angle between the shielding current and the tetragonal axis in a single crystal of tin. It had been shown by Frazer and Shoenberg (*41*) that, on the basis of the London theory (see Section III), the variation of λ should be as $\cos^2 \theta$; the experimental

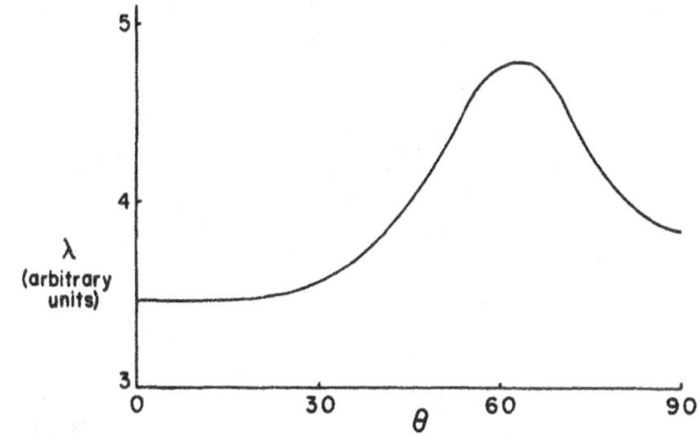

Fig. 6. Crystalline anisotropy of penetration depth λ in tin; θ is the angle in degrees between the tetragonal axis and the current. [After Pippard (*40*).]

results are clearly contradictory to this expectation. Indeed, even the magnitudes of the measured penetration depths are several times larger than predicted by the London theory. A word may be said here about the law of penetration of the magnetic field. The London theory predicts $H \sim \exp(-x/\lambda)$, where x is the distance into the superconductor from its surface. This is in general an adequate description of the actual situation, provided one takes into account a dependence of λ upon sample dimensions for microscopic specimens (dimensions $\gtrsim 10\lambda$), as further discussed in Section III.

b. *Temperature dependence of λ*

It has been found experimentally (*37,42*) that the variation of λ with the reduced temperature $t = T/T_c$ is very closely described, for all superconductors studied, by

$$\lambda(t) = \lambda(0)\,(1 - t^4)^{-1/2} \tag{5}$$

Small deviations from this relation have been observed at very low reduced temperatures, and these can be accounted for by the BCS theory (*43*).

c. *Mean free path dependence of λ*

The observation by Pippard (*44*) that the penetration depth of tin with 3% indium dissolved in it was *twice* that of pure tin, although the thermodynamic quantities like H_c and T_c were hardly affected, was an important landmark. It showed up a fundamental inadequacy in the London theory and led to its reformulation by Pippard, as will be described in Section III. This point illustrates our earlier observation on the role played by the studies of the penetration depth in developing our understanding of superconductivity. Pippard also observed that the crystalline anisotropy of λ, seen earlier in pure tin, disappeared in the alloy.

d. *Dependence of λ upon the magnetic field*

The crucial experiment here was again done by Pippard (*45*), and his results for tin are shown in Fig. 7. The remarkable point is that, even close to $t = 1$, λ changes only by about 2% between $H_a = 0$ and $H_a = H_c$. By using the thermodynamic relation $(\partial S/\partial H)_T = (\partial I/\partial T)_H$, along with the Meissner effect and the temperature variation of λ, one can compute the change in entropy of the sample as the applied field is increased to H_c. If this change is now assumed to take place only in the region where the field has penetrated, one then gets the incredible result that, near $t = 1$, the change in entropy density is as much as one-quarter of the total entropy difference between the normal and superconducting phases, even though λ changes only by a couple of per cent. It seemed more plausible to assume that the total entropy change is distributed in a region considerably larger than the penetration layer.

3. Samples of Dimensions $\sim \lambda$

It is clear that the electromagnetic properties of samples with at least one dimension comparable to λ will be different from those of bulk samples. The extensive work done up to the early 1950s in this area has been summarized by Shoenberg (*46*). The salient results are summarized here. As already mentioned,

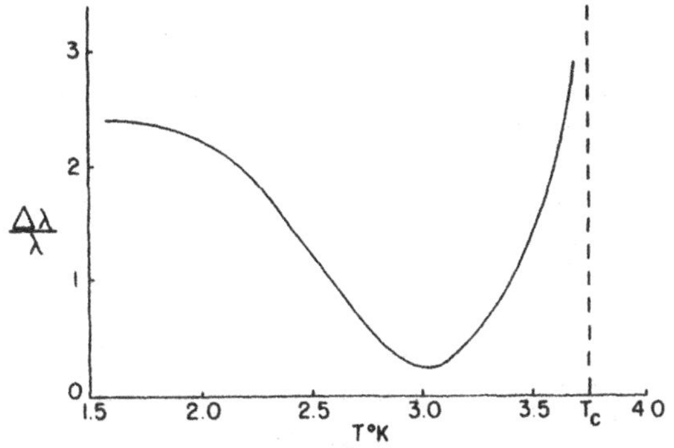

Fig. 7. Percentage change in the penetration depth in tin between zero and H_c, as a function of temperature. Note that the graduations along the $\Delta\lambda/\lambda$ axis stand for 1, 2, and 3%, respectively. [After Pippard (*45*).]

the magnetization of thin films, wires, and colloids was used to determine $\lambda(T)$. For well-annealed thin films of thickness d, it was found that the critical field for restoration of normal resistance, H_f, was given by

$$H_f/H_c = 1 + (b/d) \tag{6}$$

where b was a few times 10^{-5} cm, and showed a dependence upon temperature similar to that of λ. Typical of recent work is that of Khukhareva (*47*), who found for thin mercury films that the critical field varied inversely as d, and as $(1-t)^{1/2}$ for $t \lesssim 1$. Note that this is the approximate temperature dependence one would expect if b has the same temperature dependence as λ. The measurement of critical currents in thin films presents formidable experimental problems, because of heating effects at the contacts and nonuniform distribution of current over the cross section of the film. It became evident early, from the work of Shalnikov (*48*), that the critical currents were much smaller than those expected from the Silsbee criterion for H_f. A clue as to why this should be so had been given by H. London (*31*), who showed that there was a critical current density $J_c = cH_c/4\pi\lambda$ for thin films ($d \ll \lambda$), based on thermodynamic arguments. The recent microscopic theory (*49*) of critical currents in thin films confirms this result in order of magnitude, and in particular gives the temperature dependence of J_c near $t = 1$ as $(1-t^2)^{3/2}$. Again, one notes that the London expression leads to this temperature dependence if the experimentally observed $H_c(t)$ and

$\lambda(t)$ are used. The variation of J_c as $(1 - t^2)^{3/2}$ has been confirmed by Glover and Coffey (*50*).

The experimental work on the magnetization of small specimens has been reviewed by Shoenberg (*46*). The work of Lock (*38*) on films of tin, indium, and lead is typical of the results obtained. Figure 8 shows schematically the

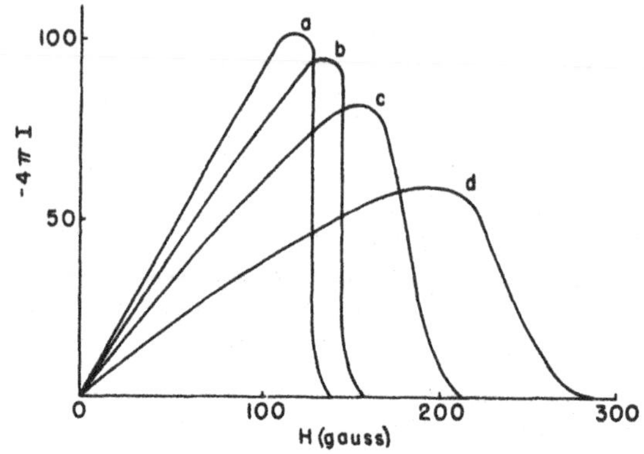

Fig. 8. Magnetization curves of thin tin films, various thicknesses at 3°K: (a) 7900 Å; (b) 5500 Å; (c) 3000 Å; (d) 2320 Å. [After Lock (*38*).]

magnetization of tin films for different thicknesses. The points to be noted are the increasing critical field with decreasing thickness, and the "rounding" of the magnetization curves. The variation of susceptibility and critical field with thickness are interpretable thermodynamically, but the rounding of the I vs. H curves has not been unequivocally explained (*51*).

The fact that thin filaments of superconductors have critical fields much higher than H_c was used by Mendelssohn (*52*) to interpret the observations by him and others on the magnetic, resistive, and thermal properties of alloy superconductors. He suggested that a filamentary superconducting mesh, a "sponge," remained superconducting up to high fields, after the bulk of the material had gone normal. Such a sponge has recently been synthesized (*53*) by pressing lead into porous Vycor glass, and shown to have many of the expected properties.

4. Persistent and Critical Currents

Mention has been made in Section I of the observation (*4*) of persistent current in a thin film loop in order to set a limit on the electrical resistance in the superconducting state. Unfortunately, in the light of subsequent knowledge, we have to modify somewhat this oft-quoted proof of vanishing resistance. The stability of the current in a ring is related to the quantization of the flux enclosed, as will be discussed in detail elsewhere. The behavior of a superconducting wire carrying a current in an otherwise normal circuit does not involve the quantization of flux, and is thus physically a quite different situation from the current-carrying ring.

It was the Londons, Fritz and Heinz, who realized (*30*) that the Meissner effect contained the key to the supercurrent; just as shielding currents uniquely related to the applied magnetic field are set up in an isolated superconductor because of the Meissner effect, similarly, the supercurrent which is fed into a wire is maintained by the magnetic field produced by it. *The supercurrent is always determined by the local magnetic field.*

A really careful verification of zero resistance in a superconductor therefore requires either a sensitive repetition of Onnes' original experiment, or a study of the stability of the shielding currents in a simply connected superconductor. One might use, for the former experiment, a sensitive voltmeter based on the Josephson effect (*54*)‡: for the latter, a stable field produced by the trapped flux in a superconducting ring. The present experimental situation, then, seems to be that there has been no careful determination of how "zero" is the zero resistance of a superconducting wire, although there is no doubt of course that it is extremely small, and the decay of a persistent current in a loop can be described as less than would be due to an effective resistivity of $\sim 10^{-23}$ Ω–cm in the superconductive state. All the preceding refers, of course, to type I superconductors. Persistent currents in type II superconducting loops *can* decay; this and other resistive properties of such superconductors will be dealt with elsewhere in the book.

The *critical current*, if it is defined operationally as the current at which the first measurable resistance appears, has a significance for bulk samples in the Silsbee sense, i.e., as the current which produces a field H_c at the surface of the sample. For thin films the critical current is that which is needed to quench the superconductivity completely, and has a more fundamental significance, as discussed in Section II.C.3.

5. Behavior in Alternating Fields

The response of a superconductor to alternating electromagnetic fields has provided one of the richest sources of clues leading to our present understanding of superconductivity. The phenomena are complex, and cannot be dealt with completely in this chapter. We shall, however, consider some of the important results. The extremes of the frequency spectrum are easily disposed of; at very low frequencies, the superconducting properties are practically unchanged from zero frequency, while at near-infrared (wavelength $\lesssim 10\ \mu$) and optical frequen-

‡ Note added in proof: This experiment has now been done by T. A. Fulton [*Bull. Am. Phys. Soc.* **13**, 75 (1968)], who has set an upper limit of 5×10^{-21} Ω–cm for the resistivity of a superconducting lead film at 4.2°K.

cies, the optical properties are unchanged on going from the normal to superconducting state (*55–57*). It had, in fact, been pointed out by H. London (*58*) that the two-fluid model would lead to a dissipation in alternating fields, which should become significant when the rf penetration depth δ becomes comparable to λ. This occurs at frequencies of the order of 10^{10} Hz. Pippard (*44*) used his measurements on the absorption of microwaves at 9.4 kMHz in alloys of tin to obtain the variation of λ with mean free path, and was led to his important generalization of the London theory and introduction of the *coherence length.*

For still higher frequencies, in the range of millimeter microwaves and far-infrared frequencies (*59–61*), absorption measurements gave convincing and perhaps the first direct evidence for the existence of an *energy gap* in the excitation spectrum of a superconductor. This topic is treated in detail in Chapter 4.

6. Interphase Surface Energy

We have at several points remarked that there must be an energy associated with normal–superconducting boundaries. This is an appropriate point to summarize the observations which pointed to the existence of such a surface energy.

The very existence of the Meissner effect, as was recognized by H. London (*31*), requires the existence of a positive surface energy; otherwise the sample would tend to break up into a mixed state of normal and superconducting layers of thickness d_n and d_s, respectively, such that $d_n \ll d_s \gtrsim \lambda$, and still have a lower energy than the Meissner state. The maximum energy per unit area that a superconducting lamina of thickness λ can gain by allowing the applied field to penetrate is $\lambda H_c^2/8\pi$. If this is to be offset by an additional surface energy α, it is obvious that one can define α in terms of a length β', such that $\alpha = \beta' H_c^2/8\pi$ and $\beta' > \lambda/2$. A more exact treatment by F. London (*30*) leads to the condition $\beta' > \lambda$ for a positive surface energy, and hence a Meissner effect.

The coherence length ξ introduced by Pippard (*45*) as "the distance [in the superconductor] over which the effects of a perturbing force are appreciable" offers a natural explanation for the surface energy. The presence of an N–S interface contributes to the surface energy an amount $\xi H_c^2/8\pi$, so that the net surface energy is $(\xi - \lambda)\, H_c^2/8\pi$. The condition for positive surface energy, leading to a Meissner effect, is therefore $\beta \equiv (\xi - \lambda) > 0$. It is, of course, now well known that if $\beta < 0$, we get type II behavior.

The phenomenon, rather inappropriately named supercooling, which is the transition into the superconducting state at fields less than H_c when the magnetic field is reduced isothermally, has been used to obtain information about the interface energy (*62*). The supercooling at low temperatures is influenced by metallurgical imperfections in the samples, but close to T_c, β becomes larger than the characteristic dimensions of such imperfections, and supercooling data in this region can be used to obtain a parameter κ, introduced by Ginzburg and Landau (*63*), which is related to ξ. It should be remarked that supercooling is possible only in a magnetic field, when the transition is a first-order one. In zero field, the transition is of second order, and no supercooling or superheating is possible.

Another direct indication for the surface energy is the observation that, in the intermediate state, the domain size is macroscopic; the interphase area is minimized so as to reduce the contribution from a positive surface energy. This factor is taken into account in the Landau theory of the intermediate state, discussed in Chapter 21.

III. PHENOMENOLOGICAL THEORIES

A. Thermodynamics of the Transition

1. Gibbs Potential

The implication, from the Meissner effect, that the superconducting–normal transition is reversible at once permits the application of thermodynamics to derive various equilibrium properties associated with the transition (*27*). We assume that the Gibbs potential G_n in the normal phase is independent of the applied magnetic field H_a—a justifiable assumption as long as we are comparing it with the Gibbs potential $G_s(H_a)$ in the superconducting phase, since the magnetic susceptibility in the normal phase is no more than 10^{-3} times that in the superconductor. We shall see later in this book that this assumption may have to be abandoned for superconductors which have very high critical fields. If $G_s(0)$ is the Gibbs potential in the superconducting phase in zero magnetic field, then in a field H_a we have, if the temperature T and stress σ_i are constant,

$$G_s(H_a, \sigma_i, T) - G_s(0, \sigma_i, T) = -V\int_0^{H_a} I\, dH \tag{7}$$

where the volume V of the superconductor is assumed to be independent of the field; i.e., the strain e_i is infinitesimal. If we now consider a long thin cylinder in a field parallel to its axis, demagnetization effects are negligible, and $I = -H_a/4\pi$. This gives

$$G_s(H_a, \sigma_i, T) - G_s(0, \sigma_i, T) = VH_a^2/8\pi \tag{8}$$

At $H_a = H_c$, we conclude from the Meissner effect that the two phases are in equilibrium, so that $G_s = G_n$ and

$$\Delta G \equiv G_n(\sigma_i, T) - G_s(0, \sigma_i, T) = VH_c^2/8\pi \tag{9}$$

Thus the free energy difference between the normal and superconducting states in zero field and at given temperature and stress is $VH_c^2/8\pi$; this result is clearly independent of sample shape and consequent demagnetization (and intermediate state) effects. Indeed, we can conclude that the area under the curve of intensity of magnetization vs. field, if it is reversible, is an invariant for all sample shapes:

$$-\int_0^{\text{normal}} I\, dH = H_c^2/8\pi \tag{10}$$

A few qualitative remarks regarding Eqs. (7)–(10) may be worthwhile. The critical field H_c is a measure of the lowering of the energy of the system when it goes superconducting, and, when this is compensated by the work done by the system in excluding an applied field, it goes normal. As we shall see later, if there are systems such as thin films which do not completely exclude the applied field, then they would stay superconducting to fields higher than H_c. Thus, even though we introduced H_c as a field which could be directly determined by measuring the transition of a suitably shaped sample in a magnetic field, it may often be more helpful to regard Eq. (9) as defining H_c.

We can now proceed to derive from Eqs. (8) and (9), by standard thermodynamic techniques, the variation of the equilibrium mechanical, thermal, and magnetic properties in the superconducting phase and at the transition.

2. Mechanical Properties

The strain tensor e_i is defined by

$$e_i = -\frac{1}{V}\left(\frac{\partial G}{\partial \sigma_i}\right)_{T,H} \tag{11}$$

and so the magnetostrictive strain $e_i^{(s)}$ in the superconductive state, and the relative strain between the phases in zero field, are, respectively,

$$e_i^{(s)}(H_a) = -\frac{H_a^2}{8\pi V}\left(\frac{\partial V}{\partial \sigma_i}\right)_{T,H} \tag{12}$$

$$\begin{aligned} e_i^n(0) - e_i^s(0) &= -\frac{H_c^2}{8\pi V}\left(\frac{\partial V}{\partial \sigma_i}\right)_{T,H} - \frac{H_c}{4\pi}\left(\frac{\partial H_c}{\partial \sigma_i}\right)_T \\ &= e_i^{(s)}(H_c) - \frac{H_c}{4\pi}\left(\frac{\partial H_c}{\partial \sigma_i}\right)_T \end{aligned} \tag{13}$$

We note here that the stress and strain tensors are written in the usual six-component form. From Eqs. (12) and (13) we see that the change in the strain at H_c, when the sample goes normal, is

$$e_i^n(H_c) - e_i^s(H_c) = \Delta e_i = -\frac{H_c}{4\pi}\left(\frac{\partial H_c}{\partial \sigma_i}\right)_T \tag{14}$$

Since a hydrostatic pressure p corresponds to the stress tensor $-p\mathbf{1}$, where $\mathbf{1}$ is the unit matrix, we get from Eq. (14) for the volume change during the transition,

$$V_n(H_c) - V_s(H_c) = \frac{VH_c}{4\pi}\left(\frac{\partial H_c}{\partial p}\right)_T \tag{15}$$

Also from Eq. (14), we find for the change in the elastic compliance tensor S_{ij},

$$\Delta S_{ij} = \Delta \frac{\partial e_i}{\partial \sigma_j} = -\frac{1}{4\pi}\left[\frac{\partial H_c}{\partial \sigma_i}\frac{\partial H_c}{\partial \sigma_j} + H_c \frac{\partial^2 H_c}{\partial \sigma_j \,\partial \sigma_i}\right] \tag{16}$$

In terms of the bulk modulus K, for a cubic crystal, we have $K = \frac{1}{3}(S_{11} + 2S_{12})$, and so

$$\Delta K = \frac{K^2}{4\pi}\left[\left(\frac{\partial H_c}{\partial p}\right)^2 + H_c \frac{\partial^2 H_c}{\partial p^2}\right] \tag{17}$$

As noted in Section II.A, all the above mechanical effects have been experimentally observed. An extensive review of the theory, with references to recent experiments, has been given by Seraphim and Marcus (*12*).

3. Thermal Properties

By successive differentiation of Eq. (9) with respect to T, we obtain the differences in entropy S and specific heat C between the normal and superconducting phases:

$$\begin{aligned}\Delta S \equiv S_n - S_s &= -\frac{\partial}{\partial T}\Delta G \\ &= -\frac{VH_c}{4\pi}\frac{\partial H_c}{\partial T}\end{aligned} \tag{18}$$

$$\begin{aligned}\Delta C = C_n - C_s &= T\frac{\partial}{\partial T}\Delta S \\ &= -\frac{VT}{4\pi}\left[\left(\frac{\partial H_c}{\partial T}\right)^2 + H_c \frac{\partial^2 H_c}{\partial T^2}\right]\end{aligned} \tag{19}$$

We note the following features of Eq. (18): ΔS vanishes at T_c (where $H_c = 0$) and also at $T = 0°K$ (from the third law of thermodynamics). We conclude that $\partial H_c/\partial T$ vanishes at $T = 0°K$. Since we know experimentally that $\partial H_c/\partial T < 0$ for finite T, we have that $S_s < S_n$. The superconducting state is a more ordered state, therefore, than the normal one. We shall discuss later several phenomenological theories based on an order parameter. The precise nature of the ordering was revealed by BCS theory. We also note from Eq. (18) that there is a latent heat

$T\,\Delta S$ associated with the transition; an amount of heat

$$\frac{VTH_c}{4\pi}\frac{\partial H_c}{\partial T}$$

is absorbed when the superconductor is driven normal at temperature T by a magnetic field. Turning now to Eq. (19), we remark that there is a jump in the specific heat at $T = T_c$ given by

$$\Delta C(T_c) = -\frac{VT_c}{4\pi}\left(\frac{\partial H_c}{\partial T}\right)^2_{T=T_c} \tag{20}$$

with the superconducting specific heat being higher than the normal specific heat. Equation (20) is known as Rutgers' formula.

4. Magnetic Properties

We saw from the discussion in Section III.A.1 that, while the free energy difference between the superconducting and normal states in zero field is $VH_c^2/8\pi$ independently of sample shape, a sharp magnetic transition at field H_c occurs only for a long thin cylinder in a parallel field. For a sample with a nonzero demagnetizing factor, as pointed out in Section II, a magnetic domain structure, called the *intermediate state*, is set up when the applied field H_a reaches some value less than H_c. The thermodynamics which we have developed so far cannot give a complete quantitative account of the intermediate state, because we have not included the effect of the normal–superconducting interface on the free energy. We can, however, following F. London (*30*), derive some thermodynamic results neglecting the interphase energy. These may be applied to macroscopic specimens in which this energy plays a relatively small role, because of the coarse scale of the domain structure.

Consider an ellipsoid in the intermediate state. Let h be the local magnetic field, so that we have $h \doteq 0$ in the superconducting regions and $h = H_c$ in the normal regions. Let a volume fraction ξ of the sample be superconducting, at an applied field H_a. Then the coarse-grained mean value of the magnetization I, over a unit volume containing many domains, is

$$I = -\xi\frac{H_c}{4\pi}$$

The contributions to the Helmholtz free energy per unit volume from the normal and superconducting regions, F_n and F_s, respectively, can be written as

$$F_n = (1-\xi)\left(\frac{H_c^2}{8\pi} - \frac{H_a^2}{8\pi}\right)$$

$$F_s = -\xi\frac{H_c^2}{8\pi} - \xi\frac{H_a^2}{8\pi}$$

so that the total free energy F is

$$F = (1 - 2\xi)\frac{H_c^2}{8\pi} - \frac{H_a^2}{8\pi}$$
$$= \frac{H_c^2}{8\pi} + H_c I - \frac{H_a^2}{8\pi}$$

from which we get, since $H = \partial F/\partial I$, that

$$H = H_c$$

This important result states that if we can assume a uniform magnetization for the intermediate state, then we can use standard results from magnetostatics, with $H = H_c$, to describe the magnetic behavior. We shall consider, as an illustration, an ellipsoid with demagnetizing coefficient $4\pi n$ in a magnetic field. Then the following relations hold in the pure superconducting state, in an applied field H_a: $B = 0$, $H = H_a/(1 - n)$, and $I = - H_a/4\pi(1 - n)$. If $H_a > H_c(1 - n)$, the sample breaks up into the intermediate state, and the appropriate equations now become

$$B = H_c - (H_c - H_a)/n \qquad I = (H_a - H_c)/4\pi n$$

These relations have been experimentally verified, and they also satisfy the thermodynamic requirement that

$$\int_0^{H_c} I\, dH_a = - H_c^2/8\pi \tag{21}$$

Certain magnetic properties of samples having at least one dimension comparable to the penetration depth λ can be treated thermodynamically. Consider a plate of thickness $2a$, in a field H_a parallel to its surface. Then one can show, by requiring that Eq. (21) be satisfied and assuming that I is linear in H_a, that

$$\frac{\chi}{\chi_0} = 1 - \frac{\lambda}{a} \qquad \text{if } a \gg \lambda \tag{22a}$$

$$\frac{\chi}{\chi_0} = \beta\left(\frac{a}{\lambda}\right)^2 \qquad \text{if } a \ll \lambda \tag{22b}$$

$$\frac{H_f}{H_c} = \left(\frac{\chi}{\chi_0}\right)^{-1/2} \qquad \text{for all } a \tag{22c}$$

where χ is the susceptibility ($\equiv I/H_a$) of the plate, $\chi_0 = - 1/4\pi$ is the susceptibility if λ is assumed to be zero, β is a parameter determined by the actual field penetration law ($\beta = \frac{1}{3}$ for an exponential penetration law), and H_f is the critical field for the plate. One thus finds the dependence of critical field of a film on its thickness $2a$:

$$\frac{H_f}{H_c} = 1 + \frac{\lambda}{2a} \qquad \text{if } a \gg \lambda \tag{23a}$$

$$\frac{H_f}{H_c} = \beta^{-1/2}\frac{\lambda}{a} \qquad \text{if } a \ll \lambda \tag{23b}$$

These relations are qualitatively correct. The discrepancies are to be attributed to the neglect of interphase surface energy and to the rounding of the experimental *I–H* curves for thin films.

The properties of the intermediate state are discussed in detail in Chapter 21.

B. Two-Fluid and Energy Gap Models

1. Two-Fluid Models

It is an interesting historical fact that two-fluid models have played a prominent role in the development of our understanding of both quantum liquids: the electron fluid in superconductors and liquid helium II. They have in both cases stimulated experimental work and provided useful models for the organization of the results, and have to some measure been reinterpreted in terms of the microscopic theories.

In the case of superconductors, the thermodynamics of a two-fluid model was worked out by Gorter and Casimir (*64*) and the electrodynamics by F. and H. London (*65,66*). We shall first describe the Gorter–Casimir two-fluid model. Gorter and Casimir realized (*67*) that the superconducting transition, because it is a transition of the second order, could be described by introducing "an internal parameter indicating a degree of superconductivity," which with rising temperature would decrease gradually until it disappeared at the transition temperature. Let this parameter be x, equal to unity at $T = 0$ and decreasing continuously to zero at $T = T_c$. It seems natural to associate x with the fraction of electrons which are superconducting. If now we assume that the total Helmholtz potential of a superconductor can be obtained by summing the contributions from the superconducting and normal fractions of the conduction electrons, we might be tempted to write

$$F(T) = xF_s(T) + (1-x)F_n(T) \tag{24}$$

where $F_s = -H_0^2/8\pi$ and $F_n = -\frac{1}{2}\gamma T^2$, where γ is the usual coefficient defining the electronic specific heat. But Eq. (24) does not permit $x(T)$ to be determined from the equilibrium condition $(\partial F/\partial x)_T = 0$; in fact, it uniquely specifies the equilibrium temperature at which x can take any value between zero and 1. At higher temperatures $x = 1$, and at lower temperatures $x = 0$. This, of course, is just the result we would expect from the application of the Gibbs phase rule to a one-component system with two independent phases. We get out of this difficulty by modifying Eq. (24) such that the equilibrium condition will yield a relation connecting x, F_s, and F_n, from which $x(T)$ can be computed; this is the same as saying that the two phases are not independent, so that their relative proportion in the superconductor is determined by their respective free energies. Gorter and Casimir chose the simplest way of doing this as follows:

$$F(T) = xF_s(T) + (1 - x)^{1/2}F_n(T) \tag{25}$$

The coefficients of F_s and F_n were chosen to give the measured T^3 dependence of the specific heat, as known at that time. Minimizing the free energy with respect to x, and using the condition that $x = 0$ at $T = T_c$, one gets the following relations:

$$\gamma = H_0^2/2\pi T_c^2 \tag{26a}$$

$$x(t) = 1 - t^4 \tag{26b}$$

$$C_s(t) = 3\gamma T_c t^3 \tag{26c}$$

As mentioned in Section II.B, the t^3-dependence of C_s leads to the parabolic dependence of h on t. Thus the Gorter–Casimir model made an assumption about the free energy for which there was no clear physical justification in order to give fairly good agreement with the early experimental results on the thermal properties. But what was remarkable was the fact, pointed out by Daunt (*68*), that the temperature variation of the penetration depth was correctly given by using $x(t)$ from Eqs. (26) for the relative concentration of superconducting electrons in the Londons' theory; this provided an unexpected, but subsequently unexplored, link between the Gorter–Casimir and London models.

There were later attempts to develop general formulations of a two-fluid model, such as the one by Marcus and Maxwell (*69*). The greatest use of a phenomenological model, such as the two-fluid model, is to reveal experimentally verifiable interrelations among physical properties, such as the temperature dependence of λ or the expression for γ [see Eqs. (26)], which it is then the task of the microscopic theory to justify. Such a model can also help in a qualitative visualization of experimental phenomena. The two-fluid model in superconductivity has had only limited success in these respects, particularly when compared with the corresponding model in liquid helium (*70*). The point here is that the Gorter–Casimir and related two-fluid models were set up so as to have the observed thermodynamic properties. But the models gave no information on, and indeed were not intended to have anything to do with, the hydrodynamic or electrodynamic aspects of the "two fluids." Further discussion of the comparison of the two-fluid models in liquid helium and superconductivity have been given by Gorter (*70*) and by Bardeen and Schrieffer (*71*).

2. Energy Gap Models

The possibility of a gap in the electronic energy spectrum of a superconductor was first suggested by F. and H. London (*65,66*) in 1935. Daunt and Mendelssohn (*72*) concluded from their observation that the Thomson heat in a superconductor is zero—that such a gap existed. Several early attempts at a microscopic theoretical model involved, or could be interpreted to give, an energy gap. Among these were Welker's (*73*) theory (involving the magnetic exchange interaction), and Heisenberg's (*74*) theory (based on the long-range part of the Coulomb interaction), as extended by Koppe (*75*). Goodman (*25*) attempted to interpret his thermal conductivity measurements on the basis of Koppe's theory, and was led to suggest an energy gap varying from $\sqrt{\pi^2/6}\, kT_c$ at 0°K to zero at T_c. There was also a thermodynamic two-fluid model with an energy gap proposed by Ginzburg (*76*), somewhat similar to the Gorter–Casimir model. Bardeen (*77*) showed that the energy gap model could lead to a relation between current and magnetic field similar to one derived by Pippard (see Section III.C). This was an important step in linking the thermal and electromagnetic properties of superconductors.

3. Discussion

The models mentioned above each suffered from one or more drawbacks which rather seriously limited their usefulness and/or longevity. The attempts to relate them to microscopic interactions, although well motivated, were unsuccessful mainly because of the lack then of a suitable apparatus for handling the many-body problem. The assumptions made in some of the models were difficult, if not impossible, to justify physically. None of them [with the exception of Bardeen's (*77*) work referred to above] gave a successful interpretation of the electrodynamic properties of a superconductor. These models nevertheless played a very useful role in the history of superconductivity, by providing a framework for the systematization of a great body of experimental data. They also served to emphasize the energy gap as a necessary consequence of a satisfactory theory of superconductivity, and the role of an order parameter in such a theory. In fact, Bardeen (*78*), after reviewing these theories in 1956, conjectured that "a possibility would be to take the relative value of the energy gap as an order parameter," a conjecture which was proved right later by Gor'kov (*79*) on the basis of the BCS theory.

It only remains now to describe two more phenomenological theories to complete this account of the pre-BCS era of superconductivity. These are, of course, the theory of F. and H. London with its important generalization by Pippard and the Ginzburg–Landau theory, discussed in the following two sections.

C. London–London–Pippard Theory

The title of this subsection, perhaps clumsily, has been chosen to emphasize the contributions of F. London, H. London, and A. B. Pippard to the development of an electrodynamics of superconductors on a phenomenological basis, which gave a consistent description of essentially all the electromagnetic properties of superconductors, thus indicating one set of results which might be expected from a microscopic theory. One might remark here that the models to be discussed in this section, as well as the Ginzburg–Landau theory to be considered next, addressed themselves primarily to the electrodynamics, and not to the thermal properties of superconductors. The LLP or GL equations did not lead to the GC relations, or vice versa. It was left to BCS to develop an inclusive description of both sets of properties.

1. London Equations

The theory of a perfect conductor due to Becker et al. (*80*), and its relation to the London electromagnetic equations, has been described by Shoenberg (*46*). The account here is based on the one given by F. London himself (*30*).

The central point of the London theory is that the supercurrent is always determined by the local magnetic field. The Meissner effect is interpreted to mean that the *diamagnetic* currents induced when a simply connected superconductor is cooled through T_c in an applied field are identical with the *supercurrents* induced when a field is applied after cooling through T_c in zero field. A persistent current has the same physical relation to the magnetic field it produces as an externally applied supercurrent in a straight wire does to its field; in both cases the magnetic field sustains the current. It is inadequate to consider a superconductor just as a perfect conductor (which would lead, in F. London's words, to "a continuum of non-equilibrium states to which one must attribute persistent currents of a truly acrobatic stability"), or as a perfect diamagnet, for this would not describe correctly the persistent current in a ring.

The London theory goes beyond the observation of the Meissner effect, namely, that $B = 0$, to a set of microscopic field equations‡ which imply definite hypotheses regarding the nature of the fine structure of the fields and currents, which one would hope to justify from a microscopic theory. In this sense F. London preferred to call the theory a *macroscopic*, rather than a *phenomenological*, theory of superconductivity.

The London theory assumes that the current in a superconductor, j, is composed of a supercurrent j_s and a normal current j_n. The electrodynamics of the

‡ See Chapter 1 of (*81*) for an excellent discussion of microscopic electromagnetic variables.

superconductor are then embodied in the following equations:

$$\operatorname{curl}(\Lambda \mathbf{j}_s) = -\frac{\mathbf{h}}{c} \tag{27}$$

$$\frac{\partial}{\partial t}(\Lambda \mathbf{j}_s) = \mathbf{e} \tag{28}$$

$$\mathbf{j} = \mathbf{j}_s + \mathbf{j}_n \tag{29}$$

$$\mathbf{j}_n = \sigma \mathbf{e} \tag{30}$$

along with

$$\operatorname{curl} \mathbf{h} = \frac{4\pi}{c}\mathbf{j} + \frac{1}{c}\frac{\partial \mathbf{e}}{\partial t} \tag{31}$$

$$\operatorname{curl} \mathbf{e} = -\frac{1}{c}\frac{\partial \mathbf{h}}{\partial t} \tag{32}$$

$$\operatorname{div} \mathbf{h} = 0 \tag{33}$$

$$\operatorname{div} \mathbf{e} = 4\pi\rho \tag{34}$$

where Λ is a quantity yet to be determined (it will turn out to be $\sim 10^{-31}$ sec^2), σ is the normal conductivity, ρ is the electric charge density, and **e** and **h** are the local fields in the Lorentz sense. We note that Eqs. (31)–(34) are the Maxwell–Lorentz field equations; Eqs. (27)–(30) contain the new assumptions of the London theory. We shall now show that for the quasi-static case, these equations lead to perfect conductivity and the Meissner effect. First, by combining Eqs. (27)–(30) we get

$$-c \operatorname{curl} \Lambda \mathbf{j} = \mathbf{h} + \sigma\Lambda \frac{\partial \mathbf{h}}{\partial t} \tag{35}$$

$$\frac{\partial}{\partial t}(\Lambda \mathbf{j}) = \mathbf{e} + \sigma\Lambda \frac{\partial \mathbf{e}}{\partial t} \tag{36}$$

We can now successively eliminate three of the four variables **e**, **h**, **j**, and ρ from Eqs. (31)–(36) and obtain

$$c^2 \operatorname{curl}\operatorname{curl} \mathbf{h} + \frac{4\pi}{\Lambda}\mathbf{h} + 4\pi\sigma\frac{\partial \mathbf{h}}{\partial t} + \frac{\partial^2 \mathbf{h}}{\partial t^2} = 0 \tag{37}$$

with identical equations for **e** and **j** obtained by replacing **h** by **e** or **j** in Eq. (37), and

$$\frac{4\pi}{\Lambda}\rho + 4\pi\sigma\frac{\partial \rho}{\partial t} + \frac{\partial^2 \rho}{\partial t^2} = 0 \tag{38}$$

We shall simply assume that there is total charge neutrality, leading to the vanishing of ρ, $\partial\rho/\partial t$, and div $\mathbf{j}$, and static conditions, so that all time derivatives vanish. We then have, using the Cartesian identity

$$\text{curl curl} = \text{grad div} - \nabla^2$$

along with Eqs. (31)–(34),

$$\nabla^2 \mathbf{h} = (4\pi/\Lambda c^2)\,\mathbf{h} \tag{39}$$

and similar equations for $\mathbf{e}$ and $\mathbf{j}$. We first note that curl $\mathbf{e} = 0$ from Eq. (32), and so, from Eq. (37) for $\mathbf{e}$ we get $\mathbf{e} = 0$. Thus the electric field in a superconductor vanishes, even though the electric current may be nonzero. This is, of course, because $\mathbf{j}$ is connected with the magnetic field via the London equation [Eq. (27)]. Thus the London equations imply perfect conductivity.

The vanishing of the field $\mathbf{h}$ inside a macroscopic superconductor is contained in Eq. (39). That this is so may be physically obvious for simple cases. It was shown quite generally by von Laue‡ that the magnitude $|h|$ of the magnetic field at a point inside the superconductor at a distance D from the surface is smaller than its mean value at the surface by a factor of the order $(D/\lambda)/\sinh(D/\lambda)$, where $\lambda = c\sqrt{\Lambda/4\pi}$, and is called the *penetration depth*. Comparison with the theory of a perfect conductor (*30*) leads to the relations

$$\lambda = (mc^2/4\pi n_s e^2)^{1/2} \tag{40}$$

$$\Lambda = m/n_s e^2 \tag{41}$$

We thus see that for large bodies, of dimensions large compared to λ, the magnetic field $B = \langle \mathbf{h} \rangle_{\text{av}} = 0$. The important new point here is that the field does not vanish discontinuously at the surface, but dies out gradually over the penetration depth λ "exponentially." Thus the London equations cover not only the Meissner effect in bulk superconductors, but also the magnetic properties of samples with dimensions comparable to λ (thin films, colloids). Finally, the conditions to be satisfied at surfaces where the material characteristics change (superconductor–superconductor, S–N, or S–vacuum) are: $e_\perp$ may be discontinuous, but $h_\parallel$, $e_\parallel$ and $h_\perp$, $j_\perp$ have to be continuous; in particular, for the S–S boundary, $(\Lambda j_s)_\parallel$ must be continuous. The subscripts here denote the parallel or perpendicular components.

2. Solutions of the London Equations

a. *Plane superconductor–vacuum interface*

Let the plane be at right angles to the x-axis and positive x be in the superconductor. Let $h_z = h_0$ at $x = 0$ and $h_x = h_y = 0$. Then the physically meaningful

‡ See proof in (*30*), p. 51.

($h_z \to 0$ as $x \to \infty$) solution of Eq. (39) is

$$h_z = h_0 \exp(-x/\lambda) \tag{42}$$

showing an exponential decay of the field with a characteristic distance λ. One can similarly show that the current is in the y-direction, and given by

$$j_y = -\frac{ch_0}{4\pi\lambda} \exp(-x/\lambda) \tag{43}$$

b. *Flat slab in parallel field*

Let the slab be of thickness $2d$ along the x-axis, $h_z = h_0$ at $x = \pm d$, and $h_y = 0 = h_x$. One then gets the solution

$$h_z = h_0 \frac{\cosh(x/\lambda)}{\cosh(d/\lambda)} \tag{44}$$

The specific magnetization I is therefore

$$\begin{aligned} I &= \frac{1}{8\pi d} \int_{-d}^{d} (h_z - h_0)\, dx \\ &= -\frac{h_0}{4\pi}\left(1 - \frac{\lambda}{d}\tanh\frac{d}{\lambda}\right) \end{aligned} \tag{45}$$

Equation (44) may be used to evaluate the critical field H_{cf} for a thin film by equating the total work done on the sample in raising the applied field to H_{cf} to the free energy difference $H_c^2/8\pi$ between the normal and superconducting phases, leading to

$$\left(\frac{H_{cf}}{H_c}\right)^2 = \left(1 - \frac{\lambda}{d}\tanh\frac{d}{\lambda}\right)^{-1} \tag{46}$$

For the limiting cases of $d \gg \lambda$ and $d \ll \lambda$, we get from the above equation

$$\frac{H_{cf}}{H_c} = 1 + \frac{\lambda}{2d} \qquad \text{for } d \gg \lambda \tag{47a}$$

$$\frac{H_{cf}}{H_c} = \sqrt{3}\,\frac{\lambda}{d} \qquad \text{for } d \ll \lambda \tag{47b}$$

in agreement with Eqs. (23a) and (23b).

c. *Sphere and circular cylinder*

By writing the London equations in appropriate coordinates, one can show that for a cylinder of radius r_0, in an applied field h_0 parallel to its axis,

$$h(r) = h_0 \frac{I_0(r/\lambda)}{I_0(a/\lambda)} \tag{48}$$

where I_0 is the Bessel function of imaginary argument. The expressions for a sphere (*30*) are more cumbersome, but we note here that, in the limiting cases of the radius of the sphere R being much greater than λ, the sphere behaves like a perfectly diamagnetic sphere of radius $R-\lambda$, and for $R \ll \lambda$ its magnetization is given by $-H_a R^5/30\lambda^2$. The latter case may hold in measurements on colloids, such as those of Shoenberg (*46*).

An interesting solution of the London equations has been considered recently by Friedel et al. (*82*). Viewing an Abrikosov line as a normal core of radius ξ surrounded by a London-type superconductor, with a field h_0 at the boundary between the two, one gets the field h at a distance $r > \xi$ from the axis of the core as

$$h = h_0 \frac{K_0(r/\lambda)}{K_0(\xi/\lambda)} \tag{49}$$

where $K_0(x)$ is the modified Bessel function of the second kind.

3. Superconducting Ring; Fluxoid Conservation

The London equations say, in agreement with experiment, that the magnetic flux inside a superconductor is zero; the shielding currents in the penetration depth which are supported by the field ensure this. This is true in the material of a multiply connected superconductor also, such as a superconductor with a hole (Fig. 9). But the hole itself can presumably contain flux, and we shall now see how the London equations apply to such a system.

Let C be a closed curve, everywhere in the superconductor, but surrounding the hole. Any surface S bounded by C will be partly in the superconductor and partly in the hole. Using Stokes' theorem to integrate Eq. (32) over S, with Eq. (28) applicable in the superconducting region, leads to

$$\frac{d}{dt}\left\{\iint_S \mathbf{h}\cdot d\mathbf{S} + c\oint_C \Lambda \mathbf{j}_s \cdot d\mathbf{s}\right\} = 0$$

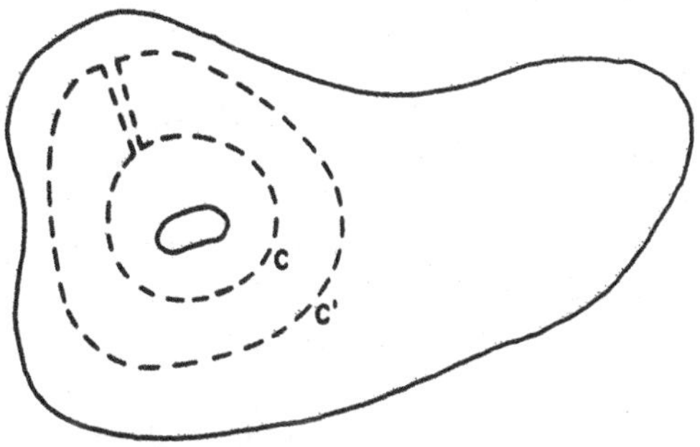

Fig. 9. Multiply connected superconductor, to illustrate fluxoid conservation. The three contours referred to in the text are C alone, C' alone, and C and C' taken together with the vanishingly narrow bridge shown to the "northwest" of the hole.

The quantity within braces was called by London the *fluxoid* ϕ_c through the contour C. Thus we have

$$\phi_c = \iint_S \mathbf{h}\cdot d\mathbf{S} + c\oint_C \Lambda \mathbf{j}_s \cdot d\mathbf{s} \tag{50}$$

$$d\phi_c/dt = 0 \tag{51}$$

Thus the fluxoid so defined is a constant in time. We note that the fluxoid is not quite the same as the magnetic flux ($=\iint_S \mathbf{h}\cdot d\mathbf{S}$) through the curve. If the curve C were placed very far from the hole, then ϕ_c would be approximated very closely by the flux through the curve.

If the curve C encloses material which is entirely superconducting, then we can integrate the London equation

$$c \operatorname{curl} \Lambda \mathbf{j}_s = -\mathbf{h}$$

and find

$$\phi_c = 0 \tag{52}$$

The fluxoid therefore vanishes if the contour encloses no hole.

Next, by considering a contour composed of two curves such as C and C' in Fig. 9, each enclosing the same hole but the new composite contour lying entirely in the superconductor, we can show easily that the fluxoid ϕ_c is identically equal to the fluxoid $\phi_{c'}$; thus the fluxoid turns out to be a property of the *hole*, and is independent of the particular contour surrounding it which is used to evaluate it.

To summarize the properties of a fluxoid: It is zero for a contour enclosing only a superconducting region, and if there is a hole in the contour, it is invariant with time and independent of the shape of the contour, and is in fact equal to the total flux through the hole and the thin penetration region surrounding it.

4. Vector Potential

The first London equation takes a particularly simple form, as F. London showed, when one introduces a vector potential $\mathbf{A}$ for the magnetic field $\mathbf{h}$:

$$\mathbf{h} = \operatorname{curl} \mathbf{A} \tag{53}$$

Then Eq. (27) is satisfied by

$$\mathbf{j}_s = -(1/\Lambda c)\,\mathbf{A} \tag{54}$$

For the static case, with $\mathbf{e} = 0$, one needs additional restrictions on $\mathbf{A}$. First, to ensure no charge buildup, $\operatorname{div} \mathbf{j}_s = 0$ and so

$$\operatorname{div} \mathbf{A} = 0 \tag{55}$$

Equation (53) is still satisfied if we add to **A** the gradient of a scalar function ϕ, which obeys, because of Eq. (55),

$$\nabla^2 \phi = 0 \tag{56}$$

For an infinite superconductor, Eq. (56) requires ϕ = constant, and so grad $\phi = 0$ and **A** is uniquely determined. For a finite superconductor, we have to consider two cases. For a simply connected superconductor, if

$$\mathbf{A} = \mathbf{A}' + \operatorname{grad} \phi \tag{57}$$

then $\mathbf{A}'$ and ϕ have to be so chosen that div$\mathbf{A} = 0$, and $A_\perp = 0$, where $A_\perp$ is the normal component of **A** at the surface of the superconductor. For a multiply connected superconductor, ϕ is further specified so that the application of Stokes' theorem to the field **h** around a contour enclosing a hole gives the fluxoid threading that hole. We thus see that, for each of the cases above, specific choice of gauge, referred to sometimes as the London gauge, enables us to express the first London relation in the form of Eq. (54).

If we define the canonical momentum field $\mathbf{p}_s$ for the superfluid as

$$\mathbf{p}_s = m\mathbf{v}_s + (e\mathbf{A}/c) \tag{58}$$

then it can be shown that the London equations can be rewritten as

$$\operatorname{curl} \mathbf{p}_s = 0 \tag{59}$$

$$\partial \mathbf{p}_s/\partial t = -e \operatorname{grad} \phi \tag{60}$$

$$\operatorname{div} \mathbf{p}_s = 0 \tag{61}$$

$$p_\perp = e\Lambda j_\perp \tag{62}$$

at the boundary, where $j_\perp$ is the normal component of the current at the boundary (i.e., current fed from external source). For an isolated superconductor, $j_\perp = 0$, and the solution to Eqs. (59)–(61) is

$$\mathbf{p}_s = 0 \tag{63}$$

which, of course, is identical to Eq. (54).

5. London Kernel

In solving problems using a given theoretical model, we shall often be interested in evaluating quantities such as the current density $\mathbf{j}_s$ given the vector potential **A**. It turns out that it is more convenient to obtain the response of the system to a single Fourier component of the influence, and then obtain the total response by suitable integration. Consider the London relation between $\mathbf{j}_s$ and **A** in the London gauge:

$$\mathbf{A}(\mathbf{r}) = -c\Lambda \mathbf{j}_s(\mathbf{r}) \tag{64}$$

The **q**th Fourier components of **A** and $\mathbf{j}_s$ are therefore related by

$$\mathbf{j}_s(\mathbf{q}) = -(1/c\Lambda)\,\mathbf{A}(\mathbf{q}) \tag{65}$$

Suppose we now wish to obtain the magnetic field as a function of position, taking into account the distribution of shielding currents, for a superconductor with a plane infinite surface in the yz-plane and extending in the x-direction. The simplest mathematical model is to consider the field h_a as produced by an external current sheet $\mathbf{j}_e$ in an infinite superconducting medium in the yz-plane and along the z-direction, producing a magnetic field h_a in the y-direction; let $\mathbf{j}_s$ be the supercurrent response. We can write the Maxwell equation

$$\text{curl}\,\mathbf{h} = (4\pi/c)\,\mathbf{j}$$

for this case as

$$\nabla^2\mathbf{A} = -(4\pi/c)\,(\mathbf{j}_e + \mathbf{j}_s) \tag{66}$$

Then **A** and $\mathbf{j}_s$ will be in the z-direction. The quantities **A**, $\mathbf{j}_e$, and $\mathbf{j}_s$ all vary along the x-axis, and each may be expressed as a Fourier integral like

$$\mathbf{A}(x) = \int_{-\infty}^{\infty} \mathbf{A}(q)\exp(iqx)\,dq \tag{67}$$

Equation (66) may therefore be rewritten

$$-q^2\mathbf{A}(q) = -(4\pi/c)\,\mathbf{j}_e(q) + K(q)\mathbf{A}(q) \tag{68}$$

where we have defined $K(q)$ by

$$\mathbf{j}_s(\mathbf{q}) = -(c/4\pi)\,K(\mathbf{q})\mathbf{A}(\mathbf{q}) \tag{69}$$

Note that for the London case, by comparing Eqs. (69) and (65), we get $K_L(q)$, the *London kernel*, as

$$K_L(q) = 4\pi/\Lambda c^2 \tag{70}$$

Thus the London kernel, as defined above, is independent of q. We shall see later that a more realistic description of the electrodynamic properties of superconductors is obtained by using the Pippard kernel, $K_P(q)$, which is a function of q. Leaving $K(q)$ for the time being as unspecified, one can show (*83*) that

$$\mathbf{h}(x) = \frac{2\mathbf{h}_a}{\pi}\int_0^{\infty}\frac{q\sin qx\,dq}{K(q)+q^2} \tag{71}$$

and the effective penetration depth λ_{eff}, defined by

$$\mathbf{h}_a\lambda_{\text{eff}} = \int_0^{\infty}\mathbf{h}(x)\,dx$$

is given by

$$\lambda_{\text{eff}} = \frac{2}{\pi}\int_0^\infty \frac{dq}{K(q) + q^2} \tag{72}$$

for specular scattering of electrons at the surface; specular, because the symmetry of the model makes electrons which pass through the current sheet have the same previous history as if they had been specularly reflected at the sheet.

The case of diffuse scattering can be treated in a manner similar to the Reuter–Sondheimer theory of the anomalous skin effect (*84*), and leads to

$$\lambda_{\text{eff}} = \pi \Big/ \int_0^\infty \ln\left[1 + \frac{K(q)}{q^2}\right] dq \tag{73}$$

Inserting the London kernel [Eq. (71)] into Eq. (72) indeed gives

$$\lambda_{\text{eff}} \equiv \lambda_L = c\sqrt{\frac{\Lambda}{4\pi}} = \sqrt{\frac{mc^2}{4\pi n_s e^2}} \tag{74}$$

as before.

6. Some Other London Topics

The reader is referred to F. London's book (see General References) for a discussion of several other aspects of the theory. We shall mention here two further points: the uniqueness of the solutions, and energy–momentum considerations. The implication of the "perfect conductor" theory, that a given set of external parameters led to the existence of an infinite number of current-carrying states, offended, in F. London's words, "the good taste of most physicists." He was able to show that the London equations completely and uniquely determine the currents and fields, given the field at the surface (or infinity) and the fluxoids connected with holes (if any).

London also showed that there is a kinetic energy density $\frac{1}{2}\Lambda \mathbf{j}_s^2$ associated with the supercurrent, in addition to the magnetic energy. For macroscopic superconductors, this energy is small compared to the total magnetic energy, and is usually neglected, but for thin films carrying persistent currents, it may have to be taken into account (*85*). One can further show that the Maxwell stress tensor T, defined by

$$T_{ik} = \frac{1}{8\pi}\left[(\mathbf{e}^2 + \mathbf{h}^2)\,\delta_{ik} - 2(e_i e_k + h_i h_k)\right] \tag{75}$$

is now supplemented by the London stress tensor S, where

$$S_{ik} = \Lambda(j_{si} j_{sk} - \tfrac{1}{2}\mathbf{j}_s^2 \delta_{ik}) \tag{76}$$

and under stationary conditions

$$\operatorname{div}(T + S) = 0 \tag{77}$$

Thus there are no volume forces acting on the superconductor. But at the surface, while T is continuous, S drops discontinuously to zero. To conserve momentum, therefore, we must have surface forces equal to the normal component of S at the surface per unit area acting on the superconductor. The seat of the mechanical forces is strictly at the surface, and this is because the only coupling between the supercurrent and the lattice is that which forces the electrons to occupy the same volume as the lattice. Further, Eq. (77) also implies that the Lorentz force on the superfluid is exactly balanced by the inertial force $-\Lambda[\mathbf{j}_s \times \operatorname{curl} \mathbf{j}_s]$, so that there will be *no Hall effect* in a superconductor.

7. London Equations and Quantum Theory

Introducing as before the local mean value $\bar{\mathbf{p}}_s$ of the canonical momentum of the superelectrons, we have, in the London gauge,

$$\bar{\mathbf{p}}_s = 0 \tag{78}$$

for a simply connected isolated superconductor, and

$$\bar{\mathbf{p}}_s = \text{const.} \tag{79}$$

for a straight cylindrical wire of constant cross section carrying an externally fed current. In contrast, a normal isolated wire in an external magnetic field has no current, and so $\mathbf{v} = 0$, or

$$\bar{\mathbf{p}} = -(e/c)\,\mathbf{A} \tag{80}$$

Thus, in a normal wire, $\bar{\mathbf{p}}$ adjusts itself to the local value of the vector potential, while in a superconductor, *something holds* $\bar{\mathbf{p}}$ *rigid* over distances equal to the dimensions of the superconductor, and does not allow it to vary with $\mathbf{A}$. F. London emphasized this *long-range order in momentum* as a fundamental characteristic of superconductivity which it was the task of the electronic theory to explain. He suggested what the wave functions in the superconducting state must do in a magnetic field from the following consideration. The current density $\mathbf{j}(\mathbf{r})$ due to N particles in a state $\psi(\mathbf{r}_1, \mathbf{r}_2, \ldots, \mathbf{r}_N)$ in a field due to the vector potential $\mathbf{A}(\mathbf{r}_j)$ is

$$\mathbf{j}(\mathbf{r}) = \sum_{j=1}^{N} \int \left\{ \frac{e\hbar}{2im} [\psi^* \nabla_j \psi - \psi \nabla_j \psi^*] - \frac{e^2}{mc} \mathbf{A}(\mathbf{r}_j)\, \psi^* \psi \right\} \times \delta(\mathbf{r} - \mathbf{r}_j)\, d\tau_1\, d\tau_2 \cdots d\tau_N \tag{81}$$

In zero field, $\mathbf{A} = 0$ and $\mathbf{j}(\mathbf{r}) = 0$, and $\psi = \psi_0$. If we now assume that the field has no effect on the wave function, we get

$$\mathbf{j}(\mathbf{r}) = -\frac{\rho(\mathbf{r})\, e^2}{mc} \mathbf{A}(\mathbf{r}) = -\frac{1}{c\Lambda} \mathbf{A}(\mathbf{r}) \tag{82}$$

where $\rho(\mathbf{r})$ is the particle density; thus the first London equation follows if the wave function is rigid in the presence of a magnetic field. London used the phrase "quantum structure on a macroscopic scale" to describe the state of this system. This is a point of view, applied not only to superconductors but also to liquid helium II, which has led recently to some very beautiful experiments described later in this book. F. London remarked that this quantum-mechanical interpretation, applied to a fluxoid, led to a quantum of fluxoid Φ_1 given by

$$\Phi_1 = hc/e \simeq 4 \times 10^{-7} \text{ G–cm}^2 \tag{83}$$

which, apart from a factor of 2 to be explained by the microscopic theory, has since been experimentally found.

8. Comments on the London Theory

The London theory basically defined the goals for the microscopic theory of superconductivity, by emphasing the diamagnetic approach, the long-range order in momentum space, the rigidity of the wave function in the superconducting state, and the relation between $\mathbf{j}_s$ and $\mathbf{A}$. In combination with the energy gap model, it provided a framework for organizing most of the experimental data on superconductors. It turned out that detailed comparison between the London theory and experimental results showed several serious discrepancies. These discrepancies stimulated the development of models by Pippard and by Ginzburg and Landau, which represented the culmination of the pre-BCS era.

9. Pippard Modification of the London Theory

a. *Recapitulation of some anomalies*

We have mentioned at appropriate points earlier in this chapter various anomalous results which do not fit into the theoretical picture developed so far. We shall summarize them here for easy reference: (1) the impossibility of a Meissner effect without introducing a negative surface energy between the normal and superconducting phases; (2) the variation of λ with orientation in a single crystal; (3) the near-independence of λ with respect to H_a near T_c, indicating an enormous change in entropy density in the penetration layer between the normal and superconducting states; and (4) the increase in λ for a Sn–3% In alloy to nearly twice its value for pure Sn, even though all thermodynamic properties changed by only a few per cent upon alloying. Pippard, who had himself obtained experimental evidence relating to points (2) and (3), had come to the realization that "the behavior of a superconductor is in some respects controlled by an interaction of rather long range (called the "range of coherence") within the electron assembly, which restricts the rate of change from point to point of the parameters describing the local properties of the assembly" (*44*).

b. *Pippard coherence length and nonlocal relation*

To study the influence of impurity atoms (acting as scattering centers for the electrons) upon the coherence length, Pippard measured λ as a function of electronic mean free path and obtained the results shown in Fig. 10. The results and their interpretation were presented with impressive lucidity in an important paper by Pippard (*44*), to which the reader is referred for details; what follows is a summary of Pippard's arguments.

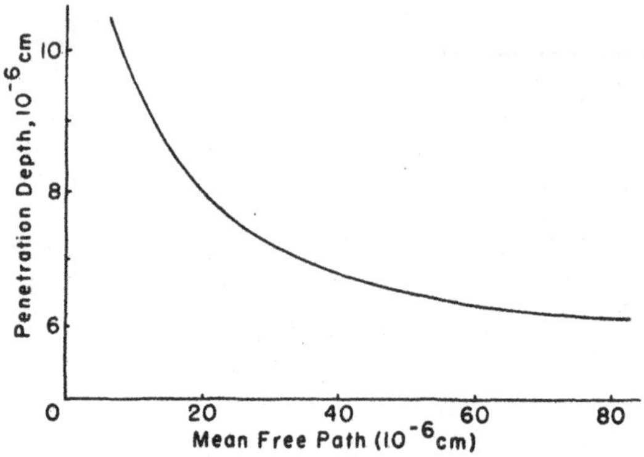

Fig. 10. Variation of penetration depth with mean free path in tin dilutely alloyed with indium. [After Pippard (*44*).]

The London relation for the penetration depth, $\lambda_L = (mc^2/4\pi n_s e^2)^{1/2}$, gives no indication at all that dilute alloying should give a significant change of λ, since m and n_s should change only slightly. We note that the rapid variation of λ begins at about where it is comparable to l, the mean free path. In a somewhat analogous situation in normal metals, with the anomalous skin effect, Chambers had shown that the local relation

$$\mathbf{j}(0) = \sigma \mathbf{E}(0) \tag{84}$$

between current $\mathbf{j}$ and electric field $\mathbf{E}$ should be replaced by a nonlocal relation

$$\mathbf{j}(0) = \frac{3\sigma}{4\pi l} \int \frac{\mathbf{r}(\mathbf{r}\cdot\mathbf{E}) \exp(-r/l)}{r^4} \, d\tau \tag{85}$$

Equation (85) allows for the possibility that the electrons can see a spatially varying field between collisions. The current at $\mathbf{r} = 0$, $\mathbf{j}(0)$, now becomes a function of $\mathbf{E}$ over the entire sample, hence the term "nonlocal." Chambers' expression for $\mathbf{j}(0)$ had been experimentally verified for the variation with l of the skin depth, showing an initial decrease with eventual leveling off as l increased, reminiscent of the behavior of λ. This fact, along with the formal similarity between Eq. (84) and the first London equation

$$\mathbf{j}_s(0) = -(1/c\Lambda)\,\mathbf{A}(0) \tag{86}$$

led Pippard to write down a modification of Eq. (86). But even before going to the nonlocal form of Eq. (86), we note that while σ depends on l, Λ is a constant depending only on the metal; and so, with Pippard, we rewrite Eq. (86) as

$$\mathbf{j}_s(0) = -(\xi/c\Lambda\xi_0)\,\mathbf{A}(0) \tag{87}$$

where ξ_0 is a constant for the metal and ξ depends upon the mean free path in some manner to be determined later, and they are both of the dimensions of length. We now write down the famous Pippard nonlocal relation between current and vector potential:

$$\mathbf{j}_s(0) = -\frac{3}{4\pi c\xi_0\Lambda}\int\frac{\mathbf{r}(\mathbf{r}\cdot\mathbf{A})\exp(-r/\xi)}{r^4}\,d\tau \tag{88}$$

One may remark here that, while the nonlocality of the current–potential relationship in Eq. (88) is very apparent, it is not quite correct, as has sometimes been done, to call the London equation [Eq. (86)] a *local* relation. The London gauge with boundary conditions implies that $\mathbf{A}$ is specified at more than just the point at which $\mathbf{j}_s$ is being measured, and hence already involves a nonlocality; the Pippard relation only carries this further, and introduces a specific nonlocality through the parameters ξ and ξ_0.

c. *Pippard kernel $K_P(q)$*

Consider a semiinfinite superconductor bounded by the yz-plane, with a current j_S flowing in the z-direction, h in the y-direction, and A in the z-direction. Then, in terms of the qth Fourier components of the field quantities j_s and A, we have, from Eq. (88),

$$j_s(q) = -\frac{3A(q)}{4\pi c\xi_0\Lambda}\int\frac{z^2\exp(-r/\xi)\exp(iqx)}{r^4}\,d\tau \tag{89}$$

The integration in spherical coordinates is easy, and leads to an expression for the kernel $K_P(q)$ defined in terms of Eq. (69):

$$K_P(q) = \frac{\xi}{\xi_0\lambda_L^2}\frac{3}{2(q\xi)^3}\left[(1+q^2\xi^2)\tan^{-1}q\xi - q\xi\right]\cdots \tag{90}$$

From this expression for $K_P(q)$, we get the following limiting forms for $j_s(q)$:

$$j_s(q) \to -\frac{\xi}{\xi_0}\frac{A(q)}{cA}\left\{1 - \frac{q^2\xi^2}{5} + \cdots\right\} \qquad \text{as } q \to 0 \tag{91a}$$

$$j_s(q) \to -\frac{3\pi}{4q\xi_0}\frac{A(q)}{cA}\left\{1 - \frac{4}{\pi q \xi} + \cdots\right\} \qquad \text{as } q \to \infty \tag{91b}$$

Using Eq. (90) we can also evaluate the penetration depth, and for diffuse scattering we get from Eq. (73),

$$\lambda = \lambda_L\left(\frac{\xi_0}{\xi}\right)^{1/2} \qquad \text{if } \xi^3 \ll \xi_0\lambda_L^2 \tag{92a}$$

and

$$\lambda \equiv \lambda_\infty = \lambda_L \frac{3^{1/6}}{(2\pi)^{1/3}}\left(\frac{\xi_0}{\lambda_L}\right)^{1/3} \qquad \text{if } \xi^3 \gg \xi_0\lambda_L^2 \tag{92b}$$

We are now in a position to be more specific about the parameters ξ_0 and $\xi(l)$. First, from Eq. (92b), using the measured value of λ and the computed λ_L for pure tin, Pippard estimated $\xi_0 \simeq 7.5 \times 10^{-5}$ cm at 0°K. A length of the order of 10^{-4} cm appears to be characteristic in superconductors, as, for example, in the treatment of the interphase surface energy. Pippard also remarked that such a length, $\xi_0 \sim hv_F/kT_c$, would emerge from an uncertainty principle argument applied to the energy kT_c, which must be the range of energies of the electrons involved in the transition. The microscopic theory later revealed that ξ_0 is actually related to the energy gap Δ, but since $\Delta \sim 1.76kT_c$, Pippard's argument still gave an approximate value for ξ_0. In fact, he found empirically that

$$\xi_0 = a(\hbar v_F/kT_c) \tag{93}$$

where v_F is the Fermi velocity and $a = 0.15$. The BCS theory gives Eq. (93) with $a = 0.18$.

We now turn to $\xi(l)$. Pippard first considered a simple proportionality, $\xi = \alpha l$, and found that his $\lambda(l)$ curve could be fitted with $\alpha = 0.93$. This was very pleasing, as indicating a range of coherence approximately equal to the mean free path. But this does allow ξ to exceed ξ_0, and indeed tend to very large values in pure superconductors. Pippard ruled this out, partly because the argument leading to Eq. (93) suggested that the maximum range of coherence is set by interactions among the electrons, independently of the impurities, and partly because of an interpretation of the nonlocal relation [Eq. (88)] to be discussed below. He chose instead a $\xi(l)$ which tended to ξ_0 as $l \to \infty$:

$$\frac{1}{\xi(l)} = \frac{1}{\xi_0} + \frac{1}{\alpha l} \tag{94}$$

which fitted his experimental results with $\xi_0 = 1.2 \times 10^{-4}$ cm and $\alpha = 0.8$.

Pippard was able to show that the model as described above accounted satisfactorily for the anomalies listed in Section III.C.9.a. Furthermore, Eq. (92b) cleared up a long-standing puzzle with λ_L, as to why measured penetration depths were always a few times larger than those computed from λ_L for pure superconductors.

10. Comments on the Pippard Theory

The Pippard theory, although it seems to be merely a modification of the London theory, represented a significant change from the earlier theory in its treatment of the problem of how an applied field influences the superconducting electrons. A basic point of the London theory was the absolute rigidity of the superconducting wave function in the presence of a field. Pippard abandoned this point, and suggested instead that a perturbing force acting at one point in the superconductor would be felt over a distance ξ, and, conversely, the response at a point due to a spatially extended perturbation would be obtained by integration over a finite region surrounding the point. Pippard emphasized the point that the similarity between Eqs. (85) and (88) is purely formal; the physical picture in the two cases is quite different.

A comparison between the London and Pippard theories can be made most conveniently in terms of the respective kernels, which we write here again for easy reference:

$$K_L(q) = \frac{1}{\lambda_L^2} \tag{95}$$

$$K_P(q) = \frac{1}{\lambda_L^2}\frac{\xi}{\xi_0}\left[1 - \frac{q^2\xi^2}{5} + \cdots\right] \qquad \text{for } q\xi \ll 1 \tag{96}$$

$$K_P(q) = \frac{1}{\lambda_L^2}\frac{\pi}{4q\xi_0}\left[1 - \frac{4}{\pi q\xi} + \cdots\right] \qquad \text{for } q\xi \gg 1 \tag{97}$$

We note here again that the essential requirement for London behavior (and an exponential decay of field) is a kernel independent of q. A comparison of Eqs. (95) and (96) shows that the Pippard kernel goes over to a London-type kernel, with a modified penetration depth, for $q\xi \ll 1$. This condition could be satisfied by $q \to 0$, corresponding to the behavior at large distances.‡ Another interesting point to observe is that, at least for the static case, Fourier components with $q \sim 1/\lambda$ are of importance, and therefore $\xi \ll \lambda$ leads to London behavior. The other limiting case of $q\xi \gg 1$ is obviously favored by $l \to \infty$, leading to $\xi \to \xi_0$ and $\xi_0 > \lambda$. We are thus led to the conclusion that there are two types of superconductors: (1) the Pippard, or type I, superconductor, with $\xi > \lambda$, exemplified by pure elemental superconductors having long mean free

‡ See (*86*), p. 10, for further discussion of this point.

paths; and (2) the London, or type II, superconductor, w'th $\xi < \lambda$, exemplified by certain transition metals which have a high T_c and therefore small ξ_0, or alloy superconductors where $\xi \simeq l < \lambda$.

The Pippard theory thus carried further the program of the London theory, by setting up in some detail the phenomenology which the microscopic theory would have to encompass. The two striking advances that the Pippard theory made were the elucidation of the nonlocal nature of the current–potential relations, and the specification of the coherence length. With the demonstration by Bardeen (*78*) that an energy gap model would lead to the Pippard nonlocal relation, theory had gone a long way toward a unified description of the thermodynamic and electrodynamic properties. But we note that so far we have still not had an explanation of the temperature dependence of λ, other than the less than satisfactory one obtained by the hybridization of the Gorter–Casimir and the London (or Pippard) theories. Nor for that matter did the isotopic variation of T_c, discovered in 1950, appear to fit into any of these theories. No serious attempt had been made to explain the peculiar magnetic behavior of alloys: spread-out magnetic transitions, high critical fields, absence of supercooling. Pippard (*44*) did suggest in 1953 that this could be due to a negative interphase surface energy, a suggestion which was worked out in detail by Goodman (*87*) in 1961.

Pippard arrived at the concept of the range of coherence partly empirically and partly on the basis of his interpretation of the nature of the superconducting state. His ξ was the shortest distance in which a significant change of electronic structure in the superconductor could occur. Meanwhile Ginzburg and Landau in 1950 developed the thermodynamics of a model in which the energy needed to produce a change in the superconducting state over *any* distance was explicitly included in the theory; their theory is the subject of the next section.

D. Ginzburg–Landau Theory

This section will serve as the briefest of introductions to a theory whose ramifications reach extensively into the rest of this treatise. The theory is based on the pioneering work of Landau (*18*) on second-order phase transitions. He made the important observation that such a transition is associated with an abrupt change in the symmetry of the system, and that the system of lower symmetry can be characterized by a parameter which is a measure of its departure from the configuration in the more symmetric phase.

In this section we shall examine the basic assumptions of the Ginzburg–Landau (GL) theory (*63*), and its relation to the Pippard theory. Many detailed applications of the theory, as well as its microscopic justification, will be dealt with elsewhere in the book.

1. Ginzburg–Landau Equations

GL noted certain inadequacies of the London theory: (1) the critical fields of thin films were correctly predicted only by assuming that λ varied with film thickness, (2) there was no explanation of the enormous positive surface energy which appeared to be associated with the normal–superconducting interface in simple superconductors, and (3) the theory did not allow a description of the critical currents in thin films, since irreversible effects were involved. It is remarkable that GL did not include, in their list of difficulties that the London theory had, the observation of Shubnikov et al. (*88*) that the magnetization of single-crystal samples of lead–indium and lead–thallium alloys showed what is now recognized as ideal type II behavior, and disagreement with the prediction of the London theory. It is now, of course, a matter of history that the most spectacular application of the GL theory has been to a description of such superconductors. To complete this historical aside, it should also be noted that it was a later remark of Landau's (*89*)—that the superconducting behavior of alloys could be due to the GL parameter κ becoming greater than $1/\sqrt{2}$—that culminated in Abrikosov's model for type II superconductors.

The detailed derivation of the GL equations will be given in Chapter 6, and need not be repeated here. We shall just note here that GL defined a parameter $\psi(r)$ which is a measure of the order in the superconducting phase, so that it is zero for $T > T_c$ and increases smoothly as T is reduced below T_c in zero field. Then the free energy density $F_{s0}(p, T, \psi)$ of the superconducting phase in zero field near T_c can be expanded as

$$F_{s0} = F_{n0} + \alpha|\psi|^2 + \beta|\psi|^4 + \cdots \tag{98}$$

where the coefficients α and β are functions of p and T. The term linear in $|\psi|$ is identically zero because of the different symmetries of the phases with $\psi = 0$ and $\psi \neq 0$, and the cubic term in $|\psi|$ vanishes to satisfy the requirement that $T = T_c$ itself represents a stable state. Minimizing the free energy at $T \lesssim T_c$, and noting that $|\psi|^2 = 0$ at $T = T_c$ and $|\psi|^2 > 0$ at $T < T_c$, we get

$$|\psi_\infty|^2 = -\frac{\alpha}{\beta} = \frac{T_c - T}{\beta(T_c)}\left(\frac{d\alpha}{dT}\right)_{T=T_c} \tag{99}$$

$$F_{s0} = F_{n0} - \frac{\alpha^2}{2\beta} \tag{100}$$

where ψ_∞ is the value in zero field (i.e., at an infinite distance from any boundary), and we have assumed that $\alpha(T) = (d\alpha/dT)_{T_c}\cdot(T_c - T)$ and $\beta(T) = \beta(T_c)$. From Eq. (100) one can get

$$H_c^2 = \frac{4\pi}{\beta(T_c)}(T_c - T)^2\left(\frac{d\alpha}{dT}\right)_{T_c}^2 \tag{101}$$

which agrees with experiment. This is the first success of the GL theory, that it predicts the right dependence of critical field upon T, at least close to T_c. This last qualification arises from the use of a finite power series in Eq. (98).

The second important step that GL took was to state that, if there was a spatial variation of ψ, as well as a magnetic field **H** derived from a potential **A**, then the energy density in the superconducting phase, F_{sH}, was given by

$$F_{sH} = F_{s0} + \frac{H^2}{8\pi} + \frac{1}{2m}|(-i\hbar\nabla - \frac{e^*}{c}\mathbf{A})\psi|^2 \qquad (102)$$

where m is the electronic mass and e^* is an effective charge. Various efforts have been made to interpret the last term on the right side of Eq. (102), none entirely satisfying. It is perhaps best to treat it as another example of the deep physical intuition of the authors of the theory.

The theory is now essentially completed by minimizing the total free energy [the volume integral of Eq. (102)] for variations in ψ^* and **A**, and establishing the boundary conditions. This leads finally to

$$\frac{1}{2m}\left|-i\hbar\nabla - \frac{e^*}{c}\mathbf{A}\right|^2 \psi + \frac{\partial F_{s0}}{\partial \psi^*} = 0 \qquad (103)$$

$$\nabla^2\mathbf{A} = -\frac{4\pi}{c}\mathbf{j}_s$$

$$= \frac{2\pi i e^*\hbar}{mc}(\psi^*\nabla\psi - \psi\nabla\psi^*) + \frac{4\pi e^{*2}}{mc^2}|\psi|^2\mathbf{A} \qquad (104)$$

with the subsidiary conditions

$$\operatorname{div}\mathbf{A} = 0 \qquad (105)$$

$$\mathbf{n}\cdot\left(-i\hbar\nabla - \frac{e^*\mathbf{A}}{c}\right)\psi = 0 \qquad (106)$$

where n is the unit vector normal to the superconducting boundary. Equations (103)–(106), taken along with Eq. (98), form the basis of the GL theory.

The application of the GL equations to a planar boundary leads to the following results. One can define parameters λ_0 (of the dimensions of length) and κ (dimensionless) such that

$$\lambda_0 = \left(\frac{mc^2}{4\pi e^{*2}\psi_\infty^2}\right)^{1/2} \qquad (107)$$

$$\kappa = \left(\frac{2e^{*2}}{\hbar^2 c^2} H_c^2 \lambda_0^4\right)^{1/2} \qquad (108)$$

so that the GL equations are formally reduced to the London equations if $\kappa \to 0$ and $\psi_\infty^2 = n_s$. Thus λ_0 becomes the penetration depth for a macroscopic sample.

But for a thin sample it turns out that the penetration depth becomes a function of the sample thickness, because the $\psi(x)$ needed to minimize the free energy depends on sample dimension.

A second important result that GL obtained was that, for $\kappa \ll 1$, the interphase surface energy density σ_{ns} between normal and superconducting phases is

$$\sigma_{ns} = \delta(H_c^2/8\pi) \tag{109}$$

where

$$\delta = 1.89\lambda_0/\kappa \tag{110}$$

thus explaining the very large positive energy ($\gg \lambda_0 H_c^2/8\pi$) needed to explain the Meissner effect, and the structure in the intermediate state. Physically, the significance of this result was that while the magnetic field decayed over a characteristic distance λ_0 to a vanishingly small value in the superconductor, $|\psi|^2$ decayed to zero toward the normal region over a much longer distance $\sim \lambda_0/\kappa$.

GL made the further observation, without pursuing it, that for $\kappa > 1/\sqrt{2}$, σ_{ns} becomes negative. This was subsequently recognized as defining the difference between type I and type II superconductors. Solutions of the GL equations for special cases have been described by de Gennes (*90*).

2. Comments on the GL Theory

As noted in the last subsection, the GL theory reduces to the London theory for $\kappa \to 0$; indeed, it is a "local" theory for all κ, and therefore distinct from the nonlocal theory of Pippard. It nevertheless leads to a satisfactory description of the interphase surface energy, a description which essentially goes over to the Pippard picture with the assumption

$$\kappa \simeq \lambda/\xi_0 \tag{111}$$

for a pure superconductor.

A second point to note is that the GL theory in its original form, as presented here, is valid only near T_c, because of the form of the expansion used for the free energy. The relaxing of the restrictions in these and other directions has been vigorously pursued on the basis of the microscopic derivation of the GL equations, and forms the subject of Chapter 6. We note here that one of the results of the microscopic theory was to show that $e^* = 2e$.

E. Prologue to the Microscopic Theory

We have now almost reached the end of our account of the essential experimental facts of superconductivity, and macroscopic–theoretical attempts to explain them. We need to mention two more points, one experimental and one theoretical, to complete the story. The experimental point is the discovery of the isotope effect in mercury *(91)* in 1950, that $T_c \sim M^{-1/2}$, where T_c is the transition temperature of an isotopically pure superconductor of isotopic mass M. This lent support to Fröhlich's suggestion *(92)* that superconductivity arises from electron–phonon interactions. The theoretical point to be mentioned is the work of Shafroth, Blatt, and Butler on a charged Fermi gas with two-particle correlations as a model for a superconductor. A discussion of this microscopic theory lies outside the scope of this chapter; reference is made to Blatt *(86)* for an account of this theory. Its relation to BCS is considered there, and also by Bardeen and Schrieffer *(93)*.

We can summarize now the features which a satisfactory microscopic theory of superconductivity was required to have, as the picture was in the mid-1950s. The superconducting state was a macroscopic quantum state, with a unique diamagnetic relation between the current and magnetic field (London); an energy gap $\sim kT_c$ was present in the excitation spectrum of the electrons; the current–field relation was nonlocal, with a range of coherence $\sim 10^{-4}$ cm (Pippard); the theory should involve an "order parameter," which went smoothly to zero as $T \rightarrow T_c$ (Ginzburg–Landau); and electron–phonon interactions should play an essential role in the phenomenon (isotope effect, Fröhlich). So the stage was set for the BCS theory.

ACKNOWLEDGMENTS

Helpful comments by P. R. Zilsel on this article, and support from the U.S. Air Force Office of Scientific Research under grant 565–66 during its preparation, are gratefully acknowledged.

GENERAL REFERENCES

The following contain further discussions of many of the topics in this chapter.

Shoenberg, D., *Superconductivity*, Cambridge, New York, 1952.

London, F., *Superfluids*, Vol. 1, Wiley, New York, 1950.

Serin, B., and J. Bardeen, articles in *Handbuch der Physik*, Vol. 15 (S. Flügge, ed.), Springer, Berlin, 1956.

Lynton, E. A., *Superconductivity*, Methuen, London, 1964.

REFERENCES

1. J. Bardeen, L. N. Cooper, and J. R. Schrieffer, *Phys. Rev.* **108**, 1175 (1957).

2. H. K. Onnes, *Leiden Comm.* 1206, 1226 (1911), *Suppl.* **34** (1913); quoted in D. Shoenberg, *Superconductivity*, Cambridge, New York, 1952, p. 1.

3. J. E. Neighbor, J. F. Cochran, and C. A. Shiffman, in *Low Temperature Physics LT9* J. G. Daunt *et al.*, eds. , Plenum Press, New York, 1965, p. 479.

4. D. J. Quinn III and W. B. Ittner III, *J. Appl. Phys.* **33**, 748 (1962).

5. W. Meissner and R. Ochsenfeld, *Naturwiss.* **21**, 787 (1933).
6. B. W. Roberts, in *Progress in Cryogenics*, Vol. 4 (K. Mendelssohn, ed.), Heywood, London, 1964, p. 159.
7. B. T. Matthias, *Progress in Low Temperature Physics*, Vol. II (C. J. Gorter, ed.), Wiley (Interscience), New York, 1955, p. 138.
8. B. T. Matthias, T. H. Geballe, and V. B. Compton, *Rev. Mod. Phys.* **35**, 1 (1963).
9. J. L. Olsen and H. Rohrer, *Helv. Phys. Acta* **30**, 49 (1957).
10. G. A. Alers and D. L. Waldorf, *IBM J. Res. Develop.* **6**, 89 (1962).
11. J. L. Olsen and H. Rohrer, *Helv. Phys. Acta* **33**, 872 (1960).
12. D. P. Seraphim and P. M. Marcus, *IBM J. Res. Develop.* **6**, 94 (1962).
13. R. W. Morse, *Progress in Cryogenics* (K. Mendelssohn, ed.), Heywood, London, 1959, p. 219.
14. M. K. Wilkinson, C. G. Shull, L. D. Roberts, and S. Bernstein, *Phys. Rev.* **97**, 889 (1955).
15. W. H. Wiedemann, P. Kienle, and F. Pobell, *The Mössbauer Effect*, Wiley, New York, 1962, p. 210.
16. W. H. Keesom and J. A. Kok, *Leiden Commun.* 221e (1932).
17. P. Ehrenfest, *Leiden Commun. Suppl. 75b* (1933).
18. L. D. Landau, *Phys. Z. Sowjet.* **11**, 545 (1937).
19. W. S. Corak, B. B. Goodman, C. B. Satterthwaite, and A. Wexler, *Phys. Rev.* **96**, 1442 (1954); W. S. Corak and C. B. Satterthwaite, *Phys. Rev.* **99**, 1660 (1954).
20. J. M. Ziman, *Electrons and Phonons*, Oxford, New York, 1960.
21. J. L. Olsen and H. M. Rosenberg, *Advan. Phys.* **2**, 28 (1953).
22. K. Mendelssohn, in *Progress in Low Temperature Physics*, Vol. 1 (C. J. Gorter, ed.), North-Holland, Amsterdam, 1955.
23. J. K. Hulm, *Proc. Roy. Soc. (London)* **A204**, 98 (1950).
24. K. Mendelssohn and J. L. Olsen, *Proc. Phys. Soc. (London)* **A63**, 2 (1950).
25. B. B. Goodman, *Proc. Phys. Soc. (London)* **A66**, 217 (1953).
26. D. E. Mapother, *IBM J. Res. Develop.* **6**, 77 (1962).
27. C. J. Gorter and H. B. G. Casimir, *Physica* **1**, 305 (1934).
28. A. G. Meshkovsky and A. I. Shalnikov, *Zh. Eksperim. i Teor. Fiz.* **17**, 851 (1947); *J. Phys. USSR* **11**, 1 (1947).
29. A. G. Meshkovsky, *Zh. Eksperim. i Teor. Fiz.* **19**, 1 (1949).
30. F. London, *Superfluids*, Vol. 1, Wiley, New York, 1950.
31. H. London, *Proc. Roy. Soc. (London)* **A152**, 650 (1935).
32. L. D. Landau, *Phys. Z. Sowjet.* **11**, 129 (1937); *J. Phys. USSR* **7**, 99 (1943).
33. D. Shoenberg, *Proc. Roy. Soc. (London)* **A175**, 49 (1940).
34. H. B. G. Casimir, *Physica* **7**, 887 (1940).
35. A. I. Shalnikov and Yu. V. Sharvin, *Zh. Eksperim. i Teor. Fiz.* **18**, 102 (1948).
36. A. B. Pippard, *Proc. Roy. Soc. (London)* **A191**, 385 (1947).
37. A. L. Schawlow, *Phys. Rev.* **109**, 1856 (1958).
38. J. M. Lock, *Proc. Roy. Soc. (London)* **A208**, 391 (1951).
39. M. S. Khaikin, *Soviet Phys. JETP* **6**, 735 (1958); *J. Exptl. Theoret. Phys. USSR* **34**, 1389 (1958).
40. A. B. Pippard, *Proc. Roy. Soc. (London)* **A203**, 98 (1950).
41. A. R. Frazer and D. Shoenberg, *Proc. Cambridge Phil. Soc.* **45**, 680 (1949).
42. A. L. Schawlow and G. E. Devlin, *Phys. Rev.* **113**, 120 (1959).
43. P. B. Miller, *Phys. Rev.* **118**, 928 (1960).
44. A. B. Pippard, *Proc. Roy. Soc. (London)* **A216**, 547 (1953).
45. A. B. Pippard, *Proc. Roy. Soc. (London)* **A203**, 210 (1950).

46. D. Shoenberg, *Superconductivity*, Cambridge, New York, 1952.
47. I. S. Khukhareva, *Zh. Eksperim. i Teor. Fiz.* **41**, 728 (1961); *Soviet Phys. JETP* **14**, 526 (1962).
48. A. I. Shalnikov, *Zh. Eksperim. i Teor. Fiz.* **10**, 630 (1940).
49. J. Bardeen, *Rev. Mod. Phys.* **34**, 667 (1962).
50. R. E. Glover III and H. T. Coffey, *Rev. Mod. Phys.* **36**, 299 (1964).
51. J. Bardeen, *Encyclopedia of Physics*, Vol. XV (S. Flügge, ed.), Springer, Berlin, 1956, p. 333.
52. K. Mendelssohn, *Proc. Roy. Soc. (London)* **A152**, 34 (1955).
53. C. P. Bean, *Rev. Mod. Phys.* **36**, 31 (1964).
54. J. Clarke, *Phil. Mag.* **8**, 115 (1966).
55. J. G. Daunt, T. C. Keeley, and K. Mendelssohn, *Phil. Mag.* **23**, 264 (1937).
56. K. G. Ramanathan, *Proc. Phys. Soc. (London)* **A65**, 532 (1952).
57. N. G. McCrum and C. A. Shiffman, *Proc. Phys. Soc. (London)* **A67**, 368 (1954).
58. H. London, *Proc. Roy. Soc. (London)* **A176**, 522 (1940).
59. M. A. Biondi and M. P. Garfunkel, *Phys. Rev. Letters* **2**, 143 (1959).
60. R. E. Glover III and M. Tinkham, *Phys. Rev.* **108**, 243 (1957).
61. P. L. Richards and M. Tinkham, *Phys. Rev.* **119**, 575 (1960).
62. T. E. Faber, *Proc. Roy. Soc. (London)* **A241**, 531 (1957).
63. V. L. Ginzburg and L. D. Landau, *Zh. Eksperim. i. Teor. Fiz.* **20**, 1064 (1950).
64. C. J. Gorter and H. B. G. Casimir, *Physik. Z.* **35**, 963 (1934); *Z. Tech. Phys.* **15**, 539 (1934).
65. F. London and H. London, *Proc. Roy. Soc. (London)* **A149**, 71 (1935).
66. F. London and H. London, *Physica* **2**, 341 (1935).
67. C. J. Gorter, *Rev. Mod. Phys.* **34**, 3 (1964).
68. J. G. Daunt, *Phys. Rev.* **72**, 89 (1947).
69. P. M. Marcus and E. Maxwell, *Phys. Rev.* **91**, 1035 (1953).
70. C. J. Gorter, *Progress in Low Temperature Physics*, Vol. I, (C. J. Gorter, ed.), North-Holland, Amsterdam, 1955, Chap. 1.
71. J. Bardeen and J. R. Schrieffer, *Progress in Low Temperature Physics*, Vol. III, (C. J. Gorter, ed.), North-Holland, Amsterdam, 1961, p. 263.
72. J. G. Daunt and K. Mendelssohn, *Proc. Roy. Soc. (London)* **A285**, 225 (1946).
73. H. Welker, *Z. Physik*, **114**, 525 (1929).
74. W. K. Heisenberg, *Z. Naturkunde* **2a**, 185 (1947).
75. H. Koppe, *Ergeb. Exakt. Naturwiss.* **23**, 283 (1950).
76. V. L. Ginzburg, *Fortschr. Physik* **1**, 101 (1953).
77. J. Bardeen, *Phys. Rev.* **97**, 1724 (1955).
78. J. Bardeen, *Encyclopedia of Physics*, Vol. XV (S. Flügge, ed.), Springer, Berlin, 1956, p. 283.
79. L. P. Gor'kov, *Zh. Eksperim. i Teor. Fiz.* **36**, 1918 (1959); *Soviet Phys. JETP* **9**, 1364 (1959).
80. R. Becker, G. Heller, and F. Sauter, *Z. Physik* **85**, 772 (1933).
81. J. H. Van Vleck, *The Theory of Electric and Magnetic Susceptibilities*, Oxford, New York, 1932.
82. J. Friedel, P. G. de Gennes, and J. Matricon, *Appl. Phys. Letters* **2**, 119 (1963). See also B. B. Goodman, *Rev. Mod. Phys.* **36**, 12 (1964).
83. M. Tinkham, in *Low Temperature Physics* (C. de Witt, B. Dreyfus, and P. G. de Gennes, eds.), Gordon & Breach, New York, 1962, p. 163.
84. G. E. H. Reuter and E. H. Sondheimer, *Proc. Roy. Soc. (London)* **A195**, 336 (1948).
85. M. Tinkham, *Rev. Mod. Phys.* **36**, 268 (1964).

86. J. M. Blatt, *Theory of Superconductivity*, Academic Press, New York, 1964.
87. B. B. Goodman, *Phys. Rev. Letters* **6**, 597 (1961).
88. L. V. Shubnikov, V. I. Khotkevich, Yu. D. Shepelev, and Yu. N. Riabinin, *Zh. Eksperim. i Teor. Fiz.* **7**, 221 (1937).
89. A. A. Abrikosov, *Zh. Eksperim. i Teor. Fiz.* **32**, 1442 (1957); *Soviet Phys. JETP* **5**, 1174 (1957).
90. P. G. de Gennes, *Superconductivity of Metals and Alloys*, Benjamin, New York, 1966.
91. E. Maxwell, *Phys. Rev.* **78**, 477 (1950); C. A. Reynolds, B. Serin, W. H. Wright, and L. B. Nesbitt, *Phys. Rev.* **78**, 487 (1950).
92. H. Fröhlich, *Phys. Rev.* **79**, 845 (1950).
93. J. Bardeen and J. R. Schrieffer, *Progress in Low Temperature Physics*, Vol. 3 (C. J. Gorter, ed.), North-Holland, Amsterdam, 1961.

2

THE THEORY OF BARDEEN, COOPER, AND SCHRIEFFER

G. Rickayzen

UNIVERSITY OF KENT AT CANTERBURY
CANTERBURY, ENGLAND

I. INTRODUCTION

The theory of superconductivity developed by Bardeen, Cooper, and Schrieffer (BCS) (*1*) was founded on a number of assumptions concerning the causes of superconductivity which, on the basis of theory and experiment, were generally agreed. Experimental evidence pointed to the fact that in the transition of a metal to the superconductive state the lattice and its properties were essentially unchanged, whereas some of the properties of the conduction electrons were changed radically. In the first instance, at least, it was reasonable to assume that the transition was caused by a change in the state of the electrons alone. Early attempts at theories suggested that the phenomenon could not arise according to the laws of classical physics, and that if the phenomenon was to be explained at all, the electrons would have to be treated by the laws of quantum mechanics. Since the theory of BCS, this has been confirmed by the direct observation of the quantization of the magnetic flux within a hollow superconductivity cylinder carrying a persistent current (*2,3*). These early attempts also suggested that a theory based upon an independent-particle model of the conduction electrons did not have within it the possibility of explaining the fundamental property of superconductors, which is their infinite conductivity. Thus the simplest model which seemed capable of explaining superconductivity was that of a gas of electrons interacting with each other through some two-particle interaction, and this was the model to which Bardeen, Cooper, and Schrieffer addressed themselves.

The question of what specific two-particle interaction is responsible for superconductivity has been considered over a long period, and for some superconductors is still a matter of controversy (*4*). When the Coulomb interaction between the electrons was little understood, it was thought that the long range of this force could bring about the long-range correlations that seemed necessary to explain superconductivity. However, the work of Bohm and Pines (*5*) showed that the long-range part of the Coulomb interaction is connected with the collective oscillations of the electrons, which, because of the high frequency of these oscillations, is hardly affected by the transition. It is now generally believed, although not conclusively proved, that the short-range repulsive Coulomb interaction alone will not give rise to superconductivity.

The difficulty in finding the right interaction is due to the small energy change between the normal and superconducting states. Measurements of the magnetic field required to destroy superconductivity near the absolute zero show that the energy difference between the two states is of the order of

$$N(0)\,(kT_c)^2$$

where $N(0)$ is the density of electronic states of one spin in the normal metal at the Fermi surface. This represents an energy of $(kT_c)^2/E_F$ per conduction electron or, more realistically, we can say that electrons within a band of energy of width kT_c around the Fermi surface change their energy by kT_c. In a typical superconductor this means that approximately 10^{-3} of the conduction electrons change their energy by approximately 10^{-3} eV. Since the energy per conduction electron due to the Coulomb interaction is of the order of 1 eV (*5*), it is clear that considerations of energy are not a good guide to the interaction. The important question is whether the interaction can lead to a qualitative change in the properties of the system, and this is difficult for one to see beforehand without solving the problem completely.

Despite this difficulty, Fröhlich (*6*) proposed the interaction which is now generally believed to give rise to superconductivity in most, if not all, known superconductors. This interaction between the electrons arises as a result of the interaction of the electrons with the possible vibrations of the lattice. Its significance in causing superconductivity is confirmed by the dependence of the critical temperature of superconductors on the isotopic mass of the lattice, a dependence which was established (*7,8*) independently and at the same time as Fröhlich put forward his proposal.

This interaction between the electrons which is mediated by the phonons can be understood from the following semiclassical picture. Consider the lattice to be a charged jelly and the electrons to be a charged gas which can move freely through the jelly. Any charge which is introduced into the medium will be screened by both the electrons and the lattice. If the foreign charge is static, it will be completely screened by the electrons and lattice and no field will be produced. If, on the other hand, the foreign charge oscillates with a frequency very much greater than the frequency of lattice vibrations (i.e., vibrations of the jelly), the lattice will be unable to follow the charge and will not contribute to the screening. The charge is screened, in this limit, by a positive dielectric constant $\epsilon_s(\mathbf{q})$, which is static (if the frequency of the motion of the charge is still much less than the plasma frequency), and which depends on the wave vector, $\mathbf{q}$, of the different spatial Fourier components of the foreign charge distribution. If the foreign charge oscillates with a frequency near a resonant frequency of the lattice, the vibrations of the lattice will become large and hence the total field produced will also be large. When the frequency of oscillation of the foreign charge passes through the resonant frequency of vibration of the lattice with the appropriate wavelength, the wavelength-dependent dielectric constant will therefore change sign and pass through zero. Hence the effect of lattice and electrons is to act as a dielectric whose dielectric constant, $\epsilon(\mathbf{q}, \omega)$, depends on the wave vector $\mathbf{q}$, and frequency ω, of the exciting charge distribution. For large ω, $\epsilon(\mathbf{q}, \omega)$ is positive. It changes sign as ω passes through the frequency $\omega_{\mathbf{q}}$ of the lattice, and it becomes infinite as ω tends to zero. At long wavelengths a reasonable approximation for $\epsilon(\mathbf{q}, \omega)$ is

$$\epsilon(\mathbf{q}, \omega) = \epsilon_s(\mathbf{q})(\omega^2 - \omega_{\mathbf{q}}^2)/\omega^2$$

Since the electrons themselves are charges in the medium, the interactions between them must be screened by this frequency-dependent dielectric constant. Hence the interaction between the electrons is dynamic and depends on their motion. This means that the methods of treating it have to be a little more subtle than those used for treating the statically screened interaction, and, in practice, the use of Green's functions (see Chapter 5) has proved fruitful. Fortunately, however, the theory is not very sensitive to the form of the interaction and can be developed formally by the replacement of the dynamic interaction by an appropriate static one. The latter can be justified by the following argument. The unscreened interaction between the electrons can be written

$$\frac{1}{2}\int d^3r\, d^3r'\, \rho(\mathbf{r})V(\mathbf{r}-\mathbf{r}')\rho(\mathbf{r}')$$

where $V(\mathbf{r}-\mathbf{r}')$ is the interaction between a charge at $\mathbf{r}$ and one at $\mathbf{r}'$, and $\rho(\mathbf{r})$ is the density of electrons at $\mathbf{r}$, that is,

$$\rho(\mathbf{r}) = \psi^*(\mathbf{r})\psi(\mathbf{r})$$

In the formalism of second quantization,

$$\psi(\mathbf{r}) = \sum_n c_n\phi_n(\mathbf{r})$$

where the ϕ_n are the basic Bloch states of the electrons. A typical term of ρ is $c_n^* c_m$, which describes the scattering of an electron from the Bloch state m to the Bloch state n. Now in the Heisenberg representation with the Hamiltonian which describes the Bloch electrons, the operator $c_n^* c_m$ oscillates with frequency $(\epsilon_m - \epsilon_n)/\hbar$. Evidently this part of the electron density should be screened by the dielectric constant at this frequency.

For a uniform superconductor the appropriate Bloch states to use are plane waves; the unscreened interaction is then

$$\frac{1}{2}\sum_{\substack{\mathbf{k},\mathbf{k}',\mathbf{q}\\ \sigma,\sigma'}} V(\mathbf{k}-\mathbf{k}')c^*_{\mathbf{k}'\sigma}c^*_{\mathbf{k}+\mathbf{q}\sigma'}c_{\mathbf{k}'+\mathbf{q}\sigma'}c_{\mathbf{k}\sigma},$$

and the screened one is

$$\frac{1}{2}\sum_{\substack{\mathbf{k},\mathbf{k}',\mathbf{q}\\ \sigma,\sigma'}} V(\mathbf{k}-\mathbf{k}')c^*_{\mathbf{k}'\sigma}c^*_{\mathbf{k}+\mathbf{q}\sigma'}c_{\mathbf{k}'+\mathbf{q}\sigma'}c_{\mathbf{k}\sigma}/\epsilon(\mathbf{k}-\mathbf{k}', \overline{\epsilon_{\mathbf{k}} - \epsilon_{\mathbf{k}'}}/\hbar) \tag{1}$$

This is often written as the sum

$$V_s + V_{\text{ph}}$$

where V_s is the Coulomb interaction screened only by the electrons and V_{ph} is referred to as the interaction mediated by the phonons. The latter is normally written in the form‡

$$\frac{1}{2}\sum_{\substack{\mathbf{k},\mathbf{k}',\mathbf{q}\\ \sigma,\sigma'}}\left[\frac{|g_{\mathbf{k}-\mathbf{k}'}|^2\hbar\,\omega_{\mathbf{k}-\mathbf{k}'}}{(\epsilon_{\mathbf{k}}-\epsilon_{\mathbf{k}'})^2-(\hbar\omega_{\mathbf{k}-\mathbf{k}'})^2}\right]c^*_{\mathbf{k}'\sigma}c^*_{\mathbf{k}+\mathbf{q}\sigma'}c_{\mathbf{k}'+\mathbf{q}\sigma'}c_{\mathbf{k}\sigma} \tag{2}$$

This interaction is attractive for small $\epsilon_{\mathbf{k}}-\epsilon_{\mathbf{k}'}$ and, as we have said, generally believed to be responsible for the occurrence of superconductivity at least in a large class of superconductors. If the interaction is calculated with a more realistic treatment of the solid, the contributions from V_s and V_{ph} do not generally cancel when $\epsilon_{\mathbf{k}}-\epsilon_{\mathbf{k}'}$ is zero.

To complete the Hamiltonian which describes the system we need to add the part which describes the independent Bloch electrons. This is

$$\mathscr{H}_0=\sum_{\mathbf{k},\sigma}\epsilon_{\mathbf{k}}c^*_{\mathbf{k}\sigma}c_{\mathbf{k}\sigma} \tag{3}$$

and the total Hamiltonian is

$$\mathscr{H}=\mathscr{H}_0+V_s+V_{ph} \tag{4}$$

Assuming that we have included the right mechanism for superconductivity, we are left with the mathematical problem of finding the eigenstates and eigenvalues of this Hamiltonian.

Although this is a mathematical problem, it is doubtful whether it could be solved without a guess based on physical intuition. We have already said that the energy of the electrons due to V_s is much greater than the condensation energy. This is still true of the small V_{ph}. In fact, the expectation value of V_{ph} in the ground state of $\mathscr{H}_0$ is of the order of

$$g^2N(0)\,(\hbar\omega_D)^2$$

where ω_D is the Debye frequency. For the common superconductors this is several orders of magnitude greater than the condensation energy. Hence the energy involved in the transition is a small part of the energy of V_{ph}. What then is this small part of the interaction which is responsible for superconductivity?

II. A NORMAL FERMI LIQUID

Before we look for this small part of the interaction, we ought to consider what role the larger part of the interaction plays in a metal. In fact, one of the

‡ This result was first obtained by Fröhlich (*9*).

surprising aspects of metals is that despite the strong interaction between the electrons in the metal, the properties of the metal in the normal state are reasonably well described by the independent-particle model. The reason for this was traced by Landau (*10*) to the fact that electrons are fermions, and to the high degeneracy temperature of the electron gas. The first of these means that the single-particle excitations of the system obey Fermi–Dirac statistics, and the second means that the number of excitations excited at normal temperatures is very small. Unless there is an essential singularity in the free energy of the system as a function of the strength of the interaction between the particles (and in metals above the superconducting transition temperature there appears to be none) the low-lying excited states of the system can be described quite simply.‡

In a uniform system the states have definite linear momentum **p**. The simplest states have properties similar to particles of momentum **p**, although they are not identical with particles; they are referred to as quasi-particles. Because of the interactions, a quasi-particle can decay into a number of other quasi-particles. Hence a state with one quasi-particle, unless it has zero excitation energy, is not an exact eigenstate. Nevertheless, provided the lifetime of the state is much greater than $\hbar/\epsilon$, where ϵ is the excitation energy, it is for many purposes sufficient to treat this state as an eigenstate. The basic bricks of Landau's theory of a normal Fermi liquid are these quasi-particles.

It is possible to have more than one quasi-particle present in a given state of the system, and the energy of this state depends only on the numbers of different quasi-particles present. The energy required to add or subtract a particular quasi-particle from this state is called the energy of the quasi-particle, and is, in this theory, a function of the numbers of other quasi-particles present. This is the essential new feature which appears in the theory of a Fermi liquid but which does not appear in an independent-particle model.

It is instructive to look at Landau's theory in the Hartree–Fock approximation, the simplest mathematical approximation for treating an interacting system of particles which brings out Landau's ideas. We shall use this also to introduce techniques that are used in the theory of superconductivity. For simplicity we consider a system of fermions interacting through a direct two-particle interaction. Such a system is described by the Hamiltonian

$$\mathscr{H} = \sum_{\mathbf{k},\sigma} \epsilon_{\mathbf{k}} c^*_{\mathbf{k}\sigma} c_{\mathbf{k}\sigma} + \frac{1}{2} \sum_{\substack{\mathbf{k},\mathbf{k}',\mathbf{q}\\ \sigma,\sigma'}} V(\mathbf{k}-\mathbf{k}')\, c^*_{\mathbf{k}'\sigma} c^*_{\mathbf{k}+\mathbf{q}\sigma'} c_{\mathbf{k}'+\mathbf{q}\sigma'} c_{\mathbf{k}}:$$

If V were zero, the eigenstates of the system would be eigenstates of the operators $c^*_{\mathbf{k}\sigma} c^*_{\mathbf{k}\sigma}$, which would, therefore, have definite values in these states. When V is

‡ For a much fuller description see (*11*).

not zero, this is not true, but provided the interaction is weak the quantum fluctuations of the operators $c^*_{\mathbf{k}\sigma}c_{\mathbf{k}\sigma}$ about their mean values $n_{\mathbf{k}\sigma}$ will be small, and the terms of the interaction which do not depend on the operators $c^*_{\mathbf{k}\sigma}c_{\mathbf{k}\sigma}$ will be negligible. Hence the Hamiltonian is approximately given by

$$\mathscr{H} = \sum_{\mathbf{k},\sigma} \epsilon_{\mathbf{k}} c^*_{\mathbf{k}\sigma}c_{\mathbf{k}\sigma} - \tfrac{1}{2} \sum_{\mathbf{k},\mathbf{k}',\sigma} V(\mathbf{k}-\mathbf{k}')\, c^*_{\mathbf{k}\sigma}c_{\mathbf{k}\sigma}c^*_{\mathbf{k}'\sigma}c_{\mathbf{k}'\sigma} + \tfrac{1}{2} \sum_{\mathbf{k},\mathbf{k}',\sigma,\sigma'} V(0)\, c^*_{\mathbf{k}\sigma}c_{\mathbf{k}\sigma}c^*_{\mathbf{k}'\sigma'}c_{\mathbf{k}'\sigma'}$$

The Hamiltonian is now diagonal in $c^*_{\mathbf{k}\sigma}c_{\mathbf{k}\sigma}$ and its eigenstates are evidently states in which the different Bloch states $\mathbf{k}, \sigma$ are either occupied or unoccupied. The total energy in one of these states is

$$E = \sum_{\mathbf{k},\sigma} \epsilon_{\mathbf{k}} n_{\mathbf{k}\sigma} - \tfrac{1}{2} \sum_{\mathbf{k},\mathbf{k}',\sigma} V(\mathbf{k}-\mathbf{k}')\, n_{\mathbf{k}\sigma}n_{\mathbf{k}'\sigma} + \tfrac{1}{2} \sum_{\mathbf{k},\mathbf{k}',\sigma,\sigma'} V(0)\, n_{\mathbf{k}\sigma}n_{\mathbf{k}'\sigma'}$$

and depends only on the occupation of the different Bloch states. The energy required to add an electron to the state $\mathbf{k}, \sigma$ (assuming that it is unoccupied) is

$$\bar{\epsilon}_{\mathbf{k}\sigma} = \epsilon_{\mathbf{k}} - \sum_{\mathbf{k}'} V(\mathbf{k}-\mathbf{k}')\, n_{\mathbf{k}'\sigma} + \sum_{\mathbf{k}'\sigma'} V(0)\, n_{\mathbf{k}'\sigma'} \tag{5}$$

The corrections to $\epsilon_{\mathbf{k}}$ arising from the interaction are referred to as the self-energy; the first is the exchange self-energy and the second the direct (or Hartree) self-energy. The results show that the energy of the quasi-particle depends on the numbers and types of other quasi-particles present.

In the Hartree–Fock approximation the quasi-particles represent good eigenstates of the system. When the terms neglected in the Hamiltonian are restored this is no longer true. The quasi-particles then have finite lifetimes which decrease with increasing excitation energy, but the quasi-particles near the Fermi surface are still good approximations to eigenstates.

Quantitatively, there are differences between the particles and quasi-particles simply because the theory of the Fermi liquid involves more parameters. Qualitatively, however, the quasi-particles have very similar properties to the original particles out of which they have been built. They have the same charge, their spin susceptibility and specific heat have the same dependence on temperature, and their transport properties are similar.

If we are trying to understand the important qualitative differences between superconductors and normal metals, it is unnecessary to include all the complications of the Fermi liquid. We can suppose in the first instance that, apart from the specific interaction which gives rise to superconductivity, the system comprises independent quasi-particles with an infinite lifetime. This was a basic assumption made by Bardeen, Cooper, and Schrieffer, and one which has turned out to be most successful. This assumption may now seem obvious, but to see it in its context one must remember that the early attempts at theories of superconductivity were based on the self-energies of quasi-particles which were either singular or led to anomalous distributions of quasi-particles in momentum space (*6*, *12*).

Many of the recent developments of the theory have been concerned with including the Fermi liquid effects in a unified treatment (see Chapters 7 and 10), the treatment most in sympathy with that of Landau being that given by Leggett (*13*). What these developments show is that, at least as far as the equilibrium properties are concerned, the assumption of BCS is very good quantitatively, and this is confirmed by experiment. As far as nonequilibrium properties are concerned, the assumption is not so good quantitatively, but it cannot be said that deviations between experiment and the theory of BCS are so great as to provide values for the unknown parameters of the Fermi liquid.

Before leaving the subject of Fermi liquids we should mention that the quasi-particles are not the only excitations of the system. There are also collective excitations of the whole system which in a charged liquid are plasma oscillations (*5*) and which in a pure neutral liquid are called zero sound (*14*). These also are ignored as a first step in the theory of BCS, although they are important for a full understanding of the validity of the theory (see Chapter 7).

III. COOPER'S PROBLEM

What then is the small part of the interaction that is responsible for superconductivity? An important clue to this was found by Cooper (*15*), who showed that with an attractive interaction between the particles, the Fermi sea is unstable to the formation of a certain kind of quasi-bound pair. Suppose we add two particles to the system and, ignoring the effect of these particles on the particles already present, consider in what state these two particles might find themselves. Since they will be interacting with each other we have a problem to solve similar to the problem of two particles interacting in free space, the main difference being that because of the exclusion principle they cannot be found in the quasi-particle states already occupied.

Subject to these limitations, the most general state in which these particles can be found is a linear superposition of pairs of quasi-particle states. In an isotropic system the total momentum of each of these pairs of states will be the same, and in the ground state we should expect this common momentum to be zero. Hence, if the total spin is also zero, we expect the ground state of the particles to be

$$|0\rangle = \sum_{\mathbf{k},\, k>k_F} \alpha_{\mathbf{k}} c^*_{\mathbf{k}\uparrow} c^*_{-\mathbf{k}\downarrow} |F\rangle \tag{6}$$

where $|F\rangle$ is the Fermi sea of particles and the coefficients $\alpha_{\mathbf{k}}$ are numbers yet to be determined. Figure 1 shows the pairs of Bloch states appearing in Cooper's state.

If there were no interaction between the particles, the $\alpha_{\mathbf{k}}$ would all be zero except for one pair of states of zero excitation energy, that is, one pair of states on the Fermi surface. The only part of the interaction which can alter this result is the part which describes the scattering of the particles from one pair of states $(\mathbf{k}\uparrow, -\mathbf{k}\downarrow)$ to another pair $(\mathbf{k}'\uparrow, -\mathbf{k}'\downarrow)$. If we retain only this part of the interaction our Hamiltonian is

$$\mathscr{H} = \sum_{\mathbf{k},\sigma} \epsilon_{\mathbf{k}} c^*_{\mathbf{k}\sigma} c_{\mathbf{k}\sigma} + \sum_{\mathbf{k},\mathbf{k}'} V_{\mathbf{k},\mathbf{k}'} c^*_{\mathbf{k}\uparrow} c^*_{-\mathbf{k}\downarrow} c_{-\mathbf{k}'\downarrow} c_{\mathbf{k}'\uparrow} \tag{7}$$

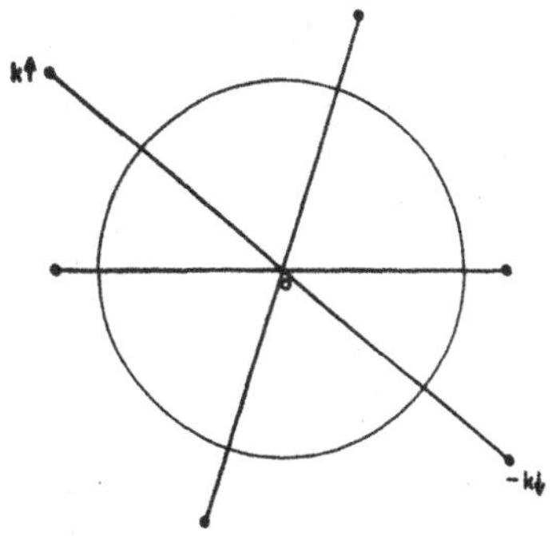

Fig. 1. Filled Fermi sea in momentum space, and some of the pairs of Bloch states superposed in Cooper's state.

This form of the Hamiltonian automatically ignores the self-energy, which, following BCS, we have decided to neglect.

The expectation value of the energy in the state $|0\rangle$ is

$$E = 2 \sum_{\mathbf{k},\, k > k_F} \epsilon_{\mathbf{k}} |\alpha_{\mathbf{k}}|^2 + \sum_{\substack{\mathbf{k},\mathbf{k}' \\ k,k' > k_F}} V_{\mathbf{k},\mathbf{k}'} \alpha^*_{\mathbf{k}} \alpha_{\mathbf{k}'} \tag{8}$$

If the state $|0\rangle$ is normalized we have

$$\sum_{\mathbf{k},\, k > k_F} |\alpha_{\mathbf{k}}|^2 = 1$$

Hence the energy is a minimum subject to this restriction when

$$(2\epsilon_{\mathbf{k}} - E)\alpha_{\mathbf{k}} + \sum_{k' > k_F} V_{\mathbf{k},\mathbf{k}'} \alpha_{\mathbf{k}'} = 0 \tag{9}$$

If Eq. (9) has a solution with E less than all the excitation energies $\epsilon_{\mathbf{k}}$, then the pair of particles will prefer to be in this coherent state rather than in independent quasi-particle states on the Fermi surface. This will mean it is favorable for electrons in the Fermi sea to come out to form such coherent states and we should expect the Fermi sea to be unstable. Hence we look to see whether such a solution exists. If we measure the excitation energies from the Fermi surface we have to seek a solution with E negative.

For a general potential $V_{\mathbf{k},\mathbf{k}'}$ we still have a complicated integral equation to solve. Fortunately with the interaction of Eq. (2), when that interaction is weak, the solution can be obtained quite simply and accurately. For, when the interaction is weak, E is small and $\alpha_{\mathbf{k}}$ is significantly different from zero only when $\epsilon_{\mathbf{k}}$ is of the order of E. The interaction between electrons with such excitation energies is essentially

$$V_{\mathbf{k},\mathbf{k}'} = -|g^2_{\mathbf{k}-\mathbf{k}'}|/\hbar\omega_{\mathbf{k}-\mathbf{k}'}$$

which is approximately independent of $\mathbf{k}$, $\mathbf{k}'$ (*16*). Hence the interaction can be taken to be a constant ($-V$), and Eq. (9) rewritten

$$(2\epsilon_{\mathbf{k}} - E)\alpha_{\mathbf{k}} - V \sum_{\mathbf{k}, k>k_F} \alpha_{\mathbf{k}} = 0$$

This has the solution

$$\alpha_{\mathbf{k}} = a/(2\epsilon_{\mathbf{k}} - E) \tag{10}$$

where a is a constant. Substituting this back into the equation, we find

$$1 = V \sum_{\mathbf{k}, k>k_F} \frac{1}{2\epsilon_{\mathbf{k}} - E} = N(0)\, V \int_0 \frac{d\epsilon}{2\epsilon - E} \tag{11}$$

where $N(0)\, d\epsilon$ is the number of Bloch states of one spin within an energy interval $d\epsilon$ near the Fermi surface.

The integral diverges logarithmically rather as a result of our approximation for the interaction than as a result of the properties of the equation. This can be seen if the form [Eq. (10)[for $\alpha_{\mathbf{k}}$ is substituted in the integral in Eq. (9). The properties of the interaction for large energy ensure the convergence of this integral.

In fact, the interaction changes sign and decreases when the energy exceeds $\hbar\omega_{\mathbf{k}-\mathbf{k}'}$. This effect must be included in the approximation, and can be taken care of by making the interaction zero when the energy is greater than $\hbar\omega$, where ω is some average phonon frequency. This method gives the correct dependence of E on the scale of the phonon frequencies. For example, if all the phonon frequencies are doubled, E will be doubled according to both the exact equation and the approximate one.

The integral of Eq. (11) will have as its upper limit $\hbar\omega$. Hence the equation can be rewritten

$$1 = \tfrac{1}{2}N(0)\, V \ln[(2\hbar\omega - E)/-E] \approx \tfrac{1}{2}N(0)\, V \ln(2\hbar\omega/-E)$$

Hence

$$E = -2\hbar\omega \exp[-2/N(0)V]$$

This solution can be obtained from the original equation in the limit of a weak interaction by somewhat grander mathematical methods (*17*). One then has specific forms for the constants $\hbar\omega$ and V in terms of the original interaction. However, since we shall not require these specific forms, this treatment is not justified here. We note that the wave function is

$$|0\rangle = a \sum_{\mathbf{k},\, k>k_F} (2\epsilon_{\mathbf{k}} - E)^{-1} c^*_{\mathbf{k}\uparrow} c^*_{-\mathbf{k}\downarrow} |F\rangle$$

and that if the band is isotropic the paired particles are in a relative *s*-state.

The fact that we have been able to find a solution of Eq. (9) shows that pairs of fermions added to the Fermi sea would like to form a state which is bound relative to the Fermi sea. This suggests that the Fermi sea is unstable, although it cannot be said to show conclusively that the system would prefer a qualitatively different state. For example, given an attractive single-particle potential one can show in the same way that an additional particle would prefer to go into a state bound relative to the Fermi sea. Nevertheless the metal containing a large number of impurities does not have a qualitatively different energy spectrum from the pure metal.

Before passing on to the question of what state replaces the filled Fermi sea when it is unstable, we should like to comment on Cooper's results. First, the result depends on the fact that the interaction is attractive. If the interaction were repulsive, V would be negative and Eq. (11) would have no solution with E negative. For an interaction which is partly attractive and partly repulsive the integral equation has to be solved with more care.

Second, we have considered only the possibility of the particles being in a singlet spin state. One can also consider the possibility of the formation of a triplet spin state. This can occur only if the interaction depends on angle (*16*). For example, if $V_{\mathbf{k},\mathbf{k}'}$ is a function of the angle between $\mathbf{k}$ and $\mathbf{k}'$, only the odd spherical harmonics in the expansion of V will contribute to the formation of such a state. To obtain a given binding energy $|E|$, the strength of the interaction in the odd angular momentum states must be greater than in the *s*-state, and normally one would not expect this to be realized in practice. Nevertheless, the superconductive properties in such a case have been considered in some detail in a number of papers (*18–23*).

Last, we considered mainly the interaction in Eq. (2). It is not certain that this is always the form of the interaction; other possibilities should be mentioned. In other cases, even if the spin state is the singlet, it is possible for the largest $|E|$ to come from a pairing other than *s*-state pairing. In the case of liquid He^3 there is reason to believe (*24*) that the most likely state for the electrons to be in is a *d*-state, and this is the case which has been most extensively studied for that problem. Further discussion of this is beyond the scope of this chapter.

IV. REDUCED HAMILTONIAN

We have seen in the previous section that with an appropriate interaction between the particles, the filled Fermi sea of fermions is unstable because of the formation of bound pairs (Cooper pairs) of particles. It is clear that a few fermions can peel off from the Fermi sea to form numbers of pairs, and the holes left behind can form similar pairs. However, when the number of particles that peels off from the Fermi sea becomes macroscopic, the pairs interfere with each other, not being able to occupy the same Bloch states simultaneously. In this case, the interesting one, the pairs cannot be treated independently and we have a real many-body problem to deal with.

If Cooper's problem really does point the way to a theory of superconductivity and if the possible pairing of electrons is paramount, we can go somewhat further before settling down to the mathematics of interdependent pairs. For the only part of the interaction responsible for pairing is that shown in Eq. (7), and this should be the small part of the interaction responsible for the superconductive state. The remainder of the interaction we can hope to treat by perturbation theory. Hence, if we wish to understand, say, the difference of the free energies of the normal and superconductive states, we can hope to do this by studying the Hamiltonian [Eq. (7)], which then plays a central role in the theory of superconductivity. BCS called this the reduced Hamiltonian and it has often since been called the pairing Hamiltonian. We shall denote it by $\mathscr{H}_p$, and write

$$\mathscr{H}_p = \sum_{\mathbf{k},\sigma} \epsilon_{\mathbf{k}} c^*_{\mathbf{k}\sigma} c_{\mathbf{k}\sigma} + \sum_{\mathbf{k},\mathbf{k}'} V_{\mathbf{k},\mathbf{k}'} c^*_{\mathbf{k}\uparrow} c^*_{-\mathbf{k}\downarrow} c_{-\mathbf{k}'\downarrow} c_{\mathbf{k}'\uparrow} \tag{7}$$

The mathematical problem is to find the eigenvalues and eigenstates of this Hamiltonian.

Since the interaction $V_{\mathbf{k},\mathbf{k}'}$ is inversely proportional to the volume, for the interaction to give a significant contribution, the expectation value of each set of operators $c^*_{\mathbf{k}\uparrow} c^*_{-\mathbf{k}\downarrow} c_{-\mathbf{k}'\downarrow} c_{\mathbf{k}'\uparrow}$ must be of order unity. If, however, these expectation values are calculated treating the whole of the interaction by perturbation theory, we find that they are of order N^{-1}, where N is the total number of particles. Hence if the Hamiltonian [Eq. (7)] has a solution which is different from the unperturbed one, it cannot be obtained by perturbation theory. This remark is presaged by the solution to Cooper's problem, where the energy E cannot be expanded in a power series in V.

The Hamiltonian [Eq. (7)] is, in fact, simple enough to be solved to order N^{-1}, which, in this problem, means that exact solutions have been obtained. Such solutions have been given both for the ground state of the system (*25,26*) and for the free energy (*27*). Since the rigorous methods of treatment, while giving confidence in the result, do not throw great physical light on the solution and are not easily generalized, we choose one of the less rigorous methods, which, nevertheless, provides the exact answer.

V. SELF-CONSISTENT SOLUTION

Consider first the ground state of the system. Since the interaction is one which creates and destroys electrons in pairs of Bloch states $(\mathbf{k}\uparrow, -\mathbf{k}\downarrow)$, the ground state will be some coherent superposition of many-body states in which the Bloch states are occupied or unoccupied in pairs. Because of the large number of particles involved, the amplitude for the occupation of a particular pair of Bloch states will not depend significantly on whether a few other states are occupied or not. Hence this amplitude is essentially the same in the ground state and low-lying excited states (provided the particular pair of states is not directly involved). This means that in the interaction the operator $c_{-\mathbf{k}\downarrow}c_{\mathbf{k}\uparrow}$ is practically a c-number, $b_{\mathbf{k}}^0$, with small quantum fluctuations about this value. Terms in the interaction which depend bilinearly on the small difference $(c_{-\mathbf{k}\downarrow}c_{\mathbf{k}\uparrow} - b_{\mathbf{k}}^0)$ should, therefore, be negligible; when they are neglected the remainder is a quadratic form in the operators c. Formally, this quadratic form can be obtained by writing all the pairs of operators $c_{-\mathbf{k}\downarrow}c_{\mathbf{k}\uparrow}$ in the form

$$c_{-\mathbf{k}\downarrow}c_{\mathbf{k}\uparrow} = b_{\mathbf{k}}^0 + (c_{-\mathbf{k}\downarrow}c_{\mathbf{k}\uparrow} - b_{\mathbf{k}}^0)$$

and neglecting the terms bilinear in the parenthetical quantities. This procedure leads to the approximate Hamiltonian which we have elsewhere [Eq. (16)] called the model Hamiltonian and which we denote by $\mathscr{H}_M$. We have then that

$$\mathscr{H}_M = \sum_{\mathbf{k},\sigma} \epsilon_{\mathbf{k}} c_{\mathbf{k}\sigma}^* c_{\mathbf{k}\sigma} + \sum_{\mathbf{k},\mathbf{k}'} V_{\mathbf{k},\mathbf{k}'} [c_{\mathbf{k}\uparrow}^* c_{-\mathbf{k}\downarrow}^* b_{\mathbf{k}'}^0 + b_{\mathbf{k}}^{0*} c_{-\mathbf{k}'\downarrow} c_{\mathbf{k}'\uparrow} - b_{\mathbf{k}}^{0*} b_{\mathbf{k}'}^0] \tag{12}$$

This Hamiltonian depends on the numbers $b_{\mathbf{k}}^0$, which are so far undetermined. For consistency these are determined at the end of the problem from the condition that the average values of the operators $c_{-\mathbf{k}\downarrow}c_{\mathbf{k}\uparrow}$ in the ground state are the numbers $b_{\mathbf{k}}^0$. Thus we have a self-consistent procedure for solving the many-body problem described by the pairing Hamiltonian. We assume certain values for the average values of the operators $c_{-\mathbf{k}\downarrow}c_{\mathbf{k}\uparrow}$, and these provide us with the model Hamiltonian for the ground state and low-lying excited states. When the eigenstates of this Hamiltonian are found, one can then check that the values which we originally assumed for $b_{\mathbf{k}}^0$ were correct. In practice, the Hamiltonian is simple enough to be solved analytically for all values of $b_{\mathbf{k}}^0$, and the condition for self-consistency can be given analytically in the form of an integral equation. The original problem is then reduced to the solution of this integral equation.

For consistency, too, one should check that the neglected terms really are negligible. Once the solution of the model Hamiltonian is found this check can be carried out in a straightforward way.‡

The procedure we have just outlined for obtaining the ground state and the low-lying excited states can be employed for finding the thermodynamic properties of the system. At a finite temperature below the critical temperature the operators $c_{-\mathbf{k}\downarrow}c_{\mathbf{k}\uparrow}$ will have a finite average value which, in this case, will be the statistical and quantum-mechanical average. There will be statistical and quantum mechanical fluctuations about the average, but these fluctuations will have a negligible effect upon the interaction energy. Hence the operators are practically equal to their expectation values and the system can again be described by the model Hamiltonian. For self-consistency the constants $b_{\mathbf{k}}^0$ must be equal to the thermal and quantum-mechanical averages of the operators $c_{-\mathbf{k}\downarrow}c_{\mathbf{k}\uparrow}$. Hence these constants depend on temperature and the model Hamiltonian is different for each temperature. This means that at different temperatures the superconductive system is in different macroscopic states. The model Hamiltonian at a given temperature describes the microscopic quantum-mechanical fluctuations of the system about some average state at that temperature.

Before solving the problem of the model Hamiltonian in detail we should like to comment briefly on the step of replacing operators by *c*-numbers. This may appear new but is, in fact, an essentially old idea in physics. For example, according to the quantum theory, the electric field is an operator which can create or destroy photons. If, however, the system is in a state with a large number of photons present, the electric field operator does not differ very much from its expectation value. In this situation the expectation value becomes a classical variable which satisfies Maxwell's equations. One might expect then that there corresponds a classical field corresponding to the operators $c_{-\mathbf{k}\downarrow}c_{\mathbf{k}\uparrow}$. This appears to be true, the appropriate field being $\langle\psi_\downarrow(\mathbf{r})\psi_\uparrow(\mathbf{r})\rangle$, where $\psi_\sigma(\mathbf{r})$ is the wave operator for electrons. In a uniform system this quantity is

$$\frac{1}{\mathscr{V}}\sum_{\mathbf{k}}\langle c_{-\mathbf{k}\downarrow}c_{\mathbf{k}\uparrow}\rangle = \frac{1}{\mathscr{V}}\sum_{\mathbf{k}} b_{\mathbf{k}}^0$$

‡ The self-consistent way of treating the theory was first introduced by Anderson (*28*) by analogy with the problem of ferromagnetism. It was taken much further by Gor'kov (*29*) and Nambu (*30*). The mathematical basis of this procedure has been discussed by Coleman and Pruski (*31*).

and its value is unaffected by microscopic fluctuations in the system. This idea of a classical or macroscopic variable being associated with different superconductive states is now recognized as fundamental in the theory and is developed in other chapters of this treatise.

The electric field is similar to the pairing operators in another way. For the electric field has zero matrix elements between states with definite numbers of photons present, and when the number of photons is measured exactly the field cannot be measured. Hence, if we ascribe a classical value to the electric field, we must be considering states with indefinite numbers of photons. As long as the number of photons present is large, this indefiniteness is unimportant. In the same way, if we ascribe classical values to the operators $c_{-\mathbf{k}\downarrow}c_{\mathbf{k}\uparrow}$ we must be considering states with indefinite numbers of fermions present. So it comes about that the model Hamiltonian does not commute with the total number of fermions, although the pairing Hamiltonian does. This again is not important as long as we are dealing with a large number of fermions. The analogy between the two kinds of field can be developed further.‡ For example, the phase of the field is measured when the corresponding particles are created or destroyed. In the case of superconductivity this occurs when particles are added to or subtracted from a superconductor as in Josephson tunneling (see Chapter 9).

Because the model Hamiltonian does not commute with the total number of particles it is not possible to work in a subspace with a definite total number of particles. In considering the ensemble averages of the operators we have then to use a grand canonical ensemble with a definite chemical potential μ. Formally this is easily included by measuring the single-particle energies relative to μ. For the free electron model of the metal this means that

$$\epsilon_{\mathbf{k}} = (\hbar^2 k^2/2m) - \mu$$

The chemical potential has to be determined afterward from the condition that the average number of particles present is a given number N,

$$\langle \sum_{\mathbf{k}\sigma} c_{\mathbf{k}\sigma}^* c_{\mathbf{k}\sigma} \rangle = N$$

Alternatively, we can regard μ as given and then this equation determines the average number of particles present. This is the view we usually adopt here.

After this digression we have now come to the point where it is necessary to find the eigenvalues and eigenstates of the model Hamiltonian. We can make the Hamiltonian look simpler by writing

$$\Delta_{\mathbf{k}} = -\sum_{\mathbf{k}'} V_{\mathbf{k},\mathbf{k}'} b_{\mathbf{k}'}^0 \tag{13}$$

‡ This analogy is developed further for the macroscopic wave function of superfluid helium by Martin (*32*).

where, for reasons that will become obvious as we continue, $\Delta_{\mathbf{k}}$ is variously known as the order parameter or gap parameter. Then

$$\mathscr{H}_M = \sum_{\mathbf{k},\sigma} \epsilon_{\mathbf{k}} c^*_{\mathbf{k}\sigma} c_{\mathbf{k}\sigma} - \sum_{\mathbf{k}} (\Delta_{\mathbf{k}} c^*_{\mathbf{k}\uparrow} c^*_{-\mathbf{k}\downarrow} + \Delta^*_{\mathbf{k}} c_{-\mathbf{k}\downarrow} c_{\mathbf{k}\uparrow} - \Delta^*_{\mathbf{k}} b^0_{\mathbf{k}}) \tag{14}$$

Since this Hamiltonian is bilinear in the creation and annihilation operators it can be diagonalized by means of a linear canonical transformation of these operators. Further, since the Bloch state $\mathbf{k}\uparrow$ is linked only to the Bloch state $-\mathbf{k}\downarrow$, one can work in the subspace of just these two states. Then the appropriate transformation‡ is of the form

$$\begin{aligned} c_{\mathbf{k}\uparrow} &= u_{\mathbf{k}} \gamma_{\mathbf{k}0} + v^*_{\mathbf{k}} \gamma^*_{\mathbf{k}1} \\ c^*_{-\mathbf{k}\downarrow} &= - v_{\mathbf{k}} \gamma_{\mathbf{k}0} + u^*_{\mathbf{k}} \gamma^*_{\mathbf{k}1} \end{aligned} \tag{15}$$

where $\gamma_{\mathbf{k}0}$, $\gamma_{\mathbf{k}1}$ are Fermi–Dirac annihilation operators, and where the coefficients $u_{\mathbf{k}}$, $v_{\mathbf{k}}$, are chosen to make the Hamiltonian diagonal; that is, they are chosen to make the coefficients of $\gamma^*_{\mathbf{k}0}\gamma^*_{\mathbf{k}1}$ and $\gamma_{\mathbf{k}1}\gamma_{\mathbf{k}0}$ in the Hamiltonian vanish. This latter condition requires that

$$2\epsilon_{\mathbf{k}} u_{\mathbf{k}} v_{\mathbf{k}} + \Delta_{\mathbf{k}} v^2_{\mathbf{k}} - \Delta^*_{\mathbf{k}} u^2_{\mathbf{k}} = 0 \tag{16}$$

Since the Eqs. (15) represent a canonical transformation, the commutation relations between the γ's being the same as those between the c's, we have the further constraints on the constants $u_{\mathbf{k}}$, $v_{\mathbf{k}}$ that

$$|u_{\mathbf{k}}|^2 + |v_{\mathbf{k}}|^2 = 1 \tag{17}$$

Equations (16) and (17) are sufficient to determine $u_{\mathbf{k}}$ and $v_{\mathbf{k}}$ in terms of Δ. From the imaginary part of Eq. (16) we find that $\Delta v/u$ is real, and hence Eq. (16) can be written

$$2\epsilon_{\mathbf{k}} |u_{\mathbf{k}} v_{\mathbf{k}}| + |\Delta_{\mathbf{k}}| (|u_{\mathbf{k}}|^2 - |v_{\mathbf{k}}|^2) = 0$$

From this and Eq. (17) we then find

$$|u_{\mathbf{k}}|^2 = \tfrac{1}{2}\left(1 + \frac{\epsilon_{\mathbf{k}}}{E_{\mathbf{k}}}\right) \qquad |v_{\mathbf{k}}|^2 = \tfrac{1}{2}\left(1 - \frac{\epsilon_{\mathbf{k}}}{E_{\mathbf{k}}}\right) \tag{18}$$

where

$$E_{\mathbf{k}} = (\epsilon^2_{\mathbf{k}} + |\Delta_{\mathbf{k}}|^2)^{1/2} \tag{19}$$

There is from the equations alone an ambiguity in the sign of the square root. It can be shown, however, that for the least free energy the positive sign must be chosen for all $\mathbf{k}$. The relative phase of u and v is given by

$$\arg(u/v) = \arg \Delta \tag{20}$$

‡ The canonical transformation was introduced independently by Bogoliubov (*17,33*) and Valatin (*34*) and is often called the Bogoliubov–Valatin transformation.

There is still an arbitrariness in the absolute phase of $u_{\mathbf{k}}$ and $v_{\mathbf{k}}$, corresponding to the fact that changing the phase of $\gamma_{\mathbf{k}0}$ and $\gamma_{\mathbf{k}1}$ does not change the Hamiltonian. This arbitrariness has no physical significance and all the coefficients $u_{\mathbf{k}}$ can for definiteness be chosen real. We have now diagonalized the model Hamiltonian, the result being

$$\mathscr{H}_M = \sum_{\mathbf{k}} E_{\mathbf{k}}(\gamma^*_{\mathbf{k}0}\gamma_{\mathbf{k}0} + \gamma^*_{\mathbf{k}1}\gamma_{\mathbf{k}1}) + \sum_{\mathbf{k}}(\epsilon_{\mathbf{k}} - E_{\mathbf{k}} + \Delta^*_{\mathbf{k}} b^0_{\mathbf{k}}) \tag{21}$$

Since the operators $\gamma_{\mathbf{k}0}, \gamma_{\mathbf{k}1}$ obey the Fermi–Dirac commutation relations, this Hamiltonian describes independent fermions with excitation energies $E_{\mathbf{k}}$. For this reason the operators γ^* are said to create quasi-particles. The question of whether the superconductive system can be described only by a new kind of fermion is one we consider later (see Sections XI and XVI).

The Hamiltonian [Eq. (21)] determines the properties of the system in terms of the numbers $b^0_{\mathbf{k}}$, but we have yet to determine these numbers. This is where the requirement of self-consistency enters, for the numbers $b^0_{\mathbf{k}}$ are the thermal and quantum averages of the operators $c_{-\mathbf{k}\downarrow}c_{\mathbf{k}\uparrow}$ when the system is described by the model Hamiltonian. It is shown in textbooks on quantum statistical mechanics that such an average is given by

$$b^0_{\mathbf{k}} = \operatorname{Tr}[\exp(-\beta\mathscr{H}_M)c_{-\mathbf{k}\downarrow}c_{\mathbf{k}\uparrow}]/\operatorname{Tr}\exp(-\beta\mathscr{H}_M)$$

where Tr stands for the trace of the following operator in the Hilbert space of the eigenstates of $\mathscr{H}_M$. If the c's are transformed to the γ's using Eqs. (15), the trace can be evaluated by standard techniques [see, for example, (*16*)]. One has

$$\begin{aligned} b^0_{\mathbf{k}} &= \operatorname{Tr}[\exp(-\beta\mathscr{H}_M)(-v^*_{\mathbf{k}}u_{\mathbf{k}}\gamma^*_{\mathbf{k}0}\gamma_{\mathbf{k}0} + u_{\mathbf{k}}v^*_{\mathbf{k}}\gamma_{\mathbf{k}1}\gamma^*_{\mathbf{k}1})]/\operatorname{Tr}\exp(-\beta\mathscr{H}_M) \\ &= u_{\mathbf{k}}v^*_{\mathbf{k}}[1 - 2f(E_{\mathbf{k}})] \end{aligned} \tag{22}$$

where $f(E)$ is the Fermi–Dirac distribution function,

$$f(E) = [\exp(\beta E) + 1]^{-1}$$

We have used the fact that in a system of independent fermions this function gives the probability of occupation of one particular state.

The gap parameter can now be written, through the use of Eqs. (13), (18), (20), and (22), as

$$\begin{aligned} \Delta_{\mathbf{k}} &= -\sum_{\mathbf{k}'} V_{\mathbf{k},\mathbf{k}'}u_{\mathbf{k}'}v^*_{\mathbf{k}'}[1 - 2f(E_{\mathbf{k}'})] \\ &= -\sum_{\mathbf{k}'} V_{\mathbf{k},\mathbf{k}'}\Delta_{\mathbf{k}'}[1 - 2f(E_{\mathbf{k}'})]/2E_{\mathbf{k}'} \end{aligned} \tag{23}$$

This equation, first obtained by BCS (*1*), is a highly nonlinear equation for the gap parameter $\Delta_{\mathbf{k}}$. Given the interaction potential the equation can be solved;

the model Hamiltonian $\mathscr{H}_M$ and, hence the thermodynamic properties of the system, are then determined. We continue with a discussion of the solution of this equation followed by derivations of the thermodynamic properties of the system.

VI. CRITERION FOR SUPERCONDUCTIVITY

The integral equation (23) always possesses a solution

$$\Delta_{\mathbf{k}} = 0 \qquad \text{all } \mathbf{k} \tag{24}$$

For this solution the operators $\gamma_{\mathbf{k}}$ become equal to the operators $c_{\mathbf{k}}$ and we have just the solution for a normal Fermi liquid. The system will possess new properties only if $\Delta_{\mathbf{k}}$ is not zero for a macroscopic number of values of $\mathbf{k}$. We shall show later (but see Section XI) that when $\Delta_{\mathbf{k}}$ is not zero, the system does possess the properties of a superconductor, and that at low temperatures the superconductive state is the stable one. The criterion for superconductivity is, therefore, that Eq. (23) should possess a nontrivial solution. This is not a very meaningful criterion until we come to discuss particular cases of the interaction $V_{\mathbf{k},\mathbf{k}'}$.

If, following Cooper, one assumes that the interaction is important only for $|\epsilon_{\mathbf{k}}|, |\epsilon_{\mathbf{k}'}| < \hbar\omega$, and that for these energies the interaction is a constant $(-V)$, the gap parameter will be a constant Δ for $|\epsilon_{\mathbf{k}}| < \hbar\omega$ and will be zero for $|\epsilon_{\mathbf{k}}| > \hbar\omega$ The constant Δ is then a solution of

$$\Delta = \Delta V \sum_{\substack{\mathbf{k} \\ |\epsilon_k| < \hbar\omega}} \frac{1 - 2f(E_{\mathbf{k}})}{2E_{\mathbf{k}}} \tag{25}$$

Since the sum is always positive, this can have no nontrivial solution if V is negative, that is, if the interaction is repulsive. On the other hand, if V is positive the equation, as we shall show, always has a nontrivial solution for sufficiently low temperatures. Hence within this model the criterion for superconductivity is that the constant interaction be attractive. Evidently the appropriate constant is the negative of the angular average of $V_{\mathbf{k},\mathbf{k}'}$ at the Fermi surface and the criterion, as first given by BCS, is

$$\langle V_{\mathbf{k},\mathbf{k}'} \rangle_{k=k'=k_F} < 0 \tag{26}$$

This means simply that at the Fermi surface the Coulomb repulsion must be dominated by the attractive interaction mediated by the phonons.

The criterion [Eq. (26)] was first evaluated and compared with experiment by Pines (*35*) and improved upon by Morel (*36*). In evaluating this comparison it must be remembered that the interaction $V_{\mathbf{k},\mathbf{k}'}$ itself is not directly amenable to experiment, and that the assumed forms for this interaction are not soundly based. With these reservations, the calculations would seem to imply that some metals which are superconductors should not be.

Apart from this disagreement between theory and experiment there are reasons for believing that this criterion is not the correct one to apply. Although it may be correct to say that the interaction V_{ph} is large only for Bloch electrons within an energy $\hbar\omega$ of the Fermi surface, there would seem to be no justification for saying this about the Coulomb repulsion. Since the Coulomb interaction involves large components of k (of the order of k_F), we might well expect the interaction to be important for Bloch electrons with excitation energies of the order of E_F, and this does seem to be the case.

The first attempt to solve the integral equation in a way which treats the Coulomb interaction more realistically was made by Bogoliubov and coworkers (*37*). In fact, if one wishes to make such a more realistic calculation, one should also include the fact that the interaction is retarded and not static as implied by Eq. (23). This kind of calculation has been performed by Morel and Anderson (*38*) and later, with different assumptions about the interaction, by Garland (*39*). The essential results of including the effect of the Coulomb repulsion on electrons of large excitation energy is not, as one might intuitively expect, to decrease the likelihood of superconductivity but to increase it. The reason for this is that the interaction changes sign at an energy of about $\hbar\omega$, the energy gap changes sign at about this energy, and the repulsive interaction helps to decrease the interaction energy and to produce superconductivity. All the numerical results must be regarded with a skeptical eye because of the simple assumptions they make about highly excited electrons. The conclusion that the Coulomb interaction is effectively reduced is, however, surely correct and probably so is the conclusion that nearly all pure nonferromagnetic metals at sufficiently low temperatures will become superconductors.‡ Since, as we shall see, the transition temperature of a superconductor is the hardest property to predict, it cannot be said, on this point, that experiment contradicts the theory.

VII. CRITICAL TEMPERATURE

A. Magnitude

We shall see that as long as the gap parameter is not zero the state with finite Δ, the superconductive state, is the stable one. At sufficiently high temper-

‡ On the question of whether all pure nonferromagnetic metals will become superconductors at sufficiently low temperatures, see Discussion 24 of (*40*).

ature the integral equation possesses no nontrivial solution and the system is normal. At sufficiently low temperatures, provided the criterion for superconductivity is satisfied, the integral equation does possess a nontrivial solution. The critical temperature T_c is the temperature which divides one of these regimes from the other, and is to be found by letting Δ become small in Eq. (23). Thus the equation for T_c is

$$\Delta_{\mathbf{k}} = -\sum_{\mathbf{k}'} V_{\mathbf{k},\mathbf{k}'}\Delta_{\mathbf{k}'} \frac{1 - 2f(|\epsilon_{\mathbf{k}}|)}{2|\epsilon_{\mathbf{k}}|} \tag{27}$$

where the Fermi–Dirac function is taken at the temperature T_c. This equation is essentially an eigenvalue equation for T_c. There is no reason, in principle, why it should not have more than one eigenvalue, and, if it does, the critical temperature for the onset of superconductivity is the greatest.

To solve Eq. (27) for an arbitrarily given potential is evidently a formidable task suitable for a computor. Fortunately, the properties of many superconductors are well described if one uses the same approximation for $V_{\mathbf{k},\mathbf{k}'}$ as in Cooper's problem:

$$V_{\mathbf{k},\mathbf{k}'} = \begin{cases} -V & |\epsilon_{\mathbf{k}}|, |\epsilon_{\mathbf{k}'}| < \hbar\omega \\ 0 & \text{otherwise} \end{cases} \tag{28}$$

In view of our remarks on the Coulomb interaction in the previous section this may seem a surprising statement to make. However, provided V is interpreted appropriately, there is no inconsistency. Even when the Coulomb interaction is treated properly, $\Delta_{\mathbf{k}}$ is approximately constant for **k** within an energy $\hbar\omega$ of the Fermi surface. The gap parameter for higher energies can be found in terms of this constant Δ and substituted in the sum. One then finds (*38*) that the net effect is an equation of the form of Eq. (27) with an interaction of the form of Eq. (28). The constant V is not, however, the average of V_{ph} at the Fermi surface but rather has the form

$$-V = \langle V_{\text{ph}} \rangle + \alpha \langle V_s \rangle$$

where α is a constant less than unity. According to Garland (*39*), for the nontransition metals the value of α is

$$[1 + N(0)\langle V_s \rangle \ln (\xi_c/\hbar\omega_D)]^{-1}$$

where ξ_c is an energy of approximately $4E_F$ and ω_D is the Debye frequency. Since the gap parameter decreases rapidly in value for $|\epsilon_{\mathbf{k}}| > \hbar\omega$, it is reasonable to take the interaction in the form of Eq. (28). Strictly this model is valid only for the weakly coupled superconductors (e.g., aluminium and tin) for which $N(0)V$ is very much less than unity, and not for strongly coupled superconductors (e.g., lead and mercury). All the same, even the properties of the latter superconductors do not depart radically from those predicted using the interaction Eq. (28). Whether the model can be applied to transition metals is not yet definitely resolved (*41,42*).

With the model interaction of Eq. (28) we have that $\Delta_{\mathbf{k}}$ is zero for $|\epsilon_{\mathbf{k}}| > \hbar\omega$, and that the equation for T_c is

$$1 = V \sum_{\mathbf{k},\, |\epsilon_k| < \hbar\omega} \frac{1 - 2f(|\epsilon_{\mathbf{k}}|)}{2|\epsilon_{\mathbf{k}}|} = N(0)\,V \int_0^{\hbar\omega} d\epsilon \frac{1 - 2f(\epsilon)}{\epsilon} \tag{29}$$

In the last step we have used the fact that as long as the density of states in the normal state varies little within an energy $\hbar\omega$ of the chemical potential the sum is symmetric about the chemical potential. We are then able to write the integral in terms of positive values of ϵ only. As long as $N(0)\,V$ is very much less than unity, kT_c comes out to be very much less than $\hbar\omega$ and can be found analytically. In fact, if the variable of integration is changed to

$$x = \epsilon/kT_c$$

and if the integral is integrated by parts, we have

$$\frac{1}{N(0)\,V} = \ln \frac{\hbar\omega}{kT_c} [1 - 2f(\hbar\omega)] + 2\int_0^{\hbar\omega/kT_c} dx \ln x \frac{d}{dx}\left(\frac{1}{e^x + 1}\right)$$

Since the integral converges rapidly we can replace the upper limit by infinity. Then, neglecting $f(\hbar\omega)$ which is very small,

$$\frac{1}{N(0)\,V} = \ln(A\hbar\omega/kT_c)$$

or

$$kT_c = A\hbar\omega \exp[-1/N(0)\,V]$$

where the number A is given by

$$\ln A = 2\int_0^{\infty} dx \ln x \frac{d}{dx}\left(\frac{1}{e^x + 1}\right)$$

The integral is, in fact, a known one, and in terms of Euler's constant γ,

$$A = 2\gamma/\pi \approx 1.13$$

Hence

$$kT_c = 1.13\hbar\omega \exp[-1/N(0)\,V] \tag{30}$$

For most superconductors, $N(0)\,V$ is of the order of 0.2–0.4. Hence $(kT_c/\hbar\omega)$ is of the order of 0.01–0.1, and the theory accounts for the lowness of the critical temperatures of superconductors.

The critical temperature is evidently a very sensitive function of the interaction V. Since, as we have emphasized before, V is not well known theoretically, we cannot expect to be able to predict transition temperatures very well. For the known superconductors one can, however, reverse the procedure and from the known critical temperature determine the value of $N(0)V$, which can then be compared with the theoretical one. This is done in Chapter 3 and, as can be seen there, the agreement is reasonable.

B. Isotope Effect

Equation (30) also predicts a definite dependence of T_c on the average isotopic mass of the ions of the lattice, M, at least if they are all of one kind. This is because the frequencies of the lattice vibrations are all proportional to $M^{-1/2}$ (*16*). If the interaction V is independent of the ionic mass, as was originally thought to be the case, one then has

$$T_c \propto M^{-1/2}$$

In fact, if one uses the interaction V_{ph}, this result follows directly from the integral equation without further simplifying the interaction. The equation for T_c is then

$$\Delta_{\mathbf{k}} = -\frac{N(0)}{4\pi}\int d\Omega_{\mathbf{k}'}\int d\epsilon' \frac{|g_{\mathbf{k}-\mathbf{k}'}|^2 \hbar\omega_{\mathbf{k}-\mathbf{k}'}\Delta_{\mathbf{k}'}}{(\epsilon_{\mathbf{k}}-\epsilon_{\mathbf{k}'})^2-(\hbar\omega_{\mathbf{k}-\mathbf{k}'})^2}\,\frac{1-2f|\epsilon'|}{|\epsilon'|}$$

where $|g|^2$ is proportional to $M^{-1/2}$ (*16*). If we then measure all energies in terms of $\hbar\omega_D$, the isotopic mass will appear in the equation only in the Fermi–Dirac function, where it will come explicitly in the ratio $kT_c/\hbar\omega_D$. Hence this ratio must be independent of M and

$$T_c \propto M^{1/2}$$

This result is changed once the Coulomb interaction is included in a reasonable way. For example, in Garland's approximation V depends on the isotopic mass. One then has

$$\begin{aligned}\frac{\delta T_c}{T_c} &= \frac{\delta M}{M}\left(-\frac{1}{2}+\frac{M}{N(0)V^2}\frac{\partial V}{\partial M}\right)\\ &= \frac{\delta M}{M}\left(-\frac{1}{2}\right)(1-\zeta)\end{aligned} \tag{31}$$

On the basis of his model, Garland (*39*) has evaluated ζ for a number of different metals, including transition metals (for which the model is more complicated) and finds results which we show in Table I. Also shown are experimental values for ζ and it can be seen that the agreement is remarkable, especially when one considers how sensitive ζ is to the interaction.

TABLE I

Deviations ζ from the Ideal Isotope Effect [from (*39*)]

Material	Calculated	Experimental
Zn	0.17 ± 0.03	0.10 ± 0.10
Cd	0.23 ± 0.05	0.0 ± 0.2
Sn	0.09 ± 0.02	0.06 ± 0.04
Hg	0.04 ± 0.01	0.00 ± 0.06
Pb	0.03 ± 0.01	0.04 ± 0.02
Tl	0.04 ± 0.04	0.0 ± 0.2
Al	0.26 ± 0.05	
Ru	0.87 ± 0.3	1.0 ± 0.20
Os	0.55 ± 0.2	0.6 ± 0.1
Mo	0.3 ± 0.15	0.3 ± 0.15
Ir	1.03 ± 0.35	
Hf	0.8 ± 0.4	
V	0.5 ± 0.25	
Ti	0.7 ± 0.35	
Zr	0.7 ± 0.35	
Ta	0.3 ± 0.15	
Re	0.29 ± 0.1	0.22 ± 0.02
Nb_3Sn	0.5 ± 0.25	$\leqslant 0.84$
Mo_3Ir	0.4 ± 0.2	$0.34 < \zeta \lesssim 0.67$
V_3Ge	0.7 ± 0.3	
V_3Ga	1.3 ± 0.5	
V_3Si	1.0 ± 0.4	
W	0.6 ± 0.6	

C. Effect of Pressure

The parameters ω_D, $N(0)$, and V all depend on the lattice parameters and hence will change with changing pressure. As was shown by early calculations of Morel (*36*), the results are very sensitive to the form of the interaction, and this is another case where the numerical predictions of the theory cannot yet be regarded as reliable. A review of the present situation has been given by Levy and Olsen (*43*) [see also (*43a*)]. Recent theoretical work (*44*) has shown that the effect is sensitive to the change in the topology of the Fermi surface with pressure.

VIII. ENERGY GAP

We now turn to the solution of Eq. (23) for the gap parameter $\Delta_{\mathbf{k}}$. In this chapter we consider only the simplest case when the interaction can be taken in the form of Eq. (28). For $|\epsilon_{\mathbf{k}}| > \hbar\omega$ the parameter is then zero, and for $|\epsilon_{\mathbf{k}}| < \hbar\omega$ the parameter is independent of **k** and satisfies the equation

$$\Delta = \Delta N(0) V \int_0^{\hbar\omega} (d\epsilon/E)[1 - 2f(E)] \tag{32}$$

where

$$E = (\epsilon^2 + |\Delta|^2)^{1/2} \tag{33}$$

For the nontrivial solution we can cancel Δ from the equation.

Equation (32) determines only the modulus of Δ and not the phase. In fact, as we have already pointed out, the existence of the phase in a uniform superconductor which is in a stationary state is connected with the indefiniteness of the number of particles. Different phases correspond to different descriptions of the same state, and this can be exploited to obtain states with a definite number of particles (*16*). There is no virtue, however, in using states with definite number of particles, so we shall fix the phase of Δ so that Δ is real. One would expect the arbitrariness of the phase of the gap parameter not to depend on the simple interaction that we have chosen. That it does not can be seen from Eq. (23). If we have any solution $\Delta_{\mathbf{k}}$ of that equation, the function $\Delta_{\mathbf{k}} \exp(i\phi)$ with ϕ independent of **k** is also a solution. One can in that case, too, take $\Delta_{\mathbf{k}}$ to be real. [If the appropriate retarded interaction is used, the gap parameter does not have to be real for all values of its argument. See (*45*) for an example.]

Strictly the density of states $N(0)$ which appears in Eq. (32) should be the density of states at the chemical potential, which for a given density of particles may not be the same in the superconducting and normal states. It can be shown, however, that since only a few electrons are involved in the transition, the relative change in the chemical potential is of the order of magnitude of $(kT_c/\mu)^2$ and is negligible. This kind of change in the chemical potential is already ignored in the theory of the Fermi liquid.

In the weak-coupling limit of

$$\Delta, kT_c \ll \hbar\omega \tag{34}$$

the solution of Eq. (32) can be given in terms of a universal function which has to be calculated on a computor. If the integral is integrated by parts we have, if we allow $\hbar\omega$ to tend to infinity when appropriate, that

$$\frac{1}{N(0)V} = \sinh^{-1}\frac{\hbar\omega}{\Delta} + 2\int_0^{\infty} d\epsilon \, \frac{\partial f(E)}{\partial \epsilon} \sinh^{-1}\frac{\epsilon}{\Delta}$$
$$\approx \ln\frac{2\hbar\omega}{\Delta} + F\left(\frac{\Delta}{kT}\right)$$

where F is a function of the ratio Δ/kT which can be computed. With Eq. (30)

this yields

$$\Delta(T) = \frac{2}{1.13} kT_c \exp\left[F\left(\frac{\Delta}{kT}\right)\right] \tag{35}$$

If follows that, within the approximations we have used, the ratio Δ/kT_c is the same function of the reduced temperature (T/T_c) for all superconductors.

This result is usually given in a slightly different way. As long as Δ is finite, in the limit that T tends to zero the Fermi–Dirac function tends to zero, and the function F tends to zero. Hence at the absolute zero of temperature,

$$\Delta(0) = 2\hbar\omega \exp[-1/N(0)V] \tag{36}$$

We can, therefore, write

$$\Delta(T)/\Delta(0) = \exp[F(\Delta/kT)]$$

and the ratio $\Delta(T)/\Delta(0)$ is a universal function of the reduced temperature. A plot of $\Delta(T)/\Delta(0)$ against reduced temperature is shown in Fig. 2. Near T_c one has the useful result

$$\frac{\Delta(T)}{\Delta(0)} = \gamma\left[\frac{8}{7\zeta(3)}\left(1 - \frac{T}{T_c}\right)\right]^{1/2} = 1.74\left(1 - \frac{T}{T_c}\right)^{1/2}$$

It follows also (subject to the same limitations) that the ratio of the gap parameter at the absolute zero to the critical temperature is the same for all

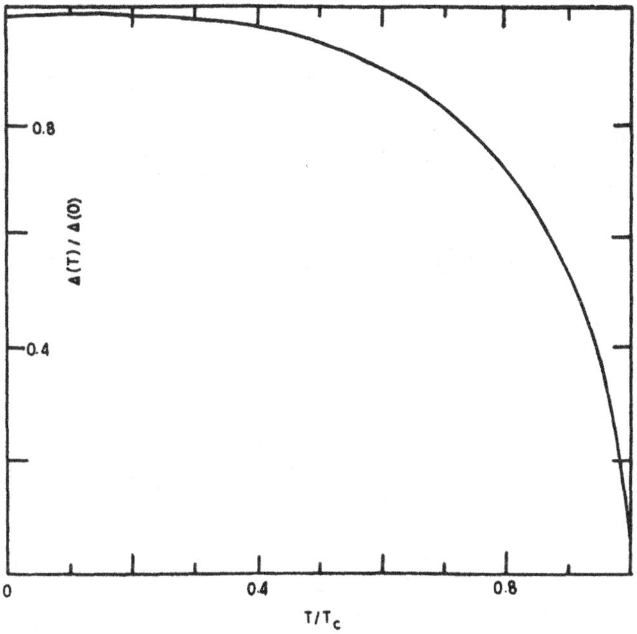

Fig. 2. Curve showing $\Delta(T)/\Delta(0)$ vs. T/T_c according to BCS.

superconductors. One finds that

$$\Delta(0)/kT_c = \pi/\gamma = 1.76 \tag{37}$$

This result is compared with experimental values in Chapter 3.

IX. EXCITATION SPECTRUM

Once we have computed Δ we can look at the energies of the quasi-particles. Since the energy of an excitation is $E_{\mathbf{k}}$, for the simple model we have been discussing there is a minimum energy required to create a new excitation, namely, Δ. Since new excitations are created in pairs, the minimum energy required to create excitations from the ground state is

$$E_G = 2\Delta$$

This gap is easily observed in absorption experiments at low temperatures. At finite temperatures excitations are always present and waves of any frequency can be absorbed, although the absorption of low-frequency waves is reduced.

The number of single-particle states will be the same in the superconductive state as in the normal state. To make up for the reduced density of states at low energy the density of states is large for energies immediately above the gap. In fact, if we write $\eta(E)$ for the density of states of one kind of excitation we have

$$\eta_s(E)/\eta_n(E) = \begin{cases} 0 & E < \Delta \\ d\epsilon/dE = E/(E^2 - \Delta^2)^{1/2} & E > \Delta \end{cases} \tag{38}$$

This spectrum is singular when E is equal to Δ. Figure 3 shows the ratio η_s/η_n plotted vs. E/Δ. As E tends to infinity the area under the curve tends to E/Δ.

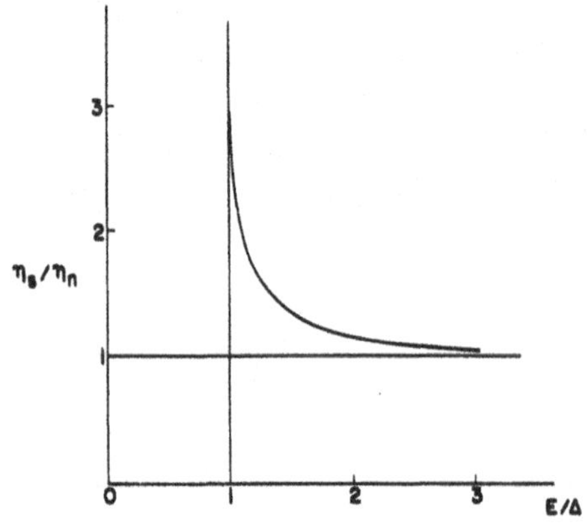

Fig. 3. Ratio of the density of states in energy in the superconductive state to that in the normal state plotted vs. E/Δ.

X. THERMODYNAMIC FUNCTIONS

A. Isotropic Superconductors

We proceed to evaluate the sum-over-states, or partition function Z, and from that the other thermodynamic functions of the superconductive state. By definition

$$\begin{aligned} Z &= \operatorname{Tr} \exp[-\beta(\mathscr{H} - \mu N)] \\ &= \operatorname{Tr} \exp(-\beta \mathscr{H}_M) \\ &= \prod_{\mathbf{k}} \operatorname{Tr} \exp(-\beta E_{\mathbf{k}} \gamma_{\mathbf{k}0}^* \gamma_{\mathbf{k}0}) \operatorname{Tr} \exp(-\beta E_{\mathbf{k}} \gamma_{\mathbf{k}1}^* \gamma_{\mathbf{k}1}) \\ &\qquad \times \exp\{-\beta(\epsilon_{\mathbf{k}} - E_{\mathbf{k}} + \Delta_{\mathbf{k}}^2[1 - 2f(E_{\mathbf{k}})]/2E_{\mathbf{k}})\} \end{aligned}$$

The traces are easily evaluated to yield

$$Z = \prod_{k} [1 + \exp(-\beta E_{\mathbf{k}})]^2 \exp\{-\beta \sum_{k} (\epsilon_{\mathbf{k}} - E_{\mathbf{k}} + \Delta_{\mathbf{k}}^2[1 - 2f(E_{\mathbf{k}})]/2E_{\mathbf{k}})\}$$

From this one calculates the thermodynamic potential for the electrons,

$$\begin{aligned} \Omega_s &= -k_B T \ln Z \\ &= -2k_B T \sum_{\mathbf{k}} \ln[1 + \exp(-\beta E_{\mathbf{k}})] + \sum_{\mathbf{k}} \{\epsilon_{\mathbf{k}} - E_{\mathbf{k}} + \Delta_{\mathbf{k}}^2[1 - 2\ (E_{\mathbf{k}})]/2E_{\mathbf{k}}\} \end{aligned} \tag{39}$$

This thermodynamic potential can be related to the critical magnetic field of the superconductor. From general thermodynamic arguments (Chapter 1), the critical magnetic field at a certain temperature can be related to the difference in the Gibbs free energy of the superconductor in the normal and superconductive states. Since this free energy change comes from the electrons and since the changes in volume and chemical potential of the electrons in the transition are negligible, the difference in the Gibbs free energy is equal to the difference in the thermodynamic potentials of the electrons. Hence we have

$$\mathscr{V} H_c^2/8\pi = \Omega_n - \Omega_s \tag{40}$$

where $\mathscr{V}$ is the volume, Ω_s is given by Eq. (39), and Ω_n is given by Eq. (39) with Δ put equal to zero.

For a general temperature, Ω_s cannot be calculated analytically. It can, however, be evaluated analytically for the absolute zero of temperature. For this case

$$\Omega_s = 2N(0) \int_0^{\hbar\omega} d\epsilon\, (\epsilon - E + \Delta^2/2E) \tag{41}$$

and, in the limit of $\hbar\omega$ very much greater than Δ,

$$\Omega_s = -\tfrac{1}{2} N(0) \Delta^2 \tag{42}$$

Hence the critical magnetic field at the absolute zero, $H_c(0)$, is given by

$$H_c(0)^2/8\pi = \tfrac{1}{2}N(0)\Delta^2 \tag{43}$$

The three quantities $H_c(0)$, $N(0)$, and $\Delta(0)$ can all be measured and the relation (43) put to the test. Near the absolute zero one can show that

$$H_c(T)/H_c(0) = 1 - 1.06(T/T_c)^2$$

Near T_c the thermodynamic potential can be evaluated (*46*) to give

$$\frac{H_c(T)}{H_c(0)} \approx \gamma\left(1 - \frac{T}{T_c}\right)\left[\frac{8}{7\zeta(3)}\right]^{1/2} \tag{44}$$

where $\zeta(x)$ is the Riemann zeta function. Hence

$$H_c(T)/H_c(0) \approx 1.74\left(1 - \frac{T}{T_c}\right)$$

Thus the slope of the critical field curve is finite at T_c, as it should be according to the thermodynamic relations (see Chapter 1).

At intermediate values of T, the critical magnetic field is calculated by computor. The result is so close to the empirical formula

$$H_c(T)/H_c(0) = 1 - (T/T_c)^2$$

that it is usual to plot the deviation from this result,

$$D(T) = \frac{H_c(T)}{H_c(0)} - [1 - (T/T_c)^2]$$

This is shown in Fig. 4, plotted vs. T/T_c; it is another universal function. These calculations confirm that as long as we can find a solution of the integral equation

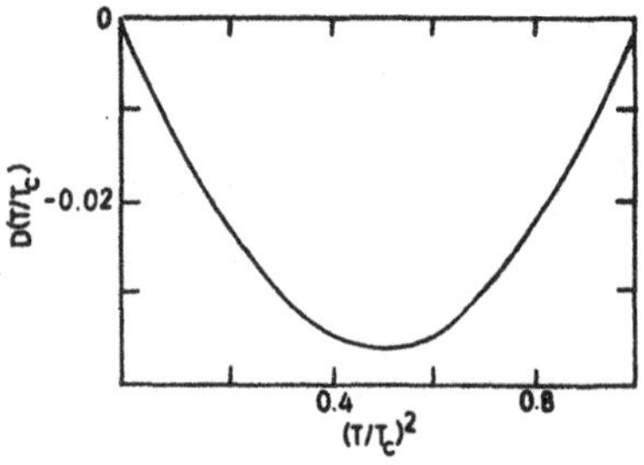

Fig. 4. Critical magnetic field of a superconductor, illustrated by a plot of the deviation $D(T/T_c)$ ($= H_c(T)/H_c(0) - [1 - (T/T_c)^2]$) from the old empirical rule versus $(T/T_c)^2$. Note the scale.

with Δ not equal to zero this solution provides a state with lower free energy than the trivial solution.

The entropy of the electrons can be calculated from the formula

$$s_s = -(\partial\Omega/\partial T)_{\mu,\mathscr{V}}$$

Using the integral equation for Δ one finds

$$s_s = -2k_B \sum_{\mathbf{k}} \{[1 - f(E_{\mathbf{k}})] \ln [1 - f(E_{\mathbf{k}})] + f(E_{\mathbf{k}}) \ln f(E_{\mathbf{k}})\} \tag{45}$$

This is the entropy of a gas of independent fermions, a result which is hardly surprising in view of the fact that the model Hamiltonian describes independent fermions.

Since the gap parameter tends to zero as T tends to T_c, the entropy is continuous at the critical temperature. There is, therefore, no latent heat evolved at the transition, and this result agrees with experiment. One can also see from Eq. (45) that as the temperature approaches the absolute zero, the entropy becomes exponentially small, being proportional to $\exp(-\Delta/kT)$.

The specific heat per unit volume of the electrons can be calculated from the formula

$$c_s = \frac{T}{\mathscr{V}}\left(\frac{\partial s_s}{\partial T}\right)_{\mu,\mathscr{V}}$$

This is strictly the electronic specific heat at constant chemical potential, but since, when the number of particles is held fixed, the change of chemical potential with temperature is negligible, we can ignore the difference between specific heats at constant μ and specific heats at constant number. This formula thus gives the measured electronic specific heat. Straightforward differentiation of Eq. (45) gives

$$\begin{aligned} c_s &= \frac{2}{\mathscr{V}} \sum_{\mathbf{k}} E_{\mathbf{k}} \frac{\partial f(E_{\mathbf{k}})}{\partial T} \\ &= -\frac{2}{\mathscr{V}T} N(0) \int_{-\infty}^{\infty} d\epsilon \left(E^2 + \frac{\beta}{2}\frac{\partial \Delta^2}{\partial \beta}\right) \frac{\partial f}{\partial E} \end{aligned} \tag{46}$$

The curve of Δ^2 against temperature has a finite slope at the critical temperature. Hence the specific heat is not continuous there, and one has

$$\left.\frac{c_s - c_n}{c_n}\right|_{T_c} = \frac{3}{2\pi^2}\beta^3 \left.\frac{\partial \Delta^2}{\partial \beta}\right|_{T=T_c} = 1.43 \tag{47}$$

At lower temperatures the specific heat decreases rapidly, and because of the gap in the spectrum of excitations it becomes proportional to $\exp[-\Delta(0)/kT]$ at temperatures below about $T_c/10$. A plot of the specific heat is shown in Fig. 5. Again, according to the theory, c_s/c_n is a universal function of the reduced temperature. In any comparison of theory with experiment it must be remembered that it is the total specific heat which is usually measured, whereas, in this section, it is the electronic specific heat which is calculated. The comparison can, therefore, be made only if other contributions to the specific heat are first subtracted from the experimental results.

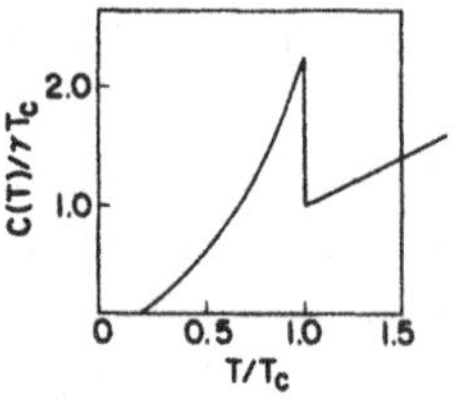

Fig. 5. Plot of the ratio of the electronic specific heat to γT_c vs. T/T_c. (The electronic specific heat in the normal state is γT.)

B. Anisotropic Superconductors

Some of the thermodynamic functions have been evaluated for anisotropic superconductors for which the gap parameter depends on angle (*47,48*). Since such superconductors have to be described by more parameters than the $N(0)\,V$ of BCS, it is not possible to give the results in such a succinct form. However, a number of inequalities have been proved (*47*) and these we give for the sake of a later comparison with experiment. From solutions of the integral equation (27) one finds that

$$E_G/kT_c \leqslant 3.53$$

$$(c_s - c_n)/c_n|_{T_c} \leqslant 1.43$$

and near the absolute zero of temperature

$$H_c(T)/H_c(0) = 1 - \chi(T/T_c)^2$$

with $\chi \geqslant 1.06$. If the dynamic coupling between the electrons is taken into account, the first of these inequalities, even in an isotropic superconductor, can be broken (*49*), but the second still holds (*50*). If, however, the superconductor possesses overlapping conduction bands, it appears that even in an isotropic superconductor it is possible to break all three inequalities (*51*).

XI. WAVE FUNCTION

We have now derived many of the thermodynamic properties of a superconductor on the basis of the original model. Considering the simplicity of the model, there is remarkable agreement with experiment, as we shall see in Chapter 3. Before we go on to discuss the effects of perturbations on the system we should like to pause to discuss the relationship of the model with the complete problem, and to make a few remarks about the wave function.

We have seen that the model Hamiltonian for the system at a given temperature describes independent Fermi quasi-particles with excitation energies $E_{\mathbf{k}}$, which depend on temperature. The ground state of the electrons is the ground state of the model Hamiltonian at the absolute zero of temperature, and satisfies the equations

$$\gamma_{\mathbf{k}0}\,|G\rangle = 0 \qquad \gamma_{\mathbf{k}1}\,|G\rangle = 0 \tag{48}$$

These equations are sufficient to define the ground state completely, and if one wants to calculate the properties of the electrons it is not necessary to know more. It is, however, quite useful to see what the ground state is like in terms of the original particle operators $c_{\mathbf{k}\sigma}$. In fact, the solution of Eq. (48) is

$$|G\rangle = \prod_{\mathbf{k}} (u_{\mathbf{k}} + v_{\mathbf{k}} c^*_{\mathbf{k}\uparrow} c^*_{-\mathbf{k}\downarrow})\,|0\rangle \tag{49}$$

where $|0\rangle$ is the vacuum state and $u_{\mathbf{k}}$, $v_{\mathbf{k}}$ are defined by Eqs. (15). This state is a coherent superposition of Slater determinants in which the pair of states $(\mathbf{k}\uparrow)$, $(-\mathbf{k}\downarrow)$ are either both occupied or both unoccupied. Bardeen, Cooper, and Schrieffer (*1*) started from the wave function [Eq. (49)] and found the coefficients $u_{\mathbf{k}}$, $v_{\mathbf{k}}$ from the variational principle.

From the wave function [Eq. (49)], which contains states of different numbers of particles, one can pick out the part with definitely N particles. Since $c^2_{\mathbf{k}\sigma} = 0$, one has

$$\begin{aligned} |G\rangle &= \prod_{\mathbf{k}} u_{\mathbf{k}} \exp\,[(v_{\mathbf{k}}/u_{\mathbf{k}})\, c^*_{\mathbf{k}\uparrow} c^*_{-\mathbf{k}\downarrow}]\,|0\rangle \\ &= \prod_{\mathbf{k}} u_{\mathbf{k}} \exp\,[\sum_{\mathbf{k}} (v_{\mathbf{k}}/u_{\mathbf{k}})\, c^*_{\mathbf{k}\uparrow} c^*_{-\mathbf{k}\downarrow}]\,|0\rangle \end{aligned}$$

The term containing exactly N particles is, therefore, proportional to

$$[\sum_{\mathbf{k}} (v_{\mathbf{k}}/u_{\mathbf{k}})\, c^*_{\mathbf{k}\uparrow} c^*_{-\mathbf{k}\downarrow}]^{N/2}\,|0\rangle \tag{50}$$

If one had a gas of fermions bound tightly together in diatomic molecules, this gas would suffer a Einstein–Bose condensation and, at the absolute zero of temperature, would have a wave function of the form of Eq. (50). Thus the formula (50) suggests that we have such a condensation. The idea of superconductivity arising from an Einstein–Bose condensation of molecular pairs was suggested before the BCS theory by Schafroth et al. (*52*). The essential difference between the simple condensation of molecular pairs and the condensation considered by BCS is that in the former the overlap between the pairs is negligible, whereas in the latter it is paramount. For example, in a superconductor such as tin, the correlation between a pair of electrons extends over a distance of the order of 10^{-5} cm, with the result that the intermediate region contains a large number of other electrons (*16*). The exact relation between the theory of BCS and the ideas of Schaforth et al. is still controversial (*53*), and since the ultimate answers do not depend on which viewpoint one holds, we will not extend the controversy here.

The excited states of the BCS theory are of the form

$$\gamma_{\mathbf{k}0}^{*}\,|G\rangle = c_{\mathbf{k}\uparrow}^{*} \prod_{\mathbf{k}'\neq\mathbf{k}} (u_{\mathbf{k}'} + v_{\mathbf{k}'}c_{\mathbf{k}'\uparrow}^{*}c_{-\mathbf{k}'\downarrow}^{*})\,|0\rangle$$

$$\gamma_{\mathbf{k}1}^{*}\,|G\rangle = c_{-\mathbf{k}\downarrow}^{*} \prod_{\mathbf{k}'\neq\mathbf{k}} (u_{\mathbf{k}'} + v_{\mathbf{k}'}c_{\mathbf{k}'\uparrow}^{*}c_{-\mathbf{k}'\downarrow}^{*})\,|0\rangle$$

In these states, one of the Bloch states is definitely occupied, the other Bloch state of the pair is definitely unoccupied, and the remaining Bloch states are occupied or unoccupied in pairs. Since the pairing energy of one pair of Bloch states has been lost, energy must have been required to have produced such an excited state. This is the origin of the gap in the spectrum of excitations. By inspection one can see that the operators $\gamma_{\mathbf{k}0}^{*}, \gamma_{-\mathbf{k}1}^{*}$ create momentum $\hbar\mathbf{k}$.

How well does the model Hamiltonian represent the pairing Hamiltonian? The difference between the two is

$$\sum_{\mathbf{k},\mathbf{k}'} V_{\mathbf{k},\mathbf{k}'}(c_{\mathbf{k}\uparrow}^{*}c_{-\mathbf{k}\downarrow}^{*} - b_{\mathbf{k}}^{0*})(c_{-\mathbf{k}'\downarrow}c_{\mathbf{k}'\uparrow} - b_{\mathbf{k}'}^{0})$$

This gives to order N^{-1} no contribution to the energy of the ground state, and to the energy of excitation of a single quasi-particle it makes only a contribution of the order of N^{-1}. Hence we have been justified in neglecting it. If, however, the numbers $b_{\mathbf{k}}^{0}$ are held fixed as quasi-particles are created, we should find that this term makes a finite contribution to the energy of a macroscopic number of quasi-particles. Since as the temperature rises macroscopic numbers of quasi-particles are created, this term as it stands makes a finite contribution to the energy. We have already made allowance for this by allowing the numbers $b_{\mathbf{k}}^{0}$ and hence $\mathscr{H}_M$ to depend on temperature. Then the contribution of the neglected term to the single-particle excitation energies about the average state of the system is indeed negligible. The state which is annihilated by all the temperature-dependent operators $\gamma_{\mathbf{k}}$ cannot be said to have any physical significance; it is a mathematical construct to help the calculation.

XII. COMPLETE HAMILTONIAN

As a first step in discussing the theory of superconductivity, we replaced the total Hamiltonian by the pairing Hamiltonian, which was believed to be enough to produce a transition in the system. The remainder was believed not to affect the existence of the transition, although it could affect some of the quantitative results. We now have to consider what effect the remainder of the Hamiltonian has. We are not here gilding the lily because the pairing Hamiltonian is not sufficient to explain the basic property of superconductors, their infinite conductivity. If the whole system described by the pairing Hamiltonian is given a push so that it has some total momentum, the pairing will not contribute to the energy and the system will be in the normal state. The current will decay. Hence to explain superconductivity we must put the rest of the Hamiltonian back.

As far as the thermodynamic properties of the system are concerned, we suppose that the model Hamiltonian contains the singular part of the interaction and that the remainder can be treated by perturbation theory. This provides us with a straightforward procedure for calculating differences between the normal and superconducting states, and has been used with great success (see Chapter 10) for improving upon the BCS theory.

The first step of obtaining the model Hamiltonian can be reformulated to fit situations more general than those we have discussed so far, for example, to problems of inhomogeneous superconductors. If the interaction is of the form

$$\sum_{\mathbf{k},\mathbf{k}'\mathbf{q}} V_{\mathbf{k},\mathbf{k}',\mathbf{q}} c^*_{\mathbf{k}\sigma} c^*_{\mathbf{k}+\mathbf{q}\sigma'} c_{\mathbf{k}'+\mathbf{q}\sigma'} c_{\mathbf{k}'\sigma}$$

this can, as a first step, be replaced by

$$\sum_{\mathbf{k},\mathbf{k}',\mathbf{q}} V_{\mathbf{k},\mathbf{k}'\mathbf{q}} \{\langle c^*_{\mathbf{k}\sigma} c^*_{\mathbf{k}+\mathbf{q}\sigma'}\rangle c_{\mathbf{k}'+\mathbf{q}\sigma'} c_{\mathbf{k}'\sigma} + c^*_{\mathbf{k}\sigma} c^*_{\mathbf{k}+\mathbf{q}\sigma'} \langle c_{\mathbf{k}'+\mathbf{q}\sigma'} c_{\mathbf{k}'\sigma}\rangle - \langle c^*_{\mathbf{k}\sigma} c^*_{\mathbf{k}+\mathbf{q}\sigma'}\rangle \langle c_{\mathbf{k}'+\mathbf{q}\sigma'} c_{\mathbf{k}'\sigma}\rangle\}$$

When this is combined with the one-particle terms of the Hamiltonian we have a bilinear model Hamiltonian which can be diagonalized and used as the zero-order Hamiltonian for perturbation theory. The constants $\langle c_{\mathbf{k}'+\mathbf{q}\sigma'} c_{\mathbf{k}'\sigma}\rangle$ have to be chosen consistently so that they are equal to the final thermal and quantum averages of the appropriate operators. For the homogeneous superconductor in thermal equilibrium only the averages $\langle c_{-\mathbf{k}\sigma} c_{\mathbf{k}\sigma'}\rangle$ are nonzero. This technique is at the basis of the method of Green's functions (Chapter 5).

We have said that once the model Hamiltonian has been separated out, the remainder of the Hamiltonian can be treated by perturbation theory. This does not mean at all that the remainder is small, only that it does not alter the essential singularity. For example, for the charged electrons the two-particle interaction includes the Coulomb interaction. To obtain reasonable values for the parameters of the superconductor we must use an approximation which properly takes into account the screening effects of this interaction. In the early sections of this chapter this was done in a phenomenological way; the sense in which this phenomenological approach is correct is now understood (see Chapters 5 and 7).

It might be thought that for a neutral Fermi liquid the perturbation would be small, but, because of the small amount of energy involved in the transition, this does not appear to be so. In fact, early calculations of a possible superfluid transition in liquid He^3 suggested critical temperatures which have now been reached without evidence of such a transition. One possible reason for this was proposed by Balian and Fredkin (*54*); according to their model, zero sound, a collective mode of the fermions, interacts weakly with the quasi-particles and can so modify their properties that the critical temperature is considerably reduced. An alternative suggestion put forward more recently (*54a*) is based on the high magnetic susceptibility of liquid He^3. In such a liquid there are collective modes, paramagnons, which are accompanied by fluctuations in the spin density and which couple to the particles. This produces another two-particle interaction, which alters the transition temperature. A similar effect can occur in transition metals and this is discussed more fully in Chapter 13. In this example the perturbation produces virtual transitions of the particles to be paired and so alters the effective pairing interaction. Thus, although the criterion for superconductivity given in Section VI is probably correct, it may be extremely difficult to find the correct pairing potential to use. In practice, then, the criterion may not be reliable.

Near the critical temperature one can expect further deviations from the simple theory. It has been assumed in the theory that some operators do not deviate substantially from their expectation values. However, near the critical temperature all these expectation values are small and one might expect the fluctuations in the values of the quantities to be as important as the quantities themselves. This is a normal feature of phase transitions (*55*), and the approach of the transition temperature is usually heralded from above and below by anomalous behavior. In fact, experimentally the normal–superconductive transition in the absence of a magnetic field is very, very sharp, no anomalous behavior being observed.

The approach of the normal metal to the transition temperature has been studied theoretically by several authors (*56,57*). In the case of a weak interaction that can lead to superconductivity they showed that near the transition temperature there is an anomalous contribution to the specific heat of the electrons. For this anomaly to be observed, however, one must be within a temperature range of the order of $(kT_c/E_F)^4\,T_c$ from T_c; that is, one must be within a range of the order of 10^{-12} °K. For all practical purposes, then, this anomaly in the specific heat is unobservable. There is a corresponding anomaly in the specific heat as the critical temperature is approached from below. This also is observable only within the same temperature range of T_c (*58*). It has, however, been suggested (*59*) that for small samples whose dimensions are of the order of $\hbar v_F/\Delta$ (of the order of 10^{-6}–10^{-4} cm for common superconductors) or smaller, the anomaly may be important at temperatures as low as 0.5°K below T_c. It has also been suggested recently (*59a*) that in amorphous superconductors one should be able to see effects due to long-range fluctuations, and there is a report that these have been observed (*59b*).

The problem of quantum fluctuations about the mean value of a macroscopic variable has a more fundamental bearing on the question of whether or not superconductivity can exist in a one-dimensional system. As a result of solving the gap equation corresponding to Eq. (27) for a one-dimensional model, Little (*60*) concluded that it should be possible to construct long organic chain molecules which would show a long-range order similar to that in a superconductor. According to Ferrell (*61*), however, because of the comparatively large number of collective excitations of low energy which a one-dimensional system possesses, the fluctuations of the system about the ordered state are so large that the order is destroyed. The same conclusion has been reached by Rice (*62*), who studied the problem in the spirit of a general theorem (*63*) which says that there can be no phase transition in a one-dimensional system with short-range forces. Using the Ginzburg–Landau theory (see Chapter 6) to describe superconductivity, Rice showed that the effect of all the fluctuations is to destroy any possible superconductivity. All the same, the discussions of Ferrell and Rice are not obviously of universal validity and Little's original idea is not yet abandoned; recently it has been suggested (*64*) that long-range order could exist in a one-dimensional system with pairing between electrons, and between electrons and holes. This suggestion is still open.‡

It has also been suggested that it is possible to have superconductivity in a two-dimensional system (*65–67*). Again, according to Rice (*62*), the fluctuations of the system about the ordered state are so large that the order is destroyed.

XIII. SPIN SUSCEPTIBILITY OF THE ELECTRONS

We turn now to the effects of static external fields on a superconductor. Of these the most interesting are the exclusion of a static magnetic field from a

‡ Note added in proof: The question of fluctuations and long-range order discussed briefly in Section XII has recently been the subject of intensive research. Applications to the theory of superconductivity are contained in the following references:

P. C. Hohenberg, *Phys. Rev.* **158**, 383 (1967).
W. A. Little, *Phys. Rev.* **156**, 396 (1967).
J. S. Langer and V. Ambegaokar, *Phys. Rev.* **164**, 498 (1968).
R. A. Ferrell and H. Schmidt, *Phys. Letters* **25A**, 544 (1967).

large specimen, the Meissner effect, and the spin susceptibility of the electrons in a magnetic field which penetrates a small specimen. To save repeating similar pieces of mathematics we will treat the former as a limiting case of the effects of varying electromagnetic fields. The latter we shall calculate in this section.

The principle of calculating any of the linear effects is the same. Since the perturbation is weak and hardly excites the superconductor, the superconductor can be described by the model Hamiltonian and the change of its state can be described by the creation and destruction of quasi-particles. The unperturbed Hamiltonian is, therefore,

$$\mathscr{H}_M = \sum_{\mathbf{k}} E_{\mathbf{k}} (\gamma^*_{\mathbf{k}0}\gamma_{\mathbf{k}0} + \gamma^*_{\mathbf{k}1}\gamma_{\mathbf{k}1})$$

where we have ignored the constant term because it does not affect the perturbation. For the spin susceptibility we consider the interaction of the spins with a static magnetic field. We can obtain the required result using simple perturbation theory if we consider the effect of a nonuniform field, and obtain the effect of a uniform field in the appropriate limit. The interaction term in the Hamiltonian is then

$$\mathscr{H}_i = -\mu_e \sum_{\mathbf{k}} H(-\mathbf{q})\,(c^*_{\mathbf{k}+\mathbf{q}\uparrow}c_{\mathbf{k}\uparrow} - c^*_{-\mathbf{k}\downarrow}c_{-\mathbf{k}-\mathbf{q}\downarrow}) + \text{c.c.}$$

where μ_e is the Bohr magneton, $H(\mathbf{q})$ is a Fourier component of the field, and c.c. denotes the complex conjugate term of the one written. In terms of the quasi-particle operators, the interaction is

$$\begin{aligned}\mathscr{H}_i &= -\mu_e \sum_{\mathbf{k}} H(-\mathbf{q}) \{(u_{\mathbf{k}+\mathbf{q}}u_{\mathbf{k}} + v_{\mathbf{k}+\mathbf{q}}v_{\mathbf{k}})(\gamma^*_{\mathbf{k}+\mathbf{q}0}\gamma_{\mathbf{k}0} - \gamma^*_{\mathbf{k}1}\gamma_{\mathbf{k}+\mathbf{q}1}) \\ &\qquad + (u_{\mathbf{k}+\mathbf{q}}v_{\mathbf{k}} - v_{\mathbf{k}+\mathbf{q}}u_{\mathbf{k}})(\gamma^*_{\mathbf{k}+\mathbf{q}0}\gamma^*_{\mathbf{k}1} + \gamma_{\mathbf{k}+\mathbf{q}1}\gamma_{\mathbf{k}0})\} + \text{c.c.} \\ &= -H(-\mathbf{q})\,M_{\text{op}}(\mathbf{q}) + \text{c.c.}\end{aligned}$$

where $M_{\text{op}}(\mathbf{q})$ is the operator for the spin magnetic moment of the electrons. From perturbation theory, the $\mathbf{q}$th Fourier component of the magnetic moment of the system is

$$M(\mathbf{q}) = \frac{\sum_{m,n} \exp(-\beta E_m) \left\{ \dfrac{\langle m|\,M_{\text{op}}(\mathbf{q})\,|n\rangle\langle n|\,\mathscr{H}_i\,|m\rangle}{E_m - E_n} + \dfrac{\langle m|\,\mathscr{H}_i\,|n\rangle\langle n|\,M_{\text{op}}(\mathbf{q})\,|m\rangle}{E_m - E_n} \right\}}{\sum_m \exp(-\beta E_m)}$$

In a uniform system only the term $H(\mathbf{q})$ of $\mathscr{H}_i$ will contribute to the sum.

Now the different operators $\gamma^*_{\mathbf{k}+\mathbf{q}0}\gamma_{\mathbf{k}0}$, $\gamma^*_{\mathbf{k}1}\gamma_{\mathbf{k},+\mathbf{q}1}$, $\gamma^*_{\mathbf{k}+\mathbf{q}0}\gamma^*_{\mathbf{k}1}$, $\gamma_{\mathbf{k}+\mathbf{q}1}\gamma_{\mathbf{k}0}$ of $M_{\text{op}}(\mathbf{q})$ will respectively scatter a quasi-particle of type 0, scatter a quasi-particle of type 1, create two quasi-particles, and destroy particles, with corresponding increases of energy $E_{\mathbf{k}+\mathbf{q}} - E_{\mathbf{k}}$, $E_{\mathbf{k}} - E_{\mathbf{k}+\mathbf{q}}$, $E_{\mathbf{k}+\mathbf{q}} + E_{\mathbf{k}}$, $-E_{\mathbf{k}+\mathbf{q}} - E_{\mathbf{k}}$. It follows that in the sum over n, the state $|n\rangle$ can differ from the state $|m\rangle$ only by the state of one quasi-particle or by having two more or two less quasi-particles. It is then simple to evaluate all the matrix elements, and if one also takes into account the thermal probability of occupation of the single-particle states, one finds for the magnetic spin susceptibility,

$$\chi(\mathbf{q}) = \frac{M(\mathbf{q})}{H(\mathbf{q})} = -2\mu_e^2 \sum_{\mathbf{k}} \left\{ (u_{\mathbf{k}+\mathbf{q}}u_{\mathbf{k}} + v_{\mathbf{k}+\mathbf{q}}v_{\mathbf{k}})^2 \frac{f_{\mathbf{k}+\mathbf{q}} - {}_{\mathbf{k}}}{E_{\mathbf{k}+\mathbf{q}} - E_{\mathbf{k}}} + (u_{\mathbf{k}+\mathbf{q}}v_{\mathbf{k}} - v_{\mathbf{k}+\mathbf{q}}u_{\mathbf{k}})^2 \frac{1 - f_{\mathbf{k}} - f_{\mathbf{k}+\mathbf{q}}}{E_{\mathbf{k}} + E_{\mathbf{k}+\mathbf{q}}} \right\}$$

The factors $u_{\mathbf{k}+\mathbf{q}}u_{\mathbf{k}} + v_{\mathbf{k}+\mathbf{q}}v_{\mathbf{k}}$ and $u_{\mathbf{k}+\mathbf{q}}v_{\mathbf{k}} - v_{\mathbf{k}+\mathbf{q}}u_{\mathbf{k}}$, which come from the matrix elements, arise because of the coherence of the single-particle states in the basic wave function. These factors are characteristic of the external interaction and the pairing and are known as "coherence factors."

In a uniform magnetic field the susceptibility is

$$\chi_s(0) = -2\mu_e^2 \sum_{\mathbf{k}} \partial f_{\mathbf{k}}/\partial E_{\mathbf{k}} \tag{51}$$

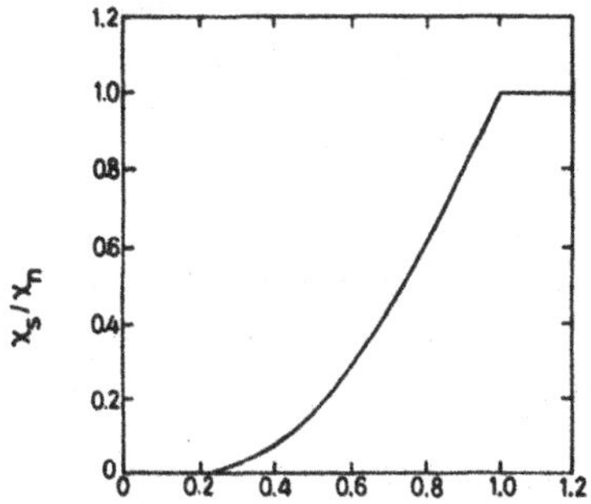

Fig. 6. Plot of the ratio χ_s/χ_n vs. T/T_c.

This result was first obtained by Yoshida (*68*). Its most striking feature is the fact that it vanishes at the absolute zero. This result is easily understood because all the electrons in the ground state are paired in such a way that the total spin of a pair is zero. To obtain a finite magnetic moment, pairs must be broken up so that some spins can be reversed. This costs a finite amount of energy which cannot be supplied by the static magnetic field. Hence the susceptibility is zero.

In a normal metal the susceptibility is

$$\chi_n = 2\mu_e^2 N(0)$$

The ratio of $\chi_s(0)$ to this is plotted against reduced temperature in Fig. 6.

XIV. EFFECTS OF TRANSVERSE ELECTROMAGNETIC FIELDS

In this section we consider the linear effects of a general transverse electromagnetic field on a superconductor. This covers a wide variety of properties of type I superconductors, and of type II superconductors below the critical magnetic field, H_{c1}. Above the critical magnetic field H_{c1} in a type II superconductor, the effect of external fields is no longer linear and will be considered elsewhere in this treatise.

Since there are strong magnetic interactions between the electrons, the field which affects any particular electron is a sum of the external field and the contributions of the fields of the other electrons. The latter fields will be subject to quantum fluctuations and should be described by operators. Fortunately, however, the quantum fluctuations in the magnetic fields are of relative order of magnitude $(v_F/c)(d/a_B)$, where v_F is the velocity of an electron, d the separation of the electrons, and a_B the Bohr radius. Since this number is of the order of 10^{-3}–10^{-2}, the quantum fluctuations are negligible; the magnetic fields of the electrons are therefore classical and can be added to the external field to form one macroscopic self-consistent field. Thus the problem is divided into two parts. In the first one finds the effect of a general self-consistent field on the superconductor. In the second, one separates the contribution of the electrons to the total field from the contribution of the external field. Hence one finds the effect on the superconductor of a given external field.

The effect of a given self-consistent field on the superconductor can be described through a classical vector potential $\mathbf{A}(\mathbf{r}, t)$. Since we are concerned only with linear effects we need only the linear term in the interaction Hamiltonian, and this is

$$\mathscr{H}_i(t) = (ieh/2mc) \int d^3r\, \psi^*(\mathbf{r})(\nabla\cdot\mathbf{A} + \mathbf{A}\cdot\nabla)\psi(\mathbf{r})$$

(The spin susceptibility of the electrons is not important for the Meissner effect and surface impedance, and is, therefore, ignored.) In terms of $\boldsymbol{a}(\mathbf{q}, t)$, the spatial Fourier transform of the vector potential, and of the quasi-particle operators, we have

$$\begin{aligned}\mathscr{H}_i(t) = -(eh/2mc) \sum_{\mathbf{k},\mathbf{q}} &[\mathbf{a}(-\mathbf{q}, t)\cdot(2\mathbf{k} + \mathbf{q})] \\ &\times [(u_{\mathbf{k}+\mathbf{q}}v_{\mathbf{k}} - v_{\mathbf{k}+\mathbf{q}}u_{\mathbf{k}})(\gamma^*_{\mathbf{k}+\mathbf{q}0}\gamma^*_{\mathbf{k}1} - \gamma_{\mathbf{k}1}\gamma_{\mathbf{k}+\mathbf{q}0}) \\ &+ (u_{\mathbf{k}}u_{\mathbf{k}+\mathbf{q}} + v_{\mathbf{k}}v_{\mathbf{k}+\mathbf{q}})(\gamma^*_{\mathbf{k}+\mathbf{q}0}\gamma_{\mathbf{k}0} - \gamma^*_{\mathbf{k}1}\gamma_{\mathbf{k}+\mathbf{q}0})]\end{aligned} \quad (52)$$

The response of the superconductor to the external field is usually given in terms of the current density. To first order in the field, the operator for the current density is

$$\begin{aligned}\mathbf{j}_{\text{op}}(\mathbf{r}) &= (e/2m)\{\psi^*(-i\hbar\nabla - e\mathbf{A}/c)\psi - [(-i\hbar\nabla + e\mathbf{A}/c)\psi^*]\psi\} \\ &= \mathbf{j}_1(\mathbf{r}) + \mathbf{j}_2(\mathbf{r})\end{aligned} \quad (53)$$

where

$$\mathbf{j}_2(\mathbf{r}) = (e^2/2mc)\,\mathbf{A}(\mathbf{r}, t)\,\psi^*\psi \tag{54}$$

and, in terms of quasi-particle operators,

$$\mathbf{j}_1(\mathbf{r}) = (e\hbar/2m\mathscr{V}) \sum_{\mathbf{k},\mathbf{q}} (2\mathbf{k} + \mathbf{q}) \exp(i\mathbf{q}\cdot\mathbf{r}) [(u_{\mathbf{k}} v_{\mathbf{k}+\mathbf{q}} - v_{\mathbf{k}} u_{\mathbf{k}+\mathbf{q}}) \times (\gamma^*_{\mathbf{k}0}\gamma^*_{\mathbf{k}+\mathbf{q}1} - \gamma_{\mathbf{k}1}\gamma_{\mathbf{k}+\mathbf{q}0}) + (u_{\mathbf{k}} u_{\mathbf{k}+\mathbf{q}} + v_{\mathbf{k}} v_{\mathbf{k}+\mathbf{q}})(\gamma^*_{\mathbf{k}0}\gamma_{\mathbf{k}+\mathbf{q}0} - \gamma^*_{\mathbf{k}+\mathbf{q}1}\gamma_{\mathbf{k}1})] \tag{55}$$

The expectation value of the current density in the external field can be found by simple perturbation theory. One has to first order in the external field (*69*) that the $\mathbf{q}$th Fourier component of this expectation value is

$$\mathbf{J}(\mathbf{q}, t) = \operatorname{Tr} \rho_M \mathbf{j}_{\rm op}(\mathbf{q}) - (i/\hbar) \int_0^t dt' \operatorname{Tr} \rho_M [\hat{\mathbf{j}}_{\rm op}(\mathbf{q}, t), \hat{\mathscr{H}}_i(t')] \tag{56}$$

where ρ_M is the density matrix for the model Hamiltonian, that is,

$$\rho_M = \exp(-\beta\mathscr{H}_M)/\operatorname{Tr}\exp(-\beta\mathscr{H}_M)$$

and $\hat{\mathbf{j}}_{\rm op}$, $\hat{\mathscr{H}}_i$ are Heisenberg operators defined by

$$\hat{\mathbf{j}}_{\rm op}(t) = \exp(i\mathscr{H}_M t/\hbar)\,\mathbf{j}_{\rm op} \exp(-i\mathscr{H}_M t/\hbar) \tag{57}$$

and a similar equation for $\hat{\mathscr{H}}_i$. It is assumed in Eq. (55) that the field is switched on at the time, $t = 0$.

Since we have so far included in the theory no dissipative mechanism, the current density given by Eq. (55) will contain transients due to the switching on of the field. In practice, dissipation would lead these transients to decay in a very short time, and the measured quantity would be the induced current density after a long time. In a rough and ready way one can take into account weak dissipation by putting a factor $\exp[-(t - t')/\tau]$ in the integrand, and looking at the current density after times long compared with τ. After such long times the different frequency components of $\mathbf{A}$ will lead to independent components of $\mathbf{J}$, of the same frequencies, which can, therefore, be calculated separately.

The component $\mathbf{j}_2$ of $\mathbf{j}_{\rm op}$ will contribute only to the first term of Eq. (55) and will give a contribution

$$\mathbf{J}_2(\mathbf{q}, \omega) = -(ne^2/mc)\,\mathbf{a}(\mathbf{q}, \omega) \tag{58}$$

where n is the density of electrons.

The component $\mathbf{j}_1$ of $\mathbf{j}_{\rm op}$ contributes only to the second term of Eq. (55) and yields a contribution

$$\mathbf{J}_1(\mathbf{q}, \omega) = \sum_m \exp(-\beta E_m) \left\{ \frac{\langle m|\mathbf{j}_1(\mathbf{q})|n\rangle\langle n|\mathscr{H}_i(\omega)|m\rangle}{E_m - E_n + \hbar\omega + i\hbar/\tau} - \frac{\langle m|\mathscr{H}_i(\omega)|n\rangle\langle n|\mathbf{j}_1(\mathbf{q})|m\rangle}{E_n - E_m + \hbar\omega + i\hbar/\tau} \right\} \times \left[\sum_m \exp(-\beta E_m)\right]^{-1} \tag{59}$$

where

$$\mathscr{H}_i(\omega) = -(\mathscr{V}/c)\sum_{\mathbf{q}} \mathbf{a}(-\mathbf{q}, \omega)\cdot\mathbf{j}_1(\mathbf{q})$$

The matrix elements here can be evaluated in the same way as for the spin susceptibility and lead to the result

$$\begin{aligned}\mathbf{J}(\mathbf{q}, \omega) = {} & \frac{e^2\hbar^2}{4m^2\mathscr{V}} \sum_{\mathbf{k}} \mathbf{a}(\mathbf{q}, \omega)\cdot(2\mathbf{k}+\mathbf{q})(2\mathbf{k}+\mathbf{q}) \\ & \times \left\{(u_{\mathbf{k}}u_{\mathbf{k}+\mathbf{q}} + v_{\mathbf{k}}v_{\mathbf{k}+\mathbf{q}})^2 \left[\frac{f_{\mathbf{k}} - f_{\mathbf{k}+\mathbf{q}}}{E_{\mathbf{k}+\mathbf{q}} - E_{\mathbf{k}} + \hbar\omega + i\hbar/\tau} + \frac{f_{\mathbf{k}+\mathbf{q}} - f_{\mathbf{k}}}{E_{\mathbf{k}} - E_{\mathbf{k}+\mathbf{q}} + \hbar\omega + i\hbar/\tau}\right]\right. \\ & \left. + (u_{\mathbf{k}}v_{\mathbf{k}+\mathbf{q}} - v_{\mathbf{k}}u_{\mathbf{k}+\mathbf{q}})^2 \left[\frac{1 - f_{\mathbf{k}} - f_{\mathbf{k}+\mathbf{q}}}{E_{\mathbf{k}+\mathbf{q}} + E_{\mathbf{k}} + \hbar\omega + i\ /\tau} + \frac{1 - f_{\mathbf{k}} - f_{\mathbf{k}+\mathbf{q}}}{E_{\mathbf{k}+\mathbf{q}} + E_{\mathbf{k}} - \hbar\omega - i\hbar/\tau}\right]\right\} \\ & - (ne^2/mc)\,\mathbf{a}(\mathbf{q}, \omega) \end{aligned} \tag{60}$$

For the weak dissipation we have assumed, this result is required in the limit of τ tending to infinity. Equation (60) is the theoretical basis of the electromagnetic properties of superconductors which we now discuss in turn.

XV. MEISSNER EFFECT AND PENETRATION DEPTH

We consider first of all the effect of a static magnetic field. For an isotropic superconductor in the static limit of Eq. (60), we have the equation

$$\mathbf{J}(\mathbf{q}) = -cK(\mathbf{q})\,\mathbf{a}(\mathbf{q})/4\pi \tag{61}$$

where

$$\begin{aligned}\frac{cK(\mathbf{q})}{4\pi} = \frac{ne^2}{mc} - \frac{e^2\hbar^2}{6m^2\mathscr{V}} \sum_{k} (2k)^2 & \left\{(u_{\mathbf{k}}u_{\mathbf{k}+\mathbf{q}} + v_{\mathbf{k}}v_{\mathbf{k}+\mathbf{q}})^2\right. \\ & \left. \times \frac{f_{\mathbf{k}} - f_{\mathbf{k}+\mathbf{q}}}{E_{\mathbf{k}+\mathbf{q}} - E_{\mathbf{k}}} + (u_{\mathbf{k}}v_{\mathbf{k}+\mathbf{q}} - v_{\mathbf{k}}u_{\mathbf{k}+\mathbf{q}})^2 \frac{1 - f_{\mathbf{k}} - f_{\mathbf{k}+\mathbf{q}}}{E_{\mathbf{k}+\mathbf{q}} + E_{\mathbf{k}}}\right\}\end{aligned} \tag{62}$$

Equation (61) expresses the current induced in the electrons by an arbitrary total macroscopic field $\boldsymbol{a}(\mathbf{q})$. The part $\boldsymbol{a}_s(\mathbf{q})$ of the total vector potential which is due to the electrons of the superconductor is given by Maxwell's equation

$$q^2\mathbf{a}_s(\mathbf{q}) = (4\pi/c)\,\mathbf{J}(\mathbf{q}) \tag{63}$$

The remainder of the vector potential $\boldsymbol{a}(\mathbf{q}) - \boldsymbol{a}_s(\mathbf{q})$ comes from the applied field. Given a particular applied field, Eqs. (61) and (63) can be solved and the solution describes the actual field inside the superconductor.

If the field varies slowly in space the solution is found easily. For in that case

$$\begin{aligned}\mathbf{J}(\mathbf{q}) &= -cK(0)\,\mathbf{a}(\mathbf{q})/4\pi \\ \mathbf{J}(\mathbf{r}) &= -cK(0)\,\mathbf{a}(\mathbf{r})/4\pi\end{aligned} \tag{64}$$

Hence, from this and Eq. (63),

$$\nabla^2 \mathbf{B}_s(\mathbf{r}) = K(0)\mathbf{B}(\mathbf{r})$$

If we have a finite superconductor with an external field imposed parallel to a plane face of the metal, inside the metal $B_s(r) = B(r)$. On the boundary, B is equal to the external field H. Therefore within the metal

$$B = H \exp(-z/\lambda_L) \tag{65}$$

where z is the distance within the conductor from the surface and λ_L is a length given by

$$\lambda_L = [K(0)]^{-1/2}$$

Hence the field penetrates only a finite distance into the superconductor if $K(0)$ exists and is positive. Since the field always varies slowly at large distances from a source, at large distances it is always proportional to $\exp(-z/\lambda_L)$ if λ_L exists. Therefore, the necessary and sufficient condition for finite penetration by the magnetic field (the Meissner effect) is that $K(0)$ should exist and be positive.

From Eq. (62) we have

$$\frac{cK(0)}{4\pi} = \frac{ne^2}{mc} - \frac{2e^2h^2}{3m^2\mathscr{V}} \sum_{\mathbf{k}} k^2 \frac{\partial f_{\mathbf{k}}}{\partial E_{\mathbf{k}}} \tag{66}$$

which, as long as Δ exists, is finite and positive. Hence the BCS theory does lead to the Meissner effect. The value of the penetration depth λ_L is for most superconductors of the order of 10^{-6}–10^{-5} cm. The length λ_L is a function of temperature being equal to

$$(mc^2/4\pi ne^2)^{1/2}$$

at the absolute zero, and tending to infinity at the critical temperature. A plot of $\lambda_L(T)$ vs. the reduced temperature is shown in Fig. 7.

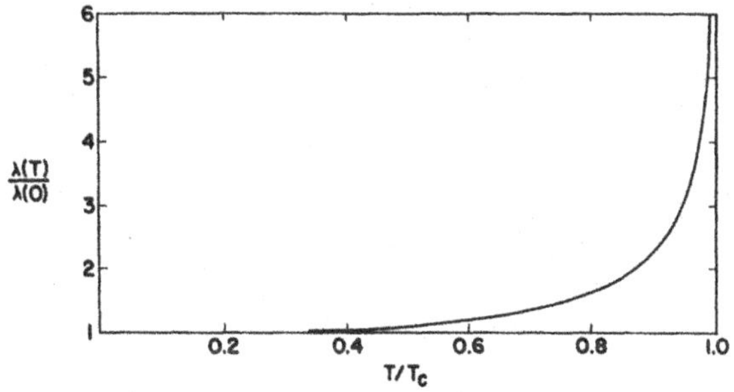

Fig. 7. Penetration depth of a pure superconductor in the London limit ($\xi_0 \ll \lambda$) as a function of temperature. The ratio $\lambda_L(T)/\lambda_L(0)$ is shown plotted vs. T/T_c.

The length λ_L is the penetration depth when the approximate relation (64) between current and field can be used instead of the exact relation (61). This will be possible when $\hbar v_F \mathbf{q}$, where v_F is the velocity of an electron at the Fermi surface, is small compared with Δ. Now the important values of q are evidently of the order of λ_L, so, for λ_L to be the penetration depth, we then require

$$\lambda_L \gg \hbar v_F / \Delta$$

It is usual to define a length called the coherence length ξ_0 by

$$\xi_0 = \hbar v_F / \pi \Delta \tag{67}$$

Then ξ_0 is the range of the Fourier transform $k(\mathbf{R})$ of $K(\mathbf{q})$ and the condition for the penetration depth to be λ_L is that this range should be much less than λ_L. Superconductors which satisfy this inequality are known as London superconductors (after F. and H. London) or type II superconductors. Pure superconductors rarely satisfy this condition, although niobium is an exception. The range of the function $k(\mathbf{R})$ is considerably reduced by impurities, with the result that many impure superconductors are type II superconductors.

When λ_L is not very much greater than ξ_0 it is necessary to use Eq. (61) in conjunction with Eq. (63) to solve the problem. In this case the relation between current and field is nonlocal and, to solve the problem of a finite superconductor, it is necessary to add boundary conditions for the electrons at the surface. Two of these have been considered in detail (*70–72*), and for the full analysis the reader is referred to the original articles. In general the penetration depth is defined in terms of the magnetic induction by

$$\lambda = \int_0^\infty B_\parallel(z)\, dz / B_\parallel(0) \tag{68}$$

where z is the distance from the surface and $B_\parallel$ is the component of the induction parallel to the surface. For the two special cases the results for λ are as follows: (1) If the electrons are specularly reflected at the surface as if at a perfectly reflecting mirror,

$$\lambda = \frac{2}{\pi} \int_0^\infty \frac{dq}{q^2 + K(q)} \tag{69}$$

(2) If the electrons are reflected randomly at the surface,

$$\lambda = \pi \left\{ \int_0^\infty dq \ln[1 + q^{-2} K(q)] \right\}^{-1} \tag{70}$$

For superconductors, known as Pippard superconductors, for which the extreme inequality $\xi_0 \gg \lambda_L$ holds, it is still possible to obtain results in an analytic form. If we denote the penetration depth in this case by λ_∞ we have that at the absolute zero

$$\lambda_\infty(0) = \left[\frac{\sqrt{3}\,\xi_0 mc^2}{8\pi^2 ne^2}\right]^{1/3} \qquad \text{(random scattering)}$$
$$\lambda_\infty(0) = \frac{8}{9}\left[\frac{\sqrt{3}\,\xi_0 mc^2}{8\pi^2 ne^2}\right]^{1/3} \qquad \text{(specular reflection)} \tag{71}$$

For random scattering we find that at an arbitrary temperature

$$\frac{\lambda_\infty(T)}{\lambda_\infty(0)} = \left[\frac{\Delta(T)\tanh\frac{1}{2}\beta\Delta(T)}{\Delta(0)}\right]^{-1/3} \tag{72}$$

A plot of this against the function $[1-(T/T_c)^4]^{-1/2}$ is almost linear. Hence the BCS theory agrees quite well with the older empirical curves.

For superconductors with parameters between the two extremes the penetration depth is a function of the temperature and of another parameter which can be taken to be $\lambda_L(0)$. No simple analytic expression can be given for the penetration depth and, so far, no tables have been published giving the penetration depth as a function of the two parameters.

To round off the discussion of penetration depth we ought to specify in more detail the effect of impurities. We have so far been able to avoid mentioning impurities because their effect (at least the effect of nonparamagnetic impurities) on the thermodynamic functions comes only through the change in the energy spectrum and this is negligible in an isotropic‡ superconductor. On the other hand, they can change the effect of a perturbation and, in particular, alter the range of the function $k(\mathbf{R})$ when the mean free path is of the order of or smaller than ξ_0. Since for the common superconductors, ξ_0 is of the order of 10^{-4} cm, the mean free path usually has an effect on the range of k.

The mathematical techniques of this chapter are too simple to be used to calculate the effects of impurities. Instead one has to use Green's functions or a similar approach. We shall, therefore, quote the result which can be given simply, and the reader is referred elsewhere for details (*16,74*). For the pure metal the relation (61) between induced current density and vector potential can be written in ordinary space as

$$\mathbf{J}(\mathbf{r}) = -(c/4\pi)\int d^3r'\,k(\mathbf{r}-\mathbf{r}')\,\mathbf{A}(\mathbf{r}')$$

In the impure metal this relation is altered to

$$\mathbf{J}(\mathbf{r}) = -(c/4\pi)\int d^3r'\,k(\mathbf{r}-\mathbf{r}')\exp[-|\mathbf{r}-\mathbf{r}'|/l]\,\mathbf{A}(\mathbf{r}') \tag{73}$$

‡ In an anisotropic superconductor the main effect is to make the superconductor appear more isotropic and so conform more to the previous theory (*73*).

where the function k is the same as in the pure metal and l is the mean free path of electrons in the normal metal.

From Eq. (73) it is clear that the current density at a particular point $\mathbf{r}$ depends on the vector potential at all points within a radius which depends on ξ_0 and l. For sufficiently impure superconductors this radius is l, and if we also have

$$\lambda_L \gg l$$

the relation between current density and vector potential is again the local one,

$$\mathbf{J}(\mathbf{r}) = -(c/4\pi)\, K(0)\, \mathbf{A}(\mathbf{r})$$

with

$$K(0) = \int d^3r'\, k(r') \exp(-r'/l) = \lambda_L^{-2}$$

Hence all sufficiently impure superconductors are London superconductors.

The penetration depth has been evaluated by Miller (*75*) for some special cases and he finds that for $l \ll \xi_0$,

$$\lambda \to \begin{cases} \lambda_L(0)\,(\xi_0/l)^{1/2} & \text{as } T \to 0 \\ \lambda_L(0)\,(2\hbar v_F k T_c/\pi l \Delta^2)^{1/2} & \text{as } T \to T_c \end{cases} \tag{74}$$

For general values of l, the penetration depth is a function of three parameters l, ξ_0, and T. There are no tabulations of λ for the most general case but some special cases have been tabulated by Miller (*75*).

Before leaving the subject of the Meissner effect and the penetration depth we must mention the question of the gauge invariance of the calculation. The essential point is that if we calculate the effect of a longitudinal vector potential by the technique used to calculate the effect of a transverse vector potential, we obtain the same formal result, Eq. (63), with, to order q/k_F, the same coefficient $K(\mathbf{q})$. This means that, since $K(\mathbf{q})$ does not vanish, a longitudinal static vector potential, according to the calculation, induces a finite current. However, we know that a longitudinal static vector potential is a mathematical construct which has no physical effects. There must be, therefore, something wrong with the calculation.

This subject was for a long time a matter of controversy, but it is now clear how the calculation is to be rectified. The author has written about this extensively elsewhere (*16*) and does not intend to go over the ground here, especially as it will be discussed in Chapter 7. The main point to be made here is that the calculation of the effect of the transverse vector potential, as given above, can be justified and the results given above are correct.

XVI. EFFECTS OF VARYING TRANSVERSE ELECTROMAGNETIC FIELDS

The title of this section covers a multiplicity of different experimental phenomena including the absorption, reflection, and transmission of different electromagnetic waves by thin films and bulk samples. Theoretically all these phenomena are covered by the two Eqs. (60) and (63), and to find the prediction for any particular phenomenon they have to be solved with the appropriate parameters and boundary conditions. We shall content ourselves by pointing out a number of special cases.

The relation (60) between current and field can also be written in the forms

$$\mathbf{J}(\mathbf{q}, \omega) = -(c/4\pi) K(\mathbf{q}, \omega) \mathbf{a}(\mathbf{q}, \omega) \tag{75}$$

$$\mathbf{J}(\mathbf{r}, \omega) = -(c/4\pi) \int d^3r' \, k(\mathbf{r} - \mathbf{r}', \omega) \mathbf{A}(\mathbf{r}', \omega) \tag{76}$$

where the function k is the spatial transform of K. For impure specimens the function $k(\mathbf{r}', \omega)$ for pure specimens is again multiplied by $\exp(-r'/l)$. Since the vector potential is related to the electric field by

$$\mathbf{a}(\mathbf{q}, \omega) = c\mathbf{E}(\mathbf{q}, \omega)/i\omega$$

the superconductor has a frequency- and wave vector–dependent complex conductivity $\sigma(\mathbf{q}, \omega)$ defined by

$$\begin{aligned} \sigma(\mathbf{q}, \omega) &= \sigma_1 + i\sigma_2 \\ &= -c^2 K(\mathbf{q}, \omega)/4\pi i\omega \end{aligned}$$

For absorption σ must have a real part, that is, $K(\mathbf{q}, \omega)$ must have an imaginary part, and this will be so if, and only if, the real parts of the denominators in Eq. (60) vanish. At the absolute zero, this requires a minimum frequency of $2\Delta/\hbar$, which is in accord with the idea that absorption takes place with the creation of pairs of quasi-particles. At high temperatures absorption can take place with the scattering of quasi-particles, and hence there is some absorption at frequencies less than $2\Delta/\hbar$. At frequencies above $2\Delta/\hbar$ the absorption increases rapidly until at very high frequencies (for example, optical frequencies) superconductivity is unimportant and the absorption is the same as in the normal state. In any particular experiment, although the observation is made at one particular frequency it is not made at one particular wave vector; the wave vectors involved are determined by the boundary conditions and the parameters of the superconductor.

The simplest special case is that when the coherence length and mean free path are much less than the penetration depth and when ω is much less than $\Delta/\hbar$ and v_F/l, the superconductor possesses a simple imaginary frequency-dependent conductivity given by

$$\sigma(\mathbf{q}, \omega) = ic^2/4\pi\omega\lambda_L^2 \tag{77}$$

Hence the field will just penetrate the superconductor to the distance λ_L but there will be no absorption. If ω is small and comparable with $v_F q \sim v_F/\lambda_L$, then the dependence of σ on wave vector becomes important again and the result is not simple.

At the other extreme, when the coherence length is much larger than the other lengths, we can expand $K(q, \omega)$ in powers of $(q\xi_0)^{-1}$ and l/ξ_0. The leading term then yields

$$\sigma(\mathbf{q}, \omega) = \left(\frac{\sigma_1}{\sigma_n} + \frac{i\sigma_2}{\sigma_n}\right)\sigma_n(\mathbf{q}) \tag{78}$$

where the ratios σ_1/σ_n, σ_2/σ_n are independent of $\mathbf{q}$ and l, and have been tabulated by Miller (*76*). Analytic forms for these ratios are

$$\frac{\sigma_1}{\sigma_n} = \frac{2}{\hbar\omega}\int_{\Delta}^{\infty} [f(E) - f(E + \hbar\omega)]\, g(E)\, dE + \frac{\theta(\hbar\omega - 2\Delta)}{\hbar\omega}\int_{\Delta - \hbar\omega}^{-\Delta} [1 - 2f(E + \hbar\omega)]\, g(E)\, dE \tag{79}$$

$$\frac{\sigma_2}{\sigma_n} = \frac{1}{\hbar\omega}\left[\int_{-\Delta}^{\Delta} dE + \theta(\hbar\omega - 2\Delta)\int_{\Delta - \hbar\omega}^{-\Delta} dE\right] \times \frac{[1 - 2f(E + \hbar\omega)]\,(E^2 + \Delta^2 + \hbar\omega\Delta)}{(\Delta^2 - E^2)^{1/2}\,[(E + \hbar\omega)^2 - \Delta^2]} \tag{80}$$

where

$$g(E) = \frac{E^2 + \Delta^2 + E\hbar\omega}{(E^2 - \Delta^2)^{1/2}[(E + \hbar\omega)^2 - \Delta^2]^{1/2}}$$

The conductivity $\sigma_n(\mathbf{q})$ is the wave vector–dependent conductivity in the normal state.

For pure bulk materials the extreme anomalous limit $ql \gg 1$ usually applies. In that case

$$\sigma_n = (3\pi/4lq)\,\sigma_{\mathrm{dc}}$$

where σ_{dc} is the dc conductivity in the normal state. The quantity which is usually determined in experiments on bulk material is the surface impedance $Z(\omega)$ defined in terms of the surface field and current by

$$Z(\omega) = -\,c^2 E_{\parallel}(0)\left\{\int_0^{\infty} j_{\parallel}(z)\, dz\right\}^{-1}$$

For the case of random scattering of electrons at the surface one finds

$$Z(\omega) = 4\pi^2 i\omega \left\{ \int_0^\infty \ln\left[1 + q^{-2}K(q, \omega)\right] dq \right\}^{-1}$$

and in the extreme anomalous limit

$$Z_s(\omega)/Z_n(\omega) = [(\sigma_1 + i\sigma_2)/\sigma_n]^{-1/3} \tag{81}$$

In situations where the mean free path is short compared with the variation of field, we can in formula (78) simply take $\sigma_n(\mathbf{q})$ to be the dc conductivity. The electrodynamic properties of the specimen in this case are described completely by a frequency–dependent complex conductivity. This result has been used in the analysis of data on thin films (*77*).

Usually the experimental conditions are such that none of the extreme formulas apply exactly. In that case it is necessary to compute the appropriate result numerically. A number of useful formulas have been given by Miller (*76*) for help in the computation.

XVII. INFINITE CONDUCTIVITY

The outstanding electromagnetic property of a superconductor is the one from which it derives its name, the superconductivity. If a superconductor is an element in a dc network, the voltage drop across it is zero (within the accuracy of the measurements). Furthermore, if a superconductor is in the form of a loop, a current can be induced in it, and this current will persist without any measurable decay. These two phenomena are evidently connected, although they are not quite equivalent. We shall discuss the former first.

The voltage drop across the superconductor is a measure of the average electric field within the superconductor. Since the voltage is zero, the average electric field is zero or, in other words, the Fourier component of the field with zero wave vector, in steady conditions, is zero. One can test whether the theory predicts such a result by considering the effect of an electric field on the superconductor. If the field produces a current which increases with time, the network will settle down to a steady state only if the field is zero. Superconductivity is implied if a given field would induce an increasing current. Since the effect exists however small the current, it is linear and we need consider only what current is induced by a uniform electric field. The induced current would be accompanied by a magnetic field, but this does not affect the issue of whether or not the metal is a superconductor.

We need, therefore, to calculate the effect of a uniform steady electric field on the superconductor. This can be obtained as the limit of the effect of a uniform sinusoidally electric field as the frequency tends to zero. In this limit the system possesses a conductivity given by Eq. (77). Hence

$$\mathbf{J}(\omega) = (ic^2/4\pi\omega\lambda_L^2)\mathbf{E}(\omega)$$

and for slowly varying uniform electric fields

$$\partial\mathbf{J}(t)/\partial t = (c^2/4\pi\lambda_L^2)\mathbf{E}(t) \tag{82}$$

This is the equation for the current induced in a system of n_s freely accelerating electrons, where

$$n_s = c^2 m/4\pi\lambda_L^2 e^2 \tag{83}$$

For this reason the density n_s is often called the density of superconducting electrons. In a pure metal at the absolute zero, n_s is equal to n, the density of electrons in the conduction band of the superconductors. In impure metals there has been some controversy (*78*) concerning what mass m should go into the formula (83); in practice only the ratio n_s/m is physically significant (*78a*), and this is well defined.

Equation (82) clearly predicts that the current induced by a uniform steady electric field does increase with time; hence the theory predicts that the metal is a superconductor. So far, the only scattering mechanism we have included in the theory is impurity scattering. However, it can be shown that this result is not affected by phonon scattering (*79*). In fact, there is clearly a close connection between the Meissner effect and the infinite conductivity; they both depend on $K(\mathbf{q}, \omega)$ as q and ω tend to zero. For the Meissner effect q tends to zero after ω is put equal to zero; for the infinite conductivity ω tends to zero after q is put equal to zero. It can be shown (*79*) that if there is any scattering which can carry away momentum, $K(0, 0)$ exists and is independent of the order in which the limits are taken. It follows that the two effects occur together as in a superconductor where $K(0, 0)$ is nonzero, or neither effect occurs as in a normal metal where $K(0, 0)$ is zero.

The mathematical calculation confirms the following simple picture of what happens in a superconductor. A uniform superconductor can be in a large number of possible metastable states, in which the Bloch states are paired $[\mathbf{k} + (\mathbf{q}/2), \uparrow]$ [where $\mathbf{k} + (\mathbf{q}/2)$ is the momentum corresponding to the state] with $[-\mathbf{k} + (\mathbf{q}/2), \downarrow]$, each value of $\mathbf{q}$ corresponding to a different metastable state. These states are metastable because any single-particle excitation requires a positive amount of energy to be excited. To change one of these states to another requires a macroscopic change of the whole system; such a macroscopic change can be and is brought about by a macroscopic field such as a uniform electric field. Scattering, however, changes only the single-particle excitations, ensuring that the system is in equilibrium in the state where the value of $\mathbf{q}$ is determined by the macroscopic field. Hence in a steady uniform electric field the current increases because the value of $\mathbf{q}$ changes with time according to

$$\frac{\partial}{\partial t}\left(\frac{\hbar\mathbf{q}}{2}\right) = e\mathbf{E} \tag{84}$$

At each instant the whole system adjusts so that it is in metastable equilibrium with the instantaneous value of $\mathbf{q}$.

In a thin loop of superconductor where a persistent current flows one has a similar picture. When the current flows the system is in a state of metastable equilibrium with a definite value of $\mathbf{q}$ and, because of the metastability, the current does not decay. The fact that we have a loop, however, adds another constraint on the system because, according to the quantum theory, only certain discrete values of $\mathbf{q}$, and hence of persistent current, are allowed. This means that the magnetic flux produced by the current within the loop is quantized. This effect is discussed more fully in Chapter 8.

There is another useful way of characterizing the different metastable states. We have seen that different states of zero current (at different temperatures) are characterized by different values of the gap parameter

$$\Delta = V\sum_{\mathbf{k}}\langle c_{-\mathbf{k}\downarrow}c_{\mathbf{k}\uparrow}\rangle$$

where, for simplicity, we take V to be independent of k. Then one can write

$$\Delta(\mathbf{r}) = V\langle\psi_{\downarrow}(\mathbf{r})\psi_{\uparrow}(\mathbf{r})\rangle$$

and the states of zero current have Δ independent of $\mathbf{r}$. Now it is easy to see that a state with a definite value of $\mathbf{q}$ has

$$\Delta(\mathbf{r}) \propto \exp(i\mathbf{q}\cdot\mathbf{r})$$

and the phase of Δ depends on position. Given the phase of Δ, the modulus of Δ is determined by the fact that the observed state will be one of metastable equilibrium. In this form the idea is easily generalized. The phase of Δ can be an arbitrary function of position (strictly speaking for this generalization it should be slowly varying) and each different phase determines a different state of metastable equilibrium. In any given problem the macroscopic boundary conditions determine the appropriate phase of Δ and hence the appropriate state of metastable equilibrium. Cases where Δ depends on position are discussed more fully elsewhere in this treatise.

XVIII. ULTRASONIC ATTENUATION

We shall consider first the attenuation of longitudinal waves in a pure superconductor. Over most of the temperature range the frequency and wave vector of the wave are related by the inequalities

$$\omega < v_F q < \Delta/\hbar$$

Under these conditions the electrons tend to follow the wave adiabatically and will screen the wave as if it were a static field. Since the screening in the superconductive state is the same as in the normal state (Chapter 10), the residual interaction is the same in the two states, and is given by

$$\mathscr{H}_i = (\chi/\mathscr{V}^{1/2}) \sum_{\mathbf{k},\mathbf{q},\sigma} \omega_{\mathbf{q}}^{1/2} c_{\mathbf{k}+\mathbf{q}\sigma}^* c_{\mathbf{k}\sigma} (b_{-\mathbf{q}}^* + b_{\mathbf{q}}) \tag{85}$$

where χ is a constant and b_q^* and b_q are, respectively, creation and destruction operators for phonons.

From the interaction (85) one can calculate the rate of absorption of phonons by the electrons using the appropriate golden rule. The probability/unit time for a transition from any possible initial state $|m\rangle$ to any state $|n\rangle$ is

$$R = (2\pi/\hbar) \sum_{m,n} \exp(-\beta E_m) |\langle n| \mathscr{H}_i |m\rangle|^2 \delta(E_m - E_n) / \sum_m \exp(-\beta E_m)$$

We take the states $|m\rangle$ and $|n\rangle$ to be states with definite numbers of phonons and quasi-particles, and we write the interaction $\mathscr{H}_i$ in terms of the quasi-particle operators:

$$\mathscr{H}_i = (\chi/\mathscr{V}^{1/2}) \sum_{\mathbf{k},\mathbf{q}} \omega_{\mathbf{q}}^{1/2} (b_{-\mathbf{q}}^* + b_{\mathbf{q}}) [(u_{\mathbf{k}+\mathbf{q}} v_{\mathbf{k}} + v_{\mathbf{k}+\mathbf{q}} u_{\mathbf{k}})(\gamma_{\mathbf{k}+\mathbf{q}0}^* \gamma_{\mathbf{k}1}^* + \gamma_{\mathbf{k}+\mathbf{q}1} \gamma_{\mathbf{k}0}) + (u_{\mathbf{k}} u_{\mathbf{k}+\mathbf{q}} - v_{\mathbf{k}} v_{\mathbf{k}+\mathbf{q}})(\gamma_{\mathbf{k}+\mathbf{q}0}^* \gamma_{\mathbf{k}0} + \gamma_{\mathbf{k}1}^* \gamma_{\mathbf{k}+\mathbf{q}1})] \tag{86}$$

Since the angular frequency is less than $\Delta/\hbar$, only the terms of the interaction which scatter quasi-particles are important.‡ Hence the probability per unit time that a phonon is absorbed is

$$R_a = (4\pi\chi^2 \omega_{\mathbf{q}} n_{\mathbf{q}}/\hbar\mathscr{V}) \sum_{\mathbf{k}} (u_{\mathbf{k}} u_{\mathbf{k}+\mathbf{q}} - v_{\mathbf{k}} v_{\mathbf{k}+\mathbf{q}})^2 f_{\mathbf{k}} (1 - f_{\mathbf{k}+\mathbf{q}}) \delta(E_{\mathbf{k}+\mathbf{q}} - E_{\mathbf{k}} - \hbar\omega_{\mathbf{q}})$$

where $n_{\mathbf{q}}$ is the number of phonons of wave vector $\mathbf{q}$ present initially, which for a macroscopic sound wave is much greater than unity. Similarly, the probability per unit time that a phonon is emitted is

$$R_e = (4\pi\chi^2 \omega_{\mathbf{q}} n_{\mathbf{q}}/\hbar\mathscr{V}) \sum_{\mathbf{k}} (u_{\mathbf{k}} u_{\mathbf{k}+\mathbf{q}} - v_{\mathbf{k}} v_{\mathbf{k}+\mathbf{q}})^2 f_{\mathbf{k}+\mathbf{q}} (1 - f_{\mathbf{k}}) \delta(E_{\mathbf{k}+\mathbf{q}} - E_{\mathbf{k}} - \hbar\omega_{\mathbf{q}})$$

‡ For fixed ω, Δ will become less than $\hbar\omega/2$ as the temperature approaches T_c and the other terms of the interaction are then important. The correction for this has been calculated by Privorotskii (*80*).

It follows that

$$\frac{dn_{\mathbf{q}}}{dt} = -\frac{4\pi\chi^2\omega_{\mathbf{q}}n_{\mathbf{q}}}{\hbar\mathscr{V}}\sum_{\mathbf{k}}(u_{\mathbf{k}}u_{\mathbf{k}+\mathbf{q}} - v_{\mathbf{k}}v_{\mathbf{k}+\mathbf{q}})^2(f_{\mathbf{k}} - f_{\mathbf{k}+\mathbf{q}})\delta(E_{\mathbf{k}+\mathbf{q}} - E_{\mathbf{k}} - \hbar\omega_{\mathbf{q}})$$

and for small $\omega_{\mathbf{q}}$ this can be written

$$\begin{aligned}\frac{dn_{\mathbf{q}}}{dt} &= \frac{4\pi\chi^2\omega_{\mathbf{q}}^2n_{\mathbf{q}}}{\mathscr{V}}\sum_{\mathbf{k}}(u_{\mathbf{k}}^2 - v_{\mathbf{k}}^2)^2\frac{\partial f_{\mathbf{k}}}{\partial E_{\mathbf{k}}}\,\delta(E_{\mathbf{k}+\mathbf{q}} - E_{\mathbf{k}})\\ &= \frac{4\pi\chi^2\omega_{\mathbf{q}}^2n_{\mathbf{q}}}{\mathscr{V}}N(0)\int d\epsilon\left(\frac{\epsilon}{E}\right)^2\frac{\partial f}{\partial E}\int\frac{d\mu}{2}\left|\frac{E}{\epsilon}\right|\delta(\epsilon_{\mathbf{k}+\mathbf{q}} - \epsilon_{\mathbf{k}})\end{aligned}$$

The attenuation coefficient is proportional to the coefficient of $n_{\mathbf{q}}$. Since this coefficient is correctly given here both for Δ finite and for Δ equal to zero, we find that

$$\frac{\alpha_s}{\alpha_n} = \int_0^\infty d\epsilon\frac{\epsilon}{E}\frac{\partial f}{\partial E}\bigg/\int_0^\infty d\epsilon\frac{\partial f}{\partial \epsilon} = 2f(\Delta) \tag{87}$$

This very simple result means that ultrasonic attenuation provides a direct measure of the gap parameter Δ.

It is worth noting that because of the conservation of energy as expressed by the delta function, the quasi-particles which contribute predominantly to the absorption are those with momentum perpendicular to the direction of motion of the acoustic wave. Hence a given acoustic wave samples only a small part of the Fermi surface (*81*). This makes ultrasonic attenuation a useful tool for finding the gap parameter in anisotropic superconductors (*82,83*).

Although the final result depends only on the gap, any test of the formula is not simply a test of a theory with a gap in its spectrum of excitations. It can be seen by tracing through the calculation that the coherence factors, which reflect an important property of the wave function, are significant in arriving at the simple result. Any test is, therefore, a partial test of the basic wave function.

We have considered ultrasonic attenuation only for the case where the mean free path is long compared with the wavelength. Other cases are also of practical importance and have been evaluated originally using Green's functions by Tsuneto (*84*) and more recently using a Boltzmann equation by a number of authors (*85–87*). Surprisingly the result [Eq. (87)] is still true in isotropic impure superconductors.

We conclude this section by mentioning the attenuation of transverse waves. In pure superconductors only the electromagnetic field accompanying the wave is important. However, except near T_c, this transverse field is effectively screened by the Meissner effect and the attenuation is zero. In impure superconductors, the electrons are dragged along with the lattice through collisions and there is a finite attenuation. For this calculation the reader is referred to the original papers (*87–89*).

XIX. THERMAL CONDUCTIVITY OF THE ELECTRONS

A. Impurity Scattering

If a small thermal gradient is applied to a superconductor, heat will be carried from the hot to the cold end by the excitations. If the gradient is held steady, the distribution of quasi-particles will be stationary but different from the equilibrium distribution. This changed distribution is what has to be calculated theoretically.

In general, the Boltzmann equation for the distribution function $f_{\mathbf{k}}(\mathbf{r})$ is

$$\frac{\partial f_{\mathbf{k}}}{\partial t} + \frac{\partial \mathbf{k}}{\partial t}\cdot\frac{\partial f_{\mathbf{k}}}{\partial \mathbf{k}} + \mathbf{v}\cdot\frac{\partial f_{\mathbf{k}}}{\partial \mathbf{r}} = \left.\frac{\partial f_{\mathbf{k}}}{\partial t}\right]_{\text{coll}} \tag{88}$$

In steady conditions with no mechanical or electromagnetic forces the first two terms on the left side are zero. In practice the thermal conductivity is measured under conditions of no net electric current. Since a thermal current tends to drag a small electric current with it, this current must be balanced by an equal and opposite current. In a normal metal, the second current is due to an electric field which is set up in the metal, and this field leads to a negligibly small correction to the thermal conductivity. In a superconductor, a small electric field would lead to infinite electric and thermal currents (*90*), and so cannot exist there. Instead, the dragged electric current is balanced by a supercurrent. As far as the thermal conductivity is concerned this supercurrent can be ignored (*91*). Hence Eq. (88) is valid and the first two terms on the left side are zero.

The last term on the left side involves the velocity of a quasi-particle of momentum $\hbar\mathbf{k}$. Since the energy of this particle is $E_{\mathbf{k}}$, we expect the velocity to be

$$\mathbf{v} = \partial E_{\mathbf{k}}/\hbar\,\partial\mathbf{k}$$

and this is borne out by more detailed considerations of the motion of wave packets (*16*). Hence

$$\mathbf{v} = \frac{\epsilon}{E}\frac{\hbar\mathbf{k}}{m} \tag{89}$$

The velocity of the quasi-particle is zero at the Fermi surface and changes sign on crossing the Fermi surface. This is in keeping with the fact that a quasi-particle has electron-like and hole-like characteristics. Above the Fermi surface it is more like an electron; below, it is more like a hole. The last term on the left side also involves the spatial derivative of the distribution function. Since this function depends on the coordinates through its dependence on temperature, it might be thought that we have to include the dependence of the gap parameter on the coordinates. A spatially dependent gap parameter, however, will not by itself produce a thermal current and so when calculating linear effects this spatial dependence can be ignored. Hence we can write

$$\frac{\partial f_{\mathbf{k}}}{\partial \mathbf{r}} = \left.\frac{\partial f_{\mathbf{k}}}{\partial \mathbf{r}}\right]_{\Delta\,\text{const}} \approx -\mathbf{v}\cdot\nabla T \frac{E}{T}\frac{\partial f_{\text{eq}}}{\partial T} \tag{90}$$

To evaluate the rate of change of the distribution function due to collisions we have to consider particular mechanisms for scattering electrons. Let us take impurity scattering first. The interaction between the electrons and a single impurity is given by

$$\mathscr{H}_i = \sum_{\mathbf{k},\mathbf{q},\sigma} U(\mathbf{q})\, c^*_{\mathbf{k}+\mathbf{q}\sigma} c_{\mathbf{k}\sigma} \tag{91}$$

where $U(\mathbf{q})$ is the Fourier transform of the potential due to the impurity. In terms of the operators for quasi-particles this can be rewritten

$$\begin{aligned}\mathscr{H}_i = \sum_{\mathbf{k},\mathbf{q}} U(\mathbf{q})[&(u_{\mathbf{k}+\mathbf{q}}v_{\mathbf{k}} + v_{\mathbf{k}+\mathbf{q}}u_{\mathbf{k}})(\gamma^*_{\mathbf{k}+\mathbf{q}0}\gamma^*_{\mathbf{k}1} + \gamma_{\mathbf{k}+\mathbf{q}1}\gamma_{\mathbf{k}0}) \\ &+ (u_{\mathbf{k}}u_{\mathbf{k}+\mathbf{q}} - v_{\mathbf{k}}v_{\mathbf{k}+\mathbf{q}})(\gamma^*_{\mathbf{k}+\mathbf{q}0}\gamma_{\mathbf{k}0} + \gamma^*_{\mathbf{k}1}\gamma_{\mathbf{k}+\mathbf{q}1})]\end{aligned}$$

Hence the probability per unit time that a quasi-particle of type 0 is scattered from $\mathbf{k}$ to $\mathbf{k}+\mathbf{q}$ is

$$R(\mathbf{k}, \mathbf{k}+\mathbf{q}) = \frac{2\pi}{\hbar}|U(\mathbf{q})|^2 (u_{\mathbf{k}}u_{\mathbf{k}+\mathbf{q}} - v_{\mathbf{k}}v_{\mathbf{k}+\mathbf{q}})^2 f_{\mathbf{k}}(1 - f_{\mathbf{k}+\mathbf{q}})\delta(E_{\mathbf{k}} - E_{\mathbf{k}+\mathbf{q}}) \tag{92}$$

It follows that the rate of increase of quasi-particles in the state $(\mathbf{k}, 0)$ is

$$\begin{aligned}\left.\frac{\partial f_{\mathbf{k}}}{\partial t}\right]_{\text{coll}} &= \sum_{\mathbf{q}} [R(\mathbf{k}+\mathbf{q}, \mathbf{k}) - R(\mathbf{k}, \mathbf{k}+\mathbf{q})] \\ &= \frac{2\pi}{\hbar}\sum_{\mathbf{q}} (u_{\mathbf{k}}u_{\mathbf{k}+\mathbf{q}} - v_{\mathbf{k}}v_{\mathbf{k}+\mathbf{q}})^2 (f_{\mathbf{k}+\mathbf{q}} - f_{\mathbf{k}})\delta(E_{\mathbf{k}} - E_{\mathbf{k}+\mathbf{q}})|U(\mathbf{q})|^2\end{aligned} \tag{93}$$

If the impurities scatter independently, this is multiplied by the density of impurities N_i for the effect of all the impurities. This completes Eq. (88) for the particles of type 0.

We can show that to first order in the temperature gradient the equation has a solution of the form

$$f_{\mathbf{k}} = f_{\text{eq},\mathbf{k}} + \mathbf{k}\cdot\nabla T\, g(E_{\mathbf{k}}) \tag{94}$$

where f_{eq} is the equilibrium distribution of quasi-particles. With the form (94) for $f_{\mathbf{k}}$ the collision term is

$$\left.\frac{\partial f_{\mathbf{k}}}{\partial T}\right]_{coll} = \frac{2\pi}{\hbar}\left(\frac{\epsilon}{E}\right)^2 g(E) N_i \sum_{\mathbf{k}} |U(\mathbf{k}-\mathbf{k}')|^2 (\mathbf{k}'-\mathbf{k})\cdot\nabla T\, \delta(E_{\mathbf{k}'}-E_{\mathbf{k}})$$

$$= -\frac{2\pi}{\hbar}\left(\frac{\epsilon}{E}\right)^2 g(E) N_i N(0)\mathbf{k}\cdot\nabla T \int d\epsilon' \frac{d\mu}{2} |U(\mathbf{k}-\mathbf{k}')|^2 (1-\mu)\,\delta(E'-E)$$

where k, k' can be put equal to k_F except in the terms involving ϵ, ϵ'. Hence

$$\left.\frac{\partial f_{\mathbf{k}}}{\partial T}\right]_{coll} = -\frac{2\pi}{\hbar}\left|\frac{\epsilon}{E}\right| g(E) N_i N(0)\mathbf{k}\cdot\nabla T \int d\mu\, |U(\mathbf{k}-\mathbf{k}')|^2 (1-\mu)$$

For the normal state the corresponding term is

$$\left.\frac{\partial f_{\mathbf{k}}}{\partial T}\right]_{coll} = -\frac{2\pi}{\hbar} g(E) N_i N(0)\mathbf{k}\cdot\nabla T \int d\mu\, |U(\mathbf{k}-\mathbf{k}')|^2 (1-\mu)$$

$$= -\mathbf{k}\cdot\nabla T\, g(E)/\tau$$

where τ is the mean free time in the normal state. Hence in the superconductive state one can write

$$\left.\frac{\partial f_{\mathbf{k}}}{\partial T}\right]_{coll} = -\left|\frac{\epsilon}{E}\right| \mathbf{k}\cdot\nabla T \frac{g(E)}{\tau}$$

where τ is still the mean free time in the normal state.

The Boltzmann equation can now be written

$$\frac{\epsilon}{E}\frac{\hbar\mathbf{k}}{m}\cdot\nabla T \frac{E}{T}\frac{\partial f}{\partial E} = \left|\frac{\epsilon}{E}\right| g(E) \frac{\mathbf{k}\cdot\nabla T}{\tau}$$

and has the solution

$$g(E) = \frac{\epsilon}{|\epsilon|}\frac{\hbar\tau}{m}\frac{E}{T}\frac{\partial f_{eq}}{\partial E} \tag{95}$$

In conjunction with (94), this provides us with the perturbed distribution of quasi-particles.

The measured quantity is the thermal conductivity, defined by

$$\kappa = \frac{\text{heat current}}{\text{thermal gradient}}$$

Since the energy of a quasi-particle is $E_{\mathbf{k}}$ and its velocity is $\mathbf{v}$, the heat current carried by one kind of quasi-particle is

$$\mathbf{W}_0 = \sum_{\mathbf{k}} E_{\mathbf{k}} \mathbf{v} f_{\mathbf{k}}$$

Since the two kinds of quasi-particle carry the same amount of heat, the total heat current is

$$\mathbf{W} = 2\sum_{\mathbf{k}} E_{\mathbf{k}} \mathbf{v}\,\mathbf{k}\cdot\nabla T \frac{\epsilon}{|\epsilon|}\frac{\hbar}{m}\frac{\tau E}{T}\frac{\partial f_{\text{eq}}}{\partial E}$$

$$= \frac{2}{3}\frac{\tau}{T} N(0) v_F^2 \nabla T \int_{-\infty}^{\infty} d\epsilon\,|\epsilon|\,E\frac{\partial f}{\partial E} \tag{96}$$

Hence

$$\kappa = \frac{4}{3}\frac{\tau}{T} N(0) v_F^2 \int_{\Delta}^{\infty} dE\,E^2\frac{\partial f}{\partial E} \tag{97}$$

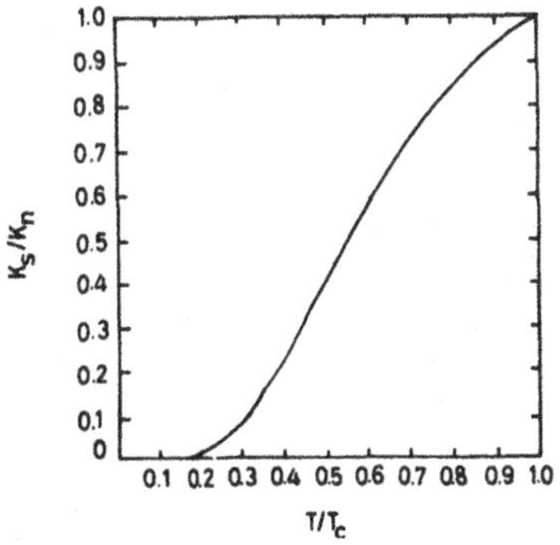

Fig. 8. Thermal conductivity of a superconductor as a function of temperature when impurity scattering is dominant *(92)*. κ_s/κ_n is shown plotted vs. T/T_c.

and the ratio of the thermal conductivities in the superconductive and normal states is

$$\frac{\kappa_s}{\kappa_n} = \frac{\int_{\Delta}^{\infty} dE\,E^2\,\partial f/\partial E}{\int_{0}^{\infty} dE\,E^2\,\partial f/\partial E} \tag{98}$$

This result is shown plotted in Fig. 8. The ratio is a universal function of the reduced temperature.

B. Phonon Scattering

If phonon scattering is important, we have the same basic Boltzmann equation but the collision term is different. Since the phonons carry energy the collisions are inelastic and more collision processes are possible. The number of quasi-particles in the state $(\mathbf{k}, 0)$ (say) is changed when a quasi-particle is scattered into or out of this state with emission or absorption of a phonon, when two quasi-particles are created by the absorption of a phonon, and when two quasi-particles are destroyed with the emission of a phonon. All these processes are described by the interaction Hamiltonian [Eq. (86)], and from it the rate of change of the distribution functions $f_{\mathbf{k}0}$, $f_{\mathbf{k}1}$ can be obtained.

The resulting Boltzmann equations for $f_{\mathbf{k}0}$, $f_{\mathbf{k}1}$ were first written down by Bardeen et al. (*92*), but their attempt to solve them by means of a variational principle failed because the trial function gave a poor result even in the normal case. Since then the equations have been solved numerically by Tewordt (*93*),

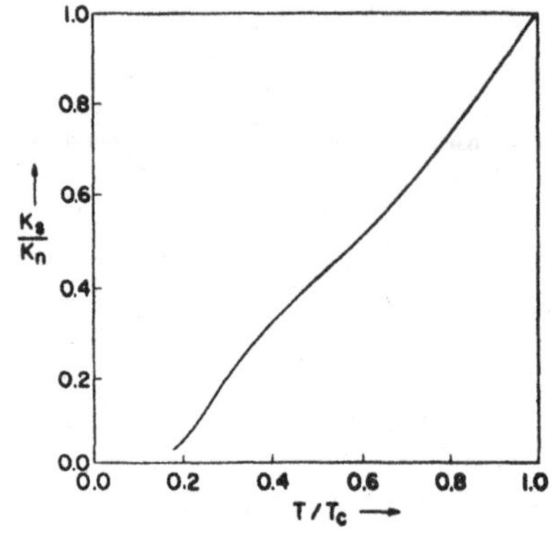

Fig. 9. Thermal conductivity of a superconductor in the weak-coupling limit as a function of temperature when phonon scattering is dominant. κ_s/κ_n is shown plotted vs. (T/T_c) (*93*).

whose result for the ratio κ_s/κ_n is plotted in Fig. 9. Near T_c this curve is strikingly different from the curve for the case when impurity scattering is dominant, and experiment can easily discriminate between the two possible scattering mechanisms.

Well below T_c, the electronic thermal conductivity is small whichever mechanism is dominant. This is purely a result of the few thermal excitations present to carry the heat, and is in accord with experiment. At these low temperatures the heat is usually carried by the phonons.

XX. TUNNELING OF ELECTRONS

If two metals are separated by a very thin strip of insulating material, electrons can tunnel quantum mechanically from one into the other, and a current can pass through the junction. For normal metals the current passing through the junction is proportional to the voltage across the junction and the resistance of the junction is ohmic. If one of the metals is superconductive or if both are, it is still possible for electrons to tunnel through the junction and for a current to pass. In this case, however, the resistance is not ohmic and, indeed, provides useful and direct information about the properties of the superconductor.

In normal circumstances it is possible to describe the tunneling process by means of a tunneling Hamiltonian (*94–96*) which describes the transfer of electrons from one metal to the other. If we use creation operators c^* for the electrons in one metal, A, and creation operators d^* for the electrons in the other, B, the tunneling Hamiltonian is

$$\mathscr{H}_T = \sum_{\mathbf{k},\mathbf{k}'\sigma} (T_{\mathbf{k},\mathbf{k}'} c_{\mathbf{k}\sigma}^* d_{\mathbf{k}'\sigma} + T_{\mathbf{k},\mathbf{k}'}^* d_{\mathbf{k}'\sigma}^* c_{\mathbf{k}\sigma}) \tag{99}$$

The first of these terms describes the transfer of an electron from B to A, the second the transfer of an electron from A to B.

If the metal A is at a potential V above that of metal B, the rate at which electrons cross from A to B is given by

$$\begin{aligned} R_{AB} = \frac{2\pi}{\hbar} \{\sum \exp[-\beta(E_{m_A} + E_{m_B})]\}^{-1} \sum_{\substack{m_A, m_B \\ n_A, n_B}} \exp[-\beta(E_{m_A} + E_{m_B})] \\ \times \{|\langle m_A, m_B| \sum T_{\mathbf{k}\mathbf{k}'}^* d_{\mathbf{k}'\sigma}^* c_{\mathbf{k}\sigma} |n_A n_B\rangle|^2 \delta(E_{m_A} + eV + E_{m_B} - E_{n_A} - E_{n_B}) \\ - |\langle m_A, m_B| \sum T_{\mathbf{k}\mathbf{k}'} c_{\mathbf{k}\sigma}^* d_{\mathbf{k}'\sigma} |n_A, n_B\rangle|^2 \delta(E_{m_A} - eV + E_{m_B} - E_{n_A} - E_{n_B})\} \end{aligned} \tag{100}$$

where the labels m_A, m_B, etc., refer to different states in the A and B metals. The current through the barrier is proportional to R_{AB}. Equation (100) describes incoherent quantum jumps of electrons from one side of the barrier to the other. If, however, both sides of the barrier are superconductive there are also coherent processes, in which the phase difference between the two sides is fixed, and which produce a current through the barrier. These processes are described in Chapters 9 and 10.

Now the interaction describes several different kinds of elementary process. In fact, in terms of the annihilation operators $\gamma_{\mathbf{k}}$, $\delta_{\mathbf{k}}$ for quasi-particles in metals A and B, respectively, we have

$$\begin{aligned} \sum_{\mathbf{k},\mathbf{k}'} T_{\mathbf{k}\mathbf{k}'}^* d_{\mathbf{k}'\sigma}^* c_{\mathbf{k}\sigma} = \sum_{\mathbf{k},\mathbf{k}'} \{T_{\mathbf{k}\mathbf{k}'}^* (u_{\mathbf{k}'B}^* \delta_{\mathbf{k}'0}^* + v_{\mathbf{k}'B} \delta_{\mathbf{k}'1})(u_{\mathbf{k}A} \gamma_{\mathbf{k}0} + v_{\mathbf{k}A}^* \gamma_{\mathbf{k}1}^*) \\ + T_{-\mathbf{k},-\mathbf{k}'}^* (-v_{\mathbf{k}'B} \delta_{\mathbf{k}'0} + u_{\mathbf{k}'B}^* \delta_{\mathbf{k}'1}^*)(-v_{\mathbf{k}A}^* \gamma_{\mathbf{k}0}^* + u_{\mathbf{k}A} \gamma_{\mathbf{k}1})\} \end{aligned} \tag{101}$$

where $u_{\mathbf{k}A}$, $v_{\mathbf{k}A}$ are probability amplitudes in metal A and $u_{\mathbf{k}B}$, $v_{\mathbf{k}B}$ are amplitudes in metal B. The interaction allows the scattering of quasi-particles from one side of the barrier to the other as well as the creation and destruction of quasi-

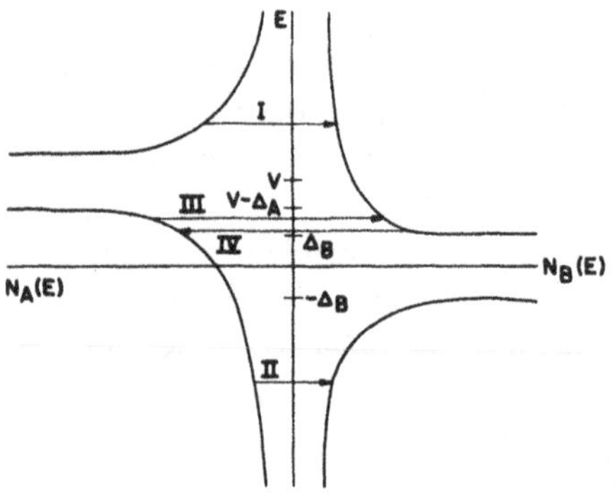

Fig. 10. Plot of density of states on two sides of barrier, and of quasi-particle processes allowed by "spin-up" terms of the Hamiltonian. The energies of quasi-particles of type 1 are conventionally taken to be negative. The arrows indicate the following processes: I, the scattering of a quasi-particle 0 from metal A to metal B; II, the scattering of a quasi-particle of type 1 from metal B to metal A; III, the creation of a quasi-particles of types 1 and 0 in metals A and B, respectively (note that an arrow pointing to a 0 curve represents the creation of a quasi-particle while an arrow pointing to a 1 curve represents the destruction of a quasi-particle); IV, the destruction of particles of type 1 and 0 in metals A and B, respectively. (For the processes allowed by the spin-down terms of the Hamiltonian we have to interchange the type 0 and type 1 particles.)

particles on the two sides of the barrier. These processes are illustrated in Fig. 10.

Since the different terms of the interaction (101) give rise to independent processes, one is led to the following form for R_{AB}:

$$\begin{aligned} R_{AB} = \frac{2\pi}{\hbar} \sum_{\mathbf{k},\mathbf{k}'} \{ & |T_{\mathbf{k},\mathbf{k}'}|^2 [|u_{\mathbf{k}A}|^2 |u_{\mathbf{k}'B}|^2 (f(E_{\mathbf{k}A})[1 - f(E_{\mathbf{k}'B})] \\ & - f(E_{\mathbf{k}'B})[1 - f(E_{\mathbf{k}A})]) \delta(E_{\mathbf{k}A} + eV - E_{\mathbf{k}'B}) \\ & + |v_{\mathbf{k}A}|^2 |v_{\mathbf{k}'B}|^2 (f(E_{\mathbf{k}'B})[1 - f(E_{\mathbf{k}A})] - f(E_{\mathbf{k}A})[1 - f(E_{\mathbf{k}'B})]) \\ & \times \delta(E_{\mathbf{k}'B} + eV - E_{\mathbf{k}A}) + |u_{\mathbf{k}'B}|^2 |v_{\mathbf{k}A}|^2 ([1 - f(E_{\mathbf{k}A})][1 - f(E_{\mathbf{k}'B})] \\ & - f(E_{\mathbf{k}A}) f(E_{\mathbf{k}'B})) \delta(eV - E_{\mathbf{k}A} - E_{\mathbf{k}'B})] \\ & + |T_{-\mathbf{k},-\mathbf{k}'}|^2 [|v_{\mathbf{k}'B}|^2 |v_{\mathbf{k}A}|^2 [f(E_{\mathbf{k}'B}) - f(E_{\mathbf{k}A})] \delta(eV + E_{\mathbf{k}'B} - E_{\mathbf{k}A}) \\ & + |u_{\mathbf{k}A}|^2 |v_{\mathbf{k}'B}|^2 [f(E_{\mathbf{k}A}) - f(E_{\mathbf{k}'B})] \delta(eV + E_{\mathbf{k}A} - E_{\mathbf{k}'B}) \\ & + |v_{\mathbf{k}A}|^2 |u_{\mathbf{k}'B}|^2 [1 - f(E_{\mathbf{k}A}) - f(E_{\mathbf{k}'B})] \delta(eV - E_{\mathbf{k}A} - E_{\mathbf{k}'B}) \end{aligned} \tag{102}$$

The important values of $\mathbf{k}$, $\mathbf{k}'$ are near the Fermi surface, and for these values $|T_{\mathbf{k},\mathbf{k}'}|^2$ can be replaced (in an isotropic superconductor) by an appropriate angular average $|T|^2$ (*97*). Further, since the sums over energy on the two sides of the barrier are independent, we can replace both $|u_A|^2$ and $|u_B|^2$ by $\frac{1}{2}$ throughout. Hence

$$R_{AB} = (\pi/\hbar)|T|^2 N_A(0) N_B(0) \int d\epsilon_A d\epsilon_B \{[f(E_A) - f(E_B)] [\delta(eV + E_A - E_B) - \delta(eV - E_A + E_B)] + [1 - f(E_A) - f(E_B)] \delta(eV - E_A - E_B)\} \quad (103)$$

One of the integrals for R_{AB} can be performed straightaway and leads to a result in which R_{AB} is a universal function of Δ_A, Δ_B, T, and V, which has to be computed numerically. So far this function has not been tabulated except for comparison with particular experiments. One special case, however, is particularly simple and is of considerable importance. This is the case where the temperature is sufficiently low for the Fermi–Dirac functions to be negligible, and where one of the metals, say B, is normal. Then

$$R_{AB} = (\pi/\hbar)|T|^2 N_A(0) N_B(0) \int d\epsilon_A d\epsilon_B \delta(eV - E_A - E_B) \quad (104)$$

where

$$E_B = |\epsilon_B|$$

Hence

$$R_{AB} = (2\pi/\hbar)|T|^2 N_A(0) N_B(0) \int d\epsilon_A \theta(eV - E_A)$$

where the θ function ensures that the integral involves only those energies for which $E_A \leqslant eV$. Since the current is proportional to R_{AB}, the conductance is proportional to R_{AB}, and the ratio of the conductance when metal A is in the superconductive state to that when the metal is in the normal state is

$$\frac{(dI/dV)_s}{(dI/dV)_n} = \frac{\int d\epsilon_A \delta(eV - E_A)}{\int d\epsilon_A \delta(eV - |\epsilon_A|)} = \frac{\eta_s(eV)}{\eta_n} = \frac{eV}{[(eV)^2 - \Delta^2]^{1/2}}$$

Hence the conductance provides a direct measure of the density of states in the superconductive state as well as of the gap. Even in models more general than that discussed by BCS this result is valid (*97*).

XXI. NUCLEAR SPIN RELAXATION TIME

According to theories of nuclear magnetic resonance (*98*) the spin-lattice relaxation time T_1 depends on the properties of the electrons solely through the tensor

$$\theta_{\alpha\beta} = Av \sum_n \langle m| S_\alpha(0) |n\rangle \langle n| S_\beta^*(0) |m\rangle \delta(E_m - E_n) \quad (106)$$

where $S(0)$ is the spin density of the electrons at a particular lattice site. In an isotropic superconductor

$$\theta_{\alpha\beta} = \theta\delta_{\alpha\beta} \tag{107}$$

$$\theta = Av\sum_n \langle m|\, S_z(0)\,|n\rangle\,\langle n|\, S_z^*(0)\,|m\rangle\,\delta(E_m - E_n) \tag{108}$$

$$R = T_1^{-1} \propto \theta$$

The ratio of relaxation times in the superconductive and normal states is, therefore, given by

$$R_s/R_n = \theta_s/\theta_n$$

Now the operator for the spin density of the electrons is

$$S_z(0) = u(0)\sum_{\mathbf{k},\mathbf{k}'} (c_{\mathbf{k}\uparrow}^* c_{\mathbf{k}'\uparrow} - c_{-\mathbf{k}'\downarrow}^* c_{-\mathbf{k}\downarrow})$$

where $u(0)$ is the value of the Bloch function at the nucleus.‡ In terms of the quasi-particle operators this is

$$S_z(0) = u(0)\sum_{\mathbf{k},\mathbf{k}'} [(u_{\mathbf{k}}u_{\mathbf{k}'} + v_{\mathbf{k}}v_{\mathbf{k}'})(\gamma_{\mathbf{k}0}^*\gamma_{\mathbf{k}'0} - \gamma_{\mathbf{k}'1}^*\gamma_{\mathbf{k}1}) + (u_{\mathbf{k}}v_{\mathbf{k}'} - u_{\mathbf{k}'}v_{\mathbf{k}})(\gamma_{\mathbf{k}0}^*\gamma_{\mathbf{k}'1}^* - \gamma_{\mathbf{k}1}\gamma_{\mathbf{k}'0})] \tag{109}$$

If the formula (109) for $S_z(0)$ is substituted in Eq. (108) one finds

$$\begin{aligned}\theta &= 2|u(0)|^2 \sum_{\mathbf{k},\mathbf{k}'} (u_{\mathbf{k}}u_{\mathbf{k}'} + v_{\mathbf{k}}v_{\mathbf{k}'})^2 f(E_{\mathbf{k}'})[1 - f(E_{\mathbf{k}})]\,\delta(E_{\mathbf{k}} - E_{\mathbf{k}'}) \\ &= |u(0)|^2 [N(0)]^2 \int d\epsilon\, d\epsilon' \left(1 + \frac{\Delta^2}{E^2}\right) f(E)[1 - f(E)]\,\delta(E - E') \\ &= |u(0)|^2 [N(0)]^2 \int_\Delta^\infty dE\, \eta_s(E)^2 (1 + \Delta^2/E^2) f(E)[1 - f(E)]\end{aligned}$$

Hence

$$R_s/R_n = -2\int_\Delta^\infty dE\,\eta_s(E)^2 (1 + \Delta^2/E^2)\,\partial f/\partial E \tag{110}$$

Unfortunately, the integral in Eq. (110) diverges logarithmically at the lower limit, a divergence which can be traced to the singular density of states in the superconductive state, $\eta_s(E)$. In their original paper, Hebel and Slichter assumed that there would be some level broadening which would remove this singularity so that the density of states of quasi-particles would go smoothly from zero to

‡ We have for most purposes assumed that the electrons can be described by plane waves. However, the relaxation of the nucleus depends strongly on the wave function at the nucleus, and to obtain correct absolute values for R it is necessary to include $u(0)$.

the BCS value as the energy increases through the value of the energy gap. It has since been pointed out (*99*) that because of the existence of thermal phonons, there is no true gap in the spectrum of single-particle excitations at a finite temperature. In fact, in Eq. (108) the states $|\alpha\rangle$ should be states of electrons and phonons. Fibich (*99*) has effectively performed this calculation for the weak coupling case and found

$$\frac{R_s}{R_n} = 2f(\Delta)\left\{1 + \frac{\Delta}{kT}[1 - f(\Delta)]\ln\left[\frac{2\Delta^{4/3}}{(\frac{1}{2}\Gamma)^{4/3}}\right]\right\}$$

where Γ depends on the coupling of the electrons to the phonons and on the phonon spectrum. If the superconductor possesses a single Debye frequency ω_D,

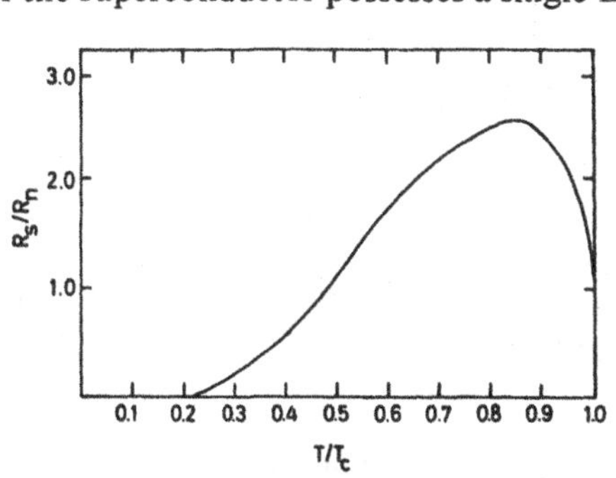

Fig. 11. Ratio R_s/R_n of the inverse spin-lattice relaxation times for aluminum in the superconductive and normal states (*99*).

and if the electron–phonon interaction changes the effective mass of an electron from m to m^* (*100*), the expression for Γ is

$$\Gamma^2(T) = (\pi^4/15)(1 - m/m^*)(kT)^4\omega_D^{-2}$$

Since R_s/R_n depends on m^* and ω_D, it is not a universal function of T/T_c. However, it depends logarithmically on m^* and ω_D and is not very sensitive to the values of these parameters. Hence a plot of R_s/R_n against temperature will have a similar shape for different metals. We show Fibich's plot for R_s/R_n for aluminum in Fig. 11.

XXII. CONCLUSIONS

The BCS theory clearly provides a basis for calculating all the properties of superconductors. We have chosen to calculate a few important ones as a guide to how to perform the calculations, and to give a feeling for the results. According to the BCS theory the superconductive system possesses an infinite number of macroscopically metastable states characterized by different values of the order parameter $\Delta(\mathbf{r})$, which can follow slowly varying external fields adiabatically. This property is responsible for superconductivity.

Dissipative and absorptive processes are caused by the quasi-particles which may exist in equilibrium or be created by external fields. The character of these processes is dominated at low temperature by the energy gap, which implies that few quasi-particles are available for absorption. Hence the absorption of low-frequency fields $\omega < 2\Delta/\hbar$ is small at low temperatures. This is a feature of ultrasonic attenuation, thermal conductivity, and nuclear spin relaxation at low temperatures. At temperatures near T_c, the gap is small and less important. The matrix elements in the form of the coherence factors play a more important part and account for the difference in the behavior of, say, ultrasonic attenuation and spin-lattice relaxation near T_c.

ACKNOWLEDGMENT

The author wishes to thank Dr. J. Garland for making available, prior to publication, the values quoted in Table I.

REFERENCES

1. J. Bardeen, L. N. Cooper, and J. R. Schrieffer, *Phys. Rev.* **108**, 1175 (1957).
2. B. S. Deaver, Jr., and W. M. Fairbank, *Phys. Rev. Letters* **7**, 43. (1961).
3. R. Doll, and M. Näbauer, *Phys. Rev. Letters* **7**, 51 (1961).
4. B. T. Matthias, T. H. Geballe, and V. B. Compton, *Rev. Mod. Phys.* **35**, 1 (1963).
5. D. Bohm and D. Pines, *Phys. Rev.* **92**, 609 (1953).
6. H. Fröhlich, *Phys. Rev.* **79**, 845 (1950).
7. E. Maxwell, *Phys. Rev.* **78**, 477 (1950).
8. C. A. Reynolds, B. Serin, W. H. Wright. and L. B. Nesbitt *Phys. Rev.* **78**, 487 (1950).
9. H. Fröhlich, *Proc. Roy. Soc. (London)* **A215**, 291 (1952).
10. L. D. Landau, *Zh. Eksperim. i Teor. Fiz.* **30**, 1058 (1956); *Soviet Phys. JETP* **3**, 920 (1957)
11. D. Pines and P. Nozières, *The Theory of Quantum Liquids*, Vol. 1, Benjamin, New York, 1966.
12. J. Bardeen, *Phys. Rev.* **59**, 928 (1941).
13. A. J. Leggett, *Phys. Rev.* **140**, A1869 (1965); **147**, 119 (1966).
14. L. D. Landau, *Zh. Eksperim. i. Teor. Fiz.* **32**, 59 (1957); *Soviet Phys. JETP* **5**, 101 (1957).
15. L. N. Cooper, *Phys. Rev.* **104**, 1189 (1956).
16. G. Rickayzen, *Theory of Superconductivity*, Wiley, New York, 1964.
17. N. N. Bogoliubov, *Nuovo Cimento* **7**, 794 (1958).
18. R. Balian and N. R. Werthamer, *Phys. Rev.* **131**, 1553 (1963).
19. I. A. Privorotskii, *Zh. Eksperim. i Teor. Fiz.* **45**, 1960 (1963); *Soviet Phys. JETP* **18**, 1346 (1964).

20. N. R. Werthamer, H. Suhl, and T. Soda, *Proc. of the Eighth Intern. Conf. on Low Temp. Physics* (R. O. Davies, ed.), Butterworth, London, 1963, p. 140.
21. S. V. Vonsovskiĭ and M. S. Svirsky, *Phys. Status Solidi* **9**, 267 (1965).
22. R. Balian, *Lectures on the Many-Body Problem*, Vol. 2, Academic Press, New York, 1964, p. 147.
23. E. G. Batyev, *Zh. Eksperim. i Teor. Fiz.* **50**, 215 (1966); *Soviet Phys. JETP* **23**, 141 (1966).
24. V. J. Emery and A. M. Sessler, *Phys. Rev.* **119**, 43 (1960).
25. J. Bardeen and G. Rickayzen, *Phys. Rev.* **118**, 936 (1960).
26. D. C. Mattis and E. Lieb, *J. Math. Phys.* **2**, 602 (1961).
27. N. N. Bogoliubov, D. N. Zubarev, and Yu. A. Tserkovnikov, *Zh. Eksperim. i Teor. Fiz.* **39**, 120 (1960); *Soviet Phys. JETP* **12**, 88 (1961).
28. P. W. Anderson, *Phys. Rev.* **112**, 1900 (1958).
29. L. P. Gor'kov, *Zh. Eksperim. i Teor. Fiz.* **34**, 735 (1958); *Soviet Phys. JETP* **7**, 505 (1958).
30. Y. Nambu, *Phys. Rev.* **117**, 648 (1960).
31. A. J. Coleman and S. Pruski, *Can. J. Phys.* **43**, 2142 (1965).
32. P. C. Martin, *Low Temperature Physics, LT9* (J. G. Daunt et al., eds.), Plenum Press, New York, 1965, p. 9.
33. N. N. Bogoliubov, *Usp. Fiz. Nauk* **67**, 549 (1959).
34. J. G. Valatin, *Nuovo Cimento* **7**, 843 (1958).
35. D. Pines, *Phys. Rev.* **109**, 280 (1958).
36. P. Morel, *J. Phys. Chem. Solids* **10**, 277 (1959).
37. N. N. Bogoliubov, V. V. Tolmachev, and D. V. Shirkov, *A New Method in the Theory of Superconductivity*, Academy of Science, Moscow, 1958, Consultants Bureau, New York, 1959.
38. P. Morel and P. W. Anderson, *Phys. Rev.* **125**, 1263 (1962).
39. J. W. Garland, Jr., *Phys. Rev. Letters* **11**, 114 (1963); *Phys. Rev.* **153**, 460 (1967).
40. B. T. Matthias et al., *Rev. Mod. Phys.* **36**, 156 (1964).
41. J. W. Garland, Jr., *Phys. Rev. Letters* **11**, 111 (1963).
42. T. H. Geballe, *Rev. Mod. Phys.* **36**, 134 (1964).
43. M. Levy and J. L. Olsen, *Physics of High Pressures and the Condensed Phase*, North-Holland, Amsterdam, 1965, p. 525.
43a. J. L. Olsen, K. Andres, and T. H. Geballe, *Phys. Letters* **26A**, 239 (1968).
44. V. I. Makarov and V. G. Baryakhatar, *Zh. Eksperim. i Teor. Fiz.* **48**, 1717 (1965); *Soviet Phys. JETP* **21**, 1151 (1965).
45. J. R. Schrieffer, D. J. Scalapino, and J. W. Wilkins, *Phys. Rev. Letters* **10**, 336 (1963).
46. B. Mühlschlegel, *Z. Physik* **155**, 313 (1959).
47. V. L. Pokrovskii and M. S. Ryvkin, *Zh. Eksperim. i Teor. Fiz.* **43**, 89 (1962); *Soviet Phys. JETP* **16**, 65 (1963).
48. B. T. Geilikman and V. Z. Kresin, *Fiz. Tverd. Tela* **5**, 3549 (1963); *Soviet Phys.-Solid State* **5**, 2605 (1964).
49. Y. Wada, *Phys. Rev.* **135**, A1481 (1964).
50. T. K. Melik-Barkhudarov, *Fiz. Tved. Tela* **7**, 1368 (1965); *Soviet Phys.-Solid State* **7**, 1103 (1965).
51. B. T. Geilikman and V. Z. Kresin, *Zh. Eksperim. i Teor. Fiz Pis'ma* **3**, 48 (1966); *Soviet Phys. JETP Letters* **3**, 28 (1966).
52. M. R. Schafroth, S. T. Butler, and J. M. Blatt, *Helv. Phys. Acta* **30**, 93 (1957).
53. J. M. Blatt, *Theory of Superconductivity*, Academic Press, New York, 1964.

54. R. Balian and D. R. Fredkin, *Phys. Rev. Letters* **15**, 480 (1965).
54a. S. Doniach and S. Engelsberg, *Phys. Rev. Letters* **17**, 750 (1966); N. F. Berk and J. R. Schrieffer, *Phys. Rev. Letters* **17**, 433 (1966).
55. R. Brout, *Phase Transitions*, Benjamin, New York, 1965.
56. D. J. Thouless, *Ann. Phys. (N.Y.)* **10**, 553 (1960).
57. E. G. Batyev, A. Z. Patashinskii, and V. L. Pokrovskii, *Zh. Eksperim. i Teor. Fiz.* **46**, 2093 (1964); *Soviet Phys. JETP* **19**, 1412 (1964).
58. V. L. Ginzburg, *Fiz. Tved. Tela* **2**, 2031 (1960); *Soviet Phys.-Solid State* **2**, 1924 (1961).
59. V. V. Shmidt, *Zh. Eksperim. i Teor. Fiz. Pis'ma* **3**, 141 (1966); *Soviet Phys. JETP Letters* **3**, 89 (1966).
59a. R. A. Ferrell and H. Schmidt, *Phys. Letters* **25A**, 54 (1967).
59b. R. E. Glover, *Phys. Letters* **25A**, 51 (1967).
60. W. A. Little, *Phys. Rev.* **134**, A1416 (1964).
61. R. A. Ferrell, *Phys. Rev. Letters* **13**, 330 (1964).
62. T. M. Rice, *Phys. Rev.* **140**, A1889 (1965).
63. L. D. Landau and E. M. Lifschitz, *Statistical Physics*, Addison-Wesley, Reading, Mass., 1958, p. 482.
64. Yu. A. Bychkov, L. P. Gor'kov, and I. E. Dzyaloshinskii, *Zh. Eksperim. i Teor. Fiz.* **50**, 738 (1966); *Soviet Phys. JETP* **23**, 489 (1966).
65. V. L. Ginzburg, *Phys. Letters* **13**, 101 (1964); *Zh. Eksperim. i Teor. Fiz.* **47**, 2318 (1964); *Soviet Phys. JETP* **20**, 1549 (1965).
66. V. L. Ginzburg and D. A. Kizhits, *Zh. Eksperim. i Teor. Fiz.* **46**, 397 (1964); *Soviet Phys. JETP* **19**, 269 (1964).
67. W. Silvert, *Phys. Rev. Letters* **14**, 951 (1965).
68. K. Yosida, *Phys. Rev.* **110**, 769 (1958).
69. R. Kubo, *J. Phys. Soc. Japan* **12**, 570 (1957).
70. G. H. Reuter and E. H. Sondheimer, *Proc. Roy. Soc. (London)* **A195**, 336 (1948).
71. R. B. Dingle, *Physica* **19**, 311 (1953).
72. A. B. Pippard, *Proc. Roy. Soc. (London)* **A216**, 547 (1953).
73. P. W. Anderson, *J. Phys. Chem. Solids* **11**, 26 (1959).
74. A. A. Abrikosov, L. P. Gor'kov, and I. E. Dzyaloshinskii, *Zh. Eksperim. i Teor. Fiz.* **36**, 900 (1959); *Soviet Phys. JETP* **9**, 636 (1959).
75. P. B. Miller, *Phys. Rev.* **113**, 1209 (1959).
76. P. B. Miller, *Phys. Rev.* **118**, 928 (1960).
77. D. M. Ginsberg and M. Tinkham, *Phys. Rev.* **118**, 990 (1960).
78. P. Nozières, *Quantum Fluids* (D.F. Brewer, ed.), North-Holland, Amsterdam, 1966, p. 8.
78a J. Bardeen, in *Quantum Theory of Atoms, Molecules, and the Solid State* (P.O. Löwdin, ed.), Academic Press, New York, 1966, p. 511
79. W. A. B. Evans and G. Rickayzen, *Ann. Phys. (N.Y.)* **33**, 275 (1965).
80. I. A. Privorotskii, *Zh. Eksperim. i Teor. Fiz.* **43**, 1331 (1962); *Soviet Phys. JETP* **16**, 945 (1963).
81. I. A. Privorotskii, *Zh. Eksperim. i Teor. Fiz.* **42**, 450 (1962); *Soviet Phys. JETP* **15**, 315 (1962).
82. R. W. Morse, T. Olsen, and J. D. Gavenda, *Phys. Rev. Letters* **3**, 15 (1959).
83. L. T. Claiborne and N. G. Einspruch, *Phys. Rev. Letters* **15**, 862 (1965).
84. T. Tsuneto, *Phys. Rev.* **121**, 402 (1961).
85. V. L. Pokrovskii and S. K. Savvinykh, *Zh. Eksperim. i Teor. Fiz.* **43**, 564 (1962); *Soviet Phys. JETP* **16**, 404 (1963).
86. M. Levy, *Phys. Rev.* **131**, 1497 (1963).

87. L. P. Kadanoff and A. B. Pippard, *Proc. Roy. Soc. (London)* **A292**, 299 (1966).
88. L. T. Claiborne, Ph. D. thesis, Brown University, Providence, R.I., 1961 (unpublished) quoted in R. W. Morse, *IBM J. Res. Develop.* **6**, 58 (1962).
89. J. R. Leibowitz, *Phys. Rev.* **136**, A22 (1964).
90. V. Ambegaokar and G. Rickayzen, *Phys. Rev.* **142**, 146 (1966).
91. J. M. Luttinger, *Phys. Rev.* **136**, A1481 (1964).
92. J. Bardeen, G. Rickayzen, and L. Tewordt, *Phys. Rev.* **113**, 982 (1959).
93. L. Tewordt, *Phys. Rev.* **129**, 657 (1963).
94. M. H. Cohen, L. M. Falicov, and J. C. Phillips, *Phys. Rev. Letters* **8**, 316 (1962).
95. R. E. Prange, *Phys. Rev.* **131**, 1083 (1963).
96. B. D. Josephson, *Advan. Phys.* **14**, 419 (1965).
97. D. J. Scalapino, J. R. Schrieffer, and J. W. Wilkins, *Phys. Rev.* **148**, 263 (1966).
98. L. C. Hebel and C. P. Slichter, *Phys. Rev.* **113**, 1504 (1959).
99. M. Fibich, *Phys. Rev. Letters* **14**, 561 (1965).
100. N. W. Ashcroft and J. W. Wilkins, *Phys. Letters* **14**, 285 (1965).

EQUILIBRIUM PROPERTIES: COMPARISON OF EXPERIMENTAL RESULTS WITH PREDICTIONS OF THE BCS THEORY

R. Meservey and B. B. Schwartz

FRANCIS BITTER NATIONAL MAGNET LABORATORY ‡
MASSACHUSETTS INSTITUTE OF TECHNOLOGY
CAMBRIDGE, MASSACHUSETTS

‡ Supported by the U.S. Air Force Office of Scientific Research.

I. INTRODUCTION

The purpose of this chapter and the next is to show to what extent the Bardeen, Cooper, and Schrieffer (BCS) (*1*) theory of superconductivity agrees with the experimental facts. The present chapter is concerned with equilibrium properties, whereas Chapter 4 discusses nonequilibrium properties. Since no rigorous distinction between these two categories is feasible we will start by listing the subjects covered in the present chapter.

1. The nature of the electron–electron interaction and pairing.
2. The density of states and the energy gap as determined by tunneling.
3. The equilibrium thermodynamic properties, including the specific heat, the critical field, and the pressure dependence of the transition temperature.
4. Equilibrium electromagnetic properties, including the penetration depth and Knight shift.

We cannot, of course discuss all the available experimental evidence, so we have chosen to emphasize those subjects which are most pertinent to the present task of comparison with the BCS theory. On a given subject we have intended to present the earliest pertinent experiments and have also tried to give the latest results. Although this method slights much important work, there have been several excellent reviews to which we can refer for more complete coverage. In addition, we have kept the theory to a minimum since this chapter is preceded by Rickayzen's on the BCS theory.

Summary of assumptions and results of the BCS theory:

1. In a metal there is an attractive force between pairs of conduction electrons caused by virtual phonon exchange and a repulsive force due to the screened Coulomb interaction.
2. When the combined effect of the two forces is attractive, the metal becomes superconducting.
3. In a superconductor, electrons are paired in the singlet state $(\mathbf{k}\uparrow, -\mathbf{k}\downarrow)$, such that each pair has the same total momentum.
4. Except for this pairing interaction, all other microscopic properties of the superconducting metal are assumed to be that of the normal metal.

In order to actually calculate measurable quantities, BCS chose a very simple model in which the interaction was assumed to be independent of electron energy, crystal direction, and phonon frequency up to some cutoff frequency, and above that frequency to be equal to zero. Since the original publication of the BCS theory, many refinements have been made, including an elegant microscopic formulation of the theory due to Gor'kov (*2*) using Green's function techniques. These later formulations will be considered as part of the BCS theory as long as the above assumptions remain as their theoretical basis.

The BCS model leads to a superconducting ground state and an excitation energy spectrum, and these in turn lead to most of the measurable properties predicted. To fully appreciate the nature of the BCS theory, however, it is important to realize that the theory was to an uncommon extent shaped to fit a wide range of empirical facts and phenomenological theories. The reason for choosing an isotropic model was based on the observation that superconductivity is not rare in metals and apparently is not associated with any particular crystal structure. The one-parameter BCS model immediately leads to a law of corresponding states for different metals, whose approximate existence had previously been well established. Lack of electrical resistance and the exponential temperature dependence of the specific heat seemed to require that the theory have an energy gap. The isotope effect very strongly hinted that the electron–electron interaction was by means of the lattice. The electrodynamics required by the experimentally based Pippard (*3*) phenomenological theory was nonlocal and involved a coherence distance. The fact that any basically simple theory could fit all these various demands of empirical knowledge is of course very impressive evidence for the theory. However, there is something particularly convincing about those predictions of a theory which are tested after the theory is complete. Because of this and because the older lines of evidence have been thoroughly presented (*1,4*) (see also Chapters 1 and 2), this chapter will emphasize recent developments such as tunneling and flux quantization.

II. BCS INTERACTION AND PAIRING

In the BCS (*1*) formulation of the theory of superconductivity, the ground state is constructed by correlating zero-momentum singlet electron pairs $(k\uparrow, -k\downarrow)$, neglecting all other electron–electron interactions which are not already present in the solution of the normal metal problem. The application of this pairing hypothesis results in the reduction of the many-body problem to a soluble problem which is believed to contain the qualitative features necessary for a description of superconductivity. This simple theoretical formulation leads to a complicated nonlinear integral equation for the (temperature– and wave number–dependent) energy gap parameter Δ_k:

$$\Delta_k = \frac{-1}{2} \sum_{k'} \frac{V_{kk'}}{E_{k'}} \Delta_{k'} \tanh\left(\beta \frac{E_{k'}}{2}\right) \tag{1a}$$

where $E_{k'} = (\epsilon_{k'}^2 + \Delta_{k'}^2)^{1/2}$, $\epsilon_k = (\hbar^2 k^2/2m) - \mu$, μ is the chemical potential, and $V_{kk'}$ is the matrix element for the scattering of a pair of electrons from state $(k\uparrow, -k\downarrow)$ to $(k'\uparrow, -k'\downarrow)$. $V_{kk'}$ is composed of two terms; the first describing the attractive electron–electron interaction mediated by phonons and the second resulting from the screened Coulomb repulsion. For realistic forms of the interaction, Eq. (1a) must be solved numerically. BCS, however, solve a model with the strong simplifying assumption that $V_{kk'}$ is isotropic and can be replaced by a constant attractive interaction $-V$ within a characteristic energy $\hbar\omega$ of the Fermi surface such that

$$V_{kk'} = \begin{cases} -V & \text{for } |\epsilon_k|, |\epsilon_{k'}| \leqslant \hbar\omega \\ 0 & \text{otherwise} \end{cases}$$

By assuming a constant density of states within an energy $\hbar\omega$ of the Fermi energy, the summation in Eq. (1a) can be replaced by an integral

$$1 = N(0)\,V \int_0^{\hbar\omega} \frac{d\epsilon}{E} \tanh\left(\frac{\beta E}{2}\right) \tag{1b}$$

where $N(0)$ is the density of electron states for one spin direction at the Fermi surface, $E = (\epsilon^2 + \Delta^2)^{1/2}$ and $\Delta_k = \Delta$ for $|\epsilon_k| \leqslant \hbar\omega$ and $\Delta_k = 0$ otherwise.

The transition temperature is simply defined as the temperature for which Δ vanishes so that E at T_c is just the normal state energy ϵ. Thus, upon integrating Eq. (1a) we obtain the famous BCS critical temperature relation,

$$kT_c = 1.13\hbar\omega \exp[-1/N(0)\,V] \tag{2a}$$

This result combined with the equation for the energy gap gives for all superconductors, when $N(0)\,V \ll 1$,

$$2\Delta(0) = 3.53kT_c \tag{3}$$

Since the most important attractive part of the interaction $V_{kk'}$ is due to phonons of high frequency, we expect $\hbar\omega$ to be on the order of $k\theta_D$. In actual practice (*5*) $\hbar\omega$ is assumed to be approximately $\frac{3}{4}k\theta_D$, so one often sees the critical temperature equation

$$T_c = 0.85\theta_D \exp[-1/N(0)\,V] \tag{2b}$$

A. Effective Potential V

In the original presentation of their model, BCS treat the interaction energy V as the one adjustable parameter which gives a correct fit for T_c. Since θ_D and $N(0)$ are measurable normal-state qualities, it is of great interest to determine V theoretically to obtain T_c. However, since T_c can be measured quite accurately and is quite sensitive to small changes in V, it is preferable to invert Eq. (2a) to obtain the experimental value of

$$[N(0)\,V]_{\text{exp}} \simeq \frac{-1}{\ln(T_c/0.85\theta_D)}$$

for comparison with theory.

The first attempt to calculate V directly from first principles was made by Pines (*6*), who, following the suggestion by BCS, averaged the screened Coulomb and phonon induced electron–electron interaction at the Fermi surface. Another approach to the problem was made by Bogoliubov et al. (*7*), who considered a more realistic interaction which was simple enough to solve yet accounted for the different energy dependence of the Coulomb interaction. In Bogoliubov's model two cutoffs were introduced: one for the attractive electron–phonon interaction $-V_1 < 0$ with an energy cutoff at $k\theta_D$ and another interaction $V_2 > 0$ representing the Coulomb repulsion with a cutoff at the Fermi energy. By making the additional assumption that $N(0)$ is not energy dependent, they obtained an equation of the same form as BCS with the interaction energy V replaced by

$$V = V_1 - \frac{V_2}{1 + N(0)\, V_2 \ln(E_F/k\theta_D)} \tag{4}$$

In 1961 an important contribution to the understanding of the electron–phonon interaction was made by Eliashberg (*8*), who showed that the energies ϵ_k appearing in the Bardeen–Pines form of the phonon-induced electron–electron interaction in a normal metal should be replaced by the quasi-particle energies E_k in the case of a superconductor. Starting from Eliashberg's retarded form of the electron phonon interaction and including a screened Coulomb repulsion with a cutoff in frequency (energy) at E_F rather than at $k\theta_D$, Morel and Anderson (*9*) improved an earlier calculation of Morel (*5*) and obtained a solution to the gap equation which led to an interaction energy identical in form to the expression found by Bogoliubov et al., except that the cutoff $k\theta_D$ and E_F appear as a result of the frequency dependence of the interactions, rather than as arbitrary cutoffs in momentum space. In Table I the values for $N(0)\ V$ obtained by Morel and Anderson for various elements are compared with the measured values. In almost all cases the agreement is good.

The calculation of Morel and Anderson predicts that all metals at sufficiently low temperatures should become superconducting. As yet one cannot be certain that this prediction is false, since it can be argued that the temperatures reached are not low enough, that magnetic impurities depress T_c, or that supercooling prevents T_c from being observed. Berk and Schrieffer (*10*) have calculated the effects of ferromagnetic spin correlations on the transition temperature. They show that the spin-induced repulsion can be many times larger than the conventional Coulomb repulsion. Thus at the beginning and end of the transition metal series, where the spin susceptibility is high, the strong ferromagnetic exchange forces are large and presumably are able to suppress superconductivity.

For strong-coupling superconductors the inversion of the microscopic equation to obtain $N(0)\ V$ from T_c and Θ_D is not correct. Recently McMillan (*10a*) has shown how to obtain the attractive V which enters into the microscopic

TABLE I

Experimental Values of $N(0)\ V$ Compared with the Theoretical Calculation of Morel and Anderson (*9*)

Element	T_c^{expt}, °K	θ_D, °K	$N(0)\ V_{calc}$	$N(0)\ V_{exptl}$
Zn	0.9	235	0.16	0.18
Cd	0.56	164	0.14	0.175
Hg	4.16	70	0.27	0.35
Al	1.2	375	0.23	0.175
In	0.4	109	0.24	0.29
Tl	2.4	100	0.23	0.27
Sn	3.75	195	0.24	0.25
Pb	7.22	96	0.30	0.39
Ti	0.4	430	0.30	0.14
Zr	0.55	265	0.26	0.16
V	4.9	338	0.35	0.24
Nb	8.8	320	0.35	0.32
Ta	4.4	230	0.34	0.25
Mo	0.92	360	0.28	0.16
U	1.1	200	0.35	0.19
	T_c^{calc}, °K			
Na	7.3×10^{-2}	160	0.13	
K	4.5×10^{-2}	100	0.13	
Cu	1.5×10^{-2}	343	0.10	
Au	6×10^{-4}	164	0.08	
Mg	2.3	342	0.20	
Ca	0.42	220	0.16	

[a] For the last six elements, superconductivity has not been observed. The value for T_c in the table represents the calculated transition temperature assuming Morel and Anderson's value for $N(0)\ V$ and the simple BCS relation $T_c \simeq 0.85\theta_D \exp[-1/N(0)\ V]$.

theory when the specific heat is known. Using this procedure one obtains a value for $N(0)\ V$ for Pb of nearly 1. Recent measurements (*10b*) on the high-transition-temperature superconductor $NbN_{0.91}$, which has a high V and low $N(0)$, indicate it more closely resembles Pb and differs from other high-transition-temperature superconductors which have high $N(0)$ and low V.

B. Decoupling of $N(0)$ and V

Since $N(0)$ can experience an appreciable variation for transition metals, a great deal of experimental attention has focused on the simple BCS exponential dependence of T_c on $N(0)\ V$. An early suggestion by Pines (*6*) indicated that in transition metals, where $N(0)$ is almost entirely determined by the high d-band density of states and V mainly by the s-s interaction, one might expect a decoupling of $N(0)$ and V with a fairly constant value of V. Experiments by Hulm and Blaugher (*11*) suggested the possibility of checking Pines' conjecture by measuring the transition temperature of two similar elements such as Nb and Mo. The early experiments (*12*) indicated that T_c decreased almost linearly with increasing Mo concentration and extrapolated to zero at a Mo concentration of about 40%. Based on the assumption that $N(0)$ varied linearly with concentration, they showed that $N(0)\ V$ and therefore V is rapidly varying function of concentration. After it was found that pure Mo indeed does become superconducting at 0.92°K (*13*), Hein et al. (*14*) reinvestigated the Mo–Nb alloy system and obtained data which contradicted the earlier results. The data suggested that superconductivity is a phenomenon common to the entire Nb–Mo alloy system, which is crucial to the constant-V conjecture of Pines. Using values of $N(0)$ obtained from specific heat data they calculated the value of V necessary to satisfy the simple BCS relation for the transition temperature. They found V increases monotonically by about 50% as one passes from pure Nb to pure Mo, whereas $N(0)\ V$ varies from 0.32 for pure Nb to about 0.12 for $Nb_{0,30}Mo_{0,70}$ back to 0.16 for pure Mo. Variations in V of about 50% are not unexpected, considering the crudeness of Pines' argument for the decoupling of $N(0)$ and V. In order to study a sample in which the number of valence electrons is constant, Bucher et al. (*15*) investigated alloys of the elements titanium and zirconium

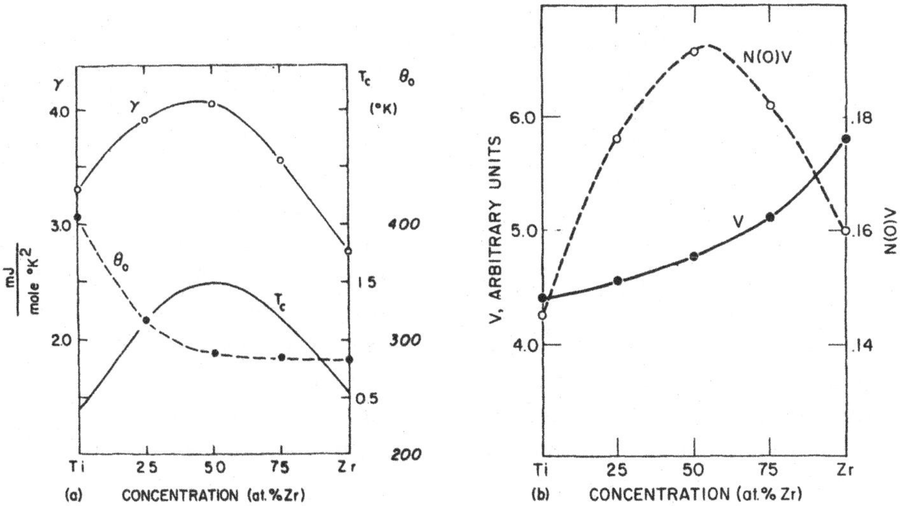

Fig. 1. (a) Experimentally determined coefficient of electronic specific heat γ, Debye temperature Θ_D, and transition temperature T_c of Ti–Zr alloys, obtained by Bucher et al. (*15*). (b) The value of $N(0)\ V$ obtained from the BCS relation $N(0)\ V = -1/\ln(T_c/0.85\Theta_D)$.

which belong to the same column in the periodic table. According to their results the transition temperature variation follows γ, the coefficient of the normal state specific heat, in good agreement with the BCS expression. If one calculates V assuming that the density of states is proportional to γ, one finds a nearly constant V in going from pure Ti to pure Zr. Their data on T_c, Θ_D, and γ are presented in Fig. 1 together with the calculated values of $N(0)\,V$ and V assuming the simple BCS relation. Superconductivity has also been investigated in alloys of the group VI transition elements tungsten and molybdenum (*16*). By assuming that the Debye temperature and density of states follow Vegard's law, a constant value of V is obtained.

Geballe (*17*) has studied binary alloys systems such as Ru–Os and Os–Ir in which data on Θ_D, T_c, and $N(0)$ seem to indicate a lack of agreement with the BCS form for T_c. For reasons which will become clearer in the discussion on the isotope effect, the interpretation of these experiments seem to indicate that another mechanism, independent of the phonon interaction, may be operative. The interpretation of the data on superconducting alloys is still open for analysis, since the experiments are difficult to perform and the idea concerning the decoupling of $N(0)$ and V is not firmly established theoretically. In addition, in these alloys, one might expect the spin-correlation effects discussed by Berk and Schrieffer (*10*) to be important.

C. Isotope Effect

Experimental clues as to the nature of the superconducting interaction were first provided by Maxwell (*18*) and Reynolds et al. (*19*), who in 1950 measured the transition temperature as a function of the mean mass of different isotopes of the same superconductor. To a good approximation they showed that $T_c \propto M^{-\beta}$, where $\beta = 0.5 \pm 0.05$. This exponential relation between T_c and $M^{-\beta}$ is called the isotope effect and indicates that the ion mass (and therefore lattice vibrations, phonons) is important in determining the transition temperature. Soon thereafter, Fröhlich (*20*), unaware of the experiments on the isotope effect, and Bardeen (*21*) separately presented a theory for the electron–phonon interaction which led naturally to models of superconductivity which were dependent on the phonon energies. The expression for T_c derived by BCS, $kT_c = 1.13\hbar\omega \exp[-1/N(0)\,V]$, had the isotope effect built in, because BCS postulated that the cutoff the attractive electron–electron interaction was at the Debye energy $k\theta_D$.

This artificial introduction of a single cutoff at approximately $k\theta_D$, leading to a value of β equal to $\frac{1}{2}$ for all superconductors, was relaxed in a calculation by Swihart (*22,23*). By making the plausible assumption that the cutoff in the Coulomb part of the electron–electron interaction is independent of the ionic mass, Swihart obtained values of β less than the BCS value of $\frac{1}{2}$. That one should no longer obtain an isotope effect with $\beta = \frac{1}{2}$ in all superconducting metals can be seen quite easily by using the form of the interaction energy investigated by Bogoliubov et al. (*7*) and later justified theoretically by Morel and Anderson (*9*):

$$V = V_1 - \frac{V_2}{1 + N(0)\, V_2 \ln(E_F/k\theta_D)}$$

If we assume $T_c \propto M^{-\beta}$ and $\theta_D \propto M^{-1/2}$, then

$$\frac{d}{dM}(\ln kT_c) = \frac{d}{dM}(\ln k\theta_D) - \frac{d}{dM}\left[\frac{1}{N(0)\, V}\right]$$

and we obtain

$$\beta = \frac{1}{2}\left\{1 - \frac{1}{[N(0)\, V]^2}\left[\frac{N(0)\, V_2}{1 + N(0)\, V_2 \ln(E_F/k\theta_D)}\right]\right\} \tag{5}$$

For metals, Morel and Anderson find that $N(0)\, V_2[1 + N(0)\, V_2 \ln(E_F/k\theta_D)]^{-1}$ is about 0.01, so that to a good approximation, Eq. (5) can be written

$$\beta = \frac{1}{2}\left[1 - \frac{0.01}{[N(0)\, V]^2}\right] \tag{6}$$

Since $N(0)\, V$ appears in the denominator in Eq. (6), the greatest deviation from $\beta = \frac{1}{2}$ should occur when the coupling is weakest. Until recently an outstanding difficulty in justifying the calculations by Swihart and those of Morel and Anderson was the fact that measurements on Zn, which is one of the weaker-coupling superconductors indicated a normal isotope effect with $\beta = 0.45$ (*24*), whereas the theoretical calculation of Morel and Anderson predicted 0.35 and Garland's (*25,26*) more exact calculation predicted 0.415. This disagreement is significant since the relevant parameters which enter into Garland's theory are very well known for Zn. A remeasurement of the isotope effect in Zn by Fassnacht and Dillinger (*27*) gives $\beta = 0.3$, and although this deviation is in the direction expected from theory, it is not clear why it should deviate so much from Garland's prediction of 0.415 ± 0.015. It would be of great interest to perform careful measurements on the isotope effect for other weak-coupling superconductors, such as Cd, to see if some of the larger disagreements between theory and earlier experiments can be resolved. Garland (*26*) calculated the isotope effect for metals by assuming an isotropic free electron model with a Coulomb energy cutoff at a few times the Fermi energy E_F. He gets somewhat better agreement with the isotope effect data, as shown in Table II. There is some question concerning the large cutoff value of the Coulomb energy; however, since this cutoff energy appears in a logarithm, the value at β does not depend critically on the cutoff energy.

TABLE II

"Best" Experimental Values for β Obtained in Fitting the Relation $T_c \propto M^{-\beta}$ Compared with the Theoretical Value of BCS (*1*), Swihart (*22,23*), Morel and Anderson (*9*), and Garland (*26*)

Element	β_{exptl}	β_{BCS}	β_{SWI}	β_{MA}	β_{GAR}	$\beta_{\text{GAR(latest)}}$
Zn	{0.45 ± 0.01; 0.30 (*27*)}	0.5	0.2	0.35	0.40	0.415 ± 0.015
Cd	0.50 ± 0.10	0.5	0.2	0.34	0.37	0.385 ± 0.025
Hg	0.50 ± 0.03	0.5	0.4	0.46	0.465	0.48 ± 0.005
Al		0.5	0.3	0.34	0.35	0.37 ± 0.025
Tl	0.50 ± 0.10	0.5	0.3	0.43	0.45	0.48 ± 0.02
Sn	0.47 ± 0.02	0.5	0.3	0.42	0.44	0.455 ± 0.01
Pb	0.48 ± 0.01	0.5	0.3	0.47	0.47	0.485 ± 0.005
Ti		0.5		0.25	0.2	0.145 ± 0.17
Zr	0.0	0.5		0.30	0.35	0.15 ± 0.17
V		0.5		0.41	0.15	0.25 ± 0.125
Ta		0.5		0.42	0.35	0.35 ± 0.075
Mo	0.33 ± 0.05	0.5	0.15	0.3	0.35	0.35 ± 0.075
Ru	0.0 ± 0.10	0.5	0.0	0.35	0.0	0.065 ± 0.15
Os	0.20 ± 0.05	0.5	0.1	0.25	0.1	0.225 ± 0.10
Ir		0.5		0.3	−0.2	−0.015 ± 0.17
Hf		0.5		0.5	0.3	0.1 ± 0.2
Re	0.39 ± 0.01	0.5		0.41	0.3	0.355 ± 0.05
U(α)	−2.2 ± 0.2	0.5	[(*32a*); see also (*198a*)].			

[a] In some cases the values for β_{SWI} and β_{MA} have been calculated by Garland using their models, respectively. (See Garland for references and (*27*) for Zn and (*32* for Zr, Mo, Re, Ru, and Os.)

It was not until 1961 that a significant departure from the $\beta = \frac{1}{2}$ dependence of the isotope effect was found. Geballe et al. (*24,28*) reported that the critical temperature of the transition elements Os and Ru (*29*) showed little or no correlation with ion mass and therefore $\beta \approx 0$. Later experiments by Hein and Gibson (*30,31*) and Bucher et al. (*32*) showed that β for Os was 0.2 and the transition temperature of Ru displayed no systematic dependence on the mean isotopic mass.

Fowler et al. (*32a*) recently measured the isotope effect in high-purity α-phase uranium isotopes 235 and 238 and obtained a value for $-\beta$ of 2. This reversal in sign of β and its large magnitude has been interpreted as proof that in α-phase uranium another mechanism besides the electron–phonon interaction is responsible for superconductivity.

To understand the occurrence of superconductivity among the elements

Matthias (*33,33a*) has divided the periodic system into regions in which three different interactions are in part responsible for the attractive electron–electron interaction. First are the usual *s-p* superconductors, in which the outer electrons are *s*- and *p*-electrons. These elements satisfy the usual BCS microscopic theory with transition temperatures of a few degrees and an isotope effect $-\beta = -0.5$ due to the electron–phonon interaction. Second, in elements with outer *d*-electrons, Matthias proposes an additional attractive interaction due to the polarization of the *d*-electron shell. The lowering of energy in a half filled *d*-shell coupled with the high lattice stability for a d^5s^1-configuration make elements in the middle of the transition series very poor superconductors. This *d*-electron interaction leads to deviation of the isotope effect from $-\beta = -0.5$ and explains the sharp dip in T_c as one goes across a row in the periodic table. Third, Matthias applies the concept of an attractive interaction caused by the creation and annihilation of a virtual *f*-electron to explain the superconductivity of La and U, which are elements at the beginning of the 4*f*- and 5*f*-series.

In narrow-band metals, such as the transition metals, Garland (*26*) proposes a simple two-band model consisting of nearly free *s*-electrons orthogonal to tightly bound *d*-electrons. Since the transition metals investigated were probably impure, pair states must be formed from linear combinations of band electrons resulting in a quasi-particle with both *s* and *d* character with a single energy gap. The effective Coulomb cutoff in such a case turns out to be on the order of one-half the *d*-subband energy. This energy dependence leads to a value for β less than $\frac{1}{2}$ and fits quite well with the experimental values. Thus Garland's calculation indicates that the lack of an isotope effect in transition metals may not be inconsistent with assuming that only electron–phonon interactions are responsible for superconductivity even in the transition metal elements. Caution must be exhibited in a comparison between Garland's theory and experiment, since the values for the parameters used in the calculation of β for the transition metals are not very well known.

D. Pairing and Flux Quantization

The BCS hypothesis of pairing electrons with zero total momentum manifests itself in the macroscopic nature of the ground-state wave function Ψ_0 for the superconducting system. Because of the gap in the single-particle energy spectrum, the response of the superconductor is rigid with respect to a weak electromagnetic perturbation. To obtain a state with a differing momentum pairing, but still preserving the pairing correlations, one can multiply the ground-state wave function by a phase factor

$$\Psi = \exp\left[i \sum_j \chi(x_j)\, \Psi_0\right] \tag{7}$$

where x_j is the position of the jth electron. This results in an increase in momentum of $\hbar\nabla\chi$ for each electron but does not change the relative internal motion of the paired electrons.

In presenting his phenomenological theory of superconductivity, London (*34*) realized that the momentum in the superconducting state appears to be held constant for each electron in some kind of long-range order. (In the BCS theory the long-range order is exhibited by the Cooper pair, not by each electron individually.) London, therefore, introduced a single coherent wave function of the form Eq. (7) to describe the behavior of all the electrons in the superconducting state. In order that the wave function Φ be single-valued, London concluded that the flux enclosed in the hole of a multiply connected superconductor must be quantized in units of $ch/e \simeq 4 \times 10^{-7}$ G–cm^2. To verify London's conjecture, experiments were performed by two groups in separate countries almost simultaneously. By measuring the flux enclosed in hollow superconducting cylinders both Deaver and Fairbank (*35*) and Doll and Näbauer (*36*) observed that the flux was indeed quantized; however, to their surprise, the unit was only one-half that predicted by London. Their data are presented in Fig. 2. Since then, many arguments have been presented leading to quantization in steps of

$$\Phi_0 = ch/2e = 2.07 \times 10^{-7} \text{ G–cm}^2$$

All the arguments are similar in that they are based on the coherence of the electron pair states which are assumed to exist according to the BCS theory of superconductivity. This half-integer quantization, when viewed in the form $\Phi_0 = ch/e^* = ch/2e$, represents a most striking verification of the existence of Cooper pairs. Parker et al. (*36a*) have measured the voltage and frequency of a Josephson tunnel junction and shown it to be $h/2e$ within the accuracy with which this constant is known from other measurements (probably about 2 parts in 10^4); since the Josephson frequency and flux quantization are different aspects of the same effect, this is very strong evidence for the correctness of the relation $\Phi_0 = ch/2e$.

We present here a simple nonrigorous argument for half-integer quantization (*37*) which is similar in spirit both to that of Onsager (*38*) and to Londons' original argument and in some sense displays more clearly the difference in behavior between a superconducting and normal metal. The velocity operator $\mathbf{v}$ is related to the canonical momentum operator $\mathbf{P}$ and the vector potential $\mathbf{A}$ by the relation

$$\mathbf{v}_j = \frac{1}{m}\left(\mathbf{P}_j + \frac{e}{c}\mathbf{A}_j\right)$$

where $-e$ and m are the charge and mass of an electron. For a multiply connected superconductor we choose an annular region deep inside the super-

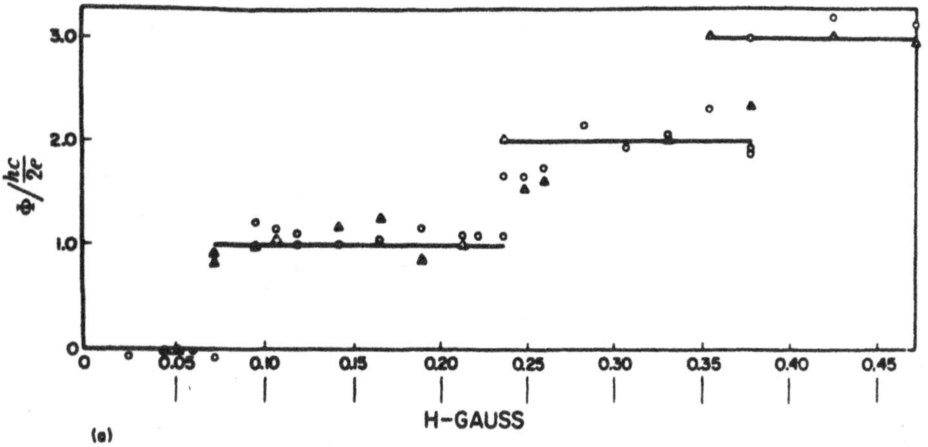

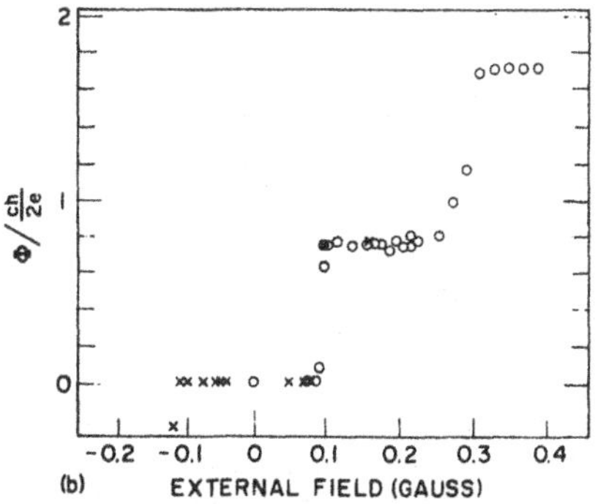

Fig. 2. Flux trapped inside a hollow superconducting cylinder as a function of the magnetic field in which the cylinder was cooled below its transition temperature. (a) Deaver and Fairbank, (*35*), (b) Doll and Näbauer. (*36*). The quantization actually observed in (b) was about 0.8 $ch/2e$.

conductor enclosing the hole and obtain the average velocity $\langle \mathbf{v} \rangle$ of all the electrons within the annular region

$$\langle \mathbf{v} \rangle = \frac{1}{m}\left[\langle \mathbf{P} \rangle + \frac{e}{c}\langle \mathbf{A} \rangle\right]$$

which can be written as the sum

$$\langle \mathbf{v} \rangle = \frac{1}{Nm}\left(\sum_{j=1}^{N} \mathbf{P}_j + \frac{e}{c}\sum_{j=1}^{N} \mathbf{A}_j\right) \tag{8}$$

where N is the total number of electrons in the annular region. Since deep inside the superconductor **J** and therefore $\langle \mathbf{v} \rangle$ is zero, upon integrating Eq. (8) around a closed path deep within the superconductor surrounding the hole, we have

$$\frac{1}{N}\sum_{j=1}^{N}\oint_c \mathbf{A}_j\cdot d\mathbf{l} = \Phi = \frac{-c}{e}\oint_c \langle\mathbf{P}\rangle\cdot d\mathbf{l}$$

where $\oint \mathbf{A}_j\cdot d\mathbf{l} = \Phi$ is the flux enclosed within the hole and is the same for all electrons. To perform the momentum summation in Eq. (8) we assume that in the superconducting state one can pick out "carriers" (presumably groups of electrons) such that *each* carrier has the *same* momentum $\mathbf{p}$. Since the momentum of the carrier is assumed to satisfy the Bohr–Sommerfeld condition, we obtain

$$\begin{aligned}\langle\mathbf{P}\rangle\cdot d\mathbf{l} &= (\text{number of carriers})\frac{1}{N}\oint \mathbf{p}\cdot d\mathbf{l}\\ &= (\text{number of carriers})\frac{\nu h}{N}\end{aligned}$$

and therefore

$$\Phi = \nu\frac{ch}{e}\left[\frac{\text{number of carriers}}{\text{number of electrons}}\right] \tag{9}$$

where $\nu = 0, \pm 1, \pm 2, \ldots$.

Since the bracketed term in Eq. (9) has been observed to be $\frac{1}{2}$, one concludes that the number of carriers is one-half the number of electrons and therefore each carrier must be composed of two electrons. These carriers can then be identified with the pairs of the BCS theory.

The quantization of flux in a superconductor is quite different from the flux behavior of a normal cylinder in a static magnetic field. In a normal metal $\langle\mathbf{v}\rangle$ is also zero; however, the momentum average $\oint\langle\mathbf{P}\rangle\cdot d\mathbf{l}$ can take on the small value $\nu h/N$ and is not restricted to values equal to $\nu h/2$ as in the case of superconductors. As a result $\langle\mathbf{P}\rangle$ and $(e/c)\langle\mathbf{A}\rangle$ can vary from point to point in such a way as to cancel one another.

It is important to emphasize that flux quantization in $ch/2e$ follows only if there is a condensed state in which a macroscopic fraction of the pairs have the *same* angular momentum. The mere existence of stable pairs is not enough (*39*). That is, in addition to pairing, the BCS interaction also restricts the pairs to the same momentum state. The condensation energy gained as a result of this restriction is the main source of the flux dependence of the energy of the system, and is responsible for the discrete nature and persistence of the currents which flow in the flux-quantized state.

For thin hollow superconducting cylinders of thickness $d < \lambda(T)$, the penetration depth, the current density is not zero, and the quantity quantized is called the fluxoid, defined by

$$\text{fluxoid} = \nu ch/2e = \oint \mathbf{A}\cdot d\mathbf{l} + \Lambda(T)\, c \oint \langle \mathbf{J} \rangle \cdot d\mathbf{l}$$

where $\Lambda(T) = 4\pi\lambda^2(T)/c^2$. Since the current which flows in the superconducting cylinder is proportional to the difference between the fluxoid and the flux, the free energy of the superconducting cylinder will be a minimum whenever the enclosed flux is a multiple of $ch/2e$. Thus the transition temperature, defined as the temperature at which the normal and superconducting free energies are equal, is a periodic function of the enclosed flux. This periodicity with magnetic field of the resistivity near the transition temperature was first observed by Little and Parks (*40,41*). The more direct measurement of T_c vs. H by Groff and Parks (*42*) on a hollow Al cylinder is presented in Fig. 3. The theoretical value for the maximum change in transition temperature for the cylinder used in the experiment is in good agreement with the measured value of 0.033°K.

Kwiram and Deaver (*43*) have observed the establishment of the flux-quantized states in times as short as 10^{-5} sec. They observed the response of a hollow tin cylinder to alternate cooling and heating through the transition temperature in a magnetic field up to a frequency of 55 Kc. A whole host of quantum-mechanical interference effects‡ based on the macroscopic nature of the BCS wave function

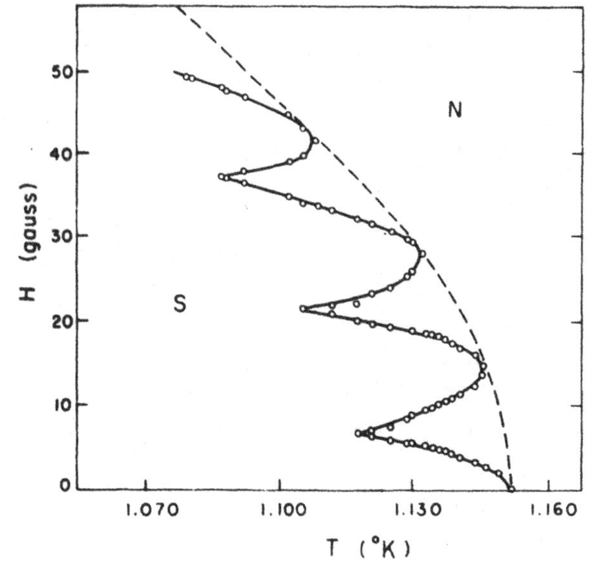

Fig. 3. Variation of the transition temperature $\Delta T/T_c$ as a function of magnetic field for a hollow 1.32-μ-diameter Al cylinder in a magnetic field as observed by Groff and Parks (*42*). The periodicity in field corresponds to a flux $ch/2e$ in the cylinder.

‡ See Chapter 8.

has further confirmed the quantization of flux in units of $ch/2e$. The experiments are performed on superconducting loops interupted by Josephson junctions or weak links. Both the theory of Abrikosov (*44*) vortices in type II superconductors and Tinkham's (*45*) model for the penetration of a magnetic field perpendicular to a thin film are based on the concept of flux quantization.

In 1962 Yang (*46*) introduced the concept of off-diagonal long-range order (ODLRO), based on earlier ideas of Penrose and Onsager (*47*). For a condensed system, superfluid helium, or superconductivity, Yang shows that the macroscopic occupation of a single quantum state results in the appearance of a term of order N in an off-diagonal component of the density matrix. For superfluid helium the single-particle density matrix exhibits ODLRO, whereas for superconductors, ODLRO is exhibited first in the two-particle density matrix, indicative of the pairing assumed in the BCS wave function.

III. DENSITY OF STATES BY TUNNELING MEASUREMENTS

The most powerful experimental method for studying the density of states in a superconductor is probably the electron tunneling technique discovered by Giaever (*48*). By this technique the tunneling current between two metals separated by a very thin insulator is measured as a function of the applied voltage, and from this the density of states as a function of energy is rather easily deduced. The simplicity of the method and the wealth of information it yields have led to many tunneling measurements, whose results constitute perhaps the most detailed comparison of the BCS theory with experiment. To allow the discussion of tunneling measurements, the semiphenomenological theory of tunneling will be briefly summarized. The purpose here is to use the summary of the theory as a means of coherently presenting results which are useful for comparison with experiments. The microscopic theory of tunneling and many specialized topics in tunneling theory are discussed at length in other chapters. The development follows Giaever and Megerle (*49*) and Shapiro et al. (*50*); and the form and presentation rely heavily on the excellent review article of Douglass and Falicov (*51*), who present other more detailed results.

A. Theory of Tunneling

The semiphenomenological theory of tunneling introduced by Giaever and Megerle (*49*) is essentially a one-dimensional model which assumes that the quasi-particles can be treated as independent Fermi–Dirac particles occupying a given state of energy ϵ with a probability $f = [1 + \exp(\beta\epsilon)]^{-1}$. The tunneling current from metal 1 to metal 2 is assumed to be proportional to the density of occupied states at a given energy in metal 1 and to the density of unoccupied states in metal 2, at the same energy. This current is also assumed to be proportional to the probability for a given quasi-particle of this energy to tunnel through the barrier. When integrated over all energies this gives

$$i_{12} = (2\pi/\hbar)\int_{-\infty}^{\infty} |M|^2 \rho_1 f_1 \rho_2 (1 - f_2)\, d\epsilon$$

where M is the matrix element between states of equal energy in the two metals. By subtracting a similar expression for the current from metal 2 to metal 1, we obtain for the net current

$$I = A\int_{-\infty}^{\infty} |M|^2 \rho_1 \rho_2 (f_1 - f_2)\, d\epsilon \tag{10}$$

Such an analysis of electron tunneling originated with Sommerfeld and Bethe (*52*) and has been applied to many problems; a presentation and extension of their work was given by Holm (*53*). More recently Esaki (*54*) has used it successfully to describe tunneling currents in degenerate semiconductor junctions.

To adapt the analysis to tunneling with superconductors the following assumptions were made:

1. The density of states in a normal metal is a constant, $\rho_n = N(0)$.
2. The density of states in a superconductor ρ_s at $T = 0°K$ is given by the BCS relation:

$$\rho_s = \begin{cases} N(0)\dfrac{|\epsilon|}{\sqrt{\epsilon^2 - \Delta^2}} & |\epsilon| \geqslant \Delta \\ 0 & |\epsilon| < \Delta \end{cases} \tag{11}$$

3. The matrix element M is independent of energy in the region of interest and is independent of whether the metals are superconducting or not.
4. The difference between the Fermi energies of the two metals equals the difference in electrical potential, V.

Measuring the energy in electron volts from the Fermi energy of metal 1 the tunneling current becomes

$$I = A\int_{-\infty}^{\infty} \rho_1(\epsilon)\,\rho_2(\epsilon + V)[f(\epsilon) - f(\epsilon + V)]\, d\epsilon \tag{12}$$

where $A = (2\pi e/\hbar)|M|^2$. When both metals are normal, Eq. (12) reduces to

$$I_{nn} = AN_1(0)\,N_2(0)\int_{-\infty}^{\infty} [f(\epsilon) - f(\epsilon + V)]\, d\epsilon$$

At low temperatures and voltages ($1 \ll \beta E_F$, $V \ll E_F$) the integral is easily evaluated and gives

$$I_{nn} = C_n V \tag{13}$$

where the constant, $C_n = AN_1(0)\,N_2(0)$, is the conductance for normal–normal tunneling. The linearity of this relation for Al–Al_2O_3–metal tunnel junctions was demonstrated by Fisher and Giaever (*55*) to hold for voltages up to about one-tenth of the oxide barrier height and covers the voltage range of present interest.

1. Superconductor–Normal Metal Tunneling

When metal 1 is superconducting and metal 2 is normal, Eq. (12) gives

$$I_{sn} = C_n \int_{-\infty}^{\infty} \rho_s [f(\epsilon) - f(\epsilon + V)]\, d\epsilon \tag{14}$$

If we differentiate this expression with respect to the applied voltage, we obtain

$$\frac{dI_{sn}}{dV} = C_n \int_{-\infty}^{\infty} \rho_s(\epsilon) \left(\frac{\beta \exp[\beta(\epsilon + V)]}{\{1 + \exp[\beta(\epsilon + V)]\}^2} \right) d\epsilon \tag{15a}$$

The second factor in this integral is a bell-shaped function which is symmetrical about its maximum, which is located at $\epsilon = -V$. The magnitude of this maximum is proportional to $1/T$ and at $T = 0$ the function degenerates into a delta function and

$$\left(\frac{dI_{sn}}{dV} \right)_{T=0} = C_n \rho_s \tag{15b}$$

This general result is independent of the density of states function ρ_s. For the BCS model at $T = 0$, Eq. (11) leads to

$$\left(\frac{dI_{sn}}{dV} \right)_{T=0} = \begin{cases} C_n \dfrac{|V|}{\sqrt{V^2 - \Delta^2}} & |V| \geqslant \Delta \\ 0 & |V| \leqslant \Delta \end{cases} \tag{15c}$$

If Δ is independent of energy these last equations can be integrated to give

$$I_{sn} = \begin{cases} C_n \sqrt{V^2 - \Delta^2} & V \geqslant \Delta \\ 0 & V \leqslant \Delta \end{cases} \tag{16}$$

Qualitatively this result can be understood from the density of states diagram of Fig. 4a and the resulting tunneling characteristic of Fig. 4b. At $T = 0$ there evidently can be no current from the superconductor until the applied voltage depresses the effective Fermi energy of the normal metal by an amount which is equal to Δ. At that time current should start to flow and at higher voltages the current should asymptotically approach the straight line of normal–normal

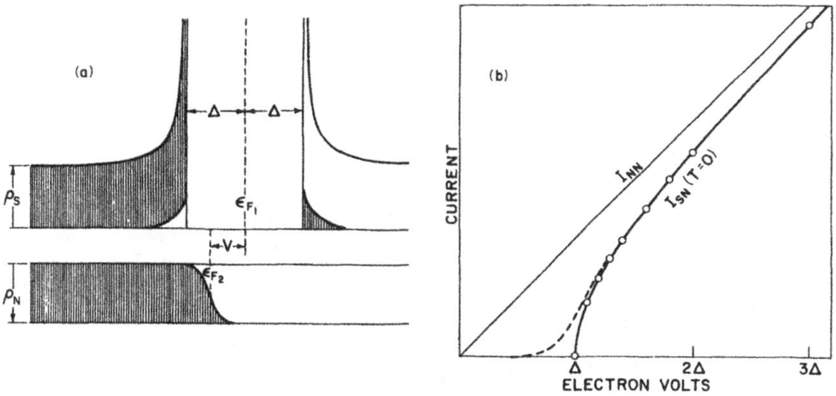

Fig. 4. (a) Schematic representation of the density of states in a superconductor–oxide–normal metal tunnel junction at a finite temperature. (b) Current as a function of voltage for such a junction at $T=0$, $T/T_c=0.2T_c$, and $T>T_c$.

tunneling. At a finite temperature well below T_c the exponential tail of the Fermi distribution always gives a finite conduction, but there remains a sudden rise for $V \approx \Delta$ as shown in Fig. 4b for $T/T_c = 0.2$.

When $T \neq 0$ the BCS conductance from Eq. (15) is

$$\frac{dI_{sn}}{dV} = C_n\left(\int_{\Delta}^{\infty} \frac{|\epsilon|}{\sqrt{\epsilon^2+\Delta^2}} \frac{\beta \exp[\beta(\epsilon+V)]}{\{1+\exp[\beta(\epsilon+V)]\}^2}\, d\epsilon + \int_{-\infty}^{-\Delta} \frac{|\epsilon|}{\sqrt{\epsilon^2+\Delta^2}} \frac{\beta \exp[\beta(\epsilon+V)]}{\{1+\exp[\beta(\epsilon+V)]\}^2}\, d\epsilon\right) \tag{17}$$

This expression for the BCS conductance has been computed numerically by Bermon (*56*) for a large range of parameters $\beta\Delta$ and βV.

The expression for the current I_{sn} in Eq. (14) has been shown by Giaever and Megerle (*49*) to be given by a series which converges rapidly for $|V| < \Delta$:

$$I_{sn} = 2C_n\Delta \sum_{m=1}^{\infty} (-1)^{m+1} K_1(m\beta\Delta) \sinh(m\beta V) \tag{18a}$$

where K_1 is the modified Bessel function of the second kind. In the low-voltage limit Eq. (18a) becomes

$$\lim_{V\to 0} I_{sn} = 2C_n\Delta \sum_{m=1}^{\infty} (-1)^{m+1} m\beta V K_1(m\beta\Delta) \tag{18b}$$

which in the low-temperature limit leads to

$$\lim_{\substack{V\to 0\\ T\to 0}} \frac{I_{sn}}{I_{nn}} = (2\pi\beta\Delta)^{1/2} \exp(-\beta\Delta) \tag{18c}$$

2. Superconductor–Superconductor Tunneling

For this case, Eqs. (11) and (12) lead to the following expression for the tunneling current:

$$I_{ss} = C_n \int_{-\infty}^{\infty} \frac{|\epsilon|}{\sqrt{\epsilon^2 - \Delta_1^2}} \frac{|\epsilon + V|}{\sqrt{(\epsilon + V)^2 - \Delta_2^2}} [f(\epsilon) - f(\epsilon + V)]\, d\epsilon \qquad (19)$$

This result was first given by Nicol et al. (*57*), who numerically calculated sample current–voltage curves and showed that at approximately $V = \pm |\Delta_2 - \Delta_1|$ there is a logarithmic singularity in the current of magnitude:

$$I_{ss} \sim \ln |V - (\Delta_2 - \Delta_1)|$$

and at $V = \Delta_1 + \Delta_2$ there is a finite discontinuity even at $T \neq 0$,

$$\Delta I_{ss} = \frac{\pi C_n \sqrt{\Delta_1 \Delta_2}}{4} \frac{\sinh[\beta(\Delta_1 + \Delta_2)/2]}{\cosh(\beta\Delta_1/2)\cosh(\beta\Delta_2/2)} \qquad (20a)$$

These features are shown in Fig. 5b and may be understood qualitatively from considering the density of states diagram of Fig. 5a and how the product of occupied and available states of the same energy changes as the voltage is changed. At $T = 0$ Eq. (19) reduces to

$$I_{ss} = C_n \int_{-V+\Delta_2}^{\Delta_1} \frac{|\epsilon|}{\sqrt{\epsilon^2 - \Delta_1^2}} \frac{|\epsilon + V|}{\sqrt{(\epsilon + V)^2 - \Delta_2^2}}\, d\epsilon \qquad (21)$$

Equation (21) can be integrated and expressed in terms of complete elliptic

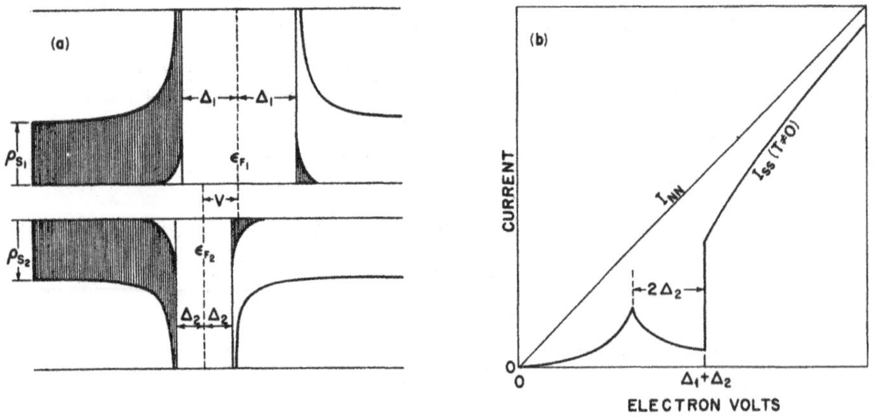

Fig. 5. (a) Schematic representation of the density of states in a superconductor–oxide–superconductor–tunnel junction at a finite temperature. (b) Current–voltage characteristic calculated from the BCS theory.

integrals (*51*). The discontinuity in the current at $V = \pm(\Delta_1 + \Delta_2)$ is at $T = 0$,

$$\Delta I = (\pi/2)\, C_n \sqrt{\Delta_1 \Delta_2} \tag{20b}$$

When $\Delta_1 = \Delta_2 = \Delta$ and $V < 2\Delta$, Taylor et al. (*58*) have computed I_{ss} numerically and have shown that a negative resistance region ($dV/dI < 0$) should appear at $T < 0.3\Delta/k$. When $T \ll \Delta/k$ and $V < 2\Delta$, the current is well approximated by

$$I_{ss} = 2C_n \exp(-\beta\Delta)\sqrt{2\Delta/(V + 2\Delta)}\,(V + \Delta) \sinh(\beta V/2)\, K_0(\beta V/2) \tag{22}$$

For $T = 0$ the discontinuity at $V = 2\Delta$ is

$$\Delta I = \pi C_n \Delta/2 \tag{20c}$$

3. Microscopic Justification of the Semiphenomenological Theory

The semiphenomenological theory as outlined above was criticized for its rather sweeping and apparently unjustified assumptions. However, from the detailed fit that Giaever and Megerle obtained with the experimental data, it appeared that for some reason these assumptions must be justified. It is therefore satisfying that the microscopic theory of tunneling, which was developed by Bardeen (*59*), Harrison (*60*), Cohen et al. (*61*), and Prange (*62*), vindicated the original assumptions. Thus, when we compare experimental results to the semiphenomenological theory, we are also comparing them to the microscopic theory. It is appropriate to mention that the careful study of the microscopic theory did bring to light the very interesting effect of Josephson tunneling; this tunneling by superconducting pairs is found between superconductors separated by a very thin tunnel barrier. Since this effect is being treated fully in Chapter 9, the experiments discussed in this chapter will be concerned almost entirely with quasi-particle tunneling, with only brief mention of pair tunneling.

B. Basic Tunneling Experiments

One of the most attractive features of Giaever's tunneling experiments is the simplicity of the experimental method. In the original experiments a thin strip of an aluminum film was evaporated on a glass substrate. The resulting film was then oxidized for a few minutes in the laboratory atmosphere and a cross strip of some other metal such as lead was then evaporated. This technique resulted in crossed thin films of aluminum and the other metal separated by a layer of aluminum oxide about 30 Å thick. Other metals which oxidize rapidly, such as magnesium, could be substituted for the aluminum; in addition lead, tin, niobium, or tantalum could be oxidized in an atmosphere of oxygen at 50–100°C for a few minutes before the second strip was evaporated. The second metal could be anything which evaporates in vacuum, and the result was a

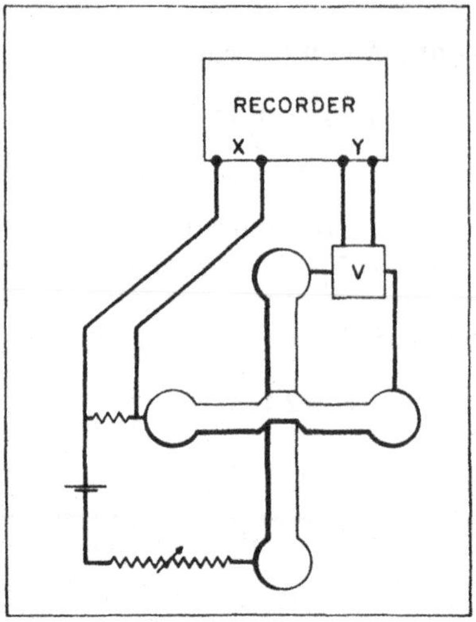

Fig. 6. Schematic diagram of a tunnel junction and the circuit used to measure current and voltage.

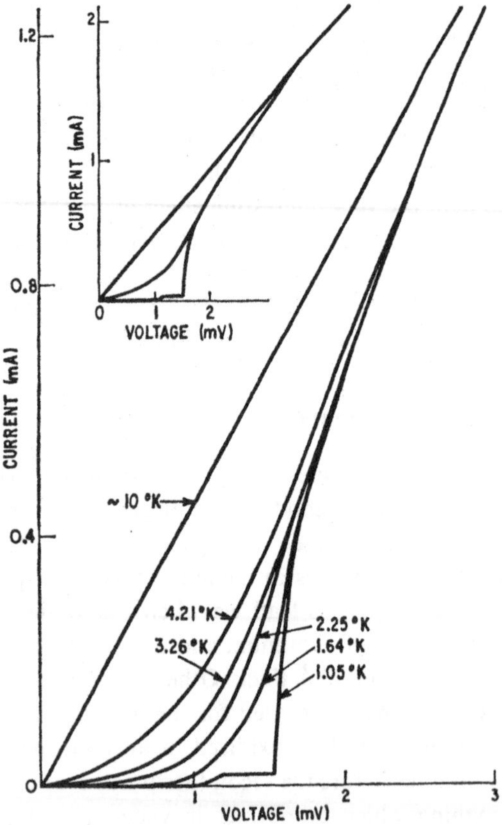

Fig. 7. Current–voltage characteristic of an Al–Al_2O_3–Sn junction at various temperatures (*49*).

four-terminal tunnel junction whose current–voltage characteristic was measured in the simple circuit shown schematically in Fig. 6.

This technique has been improved and extended in a number of ways. The oxidation can be accomplished in a more controlled way in an oxygen glow discharge. The oxide layer can also be formed by the evaporation of an insulator or the evaporation of a metal such as aluminum which can subsequently be oxidized to form the insulating layer. In addition to tunneling between thin films, it has proved possible to obtain tunneling results between bulk crystals and thin films. Levinstein and Kunzler (*63*) have developed a technique whereby a needle-like probe of niobium, tantalum, or aluminum is oxidized and used as a small area probe to obtain tunneling characteristics on a bulk sample of unoxidized metal. In the measurement of tunneling characteristics circuits have been developed to measure and plot the first and second derivatives of the current–voltage curves. These derivative plots are very useful in comparing the results with theory and have yielded a tremendous amount of detailed information. The techniques of sample preparation and circuitry have been greatly

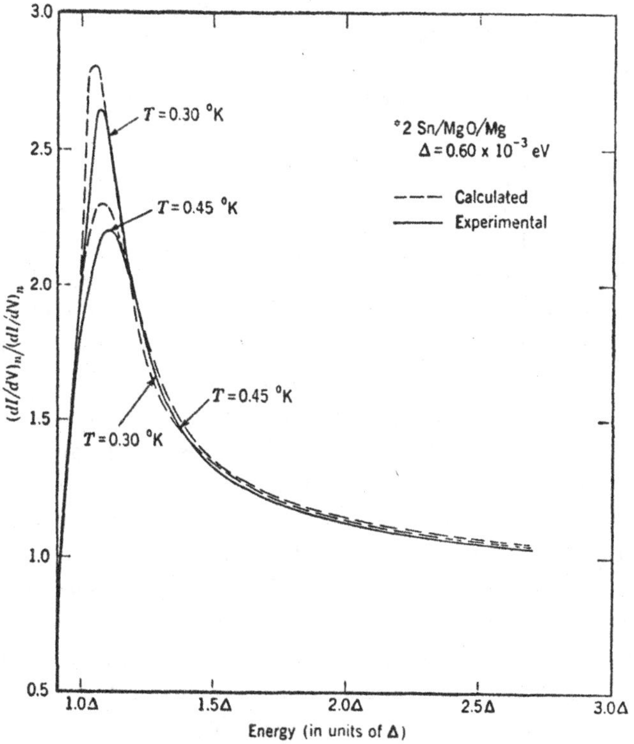

Fig. 8. Experimental and calculated relative conductance curves plotted vs. the normalized energy for two different temperatures. The BCS density of states was used in the calculation (*64*).

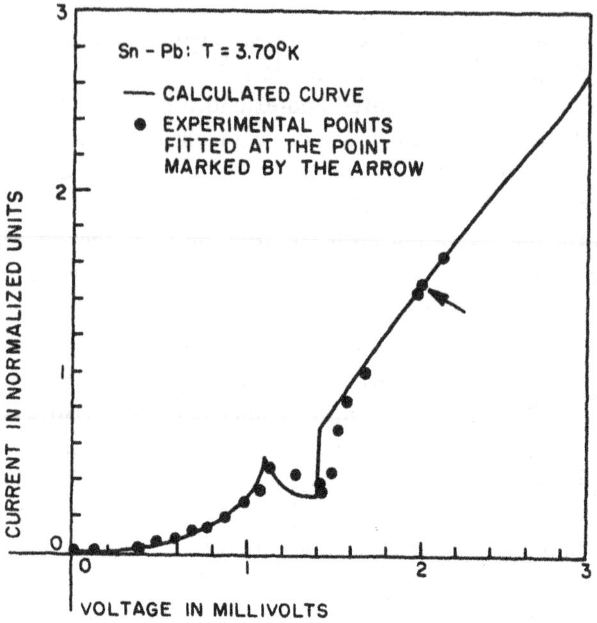

Fig. 9. Measured values of the current and voltage of a Sn–Pb tunnel junction as compared with the curve calculated from the BCS theory (*50*).

developed and are treated in detail in Chapter 11, but in spite of this development in the sophistication and power of tunneling measurements, the basic simplicity of the method remains.

Figure 7 shows one of the original tunneling curves obtained by Giaever and Megerle (*49*). The striking thing about these curves is their resemblance to what we expect from the semiphenomenological theory. At temperatures well below the transition temperature of lead but above that of aluminum the tunnel current remains low until $eV \approx \Delta_{Pb}$; here the current rises rapidly and then approaches the normal state asymptotically at higher voltages. Below the aluminum transition temperature, the curves are readily understood on the basis of the diagram of Fig. 4a, which implies: the initial rapid rise in current, the negative resistance region about equal to the gap width in aluminum, $2\Delta_{Al}$, and subsequent very rapid rise near $\Delta_{Pb} + \Delta_{Al}$. Figure 8 shows the relative conductance of a Mg–MgO–Sn junction measured by Giaever et al. (*64*) compared with the BCS theory; the agreement is excellent provided we use the measured energy gap in place of $3.53kT_c$. For superconductor–superconductor tunneling Shapiro et al., (*50*) show in Fig. 9 their measured curves for lead–tin junctions as compared with numerical calculation based on the BCS theory [Eq. (19)]. The curves are normalized at the point indicated and except for a blurring of the predicted features the general character of the curves, including the negative resistance

TABLE III

Measured Values of $2\Delta(0)/kT_c$
(BCS) theoretical value = 3.53)

Superconductor	Tunneling measurements	Ref.	Thermodynamic measurements[a]
Al	4.2 ±0.6	*(49)*	3.53
	2.5 ±0.3	*(50)*	
	2.8 – 3.6	*(68)*	
	3.37 ± 0.1	*(65)*	
Cd	3.2 ±0.1	*(65a)*	3.44
Ga			3.52, 3.50, 3.48
Hg(α)	4.6 ±0.1	*(84)*	3.95
In	3.63 ± 0.1	*(49)*	3.65
	3.45 ± 0.07	*(65)*	
	3.61	*(94)*	
La	1.65 – 3.0 (fcc)[b]	*(100)*	3.72 (fcc) (*d*-hcp)
	3.2	*(100)*	
Nb	3.84 ± 0.06	*(85)*	3.65
	3.6	*(95)*	
	3.6	*(96)*	
Pb[c]	4.29 ± 0.04	*(69)*	3.95
	4.38 ± 0.01[d]	*(74)*	
Sn	3.46 ± 0.1	*(49)*	3.61, 3.57
	3.10 ± 0.05	*(50)*	
	3.51 ± 0.18	*(85)*	
	2.8 – 4.06	*(65)*	
	3.1 – 4.3	*(87)*	
Ta	3.60 ± 0.1	*(85)*	3.63
	3.5	*(95)*	
	3.65 ± 0.1	*(97)*	
Tl	3.57 ± 0.05	*(98)*	3.63
	3.9	*(94)*	
V	3.4	*(95)*	3.50
Zn	3.2 ±0.1	*(99)*	3.44

[a] The values given here were calculated from values of $\gamma T_c^2/V_M H_c^2(0)$ assuming the equation $[2\pi V_M H_c^2(0)/3\gamma T_c^2]^{1/2} = 2\Delta(0)/kT_c$.

[b] The measured tunneling results of Edelstein and Toxen *(100)* in La are very low and widely scattered and perhaps reflect the great structure sensitivity of La. Hauser's *(100a)* later measurements are higher and less scattered.

[c] Other older measurements are collected in *(51)*.

[d] T_c assumed to be 7.193 °K.

region, was so similar to the predictions as to leave little doubt that the basic description of the phenomenon was correct.

Although the qualitative agreement between the BCS theory and early tunneling experiments was spectacular, the quantitative agreement was not so satisfactory. The BCS theory predicts [Eq. (3)] that in the weak-coupling limit all superconductors should have the same value for the normalized energy gap at $T = 0: 2\Delta(0)/kT_c = 3.53$. Table III lists the measured values and shows that, although there is rough agreement with theory, the deviations for some elements is large. In addition, it was found that the disagreement between theory and experiment was not to be found solely in the absolute value of the gap. In an effort to improve the agreement between the measured and calculated tunneling curves, Giaever et al. (*64*) attempted to fit their data on tin with a BCS density of states which had been averaged over a small fixed energy interval. With this empirical broadening width the calculated curves fitted the data very well, but not perfectly, even though the energy gap and the broadening width were both used as adjustable parameters. Zavaritskii (*65*), to avoid thermal broadening, measured the conductance of Pb, Sn, In, and Al at temperatures down to 0.1°K and so essentially reached the region where Eq. (16) should apply. However, the finite slope of the current discontinuity remained, and with it some ambiguity about the exact size of the energy gap. In the low-temperature limit near zero voltage Eq. (18c) predicts an exponential temperature dependence of the energy gap; Fig. 10 shows that the normalized experimental results have this temperature dependence and give an independent way of obtaining the energy gap. In the case of superconductor–superconductor tunneling the results of Shapiro agree fairly well with the discontinuity predicted by Eq. (20a) although the

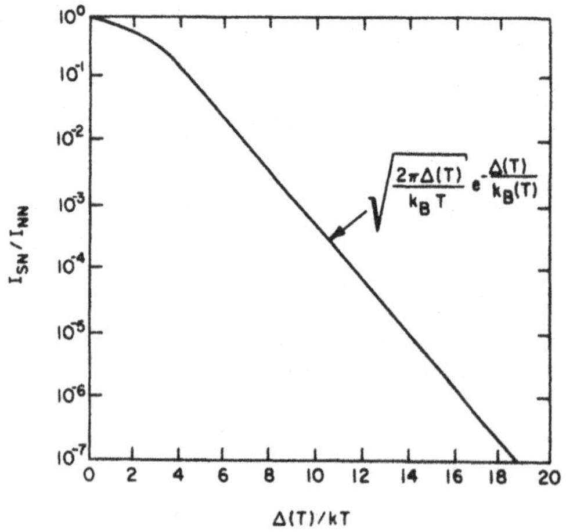

Fig. 10. Plot of the ratio I_{sn}/I_{nn} as a function of $\Delta(T)/kT$. The exponential behavior is apparent at the lower temperatures (*51*).

finite slope does not allow an exact comparison. However, even a casual inspection of the experimental curves shows that the theoretical cusp-like logarithmic singularity of Eq. (19) is not to be found on the experimental curves. In the case of identical superconductors the results of Gasparovic et al. (*66*) on a lead–lead junction are shown in Fig. 11 and compared with values numerically calculated from the BCS theory using a value of $2\Delta(T)$ adjusted to fit the experimental curve at the point shown in the figure. Again, except for the region near the gap edge, the agreement is excellent.

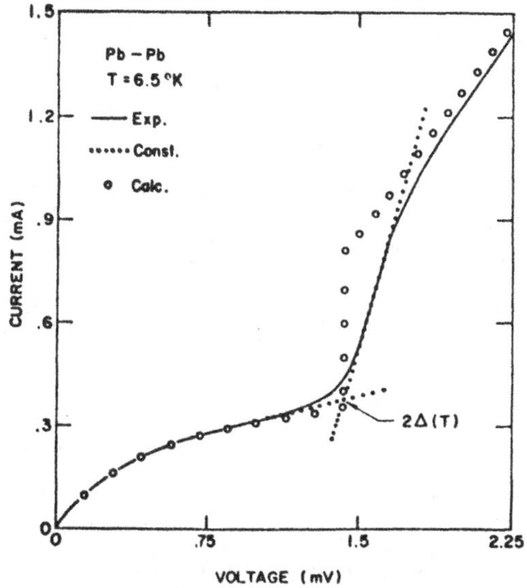

Fig. 11. Current–voltage characteristic of a lead–lead junction compared with values calculated from the strong-coupling theory for lead. Dashed lines show construction used to determine the energy gap (*66*).

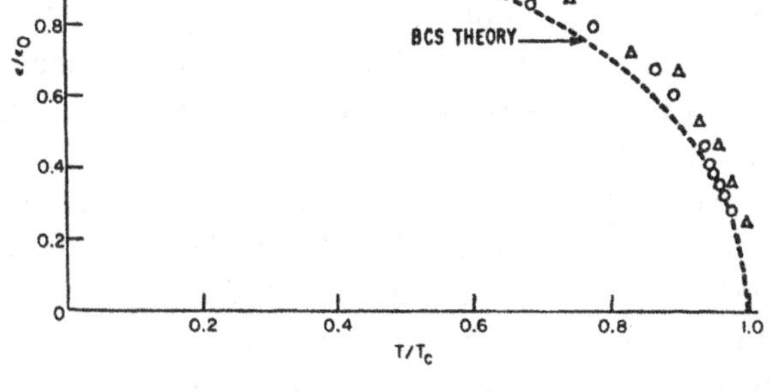

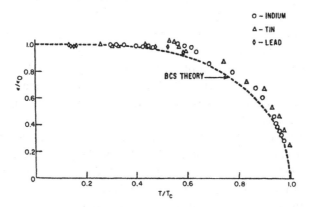

Fig. 12. Normalized measured energy gap of Pb, Sn, and In vs. temperature as compared with the BCS theory (*49*).

Figure 12 shows measurements of the temperature dependence of the energy gap in thin films of lead, tin, and indium obtained by Giaever and Megerle (*49*). Although the scatter of the measurements is large, there is good general agreement with the BCS theoretical values as calculated by Mühlschlegel (*67*) for the weak-coupling limit. Figure 13 shows later measurements of Douglass and Meservey (*68*) on aluminum thin films. Although the structure and thickness of the films were evidently important in determining the absolute value of the gap, the temperature dependence agreed with BCS up to a reduced temperature

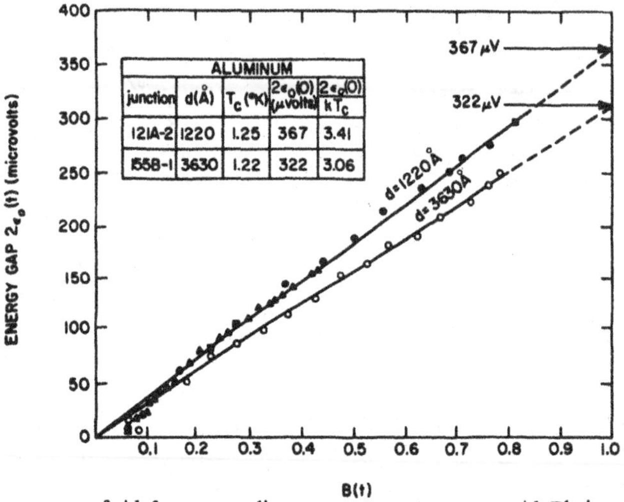

Fig. 13. Energy gap of Al from tunneling measurements on two Al–Pb junctions vs. the theoretical BCS function (*68*).

$T/T_c = 0.993$; the deviation near T_c is probably not significant because of the increasing ambiguity in obtaining the gap from experimental data as T_c is approached.

It is to be noted that quantitative agreement was found only when the results were normalized in some way. The absolute value of the energy gap deviates from the BCS model prediction of $3.53kT_c$ by as much as 30% in the case of mercury. There was even considerable scatter in the values obtained for the same metal from different thin-film samples; in the case of tin films Zavaritskii (*65*) found values of $2\Delta(0)/kT_c$ ranging from 2.8 to 4.06. In some instances the discrepancy in the energy gap between different samples has been shown to be associated with anisotropy, a subject that will be discussed in more detail in Section III.E. It should be pointed out that most of the results in Table III were obtained from thin-film samples whose properties may depend on strains caused by the substrate, on gaseous impurities, and on surface effects. Although these thin-film results are not equivalent to measurements on single crystals, the larger disagreements in Table III with the simple BCS model cannot be attributed primarily to thin-film effects.

The smearing of the predicted singularities and discontinuities is disappointing in itself; in addition, it makes the determination of the energy gap somewhat ambiguous. For large energy gaps at low reduced temperature this broadening introduces an error of perhaps only 2%. On the other hand, near the transition temperature or the critical field the value of the energy gap as judged from the tunneling characteristics becomes very uncertain. For these cases an understanding of the broadening mechanisms is essential before a precise comparison with theory is possible.

C. Effect of the Discrete Phonon Spectrum

More startling than these broadening effects was the observation by Giaever et al. (*64*) that the relative conductance curve for lead (Fig. 14) exhibits qualitative features which were not predicted by the simple BCS model. This figure

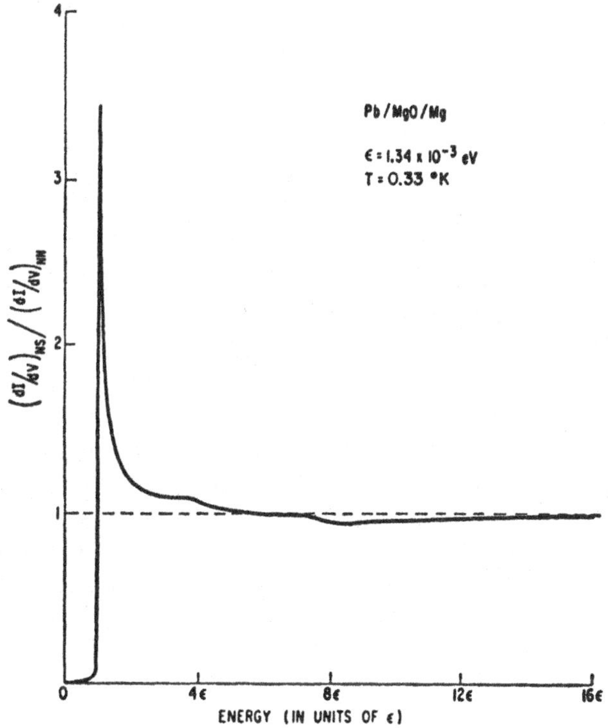

Fig. 14. Differential conductance of a Pb–Mg junction vs. voltage, showing the superconducting density of states of Pb (*64*).

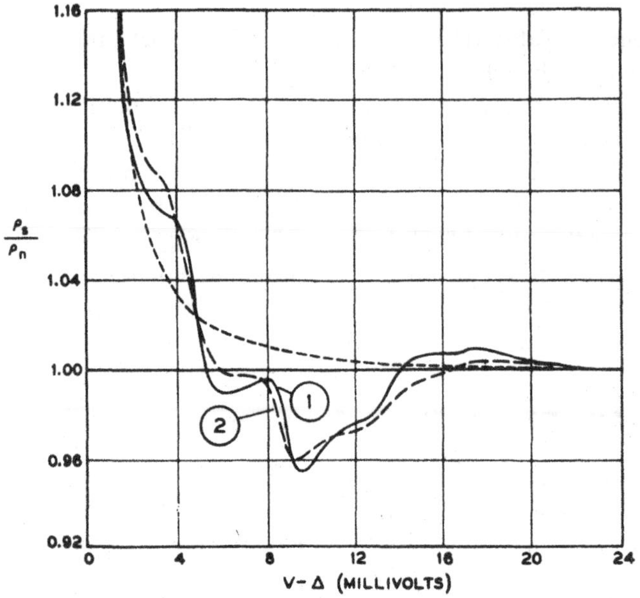

Fig. 15. Comparison of (1) the density of states vs. energy as calculated by Schrieffer et al. (solid line); (2) the measured tunneling characteristic (long-dashed line); and the BCS theory (short-dashed line) (*70*).

shows additional structure in the density of states curve and shows that the conductance in the superconducting state at high energy (or voltage) is actually less than that for the normal state. It was pointed out by the above authors that the crossover point was approximately at the Debye energy.

The discovery of structure in the density of states of lead aroused much interest because the most natural explanation was that it reflected the nature of the phonon spectrum. This was particularly plausible in lead because the Debye energy was so low that lead had been expected to deviate most strongly from the BCS constant-interaction model. Rowell et al. (*69*), using a more sensitive method of measuring dI/dV, succeeded in resolving considerable detail in the relative conductance curves in a lead–aluminum tunnel junction (Fig. 15).

The effect on the density of states of a simple but realistic phonon spectrum has been calculated by Schrieffer et al. (*70*). Using the Eliashberg (*8*) treatment of a retarded interaction, these authors showed that the density of states, as measured by tunneling experiments, could be represented by the simple expression

$$\rho_s = \rho_n \operatorname{Re}\left[\frac{\epsilon}{\sqrt{\epsilon^2 - \Delta^2(\epsilon)}}\right] \tag{23}$$

which is the same as the BCS expression except that the gap function, Δ, is complex and a function of the energy. The complex energy gap indicates that

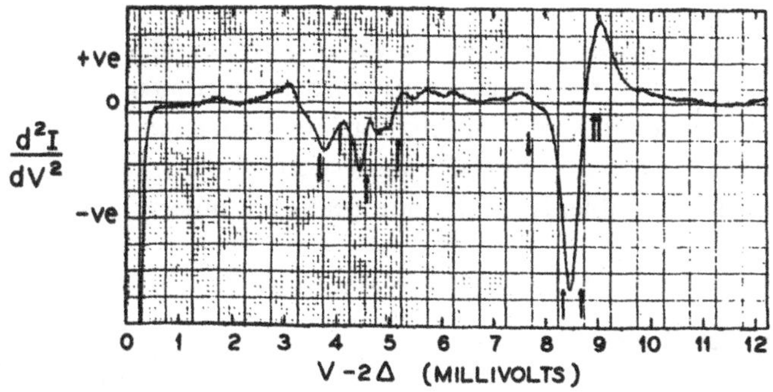

Fig. 16. (d^2I/dV^2) vs. $(V-2\Delta)$ for a Pb–Pb junction at 1.3°K. Arrows indicate the bias for Van Hove singularities expected from neutron measurements of the phonon spectrum (*74*).

in the high-energy region the damping of the quasi-particles becomes important. Using this density of states expression and a phonon spectrum consisting of two Lorentzian broadened peaks (as shown at the bottom of Fig. 15) a theoretical tunneling characteristic was calculated. The position and width of the phonon peaks were chosen to approximately fit the phonon spectrum observed in neutron scattering experiments by Brockhouse et al. (*71*), and the electron–phonon coupling constant was adjusted to give the experimental value of $\Delta(0)$. Figure 15 shows that this calculation reproduces the tunneling results not only near the phonon peaks but also at higher energies where the effect of harmonics of the phonon frequencies is observed. Discrepancies between the phonon density of states derived from neutron scattering and tunneling have been noted by Dynes et al. (*71a*) and attributed to the force constant model assumed in analyzing the tunneling data.

It was also discovered by Rowell et al. (*69*) that by observing the second derivative d^2I/dV^2 of a lead–lead junction much more structure in the effective density of states was revealed, as shown in Fig. 16. Scalapino and Anderson (*72*) attributed this structure to variations in the density of states caused by critical points (*73*) (maxima, minima, or saddle points) of the phonon spectrum, and, indeed much of the observed structure can be correlated in this way with neutron scattering measurements of the phonon spectrum. Other measurements of tin, indium, and thallium all support these general findings, although the knowledge of the phonon spectra is not complete enough to allow a complete analysis. In fact, McMillan and Rowell (*74*) have shown that this method in reverse provides a very powerful technique of determining the phonon spectrum from the observed effective density of states, and Rowell and Kopf (*75*) have applied the methods to lead, tin, indium, and thallium. The detailed correlation between the density of states and the phonon spectrum is very strong evidence that the attractive interaction between electrons is a retarded one involving phonons.

In fact, Wyatt (*76*) has argued that observation of a phonon peak in the tunneling current into tantalum proves that the attractive interaction in tantalum, a transition metal, is caused by the BCS mechanism of virtual phonon exchange and not by some other mechanism as has been suggested (*77–79*). Such a line of reasoning is, however, questionable, since phonon peaks are observed in semiconductor tunneling.

D. Strong Coupling

As mentioned previously, the measured values of $2\Delta(0)/kT_c$ for lead and mercury are considerably above the BCS model prediction of 3.53. These large values were not entirely unexpected, since the comparatively large value of the ratio T_c/Θ_D for these metals makes the validity of assuming weak coupling doubtful. Schrieffer and Wada (*80,81*) have offered a very plausible explanation of how this increase in the ratio $2\Delta(0)/kT_c$ comes about. The strong interaction causes damping of the quasi-particles and a decrease in both $\Delta(0)$ and T_c. However, since the damping increases with the temperature, T_c (determined self-consistently in the presence of thermal phonons) decreases more than $\Delta(0)$ (determined at $T_c = 0$, where there are no thermal phonons) and so the ratio $2\Delta(0)/kT_c$ increases. On the basis of this idea, Swihart et al. (*82*) have calculated the ratio $2\Delta/kT_c$ for realistic phonon spectra and obtained values of 4.33–4.40, 4.8, and 3.5 for lead, mercury, and aluminum, respectively. These calculations have been extended by Swihart (*83*) to include indium and tin. The calculated values for lead and mercury are close to the measured values and the value of 3.5 for aluminum assures us that this model reduces to the weak-coupling limit for aluminum, as it should. These authors also calculated how the BCS-model

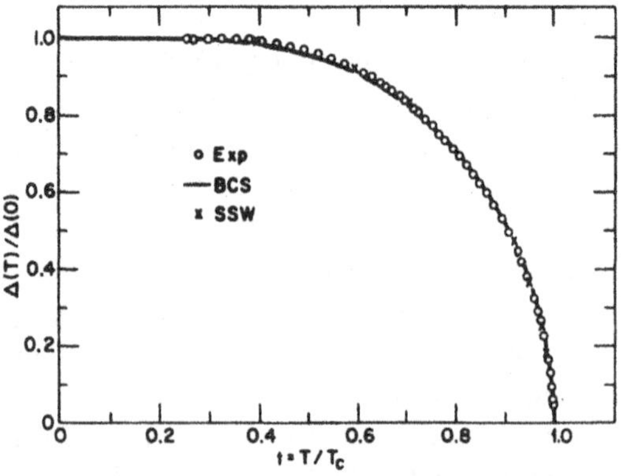

Fig. 17. Temperature dependence of the energy gap of lead vs. reduced temperature compared with the BCS model and the strong-coupling theory model of Swihart et al. [From Gasparovic et al. (*66*).]

temperature dependence of the energy gap is modified by these damping effects. Although the effects are small, they agree well with the deviations from BCS observed by Gasparovic et al. (*66*) in measurements of $\Delta(T)$ (see Fig. 17). It was noted by these workers that the data also fitted a BCS-model temperature dependence corresponding to some higher critical temperature, T_c', where $3.5kT_c' = [2\Delta(0)]$ meas. T_c' varied between 8.7 and 9.7°K as contrasted with the measured T_c of 7.3°K. This fact is pertinent to the measurements of mercury by Bermon and Ginzberg (*84*), who also found that their data fitted a BCS temperature dependence corresponding to a transition temperature about 25% higher than the measured T_c. This implies that the deviations from the simple BCS model in mercury are very similar to those in lead and perhaps both can be described by this extension of the BCS model.

E. Effects of Crystal Symmetry

It was suspected by Giaever and Megerle (*49*) that some of the lack of agreement between the BCS model and the tunneling data was to be attributed to the isotropic model assumed in the theory. An anisotropic model would give different energy gaps in different crystal directions. Even in thin films, which are a partly ordered, partly random collection of crystallites, it seemed plausible that different values of the gap should be measured in different samples and that a smearing of the gap might occur. Townsend and Sutton (*85*) observed what appeared to be more than one gap edge in thick lead films in a lead–tantalum junction. Zavaritskii (*86*) studied this problem by measuring the current–voltage characteristic of a junction formed by a tin film and a single crystal of tin; different crystal directions were observed to give energy gaps about 20% larger and smaller than the average. Later results by Zavaritskii (*87*) are shown in Fig. 18, in which the dV/dI vs. voltage clearly shows the structure at the gap edge. The complicated results of this experiment were interpreted by assuming that there are large annular regions of approximately constant Δ and sharp boundaries between the regions, corresponding presumably to singularities in the Fermi surface of tin. The results are complicated enough that no unique determination of the Fermi surface was possible, but it is interesting to note that the values of the energy gaps measured varied from 4.3 to $3.1kT_c$, just about the spread in thin-film measurements. The results are in fair agreement with the results of ultrasonic attenuation measurements and were interpreted by Zavaritskii (*87*) in terms of the anisotropic model of Pokrovskii (*88*) and later calculations by Geilikman and Kresin (*89*). These explanations are based on the assumption of anisotropy of the Fermi surface and the band structure. Bennett (*90*) has suggested that the basic anisotropy is that of the phonon spectrum, which by means of the electron–phonon interaction causes the aniso-

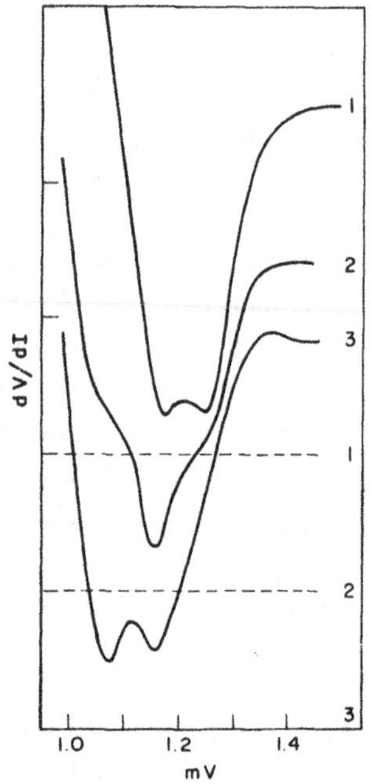

Fig. 18. dV/dI vs. V for tunneling in Sn–Sn samples of various orientations (*87*). (For clarity the origins for different curves have been displaced vertically.)

tropy in the energy gap and that the positions of the Van Hove critical points should also reflect the anisotropy and be observable in the tunneling characteristic. Rochlin and Douglass (*91*) have interpreted the structure that they and others (*92,93*) have observed in lead–lead Josephson junctions according to the Bennett theory, but as yet these very complex effects are not completely understood. McVicar and Rose (*93a*) have measured the energy gap in single-crystal niobium by tunneling and shown the anisotropy to agree fairly well with ultrasonic data of Dobbs and Perz (*229*). So far, experiments have demonstrated that the isotropic BCS model is not adequate, but the manner in which the theory should be generalized has not been positively established.

F. Magnetic Field Dependence of the Energy Gap

Giaever and Megerle (*49*) showed that a magnetic field applied parallel to the plane of a thin-film tunnel junction decreased the measured energy gap monotonically until the gap disappeared at the critical field. Douglass (*101*) realized that the effect of the magnetic field on the energy gap could be described by the Ginzburg–Landau (*102*) phenomenological theory, provided the energy gap parameter was identified with the order parameter, a result which had been obtained by Gor'kov (*2*) from the microscopic theory.

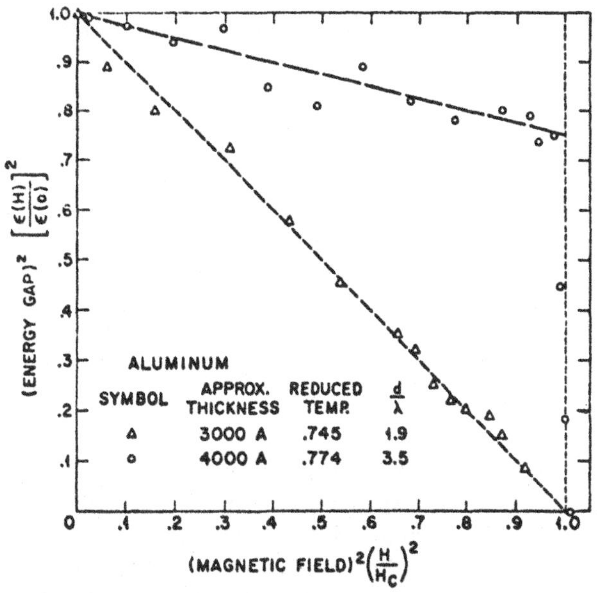

Fig. 19. Energy gap of aluminum as a function of the magnetic field from tunneling measurements on Al–Pb samples (*101*).

Figure 19 shows measurements by Douglass (*101*) of the magnetic field dependence of the energy gap on films of two different thicknesses. In the upper curve, for which the film thickness $d \gg 2\lambda$ (λ is the penetration depth at low fields), the energy gap drops almost discontinuously at H_c and corresponds to a first-order phase transition, as predicted by the Ginzburg–Landau theory. In the lower curve, for which $d < 2\lambda$, the energy gap is depressed continuously and becomes zero at H_c, and corresponds to a second-order phase transition. The Ginzburg–Landau theory predicts that the change from a first- to a second-order transition takes place at a critical value of the parameter d/λ equal to $\sqrt{5}$. More extensive measurements on aluminum films by Meservey and Douglass (*103*) confirmed the earlier results and showed that over a wide range of temperature and thickness the results could be explained by the single additional parameter d/λ, although the critical value of $(d/\lambda)_c$ was approximately 1.2 instead of the Ginzburg–Landau value of 2.23. Collier and Kamper (*104*) have made tunneling measurements of the energy gap of tin, both in a parallel field where the results agreed more quantitatively with the Ginzburg–Landau theory, and in a perpendicular field where the results could be interpreted on the basis of Tinkham's (*45*) theory for the penetration of field.

Maki (*105*) has considered in detail the effect of magnetic fields and transport currents on dirty thin-film superconductors (ones in which the coherence length is smaller than the penetration depth because of the presence of impurities) and has found that the resulting behavior of the density of states is identical to that in a "gapless" superconductor with magnetic impurities (see the next section and Chapter 16). Recently Levine (*106*) and Guyon et al. (*107*) have carried out detailed tunneling studies on dirty Sn and dirty In films and have found good agreement with Maki's theory of the field dependence of the density of states.

G. Gapless Superconductivity

Early experiments by Matthias et al. (*108*) showed a rapid decrease in the superconducting transition temperature when rare earth impurities were placed in bulk La. Similar experiments were performed by the Göttingen group (*109*), who showed that T_c decreased linearly with impurity concentration. A magnetic impurity concentration of only 1% was enough to lower the transition temperature appreciably, whereas nonmagnetic impurities had little or no effect on T_c. This sharp decrease can be explained in terms of the exchange interaction between conduction electrons and impurity atoms, resulting in spin-flip scattering which destroys the time-reversal correlation of the Cooper pairs.

In 1961 a significant paper by Abrikosov and Gor'kov (*110*) appeared in which the microscopic theory was extended to include alloys with magnetic impurities. By assuming that the impurity spin **S** is coupled to the spin of the conduction electrons **s** by an exchange interaction of the form $J(\mathbf{s}\cdot\mathbf{S})$, Abrikosov and Gor'kov obtain a scattering time τ_s for a spin-flip process. In contrast with nonmagnetic impurities, spin-exchange scattering is not time-reversal-invariant. This results in a finite lifetime for the Cooper pairs and, therefore, severely reduces the transition temperature. Furthermore, Abrikosov and Gor'kov arrived at the remarkable conclusion—that there exists a region of magnetic impurity concentration above which the gap in the excitation spectrum is zero, yet the specimen is still a superconductor in the sense of having pair correlation and a nonzero transition temperature. The possibility of a gapless superconductor was surprising, at first, since the BCS theory contains only a single parameter Δ which is both the energy gap and order parameter. At present, however, gapless superconductivity has been well established and seems to be a quite general phenomenon observable in many situations.

The expression derived by Abrikosov and Gor'kov for the transition temperature T_c of a dirty specimen with spin-flip scattering time τ_s is given by

$$\ln\left(\frac{T_{c0}}{T_c}\right) = \Psi\left(\frac{1}{2} + \frac{\hbar}{2\pi k T_c \tau_s}\right) - \Psi\left(\frac{1}{2}\right) \tag{24}$$

where T_{c0} is the transition temperature in the absence of magnetic impurities and Ψ is the di-gamma function. For low concentration of magnetic impurities $T_c = T_{c0} - \pi h/4k_B\tau_s$, which confirms the linear dependence of T_c on the impurity concentration. From Eq. (24) we also obtain the critical concentration for which T_c goes to zero, given by $\tau_{sc} = [2\hbar/\Delta(0)]$. In addition, Abrikosov and Gor'kov derived an expression for the concentration-dependent energy gap denoted as Ω_G and found that Ω_G went to zero at an impurity concentration such that

$$\frac{1}{\tau_s} = \frac{1}{2\hbar/\Delta(0)}\frac{2}{e^{\pi/4}} = \frac{0.91}{\tau_{sc}}$$

Thus the energy gap Ω_G vanishes at a concentration of only 0.91, the concentration at which superconductivity is entirely destroyed. (One often sees $\hbar/2\pi k T_c \tau_s$ written as

$$0.140\frac{\tau_{sc}}{\tau_s}\frac{T_{c0}}{T_c} = 0.140\frac{n}{n_c}\frac{T_{c0}}{T_c}$$

where n_c is the critical impurity concentration.)

The first experiments explicitly designed to test the concept of gaplessness were performed by Reif and Woolf (*111,112*) on "quenched" alloy films evaporated on substrates maintained at 4°K. This allowed them to measure the electrical resistivity to obtain the transition temperature, and the tunneling characteristics to obtain the energy gap and, in later experiments, the density of states. In Fig. 20 we compare their data for T_c/T_{c0} and $\Omega_G/\Delta(0)$ on a quenched In film as a function of Fe concentration with the theory of Abrikosov and Gor'kov as computed in detail by Skalski et al. (*113*). Since the tunneling measurements were taken at a reduced temperature of about 0.25, the data have to be compared with the more general temperature-dependent gap $\Omega_G(T)$ as determined by Skalski et al. The quantitative agreement, although not very good, is probably the best that can be expected on the basis of sample inhomogeniety and temperature broadening effects. It would be desirable to

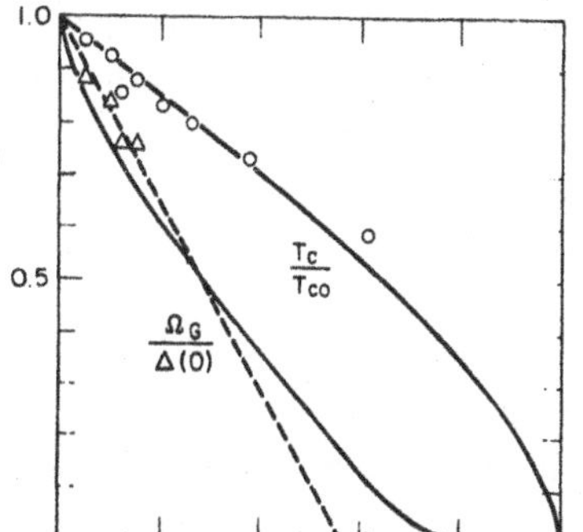

Fig. 20. Plot of T_c/T_{c0} and $\Omega_G/\Delta(0)$ at a reduced temperature of $t = 0.25$ as a function of mean spin flip scattering time τ_s as obtained by Skalski et al. (*113*). The triangle at $\tau_{sc}/\tau_s \simeq 0.6$ is about the concentration at which the experimental gap disappears.

obtain very low temperature tunneling data. Also implicit in the theory is the weak-coupling assumption of BCS.

Departures from the Abrikosov and Gor'kov temperature dependence [Eq. (24)] have been observed in lanthanum alloys containing small amounts of gadolinium impurities (*114,115*). The ratio T_c/T_{c0} experiences an anamolous increase and then a rapid decrease near a concentration corresponding to the critical scattering time τ_{sc}. This change in the temperature behavior has been interpreted by Bennemann (*116*) as being due to magnetic ordering between the magnetic impurities.

As mentioned earlier, gaplessness is not unique to superconductors with paramagnetic impurities but appears in a variety of other situations. In particular, any interaction which is not time-reversal-invariant, such as a magnetic field or a current, produces effects similar to those of paramagnetic impurities and gives rise to an effective scattering time τ_s which acts in a manner similar to the spin-flip exchange scattering time. Guertin et al. (*116a*) have recently tested many of these pair-breaking mechanisms alone and in combination by looking at the change in the transition temperature. In cases where the pair-breaking mechanisms are truly separable, their effect on the transition temperature simply adds to the argument in the di-gamma function [Eq. (24)]. Levine (*106*) has performed an experiment on the effect of a transport current on the energy gap as measured by electron tunneling. He was able to confirm the expected decrease of the energy gap with the square of the current; however, he was unable to extend his measurements to low enough temperatures, where the more exact expressions of Maki (*105*) and Fulde (*116b*) can be applied to illustrate the difference between the order parameter and energy gap. Levine has also tested the concept of gaplessness by measuring the density of states of a thin-film superconductor in a magnetic field using tunneling techniques. The density of states obtained are in excellent agreement with the Maki calculations; however, thermal smearing prevented a direct verification of the gapless region.

H. Other Tunneling Effects

1. Josephson Tunneling

A consequence of the microscopic theory of tunneling which was not included in the phenomenological theory was discovered by Josephson (*117,118*), who found that pairs could tunnel through very thin barriers at zero voltage up to a certain critical current. Above that current the observed voltage V is accompanied by a very high frequency oscillation at $\omega = 2eV/\hbar$ (*119–123*). Langenberg et al. (*124*) have measured the frequency–voltage ratio of the radiation from Josephson junctions and have been able to obtain an experimental accuracy of 6 parts in 10^6 in the value of $2e/h$. According to these authors this measurement gives a value of $2e/h$ which is more accurate than that of any other measurement. The dc and ac Josephson effects and how they modify the low-current portion of the tunneling characteristic of junctions with very low barrier heights are covered in detail in Chapter 9 and are mentioned here as an interesting example of an effect which was implicit in the BCS theory but was only discovered considerably later.

2. Multiparticle Tunneling

In experiments on current–voltage measurements on tunneling between superconductors, Taylor and Burstein (*98*) and independently Adkins (*125*) observed excess current which could not be explained within the framework of single-particle tunneling. The excess current is observed as a sharp temperature-independent jump in current at a voltage $\Delta(T)$ for identical superconductors or at $\Delta_1(T)$ and $\Delta_2(T)$ for two different superconductors. The jump is followed by an exponential rise which can be separated into a temperature-independent excess with an exponential dependence on voltage and a strongly temperature-dependent part.

To explain the onset of the current jump at $\Delta(T)$, Schrieffer and Wilkins (*126*) computed higher-order effects in which two electrons in superconductor 1 tunnel through the oxide to become two quasi-particles in superconductor 2. In this process the quasi-particle distribution in superconductor 1 is left undisturbed even though the number of pairs is decreased by 1. The conservation of energy for this process requires that the thresholds for these processes occur at Δ_1 and Δ_2. Since this higher-order-particle tunneling involves two particles, the matrix element for this process will be proportional to the square of the single-particle tunneling matrix element. Unlike single-particle tunneling, however, two-particle tunneling does not depend on the number of thermally excited quasi-particles and should therefore be temperature-independent and relatively more pronounced at low temperature where there is little single-particle tunneling.

The strongly temperature-dependent excess current observed by Taylor and Burstein for Pb–Pb junctions can be associated with single-particle tunneling in which phonons assist in exciting quasi-particles. Kleinman et al. (*127,128*) have calculated the process in which a phonon is absorbed by a ground-state pair on one side of the oxide barrier, resulting in a pair of quasi-particles, one on each side of the barrier, and the transfer of one electron across the barrier. By considering various possibilities, Kleinman was able to show that the one-phonon umklapp process dominates and yields the correct temperature and voltage dependence for the excess current in the Pb–Pb junction.

3. Photon-Assisted Electron Tunneling

Dayem and Martin (*129*) have measured the *I–V* characteristics of a tunnel junction in the presence of microwave radiation at various frequencies ν. In addition to the usual tunneling mechanism, electrons are able to absorb a photon in the process of tunneling across the oxide barrier. The experimental result shows a sharp rise in the current at a voltage $\Delta_1 + \Delta_2 - h\nu$. This rise is due to electrons near the gap edge which upon absorbing a photon can tunnel into the available states on the other side of the barrier. The sharpness of the rise is indicative of the large density of states available at the gap edges. By measuring the differential conductance dI/dV, the excess current shows up as peaks at voltage values $\Delta_1 + \Delta_2 \pm nh\nu$, where n is an integer. These peaks indicate the absorption or emission of one or more photons when the electron is tunneling across the barrier.

Tien and Gordon (*130*) have computed the change in tunneling current in the presence of microwave radiation by considering the response of the quasi-particle wave functions by simple time-dependent perturbation theory. These resulting wave functions contain components at energies $\Delta \pm nh\nu$, etc., and cause an effective change in the density of states for the superconductor. The calculated microwave power necessary to cause the measured changes in I–V seen by Dayem and Martin is larger by at least an order of magnitude than that used in the experiment.

4. Microwave-Phonon-Assisted Tunneling

Stimulated by the experiments of Dayem and Martin (*129*) with microwave photons, Abeles and Goldstein (*131,132*) and Lax and Vernon (*133*) measured the change in tunneling current ΔI_s in the presence of microwave phonons. Both experiments showed that for longitudinal phonons, ΔI_s as a function of voltage correlated closely with the d^2I/dV^2 curve of the diode without the microwave phonons. By generalizing Tien and Gordon's (*130*) expression for photons to phonons, Lax and Vernon were able to show that ΔI_s should be proportional to d^2I/dV^2. However, in the experiment reported by Abeles and Goldstein, ΔI_s for transverse phonons was different from the results for longitudinal phonons. This difference in behavior is not well understood but may be due to the lack of screening of the magnetic field associated with transverse phonons. Upon closer examination of the I–V response to longitudinal phonons, Goldstein et al. (*134*) found real deviations from the theoretical $\Delta I_s \sim d^2I/dV^2$ which were most striking in Pb–Pb junctions and less apparent in Pb–Al junctions. In addition, the magnitude of ΔI_s was an order of magnitude greater than expected from theory, indicating a similarity with the Dayem–Martin experiment. As yet, this disagreement remains unexplained.

5. Tomasch Effect

Tomasch (*135,136*) has observed oscillations in the effective tunneling density of states approximately periodic in the inverse thickness of superconducting films. McMillan and Anderson (*137*) have explained that this geometrical resonance results from interference effects between quasi-electrons and quasi-holes which are coupled by perturbations of the energy gap at or near the outer film surface. The positions of the resonances fit the theoretical energy dependence

$$\omega_n = \left[\Delta^2 + \left(\frac{n\pi\hbar v_F}{d}\right)^2\right]^{1/2}$$

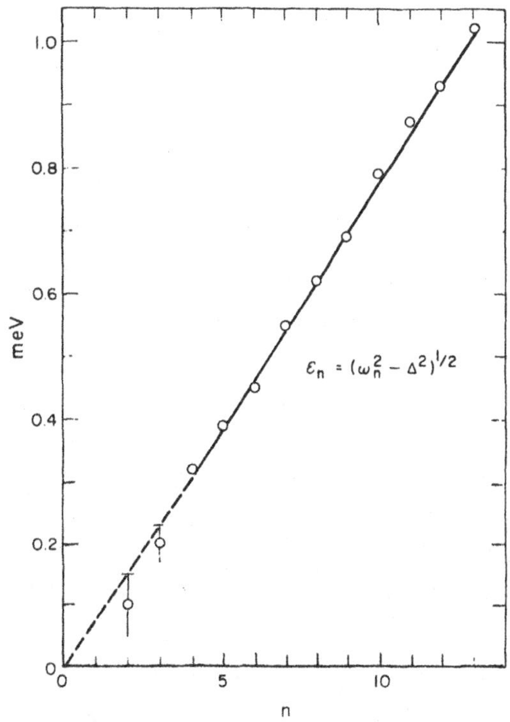

Fig. 21. Plot of data of Tomasch and Wolfram (*136*) compared with the theoretical equation $\epsilon_n = n(\hbar v_F \pi/d) = (\omega_n{}^2 - \Delta^2)^{1/2}$, where ω_n is the energy of the nth observed resonance. For In, $v_F = 1.23 \times 10^8$ cm/sec.

where v_F is the Fermi velocity, d is the thickness of the film, and n is an integer (see Fig. 21). The Tomasch effect is a demonstration of the coherent coupling between hole and electron quasi-particles in the BCS theory of superconductivity and complements the Josephson effect, which couples the phases of the coherent macroscopic functions for superconductors separated by a thin oxide layer.

IV. THERMODYNAMIC PROPERTIES

A. Classical Thermodynamics of Superconductors

Unlike tunneling experiments, thermal properties of superconductors had been measured 30 years prior to the BCS theory. As early as 1924 Keesom (*138*) had applied thermodynamics to the specific heat jump in a superconductor; but it was the reversible magnetization measurements of Meissner and Ochsenfeld (*139*) that justified the use of thermodynamics. The discovery that a superconductor acted like a perfectly diamagnetic substance implies on very general grounds that the Gibbs free energy difference between the normal and superconducting states in a magnetic field H is (per mole)

$$G_n - G_s = \left(\frac{H_c^2}{8\pi} - \frac{H^2}{8\pi}\right) V_M \tag{25}$$

where H_c is the thermodynamic critical field and V_M is the molar volume. Since $S = -(\partial G/\partial T)_H$, we obtain the entropy difference to be

$$S_n - S_s = -\frac{V_M}{4\pi} H_c \frac{dH_c}{dT} \tag{26}$$

and from $dS = C\,dT/T$ the specific heat difference is

$$C_n - C_s = -\frac{V_M T}{4\pi}\left[H_c \frac{d^2 H_c}{dT^2} + \left(\frac{dH_c}{dT}\right)^2\right] \tag{27}$$

In the absence of a magnetic field the transition from the normal state to the superconducting state occurs at T_c with no change in entropy and with a discontinuity in the specific heat of magnitude,

$$(C_n - C_s)_{T_c} = -\frac{V_M T_c}{4\pi}\left(\frac{dH_c}{dT}\right)^2_{T=T_c} \tag{28}$$

a formula first given by Rutgers (*140*). Thus the transition implied by these relations is first order with a magnetic field and second order in the absence of a magnetic field.

These classical thermodynamic equations were important in developing the London theory (*34*) and later the BCS theory. The Meissner effect, the second-order transition, and the connection between the thermal and magnetic properties became established experimental facts to which all theories had to conform. The BCS theory does reproduce these effects and in addition it gives more explicit physical meaning to the London concept of ordering in momentum space.

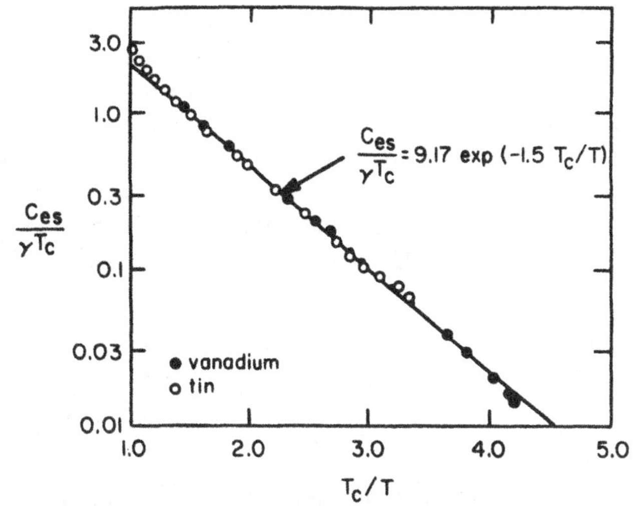

Fig. 22. Reduced electronic specific heat in superconducting vanadium and tin. [From Biondi et al., (*150*).]

The measured values of the specific heat in the superconducting state appeared to be of the form aT^3 and remained unexplained for many years. Although in 1938 the data of Keesom and Van Laer (*141*) for tin showed deviations from this law, it was not until 1953 that the measurements of Brown et al. (*142*) on niobium confirmed such a deviation. The measurements of Corak et al. (*143*) showed that the electronic specific heat of vanadium had an exponential temperature dependence at the lowest temperatures; this result and confirming measurements on tin by Corak and Satterthwaite (*144*) are shown in Fig. 22. An exponential dependence of C_{es} was strong evidence for an energy gap, a concept which had recurrently been proposed by theorists since F. London (*145*) first made the suggestion in 1935. Bardeen (*146*) has stated that it was the attempt to obtain an energy gap by an electron–phonon interaction that provided much of the motivation for developing the microscopic theory.

B. Thermodynamics According to the BCS Theory

The BCS theory leads to an expression for the free energy from which the transition temperature and other thermodynamic properties can be derived. This energy (in zero magnetic field) is

$$F_s(T) = 2k_B T \sum_k \ln[1 - f(E_k)] + \sum_k (\epsilon_k - E_k)\, v_k^2 - \sum_k (\Delta_k^2/E_k) f(E_k) \quad (29)$$

where $E_k = (\epsilon_k^2 + \Delta_k^2)^{1/2}$, $v_k^2 = \frac{1}{2}(1 - \epsilon_k/E_k)$, and the energy gap parameter Δ_k is determined from the nonlinear relation of Eq. (1a). As usual, $f(E_k) = (1 + \beta E_k)^{-1}$ is the Fermi function and v_k is the probability amplitude that the pair state $(k\uparrow, -k\downarrow)$ is occupied in the ground state. When $\Delta = 0$, the last two terms in the expression for F_s and zero and there remains the usual expression for the free energy of independent fermions.

The transition temperature is determined when the free energy difference between the superconducting and normal state is zero, a condition that in this model is equivalent to $\Delta = 0$. This condition leads to the expression for the transition temperature $T_c = 0.85\Theta_D \exp[-1/N(0)\,V]$. Since $N(0)\,V$ is typically 0.2, the order of magnitude of the transition temperatures for many superconductors is correctly predicted and an explanation is provided for why T_c is much less than Θ_D. However, as discussed in Section II, numerical agreement of the measured and calculated T_c is only roughly achieved and then only after the original BCS model has been considerably modified. A basic difficulty in calculating T_c lies in the fact that quantities which are not well known enter as exponents and as yet T_c cannot be calculated with certainty from first principles.

C. Specific Heat and Entropy

From Eq. (29) and the relations $S = -\partial F/\partial T$ we obtain the entropy (per unit volume)

$$S = -2k_B \sum_k \{[1 - f(E_k)] \ln [1 - f(E_k)] + f(E_k) \ln f(E_k)\} \tag{30}$$

This equation for the entropy is of exactly the same form as for independent fermions and is written as it is usually used for electrons and holes in normal metals. The difference is in the dispersion relation, which in the present case is $E_k = (\epsilon_k^2 + \Delta_k^2)^{1/2}$. As $T \to T_c$, Δ_k approaches zero continuously and the entropy will approach the normal-state entropy continuously, so that there will be no latent heat at T_c when there is no magnetic field, as expected from Eq. (26).

The electronic specific heat (per unit volume) is obtained from Eq. (30) and the relation $C = T\,\partial S/\partial T$:

$$C_{es} = -\frac{2}{T} \sum_k \left(E_k^2 + \frac{\beta}{2} \frac{\partial \Delta^2}{\partial \beta} \right) \frac{\partial f}{\partial E} \tag{31}$$

In the weak-coupling limit $[N(0)\,V < 1]$ Mühlschlegel (*67*) has solved Eq. (31) numerically and has tabulated values of C_{es}/C_{en} as a function of T. Useful approximations in various temperature regions have been given by Bardeen and Schrieffer; at the lowest temperatures the theoretical value approaches

$$\underset{T \to 0}{C_{es}} = 9.17\gamma T_c \exp(-1.5T_c/T) \tag{32}$$

Since the theory predicts a second-order transition, C_{es} is discontinuous at T_c. This second-order transition at T_c had been established by experiment long before the BCS theory, but with the discovery of a logarithmic singularity in the specific heat at the superfluid transition of liquid helium, it was conjectured that a similar effect might be found in the superconducting transition. Careful investigations (*147,148*) for evidence of a logarithmic singularity in C_{es} at T_c have led to negative results and the experimental evidence is that the transition is second order in zero magnetic field.

To compare the predictions of the BCS theory with specific-heat measurements, the usual method of evaluating the data has been to assume that the specific heat is the sum of an electronic contribution and a lattice contribution:

$$C_n = \gamma T + C_{ln} \tag{33}$$

$$C_s = C_{es} + C_{ls} \tag{34}$$

Here C_n and C_s are the total specific heats in the normal and superconducting state, C_{es} is the electronic specific heat in the superconducting state, and γT is the usual expression for the electronic specific heat in the normal state. The lattice specific heats in the normal and superconducting states are C_{ln} and C_{ls}, respectively, and it has usually been assumed that

$$C_{ln} = C_{ls} \tag{35}$$

This last assumption was based on measurements of the lattice constants (*149*) above and below the transition temperature, which showed that any discontinuities in the elastic properties were very small and presumably therefore any discontinuity in the lattice specific heat was also very small. At low-enough temperatures the lattice specific heat has a cubic dependence on the temperature and can be expressed as

$$C_{ls} = C_{ln} = a(T/\Theta_D)^3 \tag{36}$$

where a is a known constant and Θ_D is the Debye temperature. By subtracting C_{ls} from the total specific heat we obtain a value for C_{es}.

The measurements of C_{es} for both vanadium and tin (*150*) shown in Fig. 22 fit an exponential relationship corresponding to an energy gap of approximately $3kT_c$. Subsequent to the publication of the BCS theory there were numerous attempts to extend the measurements of C_{es} to lower temperatures, where the exponential temperature dependence of Eq. (33) was expected. Figure 23 shows data for Al, V, Zn, and Sn (*151*). The upward deviation of many of the results at low temperatures is very noticeable; in fact, some of the results fitted a T^3-law just as well as an exponential one.

Another complication was introduced by the measurements of the specific heat of indium by Bryant and Keesom (*152*) (Fig. 24), who found that after

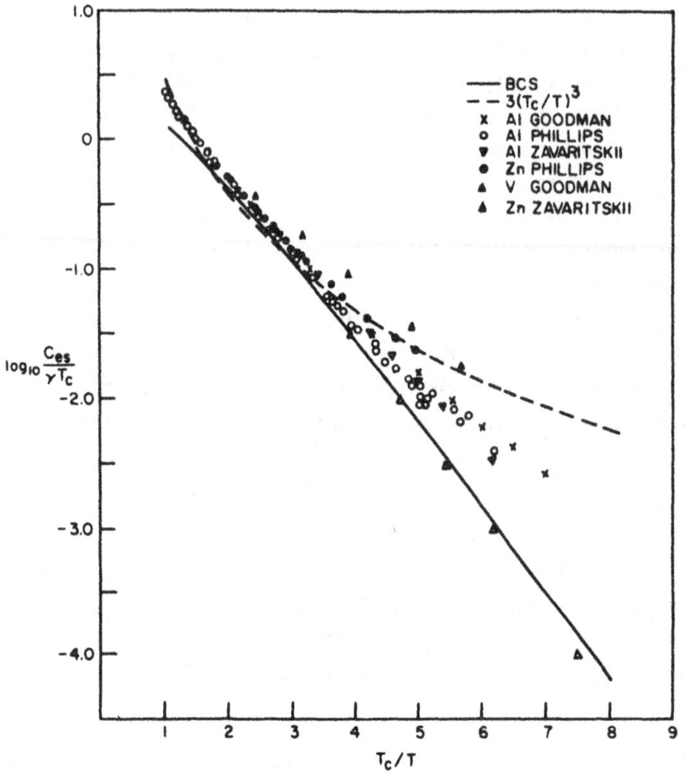

Fig. 23. Reduced superconducting electronic specific heat vs. reduced temperature for Al, Zn, and V (*151*).

subtracting the lattice part according to the assumptions of Eqs. (33)–(36), the electronic specific heat C_{es} became negative below $T/T_c = 0.8$. A similar anomaly was found in niobium by Boorse and co-workers (*153*). This result implied that the lattice specific heat constant increased below T_c. On the other hand, measurements of the elastic constants in indium by Chandrasekhar and Rayne (*154*) implied that C_{ls} and C_{ln} differ by no more than 1 part in 10^4. For niobium van der Hoeven and Keesom (*155*) showed that the apparent anomaly disappeared when the specific heat measurements were carried out at a low-enough temperature so that Eq. (36) truly applies and the correct Θ_D is obtained. This result was confirmed by Boorse and co-workers (*153*) on a single-crystal sample of niobium of higher purity. For indium, however, O'Neal and Phillips (*156*) have confirmed the values of the specific heat obtained by Bryant and Keesom (*152*), and, although the analysis of the data was slightly different, a slight anomaly still remained.

It should be mentioned that these measurements are difficult and their inter-

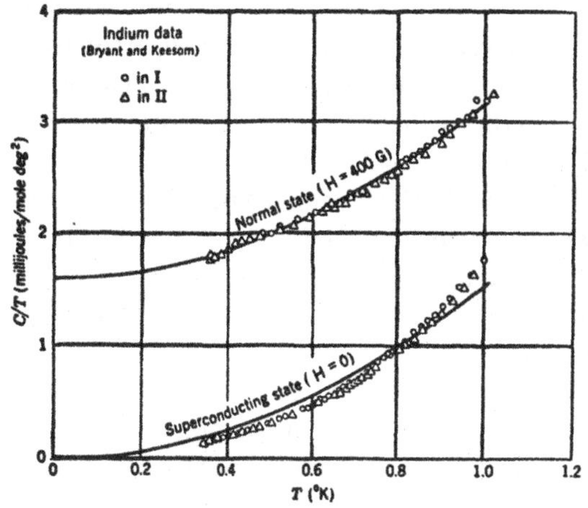

Fig. 24. Plot of C/T vs. T for indium in the normal and superconducting state. The upper line is the best fit of $\gamma + \alpha T^2$ for the normal points. The lower curve is a plot of αT^2. The experimental points for the superconductor lie below it (*152*).

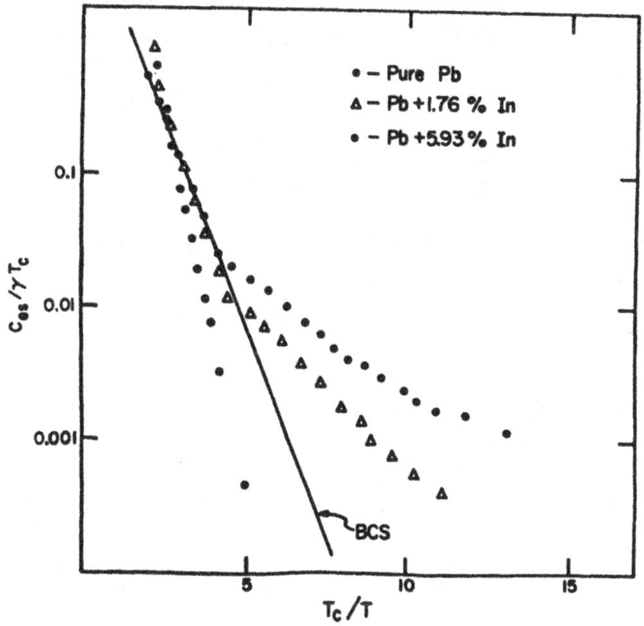

Fig. 25. Reduced electronic specific heat in the superconducting state vs. the reduced temperature for pure lead and two lead–indium alloys (*159*).

pretation complicated. For instance, for indium it is necessary to correct the normal-state specific heat by a hyperfine contribution which is proportional to T^{-2} and which is believed to be absent in the superconducting state because of a long spin-lattice relaxation time. O'Neal and Phillips (*156*) found that it was also necessary to correct the superconducting-state specific heat with a term proportional to T because of small inclusions of normal material due to locked-in magnetic flux. Phillips (*157*) has suggested that the source of the contradiction may lie in the assumption of the cubic dependence of the lattice specific heat and that actually the cubic region of C_{ls} may not yet have been reached at T_c, but as yet the anomalous behavior of indium has not been resolved.

The upward trend of the specific-heat data in the very low temperature region has been found in many metals and has received close study. Keesom and van der Hoeven (*158*) have reported measurements on lead between 0.4 and 4.2°K in which the very low temperature specific heat is proportional to T^3. More careful measurements (*159*) demonstrated two rather definite slopes on a semilog plot (Fig. 25), indicating for lead two distinct energy gaps of size $4.1kT_c$ and $1.1kT_c$. With the addition of 5.93% indium, the smaller gap at $1.1kT_c$ apparently disappeared. These facts are consistent with an anisotropic energy gap, although the amount of impurity needed to remove the anisotropy is much larger than predicted by Anderson (*160*). The value of $4.1kT_c$ is close to the value obtained in tunneling and infrared absorption measurements and to the value that has been calculated on the basis of a retarded, energy-dependent

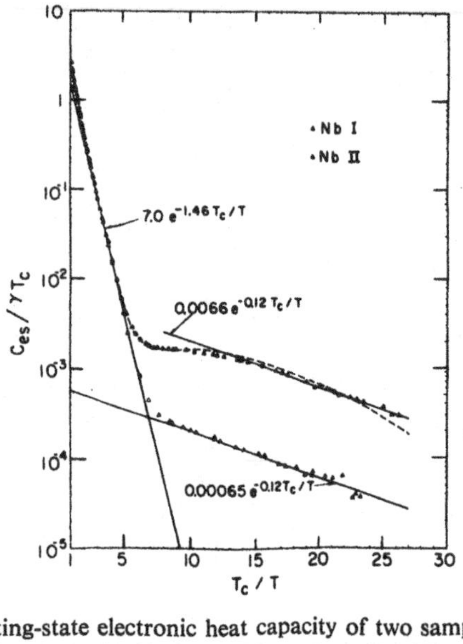

Fig. 26. Superconducting-state electronic heat capacity of two samples of Nb. The dashed curve represents the expression $C_{es}/T_c = 7.0 \exp(-1.46 T_c/T) + 0.0038 (0.25 T_c/T)^2 \exp(0.25 T_c/T) [1 + \exp(0.25 T_c/T)]^{-2}$ (*161*).

TABLE IV

Measured Values of Two Thermodynamic Functions

	$\left(\frac{C_s - C_n}{C_n}\right)_{Tc}$	Ref.	$\frac{\gamma T_c^2}{V_M H_c^2(0)^a}$	Ref.
BCS	1.43[b]	(1[b],4)	0.168	
		(166		
Al	1.29[c]–1.59[c]	(167,168)	0.171	(167)
		(148,170)		
		(171)		
	1.45[c] (av.)		0.170	(170[d])
Cd	1.32	(172)	0.177	(172)
	1.40	(173)		
Ga	1.44, 1.41	(172,174)	0.169, 0.170	(172,174)
	1.40	(175)[e]	0.173	(175)
Hg	2.37	(169,176)	0.134	(165)
		(165)		
In	1.73	(156)	0.157,0.150	(152,165)
La (hcp)	1.5	(177)		
Mo	1.28	(186)	0.182	(148)
Nb	1.87 (calorimetric)	(178)	0.157	(179)
	2.0 (magnetic)	(179)		
Pb	2.71	(180,176)	0.134	(187[d])
Sn	1.60	(147,181)	0.161	(165)
			0.164	(152)
Ta	1.59	(147,181,182)	0.161	(181,182)
Tl	1.50	(183)	0.161	(181)
U(β)	1.36, 1.52	(184)		
V	1.49	(162)	0.170	(162)
Zn	1.30	(174)	0.177	(174)
	1.24	(185)		

[a] Values of the molar volume V_M are taken from *(181)* except for La and Mo.

[b] The value 1.52 given in the original BCS paper was a numerical mistake which when corrected gives 1.43.

[c] The results from *(148)*, *(167)*, *(168)*, *(170)*, *(171)*, and other works to which they refer have a much greater scatter than can be explained by experimental error. This is perhaps a problem of sample preparation, although the magnetic measurements tend to be larger than the calorimetric ones. The average is over the 15 measurements (6 calorimetric, 9 magnetic) reported or summarized from previous work in the above references.

[d] This is the reference for $H_c(0)$. The values of γ and T_c are taken from the other references listed for the material.

[e] Value taken from *(181)*; discussion of result is given in *(175)*.

electron–phonon interaction as discussed in Section IIID. Shen et al. (*161*) have found that the specific-heat curves of high-purity Nb, Ta, and V provide very strong evidence for two energy gaps (Fig. 26). For vanadium these results have been confirmed by Radebaugh and Keesom (*162*) for a very high purity crystal. Presently it appears that some of the excess specific heat at low temperatures can be attributed to additional energy gaps. The results in lead were interpreted as the result of anisotropy (*159*); in Nb, Ta, and V the results were interpreted by Shen et al. (*161*) to result from an overlapping of the *s*- and *d*-bands at the Fermi surface as suggested by Suhl et al. (*163*). In order to fit the vanadium data Radebaugh and Keesom (*162*) find it necessary to use this two-gap theory as well as assume that the larger gap is anisotropic.

The simple BCS model predicts that in zero magnetic field the normalized discontinuity in the electronic specific heat at T_c is $[(C_{es} - C_{en})/C_{en}]_{T_c} = 1.43$. Measured values of this quantity, which is closely related to the normalized energy gap at T_c, are shown in Table IV. There is rough agreement with many superconductors, and again the most strongly coupled superconductors are the most deviant. The realistic strong-coupling model for lead [Section III.D and (*83*)] gives us much better agreement with the data and some hope that more refined models will allow this discontinuity to be calculated precisely. There remain unanswered questions and even contradictions, such as the measurements of Phillips on lead in which the anomalously large specific heat at low temperatures was absent, but in general it appears that the experimental results are converging with each other and probably with a generalized BCS theory.

D. Critical Magnetic Field

The critical magnetic field at $T = 0$ is obtained in the BCS theory by setting the diamagnetic energy of the superconductor equal to the condensation energy:

$$V_M H_c^2(0)/8\pi = \tfrac{1}{2} N(0)\,\Delta^2(0)$$

Since for a free electron model, $N(0) = 3\gamma/2\pi^2 k^2$, where γT is the molar electronic specific heat of the normal state, the above relation (for a type I superconductor) can be written entirely in terms of measurable quantities,

$$\Delta(0) = kH_c(0)\sqrt{\pi V_M/6\gamma}$$

Using the BCS value of $2\Delta/kT_c = 3.53$, we also have the result that for weakly coupled superconductors

$$\gamma T_c^2/V_M H_c^2(0) = 0.168$$

The measured values of this quantity, which are closely related to the reciprocal of the energy gap at $T = 0$, are shown in Table IV. It is seen that there is fair agreement with theory except in the case of strongly coupled superconductors, which deviate markedly from the theoretical value. However, the deviations from theory are less than in the case of the normalized specific heat discontinuity; the reason probably is related to the increase in the importance of lifetime effects near T_c, as discussed in Section III.D. For a type II superconductor H_c is not a directly measurable quantity but is defined by setting $H_c^2/8\pi$ equal to the area under a reversible magnetization curve.

It was discovered many years ago that the decrease of the critical field with temperature is approximately proportional to the square of the reduced temperature, and it has become customary to present the temperature dependence of the critical field in terms of the deviation from this empirical parabolic law,

$$D\left(\frac{T}{T_c}\right) = \frac{H_c(T)}{H_c(0)} - \left[1 - \left(\frac{T}{T_c}\right)^2\right] \tag{37}$$

The simple BCS model predicts this deviation fairly well for aluminum, a weakly coupled, isotropic superconductor (Fig. 27), but for lead and mercury the measured deviation does not agree at all, even being of the opposite sign. Again these effects may be mainly attributed to strong coupling according to Swihart et al., who have calculated the points in Fig. 27 for lead using a strong-coupling model.

With the aid of the general thermodynamic relations, Eqs. (26)–(28), we can obtain thermal quantities such as entropy and specific heat from magnetic measurements as long as the measurements are reversible. Mapother (*164*) has

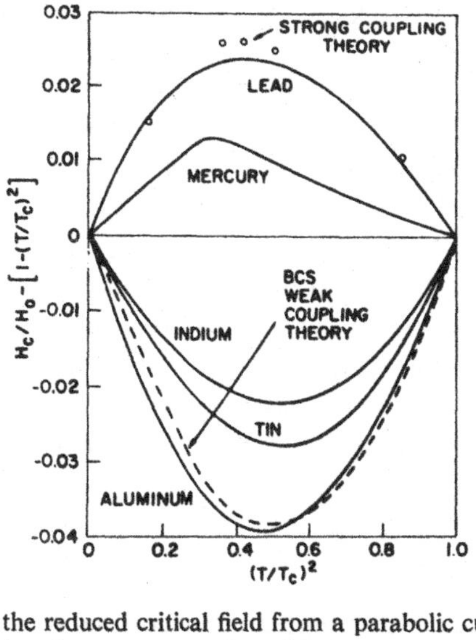

Fig. 27. Deviation of the reduced critical field from a parabolic curve as a function of the square of the reduced temperature. The solid curves are from experiment (*82*).

investigated the correspondence between magnetic and thermal measurements for tin and has found that the specific heat obtained from critical field measurements agrees with precise calorimetric measurements. The critical field measurements are usually easier and are claimed to be more precise near T_c, making them particularly useful in determining the discontinuity in C_{es}. Well below T_c the variation of H_c depends almost entirely on the specific heat in the normal state (since C_{es} decreases exponentially), and so the method becomes less useful. However, Finnemore and Mapother (*165*) claim that their critical field measurements on indium imply that there is no change in the lattice specific heat between the normal and the supercondicting states.

E. Pressure Effects

The effect of pressure on the critical field of a superconductor was measured by Sizoo et al. (*188*) in 1925. Although this and other early experiments showed that pressure was a relevant thermodynamic variable, the effects were so small that pressure has played a very minor role in the development of the theory of superconductivity until recently.

One reason for the recent increase in interest is that the technique of applying high pressures at low temperatures has been greatly developed. Another is the fact that the pressure dependence of T_c as predicted by the BCS theory is capable of analysis.

For the simple model the BCS transition temperature is

$$T_c = 0.85\Theta_D \exp[-1/N(0)\,V] \tag{2b}$$

Now Θ_D quite generally increases with pressure and so from this we might expect dT_c/dP to be positive. However, for all the common nontransition metals except thallium, dT_c/dP is negative, as shown in Table V. Actually for thallium dT_c/dP is positive only for small pressures and changes sign at higher pressures; it is believed that this low-pressure behavior is caused by anisotropy. This eliminates the view that $d\Theta_D/dP$ is the primary cause of dT_c/dP and leaves us with $N(0)$ and V. The density of states $N(0)$ can (for a free electron gas) be related to the electronic specific heat constant of the normal state: $\gamma = 2\pi^2k^2N(0)/3$.

With higher pressures it is much easier to measure the critical field than the specific heat, and in the limit as $T \to 0$, Eq. (27) reduces to

$$\gamma = \frac{\alpha}{2\pi}\,\frac{H_c^2(0)}{T_c^2} \tag{38}$$

If the critical field varied with temperature according to a parabolic law, that is, if $H_c = H_0\,(1 - t^2)$, then in Eq. (38) we would have $\alpha = 1$; furthermore, de-

TABLE V

Pressure Dependence of T_c

	Nontransition elements, $dT_c/dp \times 10^5$,[a] (°K/atm)		Transition elements, $dT_c/dp \times 10^5$,[a] °K/atm
Al	−2.9 ±0.2	La(α) (fcc)	−4.3 ±0.2
Bi II	−3.2 (*190*)	La(β) (hcp)	(−0.36 ±0.1 14(*197*)
Cd	−1.8 ±0.17 (*190*)	Mo	−0.1 ±0.1
Ga I	−1.8 ±0.03	Nb	−0.28 ±0.02
Ga II	−3 ±1	Os	−0.18 ±0.06 (*196*)
Hg(α)	−3.70 ±0.05	Re	0 ±0.1
Hg(β)	−4.8 ±0.2	Ru	0 ±0.03 (*196*)
In	−4.3 ±0.2	Ta	−0.26 ±0.01
Pb	−3.80 ±0.07	Ti	0.55[b] (*190*)
Sn	−4.6 ±0.1	Th	−1.7 ±0.2
Tl	2.3 ±0.1	U (α)	10.5[c] (*198*)
Zn	−1.6 ±0.4	U (β)	
		U (γ)	0.9 (*195*)
		V	−0.49 ±0.05
		Zr	1.5 ±0.5 (*190*) 0.9,[b] 1.4[b]

[a] Unless shown otherwise, the values are from (*189*), in which the original references are listed.

[b] High-pressure region.

[c] Low-pressure region; for higher pressures see (*198a*).

viations of α from unity can be determined from the known deviations from the parabolic law.

It is also possible to obtain $dN(0)/dP$ from measurements of the thermal expansion coefficient, but in practice data is only obtained for $P \approx 0$. Knowing $dN(0)/dP$ and assuming Eq. (2b) we should be able to calculate the pressure dependence of the interaction potential, dV/dP. However, as discussed in Section II, Eq. (2b) must be corrected to obtain realistic values of V, and one presumably must use this more refined model to calculate dV/dP or dT_c/dP. Until now the emphasis has been on the calculation of the isotope effect, and the pressure-dependence calculation remains to be done.

Brandt and Ginzburg (*190*) have assumed on the basis of an approximate model calculation of Pines (*6*) that for nontransition metals the decrease in T_c is caused essentially by the variation in V with an approximately constant value of $N(0)$. Assuming that $N(0)$ is constant and that at some critical pressure P_c, the BCS interaction parameter V goes to zero and superconductivity disappears,

Ginzburg (*191*) has derived from the theory of second-order transitions the prediction that near P_c the pressure dependence of T_c should be

$$T_c(P) = A \exp[-b/(P_{c0} - P)] \tag{39}$$

The only substance in which studies have been carried to high-enough reduced pressure to apply this formula is cadmium (Fig. 28). On the basis of these measurements and Eq. (39) Brandt and Ginzburg (*192*) predict that P_{c0} for cadmium is between 57,000 and 70,000 atm.

In the case of the transition metals dT_c/dP appears to be rather random both as to sign and magnitude, instead of being consistently negative, as with the nontransition metals. According to Morin and Maita (*193*), T_c of the transition metals is determined mainly by the density of states of the *d*-electrons. With

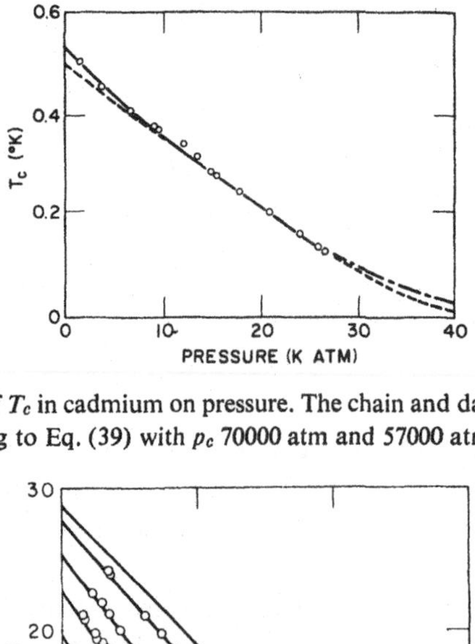

Fig. 28. Dependence of T_c in cadmium on pressure. The chain and dashed curves are plotted according to Eq. (39) with p_c 70000 atm and 57000 atm (*192*).

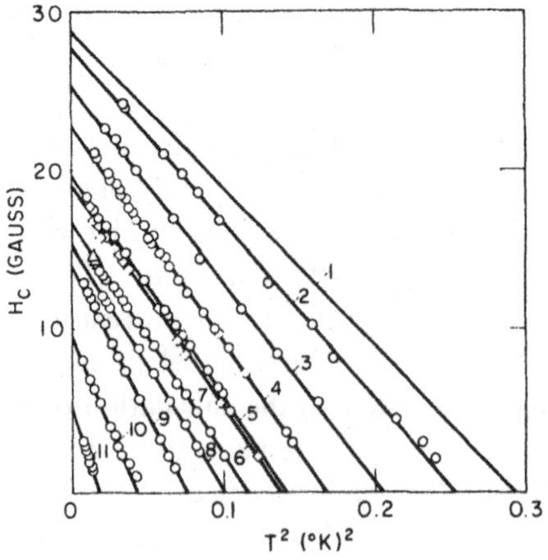

Fig. 29. Dependence of H_c on T^2 in cadmium under pressure: (1) $p = 1$ atm; (2) $p = 1.5$ katm; (3) $p = 3.7$ katm; (4) $p = 6.6$ katm; (5) $p = 9$ katm; (6) $p = 9.3$ katm; (7) $p = 12$ katm; (8) $p = 13.4$ katm; (9) $p = 15.3$ katm; (10) $p = 20.9$ katm; (11) $p = 26.4$ katm (*192*).

plausible assumptions, the sign of $dN(0)/dP$ can be correlated with the sign of dT_c/dP for a large number of elements and alloys. This result lends strong support to the view that the change in T_c with pressure in the transition metals is mainly caused by a change in $N(0)$, provided, of course, no qualitative change in the electronic structure is caused by the change in pressure (see Fig. 29). In lanthanum (*197*) and uranium (*198*) the positive pressure dependence is particularly large, and Kondo (*194*) suggested that admixtures of *f*-electron wave functions should have this effect. Kuper and others (*78*) have postulated a magnetic interaction with the *f*-electrons of an entirely different character from the BCS interaction. Present data appear to be consistent with a BCS type of interaction in which the effective value of $N(0)$ is increased by a narrow band of *f*-electrons overlapping the Fermi surface. According to Geballe et al. (*195*), application of pressure rapidly decreases the population of the 5*f*-band until it has moved far enough above the Fermi surface to no longer affect the superconducting interaction; further increase in pressure has little effect on T_c. More recently Smith and Chu (*195a*) have shown that the depression of the transition temperature on nontransition metal superconductors is proportional to the fractional change in volume over a wide range of pressures. Extropolation to $T_c = 0$ yields critical pressures for Al, Cd, and Zn of 67, 38, and 41 kbars, respectively, which is considerably lower than previously predicted.

V. ELECTRODYNAMIC PROPERTIES

A. Penetration Depth

Soon after the discovery by Meissner and Ochsenfeld (*139*) that the magnetic field is expelled from the interior of a superconductor independent of the history of the sample, it was realized that the earlier theories of Becker et al. (*199*) and de Haas–Lorentz (*200*), based on a model of freely accelerating electrons, must be modified. To account for the Meissner effect, F. and H. London (*201*) proposed the additional equation connecting the superconducting current **J** and magnetic field **H** in a superconductor

$$\nabla \times \mathbf{J}_s + (\mathbf{H}/c\Lambda) = 0$$

Λ is a constant, which for an ideal electron gas where all the electrons can accelerate freely, is equal to $\Lambda_f = m/ne^2$; here m is the electron mass and n the density of electrons. By considering the response of the electron distribution to an electric field and neglecting collisions with the lattice, one can obtain Λ in a real metal (*202*):

$$\Lambda = 3h^3/2e^2\bar{v}_F S = 3/2e^2 N(0)\,\bar{v}_F^2 \tag{40}$$

where $\bar{v}_F$ is the average Fermi velocity and S is the free area in momentum space of the Fermi surface. An experimental value for Λ can be obtained from measurements of the anomalous skin effect and the specific heat. In tin and aluminum, for example, the free electron values of Λ_f are approximately 1.0×10^{-31} sec^2 and 0.66×10^{-31} sec^2, whereas Λ is equal to 1.7×10^{-31} sec^2 and 0.35×10^{-31} sec^2, respectively (*202*).

To obtain the solution for the field penetration, we solve the Londons equation together with Maxwell's equations $c\nabla \times \mathbf{H} = 4\pi\mathbf{J}$ and $\nabla \cdot \mathbf{B} = 0$. If the applied field is in the z-direction parallel to the plane surface of a semiinfinite superconductor, the solution is

$$H_z(x) = H_z(0)\exp(-x/\lambda_L)$$

where $\lambda_L = \sqrt{\Lambda c^2/4\pi}$ is defined as the London penetration depth. For typical superconductors λ_L is about 2×10^{-6} cm.

The earliest measurements of the penetration depth were based on determining the magnetic susceptibility of small particles or thin films. By measuring the magnetization curves for colloids of mercury, Shoenberg (*203*) was able to provide evidence for an appreciable penetration depth, which increased sharply as $T \to T_c$. In 1948 Daunt et al. (*204*) suggested combining the Londons' original expression for λ with the Gorter–Casimir (*205*) two-fluid equation, where n is replaced by the temperature-dependent number of superconducting electrons $n_s(t) = n_s(0)(1 - t^4)$ and therefore

$$\lambda_L(T) = \lambda_L(0)\, y \tag{41}$$

where $y = (1 - t^4)^{-1/2}$ and $T/T_c = t$. According to this theory, a plot of the measured $\lambda(T)$ vs. y should be a straight line of slope $\lambda_L(0)$ and intercept $\lambda_L(0)$ at $y = 1$ ($t = 0$). Laurmann and Shoenberg (*206*) were able to interpret their temperature data on tin quite well using Eq. (41) and most of the quoted values for the penetration depth at $t = 0$ have been obtained from the best fit to the curve $\lambda = \lambda_0(1 - t^4)^{-1/2}$. The values obtained for λ_0 were generally several times larger than the London value $\lambda_L(0)$. Later experiments (*207–209*), however, showed that one cannot exactly equate the directly measured $\lambda(0)$ with λ_0, since the data on $\lambda(T)$ vs. y deviated slightly from a straight line for low temperatures near $y = 1$. This small deviation in the slope is predicted by the BCS theory and will be discussed later. At any rate, the discrepancy between $\lambda_L(0)$ and λ_0 remained to be explained.

Direct measurements of the penetration depth were carried out by Désirant and Shoenberg (*210*) using a composite specimen of fine mercury wires, and extensive measurements were made by Lock (*211*) on thin films of lead, tin, and indium. In 1940 Casimir (*212*) suggested obtaining λ by measuring the mutual inductance between two coils wound on a superconducting cylindrical core. As the penetration depth changes with temperature, so does the flux linkage between the coils. This method was applied successfully by Laurmann and Shoenberg (*206*) and Shalnikov and Sharvin (*213*), and more recently with some refinements by Chambers (*214*), Schawlow and Devlin (*207*), and Maxfield and McLean (*209,215*).

Many experiments concerning the penetration depth have been performed by Pippard (*216*) and his collaborators by measuring the surface impedance at microwave frequencies. At high frequencies the surface impedance of the superconductor is determined by the penetration depth, whereas in the normal metal it is determined by the skin depth δ. The actual measurement consists of recording the change in resonant frequency of a cavity containing the specimen in both the normal and superconducting state. For $T < T_c$, the shift in resonant frequency is proportional to $\delta_i(\omega) - \lambda(\omega)$, where $\delta_i(\omega)$ is the reactive part of the skin depth (defined as $\delta = \delta_r + i\delta_i = Z/4\pi\omega$, where Z is the surface impedance) and $\lambda(\omega)$ the frequency-dependent penetration depth. By making use of the Kramers–Kronig relation, one can obtain the zero-frequency value of λ from the observed changes in the resonant frequency. For the frequency used, $\lambda(\omega)$ is close to $\lambda(\omega = 0)$ and the Kramers–Kronig relations yield only a small, but important, correction.

The discrepancy between $\lambda_L(0)$ and λ_0 was clarified in 1953 by Pippard (*3*). In measurements on impure specimens he noticed that λ increased with decreasing mean free path. To explain this result he proposed that the London equation relating the current density to the vector potential is but a limiting form of the more general nonlocal relation

$$\mathbf{j}(r) = \frac{-3}{4\pi c\xi_0 \Lambda}\int \frac{\mathbf{R}[\mathbf{R}\cdot\mathbf{A}(r')]}{R^4}\exp(-R/\xi_0)\exp(-R/\alpha l)\, d^3r' \tag{42}$$

where ξ_0 was introduced as a coherence distance over which order will extend in the bulk superconductor, l is the mean free path, α is a number on the order of 1, and $R = |r - r'|$. Equation (42) is similar to the expression given by Reuter and Sondheimer (*217*) for the relation between current density and electric field for a normal metal in the anomalous limit, where the skin depth is small compared to the mean free path; this nonlocal relation reduces to the London equation when $A(r)$ varies slowly over a coherence distance ξ_0.

A nonlocal relation between j and A, similar to Pippard's expression [Eq. (42)], follows directly from the BCS theory:

$$\mathbf{j}(r) = \frac{-3}{4\pi c\xi_0 \Lambda(t)}\int \frac{\mathbf{R}[\mathbf{R}\cdot\mathbf{A}(r')]}{R^4} J(R, T)\exp(-R/l)\, d^3r' \tag{43}$$

where $\Lambda(t)$ depends on the temperature-dependent gap parameter Δ as

$$1 - \frac{n_s(t)}{n_s(0)} = 1 - \frac{\Lambda}{\Lambda(t)} = \frac{1}{2}\int_0^\infty \operatorname{sech}^2 \left\{ \frac{1}{2}\left[y^2 + \left(\frac{\Delta(T)}{kT} \right)^2 \right]^{1/2} \right\} dy \tag{44}$$

and now

$$\lambda_L(T) = c(\Lambda(t)/4\pi)^{1/2} \tag{45}$$

According to BCS and Pippard, the measured penetration depth, defined by

$$\lambda = \frac{1}{H_z(0)} \int_0^\infty H_z(x)\, dx$$

should be given by Eq. (45) when $\xi_0 \ll \lambda_L(T)$ (London limit). In the opposite limit $\xi_0 \gg \lambda_L(T)$ (Pippard limit) the expression for the penetration depth is

$$\lambda = \lambda_\infty(T) = \left[\frac{\sqrt{3}}{2\pi} \frac{\xi_0 \lambda_L^2(T)}{J(0, T)} \right]^{1/3}$$

(Both the London and Pippard limits still require $l \gg \xi_0$ and $l \gg \lambda_L$, respectively.) The kernel $J(R, T)$ is rather insensitive to temperature, and detailed calculation shows that at $T = 0$ it resembles the exponential $\exp(-R/\xi_0)$ to within 5%. In fact, the BCS coherence length, defined as $\int_0^\infty J(R, T)\, dR = \xi_0 = 0.18$ $(\hbar v_F/kT_c)$, shows that $J(0, T)$ varies from 1 to 1.33 as T varies from 0 to T_c. The nonlocal BCS or Pippard theory actually predicts a reversal in the sign of the penetrating field. This has been confirmed in thin cylindrical films by Drangeid and Sommerhalder (*218*). To determine λ for intermediate cases, BCS give a plot of the ratio of $\lambda(T)/\lambda_L(T)$ as a function of $\xi_0/\lambda_L(T)$. Since Λ and $\Lambda(t)$ are defined by Eqs. (40) and (44) and $\xi_0 = 0.18\,(\hbar v_F/kT_c)$, λ can be calculated. In Table VI we present data for $\lambda_L(0)$ and ξ_0, and compare the calculated and observed values of λ for $t = 0$. The agreement between theory and experiment is fairly good.

TABLE VI

Theoretical Penetration Depth at $t = 0$, $\lambda(0)$ compared with Experiment λ_0

Metal	$\lambda_L(0) \times 10^6$	$\xi_0 \times 10^6$	$\frac{\xi_0}{\lambda_L(0)}$	$\frac{\lambda(0)}{\lambda_L(0)}$	$\lambda(0) \times 10^6$ (random scattering)	$\lambda_0 \times 10^6$ (observed)
Sn	3.4	23	6.2	1.5	5.1	5.1
Al	1.6	160	100	3.3	5.3	4.9, 5.15
Pb	3.7	8.3	2.2	1.25	4.4	3.9
Cd	11.0	76	6.9	1.6	17.5	13
Nb	3.9	3.8	0.98	1.15	4.5	4.4

[a] See (*4*) and (*225*) for complete references and (*209*) for Nb.

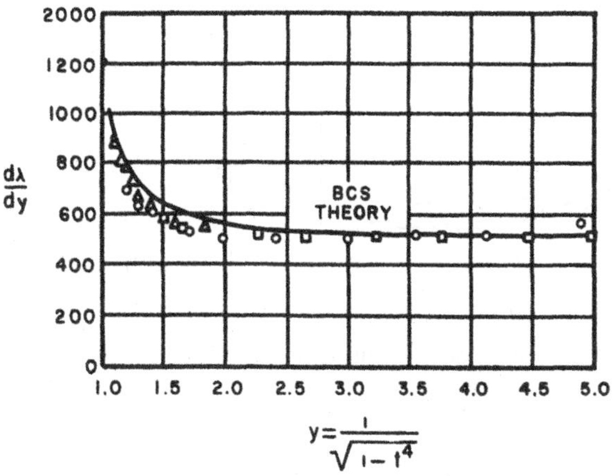

Fig. 30. Temperature dependence of $d\lambda/dy$ for tin obtained by Schawlow and Devlin (*207*) compared with the theoretical curve obtained from the BCS theory.

A more sensitive comparison between theory and experiment can be displayed by plotting $d\lambda/dy$ vs. y. In Fig. 30 we present the data of Schawlow and Devlin (*207*) on pure tin. The deviation from the straight-line prediction of the phenomenological theory is quite pronounced for values of $y < 1.5$. The deviation is barely observable in the normal plot of $\lambda(T)$ vs. y. This increase in $d\lambda/dy$ at low temperatures was first predicted by Lewis (*219*), who extended the two-fluid model of Gorter and Casimir to include a temperature-independent gap. The gap causes $\lambda(T)$ to fall more rapidly as the temperature is lowered and $d\lambda/dT$ is thus increased. The more rigorous treatment of $\Lambda(T)$ in the BCS theory takes into account the temperature-dependent gap and predicts a significant increase in $d\lambda/dy$ for $y < 2$. Eventually, however, $d\lambda/dy$ decreases to zero because of the exponential temperature dependence of the number of normal electrons. This decrease has been confirmed by Erlbach et al. (*220*). Waldram (*221,222*) has pointed out the importance of making the correction to zero frequency in the determination of $d\lambda/dy$ by surface reactance. Since Pippard thought that the zero-frequency correction to $d\lambda_0/dy$ would be negligible, he did not use the Kramers–Kronig to obtain $d\lambda(\omega = 0)/dy$ from $d\lambda(\omega)/dy$. Pippard therefore obtained the high value of $\lambda_0 = 5.6 \times 10^{-6}$ cm for pure Sn. In actuality, for low temperature, the frequency correction is almost linear in $y - 1$, so that the frequency correction to λ_0 is not zero, even at absolute zero. By taking into account the frequency correction, the agreement between the data of Schawlow and Devlin and that of Pippard and Waldram for Sn becomes excellent for $y > 1.8$ where $d\lambda/dy$ is constant and approximately equal to 5.2×10^{-6} cm. Below $y = 1.8$, however, the data of Pippard and Waldram show a smaller increase in $d\lambda/dy$ than those of Schawlow and Devlin, which, in turn, are slightly less

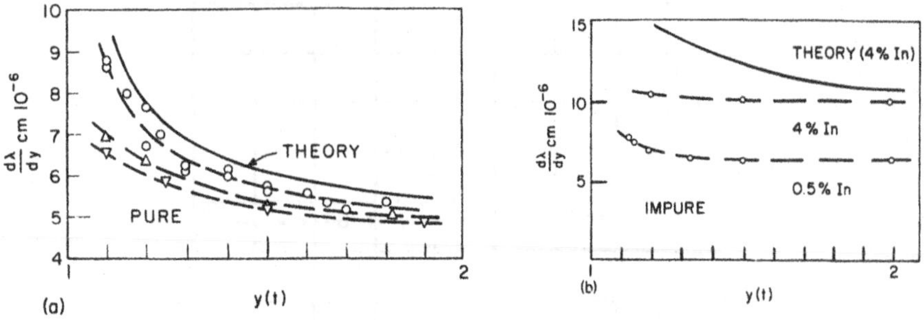

Fig. 31. Comparison of data of $d\lambda/dy$ in the temperature interval $y<2$ in both pure and impure tin. The solid curves are theoretical. In (a): ○, data by Schawlow and Devlin (*207*); △, Pippard (*216*); and ▽, Waldram (*221*): (b) Waldram (*221*).

than Miller's (*223*) prediction from the BCS theory (see Fig. 31). This disagreement between theory and experiment becomes greater with increasing impurity content.

In all measurements the data obtained actually measure $d\lambda(T)/dy$, whereas the comparison between the BCS theory and experiment is made with the temperature dependence of $\lambda_L(T)$ as defined by Eq. (45). To determine $\lambda_L(T)$ from the measured $\lambda(T)$, one must use the relation between $\lambda(T)/\lambda_L(T)$ and $\xi_0/\lambda_L(T)$ obtained by BCS. This has been done by Maxfield and McLean (*209*) for pure niobium. They find that in pure niobium $d\lambda_L(T)/dy$ increases even faster than predicted by the BCS theory for $y<1.4$. The slope $d\lambda/dy$ is actually quite sensitive to the temperature dependence of the energy gap, and therefore any deviation from the BCS temperature dependence for the gap will, as we shall see, manifest itself in the plot of $d\lambda_L/dy$ vs. y.

Measurements on the variation of $\lambda(T)$ have also been performed by Pippard and Waldram on impure specimens as a function of impurity concentration. The data indicate that for impurity content greater than about 3% (e.g., 3% In in Sn), $d\lambda/dy$ vs. y is linear within experimental error for all values of y, even between $y=1$ and 2. This is in disagreement with a calculation Miller has made of the surface impedance based on the general theory of the anomalous skin effect in normal and superconducting metals given by Mattis and Bardeen (*224*). Miller also demonstrates that in impure specimens, $d\lambda/dy$ for large y should be constant and depend on the impurity concentration. In Fig. 32 we plot the limiting slope of $d\lambda/dy$ as a function of inverse mean free path. In this limit, Miller's theory is in good agreement with experiment. Actually, as suggested by Pippard, the mean free path l produces an additional decay term in Eq. (42) and thus modifies the coherence length in the equation for the relation between current density and vector potential in such a way that $1/\xi(l) \simeq (1/\xi_0) + (1/\alpha l)$, where $\xi(l)$ can be interpreted as an effective range of coherence appearing in the relation between current density and vector potential and α is a constant

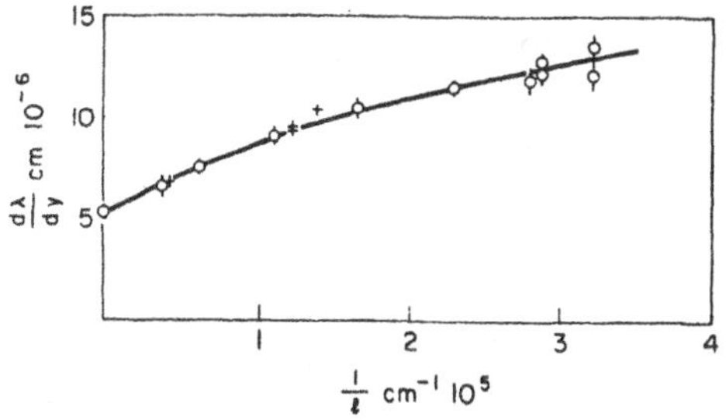

Fig. 32. Change in the limiting slope $d\lambda/dy$ as a function of mean free path. The crosses in the figure refer to monocrystalline tin. The curve is the theoretical curve calculated from Miller's curve (*223*).

on the order of 1. Note that this definition of the coherence length is proportional to l for $l \ll \xi_0$, whereas the coherence length appearing in the range of variation of the order parameter, for dirty metals, $l \ll \xi_0$, is proportional to $\sqrt{\xi_0 l}$.

Greytak and Wernick (*225*) have measured the penetration depth in type II superconductors by fitting the temperature dependence to the curve $\lambda = \lambda_0 y$. Gittleman et al. (*226*) have also measured the penetration depth in type II superconductors; however, in interpreting their data, they chose to treat λ as an adjustable parameter which gave the best fit to phenomenological theories due to Ginzburg and Landau, and Parmenter. In type II superconductors λ is appreciably greater than 10^{-6} cm and naturally satisfies the relation $\kappa \simeq \lambda/\xi > 1/\sqrt{2}$.

Erlback et al. (*220,227*) have made penetration-depth measurements on the strong-coupling superconductor Pb, based on the transmission of flux through a thin Pb film. The true temperature dependence of λ is quite complicated, and therefore they simplified their analysis by comparing their data to the theoretical variation of $\lambda_L(T)/\lambda_L(0)$ when the London limit applied and $[\lambda_\infty(T)/\lambda_\infty(0)]^{3/2}$ when the Pippard limit applied. In Pb, where the coherence length is only slightly greater than the London penetration depth, the London limit seems to fit their data better. A reasonable fit to the temperature variation of $\lambda_L(T)/\lambda_L(0)$ is obtained by using the BCS model but with the slightly higher value $2\Delta = 4.93kT_c$, which in some sense indicates the strong-coupling nature of Pb.

The experimental techniques for obtaining the penetration depth yield more accurate measurements of changes in λ than of the absolute value of λ. For example, in reactance measurements, small reflections in the transmission lines of the apparatus can cause large and undetectable errors in the zero measurement of the frequency shift. Waldram (*221*) reports that measurements of $\lambda(0)$ of different specimens of the same purity at 3000 MHz show a scatter of up to 40%. In past experiments, the value of $d\lambda/dy = \lambda_0$ (for large values of y) have been quoted as the value for $\lambda(0)$, since both the empirical relation and the

BCS theory give a constant for $(d\lambda/dy)_{y\to\infty}$. However, the microscopic theory shows that for very large values of y one always has the London limit, since $\lambda(T)$ eventually becomes much greater than ξ_0. Thus we have for large y

$$\lambda(y) = \lambda_L(y)\left[\frac{\xi_0}{\xi(l)}\right]^{1/2} = \lambda_L(y)\left(1 + \frac{\xi_0}{l}\right)^{1/2}$$

$$\left(\frac{d\lambda}{dy}\right)_{y\to\infty} = \frac{d\lambda_L(y)}{dy}\left(1 + \frac{\xi_0}{l}\right)^{1/2}$$

A recent technique for measuring the absolute value of λ due to Meservey (*228*), takes advantage of the quantum interference effects in superconducting loops interrupted by Josephson junctions. A peak in the current–voltage response of the system occurs whenever the loop encloses a quantum of flux. Since, for the specimens used by Meservey, the area of the hole of the loop was approximately $2\lambda w$, where w is the width of the two-junction system, an absolute value for λ can be obtained using the relation $\Phi_0 = \Delta H 2\lambda w$, where Φ_0 is the flux quantum $ch/2e$ and ΔH is the field interval between voltage maxima. The absolute values of λ for Sn and Al obtained in this way agree with previous measurements but the uncertainty in this technique may still be 10%.

As mentioned before, the temperature dependence of $\lambda(T)$ is quite sensitive to the temperature-dependent energy gap. Since the penetration depth is *not* a fundamental parameter of the BCS theory but a derived one, Waldram has preferred to treat the deviations in $\lambda(T)$ as a direct result of deviation of the more fundamental quantity $\Delta(T)$ from the BCS model. By using a construction explained in Fig. 33, Waldram has been able to invert the data on $\lambda(T)$ to obtain $\Delta(T)$. The resulting $\Delta(T)$ vs. T gives a slight deviation from the BCS prediction but agrees quite well with more direct measurements of $\Delta(T)$ from specific heat, acoustic attenuation, and tunneling. A similar inversion of $\lambda(T)$ for Nb to obtain $\Delta(T)$ was performed by Maxfield and McLean and again $\Delta(T)$ did not exactly agree with the BCS shape or with previous energy gap measurements from ultrasonic attenuation due to Dobbs and Perz (*229*). Leggett (*230*) has calculated the effect of Fermi liquid theory on the penetration depth and has shown that in the Pippard limit a very small change in the temperature dependence of $\lambda(T)$ should occur when T is not too close to 0 or T_c. In the London limit, Leggett concludes that Fermi liquid effects can be large and observable.

We recommend the comprehensive article by Waldram (*221*), where more experimental detail and data are presented. Waldram concludes that there is probably no large fundamental error for the behavior of $\lambda(T)$ as deduced from the BCS theory. In some situations, however, to get more exact fits it is important to make allowances for two properties which may vary from metal to

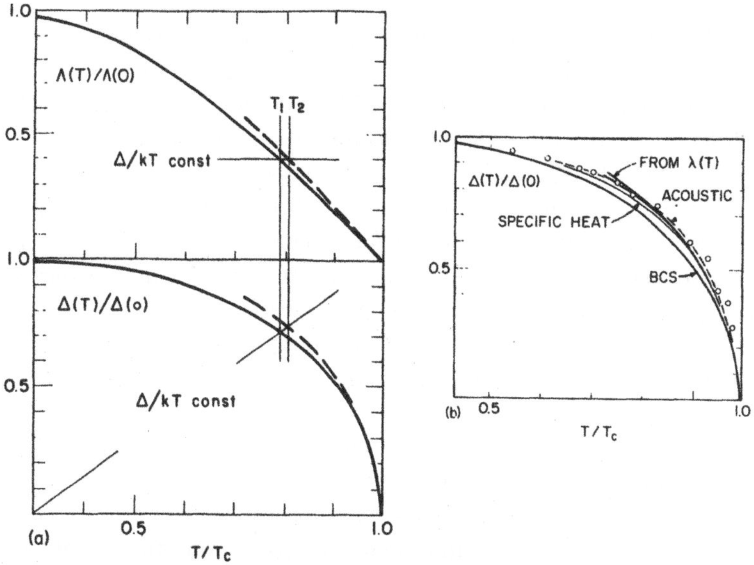

Fig. 33. (a) Construction used by Waldram (*221,222*) to obtain the value of $\Delta(T)$ which yields the observed temperature dependence of $\lambda(t)$. The solid lines are the theoretical curves obtained from the BCS $\Delta(T)$. The dashed curve in the upper figure is obtained from the data on $\lambda(t)$, and in the lower figure corresponds to a slight change in the temperature dependence of $\Delta(T)$. (b) ΔT obtained from construction in (a) compared with the result of other measurements from specific heat, Corak and Satterthwaite (*144*); acoustic attenuation, Morse and Bohm (*249*); and (circles) tunneling data of Townsend and Sutton (*85*).

metal; the complex structure of the Fermi surface leading to changes in the temperature-dependent gap due to the anisotropy and the energy dependence of the electron–phonon interaction. Therefore, for example, the slight failure of the BCS theory for $\Delta(T)$ of tin might be due to the several pieces of its Fermi surface displaying different energy gaps, even multiple gaps for the same crystal directions. It would be of great interest to obtain precise data on $\lambda(T)$ for the strong-coupling superconductors Pb and Hg.

B. Knight Shift

Soon after the appearance of the BCS theory, Yosida (*231*) calculated that the temperature-dependent spin susceptibility of the superconducting electrons, $\chi_s(T)$, should vanish exponentially as the reduced temperature T/T_c approaches zero. This behavior follows from the fact that a minimum of energy 2Δ must be expended on a pair to form excitations, which can then respond to the external magnetic field. Since the gain in Zeeman energy $2\mu_B H$ for the fields of interest in the experiment is small in comparison with 2Δ, at low temperatures only the thermally excited quasi-particles, whose number is proportional to $\exp(-T_c/T)$, contribute to the superconducting spin susceptibility.

To determine the spin paramagnetism of the superconducting electrons below T_c, the field at the atomic nuclei is measured via nuclear magnetic resonance techniques and the Knight shift is obtained. The Knight shift is defined as the fractional difference in the resonant frequency between a nucleus in a free ion and the same nucleus in a metallic medium. This shift is due, in part, to the field at the nucleus created by the paramagnetic response of the electrons, which is directly proportional to the electron spin susceptibility. To ensure that the applied field would penetrate the superconducting specimen below T_c, one of the dimensions d is made small so that the condition $d < \lambda$ (the penetration depth) is satisfied.

Although the theory predicts $\chi_s(0)$ should vanish as $T \to 0$, measurements of $\chi_s(T)$ by Reif (*232*) on superconducting mercury colloids consisting mostly of particles less than 500 Å in diameter, and Androes and Knight (*233*) on thin superconducting platelets of tin, indicated that $\chi_s(0)/\chi_n$ was approximately $\frac{2}{3}$ to $\frac{3}{4}$. Experiments by Hammond and Kelly (*234*) indicated that aluminum also displayed a Knight shift in the superconducting state about $\frac{3}{4}$ of the normal-state value; however, remeasurements by the same authors (*235*) on more carefully prepared Al specimens indicated that $\chi_s(t)/\chi_n$ extrapolates to zero at $t = 0$. Measurements were also made on vanadium by Noer and Knight (*236*), indicating no change in χ in going into the superconducting state. However, since vanadium is a transition metal, other contributions to the Knight shift, unaffected by the superconducting transition, such as orbital paramagnetism, complicate the interpretation of the data. Clogston et al. (*237*) have presented fairly detailed Knight shift measurements on V_3Ga and V_3Si, in which they have been able to separate the various contributions to the Knight shift. They conclude that a substantial part of the susceptibility of vanadium arises from the Kubo–Obata (*238*) temperature-independent orbital paramagnetism which is present in metals with partially filled degenerate bands. Since the magnitude of this orbital paramagnetism should be the same in both the normal and superconducting state, the Knight shift in vanadium should remain unchanged.

The disagreement between Rief's data on mercury and Yosida's microscopic calculation of the Knight shift soon stimulated many theoretical explanations for the observed nonzero value for the spin susceptibility in some superconductors, which are summarized below.

1. Modified Spin Pairing

The earliest attempt was made by Heine and Pippard (*239*), who suggested that the strict BCS pairing of electrons with equal and opposite momentum $(\mathbf{k}\uparrow, -\mathbf{k}\downarrow)$ should be relaxed to include pairs with differing momentum (**k**'↑, −**k**↓). This modified pairing allows the electrons to respond to a field in much the same way as a normal electron gas. Schrieffer (*240*) further argued that in small samples, where the spread in single-particle states is large compared to the momentum difference due to the change in pairing caused by the external magnetic field, one can describe the response of the superconductor by allowing a continuous change of pairing with the magnetization. The pairing, however, still occurs between two definite single-particle states. Cooper (*241*) developed their ideas further by including matrix elements connecting a pair state in an eigenstate of total momentum and a scattered pair state with a different total momentum. Cooper maintained that such transitions were allowed if the electron–electron interaction responsible for superconductivity is not spatially invariant. In this manner, he was able to construct a wave function which leads to a BCS-like condensed state, yet gives a nonzero spin susceptibility.

2. Triplet and Higher Angular Momentum State Pairing

If the pairing of electrons occurs in the triplet state, then the alignment of the spins of the electrons with the field is energetically favorable, since a down spin pair can be changed into an up spin pair without a discontinuous change in the superconducting energy. Since the overall wave function must be antisymmetric, the triplet state $s = 1$ must have odd orbital angular momentum. The theory of p-wave pairing has been treated by Balian and Werthamer (*242*) and Privorotskii (*243*). For pure metals they obtain a superconducting state with an isotropic energy gap where the spin susceptibility decreases monotonically with temperature to a limiting value of $\chi_s(0)/\chi_n = \frac{2}{3}$. Impurities, however, considerably increase the free energy of the p-wave-pair state, so that in the dirty specimens investigated, it seems unlikely that p-wave pairing is responsible for the large value of the Knight shift.

3. Spin-Orbit Coupling

Ferrell (*244*), Anderson (*245*), and Abrikosov and Gor'kov (*246*) have shown that spin-reversing scattering, arising from spin-orbit interactions at displaced surface atoms or impurities, can play an important role in determining χ_s for small specimens. In the presence of a large amount of spin-orbit scattering, the single-particle wave functions Ψ_n are no longer eigenstates of spin and in fact have no average spin component in any direction. As a result, the spin operator has off-diagonal matrix elements connecting transitions from state n to n'. Since a significant contribution to χ_n comes from states in which ϵ_n and $\epsilon_{n'}$ differ by more than the gap, χ_s is not significantly affected by superconductivity. A theoretical estimate for the ratio of the zero-temperature spin susceptibility of the superconductor to the normal-state value obtained by Anderson gives

$$\frac{\chi_s(0)}{\chi_n} = \begin{cases} 1 - \dfrac{2}{\pi}\dfrac{l_{s0}}{\xi_0} & l_{s0} \ll \xi_0 \\ \dfrac{\pi}{6}\dfrac{\xi_0}{l_{s0}} & l_{s0} \gg \xi_0 \end{cases}$$

where l_{s0} is the average distance an electron travels between spin reversing collisions and is equal to $v_F\tau_{s0}$, where τ_{s0} is the mean scattering time between spin flips. This result can be understood physically from the uncertainty principle $\delta E \approx h/\tau_{s0}$. If $\delta E \gg \Delta$, then $\hbar/\tau_{s0} \gg \hbar v_F/\pi\xi_0$, and the one-electron states are strong admixtures of spin with $\chi_s(0)/\chi_n \approx 1$, whereas if $\delta E \ll \Delta$, then spin is a good quantum number and $\chi_s(0)/\chi_n \approx 0$. A more exact expression for $\chi_s(T)/\chi_n$ has been derived by Abrikosov and Gor'kov using Green's function techniques.

In Fig. 34 the data on Hg and Sn are compared with theoretical curves fitted by Abrikosov and Gor'kov. For the heavier metals Sn and Hg, reasonable values of l_{s0} ($\xi_0/l_{s0} = 2.16$ for Sn and 1.20 for Hg) give fair agreement with the experimental data; however, the experimentally observed decrease in $\chi_s(t)/\chi_n$ for $t \lesssim 1$ is much sharper. Since l (the mean free path) is on the order of the sample size, these values of the spin flip scattering length lead to a value of l_{s0}/l on the order of $\frac{1}{10}$. This corresponds to a spin flip approximately once every 10 collisions with the surface.

The initial observation of a nonzero Knight shift in Al, however, did present a theoretical problem. In Al, one has a light metal, and spin-orbit effects were expected to be less important than in Sn and Hg. To account for the large observed residual Knight shift in Al seen by Hammond and Kelly in their first

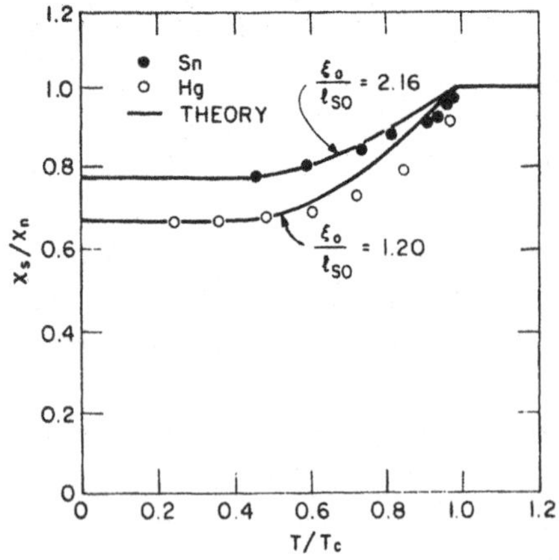

Fig. 34. Comparison between the Knight shift data on Sn (*233*) and Hg (*232*) with the theoretical predictions of Abrikosov and Gor'kov (*246*) with $\xi_0/l_{s0} = 2.16$ for Sn and 1.20 for Hg. Note that the falloff of the experimental points at $t = 1$ is more rapid than predicted from theory.

experiment (*234*), Matthias suggested that a paramagnetic oxide layer might exist on the Al surface. With each scattering at the surface, an exchange interaction can occur, mixing spin states in a manner similar to the spin-orbit interaction. If, in fact, this is the case, the exchange scattering (which, unlike the spin-orbit interaction, is not time-reversal-invariant) should also affect the transition temperature. At the time, however, the experiments on the transition temperature were not definitive on this point. Therefore, to test these ideas, Hammond and Kelly carefully prepared a thin film of Al so as to remove all surface effects. The Knight shift was observed in the superconducting state down to a reduced of temperature of $t = 0.3$. Although the data do not exactly fit the BCS temperature expression, they do extrapolate to $\chi_s/\chi_n = 0$ at $t = 0$. The disagreement of $\chi_s(t)$ with Yosida's theoretical temperature dependence may be due to depairing effects arising from the strong magnetic field.

A recent suggestion by Ferrell concerning the spin-orbit coupling force arising from the periodic crystal field was developed theoretically by Appel (*247*). This interaction can cause a virtual transition from the ground state to the excited state with energy 2Δ and therefore produces a significant contribution to the spin susceptibility. Appel concludes that in Sn and Hg this effect can be as important as the usual spin-orbit scattering at the surface. These two different spin-orbit effects should be separable because τ_{s0} (surface) is size-dependent. Gor'kov (*248*) calculated the same spin-orbit interaction with the lattice and apparently finds that it is small and cannot account for the large value of the Knight shift in superconductors. Spin-orbit coupling seems most likely to account for a substantial portion of the Knight shift in Sn and Hg. The measurements by Hammond and Kelly (*235*) on Al are the first convincing evidence that in a metal where spin-orbit effects are small, the Knight shift extrapolates to zero as $t \to 0$. This tends to remove another outstanding discrepancy between the simple BCS theory and experiments.

VI. CONCLUSIONS

In this chapter we have considered some of the experimental evidence for the validity of the BCS theory, a term which we have used to include generalizations of the original theory which retain essentially the same physical assumptions for their basis. The experimental support for the theory provided by the work surveyed in this chapter is very strong indeed. It is true that quantitative agreement is not always precise, but the range of the explained phenomena is so extensive that one can hardly doubt the soundness of most of the theoretical structure. New phenomena not originally proposed by BCS, such as tunneling and flux quantization, have fitted neatly into the existing structure. Major conflicts such as with the measurements of the Knight shift, the specific heat at low temperature, the isotope effect, and the behavior of strongly coupled superconductors are apparently being slowly resolved.

In spite of these successes, we still cannot predict the superconducting transition temperature of a metal. More generally the calculation of the absolute value of many superconducting properties is rather uncertain because of the uncertainty in the properties of the normal metal. In fact, there are situations in which one prefers to trust the theoretical foundations of the BCS theory and use experimental data on superconductors to obtain detailed information about the normal metal. The most spectacular example is the inversion of tunneling data for strongly coupled superconductors to obtain the phonon spectrum, but other examples are appearing and we can probably expect this interaction to gradually lead to establishing the BCS theory in more detail and at the same time fill in gaps in our knowledge of other low-temperature phenomena.

Finally, there remains the question of whether the BCS theory is sufficient. Are there mechanisms besides the electron–phonon interaction which lead to superconductivity? Although no such mechanism has been clearly demonstrated to cause superconductivity, there are interactions not involving phonons which can lead to an attractive interaction and can in principle cause superconductivity. The existence of such superconductors is still an open question.

ACKNOWLEDGMENTS

Suggestions from many people have been included in this review and, although we retain responsibility for all errors, we wish to acknowledge comments and help from the following persons: J. Appel, C.E. Chase, L.N. Cooper, L.M. Falicov, R.A. Ferrell, L.W. Gruenberg, P.W. Keesom, W.L. McLean, N.E. Phillips, J.M. Rowell, Y. Shapiro, J.C. Swihart, and J.R. Waldram. One of the authors (R.M.) gratefully acknowledges support from the Electronics Research Center under NASA Contract NAS 12-101 during the preparation of this chapter.

REFERENCES

1. J. Bardeen, L. N. Cooper, and J. R. Schrieffer, *Phys. Rev.* **108**, 1175 (1957).
2. L. P. Gor'kov, *Zh. Eksperim. i Teor. Fiz.* **36**, 1918 (1959); *Soviet Phys. JETP* **9**, 1364 (1959).
3. A. B. Pippard, *Proc. Roy. Soc. (London)* **A216**, 547 (1953).
4. J. Bardeen and J. R. Schrieffer, *Progress in Low Temperature Physics*, Vol. 3 (C. J. Gorter, ed.), North-Holland, Amsterdam, 1961, p. 170.
5. P. Morel, *J. Phys. Chem. Solids 10*, 277 (1959); B. B. Goodman, J. Hillairet, J. J. Veyssie, and L. Weil, *Proc. of the Seventh Intern. Conf. on Low Temp. Physics* (G. M. Graham and A. C. Hollis Hallett, eds.), Univ. Toronto Press, Toronto, 1961, p. 350.

6. D. Pines, *Phys. Rev.* **109**, 280 (1958).
7. N. N. Bogoliubov, V. V. Tolmachev, and D. V. Shirkov, *A New Method in the Theory of Superconductivity*, Academy of Science, Moscow, 1958, translation Consultants Bureau, New York, 1959.
8. G. M. Eliashberg, *Zh. Eksperim. i Teor. Fiz.* **38**, 966 (1960); *Soviet Phys. JETP* **11**, 696 (1960).
9. P. Morel and P. W. Anderson, *Phys. Rev.* **125**, 1263 (1962).
10. N. F. Berk and J. R. Schrieffer, *Phys. Rev. Letters* **17**, 433 (1966); *Proc. of the Tenth Intern. Conf. on Low Temp. Physics, Moscow, 1966*, Vol. II.A (M. P. Malkov, ed. in chief), Viniti, Moscow, 1967, p. 150; K. Andres and M. A. Jensen, *Phys. Rev.* **165**, 533 (1968); M. A. Jensen and K. Andres, *Phys. Rev.* **165**, 545 (1968).
10a. W. L. McMillan, *Phys. Rev.* **167**, 331 (1968).
10b. T. H. Geballe, B. T. Matthias, J. P. Remeika, A. M. Clogston, V. B. Compton, J. P. Maita, and H. J. Williams, *Physics* **2**, 293 (1966).
11. J. K. Hulm and R. D. Blaugher, *Phys. Rev.* **123**, 1569 (1961).
12. J. K. Hulm, R. D. Blaugher, T. H. Geballe, and B. T. Matthias, *Phys. Rev. Letters* **7**, 302 (1961).
13. T. H. Geballe, B. T. Matthias, E. Corenzwit, and G. W. Hull, Jr., *Phys. Rev. Letters* **8**, 313 (1962).
14. R. A. Hein, J. W. Gibson, and R. D. Blaugher, *Rev. Mod. Phys.* **36**, 149 (1964).
15. E. Bucher, F. Heiniger, J. Muheim, and J. Muller, *Rev. Mod. Phys.* **36**, 146 (1964).
16. J. K. Hulm, C. K. Jones, and A. Taylor, *Bull. Am. Phys. Soc.* **11**, 710 (1966).
17. T. H. Geballe, *Rev. Mod. Phys.* **36**, 134 (1964).
18. E. Maxwell, *Phys. Rev.* **78**, 477 (1950).
19. C. A. Reynolds, B. Serin, W. H. Wright, and L. B. Nesbitt, *Phys. Rev.* **78**, 487 (1950).
20. H. Fröhlich, *Phys. Rev.* **79**, 845 (1950).
21. J. Bardeen, *Phys. Rev.* **79**, 167 (1950).
22. J. C. Swihart, *Phys. Rev.* **116**, 45 (1959).
23. J. C. Swihart, *IBM J. Res. Develop.* **6**, 14 (1962).
24. T. H. Geballe and B. T. Matthias, *IBM J. Res. Develop.* **6**, 256 (1962).
25. J. W. Garland, Jr., *Phys. Rev. Letters* **11**, 111 (1963).
26. J. W. Garland, Jr., *Phys. Rev. Letters* **11**, 114 (1963); *Phys. Rev.* **153**, 460 (1967).
27. R. E. Fassnacht and J. R. Dillinger, *Phys. Rev. Letters* **17**, 255 (1966).
28. T. H. Geballe, B. T. Matthias, G. W. Hull, Jr., and E. Corenzwit, *Phys. Rev. Letters* **6**, 275 (1961).
29. D. K. Finnemore and D. E. Mapother, *Phys. Rev. Letters* **9**, 288 (1962).
30. R. A. Hein and J. W. Gibson, *Phys. Rev.* **131**, 1105 (1963).
31. J. W. Gibson and R. A. Hein, *Phys. Rev.* **141**, 407 (1966).
32. E. Bucher, J. Muller, J. L. Olsen, and C. Palmy, *Phys. Letters* **15**, 303 (1965).
32a. R. D. Fowler, J. D. G. Lindsay, R. W. White, H. H. Hill, and B. T. Matthias, *Phys. Rev. Letters* **19**, 892 (1967).
33. B. T. Matthias, M. Peter, H. J. Williams, A. M. Clogston, E. Corenzwit, and R. C. Sherwood, *Phys. Rev. Letters* **5**, 542 (1960).
33a. J. J. Engelhardt, G. W. Webb, and B. T. Matthias, *Science* **155**, 191 (1967).
34. F. London, *Superfluids*, Vol. I, Wiley, New York, 1950.
35. B. S. Deaver, Jr., and W. M. Fairbank, *Phys. Rev. Letters* **7**, 43 (1961).
36. R. Doll and M. Näbauer, *Phys. Rev. Letters* **7**, 51 (1961).
36a. W. H. Parker, B. N. Taylor, and D. N. Langenberg, *Phys. Rev. Letters* **18**, 284 (1967).
37. B. B. Schwartz and L. N. Cooper, *Phys. Rev.* **137**, A829 (1965).

38. L. Onsager, *Phys. Rev. Letters* **7**, 50 (1961).
39. D. Bohm, *Proc. of the Eighth Intern. Conf. on Low Temp. Physics* (R. O. Davies, ed.), Butterworth, London, 1963, p. 109.
40. W. A. Little and R. D. Parks, *Phys. Rev. Letters* **9**, 9 (1962).
41. R. D. Parks and W. A. Little, *Phys. Rev.* **133**, A97 (1964).
42. R. P. Groff and R. D. Parks, *Proc. of the Tenth Intern. Conf. on Low Temp. Physics, Moscow, 1966*, Vol. II.A (M. P. Malkov, ed. in chief), Viniti, Moscow, 1967, p. 253; *Phys Rev.*, to be published.
43. A. L. Kwiram and B. S. Deaver, Jr., *Phys. Rev. Letters* **13**, 189 (1964).
44. A. A. Abrikosov, *Zh. Eksperim. i Teor. Fiz.* **32**, 1442 (1957); *Soviet Phys. JETP* **5**, 1174 (1957).
45. M. Tinkham, *Phys. Rev.* **129**, 2413 (1963).
46. C. N. Yang, *Rev. Mod. Phys.* **34**, 694 (1962).
47. O. Penrose and L. Onsager, *Phys. Rev.* **104**, 576 (1956).
48. I. Giaever, *Phys. Rev. Letters* **5**, 147, 464 (1960).
49. I. Giaever and K Megerle, *Phys Rev* **122**, 1101 (1961)
50 S Shapiro, P. H. Smith, J. Nicol, J. L. Miles, and P. F. Strong, *IBM J. Res. Develop.* **6**, 34 (1962).
51. D. H. Douglass, Jr., and L. M. Falicov, *Progress in Low Temperature Physics*, Vol. 4 (C. J. Gorter, ed.), North-Holland, Amsterdam, 1964, p. 97.
52. A. Sommerfeld and H. A. Bethe, *Handbuch der Physik*, Vol. 24 (S. Flügge, ed.), Springer, Berlin, 1933, pp. 333–450.
53. R. Holm, *J. Appl. Phys.* **22**, 569 (1951).
54. L. Esaki, *Phys. Rev.* **109**, 603 (1958).
55. J. C. Fisher and I. Giaever, *J. Appl. Phys.* **32**, 172 (1961).
56. S. Bermon, *Tech. Rep. 1, NSF-GP1100*, University of Illinois, Urbana, 1964.
57. J. Nicol, S. Shapiro, and P. H. Smith, *Phys. Rev. Letters* **5**, 461 (1960).
58. B. N. Taylor, E. Burstein, and D. N. Langenberg, *Bull. Am. Phys. Soc.* **7**, 190 (1962).
59. J. Bardeen, *Phys. Rev. Letters* **6**, 57 (1961); **9**, 147 (1962).
60. W. A. Harrison, *Phys. Rev.* **123**, 85 (1961).
61. M. H. Cohen, L. M. Falicov, and J. C. Phillips, *Proc. of the Eighth Intern. Conf. on Low Tempt. Physics* (R. O. Davies, ed.), Butterworth, London, 1963, p. 178; *Phys. Rev. Letters* **8**, 316 (1962).
62. R. E. Prange, *Phys. Rev.* **131**, 1083 (1963).
63. H. J. Levinstein and J. E. Kunzler, *Phys. Letters* **20**, 581 (1966).
64. I. Giaever, H. R. Hart, Jr., and K. Megerle, *Phys. Rev.* **126**, 941 (1962).
65. N. V. Zavaritskii, *Zh. Eksperim. i Teor. Fiz.* **41**, 657 (1961); *Soviet Phys. JETP* **14**, 470 (1961).
65a. P. Kumbare, P. M. Tedrow, and D. M. Lee, *Bull. Am. Phys. Soc.* **12**, 77 (1967).
66. R. F. Gasparovic, B. N. Taylor, and R. E. Eck, *Solid State Commun.* **4**, 59 (1966).
67. B. Mühlschlegel, *Z. Physik* **155**, 313 (1959).
68. D. H. Douglass, Jr., and R. Meservey, *Phys. Rev.* **135**, A19 (1964).
69. J. M. Rowell, P. W. Anderson, and D. E. Thomas, *Phys. Rev. Letters* **10**, 334 (1963).
70. J. R. Schrieffer, D. J. Scalapino, and J. W. Wilkins, *Phys. Rev. Letters* **10**, 336 (1963).
71. B. N. Brockhouse, T. Arase, G. Caglioti, K. R. Rao, and A. D. B. Woods, *Phys. Rev.* **128**, 1099 (1962).
71a. R. C. Dynes, J. P. Carbotte, and E. J. Woll, Jr., *Solid State Commun.* **6**, 101 (1968).
72. D. J. Scalapino and P. W. Anderson, *Phys. Rev.* **133**, A921 (1964).
73. L. Van Hove, *Phys. Rev.* **89**, 1189 (1953).

74. W. L. McMillan and J. M. Rowell, *Phys. Rev. Letters* **14**, 108 (1965).
75. J. M. Rowell and L. Kopf, *Phys. Rev.* **137**, A907 (1965).
76. A. F. G. Wyatt, *Phys. Rev. Letters* **13**, 160 (1964).
77. D. C. Hamilton and M. A. Jensen, *Phys. Rev. Letters* **11**, 205 (1963).
78. C. G. Kuper, M. A. Jensen, and D. C. Hamilton, *Phys. Rev.* **134**, A15 (1964).
79. B. T. Matthias, *Science* **144**, 378 (1964).
80. J. R. Schrieffer and Y. Wada, *Bull. Am. Phys. Soc.* **8**, 307 (1963).
81. Y. Wada, *Rev. Mod. Phys.* **36**, 253 (1964).
82. J. C. Swihart, D. J. Scalapino, and Y. Wada, *Phys. Rev. Letters* **14**, 106 (1965).
83. J. C. Swihart, *Proc. of the Tenth Intern. Conf. on Low Temp. Physics, Moscow, 1966*, Vol. II.B (M. P. Malkov, ed. in chief), Viniti, Moscow, 1967, p. 275.
84. S. Bermon and D. M. Ginsberg, *Phys. Rev.* **135**, A306 (1964).
85. P. Townsend and J. Sutton, *Phys. Rev.* **128**, 591 (1962); *Proc. of the Eighth Intern. Conf. on Low Temp. Physics* (P. O. Davies, ed.), Butterworth, London, 1963, p. 182.
86. N. V. Zavaritskii, *Zh. Eksperim. i Teor. Fiz.* **43**, 1123 (1962); *Soviet Phys. JETP* **16**, 793 (1963).
87. N. V. Zavaritskii, *Zh. Eksperim. i Teor. Fiz.* **48**, 837 (1965); *Soviet Phys. JETP* **21**, 557 (1965).
88. V. L. Pokrovskii, *Zh. Eksperim. i Teor. Fiz.* **40**, 641 (1961); *Soviet Phys. JETP* **13**, 447 (1961).
89. B. T. Geilikman and V. Z. Kresin, *Zh. Eksperim. i Teor. Fiz.* **40**, 970 (1961); *Soviet Phys. JETP* **13**, 677 (1961).
90. A. J. Bennett, *Phys. Rev.* **140**, A1902 (1965).
91. G. I. Rochlin and D. H. Douglass, *Phys. Rev. Letters* **16**, 359 (1966).
92. S. M. Marcus, *Phys. Letters* **19**, 623 (1966).
93. I. K. Yanson, V. M. Svistunov, and I. M. Dmitrenko, *Zh. Eksperim. i Teor. Fiz.* **47**, 2091 (1964); *Soviet Phys. JETP* **20**, 1404 (1965).
93a. M. L. A. McVicar and R. M. Rose, *Proceedings of the Symposium on the Physics of Superconducting Devices, Charlottesville, Va, 1967.*
94. J. M. Rowell and W. L. McMillan, *Proc. of the Tenth Intern. Conf. on Low Temp. Physics, Moscow, 1966*, Vol. II.B (M. P. Malkov, ed. in chief), Viniti, Moscow, 1967, p. 296.
95. I. Giaever, *Proc. of the Eighth Intern. Conf. on Low Temp. Physics* (P. O. Davies, ed.), Butterworth, London, 1963, p. 171.
96. M. D. Sherrill and H. H. Edwards, *Phys. Rev. Letters* **6**, 460 (1961).
97. I. Dietrich, *Proc. of the Eighth Intern. Conf. on Low Temp. Physics* (P. O. Davies, ed.), Butterworth, London, 1963, p. 173.
98. B. N. Taylor and E. Burstein, *Phys. Rev. Letters* **10**, 14 (1963).
99. G. B. Donaldson, *Proc. of the Tenth Intern. Conf. on Low Temp. Physics, Moscow, 1966*, Vol. II.B (M. P. Malkov, ed. in chief), Viniti, Moscow, 1967, p. 291.
100. A. S. Edelstein and A. M. Toxen, *Proc. of the Tenth Intern. Conf. on Low Temp. Physics, Moscow, 1966*, Vol. II.B (M. P. Malkov, ed. in chief), Viniti, Moscow, 1967, p. 270.
100a. J. J. Hauser, *Phys. Rev. Letters* **17**, 921 (1966).
101. D. H. Douglass, Jr., *Phys. Rev. Letters* **7**, 14 (1961).
102. V. L. Ginzburg and L. D. Landau, *Zh. Eksperim. i Teor. Fiz.* **20**, 1064 (1950).
103. R. Meservey and D. H. Douglass, Jr., *Phys. Rev.* **135**, A24 (1964).
104. R. S. Collier and R. A. Kamper, *Phys. Rev.* **143**, 323 (1966).
105. K. Maki, *Progr. Theoret. Phys. (Kyoto)* **31**, 731 (1964).
106. J. L. Levine, *Phys. Rev.* **155**, 373 (1967).

107. E. Guyon, F. Meunier, and R. S. Thompson, *Phys. Rev.* **156**, 452 (1967).
108. B. T. Matthias, H. Suhl, and E. Corenzwit, *Phys. Rev. Letters* **1**, 92 (1958); *J. Phys. Chem. Solids* **13**, 156 (1960).
109. K. Schwidtal, *Z. Physik* **158**, 563 (1960).
110. A. A. Abrikosov and L. P. Gor'kov, *Zh. Eksperim. i Teor. Fiz.* **39**, 1781 (1960); *Soviet Phys. JETP* **12**, 1243 (1961).
111. F. Reif and M. A. Woolf, *Phys. Rev. Letters* **9**, 315 (1962).
112. M. A. Woolf and F. Reif, *Phys. Rev.* **137**, A557 (1965).
113. S. Skalski, O. Betbeder-Matibet, and P. R. Weiss, *Phys. Rev.* **136**, A1500 (1964).
114. R. A. Hein, R. L. Falge, Jr., B. T. Matthias, and E. Corenzwit, *Phys. Rev. Letters* **2**, 500 (1959).
115. J. E. Crow and R. D. Parks, *Phys. Letters* **21**, 378 (1966); J. E. Crow, R. P. Guertin, and R. D. Parks, *Proc. of the Tenth Intern. Conf. on Low Temp. Physics, Moscow, 1966*, Vol. II.A (M. P. Malkov, ed. in chief), Viniti, Moscow, 1967, p. 301.
116. K. H. Bennemann, *Phys. Rev. Letters* **17**, 438 (1966).
116a. R. P. Guertin, W. E. Masker, T. W. Mihalisin, R. P. Groff, and R. D. Parks, *Phys. Rev. Letters* **20**, 387 (1968).
116b. P. Fulde, *Phys. Rev.* **137**, A783 (1965).
117. B. D. Josephson, *Phys. Letters* **1**, 251 (1962); *Rev. Mod. Phys.* **36**, 216 (1964).
118. B. D. Josephson, *Advan. Phys.* **14**, 419 (1965).
119. S. Shapiro, *Phys. Rev. Letters* **11**, 80 (1963).
120. S. Shapiro, A. R. Janus, and S. Holly, *Rev. Mod. Phys.* **36**, 223 (1964).
121. I. K. Yanson, V. M. Svistunov, and I. M. Dmitrenko, *Zh. Eksperim. i Teor. Fiz.* **48**, 976 (1965); *Soviet Phys. JETP* **21**, 650 (1965).
122. D. N. Langenberg, D. J. Scalapino, B. N. Taylor, and R. E. Eck, *Phys. Rev. Letters* **15**, 294 (1965); errata **15**, 842 (1965).
123. D. N. Langenberg, D. J. Scalapino, and B. N. Taylor, *Proc. IEEE* **54**, 560 (1966).
124. D. N. Langenberg, W. H. Parker, and B. N. Taylor, *Phys. Rev.* **150**, 186 (1966); *Phys. Rev. Letters* **18**, 287 (1967).
125. C. J. Adkins, *Rev. Mod. Phys.* **36**, 211 (1964).
126. J. R. Schrieffer and J. W. Wilkins, *Phys. Rev. Letters* **10**, 17 (1963).
127. L. Kleinman, *Phys. Rev.* **132**, 2484 (1963).
128. L. Kleinman, B. N. Taylor, and E. Burstein, *Rev. Mod. Phys.* **36**, 208 (1964).
129. A. H. Dayem and R. J. Martin, *Phys. Rev. Letters* **8**, 246 (1962).
130. P. K. Tien and J. P. Gordon, *Phys. Rev.* **129**, 647 (1963).
131. B. Abeles and Y. Goldstein, *Phys. Rev. Letters* **14**, 595 (1965).
132. Y. Goldstein and B. Abeles, *Phys. Letters* **14**, 78 (1965).
133. E. Lax and F. L. Vernon, Jr., *Phys. Rev. Letters* **14**, 256 (1965).
134. Y. Goldstein, B. Abeles, and R. W. Cohen, *Phys. Rev.* **151**, 349 (1966).
135. W. J. Tomasch, *Phys. Rev. Letters* **15**, 672 (1965); **16**, 16 (1966).
136. W. J. Tomasch and T. Wolfram, *Phys. Rev. Letters* **16**, 352 (1966).
137. W. L. McMillan and P. W. Anderson, *Phys. Rev. Letters* **16**, 85 (1966).
138. W. H. Keesom, *Report of 4th Solvay Congress of Physics, 1924*, p. 288.
139. W. Meissner and R. Ochsenfeld, *Naturwiss.* **21**, 787 (1933).
140. P. Ehrenfest, *Leiden Commun. Suppl. 75b* (1933).
141. W. H. Keesom and P. H. Van Laer, *Physica* **5**, 193 (1938).
142. A. Brown, M. W. Zemansky, and H. A. Boorse, *Phys. Rev.* **92**, 52 (1953).
143. W. S. Corak, B. B. Goodman, C. B. Satterthwaite, and A. Wexler, *Phys. Rev.* **96**, 1442 (1954).

144. W. S. Corak and C. B. Satterthwaite, *Phys. Rev.* **99**, 1660 (1955).
145. F. London, *Proc. Roy. Soc. (London)* **A152**, 24 (1935).
146. J. Bardeen, *Proc. of the Eighth Intern. Conf. on Low Temp. Physics* (R. O. Davies, ed.), Butterworth, London, 1963, p. 3.
147. J. F. Cochran, *Ann. Phys. (N.Y.)* **19**, 186 (1962).
148. D. C. Rorer, H. Meyer, and R. C. Richardson, *Z. Naturforsch.* **18a**, 130 (1963).
149. W. H. Keesom and H. Kamerlingh Onnes, *Leiden Commun. 174b* (1924);
M. K. Wilkinson, C. G. Shull, L. D. Roberts, and S. Bernstein, *Phys. Rev.* **97**, 889 (1955).
150. M. A. Biondi, A. T. Forrester, M. P. Garfunkel, and C. B. Satterthwaite, *Rev. Mod. Phys.* **30**, 1109 (1958).
151. H. A. Boorse, *Phys. Rev. Letters* **2**, 391 (1959).
152. C. A. Bryant and P. H. Keesom, *Phys. Rev. Letters* **4**, 460 (1960).
153. H. A. Boorse, A. T. Hirschfeld, and H. A. Leupold, *Phys. Rev. Letters* **5**, 246 (1960).
154. B. S. Chandrasekhar and J. A. Rayne, *Phys. Rev. Letters* **6**, 3 (1961).
155. B. J. C. van der Hoeven, Jr., and P. H. Keesom, *Phys. Rev.* **134**, A1320 (1964).
156. H. R. O'Neal and N. E. Phillips, *Phys. Rev.* **137**, A748 (1965).
157. N. E. Phillips, private communication, 1967.
158. P. H. Keesom and B. J. C. van der Hoeven, Jr., *Phys. Letters* **3**, 360 (1963).
159. B. J. C. van der Hoeven, Jr., and P. H. Keesom, *Phys. Rev.* **137**, A103 (1965).
160. P. W. Anderson, *J. Phys. Chem. Solids* **11**, 26 (1959).
161. L. Y. L. Shen, N. M. Senozan, and N. E. Phillips, *Phys. Rev. Letters* **14**, 1025 (1965).
162. R. Radebaugh and P. H. Keesom, *Phys. Rev.* **149**, 209 (1966).
163. H. Suhl, B. T. Matthias, and L. R. Walker, *Phys. Rev. Letters* **3**, 552 (1959).
164. D. E. Mapother, *Phys. Rev.* **126**, 2021 (1962).
165. D. K. Finnemore and D. E. Mapother, *Phys. Rev.* **140**, A507 (1965).
166. T. E. Faber, *Proc. Roy. Soc. (London)* **A231**, 353 (1955).
167. N. E. Phillips, *Phys. Rev.* **114**, 676 (1959).
168. N. V. Zavaritskii, *Zh. Eksperim. i Teor. Fiz.* **37**, 1506 (1959); *Soviet Phys. JETP* **10**, 1069 (1960).
169. J. F. Cochran, C. A. Shiffman, J. E. Neighbor, *Rev. Sci. Instr.* **37**, 499 (1966).
170. S. Caplan and G. Chanin, *Phys. Rev.* **138**, A1428 (1965).
171. J. F. Cochran and D. E. Mapother, *Phys. Rev.* **111**, 132 (1958).
172. N. E. Phillips, *Phys. Rev.* **134**, A385 (1964).
173. N. V. Zavaritskii, *Zh. Eksperim. i Teor. Fiz.* **34**, 1116 (1958); *Soviet Phys. JETP* **7**, 773 (1958).
174. G. Seidel and P. H. Keesom, *Phys. Rev.* **112**, 1083 (1958).
175. T. P. Sheahen, J. F. Cochran, and W. D. Gregory, to be published (value taken from Ref. *181*).
176. N. E. Phillips, M. H. Lambert, and W. R. Gardner, *Rev. Mod. Phys.* **36**, 131 (1964).
177. D. K. Finnemore, D. L. Johnson, J. E. Ostenson, F. H. Spedding, and B. J. Beaudry, *Phys. Rev.* **137**, A550 (1965).
178. H. A. Leupold and H. A. Boorse, *Phys. Rev.* **134**, A1322 (1964).
179. D. K. Finnemore, T. F. Stromberg, and C. A. Swenson, *Phys. Rev.* **149**, 231 (1966).
180. J. Neighbor, J. F. Cochran, and C. A. Shiffman, *Phys. Rev.* **155**, 384 (1967).
181. T. P. Sheahan, *Phys. Rev.* **149**, 370 (1966).
182. D. White, C. Chou, and H. L. Johnston, *Phys. Rev.* **109**, 797 (1958).
183. B. J. C. van der Hoeven, Jr., and P. H. Keesom, *Phys. Rev.* **135**, A631 (1964).
184. B. T. Matthias, T. H. Geballe, E. Corenzwit, K. Andres, G. W. Hull, Jr., J. C. Ho, N. E. Phillips, and D. K. Wohlleben, *Science* **151**, 985 (1966).

185. N. E. Phillips, *Phys. Rev. Letters* **1**, 363 (1958).

186. D. C. Rorer, D. G. Onn, and H. Meyer, *Phys. Rev.* **138**, A1661 (1965).

187. D. L. Decker, D. E. Mapother, and R. W. Shaw, *Phys. Rev.* **112**, 1888 (1958).

188. G. J. Sizoo and H. Kamerlingh Onnes, *Leiden Commun. 180b* (1925); G. J. Sizoo, W. J. de Haas, and H. Kamerlingh Onnes, *Leiden Commun. 180c* (1925).

189. M. Levy and J. L. Olsen, *Physics of High Pressures and the Condensed Phase* (A. van Itterbeek, ed.), North-Holland, Amsterdam, 1965, p. 525.

190. N. B. Brandt and N. I. Ginzburg, *Usp. Fiz. Nauk* **85**, 485 (1965); *Soviet Phys. Usp.* **8**, 202 (1965).

191. V. L. Ginzburg, *Zh. Eksperim. i Teor. Fiz.* **44**, 2104 (1963); *Soviet Phys. JETP* **17**, 1415 (1963).

192. N. B. Brandt and N. I. Ginzburg, *Zh. Eksperim. i Teor. Fiz.* **44**, 1876 (1963); *Soviet Phys. JETP*, **17**, 1262 (1963).

193. F. J. Morin and J. P. Maita, *Phys. Rev.* **129**, 1115 (1963).

194. J. Kondo, *Progr. Theoret. Phys. (Kyoto)* **29**, 1 (1963).

195. T. H. Geballe, B. T. Matthias, K. Andres, E. S. Fisher, T. F. Smith, and W. H. Zachariasen, *Science* **152**, 755 (1966).

195a. T. F. Smith and C. W. Chu, *Phys. Rev.* **159**, 353 (1967).

196. E. Bucher, J. Müller, J. L. Olsen, and C. Palmy, *Cryogenics* **5**, 283 (1965).

197. W. E. Gardner and T. F. Smith, *Phys. Rev.* **138**, A484 (1965).

198. T. F. Smith and W. E. Gardner, *Phys. Rev.* **140**, A1620 (1965).

198a. W. E. Gardner and T. F. Smith, *Phys. Rev.* **154**, 309 (1967).

199. R. Becker, G. Heller, and F. Sauter, *Z. Physik* **85**, 772 (1933).

200. G. L. de Haas-Lorentz, *Physica* **5**, 384 (1925).

201. F. London and H. London, *Proc. Roy. Soc. (London)* **A149**, 71 (1935).

202. T. E. Faber and A. B. Pippard, *Proc. Roy. Soc. (London)* **A231**, 336 (1955).

203. D. Shoenberg, *Proc. Roy. Soc. (London)* **A175**, 49 (1940).

204. J. G. Daunt, A. R. Miller, A. B. Pippard, and D. Shoenberg, *Phys. Rev.* **74**, 842 (1948).

205. C. J. Gorter and H. B. G. Casimir, *Physik Z.* **35**, 963 (1934).

206. E. Laurmann and D. Shoenberg, *Proc. Roy. Soc. (London)* **A198**, 560 (1949).

207. A. L. Schawlow and G. E. Devlin, *Phys. Rev.* **113**, 120 (1959).

208. P. N. Dheer, *Proc. Roy. Soc. (London)* **A260**, 333 (1961).

209. B. W. Maxfield and W. L. McLean, *Phys. Rev.* **139**, A1515 (1965).

210. M. Désirant and D. Shoenberg, *Proc. Phys. Soc. (London)* **60**, 413 (1948).

211. J. M. Lock, *Proc. Roy. Soc. (London)* **A208**, 391 (1951).

212. H. B. G. Casimir, *Physica* **7**, 887 (1940).

213. A. I. Shalnikov and Yu. V. Sharvin, *Izv. Akad. Nauk SSSR* **12**, 195 (1948).

214. R. G. Chambers, *Proc. Roy. Soc. (London)* **A215**, 481 (1952).

215. W. L. McLean, *Proc. of the Seventh Intern. Conf. on Low Temp. Physics* (G. M. Graham and A. C. Hollis Hallett, eds.), Univ. Toronto Press, Toronto, 1961, p. 330.

216. A. B. Pippard, *Proc. Roy. Soc. (London)* **A203**, 210 (1950).

217. G. E. H. Reuter and E. H. Sondheimer, *Proc. Roy. Soc. (London)* **A195**, 336 (1948).

218. K. E. Drangeid and R. Sommerhalder, *Phys. Rev. Letters* **8**, 467 (1962).

219. H. W. Lewis, *Phys. Rev.* **102**, 1508 (1956).

220. E. Erlbach, R. L. Garwin, and M. P. Sarachik, *IBM J. Res. Develop.* **4**, 107 (1960).

221. J. R. Waldram, *Advan. Phys.* **13**, 1 (1964).

222. J. R. Waldram, *Rev. Mod. Phys.* **36**, 187 (1964).

223. P. B. Miller, *Phys. Rev.* **113**, 1209 (1959); **118**, 928 (1960).

224. D. C. Mattis and J. Bardeen, *Phys. Rev.* **111**, 412 (1958).

225. T. J. Greytak and J. H. Wernick, *J. Phys. Chem. Solids* **25**, 535 (1964).
226. J. I. Gittleman, B. Rosenblum, T. E. Seidel, and A. W. Wicklund, *Phys. Rev.* **137**, A527 (1965).
227. M. P. Sarachik, R. L. Garwin, and E. Erlbach, *Phys. Rev. Letters* **4**, 52 (1960).
228. R. Meservey, *Low Temperature Physics, LT9* (J. G. Daunt et al., eds.), Plenum Press, New York, 1965, Part A, p. 455.
229. E. R. Dobbs and J. M. Perz, *Rev. Mod. Phys.* **36**, 257 (1964).
230. A. J. Leggett, *Phys. Rev.* **140**, A1869 (1965).
231. K. Yosida, *Phys. Rev.* **110**, 769 (1958).
232. F. Reif, *Phys. Rev.* **106**, 208 (1957).
233. G. M. Androes and W. D. Knight, *Phys. Rev.* **121**, 779 (1961).
234. R. H. Hammond and G. M. Kelly, *Rev. Mod. Phys.* **36**, 185 (1964).
235. R. H. Hammond and G. M. Kelly, *Phys. Rev. Letters* **18**, 156 (1967).
236. R. J. Noer and W. D. Knight, *Rev. Mod. Phys.* **36**, 177 (1964).
237. A. M. Clogston, A. C. Gossard, V. Jaccarino, and Y. Yafet, *Phys. Rev. Letters* **9**, 262 (1962).
238. R. Kubo and Y. Obata, *J. Phys. Soc. Japan* **11**, 547 (1956).
239. V. Heine and A. B. Pippard, *Phil. Mag.* **3**, 1046 (1958).
240. J. R. Schrieffer, *Phys. Rev. Letters* **3**, 323 (1959).
241. L. N. Cooper, *Phys. Rev. Letters* **8**, 367 (1962); L. N. Cooper, H. J. Lee, B. B. Schwartz, and W. Silvert, *Proc. of the Eighth Intern. Conf. on Low Temp. Physics* (R. O. Davies, ed.), Butterworth, London, 1963, p. 126.
242. R. Balian and N. R. Werthamer, *Phys. Rev.* **131**, 1553 (1963).
243. I. A. Privorotskii, *Zh. Eksperim. i Teor. Fiz.* **45**, 1961 (1963); *Soviet Phys. JETP* **18**, 1346 (1964).
244. R. A. Ferrell, *Phys. Rev. Letters* **3**, 262 (1959).
245. P. W. Anderson, *Phys. Rev. Letters* **3**, 325 (1959).
246. A. A. Abrikosov and L. P. Gor'kov, *Zh. Eksperim. i Teor. Fiz.* **42**, 1088 (1962); *Soviet Phys. JETP* **15**, 752 (1962).
247. J. Appel, *Phys. Rev.* **139**, A1536 (1965).
248. L. P. Gor'kov, *Zh. Eksperim. i Teor. Fiz.* **48**, 1772 (1965); *Soviet Phys. JETP* **21**, 1186 (1965).
249. R. W. Morse and H. V. Bohm, *Phys. Rev.* **108**, 1094 (1957).

4

NONEQUILIBRIUM PROPERTIES: COMPARISON OF EXPERIMENTAL RESULTS WITH PREDICTIONS OF THE BCS THEORY

D. M. Ginsberg

DEPARTMENT OF PHYSICS AND MATERIALS RESEARCH LABORATORY
UNIVERSITY OF ILLINOIS
URBANA, ILLINOIS

L. C. Hebel

BELL TELEPHONE LABORATORIES, INCORPORATED
WHIPPANY, NEW JERSEY

I. INTRODUCTION

At the same time that the BCS theory was being formulated, measurements of nonequilibrium properties of superconductors were being carried out which showed important new features characteristic of superconductivity. The four most important types of measurements were of infrared and microwave absorption, ultrasonic attenuation, nuclear spin-lattice relaxation, and thermal conductivity. Evidence was found for an energy gap in the conduction electron excitation spectrum and, of equal importance, for the correlation of electron spin and momentum in the wave function describing the superconducting state. The purpose of this chapter is to bring together the experimental and theoretical work concerning these four types of experiments, which show considerable departure from predictions of the older two-fluid model and excellent confirmation of the spin-momentum pairing introduced in the BCS theory.

Very direct evidence for the existence of an energy gap in the conduction electron excitation spectrum is shown in the infrared transmission and absorption measurements as well as in the microwave measurements of surface impedance in superconductors. At low temperatures there is no energy absorption as the frequency is increased until the frequency exceeds a critical value, above which the absorption rises rapidly to that found in the normal metal. The gap in the excitation spectrum is found to decrease from its maximum value at very low temperature to zero at the transition temperature.

Conduction electrons also provide the major source of attenuation of ultrasonic waves in metals at low temperatures. The interaction is simplest in the case of longitudinal acoustic waves. The measurements in superconductors show a rapid drop in attenuation for temperatures below the critical temperature, which was originally interpreted as the result of a diminishing number of "normal" electrons. However, the drop in the attenuation is found to be so rapid that it is impossible to get agreement with the predictions of the two-fluid model.

In addition, the nuclear spin system in a metal is strongly influenced by the conduction electron spins. Energy exchanges by nuclear spin–electron spin scattering with the conduction electrons at the top of the Fermi distribution usually provide the quickest means for the nuclear spins in a metal to come into thermal equilibrium with their surroundings. Measurements showed that the nuclear spin-lattice relaxation time has a characteristically different temperature dependence in the superconducting and normal phases of the metal. Unlike ultrasonic attenuation, the nuclear spin-lattice relaxation rate at first increases for temperatures below the critical temperature and then finally decreases rapidly as the temperature is lowered further. Satisfactory comparison of the nuclear spin-lattice relaxation rate with the ultrasonic attenuation seemed to be impossible on the basis of two-fluid theories of superconductivity.

The interpretation of thermal conductivity measurements is complicated by the fact that heat can be conducted by both phonons and conduction electrons, and in addition several scattering mechanisms are possible in each case. Examples can be found of metals which fit each situation, so that the results for each metal must be interpreted individually. Some metals, notably tin, provided early results characteristic of an exponential temperature dependence of thermal conductivity in the superconducting phase. This was interpreted as evidence of an energy gap.

The BCS theory of superconductivity has as its main feature the spin-momentum pairing of conduction electrons with energies near the Fermi energy, as is discussed in detail in Chapter 2. It was shown there that the conduction electron excitation spectrum displays an energy gap for temperatures below the critical temperature; in addition, the pairing of electrons in a superconductor results in electron scattering and energy absorption in a superconductor which is characteristically different from that in the normal metal. The remainder of this chapter will provide a comparison of experimental results with calculations based on the BCS approach in which the correlation in scattering by paired electrons will play a very crucial role in the theory. In the course of the discussion it will be shown that no one-electron theory can simultaneously explain the results of the nuclear relaxation rate and infrared absorption, on the one hand, and the results concerning ultrasonic absorption measurements, on the other. Consequently, comparison of the two groups of measurements provides important verification of paired electron spin-momentum correlation, which is the basis of the BCS theory.

II. BASIC INTERACTIONS

In this section we will consider the interactions appropriate to electromagnetic absorption, ultrasonic attenuation, and nuclear spin-lattice relaxation. There is no single interaction characteristic of the more complicated case of thermal conductivity, and discussion of it is reserved for Section VI. The discussion here will lead to the comparison of matrix elements and transition rates for the normal and superconducting phases; further detail necessary to discuss particular experiments will be found in the sections to follow. In each case it is particularly convenient to treat the conduction electrons by using the formalism of second quantization employed in Chapter 2 and to treat the electromagnetic field (or vector potential), the phonon field, and the nuclear spins using the ordinary quantum formalism. Thus the electron spin and momentum transitions are calculated using destruction operators $c_{\mathbf{k},\sigma}$ and creation operators $c^*_{\mathbf{k},\sigma}$ for the Bloch states labeled by wave vector **k** and spin σ with the one-electron Bloch function denoted by $U_{\mathbf{k},\sigma}(r)$. Both operators obey the usual anticommutation relations.

A. Electromagnetic Absorption

Electromagnetic interactions give rise to one of the most striking features of superconductivity, the Meissner effect. As shown in Chapter 2, the exclusion of fields at low frequencies in superconductors comes about because of a cancellation in the "paramagnetic" contribution to the current operator, that is, the term proportional to $\mathbf{p}\cdot\mathbf{A}(\mathbf{r})$, where **p** is the electron momentum operator and $\mathbf{A}(\mathbf{r})$ is the electromagnetic vector potential. It is the same paramagnetic portion which is involved in the absorption of electromagnetic energies at microwave and infrared frequencies. Using second quantization for the electrons and treating $\mathbf{A}(\mathbf{r})$ classically, the interaction Hamiltonian has the form in the London (Coulomb) gauge,

$$\mathscr{H}_1 = \left(\psi(\mathbf{r}), \frac{e}{mc}\,\mathbf{p}\cdot\mathbf{A}(\mathbf{r})\,\psi(\mathbf{r})\right)$$
$$= \frac{e\hbar}{mc}\sum_{\mathbf{k},\sigma}(\mathbf{k}\cdot\mathbf{A}_{\mathbf{q}})\,c^*_{\mathbf{k}+\mathbf{q},\sigma}c_{\mathbf{k},\sigma} \qquad (1)$$

In most metals the conduction electron plasma frequency lies at optical wavelengths and is much higher than the frequencies of interest to study superconductivity, of order $kT_c/\hbar$. Thus the metal would be totally reflecting in the absence of electromagnetic dissipation in the metal. The presence of some dissipation results in electric and magnetic rf fields in the metal which are not quite 90° out of phase, and hence some absorption or transmission takes place.

The exact amount is determined by solution of the boundary-value problem in which a nonlocal field–current relationship is necessary to treat the situation in most metals, giving rise to the anomalous skin effect. A similar type of nonlocal field–current relationship is inherent to superconductivity as well, as is discussed in Chapter 2. A treatment of such effects is necessary to interpret properly the experiments and will be discussed in more detail in Section IV. Nevertheless much can be learned merely by comparing the absorption rate expressions for normal and superconducting states, as will be done in Section III.

B. Ultrasonic Attenuation

Electron–phonon interactions lead to the ordinary electrical resistivity of the metal and also provide the dominant source of attenuation of ultrasonic waves in metals at low temperatures. Studies of both shear and longitudinal wave absorption have been carried out, but only the longitudinal case has a simple interpretation. The usual formalism for treating such interactions involves the concept of the deformation potential, in which a dilatation $\Xi(\mathbf{r})$ of the lattice is involved as a longitudinal phonon is created or absorbed with a corresponding change in momentum of the electron. For acoustic phonons of long wavelength the change in energy for a spherical energy surface would be given by

$$\epsilon(\mathbf{k}, \mathbf{r}) = \epsilon_0(\mathbf{k}) + c_1 \Xi(\mathbf{r}) \tag{2}$$

so that the interaction Hamiltonian has the form

$$\begin{aligned} \mathscr{H}_1 &= (\psi(\mathbf{r}), c_1 \Xi(\mathbf{r}) \psi(\mathbf{r})) = \sum_{\mathbf{k}, \mathbf{k}', \sigma} \langle \mathbf{k}'\sigma | c_1 \Xi(\mathbf{r}) | \mathbf{k}\sigma \rangle c^*_{\mathbf{k}', \sigma} c_{\mathbf{k}, \sigma} \\ &= i c_1 \sum_{\mathbf{q}} (\tfrac{1}{2}\rho\omega_{\mathbf{q}})^{1/2} |\mathbf{q}| \sum_{\mathbf{k}, \sigma} (a_{\mathbf{q}} - a^*_{-\mathbf{q}}) c^*_{\mathbf{k}+\mathbf{q}, \sigma} c_{\mathbf{k}, \sigma} \end{aligned} \tag{3}$$

where $a^*_{\mathbf{q}}$ and $a_{\mathbf{q}}$ refer to longitudinal phonons of wave vector $\mathbf{q}$, and the selection rule $\mathbf{k} - \mathbf{k}' = \mathbf{q} \simeq 0$ would be used with the long-wavelength phonon limit. A detailed discussion of the ultrasonic interaction is reserved for Section V, including the more complicated shear wave case. However, the main features for longitudinal waves can be seen by comparing absorption rates for superconducting and normal phases, to be considered in Section III.

C. Nuclear Spin-Lattice Relaxation

For relaxation of nuclear spins into thermal equilibrium with the lattice, the most important interaction in most metals is the nuclear spin–electron spin scattering that results from the contact hyperfine interaction. It has the form of the dot product between the nuclear spin operator $\mathbf{I}$ and electron spin operator $\mathbf{S}$. When second quantization is used for the electrons, the interaction is given by

$$\mathscr{H}_{1m} = (8\pi/3)\,\gamma_e\gamma_n\hbar^2 \sum_{i,j} \mathbf{I}_i\cdot(\psi(\mathbf{r}), \mathbf{S}_j\delta(\mathbf{r}_j - \mathbf{R}_i)\,\psi(\mathbf{r}))$$
$$= (8\pi/3)\,\gamma_e\gamma_n\hbar^2\,|U(0)|^2 \sum_i \mathbf{I}_i\cdot \sum_{\substack{\mathbf{kk}'\\ \sigma\sigma'}} \mathbf{S}_{\sigma\sigma'} c^*_{\mathbf{k}'\sigma'} c_{\mathbf{k}\sigma} \tag{4}$$

where the gammas are gyromagnetic ratios and $U(0)$ is the Bloch function evaluated at the nucleus.

Since $\mathbf{I}\cdot\mathbf{S} = I_z S_z + \frac{1}{2}(I_+S_- + I_-S_+)$, where $I_\pm = I_x \pm iI_y$ and $S_\pm = S_x \pm iS_y$ with $(\pm\frac{1}{2}|S_z|\pm\frac{1}{2}) = \pm\frac{1}{2}$ and $(+\frac{1}{2}|S_+|-\frac{1}{2}) = (-\frac{1}{2}|S_-|+\frac{1}{2}) = 1$, it is convenient to break Eq. (4) into three parts:

$$\mathscr{H}_{1m} = (4\pi/3)\,\gamma_e\gamma_n\hbar^2\,|U(0)|^2 \sum_i \sum_{\mathbf{k},\mathbf{k}'} \begin{cases} I_{zi}(c^*_{\mathbf{k}'+}c_{\mathbf{k}+} - c^*_{\mathbf{k}'-}c_{\mathbf{k}-}) & m = z \\ I_{+i}c^*_{\mathbf{k}'-}c_{\mathbf{k}+} & m = + \\ I_{-i}c^*_{\mathbf{k}'+}c_{\mathbf{k}-} & m = - \end{cases} \tag{5}$$

In the normal state the calculation of the relaxation rate for nuclear spin involves approximations, because the nuclear states are not accurately known in the presence of local dipolar fields which couple the nuclei. A more complete discussion is given in Section VII, where the experiments are discussed. However, the temperature dependence comes entirely from the electron-state occupation factors; thus the essential features can be obtained from a comparison of the relaxation rates in the normal and superconducting phases, to be considered in Section III.

III. COHERENCE EFFECTS IN TRANSITION PROBABILITIES, ABSORPTION COEFFICIENTS, AND RELAXATION RATES

In calculating the matric elements between the conduction electron states in a superconductor, one finds a striking difference by comparison with the situation in the normal state. Since the spin and momentum states in the superconductor are occupied in pairs, one finds that the scattering interactions contain terms corresponding to a coherence in the scattering between the pair electrons. Thus when the matrix elements are squared, the two contributions add coherently, and the sign between them gives rise to interference effects which would be completely absent in the normal state. The situation is best seen by means of an example.

As discussed in Chapter 2, at temperatures between absolute zero and the transition temperature there are three groups of electrons which can be found in the metal. First, there are ground-state pairs which are peculiar to the superconducting state. In addition, there are single and pairwise occupations of states which are thermally excited, similar to the situation in the normal metal; the main difference is that the pairwise excitations must be orthogonal to the ground-state pair occupations below the critical temperature. As shown in Chapter 2, the BCS theory introduces the conduction electron wave function Ψ obtained from the vacuum function, denoted by Φ_0, in the following way:

$$\Psi = \prod_{\mathbf{k}} [(1 - h_{\mathbf{k}})^{1/2} + h_{\mathbf{k}}^{1/2} b_{\mathbf{k}}^*] \prod_{\mathbf{k}'} [(1 - h_{\mathbf{k}'})^{1/2} b_{\mathbf{k}'}^* - h_{\mathbf{k}'}^{1/2}] \prod_{\mathbf{k}''} c_{\mathbf{k}''}^* \Phi_0 \qquad (6)$$

In this equation $b_{\mathbf{k}}^* = c_{\mathbf{k},+}^* c_{-\mathbf{k},-}^*$ is the pair creation operator for states $\mathbf{k},+$ and $-\mathbf{k},-$, where $\pm\mathbf{k}$ labels the momentum and the second index $(\pm)$ labels the spin direction of the Bloch states. In Eq. (6) $\mathbf{k}$, $\mathbf{k}'$, $\mathbf{k}''$ label the states occupied by ground-state pairs, thermally excited pairs, and singles, respectively. The parameter $h_{\mathbf{k}}$ is given by

$$h_{\mathbf{k}} = \tfrac{1}{2}(1 - \epsilon_{\mathbf{k}}/E_{\mathbf{k}}) \qquad (7)$$

where $\epsilon_{\mathbf{k}}$ is the normal Block state energy relative to the Fermi energy. $E_{\mathbf{k}}$ is often called the quasi-particle energy and is given by

$$E_{\mathbf{k}} = \sqrt{\epsilon_{\mathbf{k}}^2 + \Delta^2} \qquad (8)$$

where $\Delta(T)$ is the superconducting energy gap parameter.

In calculating the matrix elements and determining the allowed transitions the electron quasi-particle energy $E_{\mathbf{k}}$ [Eq. (8)] plays an important role. Let us consider two excited states labeled 1 and 2. They will differ because of the assignment of states among the three types of terms: single excitations, thermal pair excitations, or ground-state pairs. It was shown in Chapter 2 that the total energy W_1 of the electrons in state 1 will differ from W_2, the total energy in state 2, by the expression

$$W_1 - W_2 = \sum_1 E_{\mathbf{k}} - \sum_2 E_{\mathbf{k}} \qquad (9)$$

where we include in the sum the value of $E_{\mathbf{k}}$ for each electron that is found in a single excitation or a thermally excited pair excitation, with no term from the ground-state pairs. We note that if $\mathbf{k}$ is used for a single electron, then $E_{\mathbf{k}}$ appears once; if it is used for an excited pair, $E_{\mathbf{k}}$ appears twice. Consequently, we do not need to distinguish between excited pairs or singles but can merely compute the probability of occupancy and energy as though any single Bloch state $\mathbf{k}$ had an energy $E_{\mathbf{k}}$ and a probability of occupation given by the Fermi function $f(E_{\mathbf{k}}, T)$. That is, if $S(T)$ labels the statistical weights of the singles, $P(T)$ labels the statistical weights of the thermally excited pairs, and $G(T)$ labels the statistical weights of the ground-state pairs, then the probabilities are given by the relations

$$S(T) = 2f(1 - f) \qquad (10a)$$

$$P(T) = f^2 \qquad (10b)$$

as in the normal metal with

$$G(T) = 1 - S(T) - P(T) = (1 - f)^2 \tag{10c}$$

Thus energy considerations restrict the kinds of matrix elements to three basic types

1. Single + single ↔ excited pair + ground-state pair.
2. Single + excited pair ↔ excited pair + single.
3. Single + ground-state pair ↔ ground-state pair + single, which will be called types 1, 2, and 3, respectively.

Let us consider a typical example, a type 1 transition, which shows how the coherence effects arise in the matrix element for the superconducting case. At first it looks peculiar to see such a transition which does not appear to conserve particles. However, such is not the case, since each pair state is partly occupied and partly empty. One uses the empty excited pair with the occupied ground-state pair or conversely. Using Eq. (6) the initial and final states for the transition of type 1 are given by

$$\begin{aligned} \Psi_i &= c_{\mathbf{k}+}^* c_{-\mathbf{k}'-}^* \Phi \\ \Psi_f &= [(1 - h_{\mathbf{k}})^{1/2} + h_{\mathbf{k}}^{1/2} c^*_{\mathbf{k},+} c^*_{-\mathbf{k},-}]\,[(1 - h_{\mathbf{k}'})^{1/2} c^*_{\mathbf{k}',+} c^*_{-\mathbf{k}',-} - h_{\mathbf{k}'}^{1/2}]\,\Phi \end{aligned} \tag{11}$$

Here Φ specifies the remaining states, not affected in the transitions in question. Now each of the four interactions considered in Section II introduces a perturbation Hamiltonian of the form

$$\mathscr{H}_1 = \sum_{\mathbf{k},\,\mathbf{k}'\sigma,\,\sigma'} B_{\mathbf{k}\sigma,\,\mathbf{k}'\sigma'} c^*_{\mathbf{k}',\,\sigma'} c_{\mathbf{k},\,\sigma} \tag{12}$$

in which B is the matrix element for the scattering from the state $\mathbf{k},\sigma$ to $\mathbf{k}',\sigma'$. If one then computes the matrix element for the type 1 example, one sees that the interaction connects two different portions of the initial and final state. That is, the operator $\mathscr{H}_1$ has a term in the sum labeled by $c^*_{\mathbf{k}',+} c_{\mathbf{k},+}$, corresponding to scattering the electron in Bloch state $\mathbf{k},+$, as well as a term labeled by $c^*_{-\mathbf{k},-} c_{\mathbf{k}',-}$, corresponding to scattering the electron in the Bloch state $-\mathbf{k}',-$.

Now, depending on the nature of the interaction, the two contributions shown above, which are typical for all three types of possible transitions, may add either constructively or destructively when the matrix element is squared. In general, $B_{\mathbf{k}\sigma,\mathbf{k}'\sigma'}$ has the same magnitude for the transitions involving $\mathbf{k},+$ as it does for those involving $-\mathbf{k},-$, because the wave-vector differences are the same in magnitude although they differ in sign. There are two general cases of symmetry for the interactions; using the notation of BCS they are

$$B_{\mathbf{k}'\sigma',\,\mathbf{k}\sigma} = \theta_{\sigma\sigma'} B_{-\mathbf{k},\,-\sigma,\,-\mathbf{k}',\,-\sigma'} \qquad \text{(case 1)}$$

$$B_{\mathbf{k}'\sigma',\,\mathbf{k}\sigma} = -\,\theta_{\sigma\sigma'} B_{-\mathbf{k},\,-\sigma,\,-\mathbf{k}',\,-\sigma'} \qquad \text{(case 2)}$$

where

$$\theta_{\sigma\sigma'} = \pm 1 \qquad \text{for } \sigma = \pm \sigma'$$

The first case applies to a scalar potential such as that involved in calculating the absorption of longitudinal ultrasonic waves. Since the interaction between the electrons and ultrasonic waves depends only on the magnitude of the electron momentum $\hbar k$ and not its direction, as seen from Eq. (3), the symmetry sign for such an interaction is positive; it belongs to case 1. In contrast, the electromagnetic interaction depends on both the direction and magnitude of electron momentum relative to the electric field. Similarly, the hyperfine interaction depends upon the direction of the electron spin relative to the nuclear spin which is being relaxed. As a consequence, the matrix element $B_{\mathbf{k}\sigma,\mathbf{k}'\sigma'}$ for these two interactions has the negative sign directly traceable to the directionality involved, so they belong to case 2.

When one calculates the matrix elements based on such interactions for the three types of elements that are possible, one finds that the contributions from each of the interfering parts give terms involving the same matrix elements as in the normal metal along with the factors $h_{\mathbf{k}}$ [Eq. (7)] which are characteristic of the superconducting state. After carrying through the algebra one finds that the two contributions from each type of transition add constructively or destructively, depending upon whether one is dealing with case 1 or case 2; the contributions involve a common factor, in addition to the normal-state matrix elements, given by the expression

$$(1 - h_{\mathbf{k}})^{1/2}(1 - h_{\mathbf{k}'})^{1/2} \mp h_{\mathbf{k}}^{1/2} h_{\mathbf{k}'}^{1/2} \tag{13}$$

where the minus sign is for case 1 and the plus sign for case 2.

Now the transition probability from a state of energy E_i to a state $E_f = E_i \pm \hbar\omega$ introduced by an interaction is proportional to the square of the matrix element which has been calculated as well as to the density of final states $N(E_f)$. To calculate the net rate of the absorption of energy one must then take the difference between direct absorption and induced emission and sum over the initial states. Strictly speaking, one should begin with the many-particle wave function and density of states involving the total electronic energy. However, the change in energy involves only the states which are being scattered, so that the absorption coefficient is expressible in terms of densities of states, Fermi factors, and matrix element factors involving only those states. (See the references cited in Section IV.)

The resulting absorption coefficient, α_s, in the superconducting phase can be simply expressed relative to that for the normal phase in terms of the matrix elements which have been calculated. If the matrix elements and density of states are taken to be independent of the angle between $\mathbf{k}$ and $\mathbf{k}'$, and the interaction $B_{\mathbf{k}'\sigma',\mathbf{k}\sigma}$ is assumed to be independent of energy in the important region of integration, one obtains

$$\alpha_s \propto \int_0^\infty d\epsilon \int_0^\infty d\epsilon' \, N_0(\epsilon') \, N_0(\epsilon) \, [f(\epsilon, T) - f(\epsilon', T)] \times [(1 - h_{\mathbf{k}})^{1/2} (1 - h_{\mathbf{k}'})^{1/2} \mp h_{\mathbf{k}}^{1/2} h_{\mathbf{k}'}^{1/2}] \tag{14}$$

where N_0 is the normal-state density of states.

The result can be considerably simplified using the quasi-particle energy $E_{\mathbf{k}} = +\sqrt{\epsilon_{\mathbf{k}}^2 + \Delta^2}$ instead of the Bloch energy $\epsilon_{\mathbf{k}}$. Defining a density of electronic states in the superconductor $N_s(E)$, given by

$$N_0(\epsilon) \, d\epsilon = N_s(E) \, dE \tag{15}$$

one finds that

$$N_s = N_0 E/(E^2 - \Delta^2)^{1/2} \tag{16}$$

which is highly peaked near the gap at $E \simeq \Delta$. Then one finds

$$\alpha_s \propto \int_\Delta^\infty \int_\Delta^\infty dE \, dE' \, (1 \mp \Delta^2/EE') \, [f(E, T) - f(E', T)] \, N_s(E) \, N_s(E') \tag{17}$$

where $f(E, T)$ is the usual Fermi function; the coherence factor $(1 \mp \Delta^2/EE')$ is all that remains of the interfering contributions from the case 1 or 2 matrix elements.

The expression for the spin-lattice relaxation rate is very similar. One wants the probability that a nucleus be flipped; because of the Pauli exclusion principle this is proportional to the number of electrons in the initial energy interval, $N_s(E_i) \, f(E_i) \, dE_i$, times the number of vacant states into which electrons can be scattered, $N_s(E_f) \, [1 - f(E_f)] \, dE_f$. (See the references cited in Section VII for more detail.) Again using the quasi-particle energy $E_{\mathbf{k}} = \sqrt{\epsilon_{\mathbf{k}}^2 + \Delta^2}$ and the corresponding superconducting density of states $N_s(E)$, the relaxation is given by

$$R_s \propto \int_\Delta^\infty \int_\Delta^\infty dE_i \, dE_f \, (1 \mp \Delta^2/EE') \, f(E_i) \, [1 - f(E_f)] \, N_s(E_i) \, N_s(E_f) \tag{18}$$

The expressions for the normal state can be obtained merely by letting the energy gap parameter $\Delta \to 0$ in both Eqs. (17) and (18). Then the coherence factor $(1 \mp \Delta^2/EE') \to 1$ and $N_s \to N_0$, and all the normal-state temperature dependences come from the integrals over the Fermi factors. Thus the comparison between normal- and superconducting-state absorption coefficients and relaxation rates can be done from Eqs. (17) and (18) without getting involved with the normal-state matrix elements. Two observations of importance can be made.

First, the presence of the energy gap can readily be seen in the Δ factor appearing both in N_s and in the limits of integration. Second, the coherence factor $(1 \mp \Delta^2/EE')$ has a profound effect on the results, depending on the sign: The minus sign applies to case 1, ultrasonic attenuation, where the coherence factor cancels the peaking which occurs in N_s for $|E - E'| \ll \Delta$ and E close to Δ; the plus sign applies to case 2; electromagnetic absorption and nuclear spin-lattice relaxation, where the coherence factor enhances the peaking in N_s. When the Fermi factors are taken into account, this difference results in a rapid drop for ultrasonic absorption as T is lowered below T_c, whereas nuclear spin-lattice relaxation at first *increases* and then rapidly decreases as T is lowered. The contrast is impossible to explain with a two-fluid model but is a natural result of the spin-momentum pairing concept inherent to the BCS theory.

IV. ELECTROMAGNETIC ABSORPTION

A. Introduction

In this section we will first discuss the qualitative features of the dependence of electromagnetic absorption on temperature and frequency, and the effect of the coherence factors which were treated in Section III. We will then consider the quantitative comparisons which have been made with the theory. Measurements of the superconducting energy gap will be listed. We will describe the effects of anisotropy and an applied dc magnetic field. Finally, we will discuss the possible existence of absorption peaks below the main gap edge (precursors). We will not deal with experiments on type II superconductors, which are included in Chapter 15, and we will restrict our discussion to calculations and experiments which seem to us to have a direct and simple bearing on the topic of this chapter, which is a comparison of data with the BCS theory (*1*). Other work is described in several excellent review articles (*2–6*).

As we have seen, electromagnetic absorption belongs to case 2, with respect to coherence effects, as does nuclear spin-lattice relaxation. The absorption of electromagnetic energy in a superconductor results from currents which are in phase with the electric field, and these are proportional to σ_1, the real part of the conductivity. This in turn should be calculated according to Eq. (17). As discussed in Section I, this equation indicates that the function σ_1 should rise and then fall as the temperature is reduced from the critical temperature T_c if the difference between E and E' is small compared with Δ, which is the case for photon energies $\hbar\omega \ll \Delta$. However, the observed absorption at these frequencies does not have such a rise and fall, but rather decreases monotonically as the temperature decreases below T_c. The explanation of this paradox is that the energy absorption is proportional to the electric field as well as to σ_1, and the electric field in the metal is limited by the mismatch in the electromagnetic impedance at the surface of the metal. This mismatch results from currents which are out of phase with the electric field and from those which are in phase. It is therefore impossible to say whether the energy absorption should initially increase or decrease until one performs a calculation. This will be described below. It is found that a monotonic decrease is actually expected as the temperature falls. One can understand this physically in terms of a decrease in both the penetration depth and the density of quasiparticles as the temperature is lowered.

The experiments which have been performed at low frequencies ($\hbar\omega \ll \Delta$) are in the microwave region. [The low-frequency measurements which we discuss are all on bulk samples, since no reliable and relevant measurements seem to have been made on films except on some (*7*) which had extremely strange electrical properties. These indicated that conduction occurred by tunneling between small particles, a mechanism which we are not considering here.] The earliest microwave measurements were made by H. London in 1940 (*8*). After the war, microwave techniques were developed by Pippard (*9*) and then by his coworkers and others (*2,3,5,10–24*). The measurements consist of observations of (1) the width and frequency of the resonance of a microwave cavity which contains the sample or which is itself the sample, or (2) attenuation by a waveguide sample, or (3) the heat absorbed by the sample in a wave guide or a cavity. It is found in all cases that the energy absorption curves tend to decrease with temperature, although there are apparent small wiggles on some of the early curves and in some recent ones (*24*). In one of these cases, the wiggles have been annihilated by improved instrumentation (*25–27*), and might therefore be presumed spurious in the other case as well. Some of the early microwave data (*10*) were thought to indicate that the rate of energy absorption did not begin to decrease until the temperature had dropped by a small finite amount, δT, below T_c, and that δT increased with frequency. However, this effect was comparable to the random scatter of the data, and was not supported by later, more precise data. The decrease of energy absorption is found to be less abrupt, the higher the frequency. This is shown for aluminum in Fig. 1. At zero frequency, the decrease occurs in a temperature range of a few millidegrees in good samples; it is thought that the transition would be even sharper if the samples were larger and more homogeneous (*28*). For photon energies very large compared with the gap width 2Δ (at near-infrared or optical frequencies), the decrease of energy absorption is too small to be observable (*29–33*). The steepness of the absorption curve changes most rapidly with photon energies in the region near $\hbar\omega \approx 2\Delta$, as one would qualitatively expect.

According to Eq. (17), the rate of energy absorption should approach zero at low temperature, because the Fermi functions approach zero. The absorption does in fact fall to less than 1% of the normal-state value for samples which are

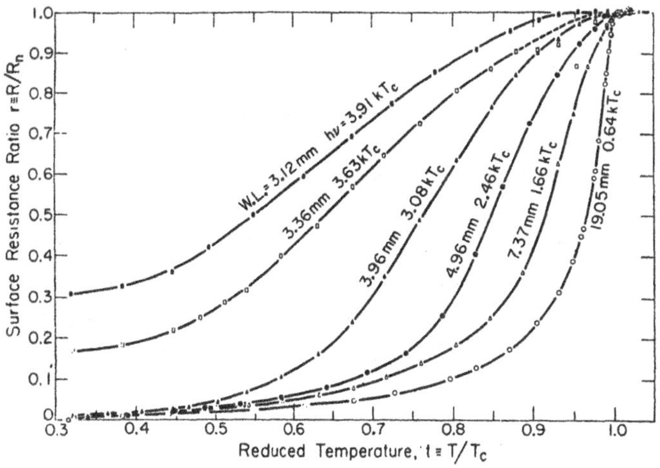

Fig. 1. Measured values of the surface resistance ratio r of superconducting aluminum as a function of the reduced temperature t at several representative wavelengths. The wavelengths and corresponding photon energies are indicated on the curves [After Biondi and Garfunkel (*15*).]

homogeneous and carefully electropolished. Surfaces which have been mechanically polished seem to have mechanisms for energy absorption which are not considered in the theoretical calculations. There may, for example, be some trapping of magnetic flux from the microwave field, creating normal regions which can absorb energy. Such an effect is beyond the scope of this chapter. Solder joints which have been present in some of the samples have also possibly been responsible for residual energy absorption at low temperatures.

B. Calculation of the Electromagnetic Response

We now turn to the problem of quantitatively comparing the observed absorption rate with theory. Any calculation of this rate must take into account the following complications:

1. The impedance mismatch at the surface of the metal is very important, as discussed above.

2. The electromagnetic wave is attenuated in the metal. The important Fourier components of the resulting wave have wave numbers q which range from zero up to and somewhat beyond the reciprocal of the penetration depth.

3. The electromagnetic kernel, which is the ratio of current density to vector potential, is a function of the wave number q, because of the nonlocal response (see Chapter 2).

4. The electron mean free path limits the range in space of this kernel, and must be taken into account.

This list shows that the calculation of energy absorption is a transport problem which must be solved for a self-consistant electromagnetic field. Although Eq. (17) is expected to be valid for each value of q, the ratio of total attenuation in the superconducting state to that in the normal state is more complicated. The nonlocal nature of the phenomenon provides in fact the main obstacle in the calculation. The theory was developed by Mattis and Bardeen (*34*), and independently by Abrikosov et al. (*35*).

By performing a perturbation calculation, using the coherence factor which appears in Eq. (17), Mattis and Bardeen derived the following expression for the current density in the presence of a vector potential **A** which has an angular frequency ω:

$$\mathbf{j}(\mathbf{r}, t) = \frac{e^2 N(0)\, v_0}{2\pi^2 \hbar c} \exp(i\omega t) \int \frac{\mathbf{R}\,[\mathbf{R}\cdot\mathbf{A}(\mathbf{r}')]\, I(\omega, R, T) \exp(-R/l)}{R^4}\, d\mathbf{r}' \tag{19}$$

where $N(0)$ is the density of normal states at the Fermi surface for one direction of spin, v_0 is the Fermi velocity, l is the electron mean path, and $\mathbf{R} = \mathbf{r} - \mathbf{r}'$. The range of the nonlocal effect of the electromagnetic field on the current density is determined by $I(\omega, R, T)$ and $\exp(-R/l)$. The range of $I(\omega, R, T)$ is approximately the coherence length $\xi_0 = \hbar v_0/\pi\Delta$. The function I is given by‡

$$\begin{aligned} I(\omega, R, T) = {} & -\pi i \int_{\Delta-\hbar\omega}^{\Delta} [1 - 2f(E + \hbar\omega)]\,[g(E)\cos(\alpha\epsilon_2) - i\sin(\alpha\epsilon_2)] \\ & \times \exp(i\alpha\epsilon_1)\, dE - \pi i \int_{\Delta}^{\infty} \{[1 - 2f(E + \hbar\omega)] \\ & [g(E)\cos(\alpha\epsilon_2) - i\sin(\alpha\epsilon_2)] \exp(i\alpha\epsilon_1) - [1 - 2f(E)] \\ & \times [g(E)\cos(\alpha\epsilon_1) + i\sin(\alpha\epsilon_1)] \exp(-i\alpha\epsilon_2)\}\, dE \end{aligned} \tag{20}$$

where ϵ_1 and ϵ_2 are the normal-state energies corresponding to E and $E + \hbar\omega$, respectively:

$$\epsilon_1 = (E^2 - \Delta^2)^{1/2} \qquad \epsilon_2 = [(E + \hbar\omega)^2 - \Delta^2]^{1/2} \tag{21}$$

$$g(E) = (E^2 + \Delta^2 + \hbar\omega E)/\epsilon_1\epsilon_2 \tag{22}$$

$$\alpha = R/\hbar v_0 \tag{23}$$

In the normal state, $\Delta = 0$, and Eq. (19) reduces to the well-known integral formula of Chambers (*36*). At very low frequencies, the expression for I in Eq. (20) is very nearly proportional to the function $\exp(-R/\xi_0)$, which was introduced to treat nonlocal effects empirically by Pippard (*37*) prior to the BCS theory.

‡ In this chaper we consider plane waves, $\exp[i(\omega t - \mathbf{q}\cdot\mathbf{r})]$, and take $\sigma = \sigma_1 - i\sigma_2$. This is the convention most often followed in the literature. In Chapter 2 the convention is to use $\exp[-i(\omega t - \mathbf{q}\cdot\mathbf{r})]$ and $\sigma_1 + i\sigma_2$.

The calculation of electromagnetic absorption is enormously simplified in the limit $l \ll \xi_0$ (the local limit) or in the limit where the skin depth $\lambda \ll \xi_0$ (the Pippard or extreme anomalous limit). In either limit, the function $I(\omega, R, T)$ varies very slowly in space compared with the rest of the integrand, so $I(\omega, R, T)$ can be replaced by $I(\omega, 0, T)$. Then the current density in the superconducting state and that in the normal state have the same spatial dependence, and the ratio of the superconducting conductivity $\sigma_s = \sigma_1 - i\sigma_2$ to the normal conductivity σ_n is independent of wave number q:

$$(\sigma_1 - i\sigma_2)/\sigma_n = I(\omega, 0, T)/-\pi i\hbar\omega \tag{24}$$

In the extreme local limit, σ_s and σ_n are *each* independent of wave number. The expressions for σ_1/σ_n and σ_2/σ_n derived by Mattis and Bardeen, which are valid in either limit, are

$$\frac{\sigma_1}{\sigma_n} = \frac{2}{\hbar\omega}\int_{\Delta}^{\infty} [f(E) - f(E + \hbar\omega)]\, g(E)\, dE + \frac{1}{\hbar\omega} \times \int_{\Delta-\hbar\omega}^{-\Delta} [1 - 2f(E + \hbar\omega)]\, g(E)\, dE \tag{25}$$

$$\frac{\sigma_2}{\sigma_n} = \frac{1}{\hbar\omega}\int_{\Delta-\hbar\omega,\, -\Delta}^{\Delta} \frac{[1 - 2f(E + \hbar\omega)]\,(E^2 + \Delta^2 + \hbar\omega E)}{(\Delta^2 - E^2)^{1/2}\,[(E + \hbar\omega)^2 - \Delta^2]^{1/2}}\, dE \tag{26}$$

The second term in Eq. (25) is to be included only if $\hbar\omega > 2\Delta$, and the lower limit of the integral in Eq. (26) is to be taken as $-\Delta$ instead of $\Delta - \hbar\omega$ if $\hbar\omega > 2\Delta$. The integrals in Eqs. (25) and (26) can be expressed in closed form (*34*) in the limit $T \to 0$.

C. Theory and Experiment for Thin Films

The experiments which can be interpreted most easily are those on films which are very thin compared with both the penetration depth and the coherence length ξ_0. Equations (25) and (26) can then be used, and also the electric field in the film is independent of position, if the electromagnetic wave is incident in a direction normal to the film's surface. The rate at which energy is reflected, absorbed, or transmitted by the film can then be very easily calculated (*38,39*). At a given frequency, the experimental data can be used to find directly σ_1/σ_n and σ_2/σ_n if two of these three rates are measured. If only one of them is measured, then σ_1/σ_n and σ_2/σ_n are not uniquely determined. However, some information can still be extracted if one assumes that the theoretical expression for σ_1/σ_n is approximately correct, and if one restricts σ_1/σ_n and σ_2/σ_n to satisfy the Kramers–Kronig relations (*40*) and the Ferrell–Glover–

Tinkham (*41,42*) sum rule. The Kramers–Kronig relations, which are valid for any linear response function obeying causality, are

$$\sigma_1(\omega) = \frac{1}{\pi}\int_{-\infty}^{\infty} \frac{\omega_1 \sigma_2(\omega_1)\, d\omega_1}{\omega_1^2 - \omega^2} + \text{const.} \tag{27}$$

$$\sigma_2(\omega) = \frac{-\omega}{\pi}\int_{-\infty}^{\infty} \frac{\sigma_1(\omega_1)\, d\omega_1}{\omega_1^2 - \omega^2} \tag{28}$$

The sum rule can be derived from Eq. (28) by taking ω to be an angular frequency which is so high compared with $2\Delta/\hbar$ that the superconducting and normal states must have approximately the same value for $\sigma_2(\omega)$. Then one finds directly that

$$\int_{-\infty}^{\infty} \sigma_{1s}(\omega)\, d\omega = \int_{-\infty}^{\infty} \sigma_{1n}(\omega)\, d\omega \tag{29}$$

The area under the σ_1/σ_n curve, which is removed at frequencies greater than zero by the superconducting transition, appears in a delta function at zero frequency. Physically, this accounts for the energy which can be stored in the superconductor in the form of persistent currents.

In the early far-infrared experiments on thin films by Tinkham with Glover (*38*) and Ginsberg (*43*), only the transmitted energy was measured, and a "reasonable" function σ_1/σ_n was calculated which was consistent with the sum rule, and which, with its Kramers–Kronig transform σ_2/σ_n, accounted for the data. (In these experiments, σ_n could be assumed real and independent of frequency because of the rapid scattering of electrons in the samples.) If there is an energy gap of width 2Δ, then it is expected that at $T=0$, the function σ_1/σ_n will be zero at finite frequencies up to the gap frequency $2\Delta/\hbar$, since photons of energy less than 2Δ cannot be absorbed. At finite temperatures, thermally excited quasi-particles are present, and can absorb photons of arbitrary energy, so σ_1/σ_n will be nonzero at all frequencies. If $\exp(-\Delta/kT) \ll 1$, then σ_1/σ_n should be very small below the gap frequency. The Kramers–Kronig transform of σ_1, σ_2, is found to decrease with frequency, becoming small near the gap frequency and above it. Near the gap frequency, both σ_1 and σ_2 are small. As a result, absorption and reflection of energy are both small, and the transmissivity of the film has a fairly sharp maximum.

The results of experiments (*43*) an annealed films of tin and lead indicated that $\sigma_1(\omega)$ rises above the gap edge somewhat more rapidly than expected. The data for indium agreed well with the theory. There seemed to be a peak in $\sigma_1(\omega)$ slightly below the main gap edge in lead (a "precursor"). It should be noted that this effect was not large compared with the uncertainty in the data, particularly if one allows a margin of error associated with necessity of finding both σ_1 and σ_2 from transmission data alone.

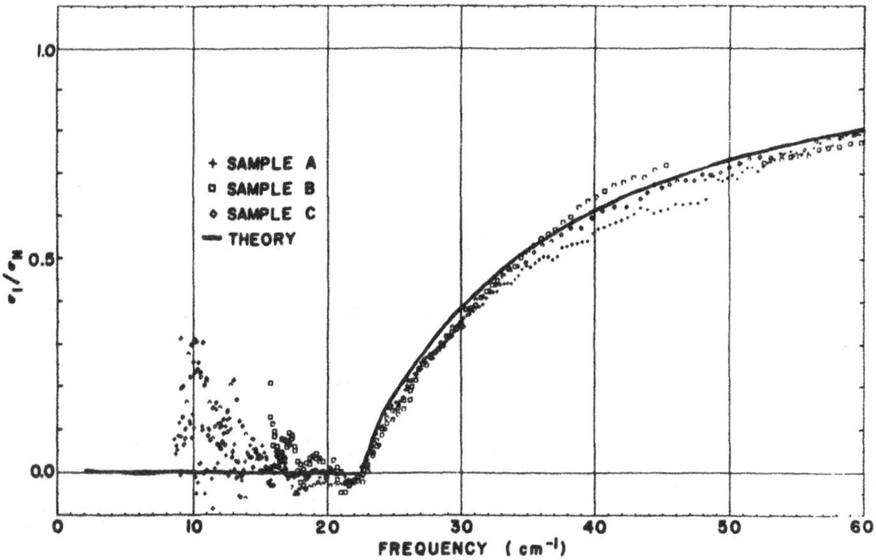

Fig. 2. Measured and calculated values of the conductivity ratio σ_1/σ_n of three superconducting lead films as a function of the photon frequency. [After Palmer (*39*).]

In recent experiments on unannealed lead films, Palmer (*39*) and Tinkham have measured both transmission and reflection. The data then yield σ_1/σ_n and σ_2/σ_n directly. The values for σ_1/σ_n are shown in Fig. 2. Tinkham has noted (*4*) that the general shape of the σ_1/σ_n curve is completely inconsistent with the shape one would calculate if electromagnetic absorption were a process involving coherence of type 1. He pointed out that this would predict a discontinuous rise in σ_1/σ_n to a value greater than unity at the gap frequency $\omega = 2\Delta/\hbar$. This theoretical discontinuous rise had been noted earlier by Pokrovskii (*44,45*), and detailed calculations by Bobetic (*46*) on very high frequency ultrasonic attenuation bear this out. It will be noted that the fit to the theoretical curves is rather good, but not perfect. There is a definite tendency for the points in Fig. 2 to lie above the theoretical curve well below the gap edge (where the scatter is large), but by a much smaller amount near the gap edge (where the scatter is small) than indicated by the analysis of the earlier data on annealed lead films.

The transmission data of Palmer and Tinkham agree well with theory above the gap edge, but poorly below it, where the observed transmission is higher than predicted. The discrepancy is on the order of 25% for sample A, and Fig. 2 shows that the other two samples deviate even more from theory. If σ_1/σ_n does indeed agree with theory, then this shows that σ_2/σ_n must disagree with it. One of Ginsberg and Tinkham's two lead films yielded transmission data which agree with the theory better than this below the gap edge; the agreement for the other one is worse. Of the three lead films which we are discussing, sample A of Palmer and Tinkham had a peak transmission ratio T_s/T_n which was about 7% higher than theoretically expected, while for Ginsberg and Tinkham's two films, one peak was 5% lower than expected, and the other was 14% higher than expected.

It should be clear that the experimental situation for lead films is unsettled at the present time. To settle conclusively the question of whether there is any absorption of energy below the main gap edge at low temperature, the absorption should be measured directly. If there is no absorption, then $\sigma_1 = 0$. (There have been absorption measurements in very thick films; we classify these as bulk samples, and they are discussed in Section G.)

D. Theory and Experiment for Bulk Samples

In comparing data on bulk samples with theory, the rate of energy absorption must be calculated without neglecting nonlocal effects. Neither the extreme anomalous limit nor the extreme local limit is well satisfied in any of the samples which have been investigated, although these limits are useful to consider for qualitative purposes. To compare the theory with the experimental data, one computes the surface impedance, defined by

$$Z = E(0) \left[\int_0^\infty j(z)\, dz \right]^{-1} \equiv R + iX \qquad (30)$$

where $E(0)$ is the electric field at the surface of the metal, located at the $z = 0$ plane, and $j(z)$ is the current density in the metal. The rate at which energy is absorbed in the metal is proportional to the surface resistance R.‡ The surface impedance can be calculated directly from the conductivity $\sigma(\omega, q)$ according to equations which were derived by Reuter and Sondheimer (*47*), and which were put into convenient form by Dingle (*48*). For the case where the electrons are specularly reflected from the surface of the metal,

$$Z(\omega) = \frac{2}{\pi} \int_0^\infty \left[\sigma(\omega, q) - \frac{iq^2c^2}{4\pi\omega} \right]^{-1} dq \qquad (31)$$

where c is the speed of light. If the electrons are diffusely scattered at the metal surface,

$$Z(\omega) = \frac{4\pi^2 i\omega}{c^2} \left\{ \int_0^\infty \ln\left[1 + \frac{4\pi\sigma(\omega, q)}{q^2c^2} \right] dq \right\}^{-1} \qquad (32)$$

The actual surface impedance should be close to the diffuse-scattering case of Eq. (32) (*49,50*).

‡ This proportionality holds only when the power absorbed is much less than the incident power. This is the usual case for metals.

In the extreme anomalous limit, these equations reduce to a very simple one:

$$Z_s/Z_n = (\sigma_n/\sigma_s)^{1/3} \tag{33}$$

where Z_n is the normal-state surface impedance, given by

$$Z_n = (1 + i\sqrt{3})\, R_n \tag{34}$$

In the extreme local limit, another simple expression is valid:

$$Z_s/Z_n = (\sigma_n/\sigma_s)^{1/2} \tag{35}$$

$$Z_n = (1 + i)\, R_n \tag{36}$$

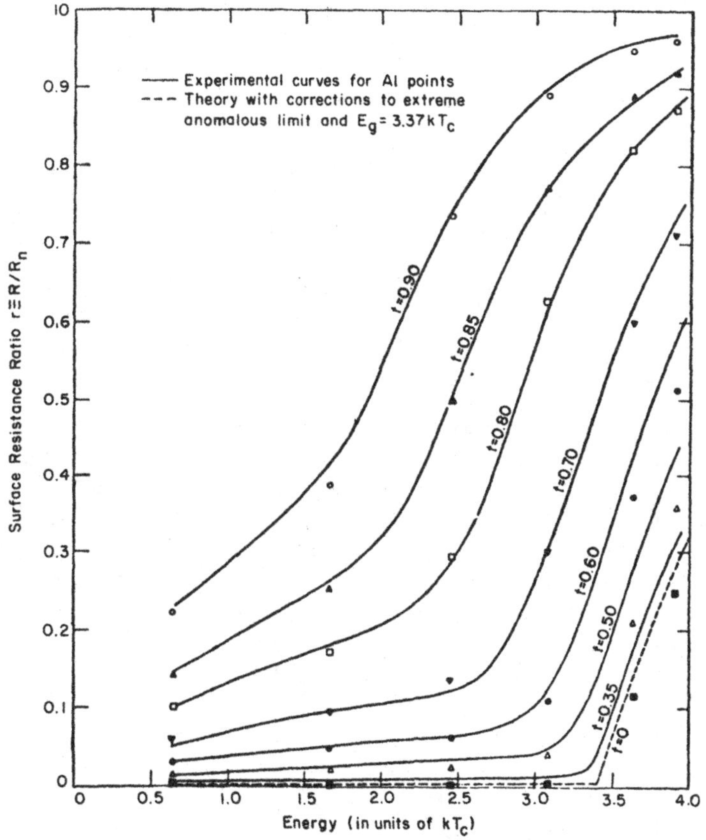

Fig. 3. Experimental curves of the resistance ratio r of superconducting aluminum as a function of reduced frequency for various values of reduced temperature t [from Biondi and Garfunkel (*15*)], with a dashed curve showing the data extrapolated to $T = 0$, compared with theoretical points, which have been calculated by assuming an energy gap at $T = 0$ of $3.37kT_c$. There are no theoretical points shown to correspond to the dashed curve at $T = 0$. [After Miller (*51*).]

Miller (*51*) has calculated the surface impedance of pure aluminum and tin. He found $\sigma(\omega, q)$ by integrating Eq. (19) to find the current density which would result from a plane electromagnetic wave in the metal. He used Eq. (32), the diffuse-reflection formula, to calculate the surface impedance. The work was simplified by restricting the expressions to the limit of infinite mean free path. This should give results very close to those for the actual mean free paths, which were very long compared with the penetration depth λ and the coherence length ξ_0. He did not assume that $\lambda \ll \xi_0$, as one would in the extreme anomalous limit. His results show that the corrections to that limit are quite sizable, particularly for tin, since λ/ξ_0 is larger for tin than for aluminum. One stage in the calculation involved graphical interpolation of $\sigma(\omega, q)$ between very low q and very high q. As Miller notes, this introduced an uncertainty which is difficult to estimate. The agreement of his results with the data of Biondi and Garfunkel (*15*) for aluminum is exceedingly good, as Fig. 3 shows.

In the case of tin, the data considered by Miller lie well below the energy gap edge, and he found it expedient to compare his results with the empirical form discovered by Pippard (*52*):

$$R_s/R_n = \phi(T/T_c)\,A(\nu) \tag{37}$$

where

$$\phi(t) = t^4(1 - t^2)/(1 - t^4)^2 \tag{38}$$

and ν is the frequency of the radiation. Eq. (37) fits well with the data for the temperature range lying between $T/T_c = 0.4$ and 0.8. The results of Miller's

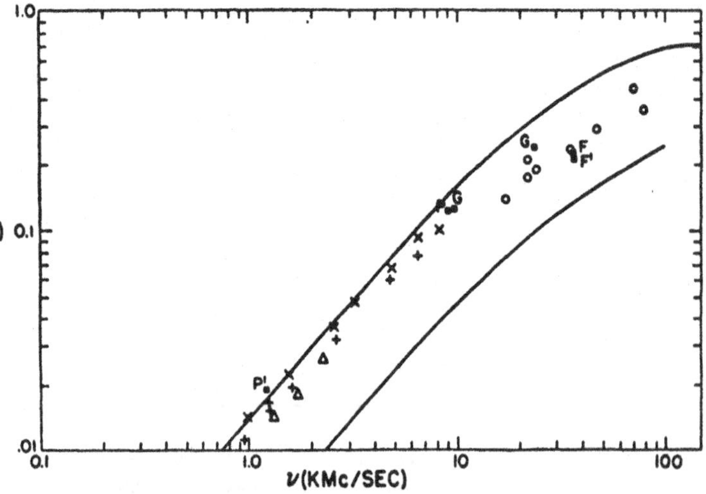

Fig. 4. $A(\nu)$, the frequency-dependent part of the surface resistance ratio r, as a function of frequency. The points show data of various investigators. The upper curve shows the calculation of Miller; the lower curve shows the extreme anomalous limit theory. [After Miller (*51*).]

calculation agree with Eq. (37) to within about 10% in the temperature range from $T/T_c = 0.4$ to 0.7 if $\hbar\omega/kT_c$ lies between 0.01 and 2. The comparison of the theoretical behavior of $A(\nu)$ with the data is shown in Fig. 4, and is seen to be satisfactory, considering the scatter of the data and the fact that the large anisotropy of tin was neglected.

Waldram (*5*) has been able to extend Miller's results semiquantitatively to tin alloys. He concludes that the temperature variation of the surface resistance in the superconducting state R_s is well described by the theory, but the magnitude of R_s is underestimated. He accounts for this by an error in the theoretical penetration depth. As related in Chapter 3, this can be explained in terms of an energy gap width which is in slight disagreement with that of the theory. It should be noted that the surface resistance is not independent of the penetration depth, because the two are related by a Kramers–Kronig type of relation (*2*).

To decrease possible effects of anisotropy, it is useful to compare theory and experiment for a rather highly doped alloy. Anderson's theory of dirty superconductors (*53*) indicates that the short electron mean free path l in such an alloy should largely wash out the anisotropy if $l \ll \xi_0$. Miller's equations, which assume that $l \gg \xi_0$, cannot be used here. Ginsberg has carried out numerical calculations (*54*) of the surface resistance, and has used an equation derived by Leplae (*55*) which is valid if $\exp(-\Delta/kT) \ll 1$:

$$\sigma_{1s}(\omega, q) = \frac{1}{2\omega}\int_{\Delta}^{\omega-\Delta} \{[g'(E, E') - 1]\,\sigma_{1n}(|\epsilon'| - |\epsilon|, q) + [g'(E, E') + 1]\,\sigma_{1n}(|\epsilon'| + |\epsilon|, q)\}\, dE \tag{39}$$

where

$$g'(E, E') = (EE' - \Delta^2)/|\epsilon\epsilon'| \tag{40}$$

In these equations, $E' = \hbar\omega - E$. E and E' are quasi-particle energies corresponding to normal-state energies ϵ and ϵ', which are measured relative to the Fermi level:

$$\epsilon^2 = E^2 - \Delta^2 \qquad \epsilon'^2 = E'^2 - \Delta^2 \tag{41}$$

The normal-state conductivity $\sigma_n = \sigma_{1n} - i\sigma_{2n}$ is calculated by the equation (*47*)

$$\sigma_n(\omega, q) = \frac{3ne^2}{4imv_0 q}\left[(S^2 - 1)\ln\left(\frac{S-1}{S+1}\right) + 2S\right] \tag{42}$$

where n is the electron number density and

$$S = \frac{\omega}{qv_0} - \frac{i}{ql} \tag{43}$$

To find $Z(\omega)$, one needs σ_2 as well as σ_1. This is calculated by a Kramers–Kronig transformation of σ_1.

Using the calculated values of $\sigma_s(\omega, q)$ and $\sigma_n(\omega, q)$, the surface impedance is determined by Eqs. (30) and (31). It should be noted that the only adjustable parameter is the gap width 2Δ. The only data on concentrated alloys with which such a calculation can be compared are the far-infrared data of Leslie and Ginsberg on lead alloys (*56*). These are shown in Fig. 5 for lead with 6% bismuth. The calculated values for specular electron scattering at the surface are too close to the values for diffuse scattering to show them separately. The figure shows that the extreme local limit is not well satisfied, even for a rather concentrated alloy. The data agree better with the numerical calculation than with the extreme anomalous limit. This is expected from Miller's results and was noted by Tinkham in a semiquantitative argument (*6*). For samples of lead with 1% thallium and 3% thallium, the results and the agreement between the numerical calculations and the data are similar, although the curves for the extreme anomalous limit and the extreme local limit are in considerably worse accord with the data.

The agreement with the calculated results is reasonably good, although the discrepancy may be significant. This may arise from an error in the method (*57*) of interpreting the data, but this seems unlikely. The discrepancy may arise from the strong electron–phonon coupling in lead, which usually results in deviations (*58,59*) from the BCS theory, a weak-coupling theory. Nam has shown (*60*) that this strong coupling would have a negligible effect for very large wave numbers q, but it might have an effect at some of the wave numbers

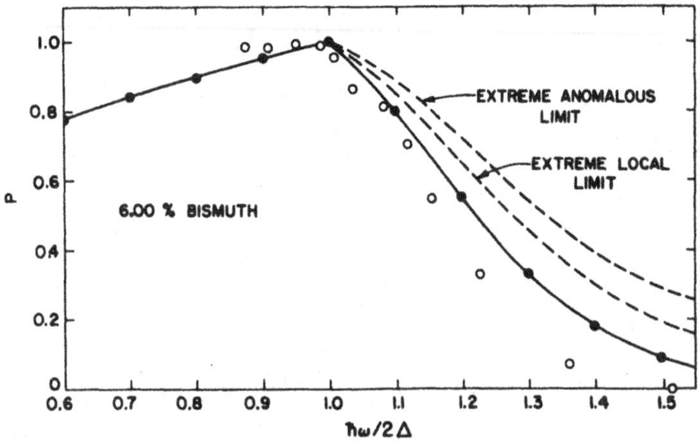

Fig. 5. Absorption curve for a lead–bismuth alloy. The open circles are the experimental data of Leslie and Ginsberg (*56*). P is the normalized value of the difference between the normal-state surface resistance R_n and the superconducting-state surface resistance R_s. The solid circles are the theoretical points for either diffuse or specular scattering of the electrons at the surface of the sample. The extreme anomalous limit and the extreme local limit are explained in the text; they are shown only at frequencies above the gap edge to avoid confusion. [After Ginsberg (*54*).]

which were important in the experiment. Calculating for the extreme anomalous limit, Nam has made other predictions about the electromagnetic properties of strong-coupling superconductors, but none of the experiments which have been done so far are sensitive enough to test the small predicted effects of strong coupling.

E. Measurements of Energy Gap Widths

We turn now to discuss briefly the determination of the energy gap width by electromagnetic absorption. The gap edge can be observed directly as a well-defined absorption edge in experiments on bulk homogeneous samples at temperatures low enough so that there are few thermally excited quasi-particles, i.e., if $\exp(-\Delta/kT) \ll 1$. At higher temperatures, there is no sharp onset of absorption, and the gap width can be determined with reasonable precision only if the data are fitted to theoretical curves. This has been done in the cases of aluminum and tin by Miller (*51*).

Anderson's theory of dirty superconductors (*53*) indicates that the introduction of impurities into a superconductor should make the gap width more isotropic, and therefore should steepen the absorption edge. This has been observed by Richards (*61*), as shown in Fig. 6. The bumps in the curves will be discussed below. [A sharp electromagnetic absorption edge in a concentrated alloy was first observed by Ginsberg and Leslie (*62*).]

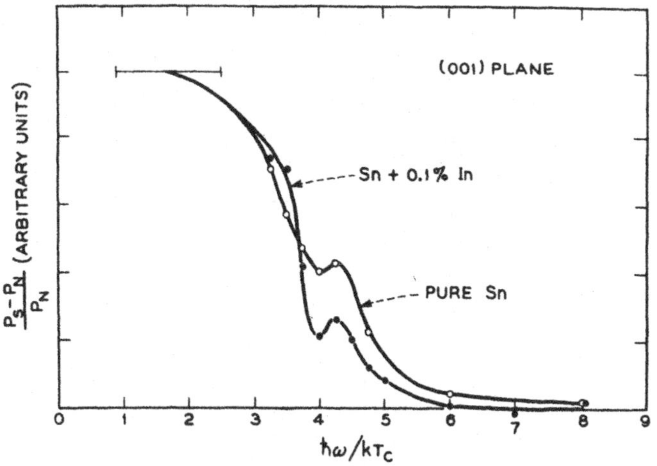

Fig. 6. Fractional difference between the power reaching the bolometer in the superconducting and normal states, as a function of frequency. This fractional difference should be approximately proportional to the difference between the normal-state surface resistance R_n and the superconducting-state surface resistance R_s. The horizontal bar indicates the approximate bandwidth of the lowest frequency point. The bandwidth for the other points was approximately 10%. [After Richards (*61*).]

TABLE I

Reduced Energy Gap Widths $2\Delta(0)/kT_c$ from Electromagnetic Absorption

Al	In	La	Pb	Hg	Nb	Ta	Sn	V	Ref.
							~ 3.5		*(26)*
3.37									*(15,51)*
	4.1 ± 0.2		4.1 ± 0.2	4.6 ± 0.2	2.8 ± 0.3	≤ 3.0	3.6 ± 0.2	3.4 ± 0.2	*(57)*
	3.9 ± 0.3		4.0 ± 0.5				3.3 ± 0.2		*(43)*
			4.37 ± 0.1						*(56)*
		2.85 ± 0.24							*(63)*

In Table I we list the gap widths that have been measured (*15,26,43,56,57,63*) by electromagnetic absorption.‡ These may be compared with values listed in Chapter 3, as measured by tunneling or specific heats, for example (*65*). The values listed here for niobium and tantalum seem low by comparison, possibly because of poor surface conditions. The values for the niobium alloys may be less sensitive to such an effect, since the waves penetrate more deeply into the samples. The agreement with the BCS value $2\Delta/kT_c = 3.53$ is qualitatively good.

The effect of crystal anisotropy is clearly seen in microwave experiments on tin (*5*) and aluminum (*20*), and in far-infrared experiments on tin (*61*). All these are close to the extreme anomalous limit, and the important electrons are those which spend a significant amount of time in the region close to the metal surface, where the electromagnetic field is to be found. In other words, almost all the electromagnetic energy is absorbed by electrons which have velocities which are nearly parallel to the surface plane. The measured surface impedance is then determined by an average over these electrons.§ This is in contrast to tunneling measurements, which investigate electrons with velocities nearly perpendicular to the surface (*66*) if the surface is sufficiently smooth. It is clear from this that tunneling is a more powerful tool than electromagnetic absorption for investigating energy gap anisotropy. Bennett (*67*) has considered the extensive tunneling data of Zavaritskii (*68*) on tin, and has concluded that they are qualitatively consistent with the far-infrared data of Richards. A more detailed calculation and comparison would be necessary for a semiquantitative comparison. It is interesting that Bennett finds theoretically that a major contribution to the anisotropy of the energy gap comes from the anisotropy of the phonon spectrum, with the anisotropy of the electron band structure contributing significantly only near the Brillouin zone edge.

F. Effect of a dc Magnetic Field

We turn now to a consideration of the effect of a dc magnetic field on the surface impedance of a type I superconductor. There have been several microwave experiments performed at frequencies much less than $2\Delta/h$, all on tin or aluminum (*17–19,21,22,52*). Some of the samples were very thin wires, and others were flat plates. It has been observed that the surface impedance R and the surface reactance X may either increase or decrease as the dc field is increased. The observed shifts are very small, on the order of a few per cent, as a result of the long-range order which characterizes a superconductor; the region of the superconductor which lies beyond the penetration depth has an

‡ Gap widths have also been measured in lead alloys (*56*) and in niobium alloys (*64*).

§ A cosine factor comes into this average from the vector dot product in Eq. (19).

effect on the electrons near the surface, and prevents the energy gap there from being strongly perturbed by the magnetic field. The variations of R and X depend qualitatively on the microwave frequency, the angle between the microwave magnetic field and the dc magnetic field, the crystal orientation, the temperature, and the electron mean free path. Any calculation which does not take all these factors into account cannot explain all the data.

Dresselhaus and Dresselhaus were the first to try to calculate (*69*) the effect. They used a two-fluid model, in which the superconducting electrons obeyed London's equations rather than having a nonlocal dependence on the electromagnetic field. The crystal anisotropy was ignored, and specular reflection of the electrons at the surface was assumed. The calculation explained some of the results of Spiewak (*17*) on tin, but later data at a somewhat higher frequency by Richards (*19*) disagreed qualitatively with the predictions. It is not unlikely that the main defect of the calculation lies in its failure to take into account the nonlocal response of a superconductor to a perturbation, which is indicated by the BCS theory.

A calculation which should be more closely related to the BCS theory has been performed by Maki (*70*). His treatment is limited to pure metals which are close to the extreme anomalous limit and to frequencies $\omega \ll \Delta(T)/\hbar$. The central idea is the well-known fact that the persistent currents which are induced by the dc magnetic field introduce an electron drift velocity $\mathbf{v}_D$, and that the kinetic energies of the normal-state electron wave functions are therefore shifted by an amount $\mathbf{p}\cdot\mathbf{v}_D$, where $\mathbf{p}$ is the electron momentum. The superconducting state is then formed from these perturbed states. Maki uses the powerful Green's function techniques of Abrikosov et al. (*35*). The results of the calculation indicate that the change in surface reactance induced by the dc field should be positive at all temperatures, and that the change in surface resistance should always be positive at low temperatures. This is in disagreement with some of the experimental data. In the calculation, the vector potential is taken as uniform close to the surface, and this is known not to be the case. Although Maki expects this not to make much difference, this is not explicitly proved. Crystal anisotropy is not taken into account at all. It is clear that no firm conclusion can be drawn about the relation of these phenomena to the BCS theory at the present time.

Budzinski and Garfunkel (*23*) have observed the effect of a dc magnetic field on the absorption of microwave energy at much higher frequencies, for which $\hbar\omega$ reached values greater than the gap width. The results are shown in Fig. 7. It is seen that the absorption edge, at a frequency of about 80 GHz, is smeared out by the dc field. They interpret this in terms of the $\mathbf{p}\cdot\mathbf{v}_D$ shift in energy which was mentioned above. One might expect the resulting effect to be small, for the following reason. The absorption of a photon, which typically has

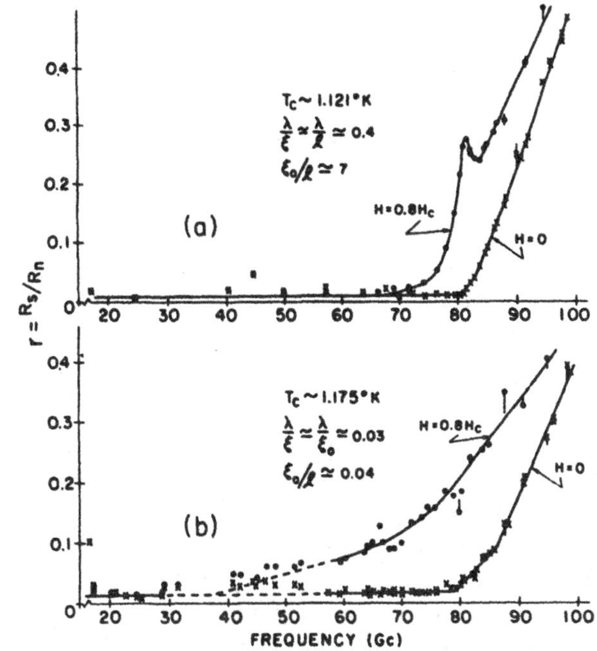

Fig. 7. Surface resistance ratio as a function of frequency for (a) silver-doped aluminum and (b) pure aluminum. In both cases, the dc magnetic field H is parallel to the radiofrequency magnetic field. Crosses are for $H=0$; circles are for $H=0.8\,H_c$. The data were taken for $T=0.34\,T_c$. [After Budzinski and Garfunkel (*23*).]

momentum on the order of $\hbar$ times the inverse of the penetration depth, is much less than p, the electron momentum. The pair of quasi-particles which are excited from the ground state of the superconductor must therefore have momenta which are almost exactly opposite in direction, so the shift in the energy of one of the quasi-particles should be almost exactly canceled by the shift in the energy of the other. However, if the electrons scatter diffusely from the surface of the metal, then this can contribute momentum parallel to the surface during the photon absorption, and the near cancellation may therefore be destroyed. The upper part of the figure shows that a doping with 0.2% silver makes the gap sharp again in the presence of the dc magnetic field, presumably because this doping level is high enough to wash out the gap anisotropy created by the $\mathbf{p}\cdot\mathbf{v}_D$ perturbation.

A semiquantitative calculation of this effect has been performed by Pincus (*71*). He points out that the perturbation introduced by the dc field is confined to a region near the sample surface, because of the Meissner effect. It is found that excited electron states may exist which are spatially bound to the surface, and which have energies that are less than the zero-field gap width.

He also considers the role of thermally excited surface-bound states. He estimates that transitions from these states to the continuum of states above the gap may have a dominating influence at temperatures which are comparable to T_c. An order of magnitude estimate then indicates that the observed effect may be consistent with this picture. He also speculates that transitions from one bound state to another may be important in explaining the low-frequency data which we considered above.

G. Precursors

We turn finally to the possibility of a singularity which appears to be a small amount of electromagnetic absorption below the main gap edge, the "precursors" in bulk samples. (The possibility of a precursor for very thin lead films was considered in Section C.) It has been thought by Richards and Tinkham and by Leslie and Ginsberg that such precursors are present for lead (*56,57*), lead alloys (*56*), and mercury (*57*). However, recent data of Norman and Douglass (*72*) indicate that the gap edge in lead is well defined, and that there is very little if any absorption below the gap edge in lead. This observation has been made by measurements of electromagnetic absorption, and therefore bears directly on the question, whereas the other measurements under discussion were of reflection. The recent data of Norman and Douglass have very little scatter, and only a systematic error make them consistent with a sizable absorption precursor. Their measurements were made on films of thickness 1400, 6000, and 54,000 Å. These are much larger than the penetration depth and the coherence length. The samples should therefore probably be considered as bulk samples, although they may not be similar to bulk samples in all respects. One cannot rule out the possibility that the precursor, if it exists, is associated with crystalline anisotropy, and that its absence in films indicates that the electron mean free path close to the films' surfaces is short enough to wash out the anisotropy. It is also not known whether the precursor might be affected by the strains which must have been introduced by differential thermal contraction during the cooling of the films of Norman and Douglass. These authors speculate that impurity of radiation may account for a spurious precursor, but this explanation is inconsistent with the observation by Richards and Tinkham that the precursor appears at a frequency which is temperature-dependent. It is not known whether the observation by Richards (*61*) of an absorption peak above the gap in tin, shown in Fig. 6, is due to anisotropy. The evidence that the peak is stronger for some crystalline orientations than for others may point to an anisotropy effect. It is probably premature to declare the question settled. Some of the theoretical implications of a precursor are discussed in Chapter 7.

V. ULTRASONIC ATTENUATION

A. Introduction

In one respect, ultrasonic attenuation is simpler than electromagnetic absorption; if the ultrasonic wave has a single frequency, then its wavelength is well defined, because the wave is attenuated appreciably only in distances which are long compared with the wavelength. In other respects, however, this phenomenon is more complicated. The wave may be either transverse, longitudinal, or (if the propagation is not along a symmetry direction) a mixture of the two. In the case of transverse waves, several mechanisms for absorption are important, some of which are case 1 and some case 2, with respect to the coherence factor. The situation is simpler for longitudinal waves, and we consider these first. After considering transverse waves, we will discuss the contribution of dislocation damping to acoustic attenuation. An unexplained anomaly in the data for lead and mercury will then be described, and finally we will discuss measurements of the energy gap width in various materials. We will assume that the phonon energy $\hbar\omega$ is much less than 2Δ. Although some calculations have been performed (*44–46*) for the case $\hbar\omega > 2\Delta$, no data are available for comparison. The early work in this field has been reviewed by Morse (*73*). The relevant theory for normal metals was developed largely by Pippard (*74–76*), Steinberg (*77*), Blount (*78*), Holstein (*79*), and by Cohen et al. (*80*).

B. Longitudinal Waves

For longitudinal ultrasonic waves, the only important interaction between the electrons and the waves is due to the change in the crystal potential which accompanies the wave. (This may be treated by means of a deformation potential.) The interaction is not changed by time reversal, so the relevant coherence is case 1. The attenuation should fall as the temperature T is lowered below the critical temperature T_c, and should be proportional to $\exp(-\Delta/kT)$ if $T \ll T_c$, according to Eq. (17). The physical reason for this is that the phonons associated with an ultrasonic wave have energy $\hbar\omega < 2\Delta$, and can be absorbed only by thermally excited quasi-particles.

A calculation of the expected ratio of the attenuation constant α_s in the superconducting state to that in the normal state α_n was first performed by Bardeen, Cooper, and Schrieffer for the case $ql \gg 1$, where q is the wave number of the acoustic wave and l is the electron mean free path. Their result was

$$\alpha_s/\alpha_n = 2f(\Delta) \tag{44}$$

where f is the Fermi function, as before. Kresin (*81*) extended the validity of

Eq. (44) to the case where $ql \ll 1$, and Tsuneto (*82*) later showed it to be valid for all values of ql. As the temperature decreases below T_c, Eq. (44) indicates there should be a rapid decrease of α_s/α_n. This was first observed by Bommel (*83*) and by Mackinnon (*84*). Not all the experiments which have been performed indicate that α_s/α_n approaches zero at low temperature. This is presumably because of one or more attenuation mechanisms which do not involve the conduction electrons. In general, the experimental curves tend to fall off somewhat more rapidly than expected near T_c. A typical case, the data of Morse and Bohm (*85*), is shown in Fig. 8. The discrepancy between the theory and the data can be understood in terms of the effects of dislocations (see below) or in terms of an energy gap function which, compared with its value at $T=0$, rises slightly too rapidly as the temperature decreases just below T_c. This is the same feature which was mentioned in Chapter 3, and which has been indicated by measurements of tunneling and penetration depths. It may be a simple result of anisotropy. If the energy gap varies with the direction of the electron velocity, then the smaller energy gap regions will contribute most heavily at low temperatures (*44*). The apparent energy gap will then be smaller when $T \ll T_c$ than would be expected from its value when $T \lesssim T_c$. The result would be a discrepancy of the observed type. The quasi-particles which are effective in attenuating an acoustic wave are those which travel along in phase with the wave, because

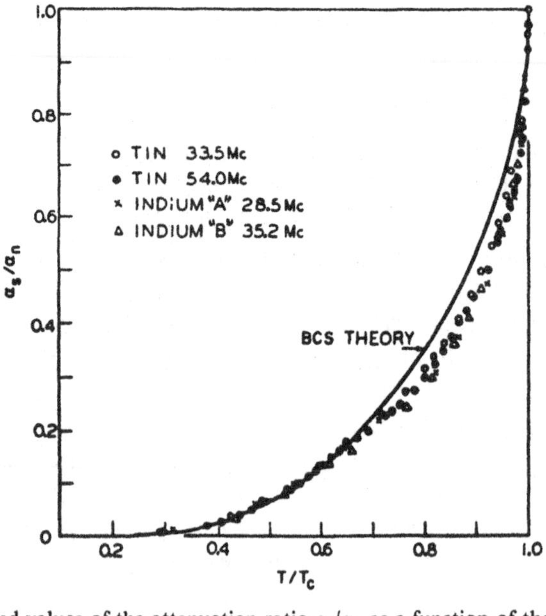

Fig. 8. Measured values of the attenuation ratio α_s/α_n as a function of the reduced temperature, compared with the theoretical BCS variation, assuming $\Delta(0) = 3.5kT_c$. [After Morse and Bohm (*85*).]

of the requirements of energy and momentum conservation. Because the electron velocity is so large compared with the sound velocity, these electrons have a velocity vector which is almost orthogonal to the acoustic propagation direction. The observed attenuation is then determined by an average over these electrons. If the gap varies with azimuthal direction, then an average energy gap is effectively at work. If ql is not $\gg 1$, then the average is over a larger region of momentum space, as a result of the uncertainty in the electron momentum (*86*). The effect of this has been seen in niobium by Dobbs and Perz (*87*). In principle, electron tunneling provides a much more powerful tool than ultrasonic attenuation for investigating the anisotropy of the energy gap, because the only electrons which contribute in a tunneling experiment are those with velocities very nearly perpendicular to the surface, if it is sufficiently smooth (*66*).

C. Shear Waves

The theoretical treatment of ultrasonic shear waves is far more complex. Because the electrons do not move exactly in phase with the ion cores, a transverse electromagnetic wave accompanies the ultrasonic wave, and the absorption processes are not all of case 1 coherence or case 2 coherence. The electrons which absorb energy are more highly selected than for longitudinal waves, because the electrons with velocities in the direction of the wave polarization contribute more strongly than those with velocities which depart significantly from this direction. As for longitudinal waves, only electrons with velocities nearly perpendicular to the wave propagation direction contribute if $ql \gg 1$. For shear waves, three separate mechanisms for energy absorption may be important. These are usually called the electrodynamic, collision drag, and shear deformation mechanisms. If two or more of the three effects are present, they are not expected to be additive (*88*). We will briefly describe each one, and the most relevant experiments.

Energy may be transmitted to the electrons by the electromagnetic wave which accompanies the acoustic wave. In superconductors, these so-called electrodynamic losses are important only near the critical temperature. At low temperature, the lossless part of the electrical conductivity of the superconductor screens the field very effectively, and the thermally excited quasiparticles therefore do not absorb much energy. As a result, the ultrasonic attenuation of shear waves decreases very rapidly with temperature just below T_c. Morse and Bohm (*89*) first observed this in aluminum as an apparently discontinuous drop in the attenuation at the critical temperature. If the earth's magnetic field is canceled out, this sudden decrease takes place in a temperature region of finite width δT, as first observed by Claiborne and Morse in alumi-

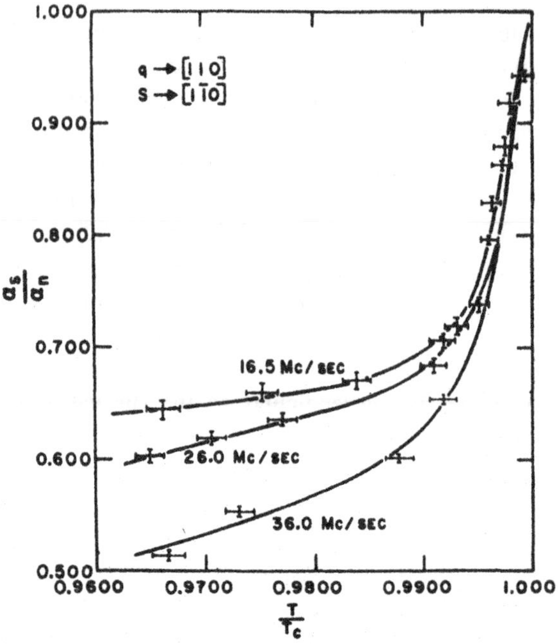

Fig. 9. Measured values of the attenuation ratio α_s/α_n as a function of the reduced temperature. The crystal orientation is [110], and the phonon propagation is in the [110] direction. [After Claiborne and Morse (*90*).]

num. (*90*). Their data are shown in Fig. 9. [The electrodynamic losses became negligible below about the temperature for which the penetration depth equals the wavelength of the acoustic wave (*91*).] They performed a calculation of this effect, using a Boltzmann equation approach, and using the London equations for the superconductor, as is appropriate, since $q\xi_0 \ll 1$. The calculated width δT agreed well with the data.

The second mechanism for attenuating transverse waves is collision drag. In the presence of a sound wave, the lattice velocity at any given location oscillates. The electron velocities relax toward the local, instantaneous lattice velocity by means of collisions with impurities, and these collisions remove energy from the sound wave. This process can be important for transverse wave attenuation if the electron mean free path is not very large. In normal metals, attenuation by collision drag is negligible compared with the electromagnetic damping, except at very high frequencies ($\gtrsim 10^9$ Hz). It was first observed in the superconducting attenuation by Claiborne and Morse (*90*) in the temperature region below that at which the electrodynamic contribution disappeared. Levy (*92*) showed theoretically that Eq. (44) should be valid for this contribution to the attenuation if $ql \ll 1$. Recently Yokota et al. (*88*) have treated the

$ql \gg 1$ case and showed that

$$\alpha_s/\alpha_n = 2f(\Delta)/ql \tag{45}$$

This result agrees well with the appropriate limit of the semiclassical calculation of Claiborne and Morse, and with their data for aluminum. The temperature dependence indicated by Eq. (45) is the same as in Eq. (44), because this process has the same coherence factor, but now a dependence on wave number and mean free path has appeared.

The third of the three attenuation mechanisms which are important for shear waves is the shear deformation contribution and is the most difficult to treat. A calculation of this effect can be made only if the topology of the Fermi surface is known; it would be absent if the Fermi surface were spherical. It is a result of the distortion of the shape of the Fermi surface which results from the lattice strains which accompany an acoustic wave. The electrons relax toward the Fermi surface which is appropriate for the local state of the lattice. The attenuation created by this mechanism is proportional to q^2, and it is more important than collision drag at high frequencies. Leibowitz (*93*) first observed it in a superconductor, tin, which has a highly nonspherical Fermi surface. At temperatures below that at which the electromagnetic damping becomes small, the attenuation ratio had a temperature dependence which followed Eq. (44) rather closely. His claim to the observation of a shear deformation effect in superconducting tin rested on the fact that the observed attenuation was larger than expected from collision drag alone, and had a different dependence on ql.

From what we have said, it should be evident that the acoustic attenuation of shear waves should have the temperature dependence described by Eq. (44), if the temperature is not too near T_c. The experiments indicate that the attenuation falls slightly more steeply than expected as the temperature diminishes. This is similar to the case for longitudinal waves, and the interpretations which were described above may presumably also be valid for transverse waves. Recently Yokota et al. (*88*) have calculated the acoustic attenuation of shear waves in aluminum, using the known atomic pseudo-potential. Their results agree reasonably well with the data of Claiborne and Morse.

D. Anisotropy

Kadanoff and Pippard (*91*) have given a unified treatment of ultrasonic attenuation in normal and superconducting metals for the case where the electron scattering is elastic scattering from impurities. Their conclusions are consistent with the sketch which we have presented here. In principal, they have taken anisotropy into account, as have other theorists (*44,66,67,88,94,95*), but in fact there has never been a definitive comparison of anisotropy effects with theory, because the required parameters have not been sufficiently well known. Anisotropy was first observed in tin by Bezuglyi et al. (*96*) and by Morse et al. (*97*). Theory (*44*) indicates that, for any given directions of propagation and polarization, $\Delta(T)/\Delta(0)$ should agree well with the BCS curve for this function. The decrease in anisotropy which is expected when the electron mean free path is shortened by impurities has been observed by Claiborne and Einspruch (*98*) for indium-doped tin.

E. Amplitude Dependence

We turn now to a brief description of a dependence of the ultrasonic attenuation ratio α_s/α_n on the amplitude of the acoustic wave. This dependence was first observed by Love et al. in lead (*99,100*). They found that the temperature dependence of the attenuation agrees rather well with the BCS expression [Eq. (44)] if the amplitude of the wave is small. However, the attenuation in the superconducting state at a fixed temperature increases as the amplitude is increased, both for transverse and longitudinal waves. For very high amplitudes the attenuation decreases again but remains larger than expected. This is shown in Fig. 10. Love and Shaw proposed two possible explanations of this. One was that the effect might be connected with the unusually strong electron–phonon coupling in lead. This explanation is apparently not correct, because the effect is weak or absent for mercury (*101*), in which the coupling is also strong, but present for indium (*102*), in which the coupling is weak. The other proposed explanation was that the effect is due to dislocation damping. This is the presently accepted view. Mason (*103,104*) showed that the magnitude of the effect is consistent with the theory of dislocation damping introduced by Granato and Lücke (*105*). This theory treats the process in which an acoustic wave of sufficiently large amplitude breaks dislocations away from pinning sites. This process requires energy, and it attenuates the ultrasonic wave. Mason found that the effect of the heavy damping of dislocations in the normal state makes the amplitude dependence much smaller than in the superconducting state. This explains why the effect may be too small to be seen above the critical temperature. One expects the amplitude dependence in the superconducting state to be strong in metals for which the pinning force is weak, as in lead. This interpretation of the effect was also put forward independently by Bezuglyi et al. (*102*) and by Tittmann and Bommel (*106*). The latter investigators showed that in lead the amplitude dependence diminishes with increasing impurity concentration but is approximately independent of frequency, as expected at the frequencies which were used. One expects the amplitude dependence to decrease for very large amplitudes, as observed. This occurs for amplitudes which are large enough so that the dislocations are broken away from a large fraction of the pinning sites.

This phenomenon seems to have great promise in the study of dislocations. It is consistent with the BCS theory but evidently does not provide a stringent test for it. Of course, measurements of the energy gap by ultrasonic attenuation must be carried out at low amplitude if Eq. (44) is to be used. Even if the amplitude of the acoustic wave is small and the dislocations are not unpinned

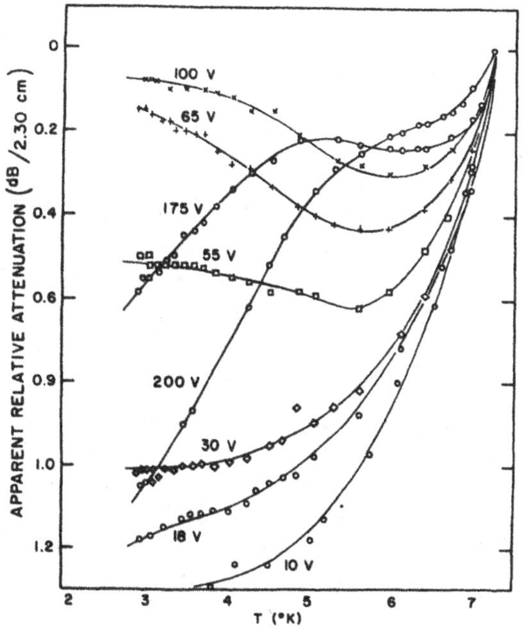

Fig. 10. Measured relative attenuation in the superconducting state as a function of temperature for various voltages across the transducer. [After Love and Shaw (*99*).]

by the wave, interactions between the electrons and the dislocations may contribute to the attenuation. This was pointed out by Mason (*104*). He also has shown that the observed deviation of α_s/α_n from the BCS behavior of Eq. (44), as shown in Fig. 10, may be due to this effect, and that the estimated energy gap values are very sensitive to it, because the contribution of dislocation damping to α_s/α_n is temperature-dependent. The damping due to dislocations would be smaller if the experiments were performed at much higher frequencies. Dislocation damping for waves of small amplitude can also be decreased by exposing the samples to a very slight amount of radiation damage, which can pin the dislocations (*107,108*).

F. Anomalous Behavior of Lead and Mercury

When the amplitude dependence is absent, the attenuation of ultrasonic waves in lead (*109*) and mercury (*101*) is in good agreement with Eq. (44) at low frequencies, where $ql \gtrsim 1$, but is in poor agreement when $ql \gg 1$. This was discovered by Deaton in lead, and is shown in Fig. 11. The reason for it is not known. A calculation by Ambegaokar (*110*), taking into account the strong electron–phonon coupling in lead, indicates that the effect theoretically does not exist. It is the only simple phenomenon in ultrasonic attenuation which is in qualitative disagreement with calculations based on the BCS theory at the present time.

G. Measurements of Energy Gap Widths

The energy gap has been measured in various substances, using Eq. (44) to interpret the results. In determining the gap width, extremely high precision is required in the measurements. The uncertainty in the result must include a contribution from the estimate of other sources of attenuation if they are present (*104*). The results are a function of the direction of propagation and polarization, and depend on the electron mean free path. We will therefore not attempt to tabulate the data here, but will refer the reader to the literature for

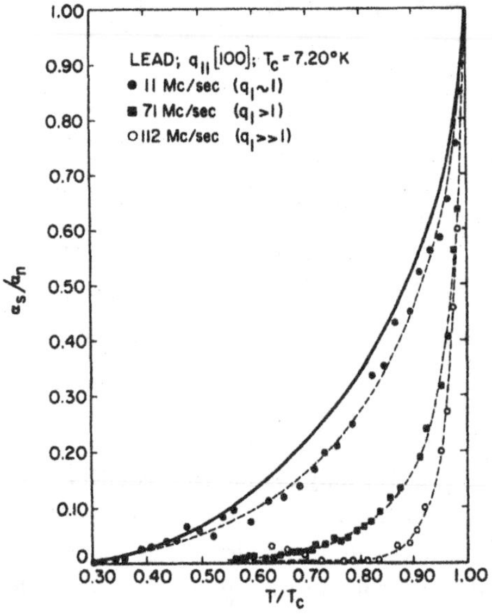

Fig. 11. Normalized longitudinal wave attenuation in lead along the [110] direction for three ultrasonic frequencies. The solid curve represents the BCS expression, assuming $2\Delta(0) = 3.5\ kT_c$. [After Deaton (*109*).]

the relevant experiments on aluminum (*89,90,111–114*), indium (*85,102*), lanthanum (*115*), lead (*83,99,100,106,109*), mercury (*101,116,117*), molybdenum (*118*), niobium (*87,119–124*), tantalum (*119,125*), tin (*84,85,93,96,97,126–132*), thallium (*133,134*), vanadium (*119,135,136*), zinc (*135*), niobium–zirconium alloys (*137–139*), and tin–indium alloys (*98*). A tabulation has been published (*65*) of some of the data which were obtained prior to 1963. The results indicate that the reduced gap width $2\Delta/kT_c$ lies in the range from about 3.1 to 4.3. This should be compared with the BCS value of 3.5.

VI. THERMAL CONDUCTIVITY

A. Introduction

The thermal conductivity κ of a metal is the sum of two parts; one is the contribution κ_e of the electrons, and the other is the contribution κ_g of the phonons:

$$\kappa = \kappa_e + \kappa_g \tag{46}$$

In the normal state, the electronic term κ_{en} is given by

$$1/\kappa_{en} = \alpha T^2 + (\beta/T) \tag{47}$$

The first and second terms arise from the scattering of electrons by phonons and by static imperfections, respectively. Also in the normal state, the phonon term κ_{gn} is given by

$$1/\kappa_{gn} = (A/T^2) + (B/T^3) \tag{48}$$

The A and B terms arise from the scattering of phonons by electrons and by static imperfections (grain boundaries and the crystal boundaries), respectively. If there is also appreciable scattering of the phonons by dislocations or point defects, a more complicated temperature dependence is expected. In the superconducting state, the same processes are at work, but their temperature dependences, except for the scattering of phonons by static imperfections, are different from those found in the normal state (*140*). It has been customary to assume that the phonon conductivity is not limited appreciably by scattering from dislocations in either the normal or the superconducting state, but this may not be true in all cases (*104,105,141*). The most definitive information about superconductivity is obtained in experiments where either the electronic contribution κ_{es} or the phonon contribution κ_{gs} is dominant, and where only one of the two types of scattering centers for the dominant carrier is important. We will restrict our attention to these cases, and refer the reader to the literature, including review articles (*141–146*), for the more complex situations. We will cite only a small fraction of the articles in this field, which abounds with a huge number of publications.

B. Phonon Conduction

Let us begin by considering a superconductor in which $\kappa_{gs} \gg \kappa_{es}$. At temperatures for which $\exp(-\Delta/kT) \ll 1$, there are very few quasi-particles to scatter the phonons, so one expects κ_{gs} to have the form T^3/B. This has been seen in many metals, for instance by Olsen and Renton (*147*) in lead, by Mendelssohn and Renton (*148*) in tin and indium, by Goodman (*149*) and Laredo (*150*) in tin. The observed thermal conductivity has the required dependence on sample dimensions, and is approximately independent of impurity content, as expected. In cases where κ_{gs} is mainly limited by the scattering of phonons by electrons, the phonons do not dominate in the heat conduction, so there are no examples in which the A/T^2 term in Eq. (48) is the only important one.

C. Electron Conduction

For samples in which $\kappa_{es} \gg \kappa_{gs}$, the heat is carried in the superconductor by the excited electrons (quasi-particles). These can be scattered almost entirely by either static imperfections or by phonons. The former case is by far the easier one to treat theoretically; this was done by Bardeen et al. (*151*), and by Geilikman and Kresin (*152*). We will consider this first, and then discuss the latter case.

When heat is being transported mainly by the quasi-particles and these are scattered by static imperfections, such as impurities, lattice defects, and grain boundaries, the scattering process is invariant under time reversal, so the case 1 coherence factor is appropriate. This has been used (*151*) to derive the transition probabilities. These are put into the Boltzmann transport equation, which is then solved. The result can be expressed in closed form:

$$\frac{\kappa_{es}}{\kappa_{en}} = \frac{2F_1(-y) + 2y\ln(1+e^{-y}) + y^2/(1+e^y)}{2F_1(0)} \tag{49}$$

where

$$y = \Delta/kT \tag{50}$$

and

$$F_1(-y) = \int_0^\infty z\,dz/(1+e^{z+y}) \tag{51}$$

The experimental data for aluminum (*146,153*), indium (*140,154*), tin (*140*), and zinc (*146*) fit well (*155*) to Eq. (49). (The conduction is expected to be more purely electronic in aluminum and zinc than in indium and tin.) In Fig. 12 are shown the data of Satterthwaite (*153*) on aluminum. The theoretical curve with the reduced gap width $2\Delta(0)/kT_c = 3.53$ (the BCS value) is a good fit to the data. The actual reduced gap width is not accurately known (*65*), but may be not far from this value. At very low temperatures, Eq. (49) indicates that κ_{es}/κ_{en} should be proportional to $\exp(-\Delta/kT)$, as we expect from Eq. (17).

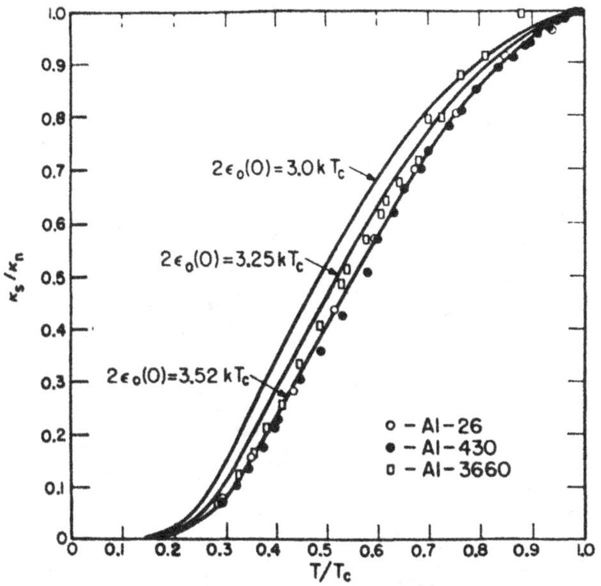

Fig. 12. Measured ratio of the superconducting to the normal thermal conductivity for aluminium. The theoretical curves are shown for the indicated values of the gap width $2\Delta(0)$ at $T=0$. [After Satterthwaite (*153*).]

This type of temperature dependence was first noticed by Goodman (*149*) for tin, and provided some of the earliest evidence for an energy gap. A similar temperature dependence has since been seen in many superconductors. Of course, if the temperature is too low, then the conduction of heat is due mainly to the phonons, and κ_{es} becomes less certain.

We turn now to the case where $\kappa_{es} \gg \kappa_{gs}$, and where the quasi-particles are scattered mainly by the phonons. This occurs near the critical temperature T_c in lead (*156,157*) and mercury (*140,157*), because they have very low Debye temperatures. The temperature dependence of κ_{es}/κ_{en} is quite different from that for impurity-limited conduction, as one can see in Fig. 13. In the present case, as T diminishes below T_c, κ_{es}/κ_{en} immediately falls off at a large rate. A successful theoretical calculation of this phenomenon has not been performed until recently. Ambegaokar, with Tewordt (*158*) and with Woo (*159*), has calculated it for lead. The calculation involves such real-metal effects as the actual phonon spectrum in lead and the dependence of the quasi-particle energy and lifetime on momentum and temperature. The calculation is far too complicated to be described here. The solution of this problem has been achieved by a succession of theorists, building on the BCS theory. By using the results of

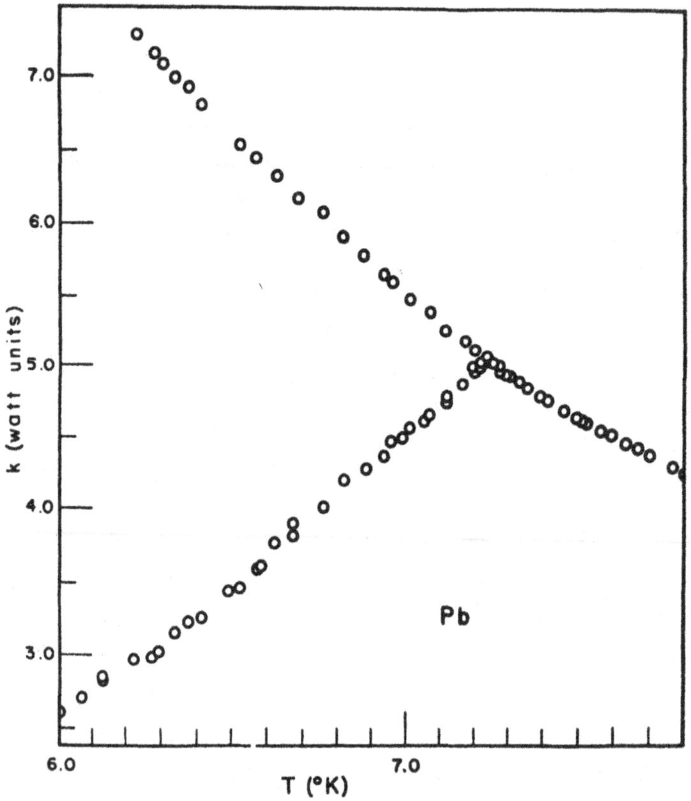

Fig. 13. Measured values of the thermal conductivity of lead as a function of temperature. Below the critical temperature, the upper points are for the normal state (in a magnetic field), and the lower points are for the superconducting state. [After Watson and Graham (*157*).]

tunneling experiments, they have been able to introduce a model which is apparently more realistic, and unfortunately much more complicated, than that of BCS.

In principle, effects of crystal anisotropy should be detectable in measurements of thermal conductivity. Such effects have actually been seen, for instance by Zavaritskii (*160*) in gallium and by Guenault (*161*) in tin. Although

Khalatnikov (*162*) has considered the theory of anisotropy effects at very low reduced temperature, there is apparently no way of rigorously comparing the theory with the data. A quantitative phenomenological solution of this problem would probably require knowledge of a mind-boggling number of parameters.

VII. NUCLEAR SPIN RELAXATION

A. Introduction

Conduction electrons strongly influence the nuclear magnetic resonance in a normal metal through the interaction of their magnetic moments with those of the nuclei. Usually, the most important term is the contact hyperfine interaction, which gives rise to two types of effects. First, the electron magnetic moment alters the static magnetic field seen by the nuclei, the resulting change in NMR field being referred to as the Knight shift. Second, energy exchanges resulting from spin–spin collisions with conduction electrons at the top of the Fermi distribution usually provide the quickest means for the nuclear spins to come into thermal equilibrium with their surroundings, a process characterized by the nuclear spin-lattice relaxation time T_1.

The Meissner effect makes it difficult to carry out conventional magnetic resonance experiments in superconductors since the necessary magnetic fields are excluded except for a thin region approximately 10^{-5}–10^{-6} cm thick. However, by using a field cycling method, Hebel and Slichter (*163*), and also Redfield (*164*) and Masuda (*165–167*), were able to make measurements of nuclear spin-lattice relaxation in superconducting metals at low temperatures. To obtain the T_1 characteristic of the superconducting phase the nuclear spin resonance was observed in the normal phase but the nuclei were allowed to relax in the superconducting state. The following cycle was used. First, the nuclear spins were allowed to come to thermal equilibrium in a static magnetic field strong enough so that the metal was in the normal phase. Typically, 500–5000 G was used. To begin the measurements the magnetic field was quickly turned to zero, the metal becoming superconducting. The nuclear spin relaxation was then allowed to take place during the time t that the field was off. Next, the field was turned on quickly, the metal returning to the normal phase, and the nuclear resonance signal was observed "on the run" as the magnetic field passed through the nuclear resonance value.

The nuclear spin-lattice relaxation time characteristic of the superconducting phase could thus be obtained by studying the height of resonance signals so obtained vs. t, because the height of the resonant signal was a direct measure of the amount of nuclear magnetization left after relaxation. To discuss the comparison of such resonance signals observed at zero field in both the normal and superconducting phases of the metal at low temperatures, it is necessary to examine the spin-lattice relaxation process (*168*) in some detail. This will lead to a theoretical expression from which one can derive the ratio of normal to superconducting relaxation times. The final expression will turn out to be that discussed in Section III, so that a direct comparison can be made of the experimental results with theoretical calculations.

B. Analysis of the Field Cycle Procedure to Measure T_1

The nuclear magnetization is usually the convenient variable for characterizing nuclear spin-lattice relaxation processes, because under normal circumstances the height of a nuclear magnetic resonance signal is directly proportional to the nuclear magnetization present. In thermal equilibrium the magnetization M is given by Curie's law:

$$M_0 = CH/\Theta_L \tag{52}$$

where M_0 is the thermal equilibrium nuclear magnetization, H is the applied static magnetic field, Θ_L is the lattice temperature, and C is the nuclear Curie constant. When the nuclear magnetization is not in thermal equilibrium, energy is exchanged between the spin system and the lattice; the magnetization relaxes toward M_0 following the equation

$$dM/dt = (M_0 - M)/T_1 \tag{53}$$

where T_1 is the spin-lattice relaxation time characteristic of energy transfer.

In looking at Eq. (52) one can see that there is a problem in applying such a method to the case of a superconductor since the relaxation will be taking place at zero magnetic field where the magnetization is characteristically zero in thermal equilibrium at all values of temperature. Such is also the case for the nonthermal equilibrium, as will be discussed later in this section and in (*168*). Thus it is necessary to use an alternative method of treatment in terms of the nuclear spin temperature. Under conditions of thermal equilibrium the nuclear spin system is characterized by the lattice temperature; that is, the energy levels of the nuclear spin system are populated according to Boltzmann statistics given by the expression

$$p_n^0/p_m^0 = \exp\left[(E_m - E_n)/k\Theta_L\right] \tag{54}$$

where p_n^0 and p_m^0 are the thermal equilibrium populations in the nth and mth nuclear energy levels with energies E_n and E_m, respectively. When not at thermal equilibrium with the lattice, the nuclear spin system is still often characterized by a temperature different from that of the lattice temperature. That is, the populations are given by the equation

$$p_n/p_m = \exp[(E_m - E_n)/k\Theta_s] \tag{55}$$

where Θ_s is the spin temperature. When a magnetic field H is present, an alternative and equivalent definition of Θ_s can be given in terms of Curie's law. That is,

$$\Theta_s = CH/M \tag{56}$$

Comparing the equations we see that in the presence of the field

$$\frac{d}{dt}\frac{1}{\Theta_s} = \left(\frac{1}{\Theta_L} - \frac{1}{\Theta_s}\right)\frac{1}{T_1} \tag{57}$$

Thus instead of characterizing nuclear spin-lattice relaxation in terms of the growth or decay of magnetization, one can use an alternative treatment in terms of the warming up or cooling down of the spin temperature as it approaches that of the lattice. From the two equations one can see that the same T_1 characterizes either treatment. It is the second form, the one characterized by spin temperature, which we can apply at zero field and is the natural form for treating nuclear spin relaxation in a superconductor.

In a strong magnetic field nuclear spin temperature measures the extent to which nuclear moments line up along rather than opposed to the external field. In zero external field the nuclei still experience magnetic fields due to their neighbors. The concept of spin temperature thus describes the degree of alignment of nuclei in their individual local fields. The fact that the local fields at various parts of the sample have basically random orientations causes the bulk nuclear magnetization, M, to be zero. It is easy to have a spin system at zero field characterized by a spin temperature which differs from Θ_L. Such a situation could be obtained, for example, as a result of adiabatic demagnetization from a strong field to zero field carried out in a time short compared with T_1 and is a part of our field cycling procedure for measuring T_1, as mentioned in the introduction. A full description of relaxation at zero external field would require as its starting point a solution of the energy levels and wave functions of the nuclei which are coupled together by their local dipole magnetic fields. Such a solution has never been obtained. However, the use of a spin temperature enables one to formulate the relaxation time calculation in terms of diagonal sums whose evaluation does not require solution of the exact energy levels. Thus with the aid of the spin temperature assumption one can calculate the spin-lattice relaxation time completely. In addition, relaxation at zero field is very easily pictured in terms of temperature changes.

The arbitrariness of the spin temperature assumption for this problem has been discussed at length in (*163*) and (*168*). It has been very fruitful in discussing cooling done by adiabatic demagnetization in many experiments and has been found in the case of relaxation in metals to provide an excellent description of the process at zero field as well as at higher fields. It will be used in the course of this discussion to provide a complete picture of relaxation in the normal and superconducting phases. The reader is referred to (*168*) for a more detailed discussion of the concept of nuclear spin temperature and the limitations of its application in describing relaxation processes.

To show the relationship between the nuclear magnetization measured in the normal state and the relaxation taking place in the superconducting state, it is necessary to introduce the characteristic expressions of nuclear magnetization. If the nuclear spin Hamiltonian is taken to consist of the nuclear Zeeman interaction $\mathscr{H}_z$ and the nuclear dipole–dipole interaction $\mathscr{H}_{dd}$, the spin-lattice interaction being treated as a perturbation, then the expression for the magnetization M and the specific heat at constant field C_H become

$$M = CH/\Theta_s \qquad C_H = C(H^2 + h_L^2)/\Theta_s^2 \tag{58}$$

where the nuclear Curie constant C and the local field h_L are given by

$$C = \mathrm{Tr}\,\mathscr{H}_z^2/kH^2\,\mathrm{Tr}\,1 \qquad h_L^2 = H^2\,\mathrm{Tr}\,\mathscr{H}_{dd}^2/\mathrm{Tr}\,\mathscr{H}_z^2 \tag{59}$$

where Tr stands for trace.

At the beginning of the cycle of measurement the nuclear spins are characterized by thermal equilibrium at the lattice temperature. When the field is switched to zero the nuclear spin system is adiabatically demagnetized. From thermodynamics one may use for an adiabatic process the equation

$$dQ = C_H\,d\Theta + \Theta(\partial M/\partial\Theta)_H\,dH = 0 \tag{60}$$

Integrating this equation between the limits H_i and H_f we obtain

$$\Theta_s(H_f) = \Theta_s(H_i)\,[1 + (H_f/h_L)^2]^{1/2}/[1 + (H_i/h_L)^2]^{1/2} \tag{61}$$

$$M(H_f) = M(H_i)\,H_f\,[1 + (H_i/h_L)^2]^{1/2}/\{H_i[1 + (H_f/h_L)^2]^{1/2}\} \tag{62}$$

where the previous equations have also been used. For the case of adiabatic demagnetization from an external field H at the lattice temperature Θ_L to zero field and a spin temperature Θ_{s1} one obtains

$$\Theta_{s1} = \Theta_L/[1 + (H/h_L)^2]^{1/2} \simeq \Theta_L h_L/H \tag{63}$$

$$M_f = 0 \tag{64}$$

One notes that from Curie's law the magnetization must go to zero at zero field, as shown by Eq. (58). Now at zero field the nuclear spin temperature begins to rise toward the lattice temperature as spin-lattice relaxation takes place. If Θ_{s2} is the temperature obtained during the time t in which the field is off, then using the basic equation for nuclear spin relaxation with a time constant T_{1s} characteristic of the superconducting phase one has

$$\frac{1}{\Theta_{s2}} = \frac{1}{\Theta_L} + \left(\frac{1}{\Theta_{s1}} - \frac{1}{\Theta_L}\right) \exp(-t/T_{1s}) \tag{65}$$

Next, the field is turned back to its original value and the spin system is adiabatically remagnetized, this time from Θ_{s2}. Using the above equations one can calculate the temperature Θ_s' and magnetization M' reached when the field is turned back to its high value again. These are given by the expressions

$$\Theta_s' = \Theta_{s2}[1 + (H/h_L)^2]^{1/2} \qquad M = CH/\Theta_s' \tag{66}$$

Thus, as a result of spin-lattice relaxation one finds Θ_s' is greater than Θ_L, so that M' is less than M. Combining all the above equations and using Curie's law, the final result for the magnetization after the cycle becomes

$$M' = (CH/\Theta_L)\{(\Theta_{s1}/\Theta_L) + [1 - (\Theta_{s1}/\Theta_L)] \exp(-t/T_{1s})\} \tag{67}$$

or

$$M' = A + B \exp(-t/T_{1s})$$

In general, h_L is very much less than H, so that Eq. (67) has almost an exponential form. One expects that a semilog plot of $M' - A$ vs. t will yield a straight line. This has been found to be true in a wide variety of experiments on metals at low temperatures (*168*) and is a good indication that spin temperature is an appropriate variable with which to treat the nuclear spin system in the presence of the nuclear dipole–dipole interaction.

Certainly it is implied in this discussion that the demagnetization and remagnetization are adiabatic and reversible. This means, of course, that the time constant for a change of field must be shorter than T_1, so that spin-lattice relaxation effects can be ignored during the field turn-on and turn-off. But it is also essential that the field be varied slowly enough compared to the characteristic nuclear spin–spin interaction time, T_2, so that the nuclear spin system can follow the external field quasi-statistically and act as a thermodynamic system. Let us discuss the second point in more detail.

Suppose the spin system is initially in thermal equilibrium with the lattice in a strong field, and the field is then turned precipitously to zero. Using the sudden approximation one can see that the nuclei now precess about their local fields which are randomly oriented; in a few precession periods the nuclei will be randomly oriented, and the magnetization will be zero. The characteristic time for this disorientation or dephasing is T_2, the spin–spin relaxation time, which arises from the magnetic interaction among the nuclei themselves. Usually $T_2 \ll T_1$ in metals at low temperatures. If the field is again turned on precipitously, the magnetization will be trapped at zero in the strong field since the magnetization will require a time of order T_2 to align itself following any sort of field change. Thus under these circumstances no resonance can be seen, because the magnetization cannot be established in the strong field without spin-lattice relaxation. The resonance has been lost completely and the magnetization and demagnetization have not been performed reversibly.

It is essential in this type of experiment that the field be turned off and on sufficiently slowly for the magnetization to follow the changes in field. If one neglects spin-lattice relaxation, one can talk crudely about the nucleus precessing about the instantaneous field which it happens to see at any given time. At strong fields the instantaneous field is practically equivalent to the external field. But as the external field is decreased to a value comparable to the local field, the nucleus precesses about the vector sum of H and h_L. Although the spatial alignment of the nuclear moments will change because of the random orientation of h_L, the moments will always be lined up along the total field which they experience, so that the degree of alignment is maintained.

Of course the local field really varies in time, but we may hope that the essential features of our description are maintained. That is, if the field is turned back on slowly the nucleus will continue to line up along the total field which it sees, so that when we reach the initial value of H, the nucleus will again have its strong-field spatial orientation. The most critical region of switching is when H is comparable to h_L, since it is at this condition that the total field is changing its direction from H to that of h_L. Thus we must avoid turning off the field too rapidly, and the requirement becomes the statement that the switching time τ must satisfy the relationship

$$T_2 \ll \tau \ll T_1$$

In the normal state this requirement can be satisfied in several metals at low temperatures, since the spin–spin time T_2 is of the order of 10 to 50 μsec and the spin-lattice time T_1 is of the order of 50 msec to several sec. However, the measurement in a superconductor poses a problem. Because of field penetration effects which occur as the metal changes from the normal to the superconducting phase and back again, the field drops to zero in a distance comparable with the penetration depth, which is of the order of 10^{-5} cm. Usually in a particle there are several field boundaries which sweep through the sample as the field is decreased or increased through the critical value. Thus in the case of a superconductor the decrease of field in the vicinity of any nucleus is much more rapid than the time constant of the magnet switching itself.

It has been found experimentally that in aluminum and cadmium‡ it is possible to satisfy the conditions relative both to T_1 and T_2 and regain most of the magnetization upon remagnetization of the sample if a very short time is spent in the superconducting state. Considerable effort was made with several other

‡ Measurements have recently been carried out in indium. See Section VII.F.

metals, notably tin, which is a classic superconductor, but in all other cases it was found to be impossible to satisfy simultaneously the conditions outlined above primarily because of the precipitous change of field at a nuclear site which is induced as the metal passes between normal and superconducting phases. All the experimental evidence on type I superconductors to be discussed in the course of this article will be based on the measurements in aluminum and cadmium. In addition, measurements have been made in some type II superconductors with results which are quite different from those obtained with the type I superconductors, as will be discussed in Section E.

C. Theory of Nuclear Spin-Lattice Relaxation

As previously mentioned, the exact nuclear spin energy levels and wave functions have never been obtained in the presence of both the nuclear Zeeman interaction $\mathscr{H}_z$ and the nuclear dipole–dipole interaction $\mathscr{H}_{dd}$. Nevertheless, the theory can be formulated in terms of these exact levels and their probability of occupation because the assumption of spin temperature enables one to put the end result in the form of diagonal sums which can be evaluated without obtaining the true level energies. The details of the theory are discussed in (*163*) and (*168*).

The essential points will be reviewed here to show how the lattice temperature dependence enters the theory and to make the connection with the discussion of nuclear spin-lattice relaxation time and matrix elements contained in Sections II and III.

The time derivative of the total nuclear spin energy $\bar{E}$ of the system relaxing at zero field may be written

$$\frac{d\bar{E}}{dt} = \frac{d}{dt}\sum_n p_n E_n = \sum_n \frac{dp_n}{dt} E_n = \frac{d}{dt}\bar{E}(\Theta_s) = \frac{\partial \bar{E}}{\partial \Theta_s}\frac{d\Theta_s}{dt} \tag{68}$$

where p_n snd E_n are the probability of occupation and the energy of the exact nuclear spin states. One can regard the equation as the first law of thermodynamics applied to the nuclear spins relaxing at zero field, so that the partial derivative is just a spin specific heat at constant field. Now the interaction between nuclear spins and electron spins causes transitions between the various nuclear spin states n and m. This interaction is the hyperfine interaction discussed in Section II. C, in which an electron spin and a nuclear spin simultaneously flip. In terms of W_{nm}, the probability per unit time of a transition from the nth to the mth state, the interaction results in a family of rate equations; that is,

$$dp_n/dt = \sum_m (p_m W_{mn} - p_n W_{nm}) \tag{69}$$

If one could solve the rate equation, one could substitute the resulting derivatives back into Eq. (68) for the change of energy to determine how the nuclear spins relax. In general one cannot carry through such a solution, but with the aid of the spin temperature assumption one finds that the procedure is tractable. Now the basic spin temperature assumption means that

$$p_n/p_m = \exp\left[(E_m - E_n)/k\Theta_s\right] \simeq 1 - (E_n - E_m)/k\Theta_s \tag{70}$$

And also one has

$$\sum_n \exp\left[-E_n/k\Theta_s\right] \simeq \sum_n \delta_{nn} \tag{71}$$

To guarantee equilibrium when the spin temperature equals the lattice temperature, the principle of detailed balance requires that

$$W_{nm}/W_{mn} = \exp\left[(E_n - E_m)/k\Theta_L\right] \simeq 1 - (E_m - E_n)/k\Theta_L \tag{72}$$

Substituting Eqs. (69)–(71) into Eq. (68), one obtains the basic simple relaxation form of Eq. (57); that is,

$$\frac{d}{dt}\frac{1}{\Theta_s} = \left(\frac{1}{\Theta_L} - \frac{1}{\Theta_s}\right)\frac{1}{T_1} \tag{73}$$

where the relaxation rate is given by

$$R = \frac{1}{T_1} = \frac{1}{2}\left[\sum_{n,\,m} (E_n - E_m)^2\, W_{nm}\right] \Big/ \sum_n E_n^2 \tag{74}$$

A solution has been carried through where the energy has taken to consist of the Zeeman term, the dipole–dipole term, and possible quadropole terms which could arise because of impurities. [See (*168*) and references contained therein.] However, we can see on the basis of Eq. (74) that the changes which occur in the relaxation rate when the sample becomes superconducting arise solely from the temperature-dependent terms in the transition probability W_{nm}, which were calculated in Section III. In that section the crucial role played by the coherence factors in the superconducting states was discussed; it was pointed out that the characteristic difference between relaxation in the normal and superconducting states is attributable to a combination of effects arising from the Fermi function $f(E)$ and effects arising from an energy gap Δ in the density of states $N_s(E)$ which appears in the neighborhood of the Fermi energy The expression obtained there for the nuclear spin-lattice relaxation rates is given by (*163*)

$$R \propto \int_\Delta^\infty \int_\Delta^\infty dE_i\, dE_f (1 + \Delta^2/E_i E_f)\, f(E_i)\left[1 - f(E_f)\right] N_s(E_i)\, N_s(E_f) \tag{75}$$

D. Comparison of Theory and Experiment for Type I Superconductors

Let us consider the situation first in the normal metal; in Eq. (75) one lets $\Delta \to 0$. Because of the function $f(E)[1 - f(E)]$ the limits of integration can be confined to a region close to the Fermi energy. Since the density of states in the normal metal is approximately constant over a small interval of energy, the density of states factors can be pulled out of the integral, which can then be evaluated easily. The result is thus just the area under the function $f(E)[1 - f(E)]$, which turns out to be directly proportional to the absolute temperature Θ. This temperature factor merely means that the number of states which can cause transitions, that is, the number of electrons with nearby vacant holes, is directly proportional to absolute temperature. Extensive experiments (*168*) have shown that the linear proportionality is characteristic of a wide variety of metals at both high and low temperatures and at high and low magnetic fields.

In the superconducting state something very characteristic occurs. First, consider the case for Θ just below Θ_c where the breadth of the energy gap is small compared with the energy associated with transition temperature, $k\Theta_c$. Then the detail in the vicinity of the energy gap—the gap itself and the peaks in the density of states—occur within the width $k\Theta_c$ delineated by the product of Fermi functions. Under these circumstances the peaking in the density of states which occurs for the nuclear relaxation case has a characteristic effect on the relaxation rate. When the breadth of the function $f(E)[1 - f(E)]$ is much wider than the gap Δ, one notes that because of the square of the density of states, one can obtain a faster relaxation time by removing states from one region and piling them up in another; one is increasing the number of nearly vacant holes to which electrons in these states can be scattered by the nuclear spins. Thus for temperatures just below the transition temperature, the peak in the density of states causes the nuclear relaxation rate R_s to rise in the superconducting state relative to its normal rate as the temperature is lowered.

On the other hand, the behavior at low temperatures is characteristically just the opposite. Under these circumstances the gap becomes much wider than the breadth $k\Theta$ characteristic of the product of Fermi functions. Under these circumstances the product $f(1 - f)$ has an appreciable value only within the gap where there are no states, and the relaxation becomes very similar to what one would expect to get in a semiconductor—a relaxation rate exponentially dependent on the inverse temperature with Δ as the characteristic energy.

Thus one expects two characteristic temperature regimes, one just below the transition temperature where the relaxation rate increases as the temperature is lowered and another at low temperatures where the relaxation rate rapidly and exponentially decreases as the temperature is lowered. This is just the

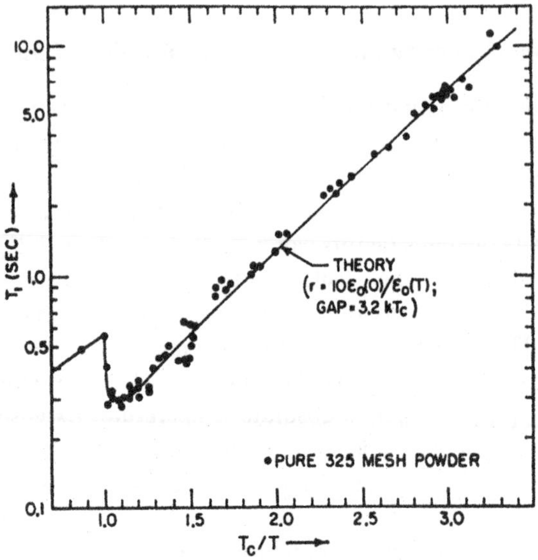

Fig. 14. Measured values of the spin-lattice relaxation time T_1 for pure aluminum as a function of reciprocal temperature. The solid line shows the theoretical curve for the indicated value of the parameter $r = \Delta(T)/\delta$. [After Masuda and Redfield (*165*).]

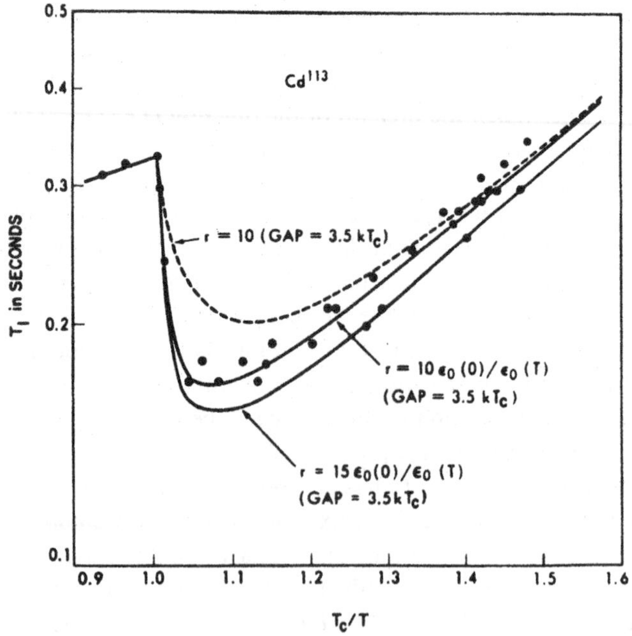

Fig. 15. Measured values of the spin-lattice relaxation time T_1 for pure cadmium as a function of reciprocal temperature. The solid lines show the theoretical curves for indicated values of the parameter $r = \Delta\,(T)/\delta$. [After Masuda (*166*).]

behavior that is found in the experiments and in the calculation as they are done on the basis of the BCS theory as shown for aluminum (*165*) in Fig. 14 and for cadmium (*166*) in Fig. 15.‡

To complete the calculation (*163*) it is necessary to take into account the fact that the states in the superconductor have a finite "breadth". The actual expression for the nuclear spin-lattice relaxation time diverges because of the peaking in density of states which occur near the gap edge. A simple rectangular breadth function is used to remove the logarithmic singularity which results from the infinitely thin energy breadth assigned to the states; in practice, either a lifetime of the state§ or an anisotropy in the energy gap could give rise to the required "breadth." That is, the BCS density of states [Eq. (16)] is replaced by

$$N_s(E) = (2\delta)^{-1}\, N_0 \int_{E-\delta}^{E+\delta} dE_1\, E_1/(E_1^2 - \Delta^2)^{1/2} \tag{76}$$

where it is convenient to express the results in terms of $r = \Delta(T)/\delta$.

One notes that, for an appropriate value of the breadth constant r, the agreement between the theory and the experiment is remarkably good. The rise in the relaxation rate is clearly found for temperatures just below the critical temperature, and the exponential limit is found at low temperatures. No one-electron theory can simultaneously explain this type of result and also explain the type of experimental results found in ultrasonic absorption (see Section V), where the absorption drops rapidly as one goes down in temperature from the critical temperature and shows no rise characteristic of the nuclear relaxation rate. Consequently, the comparison of the nuclear relaxation with the ultrasonic attenuation provides direct verification of the basic feature of spin-momentum correlation through virtual pairing in the BCS theory.

To throw light on the mechanism of the "breadth," measurements were carried out by Masuda (*167*) for aluminum containing small, controlled amounts of impurity. Results are shown in Fig. 16 for aluminum containing 0.036 at.% and 0.055 at.% zinc. Comparison with the curve for the pure metal, shown for $r = \Delta(T)/\delta = 10$, indicates that the "breadth" *decreases* as the amount of impurity is increased. Such a result is at variance with breadth from a simple scattering lifetime effect but is consistent with the theory of "dirty superconductors" proposed by Anderson (*53*).# In his formulation one-electron states are paired with their exact time-reverse states; this is a generalization of the BCS pairing of $\mathbf{k}+, -\mathbf{k}-$ and is independent of elastic impurity scatterings. Impurity scattering becomes important when the mean free path is small

‡ For aluminum alloys see (*167*).

§ A phonon scattering mechanism for the lifetime has been proposed by Fibich. See Section VII.F.

See also Chapter 5.

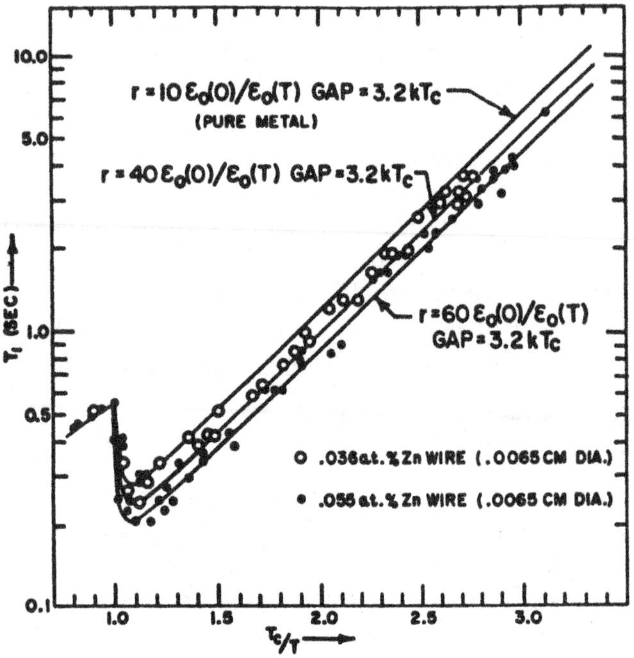

Fig. 16. Measured values of the spin-lattice relaxation time T_1 for aluminum–zinc alloys as a function of reciprocal temperature. Data are shown for two zinc concentrations. The lines are the theoretical curves for indicated values of the parameter $r = \Delta(T)/\delta$ [After Masuda (*166*).]

compared to the coherence length. Under such circumstances the time-reversed pair states are mixtures of pairs taken more or less randomly from all parts of the Fermi surface, so that impurity scattering should make the energy gap become isotropic. Masuda's measurements seem to be consistent with this interpretation.

E. Spin-Lattice Relaxation in Type II Superconductors

Relaxation of nuclear spins in vanadium has been measured by Fite and Redfield (*169*), who find characteristic differences in relaxation phenomena compared with those typically observed in type I superconductors.‡ The authors were able to understand their relaxation curves in terms of fundamental processes characteristic of type II superconductors and were even able to make conclusions concerning the vortex structure in the mixed state of vanadium from the NMR line shape.

‡ For T_1 measurements in niobium see (*170*).

The mixed state of a type II superconductor can exist in an ordered array (*172*)‡ of magnetic field lines, each surrounded by a vortex-like circulation of current. Such an array results in a periodic variation of magnetic field in the metal. Direct measurements using neutron diffraction (*172*) have shown that the vortex structure in niobium is formed from a triangular lattice. The lattice constant, typically 100–1000 Å, is determined by a flux quantum condition of the same type as that for superconducting cylinders. Using magnetic resonance methods, Fite and Redfield have independently determined from the NMR line shape that the vortex structure in the mixed state of vanadium is also based on a triangular lattice. In addition, they found that the measurements of nuclear spin-lattice relaxation time, T_1, showed separate decay contributions from nuclear spins close to vortex lines and from nuclear spins far away from vortex lines.

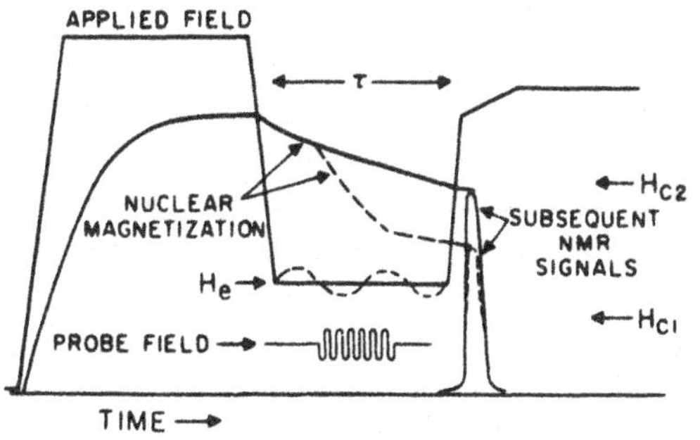

Fig. 17. Magnetic field versus time for the field cycle used in making spin-lattice relaxation time and NMR line-shape measurements in vanadium [After Fite and Redfield (*169*).]

The measurements were based on a field cycling technique shown in Fig. 17, which is similar to the cycle previously discussed for experiments in type I superconductors. First, the spins were allowed to come to thermal equilibrium in a magnetic field of 10 kG. To take the measurement the field was lowered in 30 msec to some value H_e at which the sample is superconducting. H_e may be above or below the upper critical field, H_{c2}, in the experiment. The field was kept at H_e for a time t, during which the spin magnetization relaxes with a time constant characteristic of the mixed state. Finally, the field is rapidly increased to 6 kG, and the nuclear magnetization is measured by means of standard NMR techniques. The cycle is repeated many times with varying values of t to determine the time constant T_1. In some of the measurements a small high-frequency magnetic field was applied in the direction of H_e to move the vortex structure

‡ See also Chapter 14.

in the sample relative to the spins to achieve a situation in which *each* spin in the sample relaxes at an average rate for the entire structure.

Just below the critical temperature, T_c, Fite and Redfield (*169*) did not observe the drop in T_1 relative to the normal state which is characteristic of type I superconductors. The absence of this decrease in T_1 seemed to be a real effect; while there was some evidence that it should be attributed to trapped flux phenomena, the interpretation is by no means completely clear. However, at lower temperatures a more characteristic nuclear relaxation phenomenon was seen. The measurement showed that nuclear spins far from vortex lines relax with a T_1 which is an exponential function of temperature, as predicted by the BCS theory for type I superconductors. For this portion of the spin system they found

$$T_1 = (0.8 \text{ sec} \pm 0.4 \text{ sec}) \exp(1.75\, T_c/T) \tag{77}$$

The spin magnetization at low temperatures was also found to have a portion with a faster decay time constant, roughly the same as T_1 in the normal metal. This fraction is interpreted as arising from spins within a coherence distance of a vortex. The interpretation is consistent with measurements made using the small high-frequency ac field during the decay. Under such circumstances a single average relaxation time was obtained whose value was consistent with the expected density of vortices in the material.

Direct information about the shape of the vortex structure was inferred from the NMR absorption-line-shape measurements. To obtain the line shape, the same field cycle was used repeatedly, with a fixed value of $t = 150$ msec, which

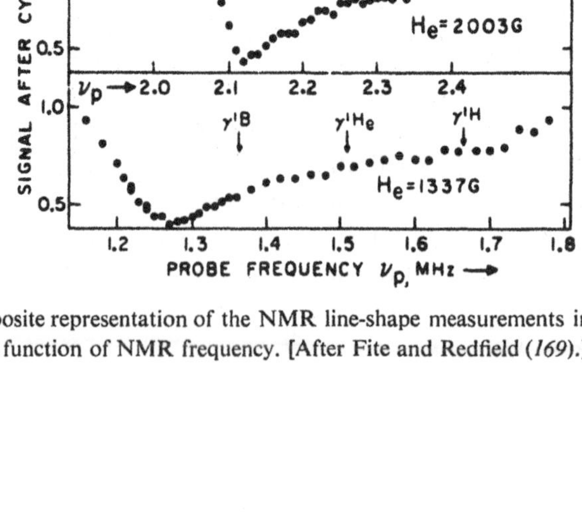

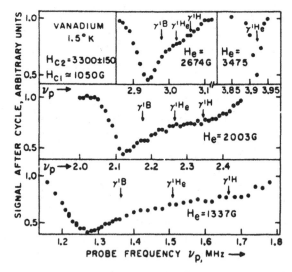

Fig. 18. Composite representation of the NMR line-shape measurements in vanadium as a function of NMR frequency. [After Fite and Redfield (*169*).]

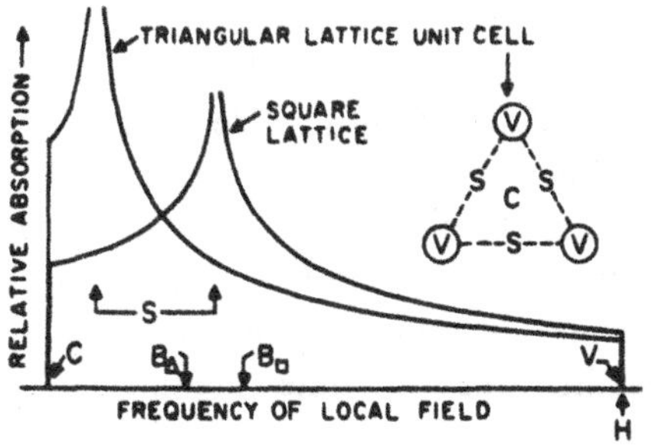

Fig. 19. Calculations of the theoretical NMR line shape expected on the basis of both triangular and square unit lattices. [After Fite and Redfield (*169*).]

was much less than T_1. During the time t, a radiofrequency field was applied *perpendicular* to H_e to come to resonance with a fraction of the spins in their local field. The line shape was inferred from the change in signal vs. frequency of the applied radiofrequency field. The results are shown in Fig. 18. If H_e is above H_{c2}, so that the sample is normal, the signal is decreased only when H_e is within about 20 G of the normal resonance conditions, as is shown in the upper right-hand trace.

If H_e is appreciably less than H_{c2}, a pronounced broadening occurs which shows that the local field becomes sufficiently inhomogeneous, so that few nuclei are close to resonance at any particular value of probe frequency. Fite and Redfield compare the line shape with what would be calculated for both square and triangular structures. The theoretical shape which they derive is shown in Fig. 19. Their experimental line shapes clearly indicate better agreement with the calculation based on a triangular lattice unit cell, in agreement with the result obtained by neutron diffraction studies in niobium.

Thus a detailed study of both T_1 and the line shape in a type II superconductor shows a two-component relaxation which arises from spins near vortices and spins far away from vortices, contributing separately at lower temperatures. The method makes possible a NMR line-shape determination which clearly shows a shape much more characteristic of the triangular lattice than of the square lattice.

F. Measurements of T_1 in Pure and Impure Indium

Measurements of the nuclear spin-lattice relaxation time T_1 in pure and impure superconducting indium have recently been made by Butterworth and MacLaughlin (*173*). The work was undertaken to provide additional experimental evidence on the questions raised in Sections VII.C and D concerning the effect on T_1 of a finite energy level breadth for BCS states. In those sections it was shown that the BCS calculation for T_1 diverges logarithmically unless the theory is modified to take into account the fact that the states in the vicinity of the gap are not infinitely sharp. Two possible mechanisms were discussed: (1) finite state lifetime leading to an energy breadth and (2) energy gap anisotropy. The work of Masuda on aluminum alloys, discussed in Section VII.D, showed that energy gap anisotropy is an important mechanism for aluminum. However, the new measurements in pure and impure superconducting indium indicate that the breadth due to state lifetime is more important for indium.

In this work solid solutions of lead in indium were studied as a function of lead concentration varying between 0.015 and 2 at.%. The measurements of the relaxation rate were made on the $(\frac{9}{2}-\frac{7}{2})$ nuclear quadrupole resonance transition of the ^{115}In nuclei at a resonant frequency close to 7.54 MHz. The quadrupole transition was saturated, and its recovery was measured to determine the relaxation rate. Butterworth and MacLaughlin (*173*) found that the growth of the signal in the normal metal after saturation was nonexponential, and the relaxation rate in the normal metal was extracted from the data.

To make the measurements in the superconductor, a field-cycling method was used. An external field $H_e > H_c(T)$ was applied to drive the sample normal during initial saturation of the nuclear quadrupole resonance transition. H_e was then quickly reduced to zero, and the quadrupole system was allowed to relax in the superconducting state for a variable time t, during which a portion of the signal would recover. The signal amplitude was then measured in the normal state after reapplication of H_e. Spin-echo techniques were used to observe the signal because of appreciably inhomogeneous broadening of the nuclear quadrupole line. The observed relaxation rate was found to be independent of H_e in the normal state and was independent of H_e in the superconducting state for $H_e > H_c(T)$.

As was the case in aluminum and cadmium, the relaxation rate in indium showed the characteristic rise in $R_s(T)/R(T_c)$ to a maximum at $T \cong 0.85T_c$ followed by a sharp drop for $T \ll T_c$ (see Fig. 14). However, the relaxation rate in indium is found to be almost independent of the impurity concentration, suggesting that energy gap anistropy plays a minor role in determining the behavior of the relaxation rate in indium as it nears its maximum. This indicates that in indium the maximum in $R_s(T)$ is limited by an effective energy breadth and hence the lifetime of the BCS states near the gap.

A mechanism for such a finite lifetime was first pointed out by Fibich (*174*), who calculated the relaxation rate in the superconductor, taking into account the scattering of thermal phonons by conduction electrons. Fibich used a retarded electron–phonon interaction in place of the constant effective potential V of the BCS theory, and he obtained the same type of complex energy-dependent gap function obtained by earlier authors. In effect the energy levels are broadened by scattering of thermal phonons, which remove the singularity in the density of states near the energy gap.

Fibich (*174*) obtained a result for aluminum which was in qualitative agreement with the data discussed in Section VII.D. However, the experimental work of Masuda previously discussed seemed to indicate that the energy gap anisotropy mechanism was the more important one for aluminum. Butterworth and MacLaughlin report a preliminary calculation for indium based on the Fibich theory which is in reasonably close agreement with their experimental results in indium, considering the approximations which are necessary to make the theoretical treatment. Their work suggests, therefore, that a more comprehensive theory of the nuclear relaxation rate is required, one which would take into account both the effects of phonons in limiting the state lifetime and also the effects of energy gap anisotropy. At the present time the work in aluminum seems to favor one mechanism, and the work in indium seems to favor the other.

VIII. NOTES ADDED IN PROOF

A. Electromagnetic Absorption

Rugheimer et al. (*175*) have extended their earlier data (*7*) on the temperature dependence of the transmission and reflection of microwaves by thin films of indium and tin. The measured characteristics of most of their samples are in good agreement with the calculation of Mattis and Bardeen (*34*). Structure of unknown origin was observed in the experimental curves for some of their films.

Palmer and Tinkham (*176*) have noticed that the sum rule (*41*, *42*) is violated by Nam's calculated curves (*60*) for the real part σ_1 and the imaginary part σ_2 of the superconducting conductivity of lead. Nam has determined that the cause of this was an error in the numerical computations which he had performed to calculate σ_2. His correction of this has brought the theoretical curve into agreement with the data below the gap edge, where there was disagreement before. At the upper end of the experimental frequency range (at about five times the gap frequency), the theoretical σ_1 curve is higher than the experimentally determined curve (*39*), and is rising more rapidly than it. However, the discrepancy is probably within the combined uncertainties of theory and the experiment. (It should be noted that the theoretical curve of greatest relevance, which we are discussing here, is that which Nam has calculated from the frequency-dependant gap parameter of lead, rather than a curve calculated by assuming an energy-independent gap parameter.) Palmer and Tinkham remark that Nam's corrected calculations show that strong coupling between electrons and phonons may account for the difference between the shapes of the calculated (*54*), and the measured (*56*) far-infrared absorption curves for lead alloys and may also explain the unusually steep absorption edge observed by Norman (*177*) for lead foils.

The far-infrared absorption data of Norman and Douglass have been extended by Norman to include foils of indium, niobium, tantalum, tin, and vanadium as well as more work on lead (*177*). He has seen evidence for multiple energy gaps in a tin foil and in two thick lead foils. This is attributed to anisotropy of the energy gap. Only one energy gap was seen in a thin lead foil. This is consistent with a washing out (*53*, *178*) of the gap anisotropy by the shortened electron mean free path in the thinner foil.

Drew and Sievers (*179*) have repeated the far-infrared absorption measurements of Norman and Douglass on lead foils and have confirmed the absence of any appreciable absorption below the gap edge. They have also measured far-infrared transmission through a wave-guide made of a rolled-up lead foil and have obtained data which are qualitatively inconsistent with theory and with all other far-infrared measurements on lead at frequencies below the gap edge. No explanation for this is available.

By measuring microwave absorption in lanthanum samples, Thompson (*180*) has shown that the ratio of energy gap width to critical temperature varies with annealing treatment. This probably accounts for the discrepancies among the values of this ratio which have been measured for lanthanum by various methods.

Koch and Pincus (*181*) and then Koch and Kuo (*182*) have determined the surface resistance R of flat samples of tin at microwave frequencies in the presence of a parallel dc magnetic field H. They found a complicated behavior of dR/dH, which varies with temperature, microwave frequency, and crystal orientation. There were peaks in dR/dH as a function of H. These disappeared if the surface was etched lightly. A qualitative interpretation was made in terms of electronic transitions from the magnetically bound surface states which Pincus discovered theoretically (*71*). Koch and Kuo have presented an alternative, qualitative, theoretical description of these states. Bennemann (*183*) has remarked that a detailed analysis of these phenomena must include the frequency dependence of the electromagnetic penetration depth.

B. Ultrasonic Attenuation

The theory (*59*) of the effect of strong electron–phonon coupling on the attenuation of longitudinal acoustic waves has been extended by Woo (*184*) to the entire range of ql values. His results show that the strong coupling has little effect on the BCS temperature dependence of α_s/α_n given in Eq. (44). This agrees with experiment, including the recent data of Tittmann and Bommel (*185*) on lead containing a small concentration of tin to remove the amplitude-dependent attenuation. Fate and Shaw (*186*) have made similar measurements on strained and impure lead and have emphasized that the temperature dependence of the electron mean free path must be included in a careful interpretation of the data for samples in which the mean free path is phonon-limited.

The amplitude-dependent attenuation in lead has been seen in the normal state as well as in the superconducting state by Tittmann and Bommel (*187*) and Hikata et al. (*188*). The former authors have also investigated the effect of strains and impurities. They have questioned the validity of Mason's theory (*104*) of the dislocation damping and have presented a derivation of their own. Ultrasonic attenuation in mercury has been measured by Ferguson and Burgess (*189*) at a frequency of 9.3 GHz. This was high enough to make the amplitude dependence negligible. Their results fit the BCS relation, Eq. (44), but with a value for the gap parameter which is somewhat smaller than that found by experiments which measure it more directly.

Fagen and Garfunkel (*190*) have determined the acoustic attenuation in aluminum at a frequency of 9.3 GHz. The energy gap width became less than the phonon energy as the temperature was increased toward T_c, and the expected (*44–46*) discontinuous increase in the attenuation was seen.

Recent measurements have been made on the attenuation of longitudinal waves in indium (*191*), doped tin (*192, 193*) niobium (*194, 195*) and vanadium (*194*). A vigorous and successful attack (*196*) has been made on an interpretation of new measurements in strained and impure tin (*197*).

Leggett has developed a theory (*198*) of the ultrasonic attenuation in a superconductor with two electron energy bands. He has predicted some interesting effects for large values of ql at low temperature.

The theory of the electrodynamic damping of acoustic shear waves (*90, 91*) has been extended by Fossheim (*199*) to all values of frequency and temperature for which $ql \gg 1$. He finds a reasonable fit of his calculations to his experimental data on indium.

C. Thermal Conductivity

Ulbrich et al. have performed a rough calculation (*200*) of the anisotropy of the electronic thermal conductivity ratio κ_{es}/κ_{en} in tin (*161*) and tin alloys (*201, 202*). Simplifying assumptions were made for the dependence of the gap parameter on direction, energy, and temperature and for the nature of the electron-impurity scattering process. Also, the scattering of electrons by phonons was neglected. The calculated anisotropy in the thermal conductivity ratio was about half as large as that which was observed. The anisotropy in this ratio was predicted to be very small below a certain value of the electron mean free path. This value agreed well with that observed in single-crystal alloys (*161*), but disagreed by an order of magnitude with that observed in polycrystalline alloys (*202*). Further theoretical work will be required to determine which of the approximations in the theory must be improved to bring about a reasonable agreement with the data.

ACKNOWLEDGMENTS

The authors wish to gratefully acknowledge helpful conversations with P.W. Anderson, J. Bardeen, M. S. Dresselhaus, A. Granato, L. P. Kadanoff, K. Maki, A. G. Redfield, C. B. Satterthwaite, C. P. Slichter, and J. A. Snow. The work of one of the authors (D. M. G.) was supported in part by the National Science Foundation and in part by the Advanced Research Projects Agency under Contract SD-131.

REFERENCES

1. J. Bardeen, L. N. Cooper, and J. R. Schrieffer, *Phys. Rev.* **108**, 1175 (1957).
2. A. B. Pippard, in *Advances in Electronics and Electron Physics*, Vol. 6 (L. Marton, ed.), Academic Press, New York, 1954, p. 1.
3. M. A. Biondi, A. T. Forrester, M. P. Garfunkel, and C. B. Satterthwaite, *Rev. Mod. Phys.* **30**, 1109 (1958).
4. M. Tinkham, in *Low Temperature Physics* (C. De Witt, B. Dreyfus, and P. G. de Gennes, eds.), Gordon & Breach, New York, 1962, p. 147.
5. J. R. Waldram, *Advan. Phys.* **13**, 1 (1964).
6. M. Tinkham, *Optical Properties and Electronic Structure of Metals and Alloys* (F. Abeles, ed.), North-Holland, Amsterdam, 1966, p. 431.
7. N. M. Rugheimer, C. V. Briscoe, A. Lehoczky, and R. E. Glovers, III, *Low Temperature Physics, LT9* (J. G. Daunt et al., eds.), Plenum Press, New York, 1965, p. 381.
8. H. London, *Proc. Roy. Soc. (London)* **A176**, 522 (1940).
9. A. B. Pippard, *Proc. Roy. Soc. (London)* **A191**, 370, 399 (1947).
10. G. S. Blevins, W. Gordy, and W. M. Fairbank, *Phys. Rev.* **100**, 1215 (1955).
11. T. E. Faber and A. B. Pippard, *Proc. Roy. Soc. (London)* **A231**, 336 (1955).
12. E. Fawcett, *Proc. Roy. Soc. (London)* **A232**, 519 (1955).
13. R. G. Chambers, *Proc. Cambridge Phil. Soc.* **52**, 363 (1956).
14. M. S. Khaikin, *Zh. Eksperim. i Teor. Fiz.* **34**, 1389 (1958); *Soviet Phys. JETP* **7**, 961 (1958).
15. M. A. Biondi and M. P. Garfunkel, *Phys. Rev.* **116**, 853, 862 (1959); *Phys. Rev. Letters* **2**, 143 (1959).
16. R. Kaplan, A. H. Nethercot, Jr., and H. A. Boorse, *Phys. Rev.* **116**, 270 (1959).
17. M. Spiewak, *Phys. Rev.* **113**, 1479 (1959).
17a. Yu. V. Sharvin and V. F. Gantmakher, *Zh. Eksperim. i Teor. Fiz.* **39**, 1242 (1960); *Soviet Phys. JETP* **12**, 866 (1961).
18. M. S. Dresselhaus, D. H. Douglass, Jr., and R. L. Kyhl, *Proc. of the Eighth Intern. Conf. on Low Temp. Physics* (R. O. Davies, ed.), Butterworth, London, 1963.

19. P. L. Richards, *Phys. Rev.* **126**, 912 (1962).
20. M. A. Biondi, M. P. Garfunkel, and W. A. Thompson, *Phys. Rev.* **136**, A1471 (1964).
21. R. Glosser and D. H. Douglass, Jr., *Low Temperature Physics, LT9* (J. G. Daunt et al., eds.), Plenum Press, New York, 1965, Part A, p. 385.
22. R. T. Lewis, *Phys. Rev.* **134**, A1 (1964).
23. W. V. Budzinski and M. P. Garfunkel, *Phys. Rev. Letters* **16**, 1100 (1966); **17**, 24 (1966).
24. S. A. Zemon and H. A. Boorse, *Phys. Rev.* **146**, 309 (1966).
25. M. A. Biondi, M. P. Garfunkel, and A. O. McCoubrey, *Phys. Rev.* **101**, 1427 (1956).
26. M. A. Biondi, M. P. Garfunkel, and A. O. McCoubrey, *Phys. Rev.* **108**, 495 (1957).
27. M. A. Biondi, A. T. Forrester, and M. P. Garfunkel, *Phys. Rev.* **108**, 497 (1957).
28. J. S. Shier and D. M. Ginsberg, *Phys. Rev.* **147**, 384 (1966).
29. E. Hirschlaff, *Proc. Cambridge Phil. Soc.* **33**, 140 (1937).
30. J. G. Daunt, T. C. Keeley, and K. Mendelssohn, *Phil. Mag.* **23**, 264 (1937).
31. A. Wexler, *Phys. Rev.* **70**, 219 (1946).
32. K. G. Ramanathan, *Proc. Phys. Soc. (London)* **A65**, 532 (1952).
33. N. G. McCrum and C. A. Shiffman, *Proc. Phys. Soc. (London)* **A67**, 386 (1954).
34. D. C. Mattis and J. Bardeen, *Phys. Rev.* **111**, 412 (1958).
35. A. A. Abrikosov, L. P. Gor'kov, and I. M. Khalatnikov, *Zh. Eksperim. i Teor. Fiz.* **35**, 265 (1958); *Soviet Phys. JETP* **8**, 182 (1959).
36. R. G. Chambers, *Proc. Phys. Soc. (London)* **65**, 458 (1952).
37. A. B. Pippard, *Proc. Roy. Soc. (London)* **A216**, 547 (1953).
38. R. E. Glover, III, and M. Tinkham, *Phys. Rev.* **108**, 243 (1957).
39. L. H. Palmer, Ph.D. thesis, University of California, Berkeley, 1966 (unpublished).
40. C. Kittel, *Elementary Statistical Physics*, Wiley, New York, 1958.
41. R. A. Ferrell and R. E. Glover, III, *Phys. Rev.* **109**, 1398 (1958).
42. M. Tinkham and R. A. Ferrell, *Phys. Rev. Letters* **2**, 331 (1959).
43. D. M. Ginsberg and M. Tinkham, *Phys. Rev.* **118**, 990 (1960).
44. V. L. Pokrovskii, *Zh. Eksperim. i Teor. Fiz.* **40**, 143, 898 (1961); *Soviet Phys. JETP* **13**, 100, 628 (1961).
45. V. L. Pokrovskii and M. S. Ryvkin, *Zh. Eksperim. i Teor. Fiz.* **40**, 1859 (1961); *Soviet Phys. JETP* **13**, 1306 (1961).
46. V. M. Bobetic, *Phys. Rev.* **136**, A1535 (1964).
47. G. E. H. Reuter and E. H. Sondheimer, *Proc. Roy. Soc. (London)* **A195**, 336 (1948).
48. R. B. Dingle, *Physica* **19**, 348 (1953).
49. R. B. Dingle, *Physica* **19**, 311 (1953).
50. R. G. Chambers, *Proc. Roy. Soc. (London)* **A215**, 481 (1952).
51. P. B. Miller, *Phys. Rev.* **118**, 928 (1960).
52. A. B. Pippard, *Proc. Roy. Soc. (London)* **A203**, 98, 210 (1950).
53. P. W. Anderson, *J. Phys. Chem. Solids* **11**, 26 (1959).
54. D. M. Ginsberg, *Phys. Rev.* **151**, 241 (1966).
55. L. Leplae, Ph.D. thesis, University of Maryland, College Park, 1962 (unpublished).
56. J. D. Leslie and D. M. Ginsberg, *Phys. Rev.* **133**, A362 (1964).
57. P. L. Richards and M. Tinkham, *Phys. Rev.* **119**, 575 (1960).
58. J. C. Swihart, D. J. Scalapino, and Y. Wada, *Phys. Rev. Letters* **14**, 106 (1965).
59. V. Ambegaokar, *Phys. Rev. Letters* **16**, 1047 (1966).
60. S. B. Nam, Ph.D thesis, University of Illinois, Urbana, 1966 (unpublished); *Phys. Rev.* **156**, 487 (1967).
61. P. L. Richards, *Phys. Rev. Letters* **7**, 412 (1961).
62. D. M. Ginsberg and J. D. Leslie, *IBM J. Res. Develop.* **6**, 55 (1962).

63. J. D. Leslie et al., *Phys. Rev.* **134**, A309 (1964).

64. R. L. Cappelletti, D. M. Ginsberg, and J. K. Hulm, *Phys. Rev.* **158**, 340 (1967).

65. D. H. Douglass, Jr., and L. M. Falicov, in *Progress in Low Temperature Physics*, Vol. IV (C. J. Gorter, ed.), North-Holland, Amsterdam, 1964, p. 97.

66. J. R. Clem, *Ann. Phys. (N.Y.)* **40**, 268 (1966).

67. A. J. Bennett, *Phys. Rev.* **140**, A1902 (1965); **153**, 482 (1967).

68. N. V. Zavaritskii, *Zh. Eksperim. i Teor. Fiz.* **48**, 837 (1965); *Soviet Phys. JETP* **21**, 557 (1965).

69. G. Dresselhaus and M. S. Dresselhaus, *Phys. Rev.* **118**, 77 (1960).

70. K. Maki, *Phys. Rev. Letters* **14**, 98 (1965).

71. P. Pincus, *Phys. Rev.* **158**, 346 (1967).

72. S. L. Norman and D. H. Douglass, Jr., *Phys. Rev. Letters* **17**, 875 (1966); **18**, 339 (1967).

73. R. W. Morse, *Progress in Cryogenics*, Vol. *I* (K. Mendelssohn, ed.), Heywood, London, 1959, p. 219.

74. A. B. Pippard, *Proc. Roy. Soc. (London)* **A257**, 165 (1960).

75. A. B. Pippard, *Phil. Mag.* **46**, 1104 (1955).

76. A. B. Pippard, *The Fermi Surface* (W. A. Harrison and M. B. Webb, eds.), Wiley, New York, 1960, p. 224.

77. M. S. Steinberg, *Phys. Rev.* **110**, 772 (1958); **111**, 525 (1958).

78. E. I. Blount, *Phys. Rev.* **114**, 418 (1959).

79. T. Holstein, *Phys. Rev.* **113**, 479 (1959).

80. M. H. Cohen, M. J. Harrison, and W. A. Harrison, *Phys. Rev.* **117**, 937 (1960).

81. V. Z. Kresin, *Zh. Eksperim. i Teor. Fiz.* **36**, 1947 (1959); *Soviet Phys. JETP* **9**, 1385 (1959).

82. T. Tsuneto, *Phys. Rev.* **121**, 402 (1961).

83. H. E. Bommel, *Phys. Rev.* **96**, 220 (1954).

84. L. Mackinnon, *Phys. Rev.* **98**, 1181 (1955).

85. R. W. Morse and H. V. Bohm, *Phys. Rev.* **108**, 1094 (1957).

86. A. B. Pippard, *J. Phys. Chem. Solids* **3**, 175 (1957).

87. E. R. Dobbs and J. M. Perz, *Rev. Mod. Phys.* **36**, 257 (1964).

88. M. Yokota, H. Kushibe, and T. Tsuneto, *Progr. Theoret. Phys. (Kyoto)* **86**, 237 (1966).

89. R. W. Morse and H. V. Bohm, *J. Accoust. Soc. Am.* **31**, 1523 (1959).

90. L. T. Claiborne, Jr., and R. W. Morse, *Phys. Rev.* **136**, A893 (1964).

91. L. P. Kadanoff and A. B. Pippard, *Proc. Roy. Soc. (London)* **A292**, 299 (1966).

92. M. Levy, *Phys. Rev.* **131**, 1497 (1963).

93. J. R. Leibowitz, *Phys. Rev.* **133**, A84 (1964).

94. J. A. Privorotskii, *Zh. Eksperim. i Teor. Fiz.* **43**, 1331 (1962); *Soviet Phys. JETP* **16**, 945 (1962).

95. J. M. Perz, *Can. J. Phys.* **44**, 1765 (1966).

96. P. A. Bezuglyi, A. A. Galkin, and A. P. Karolyuk, *Zh. Eksperim. i Teor. Fiz.* **36**, 1951 (1959); *Soviet Phys. JETP* **9**, 1388 (1959).

97. R. W. Morse, T. Olsen, and J. D. Gavenda, *Phys. Rev. Letters* **3**, 15 (1959); erratum **3**, 193 (1959).

98. L. T. Claiborne and N. G. Einspruch, *Phys. Rev. Letters* **15**, 862 (1965).

99. R. E. Love and R. W. Shaw, *Rev. Mod. Phys.* **36**, 260 (1964).

100. R. E. Love, R. W. Shaw, and W. A. Fate, *Phys. Rev.* **138**, A1453 (1965).

101. R. L. Thomas, H. C. Wu, and N. Tepley, *Phys. Rev. Letters* **17**, 22 (1966).

102. P. A. Bezuglyi, V. D. Fil', and O. A. Shevchenko, *Zh. Eksperim. i Teor. Fiz.* **49**, 1715 (1965); *Soviet Phys. JETP* **22**, 1172 (1966).

103. W. P. Mason, *Appl. Phys. Letters* **6**, 111 (1965).
104. W. P. Mason, *Phys. Rev.* **143**, 229 (1966).
105. A.V. Granato and K. Lücke, *J. Appl. Phys.* **27**, 583, 789 (1956).
106. B. R. Tittmann and H. E. Bommel, *Phys. Rev. Letters* **14**, 296 (1965).
107. D. O. Thompson and D. K. Holmes, *J. Appl. Phys.* **27**, 713 (1956).
108. A. V. Granato and K. Lücke, *Physical Acoustics*, Vol. 4A (W. P. Mason, ed.), Academic Press, New York, 1966, p. 225.
109. B. C. Deaton, *Phys. Rev. Letters* **16**, 577 (1966).
110. V. Ambegaokar, *Phys. Rev. Letters* **16**, 1047 (1966).
111. K. L. Chopra and T. S. Hutchison, *Can. J. Phys.* **36**, 805 (1958).
112. R. David and N. J. Poulis, *Proc. of the Eighth Intern. Conf. on Low Temp. Physics* (R. O. Davies, ed.), Butterworth, London, 1963, p. 193.
113. R. David, H. R. van der Laan, and N. J. Poulis, *Physica* **29**, 357 (1963).
114. R. David, *Philips Res. Rept.* **19**, 524 (1964).
115. T. Olsen, L. T. Claiborne, and N. G. Einspruch, *Low Temperature Physics, LT9* (J. G. Daunt et al., eds.), Plenum Press, New York, 1965, Part A, p. 375.
116. K. L. Chopra and T. S. Hutchison, *Can. J. Phys.* **37**, 1100 (1959).
117. L. Mackinnon and A. Myers, *Proc. Phys. Soc. (London)* **73**, 291 (1959).
118. N. H. Horwitz and H. V. Bohm, *Phys. Rev. Letters* **9**, 313 (1962).
119. M. Levy, R. Kagiwada, and I. Rudnick, *Proc. of the Eighth Intern. Conf. on Low Temp. Physics* (R. O. Davies, ed.), Butterworth, London, 1963, p. 188.
120. A. Ikushima, T. Suzuki, N. Tanaka, and S. Nakajima, *J. Phys. Soc. Japan* **19**, 2235 (1964).
121. R. Weber, *Phys. Rev.* **133**, A1487 (1964).
122. A. Ikushima, M. Fujii, and T. Suzuki, *J. Phys. Chem. Solids* **27**, 327 (1966).
123. R. Kagiwada, M. Levy, and I. Rudnick, *Phys. Letters* **22**, 29 (1966).
124. N. Tsuda, S. Koike, and T. Suzuki, *Phys. Letters* **22**, 414 (1966).
125. M. Levy and I. Rudnick, *Phys. Rev.* **132**, 1073 (1963).
126. L. Mackinnon, *Phys. Rev.* **106**, 70 (1957).
127. P. A. Bezuglyi, A. A. Galkin, and A. P. Karolyuk, *Zh. Experim. i Teor. Fiz.* **39**, 7 (1960); *Soviet Phys. JETP* **12**, 4 (1961).
128. P. A. Bezuglyi and A. A. Galkin, *Zh. Eksperim. i Teor. Fiz.* **39**, 1163 (1960); *Soviet Phys. JETP* **12**, 810 (1961).
129. K. Kamigaki, *J. Phys. Soc. Japan* **16**, 1141 (1961).
130. A. R. Mackintosh, *Proc. of the Seventh Intern. Conf. on Low Temp. Physics* (G. M. Graham and A. C. Hollis Hallett, eds.), Univ. Toronto Press, Toronto, 1961, p. 240.
131. A. G. Shepelev, *Zh. Eksperim. i Teor. Fiz.* **45**, 2076 (1963); *Soviet Phys. JETP* **18**, 1423 (1964).
132. T. Olsen, L. T. Claiborne, and N. G. Einspruch, *J. Appl. Phys.* **37**, 760 (1966).
133. G. A. Saunders and A. W. Lawson, *Phys. Rev.* **135**, A1161 (1964).
134. R. Weil and A. W. Lawson, *Phys. Rev.* **141**, 452 (1966).
135. H. V. Bohm and N. H. Horwitz, *Proc. of the Eighth Intern. Conf. on Low Temp. Physics* (R. O. Davies, ed.), Butterworth, London, 1963, p. 191.
136. J. L. Brewster, M. Levy, and I. Rudnick, *Phys. Rev.* **132**, 1062 (1963).
137. L. T. Claiborne and N. G. Einspruch, *Phys. Rev. Letters* **10**, 49 (1963).
138. L. T. Claiborne and N. G. Einspruch, *Phys. Rev.* **132**, 621 (1963).
139. L. T. Claiborne and N. G. Einspruch, *J. Appl. Phys.* **37**, 925 (1966).
140. J. K. Hulm, *Proc. Roy. Soc. (London)* **A204**, 98 (1950).

141. P. G. Klemens, *Handbuch der Physik*, Vol. 14 (S. Flügge, ed.), Springer, Berlin, 1956, p. 266.
142. J. L. Olsen and H. M. Rosenberg, *Advan. Phys.* **2**, 28 (1953).
143. H. M. Rosenberg, *Phil. Trans. Roy. Soc. (London)* **A247**, 441 (1955).
144. B. Serin, *Handbuch der Physik*, Vol. 15 (S. Flügge, ed.), Springer, Berlin, 1956 p. 261.
145. K. Mendelssohn, in *Progress in Low Temperature Physics*, Vol. 1 (C. J. Gorter, ed.), North-Holland, Amsterdam, 1955, p. 184.
146. N. V. Zavaritskii, *Zh. Eksperim. i Teor. Fiz.* **34**, 1116 (1958); *Soviet Phys. JETP* **7**, 773 (1958).
147. J. L. Olsen and C. A. Renton, *Phil. Mag.* **43**, 946 (1952).
148. K. Mendelssohn and C. A. Renton, *Phil. Mag.* **44**, 776 (1953).
149. B. B. Goodman, *Proc. Phys. Soc. (London)* **A66**, 217 (1953).
150. S. J. Laredo, *Proc. Roy. Soc. (London)* **A229**, 473 (1955).
151. J. Bardeen, G. Rickayzen, and L. Tewordt, *Phys. Rev.* **113**, 982 (1959).
152. B. T. Geilikman and V. Z. Kresin, *Dokl. Akad. Nauk SSSR* **123**, 259 (1953); *Zh. Eksperim. i Teor. Fiz.* **36**, 959 (1959); *Soviet Phys. JETP* **9**, 677 (1959).
153. C. B. Satterthwaite, *Phys. Rev.* **125**, 873 (1962).
154. R. J. Sladek, *Phys. Rev.* **97**, 902 (1955).
155. J. Bardeen and J. R. Schrieffer, *Progress in Low Temperature Physics*, Vol. 3 (C. J. Gorter, ed.), North-Holland, Amsterdam, 1961, p. 170.
156. W. J. de Haas and A. Rademakers, *Physica* **7**, 992 (1940).
157. J. H. P. Watson and G. M. Graham, *Can. J. Phys.* **41**, 1738 (1963).
158. V. Ambegaokar and L. Tewordt, *Phys. Rev.* **134**, A805 (1964).
159. V. Ambegaokar and J. Woo, *Phys. Rev.* **139**, A1818 (1965).
160. N. V. Zavaritskii, *Zh. Eksperim. i Teor. Fiz.* **37**, 1506 (1959); *Soviet Phys. JETP* **10**, 1069 (1960).
161. A. M. Guénault, *Proc. Roy. Soc. (London)* **A262**, 420 (1961).
162. I. M. Khalatnikov, *Zh. Eksperim. i Teor. Fiz.* **36**, 1818 (1959); *Soviet Phys. JETP* **9**, 1296 (1959).
163. L. C. Hebel and C. P. Slichter, *Phys. Rev.* **107**, 901 (1957); **113**, 1504 (1959); L. C. Hebel, *Phys. Rev.* **116**, 79 (1959).
164. A. G. Redfield, *Phys. Rev. Letters* **3**, 85 (1959).
165. Y. Masuda and A. G. Redfield, *Phys. Rev.* **125**, 159 (1962).
166. Y. Masuda, *IBM J. Res. Develop.* **6**, 24 (1962).
167. Y. Masuda, *Phys. Rev.* **126**, 1271 (1962).
168. L. C. Hebel, *Solid State Physics*, Vol. 15 (F. Seitz and D. Turnbull, eds.), Academic Press, New York, 1963, p. 409.
169. W. Fite, II, and A. G. Redfield, *Phys. Rev. Letters* **17**, 381 (1966).
170. K. Asayama and Y. Masuda, *J. Phys. Soc. Japan* **21**, 1459 (1966).
171. P. G. de Gennes, *Superconductivity of Metals and Alloys*, Benjamin, New York, 1966.
172. D. Cribier, B. Jacrot, L. Madhav Rao, and B. Farnoux, *Phys. Letters* **9**, 106 (1964).
173. J. Butterworth and D. E. MacLaughlin, *Phys. Rev. Letters* **20**, 265 (1968).
174. M. Fibich, *Phys. Rev. Letters* **14**, 561 (1965).
175. N. M. Rugheimer, A. Lehoczky, and C. V. Briscoe, *Phys. Rev.* **154**, 414 (1967).
176. L. H. Palmer and M. Tinkham, *Phys. Rev.* **165**, 588 (1968).
177. S. L. Norman, *Phys. Rev.* **167**, 393 (1968).
178. P. C. Hohenberg, *Zh. Eksperim. i Teor. Fiz.* **45**, 1208 (1963); *Soviet Phys. JETP* **18**, 834 (1964).

179. H. D. Drew and A. J. Sievers, *Phys. Rev. Letters* **19**, 697 (1967).
180. W. A. Thompson, *Phys. Letters* **24A**, 353 (1967).
181. J. F. Koch and P. A. Pincus, *Phys. Rev. Letters* **19**, 1044 (1967).
182. J. F. Koch and C. C. Kuo, *Phys. Rev.* **164**, 618 (1967).
183. K. H. Bennemann, *Phys. Letters* **24A**, 357 (1967).
184. J. W. F. Woo, *Phys. Rev.* **155**, 429 (1967); *Phys. Rev.* **172**, 423 (1968).
185. B. R. Tittmann and H. E. Bommel, *Phys. Rev.* **151**, 189 (1966).
186. W. A. Fate and R. W. Shaw, *Phys. Rev. Letters* **19**, 230 (1967).
187. B. R. Tittmann and H. E. Bommel, *Phys. Rev.* **151**, 178 (1966).
188. A. Hikata, R. A. Johnson, and C. Elbaum, *Phys. Rev. Letters* **17**, 916 (1966).
189. R. B. Ferguson, and J. H. Burgess, *Phys. Rev. Letters* **19**, 494 (1967).
190. E. A. Fagen and M. P. Garfunkel, *Phys. Rev. Letters* **18**, 897 (1967).
191. V. D. Fil', O. A. Shevchenko, and P. A. Bezuglyi, *Zh. Eskperim. i Teor. Fiz.* **52**, 891 (1967); *Soviet Phys. JETP* **25**, 587 (1967).
192. L. T. Claiborne and N. G. Einspruch, *Phys. Rev.* **151**, 229 (1966).
193. J. M. Perz and E. R. Dobbs, *Proc. Roy. Soc. (London)* **A297**, 408 (1967).
194. J. M. Perz and E. R. Dobbs, *Proc. Roy. Soc. (London)* **A296**, 113 (1967).
195. N. Tsuda and T. Suzuki, *J. Phys. Chem. Solids* **28**, 2487 (1967).
196. L. T. Claiborne and N. G. Einspruch, *Phys. Letters* **24A**, 589 (1967).
197. K. D. Chaudhuri and M. C. Jain, *Phys. Letters* **24A**, 84 (1967).
198. A. J. Leggett, *Prog. Theo. Phys.* **36**, 901 (1966).
199. K. Fossheim, *Phys. Rev. Letters* **19**, 81 (1967).
200. C. W. Ulbrich, D. Markowitz, R. H. Bartram, and C. A. Reynolds, *Phys. Rev.* **154**, 338 (1967).
201. G. J. Pearson et al., *Phys. Rev.* **154**, 329 (1967).
202. J. E. Gueths et al., *Phys. Rev.* **163**, 364 (1967).

5

THE GREEN'S FUNCTION METHOD

Vinay Ambegaokar

DEPARTMENT OF PHYSICS
CORNELL UNIVERSITY
ITHACA, NEW YORK

I. PROPERTIES OF GREEN'S FUNCTIONS

A. Introduction

This chapter is an introduction to the methods of quantum field theory as they apply to many-particle systems, particularly to superconductors. Although some space will be devoted to the formal structure of the methods, the emphasis will be on their utility. Several applications to the simplest model of a superconductor will be given, and the groundwork laid for more recent developments discussed in later chapters.

As a start it is appropriate to bring into focus the ideas underlying the Green's function method, and to indicate why it is particularly useful in the study of superconductivity.

A system of many particles is of necessity a system of great complexity with an extremely dense spectrum of energy levels. The details of this complexity are, however, unmeasurable and therefore uninteresting. The idea of the Green's function method is contained in the observation that in the experiments one is interested in studying the system is perturbed in relatively simple ways: by adding or removing particles, for example, or by applying electromagnetic fields, temperature gradients, sound waves, and so on. Thus it is unnecessary to attempt to calculate all the wave functions and energy levels of the system. Instead one can concentrate on the way in which it responds to simple perturbations. Even when the system one is dealing with is too complicated to permit the calculation of experimental quantities from first principles, an analysis of the sort mentioned above may allow the relation of experimental quantities to each other, or to certain parameters of the microscopic theory—calculable in principle but not in practice. This is the spirit of the Landau theory of the Fermi liquid, and also of the strong-coupling theory of superconductivity described in Chapters 10 and 11.

In essence, then, in the Green's function method one extends the basic idea of equilibrium statistical mechanics to nonthermodynamic quantities. In making this extension one can exploit to great advantage the formal similarities between the statistical operator $\exp(-\beta \mathscr{H})$ (where $\mathscr{H}$ is the Hamiltonian and β the inverse temperature in energy units) and the quantum-mechanical time-evolution operator $\exp(-i\mathscr{H}t)$. We shall see that it is extremely useful to extract once and for all certain general consequences of this formal similarity.

It is especially appropriate to study the Green's function method in connection with the theory of superconductivity, because it is in this area that the method has come into its own. There are two reasons for this. The superconductor is the first system in solid-state physics in which the interaction between electrons qualitatively changes the spectrum of excitations. It is therefore useful to have available formal methods that, as we shall see, do not at the outset conceptually prejudice one in favor of the solutions of a one-particle problem. The other reason is more subtle. The interaction that causes superconductivity is not describable by an ordinary potential: An electronic charge fluctuation with a frequency just below a characteristic frequency of the ionic lattice excites in the lattice a resonant sympathetic vibration that overcompensates for the electronic charge. As a result, part of the interaction between two electrons in the medium is a spatially short-ranged, temporally retarded *attraction*. This is the attraction that causes superconductivity, in all cases which are understood in detail. It is a cooperative interaction in which the polarization of the medium as a whole plays the essential role. But it was to treat just such interactions that the methods of quantum field theory were developed. When applied to solids, they allow ordinary potentials and dynamical interactions to be treated with equal facility, and they have the additional merit of allowing level shifts and widths to be handled in a unified and consistent way. These last remarks will not be substantiated until later chapters, but it is important to emphasize that the development which follows is not an idle formal exercise but essential for the proper phrasing of interesting questions about superconductors. For example, the remarkable developments to be described in Chapter 11—in which it is shown how superconductivity, a major mystery barely 10 years ago, is now used as a tool for studying electron–phonon interactions—depend in an essential way on analysis of the sort introduced here.

Since the task I have been given is to write an introductory chapter, pedagogic in purpose, there is no contact with very recent research. By and large, references have been given only to work that is directly related to, or supplements, the arguments of the text.

The organization of much of the material presented here is based on the elementary portions of rough lecture notes prepared for a course given at Cornell in the spring of 1963. The portion on the formal properties of Green's functions (the remainder of this section of this chapter) was in turn based on lectures given at the Brandeis Summer School in 1962 and published in *Astrophysics and the Many-Body Problem* (Benjamin, New York, 1963). I am grateful to Mr. W. A. Benjamin for permission to reuse portions of pages 341–344 of these notes. Most of the writing of the present chapter was done during a 6-week visit in the summer of 1966 to the North American Aviation Science Center in Thousand Oaks, California. I am grateful to Dr. H. Reiss, Dr. T. G. Berlincourt, and Dr. G. Lehman for extending the hospitality of this laboratory to me.

B. Formal Developments

1. Definition of Thermal Green's Functions

Consider a system of interacting fermions. For definiteness we shall restrict ourselves in this chapter to interactions describable by a sum of spin-independent two-body potentials, so that the Hamiltonian is

$$\mathscr{H} = \sum_i (p_i^2/2m) + \tfrac{1}{2} \sum_{i \neq j} v_{ij} \tag{1}$$

In Chapter 10 it will be shown that no essential complications are caused by the more realistic interaction discussed qualitatively in the last section. In the language of second quantization [an introduction to which is given in an appendix to this treatise (Volume 2)].

$$\mathscr{H} = -\frac{1}{2m}\int d^3x\, \psi_\alpha^\dagger(\mathbf{x})\, \nabla^2 \psi_\alpha(\mathbf{x})$$
$$+\frac{1}{2}\int d^3x\, d^3x'\, \psi_\alpha^\dagger(\mathbf{x})\, \psi_{\alpha'}^\dagger(\mathbf{x}')\, v(\mathbf{x}, \mathbf{x}')\, \psi_{\alpha'}(\mathbf{x}')\, \psi_\alpha(\mathbf{x}) \qquad (2)$$

(Above, and hereafter, repeated spin indices are to be summed over and $\hbar = 1$.) In Eq. (2) $\psi_\alpha(\mathbf{x})$ is the quantized field operator obeying the usual equal time anticommutation relations

$$\begin{aligned} \{\psi_\alpha(\mathbf{x}), \psi_{\alpha'}(\mathbf{x}')\} &= \{\psi_\alpha^\dagger(\mathbf{x}), \psi_{\alpha'}^\dagger(\mathbf{x}')\} = 0 \\ \{\psi_\alpha(\mathbf{x}), \psi_\beta^\dagger(\mathbf{x}')\} &= \delta_{\alpha\beta}\delta^3(\mathbf{x}-\mathbf{x}') \end{aligned} \qquad (3)$$

Let us see how to write an experimentally measurable quantity in terms of these operators. If a voltage V is applied to a superconducting tunnel junction so as to cause current to flow from left to right, one might be willing to accept on intuitive grounds that the current is

$$I \propto \int_{\mu_r}^{\mu_r + eV} d\omega\, N_r^>(\omega)\, N_l^<(\omega) \qquad (4)$$

where $N_r^>$ is the density of states available to a particle on the right, $N_l^<$ the density of states from which a particle on the left may be removed, and μ_r the chemical potential on the right. Equation (4) is indeed correct (at $T = 0$), as will be seen in a later chapter. What concern us here are the formal expressions for $N^>$ and $N^<$. Again intuitively one expects

$$N^>(\omega) = \frac{1}{V}\sum_{\mathbf{p}\alpha}\sum_n |\langle n|\, c_{\mathbf{p},\alpha}^\dagger\, |G\rangle|^2\, \delta(E_n - E_G - \omega) \qquad (5)$$

$$N^<(\omega) = \frac{1}{V}\sum_{\mathbf{p}\alpha}\sum_n |\langle n|\, c_{\mathbf{p},\alpha}\, |G\rangle|^2\, \delta(E_G - E_n - \omega) \qquad (6)$$

where G is the ground state of the system, and $c_{\mathbf{p},\alpha}$, $c_{\mathbf{p},\alpha}^\dagger$ are annihilation and creation operators for states of momentum $\mathbf{p}$ and spin α, i.e. normalizing the states in a box of volume V:

$$\psi_\alpha(\mathbf{x}) = \frac{1}{V}\sum_{\mathbf{p}} \exp(i\mathbf{p}\cdot\mathbf{x})\, c_{\mathbf{p},\alpha} \qquad (7)$$

In Eq. (5) one is asking how many states of the interacting system per unit energy interval around ω can be reached by adding a particle. Similarly, in

Eq. (6) one is asking for the spectral density of states reached by removing a particle. A simple formal transformation yields

$$N^{>}(\omega)=\frac{1}{2\pi V}\sum_{\mathbf{p}\alpha}\int_{-\infty}^{\infty}dt\,\exp(i\omega t)\,\langle G|\,c_{\mathbf{p},\alpha}(t)\,c_{\mathbf{p},\alpha}^{\dagger}(0)\,|G\rangle$$
$$N^{<}(\omega)=\frac{1}{2\pi V}\sum_{\mathbf{p}\alpha}\int_{-\infty}^{\infty}dt\,\exp(i\omega t)\,\langle G|\,c_{\mathbf{p},\alpha}^{\dagger}(0)\,c_{\mathbf{p},\alpha}(t)\,|G\rangle \tag{8}$$

where the operators are now in the Heisenberg picture, i.e., $c_{\mathbf{p},\alpha}(t)=\exp(i\mathscr{H}t)$ $c_{\mathbf{p},\alpha}(0)\exp(-i\mathscr{H}t)$. At finite temperatures one similarly expects the tunneling characteristic to involve ($T\equiv$ temperature)

$$N^{>}(\omega,T)=\frac{1}{2\pi V}\sum_{\mathbf{p},\alpha}\int_{-\infty}^{\infty}dt\,\exp(i\omega t)\,\langle c_{\mathbf{p}\alpha}(t)\,c_{\mathbf{p}\alpha}^{\dagger}(0)\rangle \tag{9}$$

and $N^{<}(\omega,T)$, similarly defined. Here the angular brackets denote an average in a thermal ensemble. One is thus led to be interested in objects like ($x\equiv\mathbf{x},\alpha$)

$$\begin{aligned}\langle\psi(x,t)\,\psi^{\dagger}(x',t')\rangle &\equiv iG^{>}(xt,x't')\\ \langle\psi^{\dagger}(x't')\,\psi(xt)\rangle &\equiv -iG^{<}(xt,x't')\end{aligned} \tag{10}$$

The single-particle thermodynamic Green's function is defined (*1–4*) as

$$G(xt,x't')\equiv\begin{cases}G^{>}(xt,x't') & t>t'\\ G^{<}(xt,x't') & t'>t\end{cases} \tag{11}$$

The usefulness of this last definition will only become apparent later but one can already note that the interesting physical quantity $N^{>}(\omega)$ is related to $G^{>}$ in a simple way. If we Fourier-transform $G^{>}$:

$$G_{\alpha\beta}^{>}(\mathbf{x}t,\mathbf{x}'t')=\frac{1}{V}\sum_{\mathbf{p}}\int_{-\infty}^{\infty}\frac{d\omega}{2\pi}\,G_{\alpha\beta}^{>}(\mathbf{p},\omega)\exp[i\mathbf{p}\cdot(\mathbf{x}-\mathbf{x}')]\exp[-i\omega(t-t')]$$

we see that

$$N^{>}(\omega)=\frac{i}{2\pi}\sum_{\mathbf{p}\alpha}G_{\alpha\alpha}^{>}(\mathbf{p},\omega) \tag{12}$$

the identical relation connecting $N^{<}$ and $-G^{<}$.

A shorthand notation for the Green's function [Eq. (11)] is

$$G(xt,x't')=-i\langle T[\psi(xt)\,\psi^{\dagger}(x't')]\rangle \tag{13}$$

where T is called Wick's time-ordering operator. It orders the operators so that the later time comes to the left, and gives a change of sign for every permutation of fermion operators. Just as the single-particle Green's function contains information about the density of states sampled by tunneling, other Green's functions contain information about the densities of states probed by more complicated measurements. Without stopping to make explicit the connection with measurable quantities we shall simply define the two-particle Green's function as

$$G(x_1t_1, x_2t_2; x_1't_1', x_2't_2') \equiv (-i)^2 \langle T[\psi(x_1t_1)\,\psi(x_2t_2)\,\psi^\dagger(x_2't_2')\,\psi^\dagger(x_1't_1')]\rangle \tag{14}$$

In general the n-particle Green's function is defined by

$$G(x_1t_1 \cdots x_nt_n; x_1't_1' \cdots x_n't_n') \equiv (-i)^n \langle T[\psi(x_1t_1)\cdots\psi(x_nt_n)\,\psi^\dagger(x_n't_n')\cdots\psi^\dagger(x_1't_1')]\rangle \tag{15}$$

The Green's functions are thus defined as expectation values of time-ordered products of field operators. They describe progressively more complicated correlations of particles in a many-body system. Most quantities whose measurement does not drive the system very far away from equilibrium are more or less directly related to low-order Green's functions. We have already seen from the example of tunneling that it is a sensible strategy to try to calculate the Green's functions rather than the complete spectrum of eigenvalues and eigenfunctions of the many-particle Hamiltonian. The latter are both more difficult to calculate and less directly related to experimental observations.

2. Theory of the Single-Particle Green's Function

Considerable information about the single-particle Green's function can be obtained just from its definition [Eq. (13)]. By studying the structure of this form we shall be able to learn the easiest ways of obtaining the information contained in it. We shall always take averages in the grand canonical ensemble, i.e.,

$$\langle 0\rangle \equiv \mathrm{Tr}\{\exp[-\beta(\mathscr{H} - \mu N)]\,0\}/\mathrm{Tr}\{\exp[-\beta(\mathscr{H} - \mu N)]\} \tag{16}$$

where $\mathscr{H}$ is the Hamiltonian, e.g., Eq. (2), and N is the total number of particles.

As a start let us consider a system of *noninteracting* particles, the Hamiltonian for which we shall call $\mathscr{H}_0$. We expand the field operators in terms of the complete set of orthonormal eigenstates of the Hamiltonian for a single particle. Then

$$\begin{aligned}\psi(x,t) &= \sum_r u_r(x)\exp(i\mathscr{H}_0 t)\,c_r\exp(-i\mathscr{H}_0 t)\\ &= \sum_r u_r(x)\exp(-i\epsilon_r t)\,c_r\end{aligned} \tag{17}$$

where ϵ_r is the energy of the single-particle state r. The single-particle Green's function is then

$$G(xt, x't') = \begin{cases} -i\sum_{rs} u_r(x)\, u_s^*(x') \exp(-i\epsilon_r t) \exp(i\epsilon_s t') \langle c_r c_s^\dagger \rangle & t < t' \\ +i\sum_{rs} u_r(x)\, u_s^*(x') \exp(-i\epsilon_r t) \exp(i\epsilon_s t') \langle c_s^\dagger c_r \rangle & t' < t \end{cases} \tag{18}$$

The expectation values are familiar from statistical mechanics:

$$\begin{aligned} \langle c_r c_s^\dagger \rangle &= \delta_{rs} \prod_{t \neq r} \{\exp[-\beta(\epsilon_t - \mu)] + 1\} / \prod_t \{\exp[-\beta(\epsilon_t - \mu)] + 1\} \\ &= \delta_{rs} \{\exp[-\beta(\epsilon_r - \mu)] + 1\}^{-1} \equiv \delta_{rs} f_r^+ \end{aligned} \tag{19}$$

$$\begin{aligned} \langle c_s^\dagger c_r \rangle &= \delta_{rs} \exp[-\beta(\epsilon_r - \mu)] / \{\exp[-\beta(\epsilon_r - \mu)] + 1\} \\ &\equiv \delta_{rs} f_r^- \end{aligned} \tag{20}$$

Substituting these results into Eq. (18) we get

$$G(xt, x't') = \begin{cases} -i\sum_r u_r(x)\, u_r^*(x') \exp[-i\epsilon_r(t - t')]\, f_r^+ & t > t' \\ i\sum_r u_r(x)\, u_r^*(x') \exp[-i\epsilon_r(t - t')]\, f_r^- & t' > t \end{cases} \tag{21}$$

In Eq. (21) we see a single oscillating time dependence associated with each single-particle level of the system. In general, the single-particle Green's function will give information about the existence and energies of single-particle-like excitations. For an interacting system the single-particle Green's function will not have any simple form like Eq. (21) but it may be that there is only a very narrow band of frequencies associated with, for example, a definite momentum. In that case one would say that the system has a "quasi-particle" excitation spectrum.

Let us now consider the general case. Since the interaction Hamiltonian occurs both in the time dependence of the operators in the Heisenberg picture and in the statistical density it is very useful to exploit the formal similarity between these two factors. In fact, if one continues the expressions for the Green's functions to complex times, the two occurrences of the interaction Hamiltonian take on very similar forms. We write, as in Eq. (11),

$$G(xt, x't') = G^>(xt, x't')\,\theta(t - t') + G^<(xt, x't')\,\theta(t' - t) \tag{22}$$

and use the cyclic property of traces

$$\mathrm{Tr}(ABC) = \mathrm{Tr}(CAB) = \mathrm{Tr}(BCA) \tag{23}$$

to get

$$G^{>}(xt, x't') = -i\frac{\mathrm{Tr}\{\exp[-\beta(\mathscr{H}-\mu N)]\exp[i\mathscr{H}(t-t')]\psi(x)\exp[-i\mathscr{H}(t-t')]\psi^{\dagger}(x')\}}{\mathrm{Tr}\{\exp[-\beta(\mathscr{H}-\mu N)]\}}$$

$$G^{<}(xt, x't') = i\frac{\mathrm{Tr}\{\exp[-\beta(\mathscr{H}-\mu N)]\psi^{\dagger}(x')\exp[i\mathscr{H}(t-t')]\psi(x)\exp[-i\mathscr{H}(t-t')]\}}{\mathrm{Tr}\{\exp[-\beta(\mathscr{H}-\mu N)]\}} \tag{24}$$

which shows explicitly [using the fact that $(\mathscr{H}, N) = 0$] that

$$G(t, t') = G(t - t') \tag{25}$$

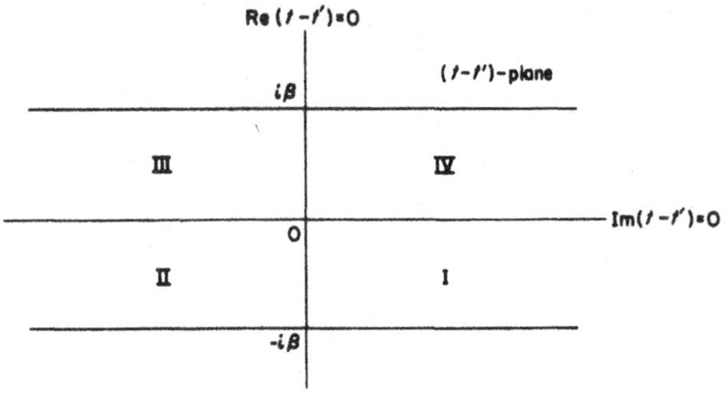

Fig. 1.

It is natural to assume that the traces in Eq. (24) converge absolutely whenever the coefficients of $\mathscr{H}$ in the exponentials have negative real parts. With this assumption we see that, with reference to Fig. 1, $G^{>}$ may be analytically continued into the regions I and II in the time difference plane, and $G^{<}$ to the regions III and IV. We therefore extend the definition of the Green's function as follows:

$$G(xt, x't') = \begin{cases} G^{>}(xt, x't') & (t-t') \text{ on positive real axis, I and II} \\ G^{<}(xt, x't') & (t-t') \text{ on negative real axis, III and IV} \end{cases} \tag{26}$$

(In the next few equations t is the time difference, i.e., $t - t' \to t$.) In the extended region of the time difference, $G^{>}$ and $G^{<}$ are not independent. For, using the relation

$$\psi^{\dagger} f(N+1) = f(N)\psi^{\dagger} \tag{27}$$

where f is any function of N, and the cyclic property [Eq. (23)], one easily sees that for t in III or IV (we omit the space-spin indices in the next several equations)

$$G^{>}(t - i\beta) = -\exp(-\beta\mu)\, G^{<}(t) \tag{28}$$

and similarly for t in I and II,

$$G^<(t+i\beta) = -\exp(\beta\mu)\, G^>(t) \tag{29}$$

The periodicity conditions [Eqs. (28) and (29)] can be used to construct a spectral representation for the Fourier transform of the Green's function for real times. Making Fourier transforms we have

$$G^>(t) = \int_{-\infty}^{\infty} \frac{d\omega}{2\pi} G^>(\omega) \exp(-i\omega t) \qquad t \text{ in I and II} \tag{30}$$

$$G^<(t) = \int_{-\infty}^{\infty} \frac{d\omega}{2\pi} G^<(\omega) \exp(-i\omega t) \qquad t \text{ in III and IV} \tag{31}$$

From Eqs. (28) and (29) it follows that

$$G^>(\omega) = -\exp[\beta(\omega-\mu)]\, G^<(\omega) \tag{32}$$

It is convenient to introduce a function called the spectral weight function, which is defined as

$$A(\omega) = i[G^>(\omega) - G^<(\omega)] \tag{33}$$

In terms of this function we have

$$G^>(\omega) = (1/i)\, A(\omega) \{\exp[-\beta(\omega-\mu)] + 1\}^{-1} \equiv (1/i)\, A(\omega) f^+(\omega) \tag{34}$$

and

$$G^<(\omega) = iA(\omega) \{\exp[\beta(\omega-\mu)] + 1\}^{-1} \equiv iA(\omega) f^-(\omega) \tag{35}$$

Now for real t we can make a Fourier transformation

$$G(t) = \int_{-\infty}^{\infty} \frac{d\omega}{2\pi} G(\omega) \exp(-i\omega t) \tag{36}$$

Using the form Eq. (11) for $G(t)$ one finds that

$$G(\omega) = i \int \frac{d\bar{\omega}}{2\pi} \left[\frac{G^>(\bar{\omega})}{\omega - \bar{\omega} + i\eta} - \frac{G^<(\bar{\omega})}{\omega - \bar{\omega} - i\eta} \right] \tag{37a}$$

where η is an infinitesimal real positive quantity. [To verify Eq. (37a) one forms $G(t)$ according to Eq. (36) and performs the ω-integral, which can be closed along an infinite semicircle in the lower half-plane for positive t and in the upper half-plane for negative t.] Substituting Eqs. (34) and (35) into Eq. (37a) one obtains the spectral representation

$$G(\omega) = \int_{-\infty}^{\infty} \frac{d\bar{\omega}}{2\pi} A(\bar{\omega}) \left[\frac{f^+(\bar{\omega})}{\omega - \bar{\omega} + i\eta} + \frac{f^-(\bar{\omega})}{\omega - \bar{\omega} - i\eta} \right] \tag{37b}$$

It is thus seen that the Green's function is completely determined by the spectral weight function. What has been achieved is a separation of the purely statistical aspects of the Green's function [contained in the brackets of Eq. (37b) from those aspects [contained in $A(\omega)$] which reflect the microscopic structure of the system one is examining. It follows from its definition that the spectral weight function can be written formally in terms of many-particle eigenstates as follows:

$$\begin{aligned} A(x, x', \omega) &= i[G^{>}(x, x', \omega) - G^{<}(x, x', \omega)] \\ &= 2\pi \sum_{nm} \rho_n [\langle n| \psi(x) |m\rangle \, \delta(E_n - E_m + \omega) \langle m| \psi^{\dagger}(x') |n\rangle \\ &\qquad + \langle n| \psi^{\dagger}(x') |m\rangle \, \delta(E_m - E_n + \omega) \langle m| \psi(x) |n\rangle] \end{aligned} \tag{38}$$

where ρ_n is the statistical weight factor

$$\rho_n = \exp[-\beta(E_n - \mu N_n)]/\mathrm{Tr}\{\exp[-\beta(\mathscr{H} - \mu N)]\} \tag{39}$$

and the subscripts m, n refer to the eigenstates of $\mathscr{H}$. For a translationally invariant system, A is a function of $\mathbf{x} - \mathbf{x}'$. Its Fourier transform is then the density of states that can be reached by adding or removing a particle of given momentum and energy. In the limit of zero temperature the ground state will in general acquire all the statistical weight. The first term in Eq. (39) will then contribute only for ω greater than the chemical potential μ, because μ is the difference in ground-state energies of two large systems which differ by the addition of one particle. Similarly, the second term in the brackets will only contribute for $\omega < \mu$. The densities of single-particle and single-hole states described by A are precisely those which we associated with tunneling.

For a noninteracting system, it is easy to see that $A(x, x', \omega)$ has the simple form

$$A(x, x', \omega) = 2\pi \sum_r u_r(x) \, u_r^*(x') \, \delta(\omega - \epsilon_r) \tag{40}$$

In particular, for a translationally invariant system with spin-independent forces, for which the wave functions are plane waves with spin projection up or down, we have

$$A_{\alpha\beta}(\mathbf{x}, \mathbf{x}', \omega) = \frac{1}{V} \sum_{\mathbf{p}} \delta_{\alpha\beta} A(\mathbf{p}, \omega) \exp[i\mathbf{p}\cdot(\mathbf{x} - \mathbf{x}')] \tag{41}$$

where

$$A(\mathbf{p}, \omega) = 2\pi\delta(\omega - \epsilon_p) \tag{42}$$

For an interacting system the spectral function $A(\mathbf{p}, \omega)$ will not have a simple peaked structure like Eq. (42), but it will obey two conditions which follow quite generally from the formal expression Eq. (38). These are

$$A(\mathbf{p}, \omega) \geqslant 0$$

$$\int_{-\infty}^{\infty} \frac{d\omega}{2\pi} A(\mathbf{p}, \omega) = \langle\{c_{\mathbf{p}}, c_{\mathbf{p}}^{\dagger}\}\rangle = 1 \tag{43}$$

This last formula is a special case, valid for a translationally invariant system with spin-independent forces, of the more general sum rule that follows from Eq. (38):

$$\int_{-\infty}^{\infty} \frac{d\omega}{2\pi} A_{\alpha\beta}(\mathbf{x}, \mathbf{x}', \omega) = \delta^3(\mathbf{x} - \mathbf{x}')\, \delta_{\alpha\beta} \tag{44}$$

We see from Eq. (42) that, for a noninteracting system, the sum rule [Eq. (43)] is exhausted by a single peak. A sharply peaked spectral function for an interacting system indicates a long-lived single-particle-like excitation. This remark can be made more precise (*5*). At $T = 0$ the propagator for a particle of momentum $\mathbf{p}$ is seen from Eq. (34) to be

$$G^{>}(\mathbf{p}, t) = \int_{\mu}^{\infty} \frac{d\omega}{2\pi} A(\mathbf{p}, \omega) \exp(-i\omega t) \tag{45}$$

If $A(\mathbf{p},\omega)$ has a pole at $\omega = E_{\mathbf{p}} - i\Gamma_{\mathbf{p}}/2$ with residue $Z_{\mathbf{p}}$and if $\Gamma_{\mathbf{p}} \ll E_{\mathbf{p}}$, it can be shown (*5*) that for $t > 1/E_{\mathbf{p}}$ one has

$$G^{>}(\mathbf{p}, t) = -iZ_{\mathbf{p}} \exp(-iE_{\mathbf{p}}t) \exp[-(\Gamma_{\mathbf{p}}/2)\, t] + O(\Gamma_{\mathbf{p}}/E_{\mathbf{p}}^2 t) \tag{46}$$

Poles of $A(\mathbf{p},\omega)$ near the real ω-axis thus describe long-lived excitations.

The spectral weight function being established as the physically significant attribute of the Green's function, the question of how best to extract it from a microscopic theory becomes merely one of strategy. A widely known strategy which is successful, direct, and not without elegance will now be described.

When the Green's function is continued to the complex t-plane as in Fig. 1, lines parallel to the imaginary time-difference axis become especially interesting because along such lines the Green's function has simple periodicity properties. For simplicity we shall consider the imaginary time-difference axis itself, between the points $i\beta$ and $-i\beta$. Time differences are restricted to this region if the two times t, t' on which the Green's function in general depends are separately restricted to the region:

$$\operatorname{Re} t = 0 \qquad -\beta < \operatorname{Im} t < 0 \tag{47}$$

[One could equally well require $\operatorname{Re} t = t_0$, where t_0 is any real number. This additional generality has been taken advantage of by Baym and Kadanoff (*2*).] The region defined in Eq. (47) will hereafter be called the restricted region.

We define

$$\tau = -i\beta \tag{48}$$

and write for the restricted region $0 < t, t' < \tau$ the inequality sign referring to the *negative imaginary part*. With these conventions the periodicity conditions

Eqs. (28) and (29) for t, t' in the restricted time region read

$$\begin{aligned} G(0, t') &= G^{<}(-t') = -\exp(\beta\mu)\, G^{>}(\tau - t') = -\exp(\beta\mu)\, G(\tau, t') \\ G(t, 0) &= G^{>}(t) = -\exp(-\beta\mu)\, G^{<}(t - \tau) = -\exp(-\beta\mu)\, G(t, \tau) \end{aligned} \tag{49}$$

It is natural to satisfy these conditions by periodically extending the Green's function in strips of width 2β parallel to the real axis, and making a Fourier series analysis (*6*):

$$G(t - t') = \frac{1}{\tau} \sum_{l} G(\zeta_l) \exp[-i\zeta_l (t - t')] \tag{50}$$

where

$$\begin{aligned} l &= 0, \pm 1, \pm 2, \dots \\ \zeta_l &= \mu + (2l + 1)\, \pi/\tau \end{aligned} \tag{51}$$

To verify that Eq. (50) does, in fact, satisfy the periodicity conditions note that

$$\begin{aligned} G(t \pm \tau) &= \frac{1}{\tau} \sum_{l} G(\zeta_l) \exp(\mp i\mu\tau) \exp[\mp (2l + 1)\, \pi i] \exp(-i\zeta_l t) \\ &= -\exp(\mp \beta\mu)\, G(t) \end{aligned} \tag{52}$$

For future reference we note here that it is sometimes convenient to measure energies with respect to the chemical potential. If one does this, the time dependence in the Heisenberg picture is governed by $\exp[i(\mathscr{H} - \mu N)t]$. The factors $\exp(\pm \beta\mu)$ then no longer occur in Eq. (49); and in the Fourier transformation Eq. (50) $\zeta_l = (2l + 1)\pi/\tau$.

The microscopic theory, as we shall see, lends itself naturally to the calculation of the Fourier series coefficients $G(\zeta_l)$. The physics is, however, contained, as we have seen, in $G(t)$ for t along the real axis, or equivalently in $A(\omega)$. Let us see how to extract this last quantity from $G(\zeta_l)$. Inverting the Fourier series ‡ [Eq. (50)]:

$$\begin{aligned} G(\mathbf{p}, \zeta_l) &= \int_0^\tau \exp(i\zeta_l t)\, G(\mathbf{p}, t)\, dt \\ &= \int_0^\tau \exp(i\zeta_l t)\, G^{>}(\mathbf{p}, t)\, dt \end{aligned} \tag{53}$$

‡ Some authors prefer to use another symbol for the time variables when these are continued to the restricted region, by writing $t = -iu$, $0 < u < \beta$. The Fourier series expansion and its inverse then read

$$G(u) = \frac{1}{\beta} \sum_{l} \exp(-i\omega_l u)\, G(\omega_l)$$

$$G(\omega_l) = \int_0^\beta du\, G(u) \exp(i\omega_l u)$$

where $i\omega_l = \zeta_l$ and

$$G(\omega_l) = iG(\zeta_l = i\omega_l)$$

Fourier-transforming $G^{>}(t)$ and substituting Eq. (34) for $G^{>}(\omega)$:

$$G(\mathbf{p}, \zeta_l) = -i\int_0^{\tau} dt \exp(i\zeta_l t)\int_{-\infty}^{\infty}\frac{d\omega}{2\pi}\exp(-i\omega t)\, A(\mathbf{p}, \omega) f^{+}(\omega) \tag{54}$$

Performing the t-integral we obtain

$$G(\mathbf{p}, \zeta_l) = \int_{-\infty}^{\infty}\frac{d\omega}{2\pi}\frac{A(\mathbf{p}, \omega)}{\zeta_l - \omega} \tag{55}$$

If $G(\mathbf{p}, \zeta_l)$ as obtained from a microscopic theory can be put in the form of Eq. (55), $A(\mathbf{p}, \omega)$ can be obtained by inspection. For the result of a microscopic calculation is to determine $G(p, \zeta_l)$ not as a set of numbers but as a function evaluated at the points ζ_l. One can, therefore, continue to the complex ζ-plane, obtaining

$$G(\mathbf{p}, \zeta) = \int_{-\infty}^{\infty}\frac{d\omega}{2\pi}\frac{A(\mathbf{p}, \omega)}{\zeta - \omega} \tag{56}$$

It is seen that $G(\mathbf{p}, \zeta)$ is analytic everywhere except for a branch cut on the real ζ-axis. The spectral weight function $A(\omega)$ is just the discontinuity across the cut. Explicitly

$$\begin{aligned} G(\zeta = \omega + i\eta) &- G(\zeta = \omega - i\eta) \\ &= \int_{-\infty}^{\infty}\frac{d\omega'}{2\pi} A(\omega')\,[(\omega - \omega' + i\eta)^{-1} - (\omega - \omega' - i\eta)^{-1}] \\ &= -i\int_{-\infty}^{\infty} d\omega'\, A(\omega')\,\delta(\omega - \omega') = -iA(\omega) \end{aligned} \tag{57}$$

There is, however, one difficulty with this procedure. A function defined on the set of points ζ_l can be continued to the complex ζ-plane in an infinite number of ways. For example, the two functions

$$G(\zeta) \qquad G(-\exp[-i\tau(\zeta - \mu)]\,\zeta) \tag{58}$$

are identical on the points ζ_l but completely different elsewhere. The continuation we are interested in has property that $G(\zeta)$ goes to zero and is analytic along any straight line off the real axis in the complex ζ-plane. It can be shown that this requirement uniquely specifies the continuation (7). We shall see later that to make the correct continuation it is often necessary to perform sums over intermediate variables ζ_l'. It is, therefore, necessary to learn how to do such sums explicitly. Consider a simple example. Suppose we are given

$$G(\mathbf{p}, \zeta_l) = (\zeta_l - \epsilon_{\mathbf{p}})^{-1} \tag{59}$$

and asked to find $G(t)$ for real t. This is, of course, a perfectly trivial example since Eq. (59) holds for a noninteracting system, and we know the answer. We can also determine the answer by inspection using the method discussed above, since we see at once that the discontinuity across the real axis is just given by Eq. (42), and $G(t)$ is then determined by Eqs. (30) and (31). However, if we did not know these formal properties we would have to sum the series

$$G(\mathbf{p}, t) = \frac{1}{\tau} \sum_{l} G(\mathbf{p}, \zeta_l) \exp(-i\zeta_l t) \tag{60}$$

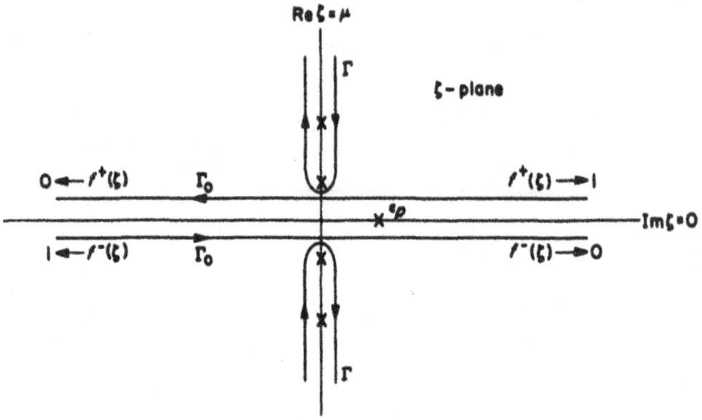

Fig. 2.

for t in the restricted region. The easiest way to do this is to transform the sum into a contour integral surrounding the poles of a function that has simple poles at the values ζ_l. The functions $f^{\pm}(\zeta)$ defined according to

$$f^{\pm}(\zeta) = \{\exp[\mp i\tau(\zeta - \mu)] + 1\}^{-1} \tag{61}$$

have the correct poles for this purpose, with residues $(\pm i\tau)^{-1}$. We can then sum the series in Eq. (60) by integrating along the contour Γ (see Fig. 2),

$$G(\mathbf{p}, t) = \frac{1}{\tau} \frac{-1}{2\pi i} (\pm i\tau) \int_{\Gamma} d\zeta \frac{f^{\pm}(\zeta) \exp(-i\zeta t)}{\zeta - \epsilon_{\mathbf{p}}} \tag{62}$$

If we choose the function (f^+ or f^-) for which the integrand is suitably convergent both to the left and to the right, we can deform the contour Γ into the contour Γ_0 without changing the value of the integral. Remembering that the inequality sign refers to the negative imaginary part and that we are considering t in the restricted time region, we see that the appropriate choice is f^+ for $t > 0$ and f^- for $t < 0$. The integral is then easily done, since the contour Γ_0 contains only a single pole at $\zeta = \epsilon_p$. We get

$$G(\mathbf{p}, t) = \begin{cases} -if^{+}(\epsilon_{\mathbf{p}}) \exp(-i\epsilon_{\mathbf{p}} t) & t > 0 \\ if^{-}(\epsilon_{\mathbf{p}}) \exp(-i\epsilon_{\mathbf{p}} t) & t < 0 \end{cases} \tag{63}$$

where t is in the restricted region. The extension to real t is now trivial, and one obtains the known result

$$G(\mathbf{p}, t) = \begin{cases} -if^{+}(\epsilon_{\mathbf{p}}) \exp(-i\epsilon_{\mathbf{p}}t) & t \text{ real and positive} \\ if^{-}(\epsilon_{\mathbf{p}}) \exp(-i\epsilon_{\mathbf{p}}t) & t \text{ real and negative} \end{cases} \tag{64}$$

The above simple example illustrates a completely general method for doing sums over ζ_l that will be of great use later.

The techniques needed for extracting $A(\omega)$ from $G(\zeta_l)$ have now been developed. Before turning to another topic, we should mention that although the above analysis was concerned specifically with the single-particle Green's function, very similar procedures can be applied to any correlation function that depends on only one time difference. Spectral representations for correlation functions depending on two and three time differences are known and have been usefully employed (*8,9*), but we shall not concern ourselves with them here.

3. Equations of Motion

The Green's functions defined at the end of Section B.1 are related by a hierarchy of coupled integrodifferential equations, which we shall now construct. These equations, together with the periodic boundary condition in the complex time plane, determine the thermal Green's functions in principle. They are also, however, extremely useful as the starting point for approximate calculations.

Under the influence of the Hamiltonian [Eq. (1)] the field operator in the Heisenberg picture is governed by the equation of motion,

$$i(\partial/\partial t)\, \psi_\alpha(\mathbf{x}, t) = [\psi_\alpha(\mathbf{x}, t), \mathscr{H}] \tag{65}$$

Working out the commutator explicitly one obtains

$$i(\partial/\partial t)\, \psi_\alpha(\mathbf{x}, t) = -(\nabla^2/2m)\, \psi_\alpha(\mathbf{x}, t) + \int d^3x'\, v(\mathbf{x}, \mathbf{x}')\, \psi^\dagger_{\alpha'}(\mathbf{x}', t)\, \psi_{\alpha'}(\mathbf{x}', t)\, \psi_\alpha(\mathbf{x}, t) \tag{66}$$

Taking the time derivative of the Green's function [Eq. (13)] one thus obtains the following equation of motion (where we now explicitly display the spin indices)

$$\begin{aligned} i(\partial/\partial t)\, G_{\alpha\alpha'}(\mathbf{x}t, \mathbf{x}'t') &= \delta(t-t')\, [\langle \psi_\alpha(\mathbf{x}t)\, \psi^\dagger_{\alpha'}(\mathbf{x}'t')\rangle + \langle \psi^\dagger_{\alpha'}(\mathbf{x}'t')\, \psi_\alpha(\mathbf{x}t)\rangle] \\ &\quad - (\nabla^2/2m)\, G_{\alpha\alpha'}(\mathbf{x}t, \mathbf{x}'t') \\ &\quad - i\int d^3x'\, v(\mathbf{x}, \mathbf{x}'')\, \langle T[\psi^\dagger_{\alpha''}(\mathbf{x}''t)\, \psi_{\alpha''}(\mathbf{x}''t)\, \psi_\alpha(\mathbf{x}t)\, \psi^\dagger_{\alpha'}(\mathbf{x}'t')]\rangle \end{aligned} \tag{67}$$

In Eq. (67) the term proportional to $\delta(t-t')$ comes from the step functions in the definition [Eq. (22)] of the Green's function. Because of the δ-function, the terms multiplying it make up an equal time anticommutator which is given by Eq. (3). The last term in Eq. (67) is clearly a special case of the two-particle Green's function [Eq. (14)]. If we add an infinitesimal positive time to the argument of the operator $\psi^{\dagger}_{x''}(\mathbf{x}'', t)$, calling the resulting argument t^+, we may permute the four field operators at will, remembering only to change the sign with each permutation, because the operator T returns the fields to the order given in Eq. (67). This equation of motion may thus be compactly written as

$$\left(i\frac{\partial}{\partial t}+\frac{\nabla^2}{2m}\right)G_{\alpha\alpha'}(\mathbf{x}t,\mathbf{x}'t')=\delta_{\alpha\alpha'}\delta^3(\mathbf{x}-\mathbf{x}')\,\delta(t-t')$$
$$-i\int d^3x''\,v(\mathbf{x},\mathbf{x}'')\,G_{\alpha\alpha'',\alpha'\alpha''}(\mathbf{x}t,\mathbf{x}''t;\mathbf{x}'t',\mathbf{x}''t^+)\tag{68}$$

Equation (68) is the first in an infinite hierarchy of equations that couple successively higher-order Green's functions. For the two-particle Green's function we obtain (combining the space and spin indices for brevity)

$$\left(i\frac{\partial}{\partial t_1}+\frac{\nabla_1^2}{2m}\right)G(x_1t_1,x_2t_2;x_3t_3,x_4t_4)=\delta(x_1-x_3)\,\delta(t_1-t_3)\,G(x_2t_2,x_4t_4)$$
$$-\delta(x_1-x_4)\,\delta(t_1-t_4)\,G(x_2t_2,x_3t_3)$$
$$-i\int dx_5\,v(x_1,x_5)\,G(x_1t_1,x_2t_2,x_5t_1;x_3t_3,x_4t_4,x_5t_1^+)\tag{69}$$

where

$$\delta(x_1-x_3)\equiv\delta^3(\mathbf{x}_1-\mathbf{x}_3)\,\delta_{\alpha\alpha'}\tag{70}$$

and $\int dx_5=\sum_{\alpha_5}\int d\mathbf{x}_5$. In deriving these integrodifferential equations the only property of the statistical average that has been used is its normalization:

$$\langle 1\rangle=1\tag{71}$$

To obtain the thermodynamic Green's functions we have to adjoin boundary conditions to the equations of motion. The appropriate boundary condition is that of periodicity along the imaginary time axis. If all times are required to lie in the restricted region discussed in the last section, it is easy to see that

$$G(t_1\cdots t_{i-1},0,t_{i+1}\cdots t_n;t_1'\cdots t_n')$$
$$=-\exp(\beta\mu)\,G(t_1\cdots t_{i-1},\tau,t_{i+1}\cdots t_n;t_1'\cdots t_n')\tag{72}$$

$$G(t_1\cdots t_n;t_1'\cdots t_{i-1}',0,t_{i+1}'\cdots t_n')$$
$$=-\exp(-\beta\mu)\,G(t_1\cdots t_n;t_1'\cdots t_{i-1}',\tau,t_{i+1}'\cdots t_n')\tag{73}$$

As a result of these periodicity conditions, the thermodynamic Green's functions may be expanded in multiple Fourier series according to

$$G(t_1 \cdots t_n; t'_1 \cdots t'_n) = \frac{1}{(\tau)^{2n}} \sum_{\{l, l'\}} \exp[i(\zeta_{l_1} t_1 + \cdots + \zeta_{l_n} t_n)]$$
$$\times \exp[-i(\zeta_{l'_1} t'_1 + \cdots + \zeta_{l'_n} t'_n)]\, G(\zeta_{l_1} \cdots \zeta_{l_n}; \zeta_{l'_1} \cdots \zeta_{l'_n}) \quad (74)$$

Because the left side of Eq. (74) is invariant under a simultaneous translation of all coordinates, the Fourier coefficients are zero unless

$$\zeta_{l_1} + \cdots + \zeta_{l_n} = \zeta_{l'_1} + \cdots + \zeta_{l'_n} \quad (75)$$

Thus the thermodynamic Green's functions are obtained in principle by continuing the equations of motion to the restricted region and then substituting Fourier expansions of the form of Eq. (74). One thus obtains a hierarchy of equations for the Fourier series coefficients. The equations in principle determine the series coefficients, and from there the thermodynamic Green's functions may be constructed according to Eq. (74) with subsequent continuation to real times. The equations of motion for the Green's functions of an noninteracting system are especially simple. If $v = 0$ in Eq. (68), this equation reduces to

$$\left(i\frac{\partial}{\partial t} + \frac{\nabla^2}{2m}\right) G^0_{\alpha\alpha'}(\mathbf{x}t, \mathbf{x}'t') = \delta_{\alpha\alpha'}\delta^3(x - x')\,\delta(t - t') \quad (76)$$

which exhibits for the first time the reason for the usage: Green's function. According to the discussion given above, the Fourier series coefficient is determined by

$$\left(\zeta_l + \frac{\nabla^2}{2m}\right) G^0_{\alpha\alpha'}(\mathbf{x}, \mathbf{x}'; \zeta_l) = \delta_{\alpha\alpha'}\delta^3(x - x') \quad (77)$$

For a translationally invariant system G^0 is a function of $\mathbf{x} - \mathbf{x}'$. The spatial Fourier transform is then

$$G_{\alpha\alpha'}(\mathbf{p}, \zeta_l) = \delta_{\alpha\alpha'}[\zeta_l - (p^2/2m)]^{-1} \quad (78)$$

which recovers a result we had previously obtained. In the general translationally invariant case, the inverse of the single-particle Green's function can always be written as

$$[G_{\alpha\alpha'}(p, \zeta_l)]^{-1} = \delta_{\alpha\alpha'}(\zeta_l - p^2/2m) - \Sigma_{\alpha\alpha'}(\mathbf{p}, \zeta_l) \quad (79)$$

which defines the so-called self-energy part Σ. The formal extension of Eq. (79) to situations that are not translationally invariant is achieved through the equation

$$\left(\zeta_l + \frac{\nabla^2}{2m}\right) G_{\alpha\alpha'}(\mathbf{x}, \mathbf{x}', \zeta_l) = \delta_{\alpha\alpha'}\delta^3(x - x')$$
$$+ \int d^3x''\, \Sigma_{\alpha\alpha''}(\mathbf{x}, \mathbf{x}'', \zeta_l)\, G_{\alpha''\alpha'}(\mathbf{x}'', \mathbf{x}', \zeta_l) \quad (80)$$

It follows from the form of Eq. (79) that $\Sigma(\mathbf{p}, \zeta)$ has the same analytic properties as $G(\mathbf{p}, \zeta)$ (*10*), so that one can write a spectral representation

$$\Sigma(\mathbf{p}, \zeta) = a_{\mathbf{p}} + \int_{-\infty}^{\infty} \frac{\Gamma(\mathbf{p}, \omega)}{\zeta - \omega} \frac{d\omega}{2\pi} \tag{81}$$

where $\Gamma(\mathbf{p}, \omega)$ is the discontinuity of Σ across the cut. From the form of Eq. (81) it follows that one can write

$$\Sigma(\mathbf{p}, \omega \pm i\eta) = \Delta(\mathbf{p}, \omega) \mp \frac{i}{2} \Gamma(\mathbf{p}, \omega) \tag{82}$$

thus defining $\Delta(\mathbf{p}, \omega)$. From the definition of $A(\mathbf{p}, \omega)$ as the discontinuity across the cut of $G(\mathbf{p}, \zeta)$ it follows that

$$A(\mathbf{p}, \omega) = \frac{\Gamma(\mathbf{p}, \omega)}{[\omega - (p^2/2m) - \Delta(\mathbf{p}, \omega)]^2 + \Gamma^2(\mathbf{p}, \omega)/4} \tag{83}$$

Since $A(\mathbf{p}, \omega)$ must be nonnegative, so must $\Gamma(\mathbf{p}, \omega)$, explaining the choice of signs in Eq. (82). Equation (83) is quite general but it of course includes the possibility that $A(\mathbf{p}, \omega)$ has simple poles in the ω-plane near the real ω-axis. It will be remembered that such poles indicate the existence of long-lived particle-like excitations [Eq. (46)].

4. Perturbation Theory and Feynman Graphs

For many applications it proves useful to make systematic expansions of the Green's functions in powers of the interaction, which in the case we are studying here is the interaction potential v. At first sight it would seem that such expansions are of no use in the theory of superconductivity, since the phenomenon involves a condensation which cannot be obtained by an expansion in the effective interaction between electrons. Perturbation theory is, however, useful in discussing the origins of the condensation. Furthermore, it is in fact possible to apply a modified form of perturbation theory to the condensed state. The theory we shall now develop will be put to both these uses later in this chapter and elsewhere in this treatise.

The basic theorem for the development of perturbation theory is the following relation for a *noninteracting* system between the n-particle Green's function and products of single-particle Green's functions:

$$G^0(1, \ldots, n; 1', \ldots, n') = \begin{vmatrix} G^0(11') \cdots G^0(1n') \\ \vdots \\ G^0(n1') \cdots G^0(n, n') \end{vmatrix} \tag{84}$$

Here we have put together space, spin, and time labels to form a single index, and the right side symbolizes the determinant of the square array. Equation (84) can be proved for the thermodynamic Green's functions by using the equations of motion and the boundary conditions. Consider first the two-particle Green's function. Its equation of motion is

$$\left(i\frac{\partial}{\partial t_1}+\frac{\nabla_1^2}{2m}\right)G^0(12,34)=\delta(13)\,G^0(24)-\delta(14)\,G^0(23) \tag{85}$$

In the restricted time region we may make Fourier series expansions as indicated in Eq. (74). We have already seen [Eq. (78)] that

$$G^0(11')=\frac{1}{\tau^2}\sum_{ll'}\exp(-i\zeta_{l_1}t_1)\exp(i\zeta_{l'_1}t'_1)\,G^0_{ll'}(x,x') \tag{86}$$

where

$$\begin{aligned}G^0_{ll'}(x,x')&=\left(\zeta_l+\frac{\nabla^2}{2m}\right)^{-1}\delta(x,x')\,\delta_{ll'}\\&=\int\frac{d^3p}{(2\pi)^3}\exp[i\mathbf{p}\cdot(\mathbf{x}-\mathbf{x}')]\cdot(\zeta_l-p^2/2m)^{-1}\,\delta_{\alpha\alpha'}\delta_{ll'}\end{aligned}$$

Thus the operator on the left of Eq. (85) when acting on a thermal Green's function is the reciprocal of the single-particle thermal Green's function.‡ Transferring this operator to the right side by matrix multiplication leads to the equation

$$G^0(12,34)=G^0(13)\,G^0(24)-G^0(14)\,G^0(23) \tag{87}$$

which is the lowest nontrivial case of Eq. (84). The general case is obtained by induction, using the equations of motion of successively higher order Green's functions.

Having seen that the Green's functions for a noninteracting system can be factored in this simple way, we now show that each Green's function for an interacting system may be expanded in a series that includes noninteracting Green's functions of arbitrarily high order. This development is very similar to that given by Dyson for relativistic quantum field theory. Because the interaction potential occurs both in the statistical factor and in the time dependence of the field operators, it is convenient to make the expansion in the restricted time region and then, if necessary, continue it to the region of real times. Following Dyson we introduce an interaction representation:

$$\psi_I(x,t)=\exp(i\mathcal{H}_0t)\,\psi(x)\exp(-i\mathcal{H}_0t) \tag{88}$$

‡ In (*4*) insufficient attention was paid to the question of giving meaning to integrating the equations of motion. The statement made there that the theorem [Eq. (84)] applies to an arbitrary normalized average is incorrect.

where the complete Hamiltonian is

$$\mathscr{H} = \mathscr{H}_0 + V$$

All the time dependence can now be expressed in terms of the U-matrix, defined as

$$U(t, t') = \exp(i\mathscr{H}_0 t) \exp[-i\mathscr{H}(t - t')] \exp(-i\mathscr{H}_0 t') \tag{89}$$

This matrix satisfies the relations

$$U(t, t'') = U(t, t')\, U(t', t'') \tag{90}$$

$$U(t, t) = 1 \tag{91}$$

The equation of motion of this operator is

$$i(\partial/\partial t)\, U(t, t') = V_I(t)\, U(t, t') \tag{92}$$

where

$$V_I(t) = \exp(i\mathscr{H}_0 t)\cdot V \exp(-i\mathscr{H}_0 t) \tag{93}$$

Integrating, we find the usual formal expression

$$\begin{aligned} U(\tau, 0) &= 1 - i\int_0^\tau dt_1\, V_I(t_1) + (-i)^2 \int_0^\tau dt_1 \int_0^{t_1} dt_2\, V_I(t_1)\, V_I(t_2) + \cdots \\ &= 1 - i\int_0^\tau dt_1\, V_I(t_1) + \frac{(-i)^2}{2!}\int_0^\tau dt_1 \int_0^\tau dt_2\, T[V_I(t_1)\, V_I(t_2)] + \cdots \\ &= T\left\{\exp\left[-i\int_0^\tau V_I(t)\, dt\right]\right\} \end{aligned} \tag{94}$$

(It may be worthwhile to remind the reader that in the above equations $\tau = -i\beta$, the times are negative imaginary and run between 0 and τ, and T orders the operators from right to left in order of increasing negative imaginary time argument.)

It is easy to see that

$$\begin{aligned} iZ_G G(xt, x't') = \operatorname{Tr}\{&\exp(\beta\mu N) \exp(-\beta\mathscr{H}_0)\, T[U(\tau, t)\, \psi_I(xt)\, U(t, t') \\ &\times \psi_I(x't')\, U(t'0)]\} \end{aligned} \tag{95}$$

where Z_G is the grand partition function

$$Z_G = \operatorname{Tr}\{\exp[-\beta(\mathscr{H} - \mu N)]\} \tag{96}$$

Let us introduce a notation for averages in the system without interactions, at temperature k/β and chemical potential μ:

$$\langle X\rangle_0 = \operatorname{Tr}\{\exp[-\beta(\mathscr{H}_0 - \mu N)]\, X\}/\operatorname{Tr}\{\exp[-\beta(\mathscr{H}_0 - \mu N)]\} \tag{97}$$

Then clearly

$$Z_G = \langle U(\tau, 0)\rangle_0 \, Z_G^0 \tag{98}$$

where Z_G^0 is obtained from Eq. (96) by the replacement $\mathscr{H} \to \mathscr{H}_0$. Furthermore, Eq. (95) becomes [the time ordering also applying to the operators in the series expansion Eq. (94)]

$$G(xt, x't') = \frac{-i\langle T[U(\tau, 0)\, \psi_I(xt)\, \psi_I^\dagger(x't')]\rangle_0}{\langle U(\tau, 0)\rangle_0} \tag{99}$$

Similarly, the two-particle Green's function may be written as

$$G(x_1t_1, x_2t_2; x_3t_3, x_4t_4) = \frac{(-i)^2 \langle T[U(\tau, 0)\, \psi_I(x_1t_1)\, \psi_I(x_2t_2)\, \psi_I^\dagger(x_4t_4)\, \psi_I^\dagger(x_3t_3)]\rangle_0}{\langle U(\tau, 0)\rangle_0} \tag{100}$$

Now if the series expansion Eq. (94) is substituted for $U(t, 0)$ in the numerators of Eq. (99) and Eq. (100) we get an expansion of the product of Z_G/Z_G^0 and the respective Green's functions. The expansion has a more symmetrical appearance if we introduce the notation

$$v(xt, x't') \equiv v(x, x')\, \delta(t - t') \tag{101}$$

Combining space spin and time into a single index, and using the form given in the last term of Eq. (2) for the interaction, one has

$$\int_0^\tau dt\, V_I(t) = \tfrac{1}{2} \int d(1) \int d(1')\, v(1, 1')\, T[\psi_I^\dagger(1^+)\, \psi_I^\dagger(1'^+)\, \psi_I(1')\, \psi_I(1)] \tag{102}$$

where

$$\int d(1) = \int_0^\tau dt_1 \int dx_1 \tag{103}$$

Substituting this form into Eq. (94) and then into Eq. (99), one obtains

$$\begin{aligned}
&(Z_G/Z_G^0)\, G(11') \\
&\quad = G^0(11') + \frac{(-i)^2}{2} \int d(2) \int d(2')\, v(2, 2') \\
&\quad \times \langle T[\psi(1)\, \psi(2)\, \psi(2')\, \psi^\dagger(2'^+)\, \psi^\dagger(2^+)\, \psi^\dagger(1')]\rangle_0 \\
&\quad + \frac{(-i)^3}{2!2^2} \int d(2) \int d(2') \int d(3) \int d(3')\, v(2, 2')\, v(3, 3') \\
&\quad \times \langle T[\psi(1)\, \psi(2)\, \psi(2')\, \psi(3)\, \psi(3')\, \psi^\dagger(3'^+)\, \psi^\dagger(3^+) \\
&\quad \psi^\dagger(2'^+)\, \psi^\dagger(2^+)\, \psi^\dagger(1')]\rangle_0 \\
&\quad + \cdots
\end{aligned} \tag{104}$$

The averages of time-ordered products are of course Green's functions for the noninteracting system, so that we have achieved the expansion

$$
\begin{aligned}
&(Z_G/Z_G^0)\, G(11') \\
&\quad = G^0(11') + \frac{i}{2}\int d(2)\int d(2')\, v(2,2')\, G^0(1,2,2';1',2^+,2'^+) \\
&\quad + \frac{i^2}{2^2 2!}\int d(2)\int d(2')\int d(3)\int d(3')\, v(2,2')\, v(3,3') \\
&\quad \times G^0(1,2,2',3,3';1',2^+,2'^+,3^+,3'^+) \\
&\quad + \cdots
\end{aligned}
\tag{105}
$$

The higher-order Green's functions for the system without interactions may now be replaced by determinants of the form Eq. (84), and a very large number of terms results. Each term is said to correspond to a certain "contraction" of the higher-order Green's function. To keep track of these terms it is convenient to introduce a "Feynman graph" for each. We represent $G^0(11')$ by a solid line directed from $1'$ to 1, and $v(22')$ by a dashed line. The term in Eq. (105) of first order in v, contains six terms which result from expanding the 3×3 determinant. The graphs that describe these terms are illustrated in Fig. 3. Figure 3 (c), for example, represents the term

$$
\frac{-i}{2}\int d(2)\int d(2')\, v(2,2')\, G^0(1,2)\, G^0(2,1')\, G^0(2',2'^+) \tag{106}
$$

This example also illustrates the role of the small times (in the direction of greater negative imaginary argument) we have introduced in Eq. (102). Evidently, these small times only play a role in equal time contractions, where they

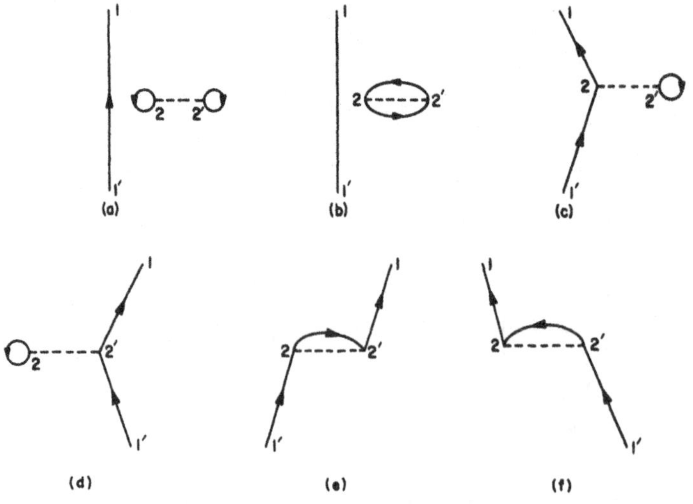

Fig. 3.

require the ordering of operators (creation operators to the left of annihilation operators) indicated in Eq. (102) and corresponding to a substraction of the interaction energy of each particle with itself (see Appendix A).

The nth-order contribution to $(Z_G/Z_G^0)\,G(1,1')$ is thus obtained by drawing n dashed lines, labeling the end points $2, 2', 3, 3', \ldots, n+1, (n+1)'$, drawing a directed solid line entering at $1'$ and leaving at 1, and connecting the end points of the dashed lines with directed solid lines (one entering and one leaving each end of a dashed line) in all possible topologically distinct ways. With each such graph is associated the multiple integral

(1) $$(i^n/2^n n!)\int d(2)\int d(2')\int d(3)\int d(3')\cdots\int d(n+1)\int d(n+1)'$$

In the integrand there is

(2) A factor $v(i, i')$ for each dashed line connecting i and i', and

(3) A factor $G^0(i,j)$ for a directed solid line running from j to i.

Finally, one has to associate a sign with each integral. A little consideration shows that the sign is

(4) $(-1)^C$, where C is the number of closed fermion loops.

For example, Fig. 3 (a) contains two closed loops; Figs. 3(b), (c), and (d) contain one; and Figs. 3(e) and (f) contain none.

Some of the graphs, for example, Figs. 3(a) and (b), that occur in the expansion of $(Z_G/Z_G^0)\,G(11')$ contain parts unconnected to the line entering at $1'$ and leaving at 1. Such parts are called unlinked parts. We shall now see that if one omits all such parts, one is making an expansion of $G(1,1')$ and getting rid of the factor Z_G/Z_G^0. Consider a linked part of order m and all possible unlinked parts. If one such graph is of order n, the order of the unlinked part then being $n-m$, the factor that goes with it is, according to rule (1) above, $(2^n n!)^{-1}$. There are, however, $n!/(n-m)!\,m!$ ways of distributing the interaction lines between the linked and unlinked parts. Furthermore, there are $(2^m)m!$ ways of permuting and interchanging the ends of the interaction lines within the linked part. Each one of these permutations corresponds to a topologically distinct labeled graph, but the integral associated with each graph is the same. This integral is the product of those associated with the linked and unlinked parts. Thus we see that by putting together the contributions of topologically distinct labeled graphs as described above, we have also achieved the correct combinatorical factor

$$\frac{1}{2^n n!}\,\frac{n!}{(n-m)!\,m!}\times 2^m m! = \frac{1}{2^{(n-m)}(n-m)!}$$

that belongs with an $(n-m)$ th-order term in the expansion of $Z_G/Z_G^0 = \langle U(\tau,0)\rangle_0$, such terms evidently being described by Feynman graphs with no external lines, or precisely what have been called unlinked parts above. Once the factor Z_G/Z_G^0 has been divided out, it is a matter of choice in the linked part whether one draws topologically distinct labeled graphs, retaining the factor $(2^m m!)^{-1}$ for an mth-order graph, or draws unlabeled graphs and omits this factor. With either choice the contribution of the sum of all linked graphs starting and ending with one line is the expansion in powers of v of the single-particle Green's function.

The perturbation expansion of other correlation functions follows in precisely the same way as the example worked out above, such correlation functions being described by Feynman graphs with more than one external line entering or leaving. An example of such an expansion will be discussed in Section II.

II. GREEN'S FUNCTIONS OF A SUPERCONDUCTOR

A. Effective Interaction

In the introduction to this chapter, and earlier in the book, it has been mentioned that, in most cases, the interaction that causes superconductivity is known to be a phonon-induced attraction between electrons. The main features of the condensation, however, depend merely on the existence of an attraction. They can be studied using a simple model, and it is with this that we shall be concerned here. The subtle and extremely interesting experimental consequences, and the theory, of the actual interaction will be the subject of a later chapter.

The effective interaction we shall use can be made plausible by a heuristic argument (*11*). Consider an interaction between two electrons in a metal in which energy and momentum are transferred from one to the other. Such an interaction is due to the Coulomb force as it is modified by the polarizability of the electron–ion system. The interesting range of momentum q, and frequency ω, transfers are

$$q \lesssim k_F \qquad \Omega_{pl} \sim \omega \ll \omega_{pl} \tag{106}$$

where k_F is the Fermi momentum, Ω_{pl} is the frequency of unscreened ionic charge fluctuations (of the order of the Debye frequency), and ω_{pl} is the electronic plasma frequency. In the range of parameters [Eq. (106)] the electron gas is capable of an essentially instantaneous screening, whereas the ionic polarizability is that of a system of free charges. The dielectric function which relates impressed and induced charge fluctuations of wave number q and frequency ω is thus approximately (k_s is the electronic screening length)

$$\kappa(q, \omega) \approx 1 + \frac{k_s^2}{q^2} - \frac{\Omega_{pl}^2}{\omega^2} \tag{107}$$

The effective interaction between electrons is then roughly given by

$$v_{\text{eff}}(q, \omega) \approx v_c(q)/\kappa(q, \omega) \tag{108}$$

where $v_c(q)$ is the Fourier transform of the instantaneous Coulomb repulsion, namely,

$$v_c(q) = 4\pi e^2/q^2 \tag{109}$$

Substituting Eqs. (107) and (109) into Eq. (108) we get

$$v_{\text{eff}}(q, \omega) \approx \frac{4\pi e^2}{q^2 + k_s^2}\left(1 + \frac{\Omega_q^2}{\omega^2 - \Omega_q^2}\right) \tag{110}$$

with

$$\Omega_q^2 \equiv q^2 \Omega_{pl}^2/(q^2 + k_s^2) \tag{111}$$

The frequency Ω_q corresponds to a zero of the dielectric function Eq. (107) and it thus is the frequency of a spontaneous charge vibration of the electron ion system, familiarly called longitudinal sound. The effective interaction Eq. (110) has been written so as to display its components of screened Coulomb repulsion and phonon-induced interaction. For $\omega \lesssim \Omega_q$ the latter is attractive and overpowers the former. The dynamical interaction Eq. (110) can be roughly simulated by a velocity-dependent potential with attractive matrix elements for transitions in which two electrons begin and end in a shell of thickness $\hbar\omega_D$ (the Debye energy) about the Fermi energy. Explicitly this model potential may be written as

$$\mathscr{H} = \tfrac{1}{2} \sum_{\substack{\mathbf{kk'K} \\ \alpha\alpha'}} V_{\mathbf{kk'}}^{\mathbf{K}} c_{\mathbf{k}+\mathbf{K}/2,\alpha}^{\dagger} c_{-\mathbf{k}+\mathbf{K}/2,\alpha'}^{\dagger} c_{-\mathbf{k'}+\mathbf{K}/2,\alpha'} c_{\mathbf{k'}+\mathbf{K}/2,\alpha} \tag{112}$$

The basic interaction vertex is illustrated in Fig. (4). It would be wrong to treat

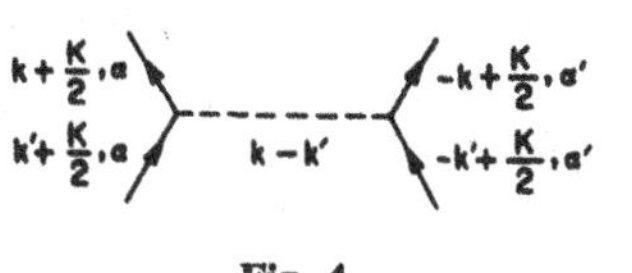

Fig. 4.

the functional form of the matrix elements $V_{\mathbf{k},\mathbf{k'}}^{\mathbf{K}}$ seriously because a retarded interaction like Eq. (110) is not strictly equivalent to a sum of velocity-dependent potentials. The main features of the interaction are simulated by taking $(\mu = k_F^2/2m)$

$$V = \begin{cases} -V & \text{for } \mu - \omega_D < \dfrac{|\pm \mathbf{k} + \mathbf{K}/2|^2}{2m} < \mu + \omega_D \\ & \text{for } \mu - \omega_D < \dfrac{|\pm \mathbf{k'} + \mathbf{K}/2|^2}{2m} < \mu + \omega_D \\ 0 & \text{for other values of } \mathbf{k}, \mathbf{k'}, \mathbf{K} \end{cases} \tag{113}$$

In other words, we assume a constant attractive matrix element when each of the momentum labels of the four field operators in Eq. (112) is in a shell around the Fermi surface. The model interaction Eqs. (112) and (113) thus takes the simple form

$$\mathscr{H} = -\frac{V}{2} \sum_{\mathbf{k}_1, \mathbf{k}_2, \mathbf{k}_3, \mathbf{k}_4}' c^\dagger_{\mathbf{k}_1\alpha} c^\dagger_{\mathbf{k}_2\alpha'} c_{\mathbf{k}_3\alpha'} c_{\mathbf{k}_4\alpha} \delta_{\mathbf{k}_1+\mathbf{k}_2, \mathbf{k}_3+\mathbf{k}_4} \tag{114}$$

where the prime indicates a sum over the shell indicated in Eq. (113). The cutoff in momentum space simulates the changeover from sympathetic to antisympathetic oscillation of the ionic lattice. If one remembers to cut off divergent momentum integrals, the interaction Eq. (114) may be written in position space as (*3*)

$$\mathscr{H}_{\text{int}} = -\frac{V}{2} \int \psi^\dagger_\alpha(x)\, \psi^\dagger_{\alpha'}(x)\, \psi_{\alpha'}(x)\, \psi_\alpha(x)\, d^3x \tag{115}$$

but it is important to remember that Eq. (115) is intended to apply only for electrons in the abovementioned momentum shells.

B. Cooper Instability

It is instructive to analyze the propagation of a pair of particles in the many-particle system under the influence of the model potential [Eq. (115)]. We shall see that a straightforward calculation (*12*) in perturbation theory indicates an instability toward a condensation of pairs.

Consider the propagator

$$L(12) \equiv (-i)^2 \langle T[\psi_\uparrow(1)\, \psi_\downarrow(1)\, \psi^\dagger_\downarrow(2)\, \psi^\dagger_\uparrow(2)] \rangle \tag{116}$$

where 1 and 2 indicate space and (imaginary) time arguments, and the arrows

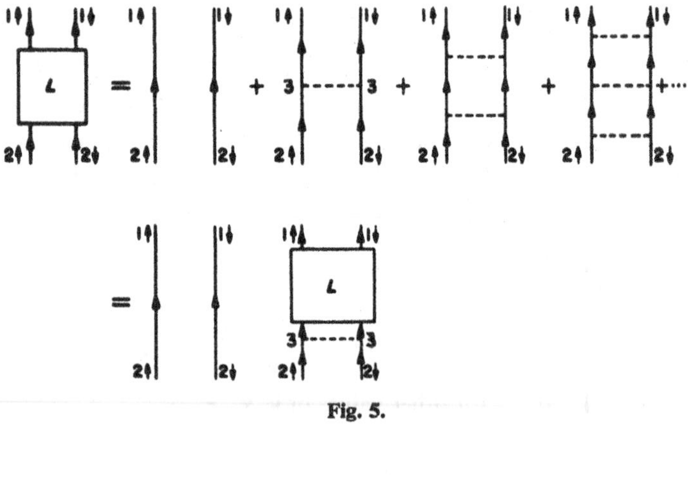

Fig. 5.

denote spin projections. One might expect the attractive interaction [Eq. (115)] to have a pronounced effect on Eq. (116). To bring this out, one can make an approximate calculation in which the initially introduced pair is allowed to sample other pair configurations under the influence of the attractive interaction. One is thus led to consider the sum of "ladder diagrams" indicated in Fig. 5. The interesting feature of the Feynman graph theory is that intermediate states corresponding to the interaction of the introduced particles with thermodynamic fluctuations are automatically included, as we shall now see. Each of the "rungs" in the series of ladders in the top right-hand part of Fig. 5 has associated with it a time which has to be integrated from 0 to $-i\beta$. If one imagines time to run over this range from the lower to the upper end of each ladder, and the position of rungs to indicate the time argument of the associated field operators, then one of the time-ordered constituents of the ladder with two rungs is as illustrated in Fig. 6. The intermediate state in this graph is one

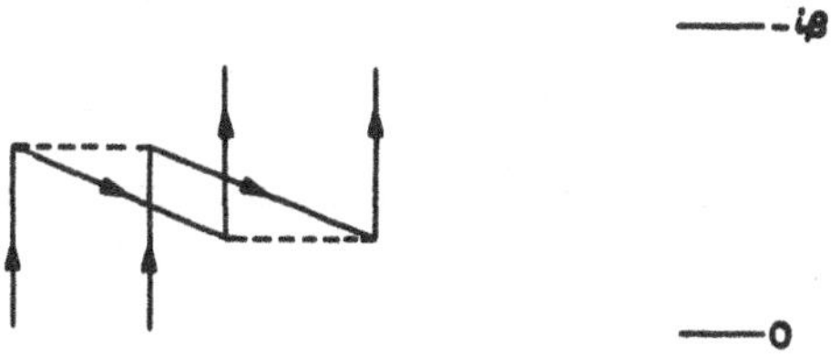

Fig. 6.

in which a thermodynamic fluctuation has occurred, the final pair configuration being occupied and an intermediate pair configuration deoccupied. The inclusion of such effects of thermodynamic fluctuations, and the other effects of finite temperature, distinguishes the calculation we are doing here from Cooper's original work (*13*).

The infinite sum of ladder graphs can be subsumed into an integral equation for L, indicated graphically in the lower part of Fig. 5. Using the rules for Feynman graphs we see that the corresponding integral equation is

$$L(12) = G^0(12)\,G^0(12) - iV\int d(3)\,L(13)\,G^0(32)\,G^0(32) \qquad (117)$$

where we have used the form of the interaction Eq. (115).

We now Fourier-transform in space, and also make a Fourier series transformation in the time arguments. Because the basic operators in Eq. (116) involve pairs of Fermi operators, there is no change of sign associated with the two parts of the time-ordered product. As a result we have a function that is periodic (instead of antiperiodic as was the single-particle Green's function) with respect to the transformation $t \to t + \tau$. A complete set of the appropriate periodic functions is $\exp(-i\tilde{\omega}_n t)$, where $\tilde{\omega}_n = 2\mu + (2\pi n/\tau)$, n being zero or a positive or negative integer. Thus we make the expansion $[(1-2) \equiv \mathbf{x}, t]$ to get

$$L(\mathbf{x}, t) = \frac{1}{\tau} \sum_n \int \frac{d^3p}{(2\pi)^3} L(\mathbf{p}, \tilde{\omega}_n) \exp[i(\mathbf{p}\cdot\mathbf{x} - \tilde{\omega}_n t)] \tag{118}$$

In terms of Fourier coefficients the integral equation (118) reads

$$L(\mathbf{p}, \tilde{\omega}_n) = \frac{1}{\tau} \sum_l \int \frac{d^3q}{(2\pi)^3} G^0(\mathbf{p}-\mathbf{q}, \tilde{\omega}_n - \zeta_l)\, G^0(\mathbf{q}, \zeta_l)\, [1 - iVL(\mathbf{p}, \tilde{\omega}_n)] \tag{119}$$

Equation (118) is thus solved by Fourier transformation:

$$L(\mathbf{p}, \tilde{\omega}_n)\left[1 + \frac{iV}{\tau} \sum_l \int \frac{d^3q}{(2\pi)^3} G^0(\mathbf{p}-\mathbf{q}, \tilde{\omega}_n - \zeta_l)\, G^0(\mathbf{q}, \zeta_l)\right]$$
$$= \frac{1}{\tau} \sum_l \int \frac{d^3q}{(2\pi)^3} G^0(\mathbf{p}-\mathbf{q}, \tilde{\omega}_n - \zeta_l)\, G^0(\mathbf{q}, \zeta_l) \tag{120}$$

Now it is not difficult to see, using the same techniques that were introduced to derive the spectral representation Eq. (55) for $G(\mathbf{p},\zeta_l)$ that $L(\mathbf{p},\tilde{\omega}_n)$ also has the spectral representation

$$L(\mathbf{p}, \tilde{\omega}_n) = \int_{-\infty}^{\infty} \frac{d\omega}{2\pi} \frac{B(\mathbf{p}, \omega)}{\tilde{\omega}_n - \omega} \tag{121}$$

where in terms of exact eigenstates [see Eq. (38)]

$$B(\mathbf{p}, \omega) = (-2\pi i) \sum_{nm} \rho_n [\langle n|\, 0_\mathbf{p}\, |m\rangle\, \delta(E_n - E_m + \omega)\, \langle m|\, 0_\mathbf{p}^\dagger\, |n\rangle$$
$$+ \langle n|\, 0_\mathbf{p}^\dagger\, |m\rangle\, \delta(E_m - E_n + \omega)\, \langle m|\, 0_\mathbf{p}\, |n\rangle]$$
$$0_\mathbf{p}^\dagger = \sum_\mathbf{q} c_{\mathbf{p}-\mathbf{q}}^\dagger c_\mathbf{q}^\dagger \tag{122}$$

The singularities of $L(\mathbf{p}, \tilde{\omega}_n \to z)$ thus determine the density of states of the interacting system that can be reached by adding or removing a pair of particles whose total momentum is $\mathbf{p}$. What are the origins of the singularities of the approximate solution for $L(\mathbf{p}, \tilde{\omega}_n)$? Referring to Eq. (120) we see that at first sight they come from the poles of the right side and the zeros of the expression in the brackets on the left. The former singularities are, however, canceled by those of the bracketed expression, which has poles at the same places. Thus the singularities are determined by the solutions of

$$\frac{-i}{\tau} \sum_l \int \frac{d^3q}{(2\pi)^3} G(\mathbf{p}-\mathbf{q}, \tilde{\omega}_n - \zeta_l)\, G(\mathbf{q}, \zeta_l) = \frac{1}{V} \tag{123}$$

the right side being positive for an attractive interaction.

The sum over l in Eq. (123) may now be performed. Introducing spectral representations for the two Green's functions according to Eq. (55) we get, using the techniques of Section II.A:

$$\frac{-i}{\tau}\sum_l \int \frac{d^3q}{(2\pi)^3}\int_{-\infty}^{\infty}\frac{d\omega}{2\pi}\int_{-\infty}^{\infty}\frac{d\omega'}{2\pi}A(\mathbf{p}+\mathbf{q},\omega)A(-\mathbf{q},\omega')\frac{1}{(\tilde{\omega}_n-\zeta_l-\omega)(\zeta_l-\omega')}$$

$$=\int \frac{d^3q}{(2\pi)^3}\int_{-\infty}^{\infty}\frac{d\omega}{2\pi}\int_{-\infty}^{\infty}\frac{d\omega'}{2\pi}A(\mathbf{p}+\mathbf{q},\omega)A(-\mathbf{q},\omega')$$

$$\times\left[\frac{f^+(\tilde{\omega}_n-\omega)}{\tilde{\omega}_n-\omega-\omega'}-\frac{f^+(\omega')}{\tilde{\omega}_n-\omega-\omega'}\right] \tag{124}$$

We notice that $\tilde{\omega}_n$ occurs in the last formula both in the denominator and in the argument of the Fermi function. If one were to make the replacement $\tilde{\omega}_n \rightarrow z$ in this formula as it stands, the resulting function of z would not have the same analytic form as the spectral representation Eq. (121). Here we run into the ambiguity regarding the continuation mentioned above Eq. (58). The resolution is clear, however. Since $\exp(-\beta\tilde{\omega}_n)=\exp(-2\beta\mu)$, we must make this replacement before continuing $\tilde{\omega}_n$ to the complex plane. Now we substitute the simple δ-function form for A given in Eq. (42), and find that Eq. (123) reduces to

$$\int \frac{d^3q}{(2\pi)^3}\frac{f^-(\tilde{\epsilon}_{\mathbf{p}-\mathbf{q}})-f^+(\tilde{\epsilon}_{\mathbf{q}})}{\tilde{\omega}_n-\tilde{\epsilon}_{\mathbf{p}-\mathbf{q}}-\tilde{\epsilon}_{\mathbf{q}}}=\frac{1}{V} \tag{125}$$

where $\tilde{\epsilon}_{\mathbf{q}}$ is the energy of a single-particle state of momentum $\mathbf{q}$. We now have to consider the region of integration in Eq. (123). Reference to the model interaction [Eq. (123)] shows that the appropriate region is such that both $\mathbf{p}-\mathbf{q}$ and $\mathbf{q}$ lie within the shell of Eq. (122). Clearly the largest region of $\mathbf{q}$ integration occurs for $\mathbf{p}=0$, and so the maximum effect of the interaction will be on pairs whose total momentum relative to the fermi sea is zero. The spectrum of zero-momentum pair states for the interacting system is thus determined by the values of $\tilde{\omega}_n$ for which

$$\frac{1}{V}=-\int' \frac{d^3q}{(2\pi)^3}\frac{\tanh[\beta(\tilde{\epsilon}_{\mathbf{q}}-\mu)/2]}{\tilde{\omega}_n-2\tilde{\epsilon}_{\mathbf{q}}} \tag{126}$$

where the prime indicates an integral over the shell about the Fermi surface.

To analyze the meaning of Eq. (126) it is convenient to consider a finite system for which the energy spectrum is discrete. Furthermore, it is convenient to measure energies with respect to the chemical potential by the replacements $\epsilon_{\mathbf{q}}\equiv\tilde{\epsilon}_{\mathbf{q}}-\mu$, $\omega_n\equiv\tilde{\omega}_n-2\mu\rightarrow z$. Equation (126) then becomes

$$\frac{1}{V}=-\sum_m \frac{\tanh(\beta\epsilon_m/2)}{z-2\epsilon_m} \tag{127}$$

where the ϵ_m's are the discrete single-particle energy levels. To find the solutions of Eq. (127) we plot both sides as a function of z, first considering z real. The result is shown in Fig. 7. The dashed vertical lines indicate the energies of pairs of noninteracting particles. The horizontal line is the graph of the left side of Eq. (127), the family of curves the graph of the right side. The intersections give the pair excitation spectrum as calculated in the ladder approximation. For $V \to 0$ the intersections approach the dashed lines, as they must. As V increases

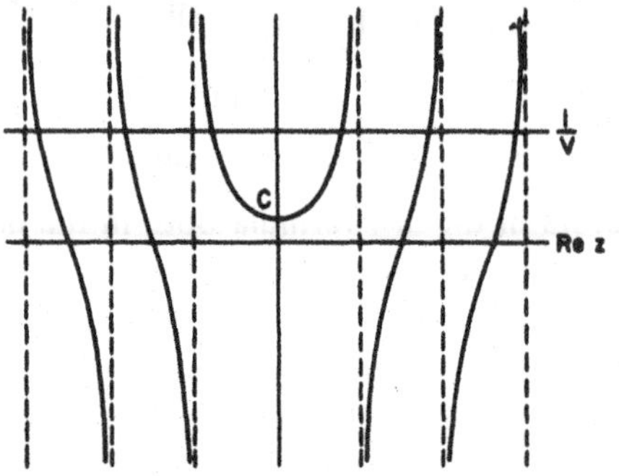

Fig. 7.

positively from zero, the calculated pair excitation spectrum changes continuously until, at C, two real roots disappear. It can be seen that the two real roots reappear at conjugate points on the imaginary z-axis. This circumstance indicates a breakdown of perturbation theory, as can be seen by considering the quantity

$$L_R(\mathbf{p}, t) \equiv \int_C dz \exp(-izt)\, L(\mathbf{p}, z) \tag{128}$$

where the contour runs from minus to plus infinity just above the real axis. Substitution of the spectral representation Eq. (121) with the explicit form of the spectral function Eq. (122) shows that $L_R(\mathbf{p}, t)$ is the retarded commutator

$$L_R(\mathbf{p}, t) = (-i)^2 \langle [0_\mathbf{p}(t), 0_\mathbf{p}^\dagger(0)] \rangle\, \theta(t) \tag{129}$$

If we trace the emergence of a pole of $L(\mathbf{p}, z)$ onto the positive imaginary z-axis we see that the contour C has to be pushed ahead of it. In other words, when a pole appears on the imaginary z-axis one is to understand that $L_R(\mathbf{p}, t)$ acquires a part that increases exponentially with time. This unphysical behavior indicates an inadequacy in the calculation, of which more will be said in the next section.

It is instructive to work out the temperature at which the instability occurs. This "critical temperature" is given by

$$\frac{1}{V} = \int' \frac{d^3q}{(2\pi)^3} \frac{\tanh(\beta_c \epsilon_\mathbf{q}/2)}{2\epsilon_\mathbf{q}} \approx N(0) \int_0^{\omega_D} d\epsilon\, \frac{\tanh(\beta_c \epsilon/2)}{\epsilon} \tag{130}$$

where $N(0)$ is the density of states for electrons of one spin at the Fermi surface, and the variation of the density of states over the region between the chemical potential and an excitation energy of ω_D has been neglected. The integral can be performed for $\beta_c\omega_D \gg 1$ and leads to the result

$$kT_c = \gamma\omega_D \exp[-1/N(0)\,V] \tag{131}$$

where γ is given by (c is Euler's constant).

$$\gamma = (2/\pi)\, e^c \sim 1.14 \tag{132}$$

Equation (131) will be recognized as the transition temperature of the BCS theory. In the present analysis, the transition has been approached from the region of higher temperatures.

C. Off-Diagonal Long-Range Order

What is one to make of the fact that the calculation described above, a calculation in the best traditions of theoretical physics, leads to results for $T < T_c$ that make no sense? Some lessons can be learned from the mishap, but it would be wrong to think that it can put right in a deductive way. A creative thought is needed, and to that we shall come momentarily.

Before the catastrophe occurred, that is, for $T > T_c$, we saw that a pair of electrons near the Fermi surface could lower its energy under the influence of the attractive interaction by sampling other pair configurations, or, in other words, acquiring a wave function that is a linear superposition of noninteracting-pair wave functions. Furthermore, we saw that the effects of the interaction were most favorable when the injected pair had the same total momentum as the electron gas as a whole. These conclusions could also be drawn from Cooper's calculation (*13*). They led Bardeen, Cooper, and Schrieffer to guess, and this is the creative thought mentioned above, a form for the wave function of the ground state. For the purpose of the Green's function theory, in which one concentrates on propagators rather than wave functions, a property [which in general has been given the name "off-diagonal long-range order" by Yang (*14*)] of the BCS wave function is of crucial importance. Here we shall extract this property from the wave function. The method to be presented is due to N. D. Mermin.

The N-particle projection of the wave function written down by BCS is (in an unnormalized form)

$$\begin{aligned} |\tilde{G}, N\rangle &= \left[\int d^3r_1\, d^3r_2\, \chi(\mathbf{r}_1 - \mathbf{r}_2)\, \psi_\uparrow^\dagger(\mathbf{r}_1)\, \psi_\downarrow^\dagger(\mathbf{r}_2)\right]^{N/2} |0\rangle \\ &= \left[\sum_{\mathbf{k}} \chi(\mathbf{k})\, c_{\mathbf{k}\uparrow}^\dagger c_{\mathbf{k}\downarrow}^\dagger\right]^{N/2} |0\rangle \end{aligned} \tag{133}$$

The last form follows from the assumption that $\chi = \chi(\mathbf{r}_1 - \mathbf{r}_2)$ and shows that the state has no total momentum, corresponding to no net current. The state also has no spin if $\chi(\mathbf{k}) = \chi(-\mathbf{k})$ because a spin rotation through π has the effect

$$c^\dagger_{\mathbf{k}\uparrow} \rightarrow c^\dagger_{\mathbf{k}\downarrow} \qquad c^\dagger_{\mathbf{k}\downarrow} \rightarrow - c^\dagger_{\mathbf{k}\uparrow} \tag{134}$$

By anticommuting the two $c^\dagger$'s and relabeling according to $\mathbf{k} \rightarrow -\mathbf{k}$ we then recover Eq. (133), showing that the state is invariant under the rotation and thus an s-state. What Eq. (133) does is put each pair into the *same* linear superposition of noninteracting pair states. In this sense it bears a loose resemblance to the ground state of a noninteracting Bose system. However, the symmetry of Eq. (133) is evidently of a Fermi character. In particular, the position and spin representative of the state is [as always, $r = (\mathbf{r}, \alpha)$]

$$\begin{aligned}\langle r_1 \cdots r_N \,|\, \tilde{G}, N\rangle &= \frac{1}{\sqrt{N!}} \langle 0|\, \psi(r_N) \cdots \psi(r_1)\,|\,\tilde{G}, N\rangle \\ &= \frac{1}{\sqrt{N!}} \sum_P (-1)^P P \chi(\mathbf{r}_1 - \mathbf{r}_2)\, \xi_{12} \chi(\mathbf{r}_3 - \mathbf{r}_4)\, \xi_{34} \cdots \\ &\qquad\qquad \chi(\mathbf{r}_{N-1} - \mathbf{r}_N)\, \xi_{N-1,N}\end{aligned} \tag{135}$$

where ξ_{12} is the spin wave function corresponding to spin up for particle 1 and spin down for particle 2.

It will now be shown that the one- and two-particle density matrices (defined below) corresponding to the state Eq. (133) have certain simple but unusual properties. Let us give the operator that is raised to the power $N/2$ in Eq. (133) the name O. Evidently

$$[\psi_\uparrow(\mathbf{r}_1), O] = \int d^3r_2\, \chi(\mathbf{r}_1, \mathbf{r}_2)\, \psi^\dagger_\downarrow(\mathbf{r}_2) \tag{136}$$

and

$$[\psi_\downarrow(\mathbf{r}_1), O] = -\int d^3r_2\, \chi(\mathbf{r}_2, \mathbf{r}_1)\, \psi^\dagger_\uparrow(\mathbf{r}_2) \tag{137}$$

Then, since $[\psi^\dagger, O] = 0$, it follows that

$$[\psi_\uparrow(\mathbf{r}_1), O^{N/2}] = \frac{N}{2} \int d^3r_2\, \chi(\mathbf{r}_1, \mathbf{r}_2)\, \psi^\dagger_\downarrow(\mathbf{r}_2)\, O^{N-2/2} \tag{138}$$

and

$$[\psi_\downarrow(\mathbf{r}_1), O^{N/2}] = -\frac{N}{2} \int d^3r_2\, \chi(\mathbf{r}_2, \mathbf{r}_1)\, \psi^\dagger_\uparrow(\mathbf{r}_2)\, O^{N-2/2} \tag{139}$$

Let $|G, N\rangle$ be the state vector Eq. (133) after normalization, i.e.,

$$|G, N\rangle \equiv \mathcal{N}_N \,|\, \tilde{G}, N\rangle \tag{140}$$

where

$$\mathcal{N}_N = [\langle N, \tilde{G} | \tilde{G}, N\rangle]^{-1/2} \tag{141}$$

The one-particle density matrix is defined as

$$\begin{aligned}\phi(\mathbf{r}_1, \mathbf{r}_1') &\equiv \langle N, G| \psi_\uparrow^\dagger(\mathbf{r}_1)\, \psi_\uparrow(\mathbf{r}_1')\, |G, N\rangle \\ &= \langle N, G| \psi_\downarrow^\dagger(\mathbf{r}_1)\, \psi_\downarrow(\mathbf{r}_1')\, |G, N\rangle\end{aligned} \tag{142}$$

Apart from a factor of i, ϕ is just $G^>$ for equal times. Because of the assumed spin rotation invariance of the state $|G, N\rangle$, ϕ evidently does not depend on the spin variables. For very large N, one also expects ϕ to be insignificantly dependent on N, a remark that will have consequences slightly later. By writing

$$|G, N\rangle = \mathcal{N}_N O^{N/2}\, |0\rangle \tag{143}$$

using $\psi|0\rangle = 0$ and the commutator Eq. (138) one obtains

$$\phi(\mathbf{r}_1, \mathbf{r}_1') = \int d^3r_2\, \hat{\chi}(\mathbf{r}_1', \mathbf{r}_2)\, \langle N, G| \psi_\uparrow^\dagger(\mathbf{r}_1)\, \psi_\downarrow^\dagger(\mathbf{r}_2)\, |G, N-2\rangle \tag{144}$$

where $\hat{\chi}$ is defined as

$$\hat{\chi}(\mathbf{r}_1, \mathbf{r}_2) = (N/2)\, \mathcal{N}_N \mathcal{N}_{N-2}^{-1} \chi(\mathbf{r}_1, \mathbf{r}_2) \tag{145}$$

The matrix element that occurs on the right of Eq. (144) plays an important role in the theory of superconductivity and it therefore merits a symbol of its own. Introducing a notation the rationale for which will become clearer in the next section, we write

$$F^>(\mathbf{r}_1, \mathbf{r}_2) \equiv -i\langle N-2, G| \psi_\uparrow(\mathbf{r}_1)\, \psi_\downarrow(\mathbf{r}_2)\, |G, N\rangle \tag{146}$$

Taking the complex conjugate

$$\begin{aligned}F^>(\mathbf{r}_1, \mathbf{r}_2)^* &= i\langle N, G| \psi_\downarrow^\dagger(\mathbf{r}_2)\, \psi_\uparrow^\dagger(\mathbf{r}_1)\, |G, N-2\rangle \\ &= -i\langle N, G| \psi_\uparrow^\dagger(\mathbf{r}_1)\, \psi_\downarrow^\dagger(\mathbf{r}_2)\, |G, N-2\rangle\end{aligned} \tag{147}$$

Equation (144) thus reads

$$\phi(\mathbf{r}_1, \mathbf{r}_1') = i\int d^3r_2\, \hat{\chi}(\mathbf{r}_1', \mathbf{r}_2)\, F^>(\mathbf{r}_1, \mathbf{r}_2)^* \tag{148}$$

By using the second expression for ϕ given in Eq. (142) and moving the $\psi_\downarrow^\dagger$ operator to the left using the adjoint of Eq. (139) one similarly obtains

$$\phi(\mathbf{r}_1, \mathbf{r}_1') = i\int d^3r_2\, \hat{\chi}(\mathbf{r}_2, \mathbf{r}_1)^*\, F^>(\mathbf{r}_2, \mathbf{r}_1') \tag{149}$$

By repeating the same simple procedure that led from Eq. (142) to Eq. (144)

one discovers that

$$F^{>}(\mathbf{r}_1, \mathbf{r}_2) = i \int d^3\mathbf{r}_2' \, \chi(\mathbf{r}_1, \mathbf{r}_1') \langle N-2, G| \psi_{\downarrow}(\mathbf{r}_2) \psi_{\downarrow}^{\dagger}(\mathbf{r}_1') |G, N-2\rangle$$

$$= i\chi(\mathbf{r}_1, \mathbf{r}_2) - i \int d^3r_2' \, \chi(\mathbf{r}_1, \mathbf{r}_1') \, \phi(\mathbf{r}_1', \mathbf{r}_2) \tag{150}$$

where in the last line we have used the canonical anticommutator of ψ and $\psi^{\dagger}$ and have also made use of the remark above Eq. (143) that ϕ is insensitive to changes in N for a system of very many particles. In a similar way one can show that

$$F^{>}(\mathbf{r}_1, \mathbf{r}_2)^* = - i\chi(\mathbf{r}_1, \mathbf{r}_2)^* + i \int d^3r_2' \, \chi(\mathbf{r}_2', \mathbf{r}_2)^* \, \phi(\mathbf{r}_1, \mathbf{r}_2') \tag{151}$$

The above analysis shows that ϕ and F are equally important amplitudes for describing the state Eq. (133).

It is well known that a Bose condensation manifests itself as an infinitely long-range behavior in the single-particle density matrix [i.e., for $|\mathbf{r}_1 - \mathbf{r}_2| \to \infty$, $\phi(\mathbf{r}_1 - \mathbf{r}_2) \neq 0$]. The choice of wave function Eq. (133) does not lead to any such behavior. It is, however, found in the two-particle density matrix, as will now be shown.

Consider the two-particle density matrix defined as

$$\phi(\mathbf{r}_1, \mathbf{r}_2; \mathbf{r}_3, \mathbf{r}_4) \equiv \langle N, G| \psi_{\uparrow}^{\dagger}(\mathbf{r}_1) \psi_{\downarrow}^{\dagger}(\mathbf{r}_2) \psi_{\downarrow}(\mathbf{r}_4) \psi_{\uparrow}(\mathbf{r}_3) |G, N\rangle \tag{152}$$

By moving the operator $\psi_{\uparrow}(\mathbf{r}_3)$ to the right and $\psi_{\downarrow}^{\dagger}(\mathbf{r}_2)$ [which anticommutes with $\psi_{\uparrow}^{\dagger}(\mathbf{r}_1)$] to the left in the manner utilized above one finds

$$\phi(\mathbf{r}_1, \mathbf{r}_2; \mathbf{r}_3, \mathbf{r}_4) = \int d^3r_3' \, d^3r_2' \, \chi(\mathbf{r}_2', \mathbf{r}_2)^* \, \chi(\mathbf{r}_3, \mathbf{r}_3')$$
$$\langle N-2, G| \psi_{\uparrow}(\mathbf{r}_2') \psi_{\uparrow}^{\dagger}(\mathbf{r}_1) \psi_{\downarrow}(\mathbf{r}_4) \psi_{\downarrow}^{\dagger}(\mathbf{r}_3') |G, N-2\rangle \tag{153}$$

If one could put the operators on the right of Eq. (153) in the proper order, one would recover the two-particle density matrix, the fact that the expectation value is in a state of $N-2$ particles again having no significance. The ordering is achieved by using the canonical anticommutation rules and one finds

$$\phi(\mathbf{r}_1, \mathbf{r}_2; \mathbf{r}_3, \mathbf{r}_4)$$
$$= [\chi(\mathbf{r}_1, \mathbf{r}_2)^* - \int d^3r_2' \, \phi(\mathbf{r}_1, \mathbf{r}_2') \, \chi(\mathbf{r}_2', \mathbf{r}_2)^*]$$
$$\times [\chi(\mathbf{r}_3, \mathbf{r}_4) - \int d^3r_3' \, \chi(\mathbf{r}_3, \mathbf{r}_3') \, \phi(\mathbf{r}_3', \mathbf{r}_4)] + \int d^3r_3' \, d^3r_2'$$
$$\times \{\chi(\mathbf{r}_2', \mathbf{r}_2)^* [\phi(\mathbf{r}_1, \mathbf{r}_3'; \mathbf{r}_2', \mathbf{r}_4) - \phi(\mathbf{r}_1, \mathbf{r}_2') \, \phi(\mathbf{r}_3', \mathbf{r}_4)] \, \chi(\mathbf{r}_3, \mathbf{r}_3')\} \tag{154}$$

Comparing the brackets in the first term on the right of Eq. (154) with Eqs. (150) and (151), we find that Eq. (154) reads

$$\phi(\mathbf{r}_1, \mathbf{r}_2; \mathbf{r}_3, \mathbf{r}_4) - F^>(\mathbf{r}_1, \mathbf{r}_2)^* F^>(\mathbf{r}_3, \mathbf{r}_4)$$
$$= \int d^3r'_3\, d^3r'_2 \{\hat{\chi}(\mathbf{r}'_2, \mathbf{r}_2)^* [\phi(\mathbf{r}_1, \mathbf{r}'_3; \mathbf{r}'_2, \mathbf{r}_4) - \phi(\mathbf{r}_1, \mathbf{r}'_2)$$
$$\times \phi(\mathbf{r}'_3, \mathbf{r}_4)]\, \hat{\chi}(\mathbf{r}_3, \mathbf{r}'_3)\} \tag{155}$$

The solution of this equation, as is verified by using Eqs. (148) and (149), is

$$\phi(\mathbf{r}_1, \mathbf{r}_2; \mathbf{r}_3, \mathbf{r}_4) = \phi(\mathbf{r}_1, \mathbf{r}_3)\, \phi(\mathbf{r}_2, \mathbf{r}_4) + F^>(\mathbf{r}_1, \mathbf{r}_2)^* F^>(\mathbf{r}_3, \mathbf{r}_4) \tag{156}$$

The meaning of Eq. (156) is perhaps more transparent when it is written out in terms of expectation values:

$$\langle \psi_\uparrow^\dagger(\mathbf{r}_1)\, \psi_\downarrow^\dagger(\mathbf{r}_2)\, \psi_\downarrow(\mathbf{r}_4)\, \psi_\uparrow(\mathbf{r}_3)\rangle = \langle \psi_\uparrow^\dagger(\mathbf{r}_1)\, \psi_\uparrow(\mathbf{r}_3)\rangle \langle \psi_\downarrow^\dagger(\mathbf{r}_2)\, \psi_\downarrow(\mathbf{r}_4)\rangle$$
$$+ \langle \psi_\uparrow^\dagger(\mathbf{r}_1)\, \psi_\downarrow^\dagger(\mathbf{r}_2)\rangle \langle \psi_\downarrow(\mathbf{r}_3)\, \psi_\uparrow(\mathbf{r}_4)\rangle \tag{157}$$

where, as we have seen, the last two expectation values represent matrix elements between the ground states of systems differing by two particles. The wave function Eq. (133) thus leads to a two-particle density matrix with the surprising property

$$\lim_{\substack{\mathbf{r}_1, \mathbf{r}_2 \to \infty \\ \mathbf{r}_3, \mathbf{r}_4 \to -\infty}} \phi(\mathbf{r}_1, \mathbf{r}_2; \mathbf{r}_3, \mathbf{r}_4) = [\lim_{\mathbf{r}_1, \mathbf{r}_2 \to \infty} F^>(\mathbf{r}_1, \mathbf{r}_2)^*] \cdot [\lim_{\mathbf{r}_3, \mathbf{r}_4 \to -\infty} F^>(\mathbf{r}_3, \mathbf{r}_4)]$$
$$\neq 0 \tag{158}$$

From the point of view taken in the present chapter this property, in the somewhat generalized from discussed below, is the central property of the condensed state.

A slightly less explicit derivation of Eq. (156) will now be given. This derivation, based on the work of Bogoliubov (*15*) and Valatin (*16*), allows one to generalize Eq. (158) form a purely algebraic property of a certain class of wave functions to a dynamical approximation for calculating the energy spectrum of superconductors. This approach also follows directly from the Eqs. (138) and (139). From these equations one readily derives:

$$c_{\mathbf{k}\uparrow}\, |N, G\rangle = \hat{\chi}(k)\, c^\dagger_{-\mathbf{k}\downarrow}\, |N-2, G\rangle \tag{159}$$

$$c_{-\mathbf{k}\downarrow}\, |N, G\rangle = -\, \hat{\chi}(k)\, c^\dagger_{\mathbf{k}\uparrow}\, |N-2, G\rangle \tag{160}$$

The states above are the normalized N and $N-2$ particle states, and $\hat{\chi}$ is defined in Eq. (145). (The property $\chi(\mathbf{k}) = \chi(-\mathbf{k})$ [see the discussion of Eq. (134)] has also been used.) Let us introduce an operator with the property

$$P^\dagger\, |N-2, G\rangle = |N, G\rangle \tag{161}$$

From Eqs. (133) and (143) we see that $P^\dagger$ has the explicit form

$$P^\dagger = \mathscr{N}_N \mathscr{N}_{N-2}^{-1} \left[\sum_{\mathbf{k}} \chi(k)\, c_{\mathbf{k}\uparrow}^\dagger c_{-\mathbf{k}\downarrow}^\dagger \right]$$
$$= \frac{2}{N} \sum_{\mathbf{k}} \hat{\chi}(k)\, c_{\mathbf{k}\uparrow}^\dagger c_{-\mathbf{k}\downarrow}^\dagger \tag{162}$$

Using this operator, Eqs. (159) and (160) may be trivially put into the suggestive form

$$[c_{\mathbf{k}\uparrow} P^\dagger - \hat{\chi}(k)\, c_{-\mathbf{k}\downarrow}^\dagger]\, |N-2, G\rangle = 0 \tag{163}$$

$$[c_{-\mathbf{k}\downarrow} P^\dagger + \hat{\chi}(k)\, c_{\mathbf{k}\uparrow}^\dagger]\, |N-2, G\rangle = 0 \tag{164}$$

As these equations show, we have found two operators that annihilate the state $|N-2, G\rangle$. It is convenient to change the normalization of these operators by writing

$$\hat{\chi}(k) = \tan \theta(k) \equiv v(k)/u(k) \qquad u^2 + v^2 = 1 \tag{165}$$

and multiplying Eqs. (163) and (164) by $u(k)$. Then we have

$$\gamma_{\mathbf{k}0}\, |N, G\rangle = \gamma_{\mathbf{k}1}\, |N, G\rangle = 0 \tag{166}$$

where

$$\gamma_{\mathbf{k}0} P^\dagger = u(k)\, c_{\mathbf{k}\uparrow} P^\dagger - v(k)\, c_{\mathbf{k}\downarrow}^\dagger \tag{167}$$

$$\gamma_{\mathbf{k}1} P^\dagger = u(k)\, c_{-\mathbf{k}\downarrow} P^\dagger + v(k)\, c_{\mathbf{k}\uparrow}^\dagger \tag{168}$$

These operators are the number-conserving generalizations of operators introduced by Bogoliubov and Valatin. They are useful because together with the adjoint operators they have anticommutation relations of a Fermi character so that they permit the construction of a spectrum of states from the single state $|N, G\rangle$. In the preceding demonstration of off-diagonal long-range order it was necessary to neglect the difference between averages taken in the N- and $(N-2)$-particle systems, which is to say that the neglect of terms of relative order $(1/N)$ was implicit in the derivation. The same sort of approximation has to be made now. Consider the commutator

$$[c_{\mathbf{k}\uparrow}, P^\dagger] = (2/N)\, \hat{\chi}(k)\, c_{-\mathbf{k}\downarrow}^\dagger \tag{169}$$

where Eq. (162) has been used. This is clearly a very small quantity for large N. We shall therefore approximate the commutators of c and $c^\dagger$ with P by zero. The operator $P^\dagger$ evidently adds a pair of particles to the condensate, and the assumption is that the effect of this operator is essentially the same on a totally paired state or one which contains a few unpaired particles. We are thus assuming that $P^\dagger$ takes almost every state of the $(N-2)$-particle system to the corresponding state of the N-particle system. Its adjoint P evidently does the reverse operation. Thus within our approximations $P^\dagger P = PP^\dagger = 1$.

Taking the adjoints of Eqs. (167) and (168) we then find the four defining equations

$$\gamma_{\mathbf{k}0} = u(\mathbf{k})\, c_{\mathbf{k}\uparrow} - v(\mathbf{k})\, c^\dagger_{-\mathbf{k}\downarrow} P \tag{170}$$

$$\gamma^\dagger_{\mathbf{k}0} = u(\mathbf{k})\, c^\dagger_{\mathbf{k}\uparrow} - v(\mathbf{k})\, c_{-\mathbf{k}\downarrow} P^\dagger \tag{171}$$

$$\gamma_{\mathbf{k}1} = u(\mathbf{k})\, c_{-\mathbf{k}\downarrow} + v(\mathbf{k})\, c^\dagger_{\mathbf{k}\uparrow} P \tag{172}$$

$$\gamma^\dagger_{\mathbf{k}1} = u(\mathbf{k})\, c^\dagger_{-\mathbf{k}\downarrow} + v(\mathbf{k})\, c_{\mathbf{k}\uparrow} P^\dagger \tag{173}$$

From these, using $u^2 + v^2 = 1$ and the properties of P, $P^\dagger$ discussed above, it is easy to see that

$$\{\gamma_{\mathbf{k}0}, \gamma^\dagger_{\mathbf{k}'0}\} = \delta_{\mathbf{k},\mathbf{k}'} \qquad \{\gamma_{\mathbf{k}1}, \gamma^\dagger_{\mathbf{k}'1}\} = \delta_{\mathbf{k},\mathbf{k}'} \tag{174}$$

all other anticommutators being zero. The γ's are thus Fermi operators that annihilate the state $|N, G\rangle$. Their algebraic properties alone can be used to derive Eq. (157). The procedure is straightforward, and the details are left as an exercise. Equations (170)–(173) may be inverted to express the c's and $c^\dagger$'s in terms of the γ's and $\gamma^\dagger$'s. The expectation value on the left of Eq. (157) may then be written as a sum of averages of γ's and $\gamma^\dagger$'s. Moving the γ's to the right and the $\gamma^\dagger$'s to the left using the commutation relations Eq. (174) allows one to reduce the left side of Eq. (157) to sums of products of averages like $\langle \gamma\gamma^\dagger \rangle$. These can be written in terms of ψ, $\psi^\dagger$ and P, $P^\dagger$. In this way the right side of Eq. (157) is recovered.

The derivation sketched above is interesting because it shows that the factorization in Eq. (157) is a special case of a Wick theorem for the Bogoliubov–Valatin operators. Let us, however, contrast this factorization with the (exact) factorization of the two-particle Green's function for a noninteracting system [Eq. (87)]. The latter factorization applied to operators in the Heisenberg picture. If one were to try to derive a particular time-ordered piece of Eq. (87) at $T = 0$ by a straightforward commutation of operators similar to that discussed in the last paragraph, it would be necessary to use the fact that for a noninteracting system the commutator of two field operators in the Heisenberg picture is an ordinary c-number. Alternatively, one has to note that for a noninteracting system the field operator is a linear combination of eigenoperators of the Hamiltonian, i.e., operators O_α with the property (E_α is a number)

$$[O_\alpha, \mathscr{H}] = E_\alpha O_\alpha \tag{175}$$

In fact, this property is all that is needed to prove Wick's theorem Eq. (84) at finite temperatures. [For an economical derivation along these lines see Gaudin (*17*).] Thus if one were to make the factorization Eq. (157) for Heisenberg representation operators, and this will in fact be done in the next section, one would implicitly be approximating the exact Hamiltonian by one for which the operators γ and $\gamma^\dagger$ are eigenoperators. This situation has some analogies with the Hartree–Fock approximation for an interacting system and a brief digression on this topic may be instructive. Whereas the factorization Eq. (87) was proved to be exact for a noninteracting system, it may be used as an approximation for the two-particle Green's function in the equation of motion for the one-particle Green's function of an interacting system. Consider the equation of motion Eq. (68). [For brevity we shall combine space and spin indices as in Eq. (69)]. When the two-particle Green's function is factored one obtains the approximate equation

$$\left[i\frac{\partial}{\partial t}+\frac{\nabla^2}{2m}-V_H(x)\right]G(xt,x't')-\int dx''\,v_F(x,x'')\,G(x''t,x't') = \delta(x,x')\,\delta(t,t') \qquad (176)$$

where

$$V_H(x)\equiv\int dx''\,v(x,x'')\,\langle\psi^\dagger(x'')\,\psi(x'')\rangle \qquad (177)$$

$$v_F(x,x'')=v(x,x'')\,\langle\psi^\dagger(x'')\,\psi(x)\rangle \qquad (178)$$

These are the Hartree and Fock potentials which are to be determined self-consistently, their value depending on the value of G obtained from Eq. (176). To be even more explicit, Eq. (176) is solved in principle by finding eigenfunctions $u_n(x)$ such that

$$\left[-\frac{\nabla^2}{2m}+V_H(x)\right]u_n(x)+\int dx''\,v_F(x,x'')\,u_n(x'')=E_n u_n(x) \qquad (179)$$

The Fourier series coefficients of G are then determined by

$$G(x,x',\zeta_l)=\sum_n\frac{u_n(x)\,u_n^*(x')}{\zeta_l-E_n} \qquad (180)$$

and the self-consistent potentials by

$$V_H(x)=-i\int d^3x''\,v(x,x'')\,G^<(x'',x'',t=0) \qquad (181)$$

$$v_F(x,x'')=-iv(x,x'')\,G^<(x,x'',t=0) \qquad (182)$$

The Hamiltonian for the interacting system has thus been approximated by

$$\mathscr{H}_{\text{eff}}=\sum_n E_n c_n^\dagger c_n \qquad (183)$$

where

$$c_n=\int dx\,u_n^*(x)\,\psi(x) \qquad (184)$$

and $\psi(x)$ is the field operator. An interesting remark is that the energies E_n are in principle temperature-dependent because the self-consistent potentials are. For a highly degenerate system like a metal this temperature dependence is of course extremely weak, the scale being determined by the degeneracy temperature.

Another remark about the above approximation is instructive. In dealing with an electron–phonon system it is reasonably realistic, as has been discussed before in this chapter, to approximate the effects of the phonons by an effective interaction between electrons that is of short range in space and retarded in time. The replacement of the two-particle Green's function in the equation of motion for the one-particle Green's function by the sum of two products of one-particle Green's function then generates an approximate one-particle excitation spectrum for the electron system which has no obvious variational character as does the Hartree–Fock solution for a system interacting through an instantaneous two-body potential. Indeed the single-particle spectrum will not correspond to a series of stable states as does Eq. (180) (E_n is *real* in the Hartree–Fock theory). However, in physical terms the approximation still corresponds to including only the average effect of the interparticle potential on the propagation of a single particle. The factorization $G_2 = G_1 G_1 - G_1 G_1$ thus has greater generality than the usual Hartree–Fock theory, its associated determinental wave functions and real energy eigenvalues.

All the above remarks have relevance when the factorization Eq. (157) is made for operators in the Heisenberg picture. On the one hand, in the approximation of instantaneous two-body interaction one recovers (as the next section will show) the results that Bardeen, Cooper, and Schrieffer obtained by a variational treatment using the wave functions $|G\rangle$, $\gamma_{k\alpha}^{\dagger}|G\rangle$, etc., as basic states. The single-particle energies so obtained are temperature-dependent for the same reasons of self-consistency as apply to the Hartree–Fock energies. However, the scale of temperatures is now governed by the critical temperature and not the degeneracy temperature.

In addition, the factorization is a more general approximation than the class of wave functions discussed in this section, and it will applied to retarded interactions in subsequent chapters. In physical terms the factorization corresponding to Eq. (157) includes in addition to the effects of Eq. (87) the average effect of the pair condensation. The approximation is, however, still of a self-consistent or "mean-field" kind. From this point of view the quantitative success of the theory which follows is perhaps surprising. Other mean field theories of phase transitions, for example, the Weiss molecular field theory of ferromagnetism and the van der Waals theory of the liquid gas transition are only qualitatively correct, their detailed predictions being notoriously unsuccessful. The reason a mean-field theory of the superconducting transition is successful would appear to be that the main correlations in a metal are governed by the extreme degeneracy of the electron gas (and the corresponding degeneracy temperature $T_c \sim 10^4\,°\mathrm{K}$). The correlations due to the pair condensation, although they have dramatic effects, are weak (being governed by the transition temperature $T_c \sim 1\text{–}10\,°\mathrm{K}$) in comparison, and may be adequately treated in an average way.

D. Gor'kov's Equations

1. Formulation and Solution

The factorization discussed in the last subsection leads via the equations of motion to a solution for the single-particle Green's function, whence physical predictions are easily extracted using the techniques developed in Section I. This formulation, at once concise, elegant, and fruitful (as much of the rest of this book will show), is due to Gor'kov (*18*).

Under the influence of the effective interaction Eq. (115) the single-particle Green's function satisfies the equation of motion

$$\left(i\frac{\partial}{\partial t}+\frac{\nabla^2}{2m}\right)G_{\uparrow\uparrow}(\mathbf{x}t,\mathbf{x}'0)=\delta^3(\mathbf{x}-\mathbf{x}')\,\delta(t) + iV\langle T[\psi_\alpha^\dagger(\mathbf{x}t)\,\psi_\alpha(\mathbf{x}t)\,\psi_\uparrow(\mathbf{x}t)\,\psi_\uparrow^\dagger(\mathbf{x}'0)]\rangle \qquad (185)$$

The average of four field operators is now factorized as discussed above. One expects that the Hartree and Fock terms will be little affected by the transition to the superconducting state and these can thus be moved to the left and combined with the single-particle Hamiltonian. The remaining term in the factorization corresponding to the assumption of off-diagonal long-range order can be written in terms of the amplitudes

$$F(\mathbf{x}t,\mathbf{x}'t')=-\,i\langle T[\psi_\uparrow(\mathbf{x}t)\,\psi_\downarrow(\mathbf{x}'t')]\rangle \qquad (186)$$

$$\bar{F}(\mathbf{x}t,\mathbf{x}'t')=-\,i\langle T[\psi_\downarrow^\dagger(\mathbf{x}t)\,\psi_\uparrow^\dagger(\mathbf{x}'t')]\rangle \qquad (187)$$

Even for $t\to t'$, F and $\bar{F}$ have a time dependence associated with their definition as matrix elements between states with different numbers of particles. Clearly

$$\bar{F}(\mathbf{x}t,\mathbf{x}'t)\propto\exp(-\,2i\mu t) \qquad (188)$$

because the energy of a condensed pair is twice the chemical potential. For the present purposes this time dependence, which can give rise to interference effects between weakly connected superconductors (see Chapter IX), is merely an inconvenience. It is removed by the device of measuring energies with respect to the chemical potential as in Section II.B [see also the discussion below Eq. (52)]. Then, Fourier-transforming in space and imaginary time one finds

$$(\zeta_l - \epsilon_k)\, G(k, \zeta_l) = 1 + \Delta F(k, \zeta_l) \tag{189}$$

where ϵ_k is the Hartree–Fock single-particle energy measured from the chemical potential and $\zeta_l = (2l+1)\pi i/\beta$. The quantity Δ is defined by

$$\Delta \equiv V\langle \psi_\uparrow(xt)\, \psi_\downarrow(xt)\rangle = \frac{iV}{\tau} \sum_l \int \frac{d^3k}{(2\pi)^3}\, F(k, \zeta_l) \tag{190}$$

The assumption that Δ is independent of position in the gauge in which **k** is the ordinary mechanical momentum is equivalent to the assumption that the condensation is in a state of zero total momentum. In Eq. (189) the spin indices on G have been omitted because the function is independent of spin projection for s-state pairing.

Another equation connecting the amplitudes G and F is obtained by considering the equation of motion of the latter. Measuring energies with respect to the chemical potential:

$$\left(-i\frac{\partial}{\partial t} + \frac{\nabla^2}{2m} + \mu\right) F(\mathbf{x}t, \mathbf{x}'0) = iV\langle T[\psi_\downarrow^\dagger(\mathbf{x}t)\, \psi_\alpha^\dagger(\mathbf{x}t)\, \psi_\alpha(\mathbf{x}t)\, \psi_\uparrow^\dagger(\mathbf{x}', 0)]\rangle \tag{191}$$

Making a factorization of the terms on the right and treating the Hartree–Fock energy in the same way as in Eq. (189) one obtains

$$(-\zeta_l - \epsilon_k)\, F(k, \zeta_l) = -\Delta^* G(k, \zeta_l) \tag{192}$$

where Δ^* is precisely the complex conjugate of Eq. (190). The solution of the pair of equations (189) and (192) is straightforward and yields

$$G(k, \zeta_l \to z) = \frac{z + \epsilon_k}{z^2 - E_k^2} \tag{193}$$

$$F(k, \zeta_l \to z) = \frac{\Delta^*}{z^2 - E_k^2} \tag{194}$$

where E_k is defined as

$$E_k = \sqrt{\epsilon_k^2 + |\Delta|^2} \tag{195}$$

Being a self-consistent parameter, Δ is temperature-dependent. Substituting Eq. (194) into Eq. (190) we obtain

$$\Delta^* = \left(-\frac{V}{\beta}\right)\left(-\frac{1}{2\pi i}\right)\left(-\beta\right)\int_{\Gamma_0} dz\, f^-(z) \int \frac{d^3k}{(2\pi)^3} \frac{\Delta^*}{z^2 - E_k^2} \tag{196}$$

where Γ_0 is the contour that surrounds the real axis (see Fig. 2), and the summation technique of Section I has been used. The poles surrounded are at $z = \pm E_k$ and a simple contour integration yields

$$1 = N(0)\, V \int_0^{\omega_D} d\epsilon \, \frac{\tanh(\beta E/2)}{E} \tag{197}$$

where we have converted the momentum sum into an integration over ϵ and approximated the density of states by its value at the Fermi surface, exactly as in Eq. (15). The condition that the matrix elements are only attractive in the shell Eq. (113) has also been imposed. Equation (197) is just the gap equation of Bardeen, Cooper, and Schrieffer, although the reason for calling Δ a "gap" will not emerge in our discussion until a little later. It determines $\Delta(T)$, which describes the average effect of the presence of condensed pairs on the propagation of particles, as discussed in the last subsection.

The transition temperature obtained from Eq. (197) is the temperature at which a solution exists for $\Delta \to 0$. In this limit the gap equation reduces to Eq. (130) and the transition temperature is given by Eq. (16).

At $T = 0$, Δ is determined by

$$1 = N(0)\, V \int_0^{\omega_D} \frac{d\epsilon}{\sqrt{\epsilon^2 + |\Delta|^2}} \tag{198}$$

which implies

$$\Delta(0) = \frac{\omega_D}{\sinh[1/N(0)\, V]} \approx 2\omega_D \exp\left[-\frac{1}{N(0)\, V}\right] \tag{199}$$

Expansions for Δ near $T = 0$ and $T = T_c$ exist (*3*), as do numerical calculations for $\Delta(T)$ (*19*). A graph of $\Delta(T)$ is given in Chapter II.

A great deal of physics is contained in this simple analysis, as we shall now proceed to show.

2. Physical Interpretation

Let us extract the single-particle excitation spectrum from the Green's function calculated above. From the general relation Eq. (57) the spectral function is

$$\begin{aligned} A(k, \omega) &= \frac{1}{i}[G(k, z \to \omega - i\eta) - G(k, z \to \omega + i\eta)] \\ &= 2\pi\left[\frac{1}{2}\left(1 + \frac{\epsilon_k}{E_k}\right)\delta(\omega - E_k) + \frac{1}{2}\left(1 - \frac{\epsilon_k}{E_k}\right)\delta(\omega + E_k)\right] \end{aligned} \tag{200}$$

At $T = 0$ the positive ω part of A describes, as we have seen, the spectrum of states available to a particle of momentum **k**. We notice that one such state is available corresponding to an excitation energy relative to the chemical potential of E_k (in absolute terms $\mu + E_k$). The interesting and unusual feature of Eq. (200) is that the available momenta are not restricted to $k > k_F$ as they would be for the normal system. One can inject a particle into a superconductor at $T = 0$ with momentum less than the Fermi momentum. However, the availability of states is given by the factor in front of $\delta(\omega - E_k)$. This factor is often called u_k^2. We see that:

(a) For $\Delta \to 0$, $E \to |\epsilon|$ and $u \to \theta(k - k_F) \equiv \begin{cases} 1 & k > k_F \\ 0 & k < k_F \end{cases}$

(b) For $\begin{array}{ll} k \gg k_F, & u \to 1 \\ k \ll k_F, & u \to 0 \end{array}$

Thus the normal-state behavior is recovered in the limit of no attractive interaction, and far away from the Fermi surface. The condensation affects the Fermi surface over a range $\Delta k \sim \Delta\,(T = 0)/\hbar v_F$, the increased kinetic energy of the ground state being repaid with interest in potential energy (as will be shown explicitly in the next section). It is as a result of this fuzzing out of the Fermi surface that particle states exist for $k < k_F$.

The spectrum of states from which a particle of momentum $\mathbf{k}$ can be removed from the system at $T = 0$ is given by the negative ω part of A. Here we again see consequences of the coherence of the ground state, namely, that we can now remove a particle with momentum $k > k_F$. The width of the region in momentum space for which this is possible is determined by the coefficient of $\delta(\omega + E_k)$, often called v_k^2. This width is again $\sim \Delta(T = 0)/\hbar v_F$. The energy of the removed particle relative to the chemical potential is seen to be $-E_k$ (in absolute terms $\mu - E_k$). The spectrum of states for particles added and removed is shown in Fig. 8. A parabolic spectrum for ϵ_k has been assumed. The important features are that an added particle must have an energy greater than $\mu + \Delta$ and a removed particle less than $\mu - \Delta$. Thus Δ, up to now just an order parameter

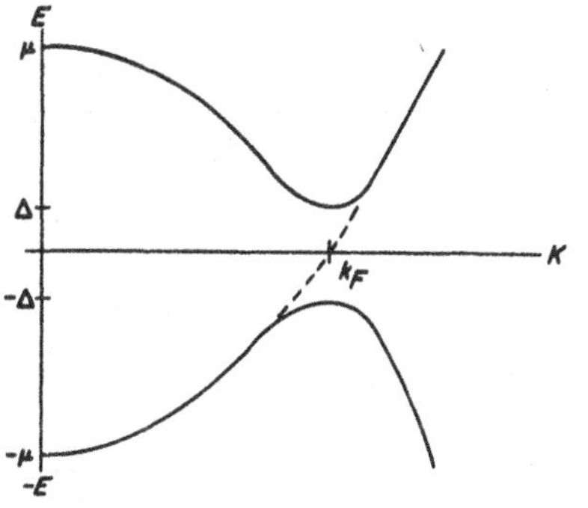

Fig. 8.

describing the condensation, is seen also to play the role of an energy gap against single-particle excitations. Physically the order parameter and the energy gap are different, and in more complicated situations the simple equality found here no longer holds. The energy gap does, however, reflect the paired nature of the condensed state. If a *pair* of particles is added it can be added to the condensate, in which case the energy required is just 2μ.

Another interesting feature is that since $E(-\epsilon) = E(\epsilon)$, there are two particle states in any direction with different momenta but the same energy. For an infinite system in which $\mathbf{k}$ is a good quantum number these are separate eigenstates. In finite samples, however, these states may be mixed, as has been experimentally demonstrated in the so-called Tomasch effect.

It is interesting to calculate the density of particle states of all momenta in the neighborhood of excitation energy ω. We consider $\omega > 0$. Then, clearly, the required density is

$$N(\omega) = \int \frac{d^3k}{(2\pi)^3} u_k^2 \delta(\omega - E_k) \tag{201}$$

Contributions to Eq. (201) come, as we have seen, from $k > k_F$ and $k < k_F$. However, on the scale of the normal density of states both these k's are extremely close to k_F. Thus we make a negligible error by approximating the momentum sum in Eq. (201) as follows:

$$N(\omega) = N(0) \int_{-\mu}^{\infty} d\epsilon\, u^2 \delta(\omega - E) \tag{202}$$

Treating positive and negative ϵ separately we get

$$\frac{N(\omega)}{N(0)} = \int_0^{\infty} d\epsilon \frac{1}{2}\left(1 + \frac{\epsilon}{E}\right) \delta(\omega - E) + \int_0^{\mu \approx \infty} d\epsilon \frac{1}{2}\left(1 - \frac{\epsilon}{E}\right) \delta(\omega - E) \tag{203}$$

where in the second term we have made the substitution $\epsilon \to -\epsilon$ and used $E(\epsilon) = E(-\epsilon)$. A cancellation occurs between the two integrals and we find

$$N(\omega) = N(0) \left.\frac{d\epsilon}{dE}\right|_{\omega = E} = N(0) \frac{\omega}{\sqrt{\omega^2 - \Delta^2}} \tag{204}$$

This is the well-known peaked density of states, first experimentally tested by Giaever in the tunneling experiment. It is exactly that of a "semiconductor" model in which only the $k > k_F$ part of the excitation spectrum is kept and u is replaced by 1. Models of this kind are misleading, however, since they gloss over the coherent nature of the ground state. They are best avoided.

III. THERMODYNAMIC PROPERTIES

A. Energy of the Ground State

In this section it will be shown that the thermodynamic functions of the model superconductor have in fact been determined by the analysis of Section II.D.1. The results obtained here, like those of the previous and next sections, were all first obtained by Bardeen, Cooper, and Schrieffer by the method described in Chapter 2.

We first consider the "binding energy," i.e., the energy difference at $T=0$ between the ordered and unordered states. This energy difference is obtained directly by the well-known trick of integrating with respect to the coupling constant. Consider the model Hamiltonian of one-particle energies and the interaction Eq. (115), where the strength of the latter is not V but λ. As an artifice we imagine being able to vary λ. The ground-state energy and wave function are then functions of λ. The complete model Hamiltonian is now

$$\mathscr{H}(\lambda) = -\frac{1}{2m}\int \psi_\alpha^\dagger(\mathbf{x})\,\nabla^2\psi_\alpha(\mathbf{x})\,d^3x - \frac{\lambda}{2}\int \psi_\alpha^\dagger(\mathbf{x})\,\psi_\beta^\dagger(\mathbf{x})\,\psi_\beta(\mathbf{x})\,\psi_\alpha(\mathbf{x})\,d^3x$$
$$\equiv \mathscr{H}_0 + \mathscr{H}_{\text{int}}(\lambda) = \mathscr{H}_0 + \lambda\mathscr{H}' \tag{205}$$

Let $E_0(\lambda)$ be the ground-state energy and $\Psi_0(\lambda)$ the normalized ground-state wave function. Then

$$E_0(\lambda) = (\Psi_0(\lambda), \mathscr{H}(\lambda)\,\Psi_0(\lambda)) \tag{206}$$

Taking the derivative of both sides with respect to λ yields

$$\partial E_0(\lambda)/\partial\lambda = E_0(\lambda)\,(\partial/\partial\lambda)\,(\Psi_0(\lambda), \Psi_0(\lambda)) + (\Psi_0(\lambda), \mathscr{H}'\Psi_0(\lambda)) \tag{207}$$

The first term on the right of this equation is zero because the norm of Ψ_0 is 1, independently of λ. Thus

$$E_0(V) - E_0(V=0) = \int_0^V \frac{d\lambda}{\lambda}\,(\Psi_0(\lambda), \mathscr{H}_{\text{int}}(\lambda)\,\Psi_0(\lambda)) \tag{208}$$

This formula is quite general and often useful. Applying it to the special case at hand by substituting for $\mathscr{H}_{\text{int}}$ from Eq. (205), we encounter the average of a product of four field operators. Again we make Gor'kov's factorization. The Hartree and Fock terms are of no interest because, as has been mentioned before, these energy shifts occur for the normal system as well and their actual value is insignificantly changed by the condensation. What remains is the pair term in the factorization, which can be written in terms of the gap parameter when λ is the value of the coupling constant. Thus one finds

$$\frac{E_s - E_n}{\text{vol.}} = -\int_0^V \frac{d\lambda}{\lambda^2}\,|\Delta(\lambda)|^2 \tag{209}$$

Now $\Delta(\lambda)$ obeys the equation [see Eq. (197) for $\beta \to \infty$]

$$\frac{1}{\lambda} = N(0) \int_0^{\omega_D} d\epsilon \frac{1}{\sqrt{\epsilon^2 + \Delta^2(\lambda)}} \tag{210}$$

from which we see that

$$d\left(\frac{1}{\lambda}\right) = N(0) \int_0^{\omega_D} d\epsilon \, d[\Delta^2(\lambda)] \frac{d}{d\Delta^2(\lambda)} \frac{1}{\sqrt{\epsilon^2 + \Delta^2(\lambda)}} \tag{211}$$

The last relation allows one to convert the integral over λ to one over Δ and we find

$$\frac{E_s - E_n}{\text{vol.}} = N(0) \int_0^{\Delta^2} dy \int_0^{\omega_D} d\epsilon \, y \frac{d}{dy} \frac{1}{\sqrt{\epsilon^2 + y}} \tag{212}$$

Doing the y integral by parts:

$$\begin{aligned} \frac{E_s - E_n}{\text{vol.}} &= N(0) \int_0^{\omega_D} d\epsilon \left\{ \frac{\Delta^2}{\sqrt{\epsilon^2 + \Delta^2}} - 2\sqrt{\epsilon^2 + \Delta^2} + 2\epsilon \right\} \\ &= N(0) \left\{ \Delta^2 \sinh^{-1} \frac{\omega_D}{\Delta} - \omega_D \sqrt{\omega_D^2 + \Delta^2} - \Delta^2 \sinh^{-1} \frac{\omega_D}{\Delta} + \omega_D^2 \right\} \\ &= -N(0)\, \omega_D^2 \{ \sqrt{1 + (\Delta/\omega_D)^2} - 1 \} \approx -\tfrac{1}{2} N(0)\, \Delta^2 (T = 0) \end{aligned} \tag{213}$$

Indeed, the next to last form is not to be taken as seriously as the last because the use of the model Hamiltonian can only be justified for $\Delta \ll \omega_D$. When this inequality is not obeyed (as for the strong-coupling superconductors, lead and mercury) one has to resort to a much more sophisticated calculation (Chapter 10) which shows that the corrections to Eq. (213) are of order $[(\Delta/\omega_D) \ln (\omega_D/\Delta)]^2$.

Equation (213) is the well-known result of Bardeen, Cooper, and Schrieffer. If for the moment we take on faith that in the model we are discussing a small applied magnetic field is expelled (Meissner effect), a simple thermodynamic argument shows that the negative of the right side of Eq. (213) is equal to $(H_0^2/8\pi)$, where H_0 is the critical field at zero temperature.

B. Thermodynamic Functions

1. Free Energy Difference—Critical Field

The difference in free energy between the normal and superconducting states can be calculated in a manner very similar to that of the last section.

Consider the grand potential Ω for the system described by Eq. (205), the artifice of a variable coupling constant being employed again so that Ω is a function of λ:

$$\Omega(\beta, \mu, \lambda) \equiv -(1/\beta) \log \text{Tr} (\exp\{-\beta[\mathscr{H}(\lambda) - \mu N]\}) \tag{214}$$

Taking the derivative with respect to λ and using the cyclic invariance of the trace to reconstruct an exponential shows that

$$\partial\Omega/\partial\lambda = \langle \mathscr{H}' \rangle \tag{215}$$

where $\mathscr{H}'$ is defined in Eq. (205) and the average is in the ensemble described by the Hamiltonian $\mathscr{H}$ of this equation. Integrating Eq. (215), factoring the product of four field operators, and including the Hartree and Fock terms in the grand potential of the normal state leads to

$$\frac{\Omega_s - \Omega_n}{\text{vol.}} = -\int_0^V \frac{d\lambda}{\lambda^2} |\Delta(\lambda, T)|^2 \tag{216}$$

This is the generalization to finite temperatures of Eq. (209). The difference in the grand potentials at fixed β, μ, and volume is on general grounds also equal to the difference in Helmholtz free energy at fixed β, N, and volume. Thus the right side of Eq. (216) is the negative of $H_c^2(T)/8\pi$, where $H_c(T)$ is the critical field. Now the integral over λ can again be converted to one over Δ, using the finite temperature gap equation [Eq. (197)] to determine Δ in an implicit way. One thus finds that

$$\frac{H_c^2(T)}{8\pi} = -N(0)\int_0^{\omega_D} d\epsilon \int_0^{\Delta^2(T)} dy\, y \frac{\partial}{\partial y}\left[\frac{\tanh \frac{1}{2}\beta\sqrt{\epsilon^2 + y}}{\sqrt{\epsilon^2 + y}}\right] \tag{217}$$

The integral over y can be done by parts leading to

$$\frac{H_c^2(T)}{8\pi} = N(0)\int_0^{\omega_D} d\epsilon \left\{\left(\frac{\Delta^2}{E} - 2E + 2\epsilon\right) - \frac{2\Delta^2}{E}[\exp(\beta E) + 1]^{-1} - \frac{4}{\beta}\ln[1 + \exp(-\beta E)] + \frac{4}{\beta}\ln[1 + \exp(-\beta\epsilon)]\right\} \tag{218}$$

Here E is, as previously, given by $E = +\sqrt{\epsilon^2 + \Delta^2}$. The first three terms in the integrand of Eq. (119) are the only ones that survive at $T = 0$, and their integration is the same as that of the last subsection. The two logarithms may be transformed by an integration by parts, and the last one may then be approximately evaluated using the Sommerfeld technique:

$$\frac{4N(0)}{\beta}\int_0^{\omega_D \approx \infty} d\epsilon \ln[1 + \exp(-\beta\epsilon)] = 4N(0)\int_0^{\infty} d\epsilon\, \epsilon[\exp(\beta\epsilon) + 1]^{-1} = N(0)\frac{\pi^2}{3\beta^2} + O(T^4) \tag{219}$$

The upper limit in Eq. (219) has been replaced by infinity because the integral is insensitive to this limit. Finally, collecting terms, we recover a BCS result:

$$\frac{H_c^2(T)}{8\pi} = N(0)\,\omega_D^2[\sqrt{1+(\Delta/\omega_D)^2} - 1]$$
$$- \frac{\pi^2}{3\beta^2} N(0)\left\{1 - \frac{6\beta^2}{\pi^2}\int_0^\infty d\epsilon\, \frac{2\epsilon^2+\Delta^2}{E}\,[\exp(\beta E)+1]^{-1}\right\} \quad (220)$$

A comparison of Eq. (210) with experiment is given in Chapter 3. As for the ground-state energy, the strong-coupling superconductors require a more subtle calculation.

2. Entropy and Specific Heat

Since the theory of Section II is a theory of independent quasi-particles of a Fermi character, the expression for the entropy may be guessed without any calculation. However, since the variational nature of the Gor'kov equation has not been shown here, it is instructive to work out the entropy from the thermodynamic potential.

Since the grand potential of the normal system is

$$\Omega_n = -N(0)\frac{\pi}{3\beta^2} + \text{const.} + O(T^4) \quad (221)$$

it follows from Eqs. (118) and (119) that

$$\Omega_s = N(0)\int_0^{\omega_D} d\epsilon\left\{\frac{\Delta^2}{E}\tanh\tfrac{1}{2}\beta E - 2E + 2\epsilon - \frac{4}{\beta}\ln[1+\exp(-\beta E)]\right\} + \text{const.} \quad (222)$$

The entropy is given by the thermodynamic relation

$$S = k\beta^2(\partial\Omega/\partial\beta)_\mu \quad (223)$$

Taking the derivative of Eq. (222) one obtains after some manipulation in which the identity $(\partial/\partial\beta)(E^2-\Delta^2)=0$ is useful:

$$S = 4kN(0)\int_0^\infty d\epsilon\,\{\ln[1+\exp(-\beta E)] + \beta E f^-(E)\} \quad (224)$$

where, as previously,

$$f^-(E) = [\exp(\beta E)+1]^{-1}$$

A more instructive form is obtained from Eq. (224) by writing

$$\beta E = \ln\left[\frac{\exp(\beta E)+1}{\exp(-\beta E)+1}\right] \quad (225)$$

Then, within the approximations that have been made in evaluating Eq. (224), one finds

$$S = -2k\sum_{\mathbf{k}}\{f^-(E_k)\ln f^-(E_k) + [1 - f^-(E_k)]\ln[1 - f^-(E_k)]\} \tag{226}$$

This is of course the form of the entropy in equilibrium for independent fermions, the factor 2 accounting for the spin.

From Eq. (226) the calculation of the specific heat proceeds in the standard way from the formula

$$C = -\beta(\partial S/\partial\beta) \tag{227}$$

The temperature dependence of the self-consistent energies $E_{\mathbf{k}}$ has, of course, to be taken into account in carrying out the derivative. A simple calculation yields

$$C = 4N(0)\int_0^\infty d\epsilon\, E\frac{\partial f^-(E)}{\partial T} \tag{228}$$

Although apparently of the form for free fermions, Eq. (228) also includes the effect of the change in condensation energy through the temperature dependence of $E = [\epsilon^2 + \Delta^2(T)]^{1/2}$. The comparison of this BCS prediction with experiment has been made in an earlier chapter.

IV. ABSORPTION OF ENERGY

The response of a many-particle system to external stimuli may be calculated in much the same spirit as the calculation of the single-particle excitation spectrum. In the case of the electronic system in a metal the stimulating forces may be due, for example, to electromagnetic fields, impressed sound waves, nuclear spin polarization, temperature gradients and so on. For weak external perturbations one is interested in the response linear in the driving force, and these "linear response functions" are often simply related to the two-particle Green's function. Two of the most striking properties of superconductors, the expulsion of a small magnetic field and the possibility of zero voltage currents, are obtained from the theory by calculating the currents that flow in response to applied magnetic and electric fields, respectively. These two properties are, however, the subject of a separate chapter. Therefore in this section we shall merely illustrate the above general remarks by calculating the absorption of energy by the model superconductor of the last two sections from a simple but fairly general class of external perturbations. The remarkable prediction of the BCS theory that the energy absorption depends markedly on the properties of the impressed perturbation under time reversal will be recovered.

To begin with we can be quite general. Consider a perturbing Hamiltonian of the form

$$\mathscr{H}'(t) = V \exp(-i\omega t) + \text{c.c.} \tag{229}$$

Suppose that the interacting many-particle system is initially in equilibrium. Then to lowest order in V, the rate of absorption of energy is given by

$$W(\omega) = 2\pi \sum_{i,f} \rho(E_i) |V_{fi}|^2 [\delta(E_f - E_i - \omega) - \delta(E_f - E_i + \omega)] \tag{230}$$

where i and f denote states of the many-particle system with energies E_i and E_f, respectively, and $\rho(E_i)$ is the statistical density corresponding to thermal equilibrium. Some simple formal transformations are suggested:

$$\begin{aligned} W(\omega) &= 2 \,\text{Im} \left\{ \left\langle V \frac{1}{\mathscr{H} - E_i - \omega - i\eta} V \right\rangle + \left\langle V \frac{1}{\mathscr{H} - E_i + \omega + i\eta} V \right\rangle \right\} \\ &= 2 \,\text{Re} \int_{-\infty}^{t} d\bar{t} \exp[i\omega(t - \bar{t})] \exp(\eta \bar{t}) \langle [V(t), V(\bar{t})] \rangle \end{aligned} \tag{231}$$

Above $V(t) = \exp(i\mathscr{H}t)\, V \exp(-i\mathscr{H}t)$, where $\mathscr{H}$ is the Hamiltonian for the system including the interactions but not the external perturbation.

Now let us specialize to perturbing potentials of the form [see Bardeen and Schrieffer (*20*)]

$$V = \sum_{\substack{\mathbf{k}\mathbf{k}' \\ \alpha\alpha'}} M_{\mathbf{k}\alpha, \mathbf{k}'\alpha'} c^{\dagger}_{\mathbf{k}\alpha} c_{\mathbf{k}'\alpha'} \tag{232}$$

This form includes a large number of interesting cases but for the present purposes one need only distinguish between two classes of matrix elements M. The first is even under time reversal and the second odd. Class 1, which includes the matrix element for the attenuation of sound, has the property

$$M_{\mathbf{k}\alpha, \mathbf{k}'\alpha'} = +\,\theta_{\alpha\alpha'} M_{-\mathbf{k}'-\alpha', -\mathbf{k}-\alpha} \tag{233}$$

where the factor $\theta_{\alpha\alpha'}$ is defined by

$$\theta_{\alpha\alpha'} = \begin{cases} 1 & \alpha = \alpha' \\ -1 & \alpha = -\alpha' \end{cases} \tag{234}$$

and takes into account the effect of rotating the spins ($\uparrow \to \downarrow$; $\downarrow \to -\uparrow$) as part of the time-reversal transformation. Class 2 includes the hyperfine interaction with nuclear spins and the interaction with electromagnetic radiation and is characterized by

$$M_{\mathbf{k}\alpha, \mathbf{k}'\alpha'} = -\,\theta_{\alpha\alpha'} M_{-\mathbf{k}'-\alpha', -\mathbf{k}-\alpha} \tag{235}$$

The importance of the difference in sign between Eqs. (233) and (235) will emerge shortly.

When Eq. (232) is substituted into Eq. (231) one again encounters the average of a product of four field operators. One's natural tendency is to factor as in Section II, but some cautionary remarks need to be made. What is being calculated in Eq. (231) is the expectation value of $V(t)$ in the states which evolve from the equilibrium distribution under the influence of V. Since V is bilinear in the field operators the calculated quantity in Eq. (231) is related to the *single-particle* Green's function for the interacting system as it is driven by the external perturbation. The external perturbation, however, affects both the free particle propagator and the self-energy part, to use the language of Section I.B.3. It can be shown (*21,22*) that what a Gor'kov factorization of Eq. (231) neglects is the change in the order parameter induced by the perturbation. It can also be shown that the major effect of the perturbation is on the phase of Δ. This effect, connected with the existence of collective oscillations of the pairs, is important for the gauge invariance of the theory (Chapter 7), but can be shown to be unimportant for long-wavelength low-frequency longitudinal fields [see (*22*) and p. 240 in (*21*)]. We shall therefore make the simple factorization of Eq. (232). Then one finds

$$\begin{aligned} W(\omega) = 2\,\mathrm{Re} \int_{-\infty}^{t} & d\bar{t} \exp[i\omega(t-\bar{t})] \exp(\eta \bar{t}) \\ & \times \sum_{\substack{\mathbf{kk}' \\ \alpha\alpha'}} \{ |M_{\mathbf{k}\alpha,\mathbf{k}'\alpha'}|^2 [G_k^<(\bar{t}-t)\, G_{k'}^>(t-\bar{t}) - G_k^>(\bar{t}-t)\, G_{k'}^<(t-\bar{t})] \\ & - M_{\mathbf{k}\alpha,\mathbf{k}'\alpha'} M_{-\mathbf{k}-\alpha,\,-\mathbf{k}'-\alpha'} \theta_{\alpha\alpha'} \\ & \times [\bar{F}_k^>(t-\bar{t})\, F_{k'}^<(\bar{t}-t) - \bar{F}_k^<(\bar{t}-t)\, F_{k'}^>(t-\bar{t})] \} \end{aligned} \tag{236}$$

where the amplitudes F and $\bar{F}$ were defined in Eqs. (186) and (187) and the superscripts denote the time-ordered constituents. (The term $\theta_{\alpha\alpha'}$ comes from the spin dependence of F and $\bar{F}$.) The Fourier transform for real times of $\bar{F}$ is obtained from the Fourier series coefficients [Eq. (194)] in the same way as was described for the normal Green's function in Section I.B.2. One has

$$\bar{F}_k^{\gtrless}(\omega) = \pm \frac{1}{i} \bar{B}_k(\omega) f^{\pm}(\omega) \tag{237}$$

where

$$\bar{B}_k(\omega) = 2\pi(\Delta^*/2E_k)\,[\delta(\omega - E_k) - \delta(\omega + E_k)] \tag{238}$$

and F has the same form with $\Delta^* \rightarrow \Delta$. Similarly, one has

$$G_k^{\gtrless}(\omega) = \pm(1/i)\, A_k(\omega) f^{\pm}(\omega) \tag{239}$$

with $A_k(\omega)$ given by Eq. (200). When the Fourier representations of $\bar{F}$, F, and

G are substituted into Eq. (236) and the integral over $\bar{t}$ performed, one finds

$$W(\omega) = 2\pi \sum_{\substack{\mathbf{kk'} \\ \alpha\alpha'}} |M_{\mathbf{k}\alpha,\,\mathbf{k'}\alpha'}|^2 \int_{-\infty}^{\infty} \frac{d\omega_1}{2\pi} \int_{-\infty}^{\infty} \frac{d\omega_2}{2\pi} [f^-(\omega_1) - f^-(\omega_2)]$$
$$\times [A_k(\omega_1) A_{k'}(\omega_2) \mp \bar{B}_k(\omega_1) B_{k'}(\omega_2)] \, \delta(\omega + \omega_1 - \omega_2) \qquad (240)$$

In the second set of brackets above the minus sign goes with matrix elements of class 1 [Eq. (233)] and the plus sign with matrix elements of class 2 [Eq. (237)]. Equation (240) can be further simplified if M depends weakly on the magnitudes of $\mathbf{k}$ and $\mathbf{k}'$ in the range $|k - k_F| \sim k_B T/\hbar v_F$. When this is so we may replace $|\mathbf{k}|$ and $|\mathbf{k}'|$ by k_F in the matrix element and average it over angles. The $\mathbf{k}$, $\mathbf{k}'$ sums are then over the spectral densities. They are performed as in Section II.D to give for the relative attenuation (s, superconductor; n, normal)

$$\frac{W_s(\omega)}{W_n(\omega)} = \frac{1}{\omega} \int_{-\infty}^{\infty} dE \, \frac{[E(E+\omega) \mp \Delta^2]}{\sqrt{E^2 - \Delta^2}\sqrt{(E+\omega)^2 - \Delta^2}} [f^-(E_1) - f^-(E_1 + \omega)] \qquad (241)$$

The integral is over the region in which the arguments of the square roots are separately positive. The upper sign (class 1) in the low-frequency limit leads to the prediction, verified for ultrasonic attenuation,

$$W_s/W_n = 2/\{\exp[\Delta(T)/kT] + 1\} \qquad (242)$$

The characteristically different behavior of the attenuation corresponding to class 2 (nuclear spin resonance) and the comparison of both predictions with experiment have been discussed in earlier chapters.

V. EFFECTS OF IMPURITIES

A. Discussion of a Normal System

An introduction to a very useful method *(23,24)* for treating the effects of dilute concentrations of impurities on the thermodynamic and transport properties of superconductors is the subject of this last section. Only the calculation of the single-particle excitation spectrum in the presence of spinless impurities will be considered, the aim being to lay the groundwork for generalizations and specific applications elsewhere in the book.

As a start let us consider a free electron gas as a model of a normal metal. If N_i impurity sites at the positions $\mathbf{R}_1 \cdots \mathbf{R}_{N_i}$ are added, the Hamiltonian for the system is of the form

$$\mathscr{H} = \mathscr{H}_0 + \mathscr{H}_{\text{imp}} \qquad (243)$$

where $\mathscr{H}_0$ is the Hamiltonian for noninteracting electrons and $\mathscr{H}_{\text{imp}}$ is

$$\mathscr{H}_{\text{imp}} = \sum_i \sum_{\alpha=1}^{N_i} u(\mathbf{x}_i - \mathbf{R}_\alpha) = \sum_{\alpha=1}^{N_i} \int dx\, \psi^\dagger(x)\, u(\mathbf{x} - \mathbf{R}_\alpha)\, \psi(x)$$
$$= \sum_{\alpha=1}^{N_i} \sum_{\mathbf{k}\mathbf{q}\sigma} c^\dagger_{\mathbf{k}+\mathbf{q},\sigma} c_{\mathbf{k},\sigma} u(q) \exp(-i\mathbf{q}\cdot\mathbf{R}_\alpha) \tag{244}$$

Above u is the impurity potential, which has been assumed to be spin-independent.

Since the interaction $\mathscr{H}_{\text{imp}}$ destroys translational invariance, the single-particle propagator $G(x, x'; \zeta_l)$ corresponding to Eq. (243) is not simply a function of $|\mathbf{x} - \mathbf{x}'|$. However, the departures from homogeneity are at a microscopic level. By averaging the propagator corresponding to Eq. (243) over impurity configurations one recovers translational invariance at a coarse level of observation.

For a given distribution of impurities, it is a matter of straightforward perturbation theory to make an expansion of the single-particle Green's function in powers of the impurity potential u. The various terms that contribute to $G(\mathbf{k}, \mathbf{k}'; \zeta_l)$, the double Fourier transform of $G(\mathbf{x}, \mathbf{x}'; \zeta_l)$ can be represented by graphs as illustrated in Fig. 9.

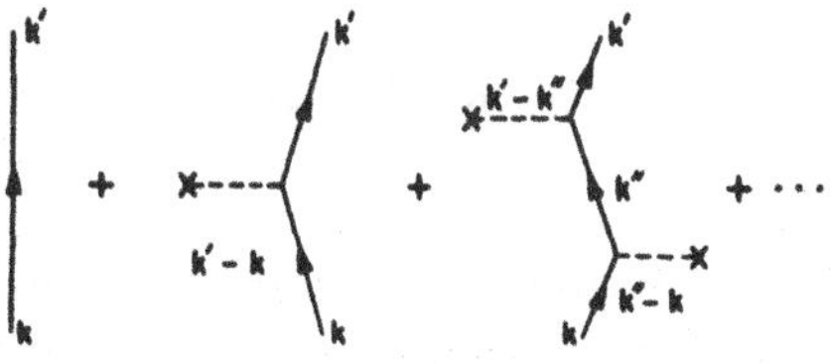

Fig. 9.

Using the techniques of Section I.B.4 one readily sees that the expansion represented by these graphs is

$$G(\mathbf{k}, \mathbf{k}'; \zeta_l) = \frac{1}{\zeta_l - \epsilon_\mathbf{k}} (2\pi)^3 \delta(\mathbf{k} - \mathbf{k}') + \frac{1}{\zeta_l - \epsilon_\mathbf{k}} U(\mathbf{k} - \mathbf{k}') \frac{1}{\zeta_l - \epsilon_{\mathbf{k}'}}$$
$$+ \int \frac{d^3k''}{(2\pi)^3} \frac{1}{\zeta_l - \epsilon_\mathbf{k}} U(\mathbf{k}'' - \mathbf{k}) \frac{1}{\zeta_l - \epsilon_{\mathbf{k}''}} U(\mathbf{k}' - \mathbf{k}'') \frac{1}{\zeta_l - \epsilon_{\mathbf{k}'}} \tag{245}$$

where U is defined as

$$U(q) = \sum_{\alpha=1}^{N_i} \exp(-i\mathbf{q}\cdot\mathbf{R}_\alpha)\, u(\mathbf{q}) \tag{246}$$

Now we average the position of each impurity over the volume of the system, so that, for example,

$$\exp[-i(\mathbf{k} - \mathbf{k}')\cdot\mathbf{R}_\alpha] \to \frac{1}{\text{vol.}} \int d^3R \exp[-i(\mathbf{k} - \mathbf{k}')\cdot\mathbf{R}] = \delta_{\mathbf{k},\mathbf{k}'} \tag{247}$$

Thus, representing the average by angular brackets,

$$\langle U(\mathbf{k}-\mathbf{k}')\rangle = n_i u(0)\,(\text{vol.})\,\delta_{\mathbf{k},\mathbf{k}'} \rightarrow (2\pi)^3\, n_i u(0)\,\delta(\mathbf{k}-\mathbf{k}') \tag{248}$$

where n_i is the density of impurities N_i/vol. Now consider the impurity average of two potentials. Here one has to distinguish between second-order scattering by a single impurity and two scatterings by differing impurities. The contribution from both these sources is

$$\begin{aligned}\langle U(\mathbf{q})\,U(\mathbf{q}')\rangle &= \langle \sum_\alpha u(\mathbf{q})\,u(\mathbf{q}')\exp[i(\mathbf{q}+\mathbf{q}')\cdot\mathbf{R}_\alpha]\rangle \\ &\qquad + \langle \sum_{\alpha\neq\alpha'} u(\mathbf{q})\,u(\mathbf{q}')\exp(i\mathbf{q}\cdot\mathbf{R}_\alpha)\exp(i\mathbf{q}'\cdot\mathbf{R}_{\alpha'})\rangle \\ &= (2\pi)^3\, n_i u(\mathbf{q})\,u(\mathbf{q}')\,\delta(\mathbf{q}+\mathbf{q}') \\ &\qquad + n_i\left(n_i - \frac{1}{\text{vol.}}\right)(2\pi)^6\, u(\mathbf{q})\,u(\mathbf{q}')\,\delta(\mathbf{q})\,\delta(\mathbf{q}')\end{aligned} \tag{249}$$

In the limit of large volume, the second term in Eq. (249) is just the square of Eq. (248). By averaging all terms of Eq. (245) in this way one obtains a prescription for calculating the self-energy part $\bar{\Sigma}$ corresponding to the averaged Green's function. These two quantities are related in the normal way according to

$$\langle G(\mathbf{k},\mathbf{k}';\zeta_l)\rangle = (2\pi)^3\,\delta(\mathbf{k}-\mathbf{k}')\,[\zeta_l - \epsilon_{\mathbf{k}} - \bar{\Sigma}(\mathbf{k},\zeta_l)]^{-1} \tag{250}$$

The terms in the expansion for $\bar{\Sigma}$ can be represented graphically by bringing together at a single cross all scatterings from the same impurity. Some terms in this graphical expansion are shown in Fig. 10. The contribution of the first-

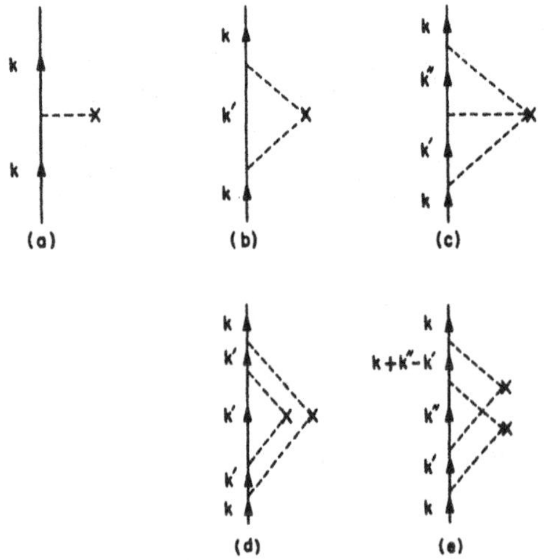

Fig. 10.

order graph is $n_i u(0)$ [Eq. (248)]. This constant shift of energy must be interpreted as a shift in the chemical potential for a system with a fixed number of electrons. Now consider the second-order term. [Notice that only the first term in Eq. (249) contributes to Σ. The other term corresponds to an iteration of the first graph in Fig. 10 and therefore makes no contribution to the self-energy part.] The part of Σ proportional to u^2 is

$$\Sigma^{(2)} = n_i \int \frac{d^3k'}{(2\pi)^3} |u(\mathbf{k}-\mathbf{k}')|^2 \frac{1}{\zeta_l - \epsilon_{\mathbf{k}'}} = \int \frac{d\omega}{2\pi} \frac{\Gamma^{(2)}(\mathbf{k}, \omega)}{\zeta_l - \omega} \tag{251}$$

where

$$\Gamma^{(2)}(\mathbf{k}, \omega) = 2\pi n_i \int \frac{d^3k'}{(2\pi)^3} |u(\mathbf{k}-\mathbf{k}')|^2 \, \delta(\omega - \epsilon_{\mathbf{k}}) \tag{252}$$

This is n_i times the relaxation rate in the Born approximation for scattering by a single impurity. To estimate the behavior of $\Gamma^{(2)}$, we may approximate the potential by a screened Coulomb potential with a screening length a_0 ($\sim k_F^{-1}$, where k_F is the Fermi momentum). Then one finds

(a) $$\Gamma^{(2)}(k_F, \mu) \sim n_i a_0^3 \mu \ll \mu \qquad \text{for } n_i a_0^3 \ll 1$$

(b) $$\frac{1}{\Gamma^{(2)}} \frac{\partial \Gamma^{(2)}}{\partial k} \sim a_0 \sim (k_F)^{-1}$$

(c) $$\frac{1}{\Gamma^{(2)}} \frac{\partial}{\partial \omega} \Gamma^{(2)} \sim \frac{1}{\mu}$$

The experimentally accessible excitations correspond to momenta and energies very close to k_F and μ, respectively, when measured on the scale of these quantities. Because of the weak dependence of Γ on k and ω in this interesting region of these variables, we may make the replacement

$$\Sigma^{(2)}(\mathbf{k}, \zeta_l \to \omega \mp i0) = \Delta \pm \frac{i\Gamma}{2}(k_F, \mu) \equiv \Delta \pm \frac{i}{2\tau} \tag{253}$$

The energy shift Δ being essentially a constant can also be absorbed into a shift of the chemical potential.

The remaining graphs in Fig. 10 have now to be considered. It is easy to see that the contribution to Γ of Fig. 10(c) is n_i times the second term in the Born series for the scattering rate by a single impurity. Summing all the graphs for Σ with only one cross thus has the effect of replacing the Born approximation matrix element $u(\mathbf{k}-\mathbf{k}')$ in Eq. (252) by the t-matrix element.

The graphs with more than one cross correspond to multiple scattering by more than one impurity. The series whose first two terms are Figs. 10(b) and (d) is easily summed. [The next term would have an insertion of two dashed lines and a cross in the line labeled $\mathbf{k}''$ in Fig. 10(d).] Clearly the effect of summing this series is to replace the δ-function in Eq. (252) by a peaked function with a spread in ω of Γ. The Ansatz $\Gamma \ll \mu$ then allows this integral equation Γ to be solved as follows. The matrix element $u(\mathbf{k}-\mathbf{k}')$ is, from our previous estimate, insensitive to changes in $\epsilon_{\mathbf{k}'}$ by Γ (if $\Gamma \ll \mu$). Taking the matrix element outside the integral over $|\mathbf{k}'|$ shows that Γ is given by Eq. (252), which is indeed much less than μ. The correction due to summing the series of Figs. 10(b), (d), etc., is therefore small, in this simple example.

The contributions of the graphs of Figs. 10(d) and (e) differ in an important way. When the external energy variable ζ_l is continued to the interesting region near $\mu \pm i0$, the major contribution to the integrals over the internal momenta comes from the neighborhood of the Fermi surface. In the contribution of Fig. 10(d) there is no restriction on the angle between $\mathbf{k}$ and $\mathbf{k}'$. In Fig. 10(e), however, $\mathbf{k}'$, $\mathbf{k}''$ and $\mathbf{k}+\mathbf{k}''-\mathbf{k}'$ must be near the Fermi surface. To estimate the reduction due to this restriction, we note that due to the scattering, the uncertainty in $\mathbf{k}$ is of the order of l^{-1}, where $l = v_F/\Gamma$. Then one finds that for fixed $\mathbf{k}$ and $\mathbf{k}'$, an angular integral over $\mathbf{k}''$ is restricted to a solid angle of the order of $(k_F l)^{-1}$. The contribution of Fig. 10(e) is thus smaller than that of Fig. 10(d) by a factor of this order.

This analysis makes it plausible that when the Born approximation is valid, the series represented by Fig. 10(b), (d), etc., gives the main contribution. Corrections to the Born approximation come from other nonoverlapping graphs.

The result [Eq. (253)] has a simple physical interpretation. In position space one has

$$\begin{aligned} G(x, \mu \pm i0) &= \int \frac{d^3k}{(2\pi)^3} \exp(i\mathbf{k}\cdot\mathbf{x}) \frac{1}{\mu - \epsilon_k \pm (i/2\tau)} \\ &= \frac{4\pi}{(2\pi)^3} \int_0^\infty dk\, k \frac{\sin kx}{x} \frac{1}{\mu - \epsilon_k \pm (i/2\tau)} \\ &\approx G^0(x, \mu \pm i0) \exp(-x/2l) \end{aligned} \tag{254}$$

where $l = v_F\tau$ and $\tau^{-1} \ll \mu$ has been used. Above G^0 is the Green's function for the pure superconductor. Since we may form the probability amplitude for a weakly excited electron or hole to travel a distance x from $G(x, \mu \pm i0)$, Eq. (254) shows explicitly that l plays the role of a mean free path.

B. Impure Superconductor

To generalize the discussion of the previous subsection to superconductors one has to remember that the superconductor is described by two propagators and that the order parameter Δ must obey a self-consistency condition. An economical way to proceed is to introduce a two-component space (*25,26*) spanned by the field operator

$$\psi(\mathbf{x}t) = \begin{pmatrix} \psi_\uparrow(\mathbf{x}t) \\ \psi_\downarrow^\dagger(\mathbf{x}t) \end{pmatrix} \tag{255}$$

The corresponding 2×2 Green's function matrix is

$$\mathscr{G}_{ij}(\mathbf{x}t, \mathbf{x}'t') = -i\langle T[\psi_i(\mathbf{x}t)\,\psi_j^\dagger(\mathbf{x}'t')]\rangle \tag{256}$$

Average values of the form $\langle\psi\psi\rangle$ and $\langle\psi^\dagger\psi^\dagger\rangle$ are to be interpreted as in Section II.c. Writing out the matrix Eq. (256) explicitly, one obtains, (in terms of previously defined propagators)

$$\mathscr{G}(\mathbf{x}t, \mathbf{x}'t') = \begin{pmatrix} G(\mathbf{x}t, \mathbf{x}'t') & F(\mathbf{x}t, \mathbf{x}'t') \\ \bar{F}(\mathbf{x}t, \mathbf{x}'t') & -G(\mathbf{x}'t', \mathbf{x}t) \end{pmatrix} \tag{257}$$

In this language, a spin-independent impurity potential is represented by

$$\begin{aligned} \sum_{i\alpha} u(\mathbf{x}_i - \mathbf{R}_\alpha) &= \sum_\alpha \int d^3x\, u(\mathbf{x} - \mathbf{R}_\alpha)\,[\psi_\uparrow^\dagger(x)\,\psi_\uparrow(x) + \psi_\downarrow^\dagger(x)\,\psi_\downarrow(x)] \\ &= \sum_\alpha \int d^3x\, u(\mathbf{x} - \mathbf{R}_\alpha)\,[\sum_{ij} \psi_i^\dagger(x)\,(\tau_3)_{ij}\,\psi_j(x) + \text{const.}] \\ &= \sum_\alpha \int d^3x\, u(\mathbf{x} - \mathbf{R}_\alpha)\,\mathrm{Tr}\,[\psi^\dagger(x)\,\tau_3\psi(x)] + \text{const.} \end{aligned} \tag{258}$$

Above τ_3 is the third Pauli matrix in its usual form. The subscripts and the trace refer to the components of Eq. (255). The constant in the last two lines comes from putting $\psi_\uparrow(x)$ to the left of $\psi_\downarrow^\dagger(x)$ and is infinite. This constant has no physical consequences, however, since equal time contractions of $\psi(x)$ and $\psi^\dagger(x)$ never occur in what follows. (In any applications where such contractions did occur, one would have to preserve the correct order.)

In the two-component language the calculation now proceeds exactly as in the last section. For the pure system, the method of Section II.D.1 leads to the following four equations:

$$(\zeta_l - \epsilon_\mathbf{k})\,G(\mathbf{k}, \zeta_l) = 1 + \Delta F(\mathbf{k}, \zeta_l) \tag{189}$$

$$(\zeta_l + \epsilon_\mathbf{k})\,\bar{F}(\mathbf{k}, \zeta_l) = \Delta^* G(\mathbf{k}, \zeta_l) \tag{192}$$

$$(\zeta_l - \epsilon_\mathbf{k})\,F(\mathbf{k}, \zeta_l) = -\Delta G(-\mathbf{k}, -\zeta_l) \tag{259}$$

$$-(\zeta_l + \epsilon_\mathbf{k})\,G(-\mathbf{k}, -\zeta_l) = 1 + \Delta^* \bar{F}(\mathbf{k}, \zeta_l) \tag{260}$$

The first two of these equations have been previously derived in Section II; the last two are obtained similarly. In terms of the matrix Eq. (257) this set of equations reads

$$\begin{pmatrix} \zeta_l - \epsilon_k & -\Delta \\ -\Delta^* & \zeta_l + \epsilon_k \end{pmatrix} \begin{pmatrix} G(\mathbf{k}, \zeta_l) & F(\mathbf{k}, \zeta_l) \\ F(\mathbf{k}, \zeta_l) & -G(-\mathbf{k}, -\zeta_l) \end{pmatrix} = \begin{pmatrix} 1 & 0 \\ 0 & 1 \end{pmatrix} \tag{261}$$

There is no loss of generality in taking Δ to be real. (In all the discussion of this chapter we have only allowed a spatially constant phase in Δ, which has no physical consequences. Since we have also implicitly been working in the gauge in which $\mathbf{k}$ is the mechanical momentum, the restriction means that we have not considered the effects of real electromagnetic fields or supercurrents.) Thus in the two-component language Eq. (261) reads

$$(\zeta_l - \epsilon_{\mathbf{k}}\tau_3 - \Delta\tau_1)\, \mathscr{G}(\mathbf{k}, \zeta_l) = 1 \tag{262}$$

In position space one thus has

$$\left[\zeta_l + \left(\frac{\nabla^2}{2m} + \mu\right)\tau_3 - \Delta\tau_1\right] \mathscr{G}(\mathbf{x}, \mathbf{x}'; \zeta_l) = \delta^{(3)}(\mathbf{x} - \mathbf{x}') \tag{263}$$

In the presence of a fixed distribution of impurities the generalization of Eq. (263) is clearly

$$\left\{\zeta_l + \left(\frac{\nabla^2}{2m} + \mu\right)\tau_3 - \left[\sum_\alpha u(\mathbf{x} - \mathbf{R}_\alpha)\right]\tau_3 - \Delta(\mathbf{x})\,\tau_1\right\} \mathscr{G}(\mathbf{x}, \mathbf{x}', \zeta_l) = \delta^{(3)}(\mathbf{x} - \mathbf{x}') \tag{264}$$

where Δ, being a self-consistent quantity, is now a function of position. After the average over impurity configurations is performed, $\langle\psi(\mathbf{x})\psi(\mathbf{x})\rangle$ will of course be independent of $\mathbf{x}$. However, the average will introduce correlations between Δ and the propagator $\mathscr{G}$ because both quantities are most strongly modified near the impurity sites. For a low density of impurities it is reasonable to neglect such correlations. Then we may replace $\Delta(\mathbf{x})$ in Eq. (263) by $\bar{\Delta}$, the average value of the order parameter, which must be determined self-consistently at the end of the calculation of $\bar{\mathscr{G}}$. Summing the diagrams corresponding to Figs. 10(b), (d), etc., one obtains

$$\bar{\Sigma}(k, \zeta_l) = n_i \int \frac{d^3q}{(2\pi)^3}\, u(\mathbf{q})^*\, \tau_3 \mathscr{G}(\mathbf{k} - \mathbf{q}, \zeta_l)\, \tau_3 u(\mathbf{q}) \tag{265}$$

This equation is solved by making an Ansatz for $\bar{\Sigma}$:

$$\bar{\Sigma}(k, z) = z - \tilde{z} - (\bar{\Delta} - \tilde{\Delta})\,\tau_1 \tag{266}$$

where $\tilde{z}$ and $\tilde{\Delta}$ are unknown functions of z. Substituting this into the right side of Eq. (265) we get

$$\bar{\Sigma}(k, z) = n_i \int \frac{d^3q}{(2\pi)^3}\, |u(\mathbf{q})|^2\, \tau_3 [\tilde{z} - \epsilon_{\mathbf{k}-\mathbf{q}}\tau_3 - \tilde{\Delta}\tau_1]^{-1}\, \tau_3 \tag{267}$$

The Green's function matrix is inverted by multiplying by one in the form

$$1 = (\tilde{z} + \epsilon_{\mathbf{k}-\mathbf{q}}\tau_3 + \tilde{\Delta}\tau_1)(\tilde{z} + \epsilon_{\mathbf{k}-\mathbf{q}}\tau_3 + \tilde{\Delta}\tau_1)^{-1} \tag{268}$$

and using the properties of the Pauli matrices. Then, equating coefficients of 1 and τ_1 on the left and right sides one obtains

$$z - \tilde{z} = n_i \int \frac{d^3q}{(2\pi)^3} |u(\mathbf{q})|^2 \frac{\tilde{z}}{\tilde{z}^2 - \epsilon_{\mathbf{k}-\mathbf{q}}^2 - \tilde{\Delta}^2} \tag{269}$$

$$\Delta - \tilde{\Delta} = n_i \int \frac{d^3q}{(2\pi)^3} |u(\mathbf{q})|^2 \frac{\tilde{\Delta}}{\tilde{z}^2 - \epsilon_{\mathbf{k}-\mathbf{q}}^2 - \tilde{\Delta}^2} \tag{270}$$

Converting the $\mathbf{q}$ integral into one over angles and one over the magnitude of $\epsilon_{\mathbf{k}-\mathbf{q}}$ and noting that the major contribution of the latter comes from near the Fermi surface, one finds

$$z - \tilde{z} = \frac{1}{2\pi\tau} \int_{-\infty}^{\infty} d\epsilon \frac{\tilde{z}}{\tilde{z}^2 - \epsilon^2 - \tilde{\Delta}^2} \tag{271}$$

where τ is the normal-state lifetime defined in Eq. (253). Performing the ϵ-integration then yields

$$z - \tilde{z} = -\frac{i}{2\tau} \frac{\tilde{z}}{\sqrt{\tilde{z}^2 - \tilde{\Delta}^2}} \tag{272}$$

Similarly, from Eq. (270),

$$\bar{\Delta} - \tilde{\Delta} = -\frac{i}{2\tau} \frac{\tilde{\Delta}}{\sqrt{\tilde{z}^2 - \tilde{\Delta}^2}} \tag{273}$$

By the square root is meant that branch of the complex function which corresponds to a positive imaginary part.

At this point one can verify that the assumption of no τ_3 component for $\bar{\Sigma}$ [Eq. (266)] is justified. The τ_3 component on the right of Eq. (265) contains $\epsilon/\tilde{z}^2 - \epsilon^2 - \tilde{\Delta}^2$. The contribution of this integral far from the Fermi surface is the same in the normal and superconducting states and shifts the chemical potential. For small ϵ the oddness of the integrand in ϵ justifies the assumption.

It is easy to see that the solution of Eqs. (272) and (273) is

$$z/\bar{\Delta} = \tilde{z}/\tilde{\Delta} \tag{274}$$

The self-energy part $\bar{\Sigma}$ is thus given by Eq. (264) with

$$\tilde{z} = z + \frac{i}{2\tau} \frac{z}{\sqrt{z^2 - \bar{\Delta}^2}} \tag{275}$$

$$\tilde{\Delta} = \bar{\Delta} + \frac{i}{2\tau} \frac{\bar{\Delta}}{\sqrt{z^2 - \bar{\Delta}^2}} \tag{276}$$

In the limit $\Delta \to 0$ we recover the normal-state result

$$\tilde{z} = z + \frac{i}{2\tau} \operatorname{sgn}(\operatorname{Im} z) \tag{277}$$

where the meaning of the square root mentioned under Eq. (273) has been used.

To determine $\bar{\Delta}$ we must impose the self-consistency condition

$$\begin{aligned}\bar{\Delta} &= V\langle \psi_\uparrow(\mathbf{x})\, \psi_\downarrow(\mathbf{x})\rangle = iV\bar{\mathscr{G}}_{12}(\mathbf{x}, \mathbf{x}; t = 0) \\ &= -\frac{V}{\beta}\sum_{\zeta_l}\int \frac{d^3p}{(2\pi)^3} \frac{\tilde{\Delta}}{\tilde{\zeta}_l^2 - \epsilon_p^2 - \tilde{\Delta}^2}\end{aligned} \tag{278}$$

Because of the approximate treatment of the interaction, the double sum is not convergent and must be cut off as discussed earlier in this chapter. It is convenient, and indeed physically most correct (because the true interaction is short-ranged in space and retarded in time), to cut off the ζ_l sum. This we shall indicate by a prime. Then, doing the momentum integral gives

$$\begin{aligned}\bar{\Delta} &= -\frac{2\pi i V}{\beta} N(0) \sum_{\zeta_l}{}' \frac{\tilde{\Delta}}{\sqrt{\tilde{\zeta}_l^2 - \tilde{\Delta}^2}} = -\frac{2\pi i V}{\beta} N(0) \sum_{\zeta_l}{}' \frac{\bar{\Delta}}{\sqrt{\zeta_l^2 - \bar{\Delta}^2}} \\ &= -\frac{V}{\beta}\sum_{\zeta_l}\int \frac{d^3p}{(2\pi)^3} \frac{\bar{\Delta}}{\zeta_l^2 - \epsilon_p^2 - \bar{\Delta}^2}\end{aligned} \tag{279}$$

In the transition between the second and third forms Eq. (274) has been used. The last form of Eq. (279) shows that $\bar{\Delta}$ obeys the same equation as the order parameter for the pure superconductor. One thus obtains the interesting result (*27*) that a dilute concentration of spinless impurities does not change the transition temperature.

Another interesting result concerns the energy spectrum. The density of single-particle states is

$$\begin{aligned}N(\omega) &= -\int \frac{d^3k}{(2\pi)^3} \frac{1}{\pi} \operatorname{Im} \mathscr{G}_{11}(k, \omega + i\eta) \\ &= -\frac{1}{\pi} \operatorname{Im} \int \frac{d^3k}{(2\pi)^3} \frac{\tilde{z} + \epsilon}{\tilde{z}^2 - \epsilon^2 - \tilde{\Delta}^2} = N(0) \operatorname{Re}\left(\frac{\tilde{z}}{\sqrt{\tilde{z}^2 - \tilde{\Delta}^2}}\right) \\ &= N(0) \operatorname{Re}\left(\frac{z}{\sqrt{z^2 - \bar{\Delta}^2}}\right)_{z = \omega + i\eta}\end{aligned} \tag{280}$$

Just above the real axis the appropriate branch ($\operatorname{Im}\sqrt{z^2 - \Delta^2} > 0$) of the complex square-root function is real and *positive for* $|z| > \bar{\Delta}$, and pure imaginary for $|z| < \bar{\Delta}$. Thus

$$\frac{N(\omega)}{N(0)} = \begin{cases} 0 & 0 < \omega < \bar{\Delta} \\ \omega/\sqrt{\omega^2 - \bar{\Delta}^2} & \omega > \bar{\Delta} \end{cases} \tag{281}$$

which again is the result we found previously for a pure isotropic system. Other properties of the impure system, such as the spectrum of states for a given momentum and the absorption properties, are of course modified. These topics are treated elsewhere in the book.

REFERENCES

1. P. C. Martin and J. Schwinger, *Phys. Rev.* **115**, 1342 (1959).
2. L. P. Kadanoff and G. Baym, *Quantum Statistical Mechanics*, Benjamin, New York, 1962.
3. A. A. Abrikosov, L. P. Gor'kov, and I. E. Dzyaloshinskii, *Methods of Quantum Field Theory in Statistical Physics*, Prentice-Hall, Englewood Cliffs, N.J., 1963.
4. V. Ambegaokar, *Astrophysics and the Many-Body Problem*, Benjamin, New York, 1963, p. 322.
5. V. M. Galitskii and A. B. Migdal, *Zh. Eksperim. i Teor. Fiz.* **34**, 139 (1958); *Soviet Phys. JETP* **7**, 96 (1958).
6. T. Matsubara, *Progr. Theoret. Phys. (Kyoto)* **14**, 351 (1955).
7. G. A. Baym and N. D. Mermin, *J. Math. Phys.* **2**, 232 (1961).
8. V. Ambegaokar and L. Tewordt, *Phys. Rev.* **134A**, 805 (1964), Appendix B.
9. G. M. Eliashberg, *Zh. Eksperim. i Teor. Fiz.* **41**, 1241 (1961); *Soviet Phys. JETP* **14**, 886 (1962).
10. J. M. Luttinger, *Phys. Rev.* **121**, 942 (1961).
11. D. Pines, *The Many-Body Problem*, Benjamin, New York, 1961.
12. L. P. Kadanoff and P. C. Martin, *Phys. Rev.* **124**, 670 (1961).
13. L. N. Cooper, *Phys. Rev.* **104**, 1189 (1956).
14. C. N. Yang, *Rev. Mod. Phys.* **34**, 694 (1962).
15. N. N. Bogoliubov, *Nuovo Cimento* **7**, 794 (1958).
16. J. G. Valatin, *Nuovo Cimento* **7**, 843 (1958).
17. M. Gaudin, *Nucl. Phys.* **15**, 89 (1960).
18. L. P. Gor'kov, *Zh. Eksperim. i Teor. Fiz.* **34**, 735 (1958); *Soviet Phys. JETP* **7**, 505 (1958).
19. B. Mühlschlegel, *Z. Physik* **155**, 313 (1959).
20. J. Bardeen and J. R. Schrieffer, *Progress in Low Temperature Physics*, Vol. III (C. J. Gorter, ed.), North-Holland, Amsterdam, 1961.
21. G. Rickayzen, *Theory of Superconductivity*, Wiley, New York, 1965.
22. V. Ambegaokar and L. P. Kadanoff, *Nuovo Cimento* **22**, 914 (1961).
23. S. F. Edwards, *Phil. Mag.* **3**, 1020 (1958).
24. A. A. Abrikosov and L. P. Gor'kov, *Zh. Eksperim. i Teor. Fiz.* **36**, 319 (1959); *Soviet Phys. JETP* **9**, 220 (1959).
25. Y. Nambu, *Phys. Rev.* **117**, 648 (1960).
26. G. M. Eliashberg, *Zh. Eksperim. i Teor. Fiz.* **38**, 966 (1960); *Soviet Phys. JETP* **11**, 696 (1960).
27. P. W. Anderson, *J. Phys. Chem. Solids* **11**, 26 (1959).

6

THE GINZBURG–LANDAU EQUATIONS AND THEIR EXTENSIONS

N. R. Werthamer

BELL TELEPHONE LABORATORIES, INCORPORATED
MURRAY HILL, NEW JERSEY

I. PHENOMENOLOGICAL APPROACH

Chapter 1 of this treatise describes in some detail the early observations of persistent currents around a superconducting ring and the magnetic flux exclusion (Meissner) effect. Chapter 1 also introduces the macroscopic electromagnetic equations of the Londons, which are able to account satisfactorily for these remarkable observations. The London equations, however, do not give a completely satisfactory macroscopic picture of all superconducting phenomena in a magnetic field, because they regard the specimen as being either entirely superconducting or entirely normal. Thus they cannot come to grips with the intermediate state, where the specimen has both normal and superconducting regions coexisting simultaneously, owing to the energetic favorability of the formation of normal–superconducting boundary layers.

These deficiencies were overcome in 1950 by Ginzburg and Landau (*1*), who proposed a phenomenological set of equations now bearing their names, allowing for spatial variations in the superconducting order due to the presence of a magnetic field. A brief discussion of these equations and their application to the intermediate state is given in Chapter 1. However, an idea of the wide variety of current topics for which these equations have been a useful tool can be gained from scanning the table of contents of this treatise; perhaps 20% or more of the whole is devoted to situations where the Ginzburg–Landau equations form the theoretical base. The ubiquitousness of the GL theory in current research is all the more remarkable upon recalling how little was known in 1950 about the microscopic origins of superconductivity. So much other speculative work of that era and earlier has passed into oblivion with the arrival of the modern BCS theory that the longevity of the GL equations must be considered a tribute to the incisiveness of the reasoning involved and the soundness of the physical intuition underlying it. Recent detailed study of the microscopic theory has served not only to confirm the essential correctness of the GL equations in their original context, but to extend their domain of validity to regimes not previously considered possible; earlier work using the equations thus has taken on a wider meaning and an increased relevancy.

Ginzburg and Landau began their argument by introducing a quantity to characterize the degree of superconductivity at various points in the material, a quantity called the "order parameter" and denoted $\Psi(\mathbf{r})$. The order parameter is defined so as to be zero for a normal region and unity for a fully superconducting region (or specimen in zero magnetic field) at zero temperature. Clearly $\Psi(\mathbf{r})$ must be closely related to the superfluid fraction in a two-fluid model, but the two quantities are not chosen to be identical. Rather, to allow for supercurrent flow, $\Psi(\mathbf{r})$ is taken as a complex function and interpreted as analogous to a "wave function" for superconductivity, so that its absolute square can be identified with the superfluid density,

$$N_s(\mathbf{r}) = |\Psi(\mathbf{r})|^2 \tag{1}$$

It should be noted that $\Psi(\mathbf{r})$ is not the system wave function for the electrons in the material (since it is defined as zero in the normal state), but further comments on what it is the wave function for are best deferred until the next section. However, pursuing the interpretation of the order parameter as a wave function, it is reasonable to write down‡ an expression for the supercurrent as

$$\mathbf{j}_s(\mathbf{r}) = \frac{e^*}{2m^*}\left[\Psi^*(\mathbf{r})\frac{\hbar}{i}\nabla\Psi(\mathbf{r}) - \Psi(\mathbf{r})\frac{\hbar}{i}\nabla\Psi^*(\mathbf{r})\right] \tag{2}$$

in the absence of a magnetic field, where e^* and m^* are the charge and mass of

‡ Because this chapter is intended as a pedagogic review, we hope to aid the nonspecialist reader by retaining explicitly all dimensional constants such as $\hbar$, c, and k_B.

the entities whose wave function is Ψ. It is equally natural to include the magnetic field via the vector potential $\mathbf{A}$:

$$\mathbf{j}_s(\mathbf{r}) = \frac{e^*}{2m^*}\left[\Psi^*(\mathbf{r})\left(\frac{\hbar}{i}\nabla - \frac{e^*}{c}\mathbf{A}(\mathbf{r})\right)\Psi(\mathbf{r}) + \Psi(\mathbf{r})\left(-\frac{\hbar}{i}\nabla - \frac{e^*}{c}\mathbf{A}(\mathbf{r})\right)\Psi^*(\mathbf{r})\right] \tag{3}$$

This expression is gauge-invariant provided Ψ indeed has a charge e^*, so that under gauge transformations

$$\mathbf{A}(\mathbf{r}) \rightarrow \mathbf{A}(\mathbf{r}) + \nabla\chi(\mathbf{r}) \tag{4}$$

Ψ transforms as

$$\Psi(\mathbf{r}) \rightarrow \exp\left[(ie^*/\hbar c)\,\chi(\mathbf{r})\right]\,\Psi(\mathbf{r}) \tag{5}$$

From the point of view of the two-fluid hydrodynamic model, the superfluid component has been regarded as a "quantum fluid" and a quantum wave function has been associated with it. Reinforcing this interpretation, a more suggestive form of Eq. (3) is

$$\mathbf{j}_s(\mathbf{r}) = e^* N_s(\mathbf{r})\,\mathbf{v}_s(\mathbf{r}) \tag{6}$$

where $\mathbf{v}_s$, the superfluid velocity, can be related to the phase of the order parameter:

$$\mathbf{v}_s(\mathbf{r}) = \frac{1}{m^*}\left[\hbar\nabla \arg\Psi(\mathbf{r}) - \frac{e^*}{c}\mathbf{A}(\mathbf{r})\right] \tag{7}$$

Furthermore, if there is no position dependence to the superconductivity and Ψ is independent of $\mathbf{r}$, then Eq. (3) reduces to the London relation

$$\mathbf{j}_s(\mathbf{r}) = -\left(e^{*2}/m^* c\right) N_s \mathbf{A}(\mathbf{r}) \tag{8}$$

as it must.

Next, a relation determining $\Psi(\mathbf{r})$ must be constructed. For this, Ginzburg and Landau focused attention on the free energy of the material, $\mathscr{F}_s$, which they wrote as a functional of Ψ and Ψ^*. The equation determining Ψ is obtained by requiring that $\mathscr{F}_s$ be a minimum with respect to variations of Ψ^* (and vice versa). The condition that $\mathscr{F}_s$ be a minimum can be shown to be equivalent to requiring that the superfluid and normal components of the two-fluid model be in stable equilibrium with respect to each other. The functional $\mathscr{F}_s$ thus plays a role analogous to a Lagrangian of Schrödinger wave mechanics, while its minimum value with respect to Ψ and Ψ^* is just the free energy of the superconducting phase in thermodynamic equilibrium with the magnetic field. Ginzburg and Landau chose the form for $\mathscr{F}_s$,

$$\mathscr{F}_s = \mathscr{F}_n + \int d^3r \left\{ -F(|\Psi(\mathbf{r})|^2) + \frac{1}{8\pi}|\mathbf{H}(\mathbf{r}) - \mathbf{H}_a|^2 + \frac{1}{2m^*}\left|\left[\frac{\hbar}{i}\nabla - \frac{e^*}{c}\mathbf{A}(\mathbf{r})\right]\Psi(\mathbf{r})\right|^2 \right\} \tag{9}$$

The term $\mathscr{F}_n$ is the free energy, at the same temperature, of the underlying normal phase, which could be obtained by increasing the magnetic field above its critical value. The function $F \geqslant 0$ represents the free energy lowering of the system due to having formed the superconducting correlation, and $F(0) = 0$ because $|\Psi|$ must go continuously to zero as the temperature is raised toward a second-order transition into the normal phase. For temperatures just below the zero field transition temperature, $T_c - T \ll T_c$, it should be sufficient to expand F in a power series and keep just the first two nonvanishing terms:

$$F(|\Psi|^2) \simeq \alpha|\Psi|^2 + \tfrac{1}{2}\beta|\Psi|^4 \tag{10}$$

The next term in Eq. (9), the magnetic field energy, represents the increase in superconducting free energy due to the expulsion of magnetic flux. The field applied externally, and which would exist in the absence of the specimen, is denoted $\mathbf{H}_a$. The final term in Eq. (9) represents the increase in superconducting free energy coming from spatial variations in the order parameter and from current flow. It could also be written, using definition (7), as

$$\frac{1}{2} N_s m^* v_s^2 + \frac{1}{2m^*} |\hbar\nabla|\Psi||^2$$

which illustrates more clearly that the term is of the form of a supercurrent kinetic energy plus a "stiffness" against rapid changes in the superfluid density. It is the latter contribution which tends to prevent the formation of superconducting–normal domain boundaries. There is no general criterion for not including higher derivatives of $|\Psi|$, except that these are likely to be small when $F(|\Psi|^2)$ is small, such as near T_c; this point will be discussed in more detail in a later section.

A further argument for the plausibility of form of these terms is that expression (3) for the supercurrent follows from a variation of the free energy functional with respect to $\mathbf{A}$. Thus minimizing $\mathscr{F}_s$ with respect to $\mathbf{A}$ is equivalent to solving Maxwell's equation,

$$\nabla \times \mathbf{H} = (4\pi/c)\,\mathbf{j}_s \tag{11}$$

for the net field $\mathbf{H}$. An analogous property holds for the Lagrangian of Schrödinger wave mechanics, and reinforces the formal similarity between Ψ and the usual Schrödinger wave function.

To recapitulate, the Ginzburg–Landau free energy functional represents a plausible extension of the London theory and two-fluid model to situations where the superconducting order is position-dependent, through the introduction of a complex order parameter, mathematically reminiscent of a Schrödinger wave function. The order parameter and magnetic field configurations are determined by the requirement that $\mathscr{F}_s$ be a minimum with respect to variations in

$\Psi(\mathbf{r})$, $\Psi^*(\mathbf{r})$, and $\mathbf{A}(\mathbf{r})$, together with the obvious requirement that $\mathscr{F}_s \leqslant \mathscr{F}_n$. Varying $\mathscr{F}_s$ leads to the set of equations

$$\left[-\frac{\partial F(|\Psi|^2)}{\partial|\Psi|^2}+\frac{1}{2m^*}\left(\frac{\hbar}{i}\nabla-\frac{e^*}{c}\mathbf{A}\right)^2\right]\Psi=0 \tag{12}$$

together with its complex conjugate, with Eqs. (11), and with the boundary condition

$$\hat{n}\cdot\left(\frac{\hbar}{i}\nabla-\frac{e^*}{c}\mathbf{A}\right)\Psi=0 \tag{13}$$

at the surface of the specimen, where $\hat{n}$ is the surface normal. The physical reason for choosing boundary condition (13) in preference to the other mathematical alternative, $\Psi=0$ at the surface, will be discussed shortly, even though both conditions reassuringly result in no supercurrent leaving the specimen. When Ψ satisfies the variational Eq. (12) and its complex conjugate, it is easily shown [by multiplying Eq. (12) by Ψ^* and subtracting the complex conjugate] that the supercurrent is source-free; $\nabla\cdot\mathbf{j}_s=0$, as expected. Furthermore, at the minimum points of $\mathscr{F}$ where both Eqs. (11) and (12) are satisfied, the superconductor sustains no net forces, in the sense of having a vanishing divergence to the stress tensor (*2*).

For temperatures just below T_c, where expansion (10) is valid, the so-far-unknown parameters α, β can easily be replaced by quantities accessible to measurement. We note that if the sample shows a perfect Meissner effect, so that $\Psi(\mathbf{r})$ is everywhere equal to its zero-field value (denoted by GL as Ψ_∞, a notation chosen to indicate a value *far* from any boundaries), then Eq. (12) gives

$$\alpha+\beta\Psi_\infty^2=0 \tag{14}$$

Also in this situation, the superconducting and normal free energies will be equal when the applied field equals the thermodynamic critical field H_c, so that

$$\alpha\Psi_\infty^2+\tfrac{1}{2}\beta\Psi_\infty^4=H_c^2/8\pi \tag{15}$$

Solving Eqs. (14) and (15) simultaneously to eliminate α and β, we find we can write

$$F(|\Psi|^2)=\frac{H_c^2}{8\pi}\left(2\frac{|\Psi|^2}{\Psi_\infty^2}-\frac{|\Psi|^4}{\Psi_\infty^4}\right) \tag{16}$$

Next, the quantity Ψ_∞ can be eliminated in favor of the weak-field penetration depth

$$\delta=(4\pi\Psi_\infty^2 e^{*2}/m^*c^2)^{-1/2} \tag{17}$$

at the same time that the Ginzburg–Landau equations (11) and (12) are reduced to a convenient dimensionless form. If all lengths are measured in units of δ;

if the magnetic field is measured in units of‡ $2^{1/2}H_c$, so that the vector potential is measured in units of $2^{1/2}H_c\delta$; and if a reduced order parameter $\Psi_0(\mathbf{r}) \equiv \Psi(\mathbf{r})/\Psi_\infty$ is introduced, then Eqs. (9), (11)–(13), and (16) become

$$\mathscr{F}_s = \mathscr{F}_n + \frac{H_c^2}{8\pi}\int d^3r\left[-2|\Psi_0|^2 + |\Psi_0|^4 + 2|\mathbf{H}-\mathbf{H}_a|^2 + \left|\left(\frac{1}{i\kappa}\nabla - \mathbf{A}\right)\Psi_0\right|^2\right] \tag{18}$$

$$\nabla \times \mathbf{H} = \frac{1}{2i}\left[\Psi_0^*\left(\frac{1}{i\kappa}\nabla - \mathbf{A}\right)\Psi_0 + \Psi_0\left(-\frac{1}{i\kappa}\nabla - \mathbf{A}\right)\Psi_0^*\right] \tag{19}$$

$$\left(1 - \Psi_0^2\right)\Psi_0 - \left(\frac{1}{i\kappa}\nabla - \mathbf{A}\right)^2 \Psi_0 = 0 \tag{20}$$

$$\hat{n}\cdot\left(\frac{1}{i\kappa}\nabla - \mathbf{A}\right)\Psi_0 = 0 \qquad \text{on surface} \tag{21}$$

The dimensionless parameter κ, now known as the Ginzburg–Landau parameter, is defined as

$$\kappa = 2^{1/2}(e^*/\hbar c)\,H_c\delta^2 \tag{22}$$

Equations (18)–(22) constitute the Ginzburg–Landau theory in its most usual form.

Even though the Ginzburg–Landau theory relies on the two pieces of experimental information H_c and δ, in fact it is just a one-parameter theory. Only the differences of κ from one material to another prevent the equations from scaling perfectly into a single "law of corresponding states" valid for all superconductors. It can be seen that whereas the magnetic field varies spatially over characteristic lengths of order δ, the reduced order parameter Ψ_0 has spatial variations with a quite different characteristic length δ/κ. The ratio of these two lengths, κ itself, thus determines the relative balance between rapidity of changes of magnetic field and of order parameter in the final solution. As for the temperature dependence of κ, it should be noted that $\delta \to \infty$ as $T \to T_c$ while $H_c \to 0$. But using the approximate Gorter–Casimir temperature dependence for H_c and δ to get a quick, even though rough, idea, we find

$$\kappa \simeq 2^{1/2}\frac{e^*H_c(0)}{\hbar c}\frac{\delta^2(0)}{1+t^2}$$

so that κ is both finite at T_c and relatively insensitive to temperature.

We are now in a position to discuss in somewhat more detail the choice of boundary condition for Ψ. As remarked earlier, the variational principle on $\mathscr{F}_s$

‡ The factor of $2^{1/2}$ might well be considered an anachronism, but in reality must be perpetuated to keep intact the standard definition of κ, Eq. (22).

plus the requirement of zero current leaving the sample leads either to Eq. (13) or to $\Psi = 0$ on the surface. Although the latter would at first sight appear more natural because it is the usual requirement on the Schrödinger wave function, it is incompatible with the experimental fact that very thin films of superconductor have very nearly the same transition temperature as the bulk material. It is possible to verify this statement by solving exactly the corresponding Ginzburg–Landau equation for a film of thickness d and by testing the boundary conditions. Such a calculation also illustrates ways in which the GL equations can be solved in special cases, and the important feature that solutions $\Psi_0(\mathbf{r})$ cannot vary more rapidly than on a scale of lengths of order δ/κ.

The appropriate one-dimensional, zero-field form of Eq. (20) is

$$\frac{1}{\kappa^2}\frac{d^2}{dx^2}\Psi_0 + \Psi_0 - \Psi_0^3 = 0 \tag{23}$$

together with the boundary conditions being tested,

$$\Psi_0(\pm\, d/2\delta) = 0 \tag{24}$$

A first integral of Eq. (23) is easily found to be

$$\left(\frac{1}{\kappa}\frac{d\Psi_0}{dx}\right)^2 = \tfrac{1}{2}(1 - \Psi_0^2)^2 + \text{constant} \tag{25}$$

The constant of integration can be determined by requiring, with no loss of generality, that $\Psi_0(x)$ is symmetric about $x = 0$, so that

$$[d\Psi_0(x)/dx]_{x=0} = 0$$

The final integration of Eq. (25) is then straightforward, with the result that the solution with the fewest nodes (and the lowest free energy) is given by the implicit relation

$$\kappa\left(\frac{d}{2\delta} - |x|\right) = \left[\frac{2}{2 - \Psi_0^2(0)}\right]^{1/2} F\left(\sin^{-1}\frac{\Psi_0(x)}{\Psi_0(0)}, \frac{\Psi_0(0)}{[2 - \Psi_0^2(0)]^{1/2}}\right) \tag{26}$$

where F is the incomplete elliptic integral of the first kind. Setting $x = 0$ yields the transcendental equation for $\Psi_0(0)$:

$$\frac{\kappa d}{2\delta} = \left[\frac{2}{2 - \Psi_0^2(0)}\right]^{1/2} F\left(\frac{\pi}{2}, \frac{\Psi_0(0)}{[2 - \Psi_0^2(0)]^{1/2}}\right) \tag{27}$$

Although Eqs. (26) and (27) could be plotted in great detail, the only observation needed for present purposes is that the right side of Eq. (27) is $\geqslant \pi/2$, and is a monotonically increasing function of $\Psi_0(0)$. This means that the film would not be superconducting at all [i.e., we find $\Psi_0(x) \equiv 0$] unless the thickness is greater than a finite critical value, $d \geqslant d_c = \pi\delta/\kappa$, a result which is in contradiction with experiment. Thus the boundary condition $\Psi_0 = 0$ at the surface must be rejected, whereas the condition $d\Psi_0/dx = 0$ at the surface produces the solution $\Psi_0(x) \equiv 1$, as desired.

II. GOR'KOV'S DERIVATION FROM MICROSCOPIC THEORY

Once the parameter κ is determined by specifying H_c and δ, no further input of physical knowledge is needed before proceeding to the purely mathematical problem of solving the Ginzburg–Landau equations in any given situation. In fact, the majority of the more interesting cases (the most prominent exception being the surface sheath solution) were worked out in the years prior to the advent of the BCS theory, and these form the substance of Chapters 14 on type II superconductivity, 16 on thin films, and 17 on proximity effects. What the microscopic theory is able to provide, however, is a deeper understanding of the fundamental meaning of the GL equations and a detailing of the circumstances under which they are or are not a valid representation of the actual situation.

The earliest and most significant progress in this direction was made by Gor'kov in a trio of classic papers during the two years following the appearance of the BCS theory. In the first of these papers (*3*), Gor'kov succeeded in recasting the original BCS variational wave function approach into the mathematical language of Green's functions, a development to which Chapter 5 has been devoted. From the present viewpoint, the crucial importance of Gor'kov's work was in giving the BCS theory such a formal simplicity and compactness that it could easily be generalized to situations of spatial inhomogeneity. The next two papers (*4*) used the coordinate-space Green's function technique in the presence of a strong static magnetic field to derive the Ginzburg–Landau equations at temperatures just below T_c, first for a pure material and then for an alloy. It is probably not controversial to claim that this derivation could never have been carried out successfully by attempting to guess the form of the electron-state pairing and the trial system wave function (or density matrix) in the BCS manner.

An appropriate starting point for presenting the GL theory derivation is the Gor'kov Green's function equations discussed in Chapter 5:

$$\left[i\omega - \frac{1}{2m}\left(\frac{\hbar}{i}\nabla - \frac{e}{c}\mathbf{A}(\mathbf{r})\right)^2 + \mu + U(\mathbf{r})\right] \times G_\omega(\mathbf{r},\mathbf{r}') + \Delta(\mathbf{r})\, F_\omega^+(\mathbf{r},\mathbf{r}') = \delta^3(\mathbf{r}-\mathbf{r}') \tag{28}$$

$$\left[-i\omega - \frac{1}{2m}\left(-\frac{\hbar}{i}\nabla - \frac{e}{c}\mathbf{A}(\mathbf{r})\right)^2 + \mu + U(\mathbf{r})\right] \times F_\omega^+(\mathbf{r},\mathbf{r}') - \Delta^*(\mathbf{r})\, G_\omega(\mathbf{r},\mathbf{r}') = 0 \tag{29}$$

together with the self-consistency relation determining Δ,

$$\Delta(\mathbf{r}) = Vk_BT\sum_{\nu} F_\omega(\mathbf{r}, \mathbf{r}) \tag{30}$$

Reviewing the notation, G_ω and F_ω are the normal and "anomolous" Green's functions dependent on discrete frequencies $\omega = (2\nu+1)\pi k_BT$, $\nu = 0, \pm 1, \pm 2, \ldots$; $U(\mathbf{r})$ is the potential due to a set of static impurities at fixed positions $\mathbf{R}_i$, each with potential $u(\mathbf{r})$,

$$U(\mathbf{r}) = \sum_i u(\mathbf{r} - \mathbf{R}_i) \tag{31}$$

$\mathbf{A}(\mathbf{r})$ is the net static vector potential coupled to electrons of the usual charge and mass, e and m, and V is the BCS coupling constant for the attraction between electrons. It is important to keep in mind the simplifying assumptions which underlie the Gor'kov equations:

1. The normal metal is regarded as a free, highly degenerate electron gas.
2. The superconductivity is regarded as arising due to the BCS model potential, an instantaneous, short-range attraction between electrons whose energy is within a narrow shell about the Fermi surface.
3. The Pauli paramagnetism of the electron spins is neglected.

Approximations 1 and 2 imply that no account is taken of band structure (other than via a possible effective mass) or of realistic electron–electron or electron–phonon interaction effects. Removing these approximations in complete generality is difficult, but some progress has been made very recently. On the other hand, it is easy to remove approximation (3), with consequences which will be discussed later in this chapter.

It is useful to convert Eqs. (28) and (29) from differential to integral form,

$$G_\omega(\mathbf{r}, \mathbf{r}') = G^n_\omega(\mathbf{r}, \mathbf{r}') - \int d^3r_1\, d^3r_2\, G^n_\omega(\mathbf{r}, \mathbf{r}_1)\, \Delta(\mathbf{r}_1)\, G^n_{-\omega}(\mathbf{r}_2, \mathbf{r}_1)\, \Delta^*(\mathbf{r}_2)\, G_\omega(\mathbf{r}_2, \mathbf{r}') \tag{32}$$

$$F^+_\omega(\mathbf{r}, \mathbf{r}') = \int d^3r_1\, G^n_{-\omega}(\mathbf{r}_1, \mathbf{r})\, \Delta^*(\mathbf{r}_1)\, G^n_\omega(\mathbf{r}_1, \mathbf{r}') - \int d^3r_1\, d^3r_2\, G^n_{-\omega}(\mathbf{r}_1, \mathbf{r})\, \Delta^*(\mathbf{r}_1)\, G^n_\omega(\mathbf{r}_1, \mathbf{r}_2)\, \Delta(\mathbf{r}_2)\, F^+_\omega(\mathbf{r}_2, \mathbf{r}') \tag{33}$$

by introducing the Green's function for the normal metal, G^n_ω, satisfying

$$\left[i\omega - \frac{1}{2m}\left(\frac{\hbar}{i}\nabla - \frac{e}{c}\mathbf{A}(\mathbf{r})\right)^2 + \mu + U(\mathbf{r})\right] G^n_\omega(\mathbf{r}, \mathbf{r}') = \delta^3(\mathbf{r} - \mathbf{r}') \tag{34}$$

Since Gor'kov considers the regime of temperatures just below T_c, the quantities $|\Delta(\mathbf{r})|$, $H(\mathbf{r})$, and the inverse penetration depth can all be regarded as small parameters; in particular, an iteration of Eqs. (32) and (33) in powers of Δ may be truncated by replacing the unknown function on the right side by the inhomogeneous term. Then the equations can be averaged over an ensemble of randomly distributed impurity configurations as discussed in Chapter 5. Just as there, it is a good approximation to regard $\Delta(\mathbf{r})$ as very nearly independent of impurity configuration. Denoting the ensemble average by angular brackets, the resulting equations become

$$\langle G_\omega(\mathbf{r},\mathbf{r}')\rangle \simeq \langle G^n_\omega(\mathbf{r},\mathbf{r}')\rangle - \int d^3r_1\, d^3r_2 \times \langle G^n_\omega(\mathbf{r},\mathbf{r}_1)\, G^n_{-\omega}(\mathbf{r}_2,\mathbf{r}_1)\, G^n_\omega(\mathbf{r}_2,\mathbf{r}')\rangle\, \Delta(\mathbf{r}_1)\, \Delta^*(\mathbf{r}_2) \tag{35}$$

$$\begin{aligned}\langle F^+_\omega(\mathbf{r},\mathbf{r}')\rangle \simeq \int d^3r_1 \langle G^n_{-\omega}(\mathbf{r}_1,\mathbf{r})\, G^n_\omega(\mathbf{r}_1,\mathbf{r}')\rangle\, \Delta^*(\mathbf{r}_1) \\ - \int d^3r_1\, d^3r_2\, d^3r_3 \langle G^n_{-\omega}(\mathbf{r}_1,\mathbf{r})\, G^n_\omega(\mathbf{r}_1,\mathbf{r}_2)\, G^n_{-\omega}(\mathbf{r}_3,\mathbf{r}_2) \\ \times G^n_\omega(\mathbf{r}_3,\mathbf{r}')\rangle\, \Delta^*(\mathbf{r}_1)\, \Delta(\mathbf{r}_2)\, \Delta^*(\mathbf{r}_3)\end{aligned} \tag{36}$$

For the moment, Eq. (35) will be ignored and attention concentrated on Eq. (36), which is substituted into Eq. (30) to give a nonlinear integral equation for $\Delta^*(\mathbf{r})$:

$$\begin{aligned}\Delta^*(\mathbf{r}) = Vk_BT\sum_\nu \Big\{ \int d^3r_1 \langle G^n_{-\omega}(\mathbf{r}_1,\mathbf{r})\, G^n_\omega(\mathbf{r}_1,\mathbf{r})\rangle\, \Delta^*(\mathbf{r}_1) \\ - \int d^3r_1\, d^3r_2\, d^3r_3 \langle G^n_{-\omega}(\mathbf{r}_1,\mathbf{r})\, G^n_\omega(\mathbf{r}_1,\mathbf{r}_2)\, G^n_{-\omega}(\mathbf{r}_3,\mathbf{r}_2)\, G^n_\omega(\mathbf{r}_3,\mathbf{r})\rangle \\ \times \Delta^*(\mathbf{r}_1)\, \Delta(\mathbf{r}_2)\, \Delta^*(\mathbf{r}_3)\Big\}\end{aligned} \tag{37}$$

We now make the Ansatz, to be confirmed shortly, that the absolute value of solutions $\Delta^*(\mathbf{r})$ to Eq. (37) will show spatial variations over a scale of lengths long compared to the range of the Green's function $G^n_\omega(\mathbf{r},\mathbf{r}')$. Since the term in Eq. (37) of third order in $|\Delta|$ is already a small correction to the term of first order, it is consistent in the third-order term to ignore both the (small) magnetic field as well as the differences in position arguments of the three factors of $|\Delta|$. No alteration can be made in the position dependence of the phase of Δ^* since responses of the phase and of the vector potential to gauge transformations must cancel exactly in order to maintain overall gauge invariance. In the first integral, again because the magnetic field is small, it is permissible to regard the correlations introduced by the impurity averaging as nearly independent of the field, so that $\langle GG\rangle$ has nearly the same field dependence as $\langle G\rangle\langle G\rangle$. Finally, an Ansatz closely related to the one above and which also will be examined in detail shortly, is that the range of $\langle G^n_\omega(\mathbf{r},\mathbf{r}')\rangle$ is short compared with the radius of an electron cyclotron orbit in the magnetic field; this permits the use of the semiclassical phase integral approximation,

$$\langle G^n_\omega(\mathbf{r}, \mathbf{r}')\rangle \simeq \langle G^n_\omega(\mathbf{r}, \mathbf{r}')\rangle_{H=0} \exp\left(\frac{ie}{\hbar c}\int_{\mathbf{r}'}^{\mathbf{r}} d\mathbf{s}\cdot\mathbf{A}(\mathbf{s})\right) \tag{38}$$

where the path of integration between $\mathbf{r}'$ and $\mathbf{r}$ is a straight line. Putting all these approximations together yields the altered version of Eq. (37),

$$\Delta^*(\mathbf{r}) \simeq VT\sum_\nu\left\{\int d^3r_1\, Q_\omega(\mathbf{r}_1-\mathbf{r})\exp\left(\frac{2ie}{\hbar c}\int_{\mathbf{r}}^{\mathbf{r}_1} d\mathbf{s}\cdot\mathbf{A}(\mathbf{s})\right)\Delta^*(\mathbf{r}_1) - R_\omega|\Delta(\mathbf{r})|^2\Delta^*(\mathbf{r})\right\} \tag{39}$$

where we introduce the notation

$$Q_\omega(\mathbf{r}_1-\mathbf{r}) \equiv \langle G^n_{-\omega}(\mathbf{r}_1, \mathbf{r})\, G^n_\omega(\mathbf{r}_1, \mathbf{r})\rangle_{H=0} \tag{40}$$

$$R_\omega \equiv \int d^3r_1\, d^3r_2\, d^3r_3\langle G^n_{-\omega}(\mathbf{r}_1, \mathbf{r})\, G^n_\omega(\mathbf{r}_1, \mathbf{r}_2')\, G^n_{-\omega}(\mathbf{r}_3, \mathbf{r}_2)\, G^n_\omega(\mathbf{r}_3, \mathbf{r})\rangle_{H=0} \tag{41}$$

A lemma which is convenient at this point, and of considerable importance for later work, is that for any functions $\mathbf{A}$ and Δ^*

$$\exp\left(\frac{2ie}{\hbar c}\int_{\mathbf{r}}^{\mathbf{r}_1} d\mathbf{s}\cdot\mathbf{A}(\mathbf{s})\right)\Delta^*(\mathbf{r}_1) = \exp[-i(\mathbf{r}_1-\mathbf{r})\cdot\mathbf{\Pi}^\dagger]\,\Delta^*(\mathbf{r}) \tag{42}$$

where we define the operators

$$\mathbf{\Pi} \equiv \frac{1}{i}\nabla - \frac{2e}{\hbar c}\mathbf{A}(\mathbf{r}) \qquad \mathbf{\Pi}^\dagger \equiv -\frac{1}{i}\nabla - \frac{2e}{\hbar c}\mathbf{A}(\mathbf{r}) \tag{43}$$

The proof begins by parameterizing explicitly the straight-line path of the $\mathbf{s}$ integral:

$$\mathbf{s} = \mathbf{r} + (\mathbf{r}_1 - \mathbf{r})\,\theta \qquad 0 \leqslant \theta \leqslant 1$$

This together with Taylor's theorem gives

$$\begin{aligned}\int_{\mathbf{r}}^{\mathbf{r}_1} d\mathbf{s}\cdot\mathbf{A}(\mathbf{s}) &= \int_0^1 d\theta\,(\mathbf{r}_1-\mathbf{r})\cdot\mathbf{A}(\mathbf{r}+\theta(\mathbf{r}_1-\mathbf{r}))\\ &= \int_0^1 d\theta\exp[\theta(\mathbf{r}_1-\mathbf{r})\cdot\nabla]\,(\mathbf{r}_1-\mathbf{r})\cdot\mathbf{A}(\mathbf{r})\exp[-\theta(\mathbf{r}_1-\mathbf{r})\cdot\nabla]\end{aligned} \tag{44}$$

where the derivatives act only on the $\mathbf{r}$ dependence of $\mathbf{A}$ and not on $\mathbf{r}_1 - \mathbf{r}$. Hence proving the desired lemma is equivalent to proving the operator identity

$$\exp\left[\int_0^1 d\theta\, e^{\theta P}Qe^{-\theta P}\right]e^P = e^{P+Q} \tag{45}$$

with P and Q being operators which do not necessarily commute, but which do satisfy

$$[e^{\theta P}Qe^{-\theta P}, e^{\theta' P}Qe^{-\theta' P}] = 0 \tag{46}$$

To demonstrate Eq. (45), we introduce an auxiliary quantity

$$V(t) \equiv e^{t(P+Q)}e^{-tP} \tag{47}$$

for which

$$\frac{dV(t)}{dt} = e^{t(P+Q)}Qe^{-tP} = V(t)\, e^{tP}Qe^{-tP} \tag{48}$$

Solving Eq. (48) for $V(t)$, we find

$$V(t) = \exp \int_0^t d\theta\, e^{\theta P}Qe^{-\theta P} \tag{49}$$

where the operator expression on the right side is well defined because of condition (46). Then using the definition of $V(t)$ and setting $t = 1$ gives the desired theorem, Eq. (45), or equivalently Eq. (42).

With Eq. (42) established, we can now appeal once more to the approximations that the field is small and the spatial variations of $|\Delta|$ slow so as to expand Eqs. (39) and (42) in powers of $(\mathbf{r}_1 - \mathbf{r})\cdot\mathbf{\Pi}^\dagger$ to lowest nontrivial order in this small quantity:

$$\Delta^*(\mathbf{r}) \simeq Vk_BT\sum_\nu \left\{ \int d^3r_1\, Q_\omega(\mathbf{r}_1)\,[1 - \tfrac{1}{2}(\mathbf{r}_1\cdot\mathbf{\Pi}^\dagger)^2] - R_\omega|\Delta(\mathbf{r})|^2 \right\} \Delta^*(\mathbf{r}) \tag{50}$$

The remainder of Gor'kov's derivation consists of evaluating Q_ω and R_ω, whereupon the approximations employed above can be validated by inspection.

The procedure for evaluating the impurity averages in Q_ω and R_ω begins by expanding the normal-state Green's function defined by Eq. (34), but with $\mathbf{A} = 0$, in powers of U, and substituting into expressions (40) and (41). Then using Eq. (31), each impurity position coordinate $\mathbf{R}_i$ is regarded as an independent random variable. The angular brackets formally denoting the impurity average take on the operational meaning

$$\langle f\{R_j\}\rangle = \prod_i \left(\Omega^{-1} \int d^3R_i\right) f(\{R_j\}) \tag{51}$$

where each integration ranges over the volume of the sample Ω. After carrying out the integrations, the resulting series expressions for Q_ω and R_ω can be placed in a one-to-one correspondence with an infinite set of diagrams specified by simple topological rules, which have already been developed in Chapter 5.

The impurity averaging diagram technique is a simple and convenient one only in the limit of small impurity concentration. It has been shown in Chapter 5 that averages can be computed to lowest order in the impurity concentration by taking only a small subset of diagrams (the "nonintersecting" ones) which can be summed easily. On the other hand, it would be desirable to have a theoretical description which could be applicable also to highly concentrated alloys. Such a description has in fact been successfully formulated by de Gennes and co-workers (5), who introduced the exact (but unknown) wave function for an electron in a fixed configuration of impurities, then noted that for a high concentration of randomly distributed impurities the electron's motion is a diffusion with a certain relaxation time, and finally removed the unknown wave functions by use of sum rules leaving only the relaxation time in final formulas. We will not go further into de Gennes formulation here, since it becomes difficult to apply in more complicated situations such as with spin-orbit scattering present, but it is important to note that this method leads to precisely the same formulas as does the dilute concentration diagramatic technique *when both are expressed in terms of a scattering time.* Throughout this chapter we will employ the diagram technique, and by appeal to de Gennes' results we will argue that our formulas are also valid for concentrated alloys when expressed in terms of scattering times rather than the concentration.

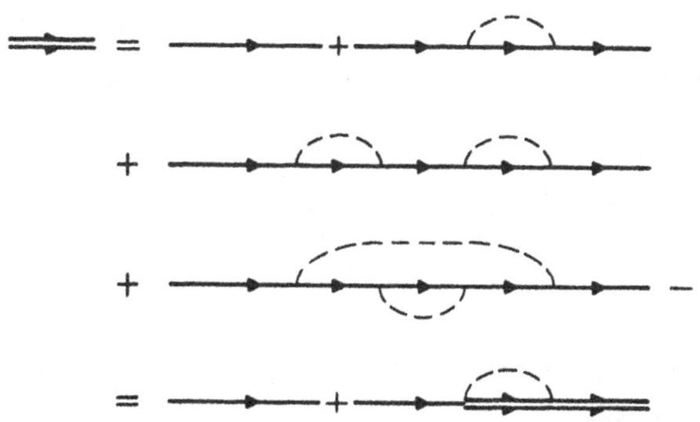

Fig. 1. Impurity averaging diagrams of self-energy type leading to $\langle G_\omega^n \rangle$.

Using the dilute concentration diagram rules of Chapter 5, the sum of self-energy graphs illustrated in Fig. 1 leads to the equation for the average of a single normal-state Green's function

$$\left(i\tilde{\omega} + \frac{\hbar^2}{2m}\nabla^2 + \mu\right)\langle G_\omega^n(\mathbf{r}-\mathbf{r}')\rangle_0 = \delta^3(\mathbf{r}-\mathbf{r}') \tag{52}$$

where

$$\tilde{\omega} \equiv \omega + \frac{\hbar}{2\tau}\,\mathrm{sgn}\,\omega$$

The inverse scattering time, τ^{-1}, is identified in the dilute limit with the product of impurity concentration, n, times the s-wave scattering probability from a single impurity; in first Born approximation, with the impurity potential Fourier transformed,

$$\frac{1}{\tau} = \frac{2\pi}{\hbar}\, n \int \frac{d^3k'}{(2\pi)^3}\, |u(\mathbf{k}-\mathbf{k}')|^2\, \delta(\epsilon(k')) \tag{53}$$

where $\mathbf{k}$ is on the Fermi surface and

$$\epsilon(k') \equiv (\hbar^2 k'^2/2m) - \mu$$

The solution of Eq. (52) is easily found to be

$$\begin{aligned}\langle G^n_\omega(\mathbf{r}-\mathbf{r}')\rangle_0 &= \int \frac{d^3k}{(2\pi)^3} \frac{\exp[i\mathbf{k}\cdot(\mathbf{r}-\mathbf{r}')]}{i\omega - \epsilon(k)} \\ &\simeq \frac{-m}{2\pi|\mathbf{r}-\mathbf{r}'|} \exp\left\{\left[ik_F \operatorname{sgn}\omega - \frac{|\tilde\omega|}{\hbar v_F}\right]|\mathbf{r}-\mathbf{r}'|\right\} \end{aligned} \tag{54}$$

in the limit $\mu \gg |\tilde\omega|$. The kernel Q_ω, involving $\langle GG\rangle$, can be expressed in terms of $\langle G\rangle\langle G\rangle$ by summing the ladder graphs of Fig. 2, which represent a vertex

Fig. 2. Impurity averaging diagrams of vertex type giving the equation satisfied by $Q_\omega = \langle G^n_\omega G^n_{-\omega}\rangle$.

correction. The resulting integral equation for Q_ω is difficult to solve with a general impurity scattering amplitude, but the equation simplifies greatly if the scattering amplitude is s-wave only. In this special case, Q_ω satisfies

$$Q_\omega(\mathbf{r}-\mathbf{r}') = Q^0_\omega(\mathbf{r}-\mathbf{r}') + [\hbar/2\pi\tau N(0)] \int d^3r_1\, Q^0_\omega(\mathbf{r}-\mathbf{r}_1)\, Q_\omega(\mathbf{r}_1-\mathbf{r}') \tag{55}$$

where, from Eq. (54),

$$\begin{aligned} Q^0_\omega(\mathbf{r}-\mathbf{r}') &\equiv \langle G^n_\omega(\mathbf{r}-\mathbf{r}')\rangle_0 \langle G^n_{-\omega}(\mathbf{r}-\mathbf{r}')\rangle_0 \\ &= \left(\frac{m}{2\pi|\mathbf{r}-\mathbf{r}'|}\right)^2 \exp\left(-\frac{2|\tilde\omega|}{\hbar v_F}|\mathbf{r}-\mathbf{r}'|\right) \\ &= \int \frac{d^3k}{(2\pi)^3} \exp[i\mathbf{k}\cdot(\mathbf{r}-\mathbf{r}')] \frac{2\pi N(0)}{\hbar v_F k} \tan^{-1}\frac{\hbar v_F k}{2|\tilde\omega|} \end{aligned} \tag{56}$$

The solution to Eq. (55) is immediate upon Fourier transformation:

$$\begin{aligned} Q_\omega(\mathbf{r}-\mathbf{r}') = \int \frac{d^3k}{(2\pi)^3} &\exp[i\mathbf{k}\cdot(\mathbf{r}-\mathbf{r}')]\, 2\pi N(0) \\ &\times \left[\left(\frac{1}{\hbar v_F k}\tan^{-1}\frac{\hbar v_F k}{2|\tilde\omega|}\right)^{-1} - \frac{\hbar}{\tau}\right]^{-1} \end{aligned} \tag{57}$$

Then by expanding the integrand of Eq. (57) in powers of k,

$$\left[\left(\frac{1}{\hbar v_F k}\tan^{-1}\frac{\hbar v_F k}{2|\tilde\omega|}\right)^{-1} - \frac{\hbar}{\tau}\right]^{-1} \simeq \frac{1}{2|\omega|} - \frac{1}{3}\left(\frac{\hbar v_F k}{2|\omega|}\right)^2 \frac{1}{2|\tilde\omega|} + O(k^4) \tag{58}$$

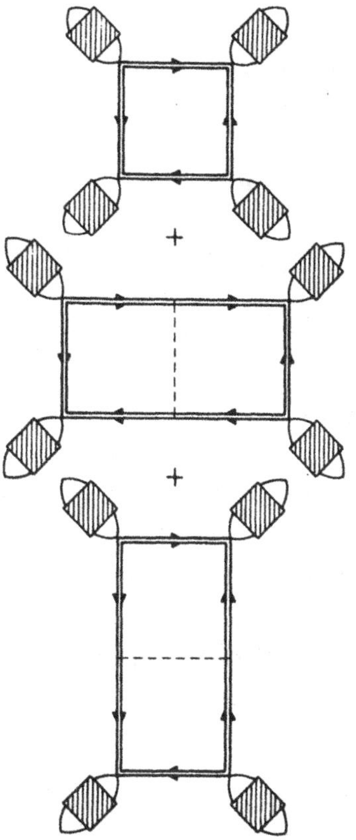

Fig. 3. Impurity averaging diagrams contributing to R_ω. Diagrams with more than one impurity line across the "box" equal exactly zero, because their contribution has an integrand with an even number of poles in the same half-plane.

values can be read off for the integrals

$$
\begin{aligned}
\int d^3r\, Q_\omega(r) &= \frac{2\pi N(0)}{2|\omega|} \\
\tfrac{1}{2}\int d^3r\, \tfrac{1}{3}r^2 Q_\omega(r) &= \frac{2\pi N(0)}{6|\tilde{\omega}|}\left(\frac{\hbar v_F}{2\omega}\right)^2
\end{aligned}
\tag{59}
$$

which occur in Eq. (50).

The quantity R_ω is evaluated in a similar manner, but is rather more intricate. All diagrams with nonintersecting impurity lines must be summed, so that in addition to the self-energy diagrams and vertex corrections contained in Fig. 1, graphs with impurity lines as in Fig. 3 must also be taken into account. Again assuming s-wave impurity scattering only, the vertex correction at each of the four vertices is

$$\int d^3r\, Q_\omega(r) / \int d^3r\, Q^0_\omega(r) = |\tilde{\omega}|/|\omega| \tag{60}$$

and hence

$$\begin{aligned}
R_\omega = {} & \left(\frac{\tilde{\omega}}{\omega}\right)^4 \int d^3r_1\, d^3r_2\, d^3r_3 \\
& \times \Big\{ \langle G^n_{-\omega}(\mathbf{r}_1, \mathbf{r})\rangle_0 \langle G^n_\omega(\mathbf{r}_1, \mathbf{r}_2)\rangle_0 \langle G^n_{-\omega}(\mathbf{r}_3, \mathbf{r}_2)\rangle_0 \langle G^n_\omega(\mathbf{r}_3, \mathbf{r})\rangle_0 \\
& + \frac{\hbar}{2\pi\tau N(0)} \int d^3r_4 \Big[\langle G^n_{-\omega}(\mathbf{r}_1, \mathbf{r})\rangle_0 \langle G_\omega(\mathbf{r}_1, \mathbf{r}_4)\rangle_0 \langle G^n_\omega(\mathbf{r}_4, \mathbf{r}_2)\rangle_0 \\
& \times \langle G^n_{-\omega}(\mathbf{r}_3, \mathbf{r}_2)\rangle_0 \langle G^n_\omega(\mathbf{r}_3, \mathbf{r}_4)\rangle_0 \langle G^n_\omega(\mathbf{r}_4, \mathbf{r}_2)\rangle_0 \\
& + \langle G^n_{-\omega}(\mathbf{r}_4, \mathbf{r})\rangle_0 \langle G^n_{-\omega}(\mathbf{r}_1, \mathbf{r}_4)\rangle_0 \langle G^n_\omega(\mathbf{r}_1, \mathbf{r}_2)\rangle_0 \\
& \times \langle G^n_{-\omega}(\mathbf{r}_4, \mathbf{r}_2)\rangle_0 \langle G^n_{-\omega}(\mathbf{r}_3, \mathbf{r}_4)\rangle_0 \langle G^n_\omega(\mathbf{r}_3, \mathbf{r})\rangle_0 \Big] \Big\} \\
= {} & \left(\frac{\tilde{\omega}}{\omega}\right)^4 \Big\{ \int \frac{d^3k}{(2\pi)^3} \frac{1}{[\tilde{\omega}^2 + \epsilon^2(k)]^2} + \frac{1}{2\pi\tau N(0)} \\
& \times \Big[\Big(\int \frac{d^3k}{(2\pi)^3} \frac{1}{[i\tilde{\omega} - \epsilon(k)]^2 [-i\tilde{\omega} - \epsilon(k)]} \Big)^2 + \text{c.c.} \Big] \Big\} \\
= {} & \frac{\pi}{2} \frac{N(0)}{|\omega|^3}
\end{aligned} \tag{61}$$

Finally substituting expressions (59) and (61) into Eq. (50), we arrive at

$$\Delta^*(\mathbf{r}) = 2\pi N(0)\, V k_B T \sum_\nu \left[\frac{1}{2|\omega|} - \frac{1}{6|\tilde{\omega}|} \left(\frac{\hbar v_F}{2\omega}\right)^2 \Pi^2 - \frac{1}{4|\omega|^3} |\Delta(\mathbf{r})|^2 \right] \Delta^*(\mathbf{r}) \tag{62}$$

The divergence in the ν summation is quickly cured by recalling from Chapter 5 that the sum is actually cut off at the Debye energy, so that

$$\sum_\nu \frac{2\pi T}{2|\omega|} = \ln\left(\frac{1.14\Theta_D}{T}\right) \qquad \frac{1}{N(0)\, V} = \ln\left(\frac{1.14\Theta_D}{T_c}\right) \tag{63}$$

Since T is very close to T_c, Eq. (62) reduces to

$$\left[\left(1 - \frac{T}{T_c}\right) - \frac{7\zeta(3)}{8(\pi k_B T_c)^2} |\Delta(\mathbf{r})|^2 - \frac{1}{6} \left(\frac{\hbar v_F}{\pi k_B T_c}\right)^2 \frac{7\zeta(3)}{8} \chi\left(\frac{\hbar}{2\pi k_B T_c \tau}\right) \Pi^2 \right] \Delta^*(\mathbf{r}) = 0 \tag{64}$$

where

$$\chi(x) \equiv \sum_{\nu=0}^{\infty} (2\nu + 1)^{-2} (2\nu + 1 + x)^{-1} / \sum_{\nu=0}^{\infty} (2\nu + 1)^{-3} \tag{65}$$

and ζ is the Riemann zeta function.

Equation (64) is the final result of the microscopic theory, as initially formalized by the Gor'kov equations (32) and (33), when simplified in the limit $T \to T_c$. Before discussing this equation in detail, however, we must go back to tidy up points passed over hastily in its derivation. First, we claimed that the semiclassical phase integral formula (38) for $\langle G^n_\omega(\mathbf{r}, \mathbf{r}')\rangle$ was adequate because the range of the Green's function was short compared to the radius of a cyclotron orbit. It is easiest to examine this in the special case of a uniform field (near T_c, where the penetration depth is long, the net field has only slow spatial variations) in the gauge $\mathbf{A}(\mathbf{r}) = \frac{1}{2}\mathbf{H} \times \mathbf{r}$, for which Eq. (38) becomes

$$\langle G^n_\omega(\mathbf{r}, \mathbf{r}')\rangle \simeq \langle G^n_\omega(\mathbf{r} - \mathbf{r}')\rangle_0 \exp\left(\frac{ie}{\hbar c}\frac{1}{2}\mathbf{H}\cdot\mathbf{r} \times \mathbf{r}'\right) \tag{66}$$

Substituting this into the exact relation [the average of Eq. (34)],

$$\left[i\tilde{\omega} - \frac{1}{2m}\left(\frac{\hbar}{i}\nabla - \frac{e}{c}\mathbf{A}(\mathbf{r})\right)^2 + \mu\right]\langle G^n_\omega(\mathbf{r}, \mathbf{r}')\rangle = \delta^3(\mathbf{r} - \mathbf{r}') \tag{67}$$

gives

$$\left[i\tilde{\omega} - \frac{\hbar^2}{2m}\left(\frac{1}{i}\nabla - \frac{e}{\hbar c}\mathbf{H} \times (\mathbf{r} - \mathbf{r}')\right)^2 + \mu\right]\langle G^n_\omega(\mathbf{r} - \mathbf{r}')\rangle_0 \simeq \delta^3(\mathbf{r} - \mathbf{r}') \tag{68}$$

From the solution in zero field, Eq. (54), the range of $\langle G^n_\omega(\mathbf{r} - \mathbf{r}')\rangle_0$ is found to be $\hbar v_F/2|\tilde{\omega}| \leqslant \hbar v_F/[2\pi k_B T + (\hbar/\tau)]$, so that the additional magnetic field terms are indeed negligible if

$$k_F \gg \frac{eH}{\hbar c}\frac{\hbar v_F}{2\pi k_B T + (\hbar/\tau)} \tag{69}$$

This is just the statement that the range is much less than the cyclotron radius, $R_c = mv_F c/eH$. The inequality can also be reexpressed as

$$\mu_B H \ll 2\pi k_B T + (\hbar/\tau) \tag{70}$$

which says that the spacing between Landau levels must be small compared to their thermal and collision broadening. The inequalities are certainly satisfied for T near T_c where H is small.

It was also claimed previously that $\Delta^*(\mathbf{r})$ had slow spatial variations compared with the range of $\langle G^n_\omega(\mathbf{r} - \mathbf{r}')\rangle_0$. This can be checked by examining Eq. (64) determining $\Delta^*(\mathbf{r})$, which indicates that solutions will have spatial variations over a scale of lengths estimated crudely as

$$\sim \frac{\hbar v_F}{2\pi k_B T_c}\left[\frac{\chi(\hbar/2\pi k_B T_c \tau)}{1 - (T/T_c)}\right]^{1/2} \sim \left[\frac{\hbar v_F}{2\pi k_B (T_c - T)}\frac{\hbar v_F}{2\pi k_B T_c + (\hbar/\tau)}\right]^{1/2} \tag{71}$$

This will certainly be long compared to the range of the Green's function provided that

$$1-(T/T_c) \ll 1+(\hbar/2\pi k_B T_c \tau) \tag{72}$$

which is just our basic assumption. Exactly the same argument justifies the equivalent approximation that the quantity $(\mathbf{r}_1 - \mathbf{r})\cdot\mathbf{\Pi}^\dagger$ appearing in Eq. (42) is a small parameter.

This completes the derivation of Eq. (64), which we can now point out is in the same form as the Ginzburg–Landau equations (10) and (12), except for a scaling of the variables. The Ginzburg–Landau current Eq. (3) also follows from the microscopic theory near T_c, using similar arguments but beginning with Eq. (35), which we had previously ignored. Since the reduction involves only methods already described in detail above, we will not take the space to give more than the final result,

$$\begin{aligned} \mathbf{j}(\mathbf{r}) &= \frac{e\hbar}{m} \lim_{\mathbf{r}'\to\mathbf{r}} (\mathbf{\Pi} + \mathbf{\Pi}^{\dagger\prime}) \langle G_\omega(\mathbf{r},\mathbf{r}')\rangle \\ &\simeq \frac{Ne\hbar}{2m} \frac{7\zeta(3)}{8(\pi k_B T_c)^2} \chi\left(\frac{\hbar}{2\pi k_B T_c \tau}\right) (\Delta^*(\mathbf{r})\,\mathbf{\Pi}\Delta(\mathbf{r}) + \text{c.c.}) \end{aligned} \tag{73}$$

where N is the total number of electrons, including both spins, per unit volume. The correspondence of Eqs. (64) and (73) with the Ginzburg–Landau equations becomes precise upon making the identifications

$$\begin{aligned} &\Psi \leftrightarrow [7\zeta(3)\,N\chi/8]^{1/2}\,(\Delta/\pi k_B T_c) \\ &e^* \leftrightarrow 2e \qquad m^* \leftrightarrow 2m \end{aligned} \tag{74}$$

Since the Gor'kov equations from which we began are equivalent to BCS theory, it is not surprising that Eqs. (64) and (73) give us precisely the BCS values (Chapter 2) for the temperature dependence (near T_c) of the fundamental quantities

$$\begin{aligned} \Delta_\infty^2 &= [8/7\zeta(3)]\,(\pi k_B T_c)^2\,(1-t) \\ \frac{H_c^2}{8\pi} &= \tfrac{1}{2}N(0)\,\frac{8(\pi k_B T_c)^2}{7\zeta(3)}\,(1-t)^2 \\ \frac{1}{\delta^2} &= \frac{4\pi N e^2}{mc^2}\,2(1-t)\,\chi\left(\frac{\hbar}{2\pi k_B T_c \tau}\right) \end{aligned} \tag{75}$$

Using these relations, we arrive at an expression for the phenomenological constant κ in terms of microscopic parameters of the normal metal,

$$\begin{aligned} \kappa &= 2^{1/2}(2eH_c/\hbar c)\,\delta^2 \\ &= \frac{1}{\chi}\left[\frac{6}{7\zeta(3)}\,\frac{mc^2}{4\pi N e^2}\right]^{1/2} \frac{2\pi k_B T_c}{\hbar v_F} \end{aligned} \tag{76}$$

As before, κ does not have any pronounced temperature dependence.

Several features of the above results deserve special attention. The most prominent is that the phenomenological charge e^* turns out to equal *twice* the electronic charge, a fact which will be of crucial significance for the discussion of flux quantization in Chapter 8. The double charge arises because the fundamental entities characteristic of superconductivity are correlated electron pairs, reflected in the nonvanishing of the "pair wave function," $\Delta(\mathbf{r}) \propto \langle\psi(\mathbf{r})\psi(\mathbf{r})\rangle$. This is also seen in the proportionality of $\Psi(\mathbf{r})$ and $\Delta(\mathbf{r})$; the Ginzburg–Landau order parameter Ψ is also the wave function (with different normalization) for the correlated electron pairs.

Expression (76) for κ shows its dependence on the normal-state mean free path, $l = v_F\tau$. For a pure specimen, $\chi \to 1$, κ is roughly the ratio of Landau penetration depth to BCS coherence distance, $\kappa \sim \delta_L/\xi_0$, so that for most pure materials, κ is substantially less than unity. Exceptions are Nb and V, which are type II superconductors (i.e., $\kappa > 2^{-1/2}$) even when pure, and the extraordi-

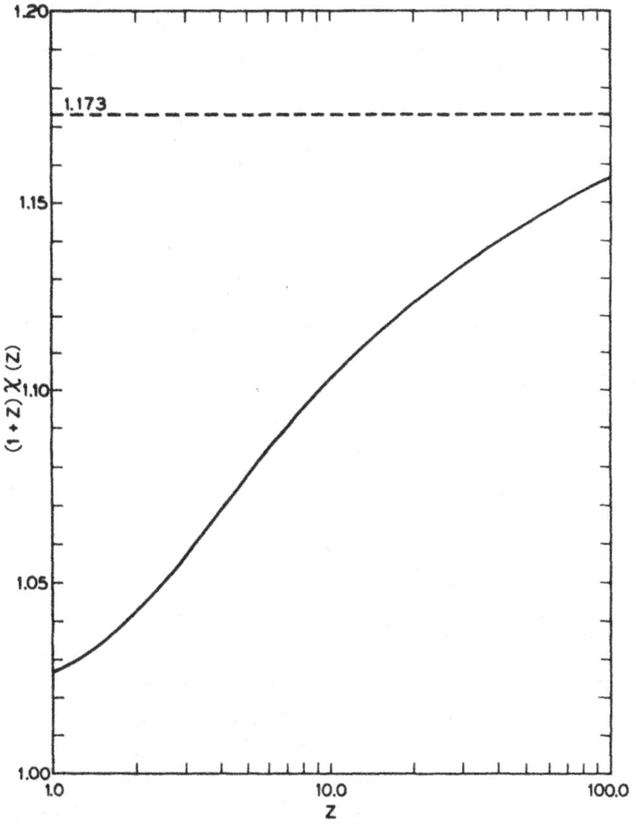

Fig. 4. Plot of $(1+z)\chi(z)$ vs. z, where $\chi(z)$ is defined by Eq. (65). Below $z = 1$, the function can be adequately described by $(1+z)\chi(z) = 1 + 0.0352z - 0.0096z^2$.

nary V_3X and Nb_3X intermetallic compounds with extremely small ξ_0 and $\kappa \gg 1$ arising from anomalies in the underlying band structure. Replacing $\chi(x)$ by the simpler form $(1 + x)^{-1}$ which roughly approximates it (Fig. 4), we find for alloys that κ increases with decreasing mean free path, approaching $\kappa \sim \delta_L/l$ in the dirty limit, $l \ll \xi_0$.

Finally, we turn to the question of the boundary conditions satisfied by $\Delta^*(\mathbf{r})$. Although clearly the derivation of the GL equations from microscopic theory cannot be considered complete until Eq. (13) has also been demonstrated, this was not carried out by Gor'kov but rather was first supplied by de Gennes (*6*). The form in which we present it is closest to, but slightly more general than, that of Abrikosov (*7*). We consider only the simplified situation of a semiinfinite pure superconductor having a plane interface with vacuum or an insulator, although interfaces between two different metals with impurities have also been treated by de Gennes (*6*) (see Chapter 17). Then returning to Eq. (39) we have $\Delta^*(\mathbf{r}) \equiv 0$ in the half-space (say) $\hat{n}\cdot\mathbf{r} \leqslant 0$, and the $\mathbf{r}_1$ integration is only over $\hat{n}\cdot\mathbf{r}_1 > 0$. The normal-state Green's function $G^n_\omega(\mathbf{r}, \mathbf{r}')$ with both $\mathbf{r}$ and $\mathbf{r}'$ defined in $\hat{n}\cdot\mathbf{r} > 0$, $\hat{n}\cdot\mathbf{r}' > 0$, must be chosen to satisfy Eq. (34) together now with the boundary conditions

$$G^n_\omega(\mathbf{r}, \mathbf{r}')\,|_{\hat{n}\cdot\mathbf{r}=0} = G^n_\omega(\mathbf{r}, \mathbf{r}')|_{\hat{n}\cdot\mathbf{r}'=0} = 0 \tag{77}$$

This is accomplished by the solution

$$G^n_\omega(\mathbf{r}, \mathbf{r}') = G^n_\omega(|\mathbf{r} - \mathbf{r}'|)\exp\left(\frac{ie}{\hbar c}\int_{\mathbf{r}'}^{\mathbf{r}} d\mathbf{s}\cdot\mathbf{A}(\mathbf{s})\right) - G^n_\omega(|\mathbf{r} - \tilde{\mathbf{r}}'|)\exp\left(\frac{ie}{\hbar c}\int_{\tilde{\mathbf{r}}'}^{\mathbf{r}} d\mathbf{s}\cdot\mathbf{A}(\mathbf{s})\right) \tag{78}$$

where $G^n(|\mathbf{r} - \mathbf{r}'|)$ on the right side denotes the solution of Eq. (34) for an infinite (and hence translationally invariant) medium, and where $\tilde{\mathbf{r}}'$ is the image point of $\mathbf{r}'$:

$$\hat{n}\cdot\tilde{\mathbf{r}}' \equiv -\hat{n}\cdot\mathbf{r}' \qquad \hat{n}\times\tilde{\mathbf{r}}' \equiv \hat{n}\times\mathbf{r}' \tag{79}$$

The expression (78) does in fact satisfy conditions (77) only if the vector potential is continued into the image space by the prescription

$$\hat{n}\cdot\mathbf{A}(\tilde{\mathbf{r}}') \equiv -\hat{n}\cdot\mathbf{A}(\mathbf{r}') \qquad \hat{n}\times\mathbf{A}(\tilde{\mathbf{r}}') = \hat{n}\times\mathbf{A}(\mathbf{r}') \tag{80}$$

which then implies the symmetry properties for the image magnetic field

$$\hat{n}\cdot\mathbf{H}(\tilde{\mathbf{r}}') = \hat{n}\cdot\mathbf{H}(\mathbf{r}') \qquad \hat{n}\times\mathbf{H}(\tilde{\mathbf{r}}') = -\hat{n}\times\mathbf{H}(\mathbf{r}') \tag{81}$$

Thus we have satisfied the boundary conditions on the normal-state electron wave functions by the method of images, such that the components of **H** *parallel* to the interface are continued into the image space *antisymmetrically*. Using the Green's function (78) in the definition of Q_ω, Eq. (39) is modified by the presence of the interface to

$$\Delta^*(\mathbf{r}) = Vk_BT \sum_{\nu} \Bigg\{ \int_{\hat{n}\cdot\mathbf{r}'>0} d^3r' \Bigg[Q_\omega(|\mathbf{r}'-\mathbf{r}|) \exp\left(\frac{2ie}{\hbar c}\int_{\mathbf{r}}^{\mathbf{r}'} d\mathbf{s}\cdot\mathbf{A}(\mathbf{s})\right) + Q_\omega(|\tilde{\mathbf{r}}'-\mathbf{r}|) \exp\left(\frac{2ie}{\hbar c}\int_{\mathbf{r}}^{\tilde{\mathbf{r}}'} d\mathbf{s}\cdot\mathbf{A}(\mathbf{s})\right) - (\text{cross terms}) \Bigg] \times \Delta^*(\mathbf{r}') - R_\omega |\Delta(\mathbf{r})|^2 \Delta^*(\mathbf{r}) \Bigg\} \tag{82}$$

The cross terms, involving products like $G^n_\omega(|\mathbf{r}'-\mathbf{r}|)G^n_{-\omega}(\tilde{\mathbf{r}}'-\mathbf{r}|)$, ensure that $\Delta^*(\mathbf{r})=0$ at the interface. However, inspection of Eq. (54) shows that these cross terms are rapidly oscillating, and hence make negligible contribution to the integral in Eq. (82), unless both $\mathbf{r}$ and $\mathbf{r}'$ are within k_F^{-1} of the interface, a region of negligible volume. Thus as soon as $\mathbf{r}$ is farther from the surface than k_F^{-1}, $\Delta^*(\mathbf{r})$ satisfies Eq. (82) with the cross terms dropped.‡ But then it is straightforward to differentiate explicitly both sides of Eq. (82) and verify directly that

$$\lim_{\hat{n}\cdot\mathbf{r}\to 0} (\hat{n}\cdot\mathbf{\Pi}^\dagger)\,\Delta^*(\mathbf{r}) = 0 \tag{83}$$

because of Eqs. (79) and (80). This confirms that the Ginzburg–Landau boundary conditions follow from the microscopic theory§ over a scale of lengths long compared with k_F^{-1}. Note that condition (83) can be incorporated directly into Eq. (82) by also continuing $\Delta^*(\mathbf{r})$ into the image space,

$$\Delta^*(\tilde{\mathbf{r}}) \equiv \Delta^*(\mathbf{r}) \tag{84}$$

Then the integration in Eq. (82) can be extended over the image space as well, and we recover Eq. (39), but with continuations (81) and (84) for $\mathbf{H}$ and Δ^*. The arguments used here are equivalent to the assumption of specular reflection; treating the case of diffuse reflection is more difficult.

III. EXTENSIONS OF THE GINZBURG–LANDAU EQUATIONS

The derivation reviewed above of the GL equations beginning from the Gor'kov Green's functions has of course relied heavily on the assumption of T being just below T_c. As we have already noted, this permitted two different kinds of expansions simultaneously: (1) the expansion in Δ/k_BT since Δ is small, and (2) the expansion in Π since Δ and H are slowly varying. It is natural to attempt to derive extensions of the GL equations which are applicable at lower temperatures, to take advantage of the conceptual and computational simplicity

‡ This behavior of $\Delta(\mathbf{r})$ very near a surface was first discussed in (8).

§ The fact that this derivation could be carried through without expanding in powers of Π the term linear in Δ indicates a contradiction with the work of Ebneth and Tewordt (9).

of differential rather than integral relations. At first sight, it would appear that neither of the two approximations, expanding in Δ and in Π, would be valid at low temperatures, and that no use could be made of either one. This is true in general; nevertheless closer inspection shows, first, that no analytical progress is possible if neither approximation is resorted to, and, second, that the two approximations do have certain limited regions of validity at lower T and are sufficiently independent that each can be made separately without invoking the other. Thus work on extending the GL equations has proceeded along two entirely distinct paths: expansion in powers of Δ with less restrictive assumptions on gradients, and expansion in powers of gradients with less restrictive assumptions on the magnitude of Δ. We will discuss each of these two approaches in turn.

A. Slow Variations

Of the two procedural paths, the first to be investigated was that of slow spatial variations with the removal of restrictions on the magnitude of Δ. This was carried out by Werthamer (*10*) for a clean material systematically to second order in Π, and independently by Tewordt (*11*), who chose to neglect certain second-order terms. Werthamer's results (*10*) were confirmed by Zumino and Uhlenbrock (*12*), and by Eilenberger (*13*). Tewordt carried the expansion further to include all terms through fourth order in Π, first (*14*) for a clean material and then (*15*) for a material with a finite impurity mean free path. A portion of Tewordt's results for the alloy had previously been obtained by Werthamer (*16*). Several authors also derived special cases of these formulas to order Π^2 in the limit $|\Delta(\mathbf{r})| \to \Delta_{BCS}$: Tsuzuki (*17*) attempted only the pure case, while Melik-Barkhudarov (*18*) and, independently, Maki (*19*) took account of a finite mean free path. A very compact method of deriving the expansion in a general form is due to Eilenberger (*20*), and we will adhere quite closely to his analysis in the following exposition.

We begin, of course, with the Gor'kov equations (28) and (29), which we now find most convenient to recast into a Nambu matrix notation (see Chapter 5). It is also best for present purposes to alter slightly the definition of the matrix Green's function, $\mathcal{G}_\omega$, to read

$$\mathcal{G}_\omega(\mathbf{r}, \mathbf{r}') \equiv \begin{pmatrix} G_\omega(\mathbf{r}, \mathbf{r}') & F_\omega(\mathbf{r}, \mathbf{r}') \\ F_\omega^+(\mathbf{r}, \mathbf{r}') & G_\omega^+(\mathbf{r}, \mathbf{r}') \end{pmatrix} \times \begin{pmatrix} \exp\left[\dfrac{ie}{\hbar c}\displaystyle\int_{\mathbf{r}'}^{\mathbf{r}} d\mathbf{s}\cdot\mathbf{A}(\mathbf{s})\right] & 0 \\ 0 & \exp\left[-\dfrac{ie}{\hbar c}\displaystyle\int_{\mathbf{r}'}^{\mathbf{r}} d\mathbf{s}\cdot\mathbf{A}(\mathbf{s})\right] \end{pmatrix} \tag{85}$$

in order to remove an otherwise irrelevant portion of the magnetic field dependence. Then $\mathscr{G}_\omega$ satisfies

$$\begin{pmatrix} i\omega-\frac{1}{2m}\left(\frac{\hbar}{i}\nabla-\theta\frac{2e}{c}\mathbf{A}(\mathbf{r})\right)^2+\mu-U(\mathbf{r}) & \Delta(\mathbf{r}) \\ -\Delta^*(\mathbf{r}) & -i\omega-\frac{1}{2m}\left(-\frac{\hbar}{i}\nabla-\theta\frac{2e}{c}\mathbf{A}(\mathbf{r})\right)^2+\mu-U(\mathbf{r}) \end{pmatrix} \times \mathscr{G}_\omega(\mathbf{r},\mathbf{r}') = \delta^3(\mathbf{r}-\mathbf{r}') \tag{86}$$

where the factor θ is defined as

$$\theta \equiv \begin{cases} 0 & \text{when acting on } G \text{ or } G^+ \\ 1 & \text{when acting on } F \text{ or } F^+ \end{cases} \tag{87}$$

As in Gor'kov's derivation of the previous section, we have neglected terms proportional to $(\mathbf{r}-\mathbf{r}')\times\mathbf{H}(\mathbf{r})$, which lead to Landau level quantization. We thus assume as before that inequalities (69) and (70) hold.

Gor'kov's derivation above arrived at approximate expressions for $\mathscr{G}_\omega(\mathbf{r},\mathbf{r}')$ which were then used in Eqs. (30) and (73) to obtain equations determining Δ, Δ^*, and $\mathbf{A}$. From these equations it was possible to construct a free energy functional whose stationary variation with respect to Δ, Δ^*, and $\mathbf{A}$ reproduced the equations, and which equaled the superconducting free energy at the stationary points. Eilenberger's more elegant and concise approach proceeds in the opposite direction. He shows *(13)* that Gor'kov's original Green's function equations (28) and (29) follow as exact consequences of a general stationary variational principle on the free energy functional

$$\mathscr{F}_s = \int d^3r\left[\frac{1}{V}|\Delta(\mathbf{r})|^2 + k_BT\sum_{\nu=-\infty}^{\infty}\int_{\omega}^{\infty\,\mathrm{sgn}\,\omega} d\omega'\,\mathrm{Tr}\{i\tau_3\mathscr{G}_{\omega'}(\mathbf{r},\mathbf{r})\} + \frac{1}{8\pi}|\mathbf{H}(\mathbf{r})-\mathbf{H}_a|^2\right] \tag{88}$$

where τ_3 is the third Pauli spin matrix. Thus any approximation can be substituted into this formula to give a functional whose stationary variation automatically leads to the same equations as would be obtained by substituting the approximate $\mathscr{G}_\omega$ into the right sides of Eqs. (30) and (73). Furthermore Eilenberger argues *(20)* that these relationships hold true even after an impurity configuration averaging provided only that $\mathscr{G}_\omega(\mathbf{r},\mathbf{r})$ in formula (88) is replaced by $\langle\mathscr{G}_\omega(\mathbf{r},\mathbf{r})\rangle$. The fact that $\mathscr{G}_\omega$ is needed for these purposes only with equal position variables shows why the phase factors in Eq. (85) were claimed to be "irrelevant."

It is now useful to consider an equation slightly more general than Eq. (86),

$$\begin{pmatrix} i\omega - \dfrac{\hbar^2}{2m}\left(\dfrac{1}{i}\nabla + \partial(\hat{\mathbf{r}})\right)^2 + \mu - U(\mathbf{r}) & \Delta(\hat{\mathbf{r}}) \\ -\Delta^*(\hat{\mathbf{r}}) & -i\omega - \dfrac{\hbar^2}{2m}\left(-\dfrac{1}{i}\nabla + \partial(\hat{\mathbf{r}})\right)^2 + \mu - U(\mathbf{r}) \end{pmatrix} \times \mathscr{G}_\omega(\mathbf{r}, \mathbf{r}'; \hat{\mathbf{r}}) = \delta^3(\mathbf{r} - \mathbf{r}') \tag{89}$$

where we define

$$\partial(\hat{\mathbf{r}}) \equiv \begin{cases} \mathbf{\Pi}(\hat{\mathbf{r}}) & \text{when acting on } \Delta(\hat{\mathbf{r}}) \\ \mathbf{\Pi}^\dagger(\hat{\mathbf{r}}) & \text{when acting on } \Delta^*(\hat{\mathbf{r}}) \end{cases} \tag{90}$$

and where $\mathscr{G}_\omega$ depends on $\hat{\mathbf{r}}$ only parametrically through its functional dependence on $\Delta(\hat{\mathbf{r}})$, $\Delta^*(\hat{\mathbf{r}})$, and $\mathbf{A}(\hat{\mathbf{r}})$. Inspection shows that the solution of Eq. (89) reduces to the solution of Eq. (86) upon setting $\hat{\mathbf{r}} = \mathbf{r}$.

To solve Eq. (89) we take its Fourier transform, after which we can set $\hat{\mathbf{r}} = \mathbf{r}$, so that

$$\begin{pmatrix} i\omega - \epsilon(k) & \Delta(\mathbf{r}) \\ -\Delta^*(\mathbf{r}) & -i\omega - \epsilon(k) \end{pmatrix} \mathscr{G}_\omega(\mathbf{k}, \mathbf{k}'; \mathbf{r}) - \int \frac{d^3q}{(2\pi)^3} U(\mathbf{k} - \mathbf{q})\, \mathscr{G}_\omega(\mathbf{q}, \mathbf{k}'; \mathbf{r})$$
$$= (2\pi)^3 \delta^3(\mathbf{k} - \mathbf{k}') + \left(\frac{\hbar}{m}\mathbf{k}\cdot\partial + \frac{\hbar^2}{2m}\partial^2\right) \mathscr{G}_\omega(\mathbf{k}, \mathbf{k}'; \mathbf{r}) \tag{91}$$

We wish to obtain $\mathscr{G}$ as a power series in $\mathbf{k}\cdot\partial$, while also carrying out the average over impurity configurations. Hence

$$\langle \mathscr{G}_\omega(\mathbf{r}, \mathbf{r}) \rangle = \int \frac{d^3k}{(2\pi)^3} \frac{d^3k'}{(2\pi)^3} \exp[i(\mathbf{k} - \mathbf{k}')\cdot\mathbf{r}] \langle \mathscr{G}_\omega(\mathbf{k}, \mathbf{k}'; \mathbf{r}) \rangle \tag{92}$$

$$\langle \mathscr{G}_\omega(\mathbf{k}, \mathbf{k}'; \mathbf{r}) \rangle = (2\pi)^3 \delta^3(\mathbf{k} - \mathbf{k}') [\mathscr{G}_\omega^{(0)}(\mathbf{k}; \mathbf{r}) + \mathscr{G}_\omega^{(1)}(\mathbf{k}; \mathbf{r}) + \mathscr{G}_\omega^{(2)}(\mathbf{k}; \mathbf{r}) + \cdots] \tag{93}$$

The lowest-order term, $\mathscr{G}_\omega^{(0)}$, is obtained by summing the same impurity diagrams (i.e., the same topologically, even though the solid lines now represent matrix propagators and arrows on the lines are no longer needed) as in Fig. 1, leading to the equation

$$\mathscr{G}_\omega^{(0)}(\mathbf{k}; \mathbf{r}) = \begin{pmatrix} i\omega - \epsilon(k) & \Delta(\mathbf{r}) \\ -\Delta^*(\mathbf{r}) & -i\omega - \epsilon(\mathbf{k}) \end{pmatrix}^{-1} \times \left[1 + n \int \frac{d^3k'}{(2\pi)^3} |u(\mathbf{k} - \mathbf{k}')|^2\, \mathscr{G}_\omega^{(0)}(\mathbf{k}'; \mathbf{r})\, \mathscr{G}_\omega^{(0)}(\mathbf{k}; \mathbf{r})\right] \tag{94}$$

It is easily verified that the solution to this equation is

$$\mathcal{G}_\omega^{(0)}(\mathbf{k};\mathbf{r}) = \frac{1}{[W_\omega(\mathbf{r})\,\eta_\omega(\mathbf{r})]^2 + \epsilon^2(k)} \begin{pmatrix} -i\omega\eta_\omega(\mathbf{r}) - \epsilon(k) & -\Delta(\mathbf{r})\,\eta_\omega(\mathbf{r}) \\ \Delta^*(\mathbf{r})\,\eta_\omega(\mathbf{r}) & i\omega\eta_\omega(\mathbf{r}) - \epsilon(k) \end{pmatrix} \tag{95}$$

$$\int \frac{d^3k}{(2\pi)^3}\,\mathcal{G}_\omega^{(0)}(\mathbf{k};\mathbf{r}) = \frac{\pi N(0)}{W_\omega(\mathbf{r})} \begin{pmatrix} -i\omega & \Delta(\mathbf{r}) \\ -\Delta^*(\mathbf{r}) & i\omega \end{pmatrix} \tag{96}$$

where in terms of the s-wave scattering probability (53),

$$\eta_\omega(\mathbf{r}) \equiv 1 + [\hbar/2\tau W_\omega(\mathbf{r})] \tag{97}$$

and we use the notation

$$W_\omega(\mathbf{r}) \equiv [\omega^2 + |\Delta(\mathbf{r})|^2]^{1/2} \tag{98}$$

The next terms in the series (93) for $\langle \mathcal{G}_\omega \rangle$ are obtained by iteration of Eq. (91):

$$\mathcal{G}_\omega^{(1)}(\mathbf{k};\mathbf{r}) = \mathcal{G}_\omega^{(0)}(\mathbf{k};\mathbf{r})\,\frac{\hbar}{m}\,\mathbf{k}\cdot\partial\,\mathcal{G}_\omega^{(0)}(\mathbf{k};\mathbf{r})\,(1 + \text{vertex corrections}) \tag{99}$$

$$\mathcal{G}_\omega^{(2)}(\mathbf{k};\mathbf{r}) = \mathcal{G}_\omega^{(0)}(\mathbf{k};\mathbf{r}) \left(\frac{\hbar}{m}\,\mathbf{k}\cdot\partial\,\mathcal{G}_\omega^{(0)}(\mathbf{k};\mathbf{r})\,\frac{\hbar}{m}\,\mathbf{k}\cdot\partial + \frac{\hbar^2}{2m}\,\partial^2 \right) \times \mathcal{D}_\omega^{(0)}(\mathbf{k};\mathbf{r})\,(1 + \text{vertex corrections}) \tag{100}$$

Because the first-order term $\mathcal{G}_\omega^{(1)}(\mathbf{k};\mathbf{r})$ is odd in $\mathbf{k}$, it makes no contribution to $\langle \mathcal{G}_\omega(\mathbf{r},\mathbf{r}) \rangle$, as can be seen from Eqs. (92) and (93). Also we could carry along the term $(\hbar^2/2m)\,\partial^2$ in $\mathcal{G}_\omega^{(2)}$, but we would find that it is odd in $\epsilon(k)$. Hence it would again make no contribution to $\langle \mathcal{G}_\omega(\mathbf{r},\mathbf{r}) \rangle$, this time because of particle-hole symmetry (density of states nearly constant in an energy shell of width $\sim W_\omega$ about the Fermi surface), and so we drop it here.

The impurity vertex corrections for the remainder of $\mathcal{G}_\omega^{(2)}(\mathbf{k};\mathbf{r})$ are calculated by summing the graphs of Fig. 5, which are reminiscent of those in Figs. 2 and 3. Since only graphs with nonintersecting impurity lines are to be included, the result for $\mathcal{G}_\omega^{(2)}(\mathbf{k};\mathbf{r})$ must be of the form [Fig. 5a]

$$\mathcal{G}_\omega^{(2)}(\mathbf{k};\mathbf{r}) = \lim_{\mathbf{r}_1,\mathbf{r}_2\to\mathbf{r}} \mathcal{G}_\omega^{(0)}(\mathbf{k};\mathbf{r})\,\frac{\hbar}{m}\,\mathbf{k}\cdot\bar{\partial}(\mathbf{r}_1)\,\mathcal{A}_\omega(\mathbf{r},\mathbf{r}_1)\,\mathcal{G}_\omega^{(0)}(\mathbf{k};\mathbf{r}_1) \times \frac{\hbar}{m}\,\mathbf{k}\cdot\partial(\mathbf{r}_2)\,\mathcal{A}_\omega(\mathbf{r}_1,\mathbf{r}_2)\,\mathcal{G}_\omega^{0}(\mathbf{k};\mathbf{r}_2)\,\mathcal{B}_\omega(\mathbf{r}_2,\mathbf{r}) \tag{101}$$

The correction factors $\mathcal{A}_\omega$ and $\mathcal{B}_\omega$ at individual vertices are given by the equations corresponding to Fig. 5b,

$$\mathbf{k}\mathcal{A}_\omega(\mathbf{r},\mathbf{r}_1) = \mathbf{k} + n\int \frac{d^3k'}{(2\pi)^3}\,|u(\mathbf{k}-\mathbf{k}')|^2\,\mathcal{G}_\omega^{(0)}(\mathbf{k}';\mathbf{r})\,\mathbf{k}'\mathcal{A}_\omega(\mathbf{r},\mathbf{r}_1)\,\mathcal{G}_\omega^{(0)}(\mathbf{k}';\mathbf{r}_1) \tag{102}$$

$$\mathcal{B}_\omega(\mathbf{r},\mathbf{r}_1) = 1 + n\int \frac{d^3k'}{(2\pi)^3}\,|u(\mathbf{k}-\mathbf{k}')|^2\,\mathcal{G}_\omega^{(0)}(\mathbf{k}';\mathbf{r})\,\mathcal{B}_\omega(\mathbf{r},\mathbf{r}_1)\,\mathcal{G}_\omega^{(0)}(\mathbf{k}';\mathbf{r}_1)$$

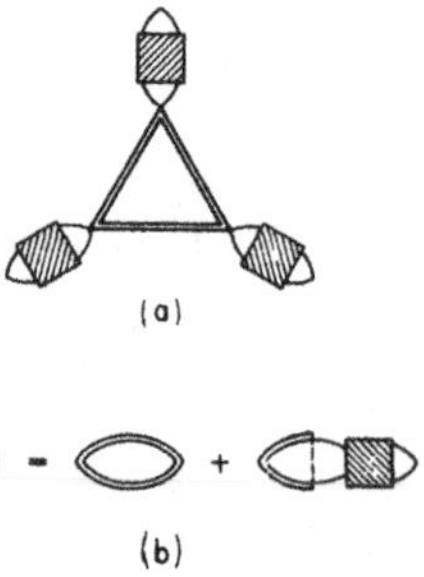

Fig. 5. Impurity averaging diagrams leading to $\mathscr{G}_\omega(\mathbf{k};\mathbf{r})$, (a), with the vertex corrections defined by diagrams (b).

Although equations (102) are the matrix analogs of the easily soluble Eq. (55), their form is such that they are not amenable to solution by straightforward matrix algebra. Proceeding naively, one would rather have to write out all four elements of each 2×2 matrix equation and solve for the elements of $\mathscr{A}_\omega$ and $\mathscr{B}_\omega$ by inversion of the resulting 4×4 matrices. A more clever method, due to Eilenberger, is to rewrite the expression (95) for $\mathscr{G}_\omega^{(0)}$ in the form of partial fractions,

$$\mathscr{G}_\omega^{(0)}(\mathbf{k};\mathbf{r}) = \frac{1}{2}\sum_{\gamma=\pm i} \frac{1+\gamma\mathscr{M}_\omega(\mathbf{r})}{\epsilon(k)+\gamma W_\omega(\mathbf{r})\,\eta_\omega(\mathbf{r})} \tag{103}$$

where we introduce

$$\mathscr{M}_\omega(\mathbf{r}) \equiv \frac{1}{W_\omega(\mathbf{r})}\begin{pmatrix} -i\omega & -\Delta(\mathbf{r}) \\ \Delta^*(\mathbf{r}) & i\omega \end{pmatrix} \tag{104}$$

The matrix $\mathscr{M}$ has the property that $\mathscr{M}^2 = -1$, leading to the key relation

$$(1+\gamma\mathscr{M})(1+\gamma'\mathscr{M}) = 2\delta_{\gamma,\gamma'}(1+\gamma\mathscr{M}) \tag{105}$$

We can solve both of equations (102) together by considering a linear combination of them with arbitrary coefficients a and b. Upon substituting Eq. (103) into this linear combination and noting that $\mathbf{k}$ is required only on the Fermi surface, the k'-integration can be carried out, yielding

$$a\mathscr{A}_\omega(\mathbf{r},\mathbf{r}_1) + b\mathscr{B}_\omega(\mathbf{r},\mathbf{r}_1) = a + b + \frac{1}{4}\sum_{\gamma,\gamma'=\pm i}\frac{1-\delta_{\gamma,\gamma'}}{W_\omega(\mathbf{r})+W_\omega(\mathbf{r}_1)+(\hbar/\tau)}$$
$$\times[1+\gamma\mathscr{M}_\omega(\mathbf{r})]\left[a\left(\frac{\hbar}{\tau}-\frac{\hbar}{\tau_{tr}}\right)\mathscr{A}_\omega(\mathbf{r},\mathbf{r}_1)+b\frac{\hbar}{\tau}\mathscr{B}_\omega(\mathbf{r},\mathbf{r}_1)\right][1+\gamma'\mathscr{M}_\omega(\mathbf{r}_1)] \tag{106}$$

The transport relaxation time, τ_{tr}, is defined in analogy with Eq. (53) as

$$\frac{1}{\tau_{tr}} = \frac{2\pi}{\hbar}n\int\frac{d^3k'}{(2\pi)^3}|u(\mathbf{k}-\mathbf{k}')|^2(1-\hat{k}\cdot\hat{k}')\,\delta(\epsilon(k')) \tag{107}$$

involving both s- and p-wave scattering. Making the Ansatz that

$$a\mathscr{A}_\omega(\mathbf{r},\mathbf{r}_1)+b\mathscr{B}_\omega(\mathbf{r},\mathbf{r}_1)=\frac{1}{4}\sum_{\gamma,\gamma'}[1+\gamma\mathscr{M}_\omega(\mathbf{r})]\,[aA_{\gamma,\gamma'}(\mathbf{r},\mathbf{r}_1)+bB_{\gamma,\gamma'}(\mathbf{r},\mathbf{r}_1)]\,[1+\gamma'\mathscr{M}_\omega(\mathbf{r}_1)] \tag{108}$$

and using the important Eq. (105), allows the reduction of Eq. (106) to the simpler relation for $aA_{\gamma,\gamma'}+bB_{\gamma,\gamma'}$,

$$aA_{\gamma,\gamma'}(\mathbf{r},\mathbf{r}_1)+bB_{\gamma,\gamma'}(\mathbf{r},\mathbf{r}_1)=a+b+\frac{1-\delta_{\gamma,\gamma'}}{W_\omega(\mathbf{r})+W_\omega(\mathbf{r}_1)+(\hbar/\tau)}\times\left[a\left(\frac{\hbar}{\tau}-\frac{\hbar}{\tau_{tr}}\right)A_{\gamma,\gamma'}(\mathbf{r},\mathbf{r}_1)+b\frac{\hbar}{\tau}B_{\gamma,\gamma'}(\mathbf{r},\mathbf{r}_1)\right] \tag{109}$$

so that

$$A_{\gamma,\gamma'}(\mathbf{r},\mathbf{r}_1)=1+\left(\frac{\hbar}{\tau}-\frac{\hbar}{\tau_{tr}}\right)\frac{1-\delta_{\gamma,\gamma'}}{W_\omega(\mathbf{r})+W_\omega(\mathbf{r}_1)+(\hbar/\tau_{tr})} \tag{110}$$

$$B_{\gamma,\gamma'}(\mathbf{r},\mathbf{r}_1)=1+\frac{\hbar}{\tau}\frac{1-\delta_{\gamma,\gamma'}}{W_\omega(\mathbf{r})+W_\omega(\mathbf{r}_1)} \tag{111}$$

Furthermore, upon substitution of Eqs. (103) and (108) into Eq. (101), a similar reduction is possible using Eq. (105); we obtain

$$\mathscr{G}^{(2)}_\omega(\mathbf{k};\mathbf{r})=\lim_{\mathbf{r}_1,\mathbf{r}_2\to\mathbf{r}}\frac{1}{8}\sum_{\gamma,\gamma_1,\gamma_2}\frac{1+\gamma\mathscr{M}_\omega(\mathbf{r})}{\epsilon(k)+\gamma W_\omega(\mathbf{r})\,\eta_\omega(\mathbf{r})}\frac{\hbar}{m}\mathbf{k}\cdot\partial(\mathbf{r}_1)\times A_{\gamma,\gamma_1}(\mathbf{r},\mathbf{r}_1)\frac{1+\gamma_1\mathscr{M}_\omega(\mathbf{r}_1)}{\epsilon(k)+\gamma_1 W_\omega(\mathbf{r}_1)\,\eta_\omega(\mathbf{r}_1)}\frac{\hbar}{m}\mathbf{k}\cdot\partial(\mathbf{r}_2)\times A_{\gamma_1,\gamma_2}(\mathbf{r}_1,\mathbf{r}_2)\frac{1+\gamma_2\mathscr{M}_\omega(\mathbf{r}_2)}{\epsilon(k)+\gamma_2 W_\omega(\mathbf{r}_2)\,\eta_\omega(\mathbf{r}_2)}B_{\gamma_2,\gamma}(\mathbf{r}_2,\mathbf{r}) \tag{112}$$

The integral of this expression over $\mathbf{k}$ can be carried out and Eqs. (110) and (111) substituted in, leading to

$$\int\frac{d^3k}{(2\pi)^3}\mathscr{G}^{(2)}_\omega(\mathbf{k};\mathbf{r})=\lim_{\mathbf{r}_1,\mathbf{r}_2\to\mathbf{r}}\frac{2\pi}{8i}N(0)\frac{v_F^2}{3}\partial(\mathbf{r}_1)\cdot\partial(\mathbf{r}_2)\times\left\{\left[\frac{[1+i\mathscr{M}_\omega(\mathbf{r})]\,[1+i\mathscr{M}_\omega(\mathbf{r}_1)]\,[1-i\mathscr{M}_\omega(\mathbf{r}_2)]}{[W_\omega(\mathbf{r})+W_\omega(\mathbf{r}_2)]\,[W_\omega(\mathbf{r}_1)+W_\omega(\mathbf{r}_2)+(\hbar/\tau_{tr})]}+\frac{[1+i\mathscr{M}_\omega(\mathbf{r})]\,[1-i\mathscr{M}_\omega(\mathbf{r}_1)]\,[1+i\mathscr{M}_\omega(\mathbf{r}_2)]}{[W_\omega(\mathbf{r})+W_\omega(\mathbf{r}_1)+(\hbar/\tau_{tr})]\,[W_\omega(\mathbf{r}_1)+W_\omega(\mathbf{r}_2)+(\hbar/\tau_{tr})]}+\frac{[1-i\mathscr{M}_\omega(\mathbf{r})]\,[1+i\mathscr{M}_\omega(\mathbf{r}_1)]\,[1+i\mathscr{M}_\omega(\mathbf{r}_2)]}{[W_\omega(\mathbf{r})+W_\omega(\mathbf{r}_1)+(\hbar/\tau_{tr})]\,[W_\omega(\mathbf{r})+W_\omega(\mathbf{r}_2)]}\right]-[i\to -i]\right\} \tag{113}$$

It is worth noting that only the transport relaxation time enters this expression; all dependence on τ has dropped out. Finally, Eqs. (113) and (96) can be substituted into formula (88) to obtain an approximate expression for $\mathscr{F}_s$ through second order in space derivatives. When the trace operation and the ω'-integration are performed, we find

$$\mathscr{F}_s = \mathscr{F}_n + \int d^3r \left\{ \frac{|\Delta(\mathbf{r})|^2}{V} + N(0)\, 2\pi k_B T \sum_{\nu=0}^{\infty} \left[-2[W_\omega(\mathbf{r}) - \omega] \right.\right.$$
$$\left.\left. + \frac{1}{3} \frac{[\hbar v_F/2W_\omega(\mathbf{r})]^2}{W_\omega(\mathbf{r}) + (\hbar/2\tau_{tr})} \left(|\mathbf{\Pi}(\mathbf{r})\, \Delta(\mathbf{r})|^2 - \frac{1}{[2W_\omega(\mathbf{r})]^2} |\nabla |\Delta(\mathbf{r})|^2|^2 \right) \right] + \frac{|\mathbf{H}(\mathbf{r}) - \mathbf{H}_a|^2}{8\pi} \right\} \tag{114}$$

This final expression for $\mathscr{F}_s$ in the slow variation approximation incorporates a number of previously known results as special cases, and so has an interesting physical interpretation. First, in the absence of any position dependence to $|\Delta|$, it reduces to the well-known results of BCS theory. In particular, the variation of $\mathscr{F}_s$ with respect to Δ leads in zero field to the familiar BCS gap equation,

$$\frac{1}{N(0)\, V} = 2\pi k_B T \sum_{\nu=0}^{\infty} \frac{1}{(\omega^2 + \Delta_T^2)^{1/2}} \tag{115}$$

where we use the notation Δ_T for the solution of this equation at temperature T. In fact, it is usually convenient to replace $1/N(0)\,V$ in Eq. (114) by the right side of Eq. (115). When Eq. (115) is satisfied, still in zero field, $\mathscr{F}_s - \mathscr{F}_n$ becomes just the space integral of $H_c^2/8\pi$, where H_c is the BCS value for the thermodynamic critical field:

$$\frac{H_c^2}{8\pi} = N(0)\, 2\pi k_B T \sum_{\nu=0}^{\infty} \left\{ \frac{\Delta_T^2}{(\omega^2 + \Delta_T^2)^{1/2}} - 2[(\omega^2 + \Delta_T^2)^{1/2} - \omega] \right\} \tag{116}$$

Also, the stationary variation of $\mathscr{F}_s$ with respect to $\mathbf{A}$ (again with $|\Delta|$ position independent) reproduces the London equation,

$$\nabla \times \mathbf{H} = -\frac{4\pi N e^2}{mc^2} \frac{\Lambda}{\Lambda_T} \left(\mathbf{A} - \frac{\hbar c}{2e} \nabla \arg \Delta \right) \tag{117}$$

The temperature dependence of the London penetration depth is indeed given by the BCS expression,

$$\frac{\Lambda}{\Lambda_T} = 2\pi k_B T \sum_{\nu=0}^{\infty} \frac{\Delta_T^2}{(\omega^2 + \Delta_T^2)\,[(\omega^2 + \Delta_T^2)^{1/2} + (\hbar/2\tau_{tr})]} \tag{118}$$

whose mean-free-path dependence agrees with that calculated by Abrikosov and Gor'kov (*21*).

Another limiting case of Eq. (114) is $T \rightarrow T_c$. It is easily verified that in this limit all results of Gor'kov presented in the previous section are reproduced. Because *p*-wave impurity scattering has been taken into account here, we learn that the definition of κ, Eq. (76), actually should contain τ_{tr} in place of τ.

We thus arrive at a physical interpretation for Eq. (114) which is quite similar to that given earlier for the phenomenological Ginzburg–Landau free energy functional. Again there are essentially four contributions to $\mathscr{F}_s - \mathscr{F}_n$: a negative energy of pairing correlation, a positive superfluid kinetic energy, a positive "quantum-mechanical pressure" arising from spatial variations in $|\Delta|$, and a positive magnetic field exclusion energy. The important generalization here, however, is that each of these contributions is assigned its value as calculated in the BCS theory, but with Δ_T everywhere replaced by the local value $\Delta(\mathbf{r})$. In other words, the slow variation approximation has led to a picture very close to a two-fluid model, in which each small element of superfluid is in local equilibrium and described by a local BCS theory.

Up to this point, the approximation of slow variation has been merely an assumption, guessed at as an appropriate and useful way to proceed, but without real justification. It must now be investigated under what circumstances the assumptions made in the derivation of $\mathscr{F}_s$, Eq. (114), are indeed satisfied at its stationary points. Strictly speaking, the only way to give a rigorous treatment is to carry the approximations to one higher order, in this case an iteration through order $(\mathbf{k}\cdot\partial)^4$, and examine the magnitude of the higher-order contribution relative to that of the lower order. As mentioned earlier, the fourth-order terms have been fully worked out by Tewordt (*15*) and confirmed by Eilenberger (*20*). Unfortunately, these terms are sufficiently complicated that it is difficult to evaluate them in any generality, and so little is to be gained from exhibiting them in detail here. What is worth noting is that since $\mathbf{\Pi}$ is the sum of a gradient and a vector potential term, the various components of $\mathbf{\Pi}$ do not commute, and hence, very schematically,

$$\int d\hat{k}\,(\mathbf{k}\cdot\mathbf{\Pi})^4 \sim k^4\,[(\Pi^2)^2 + O(eH/\hbar c)^2] \tag{119}$$

Thus the higher-order terms in $\mathbf{k}\cdot\partial$ introduce not only higher derivatives of $|\Delta|$, but also powers of v_s, H, and their derivatives. Because of this proliferation of quantities entering in successively higher orders, it becomes an intricate task to give even a qualitative discussion of the validity of the approximation with any thoroughness. Rather we will only try to present several crude indications of what we believe to be the appropriate conditions.

First, let us consider a model with vanishing magnetic field and supercurrent.

Then the position dependence of $|\Delta|$ will be determined by an equation of the schematic form

$$\frac{dF(|\Delta|^2)}{d|\Delta|^2}|\Delta| = \xi^2\nabla^2|\Delta| + O(\xi^4\nabla^4|\Delta|) \tag{120}$$

The fourth-order contribution is negligible provided $\xi^2\nabla^2|\Delta| \ll |\Delta|$, which in turn is possible only if $F'(|\Delta|^2) \ll 1$. But since

$$F'(|\Delta|^2) \sim 2\pi k_B T \sum_\nu [(\omega^2 + |\Delta|^2)^{-1/2} - (\omega^2 + \Delta_T^2)^{-1/2}] \tag{121}$$

we find $F'(|\Delta|^2) \ll 1$ only if the condition

$$\Delta_T^2 - |\Delta|^2 \ll (\pi k_B T)^2 + \Delta_T^2 \sim (\pi k_B T_c)^2 \tag{122}$$

is satisfied. Thus $|\Delta|$ is slowly varying only if its magnitude is close to the BCS value appropriate to the temperature. Looked at more qualitatively, Eq. (120) is reminiscent of an equation describing the deformation of an elastic medium, where F' plays the role of a restoring force. A deformation can spread over a large distance only if the restoring forces are very weak. In the superconducting situation, only small deviations in $|\Delta|$ from its BCS value can be maintained over large distances.

If we reintroduce the magnetic field and the supercurrents it produces, then we might consider the highly schematic model which generalizes Eq. (120),

$$\mathscr{F}_s = \mathscr{F}_n + \int d^3r \{N(0)[F(|\Delta|^2) + \xi^2((\nabla|\Delta|)^2 + (mv_s/\hbar)^2|\Delta|^2)$$
$$+ \xi^4((\nabla^2|\Delta|)^2 + (mv_s/\hbar)^4|\Delta|^2) + \xi'^4(eH/\hbar c)^2|\Delta|^2] + (H - H_a)^2/8\pi\} \tag{123}$$

where as usual we identify $m\mathbf{v}_s = \hbar\nabla \arg\Delta - (e/c)\,\mathbf{A}$. The conditions that all contributions of fourth order to determining $|\Delta|$ are negligible now become $F' \ll 1$ as before, together with

$$\xi'^4(eH/\hbar c)^2 \ll \xi^2(mv_s/\hbar)^2 \ll 1 \tag{124}$$

But in addition, the fourth-order terms make a contribution to determining H via Maxwell's equation,

$$\nabla \times \mathbf{H} \sim N(0)\frac{e}{\hbar c}|\Delta|^2\left\{\xi^2\frac{m\mathbf{v}_s}{\hbar}\left[1 + O\left(\xi\frac{mv_s}{\hbar}\right)^2\right] + O\left(\xi'^4\nabla \times \frac{e\mathbf{H}}{\hbar c}\right)\right\} \tag{125}$$

If we introduce a schematic penetration depth λ,

$$1/\lambda^2 \sim N(0)(e/\hbar c)^2|\Delta|^2\xi^2 \tag{126}$$

then the requirement that the fourth-order contribution is also negligible in Eq. (125) introduces the final condition,

$$\xi'^4/\xi^2\lambda^2 \ll 1 \tag{127}$$

This condition is not really independent of inequalities (124) because the Maxwell–London equation (125) relates H and v_s,

$$\frac{eH}{\hbar c} \sim \frac{1}{\lambda}\frac{mv_s}{\hbar} \tag{128}$$

We thus arrive at the additional conclusions that Eq. (114) for $\mathscr{F}_s$, which neglects fourth-order terms, is valid provided the magnetic field is not sufficiently strong to force $|\Delta|$ substantially away from Δ_T, and provided the electrodynamics is local via a coherence length being much less than a penetration depth. Although these conclusions have been deduced in a very sketchy, intuitive way, we believe that they contain all the essential physics of the slow variation approximation.

Since $|\Delta|$ must not differ much from Δ_T, there is no point in keeping Eq. (114) intact in its full and unnecessary generality. Expanding $|\Delta|$ about Δ_T to an appropriate order, we arrive at the somewhat simpler form,

$$\begin{aligned}\mathscr{F}_s = \mathscr{F}_{sT} + \int d^3r \Bigg\{ N(0)\, 2\pi k_B T \sum_{\nu=0}^{\infty} \Bigg[&\frac{1}{4}\frac{(|\Delta|^2 - \Delta_T^2)^2}{(\omega^2 + \Delta_T^2)^{3/2}} \\ &+ \frac{1}{12}\frac{[\hbar v_F \omega/(\omega^2 + \Delta_T^2)]^2}{(\omega^2 + \Delta_T^2)^{1/2} + (\hbar/2\tau_{tr})} |\nabla|\Delta||^2 \\ &+ \frac{1}{12}\left(1 + (|\Delta|^2 - \Delta_T^2)\frac{\partial}{\partial \Delta_T^2}\right) \\ &\times \frac{(mv_F)^2 \Delta_T^2/(\omega^2 + \Delta_T^2)}{(\omega^2 + \Delta_T^2)^{1/2} + (\hbar/2\tau_{tr})} v_s^2 \Bigg] + \frac{|\mathbf{H} - \mathbf{H}_a|^2}{8\pi} \Bigg\}\end{aligned} \tag{129}$$

where $\mathscr{F}_{sT}$ is the free energy of the BCS state in zero field. To recapitulate the domain of validity of this form, we require being close to the BCS state in the sense

$$\frac{\Delta_T - |\Delta|}{\Delta_T} \ll \left(\frac{\pi k_B T_c}{\Delta_T}\right)^2 \tag{130}$$

and also that the electrodynamics is local. Using the crude but qualitatively correct functional dependences distilled from Eq. (129) and from the fourth-order terms calculated by Tewordt (*15*) and by Eilenberger (*20*)

$$\xi^2 \sim \frac{(\hbar v_F/\pi k_B T_c)^2}{1 + (\hbar/2\pi k_B T_c \tau_{tr})} \qquad \xi'^4 \sim \frac{(\hbar v_F/\pi k_B T_c)^4}{[1 + (\hbar/2\pi k_B T_c \tau_{tr})]^3} \tag{131}$$

$$\frac{1}{\lambda^2} \sim \frac{4\pi N e^2}{mc^2}\frac{(\Delta_T/\pi k_B T_c)^2}{1 + (\hbar/2\pi k_B T_c \tau_{tr})} \tag{132}$$

the locality condition becomes

$$\frac{\xi'^4}{\xi^2\lambda^2} \sim \frac{1}{\kappa^2}\frac{(\Delta_T/\pi k_B T_c)^2}{1+(\hbar/2\pi k_B T_c \tau_{tr})} \ll 1 \tag{133}$$

It is gratifying that $\xi'^4/\xi^2 \sim$ Pippard coherence length, so that the locality condition here is just the same as the usual BCS requirement for a local London electrodynamics.

Of course, Eq. (129) reduces to the Ginzburg–Landau–Gor'kov theory as $T \to T_c$. Equation (133) informs us of a requirement on the temperature for applicability of Gor'kov's derivation even more stringent than Eq. (72):

$$1-(T/T_c) \ll \text{smaller of } \{1, \kappa^2 [1+(\hbar/2\pi k_B T_c \tau_{tr})]\} \tag{134}$$

It is amusing to note that Eq. (129) is actually very close in structure to the GL theory for all temperatures, with the slight difference that it represents an expansion about a local BCS–Meissner state rather than the normal state. Except for specifying the temperature-dependent coefficients, much of the form of this equation might have been guessed from the same kind of phenomenological reasoning as originally used by Ginzburg and Landau.

Although applications of the various generalized GL equations are not discussed in this chapter, it should be mentioned that Melik-Barkhudarov (*18*), Maki (*19*), and Werthamer (*22*) used calculations equivalent to that of Eq. (129) to show that the lower critical field, H_{c1}, of type II superconductors in the dirty, large κ limit was governed at all temperatures by a κ-parameter which extends the GLAG definition,

$$\kappa(T) = 2^{1/2} 2eH_c(T)\,\lambda^2(T)/\hbar c \tag{135}$$

where H_c and λ have the BCS temperature dependence. Tewordt and collaborators (*23–25*) have carried out extensive calculations of a number of properties of type II superconductors, using Eq. (129) together with all terms of fourth order as well, but working exclusively in the temperature region $1-t \ll 1$ where no restrictions on $|\Delta|$ or κ are required.

B. Small Order Parameter in Local Limit

The slow variations approach discussed in the previous section was found to be appropriate for fields sufficiently weak that the order parameter is close in magnitude to the zero-field, BCS value. It is helpful to visualize this domain in a two-dimensional space with coordinates $|\Delta|/\pi k_B T_c$ and T/T_c, as in region 1 of Fig. 6. A possible further restriction on the domain is the condition (133), which plays a role if $\kappa \lesssim 1$ but which we have not illustrated.

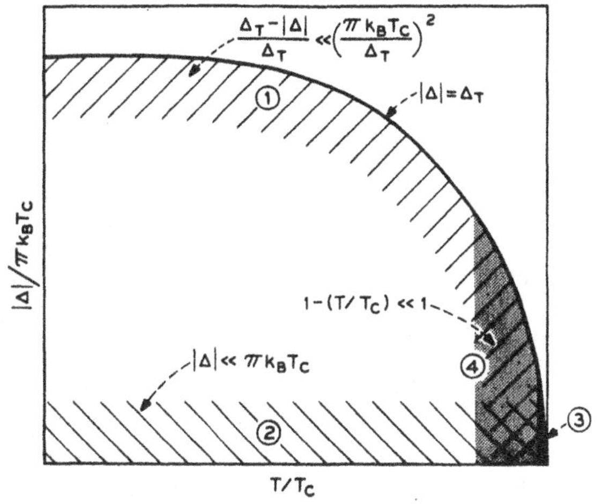

Fig. 6. Plot of the "phase plane" $|\Delta|$ vs. T, with the regions of validity of the various approximation schemes to the Gor'kov equations.

The other major approach to generalizing the GL equations is a direct iteration of the Gor'kov equations in powers of $|\Delta|/\pi k_B T_c$, which is then valid in the region labeled 2. That tiny portion of the overlap of regions 1 and 2 which satisfies $0 \leqslant |\Delta| \leqslant \Delta_T$, and which we have labeled 3, is the domain of validity of the GL theory. Tewordt's calculations (*23–25*) including all fourth-order terms are valid in region 4. The approximation of small $|\Delta|$ is appropriate when the superconductor is at or just below a second-order transition to the normal state. Unfortunately, this one approximation alone is not enough to permit in generality a reduction from an integral to a differential equation. Thus it is also usually assumed that the electrodynamics is local.

A theory of this type was first worked out by Maki (*26*) in the slightly more specialized case of a very short mean free path. Maki's results for the transition point itself were arrived at independently by de Gennes (*27*), using a somewhat different point of view. However, it was later pointed out by Caroli et al. (*28*) that Maki's calculations just below the transition contained an error, and their corrections were then confirmed by Maki (*29*). We shall present here the corrected version, while assuming only locality rather than the more severe restriction of very short mean free path. This will enable us to incorporate in our formulas the clean limit of the GLG equations near $t = 1$.

The predominant mathematical advantage to assuming a local electrodynamics is that it permits us to ignore the failure of $\mathbf{\Pi}$ to commute with itself. This circumstance has been discussed above in connection with Eqs. (119) and (125). Thus we are able to manipulate expressions containing $\mathbf{\Pi}$ as though it were a c-number. An interrelated consequence is that Π^2 itself is comparatively small:

We maintain the analog of one of inequalities (124), together with Eqs. (131),

$$\left[\frac{\hbar v_F/\pi k_B T}{1+(\hbar/2\pi k_B T\tau_{tr})}\right]^2 \frac{2eH}{\hbar c} \lesssim \left[\frac{\hbar v_F/\pi k_B T}{1+(\hbar/2\pi k_B T\tau_{tr})}\right]^2 \Pi^2 \ll 1 \qquad (136)$$

but now at temperature T rather than T_c. However, here we need no restriction to very small fields or very slow variations; we do not necessarily require the other of inequalities (124),

$$\frac{(\hbar v_F/\pi k_B T)^2}{1+(\hbar/2\pi k_B T\tau_{tr})} \Pi^2 \ll 1 \qquad (137)$$

The most convenient starting point for the derivation is the iteration of the Gor'kov equations given earlier, Eqs. (35) and (37). Now, however, the kernels cannot immediately be assumed to be short-ranged, but rather the impurity average must be carried out first. Thus we define the analogs of Eqs. (40) and (56),

$$\begin{aligned} Q_\omega(\mathbf{r}_1, \mathbf{r}) &\equiv \langle G^n_{-\omega}(\mathbf{r}_1, \mathbf{r})\, G^n_\omega(\mathbf{r}_1, \mathbf{r})\rangle \\ Q^0_\omega(\mathbf{r}_1, \mathbf{r}) &\equiv \langle G^n_{-\omega}(\mathbf{r}_1, \mathbf{r})\rangle\, \langle G^n_\omega(\mathbf{r}_1, \mathbf{r})\rangle \end{aligned} \qquad (138)$$

Here each Green's function describes electron propagation in the magnetic field, and hence contains the semiclassical phase factor of Eq. (38). The impurity averaging of Q_ω again follows from the diagrams of Fig. 2, leading to the analog of Eq. (55),

$$Q_\omega(\mathbf{r}, \mathbf{r}') = Q^0_\omega(\mathbf{r}, \mathbf{r}') + [\hbar/2\pi\tau N(0)] \int d^3r_1\, Q^0_\omega(\mathbf{r}, \mathbf{r}_1)\, Q_\omega(\mathbf{r}_1, \mathbf{r}') \qquad (139)$$

But from Eqs. (38) and (42) we have

$$\int d^3r'\, Q^0_\omega(\mathbf{r}, \mathbf{r}')\, \Delta^*(\mathbf{r}') = \int d^3r'\, [Q^0_\omega(\mathbf{r}-\mathbf{r}')]_{H=0} \exp[-i(\mathbf{r}'-\mathbf{r})\cdot\mathbf{\Pi}^\dagger]\, \Delta^*(\mathbf{r}) \qquad (140)$$

Introducing the Fourier transform of $[Q^0_\omega(\mathbf{r}-\mathbf{r}')]_{H=0}$ as per Eq. (56),

$$\tilde{Q}^0_\omega(\mathbf{k}) \equiv \frac{2\pi N(0)}{\hbar v_F k} \tan^{-1}\left(\frac{\hbar v_F k}{2|\tilde{\omega}|}\right) \qquad (141)$$

and using the locality approximation that the components of the vector operator $\mathbf{\Pi}$ commute with each other, we can solve Eq. (139) to find

$$\int d^3r'\, Q_\omega(\mathbf{r}, \mathbf{r}')\, \Delta^*(\mathbf{r}') = \left[(\tilde{Q}^0_\omega(\mathbf{\Pi}^\dagger))^{-1} - \frac{\hbar}{2\pi\tau N(0)}\right]^{-1} \Delta^*(\mathbf{r}) \qquad (142)$$

Furthermore, condition (136) is also required, so that $\tilde{Q}^0_\omega(\mathbf{\Pi}^\dagger)$ must be expanded in $\Pi^\dagger$,

$$\int d^3r'\, Q_\omega(\mathbf{r}, \mathbf{r}')\, \Delta^*(\mathbf{r}') \simeq \frac{2\pi N(0)}{2|\omega| + \frac{1}{3}[(\hbar v_F)^2/2|\tilde{\omega}|]\, \Pi^{\dagger 2}}\, \Delta^*(\mathbf{r}) \tag{143}$$

Because we are not requiring condition (137), we do not a priori continue to expand the right side in powers of $\Pi^{\dagger 2}$.

The computation of the term cubic in Δ again uses the diagrams of Fig. 3. As seen from Eq. (142), the vertex correction at each of the four vertices is now

$$\eta_\omega(\mathbf{r}) \equiv 2|\omega| \left[2|\omega| + \frac{1}{3}\frac{\hbar}{\tau}\left(\frac{\hbar v_F}{2\tilde{\omega}}\right)^2 \Pi^2(\mathbf{r}) \right]^{-1} \tag{144}$$

so that in analogy with Eqs. (41) and (61),

$$\begin{aligned}
&\int d^3r_1\, d^3r_2\, d^3r_3 \langle G^n_{-\omega}(\mathbf{r}_1, \mathbf{r}) G^n_\omega(\mathbf{r}_1, \mathbf{r}_2) G^n_{-\omega}(\mathbf{r}_3, \mathbf{r}_2) G^n_\omega(\mathbf{r}_3, \mathbf{r}) \rangle \Delta^*(\mathbf{r}_1) \Delta(\mathbf{r}_2) \Delta^*(\mathbf{r}_3) \\
&\quad = \eta_\omega(\mathbf{r}) \lim_{\mathbf{r}_1, \mathbf{r}_2, \mathbf{r}_3 \to \mathbf{r}} \eta^\dagger_\omega(\mathbf{r}_1)\, \eta_\omega(\mathbf{r}_2)\, \eta^\dagger_\omega(\mathbf{r}_3) \int d^3r'_1\, d^3r'_2\, d^3r''_3 \\
&\quad \times \exp[i(\mathbf{r}'_1 - \mathbf{r})\cdot\mathbf{\Pi}^\dagger(\mathbf{r}_1) - i(\mathbf{r}'_2 - \mathbf{r})\cdot\mathbf{\Pi}(\mathbf{r}_2) + i(\mathbf{r}'_3 - \mathbf{r})\cdot\mathbf{\Pi}^\dagger(\mathbf{r}_3)] \\
&\quad \times \Big\{ \langle G^n_{-\omega}(\mathbf{r}'_1, \mathbf{r})\rangle_0 \langle G^n_\omega(\mathbf{r}'_1, \mathbf{r}'_2)\rangle_0 \langle G^n_{-\omega}(\mathbf{r}'_3, \mathbf{r}'_2)\rangle_0 \langle G^n_\omega(\mathbf{r}'_3, \mathbf{r})\rangle_0 \\
&\quad + \frac{\hbar}{2\pi\tau N(0)} \int d^3r'_4 \Big[\langle G^n_{-\omega}(\mathbf{r}'_1, \mathbf{r})\rangle_0 \langle G^n_\omega(\mathbf{r}'_1, \mathbf{r}'_4)\rangle_0 \langle G^n_\omega(\mathbf{r}'_4, \mathbf{r}'_2)\rangle_0 \\
&\quad \times \langle G^n_{-\omega}(\mathbf{r}'_3, \mathbf{r}'_2)\rangle_0 \langle G^n_\omega(\mathbf{r}'_3, \mathbf{r}'_4)\rangle_0 \langle G^n_\omega(\mathbf{r}'_4, \mathbf{r})\rangle_0 + \langle G^n_{-\omega}(\mathbf{r}'_4, \mathbf{r})\rangle_0 \\
&\quad \times \langle G^n_{-\omega}(\mathbf{r}'_1, \mathbf{r}'_4)\rangle_0 \langle G^n_\omega(\mathbf{r}'_1, \mathbf{r}'_2)\rangle_0 \langle G^n_{-\omega}(\mathbf{r}'_4, \mathbf{r}'_2)\rangle_0 \langle G^n_{-\omega}(\mathbf{r}'_3, \mathbf{r}'_4)\rangle_0 \\
&\quad \times \langle G^n_\omega(\mathbf{r}'_3, \mathbf{r})\rangle_0 \Big] \Big\} \Delta^*(\mathbf{r}_1) \Delta(\mathbf{r}_2) \Delta^*(\mathbf{r}_3)
\end{aligned} \tag{145}$$

The evaluation of the quantity in parentheses is carried out similarly to that in Eq. (61) except that an expansion through *second* order in Π must be carried out because of conditions (136) and (137), the point which was originally overlooked by Maki. After a fairly lengthy calculation, the final result is that the right side of Eq. (145)

$$\begin{aligned}
&= \frac{\pi}{2} N(0) \lim_{\mathbf{r}_1, \mathbf{r}_2, \mathbf{r}_3 \to \mathbf{r}} \left\{ \prod_{i=1}^{4} \left[|\omega| + \frac{1}{3}\frac{\hbar}{2\tau}\left(\frac{\hbar v_F}{2\tilde{\omega}}\right)^2 \partial^2(\mathbf{r}_i) \right] \right\}^{-1} \\
&\quad \times \left[|\omega| + \frac{1}{3}\frac{\hbar}{2\tau}\left(\frac{\hbar v_F}{2\tilde{\omega}}\right)^2 \frac{1}{4}\left([\partial(\mathbf{r}_1) - \partial(\mathbf{r}_3)]^2 + [\partial(\mathbf{r}_2) - \partial(\mathbf{r}_4)]^2\right) \right] \\
&\quad \times \Delta^*(\mathbf{r}_1) \Delta(\mathbf{r}_2) \Delta^*(\mathbf{r}_3)
\end{aligned} \tag{146}$$

where we again make use of definition (90) and use the notation

$$\partial(\mathbf{r}_4) \equiv \partial(\mathbf{r}_1) - \partial(\mathbf{r}_2) + \partial(\mathbf{r}_3) \tag{147}$$

The evaluation of the current follows similar lines, in analogy with Eq. (73):

$$\mathbf{j}(\mathbf{r}) = \frac{Ne\hbar}{2m} \lim_{\mathbf{r}_1, \mathbf{r}_2 \to \mathbf{r}} [\mathbf{\Pi}^\dagger(\mathbf{r}_1) + \mathbf{\Pi}(\mathbf{r}_2)] \, \pi T \sum_{\nu=0}^{\infty} \frac{1}{|\tilde{\omega}|}$$
$$\times \left\{ \prod_{i=1}^{2} \left[|\omega| + \frac{1}{3} \frac{\hbar}{2\tau} \left(\frac{\hbar v_F}{2\tilde{\omega}} \right)^2 \partial^2(\mathbf{r}_i) \right] \right\}^{-1} \Delta^*(\mathbf{r}_1) \, \Delta(\mathbf{r}_2) \qquad (148)$$

Although the linearized version of the Maki–de Gennes theory has been used in a variety of situations, the nonlinear portion has been applied almost exclusively to bulk type II superconductivity. There it turns out that much of the complication of expression (146) is unnecessary, because Δ is such that it is one of the degenerate set of lowest eigenfunctions of

$$\Pi_a^2 \equiv \left(\frac{1}{i} \nabla - \frac{2e}{\hbar c} \mathbf{A}_a \right)^2$$
$$\Pi_a^2(\mathbf{r}) \, \Delta(\mathbf{r}) \simeq q^2 \Delta(\mathbf{r}) \qquad (149)$$
$$[\mathbf{\Pi}_a(\mathbf{r}) \, \Delta(\mathbf{r})] \cdot [\mathbf{\Pi}_a(\mathbf{r}) \, \Delta(\mathbf{r})] \simeq 0$$

with corrections of order $|\Delta|^2$. Thus in this application, Eqs. (37), (143), (146), and (148) reduce very nearly to the GLAG form

$$\mathscr{F}_s = \mathscr{F}_n + \int d^3r \Big\{ N(0) \, |\Delta(\mathbf{r})|^2 \ln t + N(0) \, 4\pi k_B T \sum_{\nu=0}^{\infty}$$
$$\times \left[\left(\frac{1}{2\omega} - \left(2\omega + \frac{1}{3} \frac{(\hbar v_F)^2}{2\tilde{\omega}} q^2 \right)^{-1} \right) |\Delta(\mathbf{r})|^2 \right.$$
$$+ \left(2\omega + \frac{1}{3} \frac{\hbar}{\tau} \left(\frac{\hbar v_F}{2\tilde{\omega}} \right)^2 q^2 \right)^{-3} |\Delta(\mathbf{r})|^4$$
$$\left. + \frac{1}{3} \frac{(\hbar v_F)^2}{2\tilde{\omega}} \left(2\omega + \frac{1}{3} \frac{(\hbar v_F)^2}{2\tilde{\omega}} q^2 \right)^{-2} (|\mathbf{\Pi}(\mathbf{r}) \, \Delta(\mathbf{r})|^2 - q^2 |\Delta(\mathbf{r})|^2) \right]$$
$$+ \frac{|\mathbf{H}(\mathbf{r}) - \mathbf{H}_a|^2}{8\pi} \Big\} \qquad (150)$$

with somewhat more formidable temperature- and field-dependent coefficients. This expression can be scaled into the GLAG dimensionless form, resulting in a κ-parameter whose temperature dependence is given by

$$\frac{\kappa(t)}{\kappa(1)} = \frac{\displaystyle\sum_{\nu} \frac{1}{2\tilde{\omega}_c} \frac{1}{(2\omega_c)^2}}{\left[\displaystyle\sum_{\nu} \frac{1}{(2\omega_c)^3} \right]^{1/2}} \, \frac{\left[\displaystyle\sum_{\nu} \left(2\omega + \frac{1}{3} \frac{\hbar}{\tau} \left(\frac{\hbar v_F}{2\tilde{\omega}} \right)^2 q^2 \right)^{-3} \right]^{1/2}}{\displaystyle\sum_{\nu} \frac{1}{2\tilde{\omega}} \left(2\omega + \frac{1}{3} \frac{(\hbar v_F)^2}{2\tilde{\omega}} q^2 \right)^{-2}} \qquad (151)$$

where ω_c is ω evaluated at $T = T_c$.

Here $\kappa(1)$ is just the quantity given by Gor'kov's derivation, Eq. (76). The explicit temperature dependence of $\kappa(t)$ cannot be given until the temperature dependence of q^2 has been worked out, which in turn requires a further input of knowledge about type II superconductivity and which we forego here. We can state that in the situation where q^2 is determined by the vanishing of the coefficient of $|\Delta(\mathbf{r})|^2$ in Eq. (150),

$$\ln\frac{1}{t} = 4\pi k_B T \sum_{\nu=0}^{\infty}\left[\frac{1}{2\omega} - \left(2\omega + \frac{1}{3}\frac{(\hbar v_F)^2}{2\tilde{\omega}}q^2\right)^{-1}\right] \tag{152}$$

that $\kappa(t)$ is only weakly dependent on temperature, rising monotonically with decreasing t such that *(28)* $\kappa(0)/\kappa(1) = 1.203$. Equations (150) and (151) are now applicable at all temperatures provided only that inequality (136) is satisfied. At lower temperatures, this requires a short mean free path.

C. Spin Paramagnetism

If a material remains superconducting in very high magnetic fields, values above say 20 kG such as are obtainable with type II superconductors or very thin films, then the Pauli paramagnetic energy of the electron spins becomes a significant fraction of the total energy of superconductivity, and must be taken into account. Since a superconductor can sustain such large magnetic fields only at lower reduced temperatures and with a gap function depressed substantially below its BCS zero-field value, the spin paramagnetism need be incorporated only into the small-order-parameter theory discussed in the previous section. Furthermore, in most situations encountered experimentally, the high critical fields are associated with a short mean free path, so that not only is the locality approximation appropriate, but also the dirty limit may be taken, $(\pi k_B T_c \tau/\hbar)\ln(\hbar/\pi k_B T_c \tau) \ll 1$. In actuality, if the locality approximation were not valid, which could be the case for a clean but high-field type II superconductor such as V_3Si, then the semiclassical phase approximation for electron propagation in the magnetic field would also break down, and Landau diamagnetism would have to be taken into account in addition to the spin paramagnetism.

When considering spin paramagnetic effects in superconductors, it is also necessary to allow for spin-orbit impurity scattering. This type of scattering can arise from two sources. The first is when an electron in a matrix with low atomic number Z, and little spin-orbit mixing in its bands, encounters a high Z impurity and is scattered through the relativistic effects of the strong Coulomb field near the impurity nucleus. The other is when an electron in a high Z matrix with substantial spin-orbit band mixing encounters a low Z impurity or crystal defect. The latter case is identical in effect to the former because the superconducting pairs in the pure matrix are still formed from time-reversal symmetric electron states; in the presence of the spin-orbit band mixing the pairs of the pure matrix are no longer exclusively in the singlet spin states of the BCS theory but rather have mixed spin. The pure system can be canonically transformed back into the spin singlet BCS representation with no change in physical properties; but then the otherwise spin-orbit-free scattering centers develop a spin-orbit component. The spin-orbit effect becomes important in magnetic fields because if an electron can flip its spin while scattering, so that the spin of a superconducting pair is no longer a good quantum number, then the spin paramagnetism of the pair will clearly be affected (*30*).

The mathematical technique of treating a spin-orbit component in the impurity scattering amplitude is due to Abrikosov and Gor'kov (*31*). The Fourier transform of the impurity scattering amplitude, denoted in formulas such as (53) or (107) by $u(\mathbf{k}-\mathbf{k}')$, is assumed to develop an additional term,

$$u(\mathbf{k}-\mathbf{k}') \rightarrow u(\mathbf{k}, \mathbf{k}') \equiv u(\mathbf{k}-\mathbf{k}') + iu_{so}(\mathbf{k}-\mathbf{k}')\, \hat{k} \times \hat{k}' \cdot \boldsymbol{\sigma} \qquad (153)$$

where $\mathbf{k}$ and $\mathbf{k}'$ are vectors on the Fermi surface. Because of the $\boldsymbol{\sigma}$ dependence in u, together with its appearance in the magnetic field Zeeman energy, $\mu_B \boldsymbol{\sigma} \cdot \mathbf{H}$, it is necessary to generalize the Gor'kov Green's functions to include the spin quantum number. This generalization is comparatively straightforward, but because it leads to a proliferation of subscripts and indices and requires a substantial amount of rather tedious algebra, we choose to give little more than the final formulas here. Details of the application to the GL equations can be found in papers by Werthamer et al. (*32*) and by Maki (*29*).

An important intermediate result is for the impurity averaged normal-state electron Green's function, which in the presence of spin paramagnetism and spin-orbit scattering becomes

$$\langle G^n_{\omega\sigma}(\mathbf{r}, \mathbf{r}')\rangle = \frac{-m}{2\pi|\mathbf{r}-\mathbf{r}'|} \exp\left\{\operatorname{sgn}\omega \left[ik_F - \frac{\tilde{\omega} + i\mu_B \sigma H}{\hbar v_F}\right]|\mathbf{r}-\mathbf{r}'|\right\} \times \exp\left(\frac{ie}{\hbar c}\int_{\mathbf{r}'}^{\mathbf{r}} d\mathbf{s}\cdot\mathbf{A}(\mathbf{s})\right) \qquad (154)$$

generalizing Eqs. (38) and (54). Here $\sigma = \pm 1$, and

$$|\tilde{\omega}| = |\omega| + (\hbar/2\tau)$$

where now

$$\frac{1}{\tau} = \frac{2\pi}{\hbar} n \int \frac{d^3k'}{(2\pi)^3} |u(\mathbf{k}-\mathbf{k}')|^2\, \delta(\epsilon(k')) + \frac{1}{\tau_{so}} \qquad (155)$$

and the spin-orbit scattering time τ_{so} is defined by

$$\frac{1}{\tau_{so}} \equiv \frac{2\pi}{\hbar} n \int \frac{d^3k'}{(2\pi)^3} |u_{so}(\mathbf{k}-\mathbf{k}')|^2\, |\hat{k} \times \hat{k}'|^2\, \delta(\epsilon(k')) \qquad (156)$$

It is expected for a spin-orbit interaction that $\tau_{so}^{-1} \sim Z^4$. Carrying through the remainder of the impurity averaging operations involved in the derivation of the local small-order-parameter theory, we arrive at a free energy functional expression which is of the same form as Eq. (150),

$$\mathscr{F}_s = \mathscr{F}_n + \int d^3r \Big\{ N(0) \Big[|\Delta(\mathbf{r})|^2 \ln t + C_1 |\Delta(\mathbf{r})|^2 + \tfrac{1}{2} C_2 |\Delta(\mathbf{r})|^4 + \tfrac{1}{3}\hbar v_F^2 \tau C_3 (|\mathbf{\Pi}(\mathbf{r})\, \Delta(\mathbf{r})|^2 - q^2 |\Delta(\mathbf{r})|^2) + C_4 |\Delta(\mathbf{r})|^2 \mu_B^2 (|\mathbf{H}(\mathbf{r})|^2 - H_a^2) \Big] + (1/8\pi)\, |\mathbf{H}(\mathbf{r}) - \mathbf{H}_a|^2 \Big\} \tag{157}$$

but with more complicated coefficients. Beginning with C_1,

$$C_1 = 4\pi k_B T \sum_{\nu=0}^{\infty} \left\{ \frac{1}{2\omega} - \left[2\omega + \frac{\hbar v_F^2 \tau}{3} q^2 + (2\mu_B H_a)^2 \times \left(2\omega + \frac{\hbar v_F^2 \tau}{3} q^2 + \frac{4}{3}\frac{\hbar}{\tau_{so}} \right)^{-1} \right]^{-1} \right\}$$

$$\equiv 2\pi k_B T \sum_{\nu=0}^{\infty} \left[\frac{1}{\omega} - \frac{\omega + a + b}{(\omega + a)^2 - b^2 + I^2} \right] \tag{158}$$

where in the notation of Maki (*29*),

$$a \equiv \frac{\hbar v_F^2 \tau}{6} q^2 + \frac{1}{3}\frac{\hbar}{\tau_{so}} \qquad b \equiv \frac{1}{3}\frac{\hbar}{\tau_{so}} \qquad I \equiv \mu_B H_a \tag{159}$$

The other coefficients are

$$C_2 = \pi k_B T \sum_{\nu=0}^{\infty} \left\{ \frac{(\omega + a + b)^3 - 3I^2(\omega + a + b)}{[(\omega + a)^2 - b^2 + I^2]^3} - 2I^2 b \frac{(\omega + a + b)^2 + I^2}{[(\omega + a)^2 - b^2 + I^2]^4} \right\}$$

$$C_3 = \pi k_B T \sum_{\nu=0}^{\infty} \frac{(\omega + a + b)^2 - I^2}{[(\omega + a)^2 - b^2 + I^2]^2} \tag{160}$$

$$C_4 = 2\pi k_B T \sum_{\nu=0}^{\infty} \frac{\omega + a + b}{[(\omega + a)^2 - b^2 + I^2]^2}$$

In deriving these results, the assumption has been made that spin flip scattering is comparatively infrequent, $\tau_{so} \gg \tau$. Formulas (157)–(161) of course do not quite reduce to those of the previous section in the limit $I \to 0$ because here the dirty limit has already been taken.

Several features of the present results deserve attention. The first, which can be seen in Eq. (158) for C_1, is that $(\mu_B H_a)^2$ occurs in sum with $\hbar v_F^2 \tau q^2$ and hence that the Pauli paramagnetism tends to lower the critical field of the superconductor. Furthermore, $(\mu_B H_a)^2$ occurs in C_1 divided by a factor containing the term $\hbar/\tau_{so}$, showing as expected that an increased rate of spin-orbit scattering tends to counteract the spin paramagnetic effect. Also noteworthy is that the term in $\mathscr{F}_s$ involving the factor C_4, which did not appear in Eq. (150), can be related to the spin magnetization energy of the superconductor. In the limit $q^2 \to 0$ and $I \to 0$ we find that

$$2N(0)\, C_4 \Delta_T^2 \mu_B^2 \to 2N(0)\, \mu_B^2 2\pi k_B T \sum_{\nu=0}^{\infty} \frac{\Delta_T^2}{\omega^2} \left(\omega + \frac{2}{3} \frac{\hbar}{\tau_{so}} \right)^{-1} \tag{161}$$

This is just the difference in spin susceptibilities, $\chi_s - \chi_n$, of the BCS superconducting and normal states (for $\Delta_T \ll \pi k_B T$) as calculated by Abrikosov and Gor'kov (*31*). Again it can be seen that an increased spin-orbit scattering brings χ_s closer to χ_n.

Although there has been a certain amount (*29,32*) of numerical computation of the coefficients $C_1, \ldots, C_4$ in several limiting cases with application to bulk type II superconductivity, an exhaustive evaluation over a complete range of the parameters t, $v_F^2 \tau$, and τ_{so} remains to be carried out.

D. Other Extensions

A number of other extensions of the GL theory have been investigated, to remove one or another of its remaining approximations or simplifying physical assumptions. Nonetheless we will only mention these extensions in passing here, because either they are restricted to a particular specialized application and should more properly be discussed in the chapter devoted to that application, or they have not yet been developed sufficiently to realize their potential usefulness.

1. Nonlocality

The approximation of a local electrodynamics has been present throughout this chapter. To a great extent it has been this approximation which has permitted the reduction of the Gor'kov integral equations to the more convenient GL differential form. Thus it should not be surprising that extensions in this direction lead to rather complicated integrodifferential relations. It is perhaps not unfair to say that the practical value of such work is determined primarily by its degree of success in solving the resulting integral equations. Because the geometry of the application is now much more closely tied to the nature of the equations to be solved, it becomes necessary to consider each geometrical configuration separately, rather than, as before, exhibiting the most general solution to a differential equation and then applying the particular boundary condition appropriate to the situation.

The most extensively developed of these approaches is that of Helfand and Werthamer (*33*) in calculating the bulk upper critical field, H_{c2}, of type II superconductors. The restriction to just the second-order transition point enables the Gor'kov equations to be linearized in Δ, leading to a linear homogeneous integral equation whose lowest eigenvalue fixes the temperature dependence of H_{c2}. The restriction to a bulk sample without finite boundaries enables the integral equation to be solved exactly. Curiously enough, the solution for $\Delta(\mathbf{r})$ turns out to be identical to that of the corresponding GLAG differential equation applicable near $t=1$. This approach has been continued to just below the transition by Eilenberger (*34*).

A careful and fairly complete nonlocal theory of the second-order transition field of a thin film has been worked out by Thompson and Baratoff and by Shapoval (*35*), following upon an extensive literature of more approximate and/or limited treatments. It is strange that the seemingly simpler problem of the surface upper critical field, H_{c3}, of a type II superconductor, involving only one surface rather than two, has been seriously neglected in comparison. There is also recent work by Schattke (*36*), reported only briefly, who obtains the nonlocal generalization of the expansion of Section III.A about the BCS–Meissner state.

2. Anisotropy of the Fermi Surface

A basic assumption of both the GL theory and of Gor'kov's equations is that the underlying normal metal Fermi surface is isotropic. Being rather unphysical, this assumption was removed from the phenomenological theory already in 1952 by Ginzburg (*37*), who made the generalization

$$\tfrac{1}{2}(\hbar^2/m)\,\Pi^2 \rightarrow \tfrac{1}{2}\hbar^2 \mathbf{\Pi}\cdot \boldsymbol{m}^{-1}\cdot\mathbf{\Pi} \tag{162}$$

where $\boldsymbol{m}^{-1}$ is some effective inverse mass tensor. This guess was confirmed from the microscopic theory by both Caroli et al. (*5*) and independently by Gor'kov and Melik-Barkhudarov (*38*), who showed that $\boldsymbol{m}^{-1}$ was related to the normal-state conductivity effective mass,

$$\boldsymbol{m}^{-1} \approx \frac{1}{N}\int \frac{d^3k}{(2\pi)^3}\,\mathbf{v}(\mathbf{k})\,\mathbf{v}(\mathbf{k})\,\delta(\epsilon(\mathbf{k})) \tag{163}$$

The two derivations (*5,38*) were not identical, however. Caroli et al. (*5*) worked entirely in the dirty limit, where correspondence (163) is precise:

$$N(0)\,\tfrac{1}{3}v_F^2\tau_{tr}\Pi^2 \rightarrow e^{-2}\mathbf{\Pi}\cdot\boldsymbol{\sigma}\cdot\mathbf{\Pi} \tag{164}$$

with $\boldsymbol{\sigma}$ being the normal-state dc conductivity tensor. Gorkov and Melik-Barkhudarov (*38*) worked entirely in the clean limit, and took account of the gap function also being anisotropic. If

$$\Delta(\mathbf{r}) \rightarrow \Delta(\mathbf{r}, \mathbf{k}) = \Delta(\mathbf{r})\,\phi(\mathbf{k}) \tag{165}$$

with $\phi(\mathbf{k})$ suitably normalized, then it was shown (*38*) that

$$\boldsymbol{m}^{-1} = \frac{1}{N}\int \frac{d^3k}{(2\pi)^3}\,\phi^2(\mathbf{k})\,\mathbf{v}(\mathbf{k})\,\mathbf{v}(\mathbf{k})\,\delta(\epsilon(\mathbf{k})) \tag{166}$$

In the dirty limit, of course, it is well known that $\phi^2 \rightarrow 1$.

The GL equations with an effective mass tensor have been applied by Tilley et al. (*39,40*) to calculate the anisotropy of H_{c2}. However, it has been pointed out by Hohenberg and Werthamer (*41*) that for cubic materials the effective mass must of necessity be isotropic, and that the observed anisotropy (*39,42*) in a material such as pure single-crystal Nb can occur only because of nonlocality. It should also be remarked that none of the above calculations has been careful in taking account of an anisotropy in the impurity scattering rate, although it can be seen from Eq. (53) that there is the possibility of $\tau \rightarrow \tau(\hat{k})$.

Finally, Eq. (164) shows for a polycrystalline specimen, where there can be no macroscopic anisotropy even with an arbitrary Fermi surface, that it is exact to manipulate the GLAG parameter κ defined by the dirty limit of Eq. (76) into the form (*4*)

$$\begin{aligned}\kappa &= \frac{7\zeta(3)}{\pi^2}\left[\frac{6}{7\zeta(3)}\,\frac{mc^2}{4\pi Ne^2}\right]^{1/2}\frac{1}{v_F\tau_{tr}}\\ &= \frac{ec\gamma^{1/2}}{\sigma k_B}\left[\frac{21\zeta(3)}{2\pi^7}\right]^{1/2}\end{aligned} \tag{167}$$

and to use experimentally measured values for the normal-state electronic specific heat coefficient γ and dc conductivity σ.

3. Landau Diamagnetism

As mentioned briefly in Section III.C, there are materials such as V_3Si which appear to be very high critical field type II superconductors even when entirely pure, as deduced by extrapolation in resistivity from the fairly high purity samples grown up to now. It can be expected that condition (69) and (70) for validity of the semiclassical phase approximation will break down for these materials, and that it will be necessary to take account of Landau level quantization in describing their high-field, low-temperature properties. A useful operational criterion is that Landau level effects are important for superconductivity in a given sample if de Haas–van Alphen oscillations or other such oscillatory phenomena can be observed in it just above the upper critical field. Theoretical work in this limit is sparse. An added mathematical complication is the necessity of using a fully nonlocal theory, again because of the low temperatures and high purities. On the other hand, the normal-state Green's function is known through its Laplace transform:

$$G_\omega^n(\mathbf{r}, \mathbf{r}') = \int_0^\infty ds \exp[(\mu - i\omega)s]\left(\frac{m}{2\pi\hbar^2 s}\right)^{3/2} \frac{\mu_B H s}{\sinh \mu_B H s}$$
$$\times \exp\left[-\frac{m}{2\hbar^2 s}\left(\frac{\mu_B H s}{\tanh \mu_B H s}|\mathbf{r} - \mathbf{r}'|_\perp^2 + |\mathbf{r} - \mathbf{r}'|_\parallel^2\right) + \frac{ie}{\hbar c}\int_{\mathbf{r}'}^{\mathbf{r}} d\mathbf{s}\cdot\mathbf{A}(\mathbf{s})\right] \tag{168}$$

In a recent note, Gunther and Gruenberg (*43*) have used this Green's function to extend the Helfand–Werthamer solution (*33*) for $H_{c2}(T)$, or more properly $T_c(H)$, and have calculated the quantum oscillations of this quantity.

IV. TIME-DEPENDENT GINZBURG–LANDAU EQUATIONS

The original GL theory and all the extensions of it discussed so far have concentrated on a time-independent thermal equilibrium situation, where the superconductor is in a static magnetic field. However, considerable attention in the last few years has been devoted to the time-dependent, nonequilibrium situation where an electric field is also present in the superconductor. The application of most importance has been to the type II superconductor with an imposed transport current. Here the superconductor develops a resistance which is associated with the driven motion of the order parameter vortex structure (see Chapter 19). It should also prove possible to treat phase-boundary propagation in films and the intermediate state by similar means.

Thus a question of interest is whether a generalization of the GL theory exists which can describe variations of the order parameter with time. Time-varying magnetic fields imply, through Maxwell's equations, that time-varying electric fields must also be considered. To describe the electric field in a gauge-invariant form, we introduce the scalar potential Φ, so that

$$\mathbf{E}(\mathbf{r}, t) = -\nabla\Phi(\mathbf{r}, t) - \frac{1}{c}\frac{\partial}{\partial t}\mathbf{A}(\mathbf{r}, t) \tag{169}$$

It also proves convenient to rescale the phase of the gap function,

$$\Delta(\mathbf{r}, t) \equiv \exp\left[\frac{2ie}{\hbar c} W(\mathbf{r}, t)\right] |\Delta(\mathbf{r}, t)| \tag{170}$$

Under the gauge transformation

$$\begin{aligned} \mathbf{A}(\mathbf{r}, t) &\to \mathbf{A}(\mathbf{r}, t) + \nabla\chi(\mathbf{r}, t) \\ \Phi(\mathbf{r}, t) &\to \Phi(\mathbf{r}, t) - \frac{1}{c}\frac{\partial}{\partial t}\chi(\mathbf{r}, t) \end{aligned} \tag{171}$$

for which **E** and **H** are invariant, $\Delta(\mathbf{r}, t)$ transforms such that

$$W(\mathbf{r}, t) \to W(\mathbf{r}, t) + \chi(\mathbf{r}, t) \tag{172}$$

Whereas in the static situation we wrote down an equation for $\Delta(\mathbf{r})$, together with an expression for $\mathbf{j}(\mathbf{r})$ to be substituted into Maxwell's equations to determine $\mathbf{A}(\mathbf{r})$, in the time-varying situation it is also necessary to exhibit an expression for the charge density $\rho(\mathbf{r}, t)$ which determines the net electric field.

Unfortunately, calculations incorporating all these generalizations prove to be elaborately intricate. Because of the requirement that there be a variation with time about thermal equilibrium, it is necessary to use a sophisticated analytic continuation of mathematical expressions in the complex frequency–temperature plane. Furthermore, the impurity averaging introduces an additional algebraic complexity. As a consequence, limitations of space restrict us to presenting no more than a brief sketch of the more significant results.

Theoretical investigation has pursued both of the avenues we have discussed for the static case: slow variations of small amplitude from the equilibrium state, and fluctuations just below the second-order transition to the normal state. The first calculations along the former of these lines were reported by Stephen and Suhl (*44*) for a pure superconductor, and their conclusions were generally confirmed from a simpler and more qualitative viewpoint by Anderson et al. (*45*). Subsequently, very detailed investigations including impurity scattering have been carried out by Stephen (*46*), Lucas (*47*), Schmid (*48*), and particularly by Abrahams and Tsuneto (*49*). The latter avenue has recently been explored by Caroli and Maki in the dirty (*50*) and clean (*51*) limits.

The goal of deriving a time-dependent GL theory really means establishing a simple *differential* equation in time and space variables for the order parameter. The schematic procedure for such a derivation is to begin with the Gor'kov equations expressed as an integral equation for the space–time dependence of the order parameter, and then to expand the Fourier transform of the integral kernel in powers of frequency Ω and wave vector Q. If such an expansion in Ω and Q *separately* is possible, then the desired differential equation will result. However, it turns out that the kernel also depends in general on the ratio

$\Omega/v_F Q$, thus precluding a differential equation being valid under all circumstances even for low frequencies and long wavelengths. A similar situation occurs for the nonequilibrium response of the normal state, where the anomalous skin effect regime requires a nonlocal integral description rather than a local differential equation. In the superconducting case, however, there are three parameter regimes where the kernel becomes independent of $\Omega/v_F Q$ and a differential equation follows: very low temperatures ($T \ll T_c$), temperatures just below the transition ($T_c - T \ll T_c$), and the dirty limit (Ω and $v_F Q \ll \tau^{-1}$). Of course the latter two regimes are just those where the electrodynamics also become local.

Another dimensionless quantity which becomes of great importance in the time-dependent case is the ratio Ω/Δ. Just as this is the dominant parameter in determining the electromagnetic response of the BCS state, so also is the form of differential equation determined by whether $\Omega/\Delta \lessgtr 1$. For $\Omega/\Delta \ll 1$ the frequency dependence is not sufficient to break up pairs and a weakly damped two-fluid hydrodynamics might be expected, whereas for $\Omega/\Delta \gg 1$ pairs are rapidly converted into single-particle excitations and a highly damped situation is likely. The assumption that $\Omega/\Delta \ll 1$ is implicit in obtaining a differential equation in the low-temperature regime, since the expansion is in small, slow variations about the BCS state. In contrast, the transition region has been studied by expanding directly in powers of Δ, thus implicitly forcing the limit $\Omega/\Delta \gg 1$. For intermediate temperatures, the interconversion of normal fluid and superfluid complicates the situation.

In the low-temperature regime, the work of Abrahams and Tsuneto (*49*) shows that $\Delta(\mathbf{r}, t)$ satisfies a wave equation. The wave velocity for the pure case (*52*) is just $v_F/\sqrt{3}$, the velocity of the collective excitations of the neutral superconductor (see Chapter 7). With a finite impurity concentration, the wave velocity is modified to $(v_F/\sqrt{3})\,[n_s(\hbar/2\tau_{tr}\Delta_0)]^{1/2}$, where n_s is a "superfluid fraction" dependent on impurity concentration:

$$n_s(x) \equiv \begin{cases} \dfrac{1}{x}\left(\dfrac{\pi}{2} - \dfrac{\cos^{-1} x}{(1-x^2)^{1/2}}\right) & x \leqslant 1 \\[2ex] \dfrac{1}{x}\left(\dfrac{\pi}{2} - \dfrac{\cosh^{-1} x}{(x^2-1)^{1/2}}\right) & x \geqslant 1 \end{cases} \tag{173}$$

The function n_s decreases from unity with increasing impurity content. Furthermore, the deviation of the charge density from uniformity is given by

$$\delta\rho(\mathbf{r}, t) = -N(0)\,e^2\left(\Phi(\mathbf{r}, t) + \frac{1}{c}\frac{\partial}{\partial t}\,W(\mathbf{r}, t)\right) \tag{174}$$

while the expression for the current becomes a time-dependent generalization of Eq. (117),

$$\mathbf{j}(\mathbf{r},t) = -\frac{Ne^2}{mc} n_s\left(\frac{\hbar}{2\tau_{tr}\Delta_0}\right)(A(\mathbf{r},t) - \nabla W(\mathbf{r},t)) \tag{175}$$

Equations (171) and (172) show that ρ and $\mathbf{j}$ are indeed gauge-invariant, while the equation of continuity,

$$\nabla\cdot\mathbf{j} + \frac{\partial\rho}{\partial t} = 0 \tag{176}$$

is satisfied because of the wave equation obeyed by Δ. This indicates that all these results follow from a variational principle on the "action"

$$S = \int d^3r \int dt \left\{\frac{1}{4} N(0)\left[\frac{(|\Delta|^2 - \Delta_0^2)^2}{\Delta_0^2} + n_s\left(\frac{\hbar}{2\tau_{tr}\Delta_0}\right)\frac{v_F^2}{3} \times \left|\left(\frac{\hbar}{i}\nabla - \frac{2e}{c}\mathbf{A}\right)\frac{\Delta}{\Delta_0}\right|^2 - \left|\left(\frac{\hbar}{i}\frac{\partial}{\partial t} + 2e\Phi\right)\frac{\Delta}{\Delta_0}\right|^2\right] + \frac{H^2 - E^2}{8\pi}\right\} \tag{177}$$

which is analogous to the free energy functional of the time-independent work. The "action" functional is to be stationary with respect to variations in Φ as well as in $\mathbf{A}$, Δ, and Δ^*.

A significant application of Eq. (174) is that it in effect represents a microscopic derivation of the London acceleration equation, which is important for completing the two-fluid hydrodynamic description of the superconductor which we expect to be valid in the slow variation limit. Taking the gradiant of Poisson's equation and using Eq. (174), we find

$$\nabla\nabla\cdot\mathbf{E} = \frac{1}{\lambda_D^2} n_s\left[\mathbf{E} + \frac{1}{c}\frac{\partial}{\partial t}(\mathbf{A} - \nabla W)\right] \tag{178}$$

where the Debye shielding length is defined by $\lambda_D^{-2} \equiv 4\pi N(0)\, e^2$. The shielding in a metal is strong, since electrons can freely rearrange themselves to neutralize an imposed point charge disturbance in a distance $\lambda_D \sim 10^{-8}$ cm, short compared with a typical length for superconductivity. Thus it is a good approximation to take $\lambda_D \to 0$ (or equivalently $\delta\rho = 0$) so that we find

$$\mathbf{E}(\mathbf{r},t) = \frac{1}{c}\frac{\partial}{\partial t}(\nabla W(\mathbf{r},t) - \mathbf{A}(\mathbf{r},t)) \tag{179}$$

But using definition (7) and its curl, we can rearrange this into

$$\frac{\partial}{\partial t}\mathbf{v}_s - \mathbf{v}_s \times \nabla \times \mathbf{v}_s = \frac{e}{m}\left(\mathbf{E} + \frac{1}{c}\mathbf{v}_s \times \mathbf{H}\right) \tag{180}$$

So far this result is not Galilean-invariant. However, it has been argued by Abrahams and Tsuneto (*49*) that an additional term, $\frac{1}{2}\nabla v_s^2$, should appear on the left side. This term arises because whenever a scalar potential, $e\Phi$, appears in an electron Hamiltonian, the added kinetic energy term $(e^2/2mc^2)A^2$ is associated with it. Because only the gauge-invariant combination $A - \nabla W$ can appear in a superconducting contribution, the $\mathbf{E}$ term in Eq. (178) should be modified,

$$e\mathbf{E} \rightarrow e\mathbf{E} - \frac{e^2}{2mc^2}\nabla(\mathbf{A} - \nabla W)^2 = e\mathbf{E} - \tfrac{1}{2}m\nabla v_s^2 \tag{181}$$

In a sense, this was already seen just below Eq. (100), a context where the term in question dropped out because it could in effect be incorporated into the chemical potential; but upon so doing it must then reappear in the charge equation. When this modification is introduced into Eq. (180), we obtain

$$\frac{d}{dt}\mathbf{v}_s \equiv \frac{\partial}{\partial t}\mathbf{v}_s + (\mathbf{v}_s\cdot\nabla)\mathbf{v}_s = \frac{e}{m}\left(\mathbf{E} + \frac{1}{c}\mathbf{v}_s \times \mathbf{H}\right) \tag{182}$$

This is the London acceleration equation, and states in a Galilean-invariant manner that superfluid motion is governed by the familiar electromagnetic forces. If we had not enforced electrical neutrality, then additional forces would appear arising from density gradients.

It is worth repeating that the above results are valid only to the extent that space–time variations are slow, requiring that $|\Delta|$ be close to the equilibrium value Δ_0; and that the temperature is very low, with correction terms of order $\exp(-|\Delta|/k_BT)$. At intermediate temperatures where a finite fraction of normal fluid is thermally excited, the two-fluid hydrodynamics must allow for the interconversion with the superfluid, and the London acceleration equation is no longer adequate.

Near the second-order transition to the normal state the situation is quite different. As we remarked, the initial expansion in powers of Δ automatically requires $\Omega/\Delta \gg 1$, and a large first time derivative term appears in the equation for Δ. This large damping term converts the equation from wave to diffusion. The result of Caroli and Maki for the dirty limit (*50*), confirmed also by Gunther (*53*), is that the linearized equation for $\Delta(\mathbf{r}, t)$ at the transition is

$$\left[\frac{\partial}{\partial t} + \frac{2ie}{\hbar}\Phi(\mathbf{r}, t) + \frac{v_F^2}{3}(\mathbf{\Pi}(\mathbf{r}, t))^2\right]\Delta(\mathbf{r}, t) = \Omega\Delta(\mathbf{r}, t) \tag{183}$$

When this equation is satisfied, the coefficient of the $|\Delta(\mathbf{r}, t)|^2$ term in the action functional reduces to the analogue of Eq. (152) with a simple dependence on the eigenvalue Ω:

$$\ln\left(\frac{T}{T_c}\right) + 4\pi k_BT\sum_{\nu=0}^{\infty}\left(\frac{1}{2\omega} - \frac{1}{2\omega + \hbar\Omega}\right) \tag{184}$$

which can also be expressed in terms of the digamma function, ψ,

$$\ln\left(\frac{T}{T_c}\right) + \psi\left(\frac{1}{2} + \frac{1}{2}\frac{\hbar\Omega}{2\pi k_B T}\right) - \psi(\tfrac{1}{2}) \tag{185}$$

The clean limit leads to rather more complicated results (*51,54*). Equation (183) has been used to show (*48,54*) that a transport current applied to a type II superconductor induces a uniform motion of the vortex structure at right angles to both the electric and magnetic static fields.

The third regime in which a simple differential equation becomes applicable is for high temperatures, $T_c - T \ll T_c$. The work of Schmid (*48*) and Abrahams and Tsuneto (*49*) shows that here $\Delta(\mathbf{r}, t)$ satisfies the equation generalizing Eq. (64),

$$\left\{1 - \frac{T}{T_c} - \frac{7\zeta(3)/8}{(\pi k_B T_c)^2}|\Delta|^2 - \frac{7\zeta(3)/8}{(\pi k_B T_c)^2}\frac{\hbar^2}{2} \times\left[\frac{v_F^2}{3}\chi\left(\frac{\hbar}{2\pi k_B T_c \tau_{tr}}\right)\Pi^2 + \left(\frac{\partial}{\partial t} + \frac{2ie}{\hbar}\Phi\right)^2\right] - \frac{\pi^2/8}{\pi k_B T_c}\hbar\left(\frac{\partial}{\partial t} + \frac{2ie}{\hbar}\Phi\right)\right\}\Delta = 0 \tag{186}$$

where χ is defined by Eq. (65). Here the wave equation terms much like those found for $T \sim 0$ are swamped by the first time derivative term similar to that of Eq. (183), and the equation becomes of diffusion form.

Lucas and Stephen (*55*) have pointed out that an even larger contribution to the relaxation rate of the order parameter than the mechanism of pair breakup into single-particle excitations, operative in Eqs. (183) and (186), comes at finite temperatures from pair breakup with the absorption of a phonon. This argument recognizes that the attractive interaction in metals which produces superconductivity is due to the exchange of phonons. Lucas and Stephen estimate the phonon absorption contribution to the coefficient of $\partial/\partial t$ in Eq. (186), but their use of the BRT Boltzmann equation might be considered suspect (*56*). In any event, this is the first occasion where the realistic phonon-exchange aspects of superconductivity have made contact with the GLG theory and its extensions, which in every other instance have found the BCS potential model to be entirely adequate. It would seem certain that the important direction for future developments in questions concerning time-dependent GL equations and relaxation to equilibrium of the order parameter is to link up with the strong-phonon-coupling theory described in Chapter 10.

ACKNOWLEDGMENTS

We would like to thank Professors Michael J. Stephen and Elihu Abrahams for numerous informative conversations and correspondence. We have also benefited from a close collaboration with our colleagues, Dr. Eugene Helfand and Dr. Pierre C. Hohenberg.

REFERENCES

1. V. L. Ginzburg and L. D. Landau, *Zh. Eksperim. i Teor. Fiz.* **20**, 1064 (1950).
2. L. Landau and E. M. Lifschitz, *The Classical Theory of Fields*, Addison-Wesley, Reading, Mass., 1951, pp. 80–81.
3. L. P. Gor'kov, *Zh. Eksperim. i Teor. Fiz.* **34**, 735 (1958); *Soviet Phys. JETP* **7**, 505 (1958).
4. L. P. Gor'kov, *Zh. Eksperim. i Teor. Fiz.* **36**, 1918 (1959); **37**, 1407 (1959); *Soviet Phys. JETP* **9**, 1364 (1959); **10**, 998 (1960).
5. C. Caroli, P. G. de Gennes, and J. Matricon, *Physik Kondensierten Materie* **1**, 176 (1963).
6. P. G. de Gennes, *Rev. Mod. Phys.* **36**, 225 (1964).
7. A. A. Abrikosov, *Zh. Eksperim. i Teor. Fiz.* **47**, 720 (1964); *Soviet Phys. JETP* **20**, 480 (1965).
8. D. S. Falk, *Phys. Rev.* **132**, 1576 (1963).
9. G. Ebneth and L. Tewordt, *Z. Physik* **185**, 421 (1965).
10. N. R. Werthamer, *Phys. Rev.* **132**, 663 (1963).
11. L. Tewordt, *Phys. Rev.* **132**, 595 (1963).
12. B. Zumino and D. A. Uhlenbrock, *Nuovo Cimento* **33**, 1446 (1964).
13. G. Eilenberger, *Z. Physik* **182**, 427 (1965).
14. L. Tewordt, *Z. Physik* **180**, 385 (1965).
15. L. Tewordt, *Phys. Rev.* **137**, A1745 (1965).
16. N. R. Werthamer, *Rev. Mod. Phys.* **36**, 292 (1964).
17. T. Tsuzuki, *Progr. Theoret. Phys. (Kyoto)* **31**, 388 (1964).
18. T. K. Melik-Barkhudarov, *Zh. Eksperim. i Teor. Fiz.* **47**, 311 (1964); *Soviet Phys. JETP* **20**, 208 (1965).
19. K. Maki, *Physics* **1**, 127 (1964).
20. G. Eilenberger, *Z. Physik* **190**, 142 (1966).
21. A. A. Abrikosov and L. P. Gor'kov, *Zh. Eksperim. i Teor. Fiz.* **35**, 1538 (1958); **36**, 319 (1959); **39**, 480 (1960); *Soviet Phys. JETP* **8**, 1090 (1959); **9**, 220 (1959); **12**, 351 (1961).
22. N. R. Werthamer, *Proc. Intern. Conf. on Physics of Type II Superconductivity, Western Reserve University*, Cleveland, Ohio, 1964.
23. L. Tewordt, *Z. Physik* **184**, 319 (1965).
24. L. Neumann and L. Tewordt, *Z. Physik* **189**, 55 (1966).
25. L. Neumann and L. Tewordt, *Z. Physik* **191**, 73 (1966).
26. K. Maki, *Physics* **1**, 21 (1964).
27. P. G. de Gennes, *Physik Kondensierten Materie* **3**, 79 (1964).
28. C. Caroli, M. Cyrot, and P. G. de Gennes, *Solid State Commun.* **4**, 17 (1966).
29. K. Maki, *Phys. Rev.* **148**, 362 (1966).
30. R. A. Ferrell, *Phys. Rev. Letters* **3**, 262 (1959); P. W. Anderson, *Phys. Rev. Letters* **3**, 325 (1959).
31. A. A. Abrikosov and L. P. Gor'kov, *Zh. Eksperim. i Teor. Fiz.* **42**, 1088 (1962); *Soviet Phys. JETP* **15**, 752 (1962).
32. N. R. Werthamer, E. Helfand, and P. C. Hohenberg, *Phys. Rev.* **147**, 295 (1966).
33. E. Helfand and N. R. Werthamer, *Phys. Rev. Letters* **13**, 686 (1964); *Phys. Rev.* **147**, 288 (1966).
34. G. Eilenberger, *Phys. Rev.* **153**, 584 (1967).
35. R. S. Thompson and A. Baratoff, *Phys. Rev. Letters* **15**, 971 (1965); *Phys Rev.* **167**, 361 (1968); E. A. Shapoval, *Zh. Eksperim. i Teor. Fiz.* **49**, 930 (1965); *Soviet Phys. JETP* **22**, 647 (1966).
36. W. Schattke, *Phys. Letters* **20**, 245 (1966).

37. V. L. Ginzburg, *Zh. Eksperim. i Teor. Fiz.* **23**, 236 (1952).
38. L. P. Gor'kov and T. K. Melik-Barkhudarov, *Zh. Eksperim. i Teor. Fiz.* **45**, 1493 (1963); *Soviet Phys. JETP* **18**, 1031 (1964).
39. D. R. Tilley, G. J. van Gurp, and C. W. Berghout, *Phys. Letters* **12**, 305 (1964).
40. D. R. Tilley, *Proc. Phys. Soc. (London)* **85**, 1177 (1965); **86**, 289, 678 (1965).
41. P. C. Hohenberg and N. R. Werthamer, *Phys. Rev.* **153**, 493 (1967).
42. W. A. Reed, E. Fawcett, P. P. M. Meincke, P. C. Hohenberg, and N. R. Werthamer, *Proc. of the Tenth Intern. Conf. on Low Temp. Physics, Moscow, 1966,* Vol. II.A (M. P. Malkov, ed. in chief), Viniti, Moscow, 1967, p. 368.
43. L. Gunther and L. W. Gruenberg, *Solid State Commun.* **4**, 329 (1966).
44. M. J. Stephen and H. Suhl, *Phys. Rev. Letters* **13**, 797 (1964).
45. P. W. Anderson, N. R. Werthamer, and J. M. Luttinger, *Phys. Rev.* **138**, A1157 (1965).
46. M. J. Stephen, *Phys. Rev.* **139**, A197 (1965).
47. G. L. Lucas, Ph.D. dissertation, Yale University, New Haven, Conn., 1966.
48. A. Schmid, *Physik Kondensierten Materie* **5**, 302 (1966).
49. E. Abrahams and T. Tsuneto, *Phys. Rev.* **152**, 416 (1966).
50. C. Caroli and K. Maki, *Phys. Rev.* **159**, 306 (1967).
51. C. Caroli and K. Maki, *Phys. Rev.* **159**, 316 (1967).
52. Similar results for the neutral pure superconductor at $T=0$ have also been obtained by M. P. Kemoklidze and L. P. Pitaevskii, *Zh. Eksperim. i Teor. Fiz.* **50**, 243 (1966); *Soviet Phys. JETP* **23**, 160 (1966).
53. L. Gunther, *Solid State Commun.* **5**, 411 (1967).
54. C. Caroli and K. Maki, *Phys. Rev.* **164**, 591 (1967).
55. G. L. Lucas and M. J. Stephen, *Phys. Rev.* **154**, 349 (1967).
56. E. Abrahams, private communication, 1967.

7

COLLECTIVE MODES IN SUPERCONDUCTORS

Paul C. Martin

LYMAN LABORATORY OF PHYSICS
HARVARD UNIVERSITY
CAMBRIDGE, MASSACHUSETTS

I. INTRODUCTION

If the space in this volume were apportioned according to the experimental importance of the phenomena this chapter would be even shorter than it is. Apart from collective modes which have absolutely nothing to do with superconductivity (phonons and plasma oscillations), there is only one phenomenon in strong-coupling superconductors which might possibly be connected with a collective mode, and probably it is not. The subject would be only slightly less academic if He^3 were to undergo a "superconducting" transition.‡ A neutral "superconductor" will generally have a characteristic longitudinal

‡ We shall include in the term superconductor all Fermi systems, charged or neutral, which have a superfluid phase (i.e., a thermodynamic phase with condensation and phase coherence).

collective mode with a phonon dispersion relation. However, in this specific superconducting system (optimistically), the experimental manifestations of the collective mode would hardly be dramatic. In the first place, the collective mode would have the same dispersion relation as ordinary hydrodynamic sound. In the second place, although the mode would be present in normal He^3 at frequencies higher than a characteristic collision frequency ($1/\tau$) only as a result of the transition, it would hardly appear abruptly. Because the interatomic forces in helium are strong, it would differ only slightly in strength and frequency from the collisionless mode (zero sound) present in the normal state. The small change in velocity and frequency would occur gradually as the temperature was reduced from T_c to $T=0$.

Pronounced effects of superfluidity on the longitudinal collective mode requires a neutral "superconductor" even more hypothetical than He^3—for example, a neutral Fermi system with weak attractive interparticle forces. In such a system there would be no collective mode with small wave number, k, in the normal state and only as a result of the superfluid transition would a collective mode develop. This mode would become well defined as the normal fluid density decreased and at zero temperature would be an infinitely long-lived phonon mode with the velocity of ordinary sound, i.e., $c^2 = dp/d\rho$.

Why then, has there been so much discussion of collective modes? There are two reasons. The first, which concerns the longitudinal mode is circuitous, pedagogical, and historical. The values of certain matrix elements in the neutral superconductor, and related properties in the charged superconductor were ambiguous in the original BCS description. Specifically, in a neutral superconductor, the matrix elements of the density $\rho_{\mathbf{k}}$ and its current (the momentum, $\mathbf{g}_{\mathbf{k}}$) did not satisfy current conservation, i.e.,

$$\left\langle 0 \left| \frac{\partial \rho_{\mathbf{k}}}{\partial t} \right| b \right\rangle = \frac{1}{i\hbar} \langle 0 | \left[\rho_{\mathbf{k}}, H \right] | b \rangle = \frac{i}{\hbar} (E_b - E_0) \langle 0 | \rho_{\mathbf{k}} | b \rangle$$
$$\neq - \langle 0 | \nabla \cdot \mathbf{g}_{\mathbf{k}} | b \rangle = - i\mathbf{k} \cdot \langle 0 | \mathbf{g}_{\mathbf{k}} | b \rangle$$

As a consequence of this failure, a perturbation calculation using BCS longitudinal current matrix elements of the momentum density, $\mathbf{g}$, when the system is given a small constant velocity, $\mathbf{v}$, did not give $\mathbf{g} = \rho \mathbf{v}$ but the nonsensical result $\mathbf{g} = \rho_n \mathbf{v}$, where ρ_n is the normal fluid density. This was particularly serious since the calculation was essentially identical with the perturbation calculation, involving transverse momentum matrix elements, which demonstrated the reduction of the moment of inertia (the analogue of the Meissner effect) for the neutral system, i.e.,

$$I = I_{\text{rigid}} (\rho_n / \rho)$$

Until the second result was obtained (the rotating-bucket demonstration of superfluidity) by a procedure which maintained particle conservation (and consequently gave $\mathbf{g} = \rho\mathbf{v}$, not $|\mathbf{g}| < \rho|\mathbf{v}|$), one had to treat the demonstration of superfluidity with some skepticism.

The second reason for the continued discussion of collective modes is a more genuine experimental one. Ever since 1959, there have been indications of a sizable absorption at very low temperatures in strongly coupled superconductors (lead and mercury) at frequencies lower than the frequency, $2\Delta/\hbar$, necessary to break pairs. Attempts have been made by increasingly sophisticated theories, to ascribe this absorption to transverse collective exciton-like modes. While these attempts have been unsuccessful and discouraging, it is not completely impossible that they could give rise to some "precursor absorption."

It was the first of these two reasons, the elucidation of the perturbative demonstration of superfluidity and the Meissner effect, which provided the historical incentive for obtaining reliable matrix elements of the density and current and resulted in the first flurry of activity on collective modes. Logically, however, the Meissner effect (an equilibrium property in a spatially varying time-independent field) can be calculated in the Landau limit, and shown to be gauge-invariant more generally by treating the magnetic field nonperturbatively. Such a nonperturbative calculation implicitly determines the sum of the squares of the matrix elements (with the magnetic field turned off) we referred to earlier without requiring one to choose between individual ambiguous matrix elements, at least some of which must be incorrect. Such a procedure, however, only defers the problem, since this ambiguity cannot be avoided in the calculation of the response to time-varying fields, e.g., accoustic attenuation.

To maintain particle conservation and avoid all such contradictions it is necessary to include correlations beyond those which lead to the BCS spectrum and matrix elements. The substance of the various complicated treatments which have ensued is that the additional correlations do not substantially modify the original BCS predictions for superconductors; in particular, they do not significantly modify the transverse current matrix elements which predict a Meissner effect, and they do not significantly modify the matrix elements of the density BCS chose to employ (instead of current matrix elements) in the calculation of ultrasonic attenuation. However, in neutral "superconductors" at wavelengths k^{-1} large compared to the coherence length $\xi \sim \hbar v/\Delta$, these additional correlations lead to a collective mode that lies in the gap. At shorter wavelengths they restore current conservation in neutral superconductions without producing a collective mode but the correction is smaller by $1/k\xi$, owing to the fact that the predicted supercurrent is smaller by that amount in what corresponds in a neutral system to the Pippard nonlocal limit.

Furthermore, in charged superconductors these same correlations restore gauge invariance and are necessary to clarify the calculation of ultrasonic absorption, although once again they do not introduce a new collective mode. Here the longitudinal collective mode of the normal state, the plasmon, is essentially unaltered. It is this circuitous reasoning which has led to the extensive and obscure literature coupling the unrelated questions of the Meissner effect, gauge invariance, and collective modes.

Apart from the longitudinal collective mode of the neutral one-component superconductor, at low temperature the effect of superconductivity on low-frequency collisionless modes (that is, modes for which the frequency ω is small, but much larger than the inverse collision time τ) is a destructive one. Modes of this type (e.g., spin waves, or transverse phonons), which may be present in strongly interacting normal Fermi fluids, depend typically on the restoring force of the normal fluid and consequently disappear when the normal component is diminished.

The modes we have just discussed (density, spin density, and current density oscillation) have their counterpart in normal systems, where they may be discussed in terms of "hole-particle" scattering and also give rise to low-frequency collective oscillations. In addition to these modes, it is in principle possible for a superconductor to have exciton-like modes with finite frequency lying in the gap. They result, as does the superconducting transition, from interactions which would be described in the normal state as "particle–particle" scattering. It is these modes and their possible connection with the observed precursor currents with which the second large and unsatisfying fraction of the literature on collective modes is concerned. In particular, the excitons for which the total momentum vanishes correspond crudely to excited states of the superconducting condensed pairs. In a weak-coupling theory these pairs would be expected to lie only slightly, if at all below the continuum, since the higher harmonics of the attractive potential which binds pairs should be quite weak compared to the s-state attraction provided by the phonons. They would also be expected to be strongly modified and raised in energy by impurities which would tend to diminish angular variations of the potential. In addition, they would not be expected to be strongly coupled to electromagnetic radiation. By contrast, the absorption reported in early experiments on strongly coupled superconductors is rather large and insensitive to impurities. The disagreement between theory and experiment would not be insurmountable if we could reject the early experiments on very impure samples in favor of more recent ones which do not show the precursor. That is to say, while theoretical refinements cannot account for the lack of variation with impurity concentration they do indicate that not very unreasonable interactions between the quasi-particles could result in the coupling to the electromagnetic radiation and the binding of excitons reported in the early experiments on pure samples. The most recent experiments on pure samples, however, suggest that the earlier reports on these samples may also have been in error. There may well be no precursor.‡

For completeness we mention two other kinds of modes, although we shall not consider them further:

1. Second sound. As a hydrodynamic mode (with $\omega\tau \ll 1$) such a mode will only exist in the neutral system away from $T=0$. In a charged system where the conductivity of the normal fluid determines ω, the frequency condition $\omega\tau \ll 1$ is always violated. Likewise, in a neutral system at $T=0$, $\tau \to \infty$, and the hydrodynamic requirement is violated. There has been some discussion about collisionless ($\omega\tau \gg 1$) second sound, a mode of relative oscillation of normal and superfluid but there appears to be difficulty even in defining these two fluid ideas when $\omega\tau \gg 1$, let alone in pursuing them quantitatively. What work has been done quantitatively argues against such a mode.

2. Collective modes connected with vortex oscillations in the inhomogeneous superconductor. These will be discussed elsewhere in this treatise.

In the remainder of this chapter we shall fill in to some extent the arguments which lead to these rather barren conclusions. In Section II we begin with the simple but academic example of the neutral superconductor with its longitudinal collective mode and then discuss briefly how the correlations responsible for this collective mode remove the objectionable features of calculations in the charged superconductor without producing a collective mode. Next we turn to the strongly coupled neutral superconductor, that is, hypothetically superfluid He^3, and indicate how the interactions modify various possible collisionless modes, low-frequency modes. In Section III we discuss possible exciton modes.

II. COLLECTIVE MOTIONS OF THE EXCITATIONS

A. Particle Conservation§

In a neutral (one-component) system it is easy to verify directly that the commutator of the momentum density

$$\mathbf{g}(\mathbf{r}) = \tfrac{1}{2}\sum_{\alpha} [\mathbf{p}_\alpha \delta(\mathbf{r}-\mathbf{r}_\alpha) + \delta(\mathbf{r}-\mathbf{r}_\alpha)\,\mathbf{p}_\alpha]$$

and the mass density

$$\rho(\mathbf{r}) = m\sum_{\alpha} \delta(\mathbf{r}-\mathbf{r}_\alpha)$$

is

$$[\mathbf{g}(\mathbf{r}), \rho(\mathbf{r}')] = i\hbar\rho(\mathbf{r})\,\nabla'\delta(\mathbf{r}-\mathbf{r}') \tag{1}$$

‡ Note added in proof. This appears to be the case; the absence of precursors has been corroborated and the apparent precursor absorption ascribed to the presence of multiples of the expected frequency.

§ The puzzle posed in this section is essentially the one raised by Buckingham (*1*) and Schafroth (*2*). The specific approach entails elaborations from (*4*) and (*11*).

Together with mass conservation,

$$\dot{\rho}(\mathbf{r}) = (1/i\hbar)[\rho(\mathbf{r}), H] = -\nabla\cdot\mathbf{g}(\mathbf{r}) \tag{2}$$

this equation implies that the Fourier transform of the expectation value of the longitudinal momentum commutator,‡

$$\chi''_{g_ig_j}(\mathbf{k}\omega) = \int_{-\infty}^{\infty} dt \int d\mathbf{r}\, \exp(-i\mathbf{k}\cdot\mathbf{r} + i\omega t)\langle[g_i(\mathbf{r}t), g_j(\mathbf{r}'t')]\rangle$$

$$\chi_{g_ig_j}(\mathbf{k}z) = \int \frac{d\omega}{\pi} \frac{\chi''_{g_ig_j}(\mathbf{k}\omega)}{\omega - z}$$

$$\chi''^{l}(k\omega) = \hat{k}_i\chi''_{g_ig_j}(\mathbf{k}\omega)\,\hat{k}_j \qquad \chi^{l}(kz) = \hat{k}_i\chi_{g_ig_j}(\mathbf{k}z)\,\hat{k}_j \tag{3}$$

satisfies the sum rule

$$\int \frac{d\omega}{\pi} \frac{\chi''^{l}(k\omega)}{\omega} = \rho \tag{4}$$

In the limit as $k \to 0$, this commutator gives the perturbative answer to the question: What is the ratio between density and velocity when the system is given a small constant velocity? So we had better obtain ρ. At vanishing temperature we may write the sum rule in the form

$$\sum_n \frac{|\langle 0|\,\mathbf{g}_\mathbf{k}\cdot\hat{k}\,|n\rangle|^2}{E_n - E_0} = \rho \tag{5}$$

We may ask a related question: What is the moment of inertia, I, when we slowly rotate the system? The answer to that question defines the normal fluid density, i.e., $I/I_{\text{rigid}} \equiv \rho_n/\rho$. It is given by

$$\lim_{k\to 0} \int \frac{\chi''^{t}(k\omega)}{\omega} \frac{d\omega}{\pi} = \rho_n \tag{6}$$

where

$$\chi''^{t}(\mathbf{k}\omega) = \tfrac{1}{2}\chi''_{g_ig_j}(\mathbf{k}\omega)[\delta_{ij} - \hat{k}_i\hat{k}_j]$$

At vanishing temperature this expression reduces to the more familiar perturbation expression

$$\lim_{k\to 0} \sum_n \frac{|\langle 0|\,\mathbf{g}_\mathbf{k}\times\hat{k}\,|n\rangle|^2}{E_n - E_0} = \rho_n \tag{7}$$

Now in a one-component system momentum conservation guarantees that when $\mathbf{k} = 0$, $\langle 0|\, g_0\, |n\rangle = 0$ unless $E_n = E_0$. We may therefore infer that as $\mathbf{k} \to 0$,

‡ This notation and the sum rules are commented upon in great detail in (*3*). This notation is useful because it involves physical quantities. It is not, however, essential for an understanding of this chapter.

$\langle 0|\, g_{\mathbf{k}}\, |n\rangle \to 0$ unless $E_n \to E_0$. If there were no such states with $E_n \sim E_0$, we could not satisfy the first sum rule [Eq. (5)], so there must be low-lying longitudinal excitations. The second sum rule [Eq. (7)] states that if there are not low-lying transverse excitations $\rho_n \to 0$. The original BCS picture gives matrix elements which vanish as $k \to 0$ between the ground state and states with quasi-particle pairs for which $E_n - E_0 \geqslant 2\Delta$; it therefore predicts that $\rho_n \to 0$ but at the expense of violating the longitudinal sum rule [Eq. (5)]. Since Eq. (5) is a consequence of Eqs. (1) and (2), the BCS theory must violate one or the other. It is straightforward to check that Eq. (1) is violated, as we stated in Section I. It is of interest to see, therefore, the nature of the states which permit Eqs. (1) and (5) to be satisfied, and to see that they do not alter Eq. (7).

B. Collective Mode and Broken Symmetry

The nature of the collective states was first clarified by Anderson (*4,5*) and Bogoliubov et al. (*6*); further refinements were made by Nambu (*7*). Indeed, collective states like them must exist in a large class of systems. This fact (*7*) was formalized by Goldstone and has since been the subject of numerous articles (*8,9*). Proceeding antihistorically we may use the Goldstone theorem to understand the collective modes from another general point of view. The Goldstone theorem states physically that if an infinitesimal uniform change of the ground state produces a new ground state degenerate with it, an infinitesimal slowly varying change in the state (which interpolates between the ground states) will result in a state of low energy provided only (as is almost always the case) that the restoring force is proportional to the gradient of the variation. Thus, for example, in an isotropic ferromagnet the fact that there are ground states for each direction of the magnetization implies that states in which the magnetization direction varies slowly in space have low energy. These are the spin waves.

In the superconductor the degeneracy is connected with the phase of the condensed pairs of fermions which characterize its ground state.

To formalize the argument, one says the following: The nonvanishing quantity $\langle \psi_\uparrow(\mathbf{r})\psi_\downarrow(\mathbf{r})\rangle$ takes on different values in the degenerate ground states obtained from one another by an infinitesimal gauge transformation (whose generator is the number operator, N). Thus under a gauge transformation

$$\langle \psi_\uparrow\psi_\downarrow\rangle \to \langle \exp(iN\phi)\, \psi_\uparrow\psi_\downarrow \exp(-iN\phi)\rangle = \langle \psi_\uparrow\psi_\downarrow\rangle \exp(2i\phi)$$

and under an infinitesimal transformation

$$\begin{aligned}\delta\langle \psi_\uparrow\psi_\downarrow\rangle &= 2i\delta\phi\,\langle \psi_\uparrow\psi_\downarrow\rangle = \delta\phi\,\langle[N, \psi_\uparrow\psi_\downarrow]\rangle \\ &= \delta\phi \int d\mathbf{r}\,\langle[n(\mathbf{r}0), \psi_\uparrow\psi_\downarrow]\rangle\end{aligned}$$

Moreover, since N is time-independent, so that $[H, N]=0$, and the state is an energy eigenstate, we may write

$$\langle[H,[N,\psi_\uparrow\psi_\downarrow]]\rangle+\langle[\psi_\uparrow\psi_\downarrow[H,N]]\rangle=0$$

We therefore obtain from the Jacobi identity,

$$0=\int d\mathbf{r}\left\langle\left[n(\mathbf{r}0),\frac{\partial}{\partial t}\psi_\uparrow\psi_\downarrow\right]\right\rangle$$

This leads to the two statements

$$0\neq\int\frac{d\omega}{\pi}\chi''_{n,\psi\psi}(0,\omega)\qquad\text{and}\qquad 0=\int\frac{d\omega}{\pi}\chi''_{n,\psi\psi}(0\omega)\,\omega^n$$

whence, as before, in the limit $k\to 0$, there must be nonvanishing matrix elements of both $n_\mathbf{k}$ and $\Sigma_\mathbf{q}\psi_{\downarrow\mathbf{q}}\psi_{\uparrow\mathbf{k}-\mathbf{q}}$ between states of arbitrarily small energy. The difficulty with the BCS theory is that it omits the coherent states in which the total momentum of the pairs varies slowly, and in which the density varies slowly.

In contrast with the momentum and density argument for low-lying excitations (which are valid for the free gas, for example), the Goldstone argument may be sharpened under certain conditions (*9*) to assert further that the excitations of long wave number at zero temperature are eigenstates. Their energy, $E(k)$, then follows immediately from the compressibility sum rule; it is $E(k)\cong ck$ with $c^2=dp/d\rho$.

A more physical, but less rigorous, understanding of the last result is obtained by arguing from the point of view of density fluctuations

$$\frac{\partial^2\delta\rho}{\partial t^2}=-\frac{\partial}{\partial t}\nabla\cdot\delta\mathbf{g}=\nabla^2\delta p=\left(\frac{\partial p}{\partial\rho}\right)\nabla^2\delta\rho$$

or, from the point of view of phase variation (*10*),

$$\frac{\partial^2\delta\mathbf{v}_s}{\partial t^2}=-\frac{\partial}{\partial t}\nabla\delta\mu=-\left(\frac{\partial\mu}{\partial\rho}\right)\nabla\frac{\partial\delta\rho}{\partial t}=\left(\frac{\partial\mu}{\partial\rho}\right)\nabla\nabla\cdot\delta\mathbf{g}$$

and, since at vanishing temperature $\mathbf{g}=\rho\mathbf{v}_s$,

$$\frac{\partial^2\delta\mathbf{v}_s}{dt^2}=\rho\left(\frac{d\mu}{d\rho}\right)\nabla^2\delta\mathbf{v}_s=\left(\frac{dp}{d\rho}\right)\nabla^2\delta\mathbf{v}_s$$

At any rate, one concludes that there are phonon excitations with velocity $c^2=dp/d\rho$ at vanishing temperature.

In a weakly interacting neutral condensed Fermi system, then, we have at zero temperature a collective mode with velocity $(dp/d\rho)^{1/2}$ which exhausts the various sum rules for the long-wavelength density fluctuation. By contrast, in a weakly interacting normal Fermi system with repulsive interactions there is only a weak collective mode (zero sound) with a velocity slightly larger than $(3dp/d\rho)^{1/2}$.

Since the collective mode is deduced from purely longitudinal gauge-invariance considerations, there is no reason to expect a modification of the transverse calculation which leads to superfluidity.

C. Consistent Calculations of Current Correlations

In this discussion we have indicated why the simple BCS excitation spectrum and matrix elements must be incorrect in some cases, and why at zero temperature a collective mode of the neutral condensed system is required by gauge invariance. Such a series of arguments, however, is still somewhat unsatisfying.

A reasonable approximation scheme for calculating correlation functions should always give results consistent with conservation laws, and if there are collective modes they should appear as a matter of course. We will now discuss such a scheme.

The flaw in the BCS treatment of response functions involving off-diagonal matrix elements of single-particle operators is one that has been frequently encountered elsewhere. The same violation of the conservation law [Eq. (2)] results from the same kind of approximation in the normal systems with impurities; there it leads to a Drude theory, i.e., a linearized Boltzmann equation with no back-scattering terms and hence violated conservation laws. It has also been encountered in nuclear physics in evaluating moments of inertia and it also leads to difficulties with Gaililean invariance in normal interacting Fermi systems when off-diagonal elements of the density and current are computed in a Hartree approximation. In each case it is necessary at least to go beyond the Hartree approximation and use either a kinetic equation, a time-dependent Hartree equation, or a random phase approximation. All these terms are synonymous. Technically, more generally, to remove all these inconsistencies in the values of off-diagonal matrix elements (vertex functions), that is, to satisfy conservation laws (Ward identities), it is necessary to solve kinetic equations (Bethe–Salpeter equations) for the off-diagonal matrix elements of the single-particle operators. There is a well-defined procedure (*11–13*), not only for calculating these matrix elements exactly but for calculating them approximately to the same level of approximation as the diagonal matrix elements are calculated in a BCS or any generalized Hartree–Fock theory, and in more complicated approximations. Needless to say, the same procedures are required for the equivalent problem—the calculation of correlation functions of single-particle operators in the equilibrium state or distribution. These procedures lead to an inhomogeneous integral equation which requires neither subsidiary or boundary conditions. The inhomogeneous term of the integral equation is the (Hartree–Fock or BCS) approximation to the matrix element.

For normal neutral systems, this integral equation is just the integral equation whose differential form is the so-called random phase approximation, or Landau kinetic equation. For the normal charged system, it is the same equation but with a self-consistent electric field. All discussions (*5–7,11,14*) of the collective modes are essentially concerned with the solution of integral equations of this type as generalized to neutral condensed and charged condensed systems. Most generally they are concerned with determining the response function $\chi_{g_i g_j}(kz)$ and the corresponding function $\chi_{\rho\rho}(kz)$ in a satisfactory manner, consistent with Eqs. (1) and (2), which are equivalent to

$$z^2\chi_{\rho\rho}(kz) = k_i k_j \chi_{g_i g_j}(kz) - \rho k^2$$

In the BCS approximation, the absorptive part, $\chi''(k\omega)$, of these functions vanishes at zero temperature except when $\omega \gtrsim 2\Delta$, since one must create quasi-particle pairs. At finite temperatures the contribution from the "forbidden" region due to scattering of thermally excited quasi-particles is proportional to the normal fluid density. By a collective mode one means a large contribution to the function $\chi''(k\omega)$ in a relatively small frequency region. Typically, such a mode can only lie in the "forbidden" region, since if it were in the "allowed" region its coupling to the quasi-particles would lead to Landau damping and prevent it from being well defined. Thus, at zero temperature, collective modes are possible in the gap, but at T_c they should only be possible outside the allowed region $\omega < kv_F$ for the normal Fermi system.

The numerous zero-temperature methods all show that for small k there is a mode

$$\chi''_{g_i g_j}(\mathbf{k}\omega) \simeq \pi\rho k_i k_j c^2 \delta(\omega^2 - c^2k^2)\,\mathrm{sgn}\,\omega$$

At finite temperature the more and less satisfactory do not differ in principle – however, they differ in practice, in the care with which terms are unimportant at $T = 0$, but dominant near $T \approx T_c$, are retained.

The Meissner effect is related to the value of $\chi(k0)$. Methods which effectively determine it directly [e.g., the discussions of Blatt (*15*) and Thouless (*16*)], determine the integral

$$\chi_{ij}(k0) = \int \frac{d\omega}{\pi} \frac{\chi''_{ij}(k\omega)}{\omega}$$

They therefore automatically give information about collective modes only when there is no continuum, e.g., for small k and $T = 0$. They do not give information about the frequency of the collective mode in general. Some, but not all, of the calculations of $\chi(k0)$ use essentially the same integral equations which determine $\chi(k\omega)$.

D. Connection between Charged and Neutral Systems

Before discussing the generalized equations in detail, let us indicate the effect of the Coulomb forces on the longitudinal correlation, at vanishing and non-vanishing temperature. Basically all the large effects of the Coulomb interaction on the charge and current correlations have nothing to do with the existence of superconductivity. The strong Coulomb forces reduce longitudinal current and charge fluctuations and therefore the transition matrix elements between low-lying states. The fact that the sum of the squared matrix elements to all intermediate states is invariant is in normal and superconducting systems maintained by the occurrence of the plasma mode, which is extraneous to all considerations. Nevertheless, the discussion of the neutral system is relevant for the following reason. To a good approximation, the response function, which is calculated in terms of matrix elements of the uncharged system, gives the dielectric constant or complex conductivity of the charged system. The relationship of this dielectric constant to the charge and current matrix elements of the system with electromagnetic forces implies drastic changes in the matrix elements of the charge or currents. The specific relation *(4–7,11,14,17)* is the following. The charge correlation function $\chi_{\rho\rho}(kz)$ in the presence of Coulomb interactions defines a function $\chi^{sc}_{\rho\rho}(kz)$:

$$\epsilon^L(kz) = [1 - \chi_{\rho\rho}(kz)/k^2]^{-1} \equiv 1 + \chi^{sc}_{\rho\rho}(kz)/k^2$$

To a good approximation $\chi^{sc}_{\rho\rho}(kz)$ is equal to the correlation function calculated, neglecting the long-range Coulomb forces between the particles, i.e., for a neutral system. (A corresponding but slightly more complicated statement holds for transverse excitations.) As a result, it is the zero of ϵ at $z = \omega_p$, which corresponds to the plasma excitation pole of $\chi_{\rho\rho}(kz)/k^2$; conversely, the poles of $\chi^{sc}_{\rho\rho}(kz)/k^2$, which would correspond to collective modes of the system with no Coulomb interaction, are poles of the dielectric constant but not natural frequencies (or collective modes) of the system with charge. The additional correlations which gave rise to collective modes in the neutral system must still be included in an improved calculation to justify the perturbative derivation of the Meissner effect; they modify and make unambiguous the dielectric constant, but they do not significantly modify the zero of the normal or BCS dielectric constant, i.e., the plasma mode. It is for this reason that the utility of the better calculation is in clarifying the calculation of properties such as accoustic attenuation, which depend on the frequency-dependent dielectric constant.

The calculations, which take into account the correlations which guarantee gauge invariance, show that the correct value of $\epsilon^L(kz)$ is approximately the result calculated from BCS density matrix elements when $(\omega/k)^2$ is much less than the electronic compressibility (i.e., $\approx \frac{1}{3} v_F^2$) and the result calculated from BCS current matrix elements in the opposite limit; thus the original BCS result is essentially correct for sound attenuation, since $(\omega/k)^2 \approx (m/M)\, v_F^2$.

E. Reconciliation with Goldstone Theorem

The absence of a low-lying collective mode should also be reconciled with our two proofs that there was one. Our first proof was based on momentum conservation—with the positive background, the charge density is conserved but the current (no longer the momentum) is not. The violation of the second proof is more subtle. There is no difficulty with the wordy description: the Goldstone theorem requires a restoring force which behaves as k^2 for small k, and the Coulomb force gives a larger restoring force, ω_p^2, as $k \to 0$. However, the way this condition comes into our proof is that we can no longer interchange infinite volume and long-wavelength limits, because the surface charge that builds up at infinity is important. This results‡ in the fact that $[H, N] \neq 0$ when $V \to \infty$. Of course, the same argument we used previously to deduce the velocity of the collective mode now gives for its frequency the plasma frequency; for example,

$$\partial \delta \mathbf{v}_s / \partial t = e(\delta \mathbf{E}/m)$$

implies that at vanishing temperature

$$(\partial^2/\partial t^2)\, \delta \mathbf{v}_s = -(e/m)\, \delta \mathbf{J} = -(e/m)\, \rho \delta \mathbf{v}_s$$

F. Collisionless Low-Frequency Modes of Strongly Coupled Superconductors

We return now to our discussion of the behavior of possible longitudinal collective modes in the neutral system at finite temperatures and of possible transverse collective modes in charged or neutral systems at finite temperatures. First let us recall the low-temperature conclusions for normal Fermi systems (*18*). There a weak attractive interaction produces no collective mode, a weak repulsive interaction gives rise to a weak collective longitudinal mode (zero sound), and a strong repulsive interaction strengthens the zero sound mode, transforming it into an ordinary phonon with velocity $c^2 = dp/d\rho$ in the strong potential limit. The Coulomb force raises its energy still higher, to the plasma frequency. Besides this longitudinal mode various other modes are possible when the interaction is sufficiently strong and angularly dependent. In particular, when the interaction is characterized by *s*- and *p*-waves, a strong *p*-wave

‡ This argument may be sharpened (*8,9*): There will be Goldstone modes, i.e., $[H, N] = 0$, when $\int v(\mathbf{r})\, d\mathbf{r} \neq \infty$.

interaction gives rise to a collective transverse mode at low temperature. This is the behavior one expects and finds above T_c in a superfluid system, and also below but near T_c, since the superfluid component is small. As the temperature is lowered these modes, with the possible exception of the longitudinal mode, in which the superfluid can partake, require stronger forces. Thus with only *s*- and *p*-waves, the existence of a transverse mode requires arbitrarily large forces at vanishing temperature (*19*). Likewise, in normal systems when the spin-dependent interaction has proper sign, spin-wave collective modes will occur for $\omega\tau \gg 1$. It is likely that these modes will be present in superconductors except at very low temperatures, although it will be more difficult to observe them without the magnetic fields used to observe (*20*) corresponding modes in the alkalis (here the spin-dependent interaction has the opposite sign).

At the same time as these modes disappear in weakly attractive systems a mode begins to appear which sharpens as the temperature decreases. This is the unique longitudinal mode of the weakly coupled neutral superconductor. Its Landau damping is proportional to the frequency and to the density of thermally excited quasi-particles. For sufficiently strong short-range forces, as we stated in Section I, the situation is less dramatic. A zero sound mode of approximately the hydrodynamic velocity alters slowly into a collective mode with the hydrodynamic velocity. For weaker repulsive forces in *s*-states, it appears that a system might have a longitudinal collective zero sound mode near to T_c, which disappears somewhat below T_c, and reappears at still somewhat lower temperatures.

The analysis which demonstrates these results has been carried out in four stages: (1) the calculation (*5–7,14*) of consistent transition matrix elements and collective modes of the BCS theory at zero temperature; (2) the calculation (*11*) of these same properties at finite temperature; (3) the generalization (*21*) of the analysis to strong interactions (i.e., Fermi liquid theory) at zero temperature; and (4) the calculation (*19*) of the temperature dependence in Fermi liquid theory. The last of these analyses (*19*), although the most complicated, is the most easily readable and comprehensible. In technical language, the first two analyses are concerned with the Bethe–Salpeter equation for the superconductor when the potential is the interaction—an equation which gives off-diagonal elements consistent with the BCS or Hartree thermodynamic theory; the last two deal with the same Bethe–Salpeter equation at long wavelengths and low frequencies that is associated with the exact gap equation. From the first and third, one deduces that there is a long-wavelength longitudinal mode which exhausts the sum rule. Its calculated velocity is with (1), the Hartree–Fock, and with (3), the exact compressibility. The second and fourth calculations tell how the thermally excited quasi-particles modify these results and merge, as $T \to T_c$, with the predictions of normal fluid kinetic theory.

III. EXCITON-LIKE MODES

As we stressed in Section I, the discussion up to this point deals fairly unambiguously with theoretical notions which have few useful consequences. The remainder of the discussion, by contrast, deals with theoretical notions which are difficult to treat adequately and their comparison with ambiguous experiments. These experiments and calculations are connected with the exciton states—the states which may be looked upon as excited bound-pair states. Such excited pairs of identical particles must either be singlet even or triplet odd. For pairs at rest these states may be classified by their angular momentum.

To get some feeling for these exciton states let us begin by noting that the Schrödinger equation for a pair of particles in free space can be written in the form

$$\left(\Omega - \frac{\mathbf{p}_1^2}{2m} - \frac{\mathbf{p}_2^2}{2m}\right)\phi = v\phi$$

where v is the interparticle potential. If we denote the total momentum by $\mathbf{P} = \mathbf{p}_1 + \mathbf{p}_2$ and the relative momentum by $\mathbf{p}_1 - \mathbf{p}_2 = 2\mathbf{p}$, the Schrödinger equation in momentum space is

$$\left[\Omega - \frac{(\frac{1}{2}\mathbf{P}+\mathbf{p})^2}{2m} - \frac{(\frac{1}{2}\mathbf{P}-\mathbf{p})^2}{2m}\right]\phi_{\mathbf{P},\Omega}(\mathbf{p}) = \int \frac{d\mathbf{p}'}{(2\pi)^3} v(\mathbf{p}-\mathbf{p}')\,\phi_{\mathbf{P},\Omega}(\mathbf{p}')$$

$$v(\mathbf{p}-\mathbf{p}') = \int d\mathbf{r}\,\exp[i(\mathbf{p}-\mathbf{p}')\cdot\mathbf{r}]\,v(\mathbf{r})$$

We may write this equation in the form

$$\phi_{\mathbf{P},\Omega}(\mathbf{p}) = \frac{1}{\Omega - \epsilon(\mathbf{p}_+) - \epsilon(\mathbf{p}_-)} \int \frac{d\mathbf{p}'}{(2\pi)^3} v(\mathbf{p}-\mathbf{p}')\,\phi_{\mathbf{P},\Omega}(\mathbf{p}')$$

and complicate it further by writing

$$\phi_{\mathbf{P},\Omega}(\mathbf{p}) = -\int \frac{d\mathbf{p}'}{(2\pi)^3} \int \frac{d\omega}{2\pi i} \frac{1}{[\omega_+ - \epsilon(\mathbf{p}_+)]} \frac{1}{[-\omega_- - \epsilon(\mathbf{p}_-)]} v(\mathbf{p}-\mathbf{p}')\,\phi_{\mathbf{P},\Omega}(\mathbf{p}')$$

where

$$\mathbf{p}_\pm = \mathbf{p} \pm \tfrac{1}{2}\mathbf{P} \qquad \omega_\pm = \omega \pm \tfrac{1}{2}\Omega \qquad \epsilon_0(\mathbf{p}) = \mathbf{p}^2/2m$$

Finally we may introduce the symbols $G_0^{-1}(\mathbf{p},\omega) \equiv \bar{G}_0^{-1}(-\mathbf{p},-\omega) \equiv \omega - \epsilon_0(\mathbf{p})$ to characterize the free-particle propagators that occur in this last form,

$$\phi_{\mathbf{P},\Omega}(\mathbf{p}) = -\int \frac{d\mathbf{p}'}{(2\pi)^3} \int \frac{d\omega}{2\pi i} G_0(\mathbf{p}_+,\omega_+)\,\bar{G}_0(\mathbf{p}_-,\omega_-)\,v(\mathbf{p}-\mathbf{p}')\,\phi_{\mathbf{P},\Omega}(\mathbf{p}')$$

The equation could be simply solved to obtain the eigenstates with $\mathbf{P} = 0$ and given angular momentum if the dependence of the potential on $(|\mathbf{p}-\mathbf{p}'|)$ of

$$v(\mathbf{p}-\mathbf{p}') = \sum_l V_l(|\mathbf{p}-\mathbf{p}'|)\,P_l(\hat{\mathbf{p}}\cdot\hat{\mathbf{p}}')\frac{2l+1}{4\pi}$$

could be neglected, since we could then write the Schrödinger eigenvalue equation as

$$0 = \left[1 + V_l \int \frac{d\mathbf{p}}{(2\pi)^3} \int \frac{d\omega}{2\pi i} G_0(\mathbf{p}_+, \omega_+)\, \bar{G}_0(\mathbf{p}_-, \omega_-)\right]$$

That is to say, the value of Ω for which the bracketed expression vanishes are the binding energies of the pairs. Since superconducting excitons are bound pairs of quasi-particles rather than free particles, even if we make all the simplifications we made above, and take the coupling to be weak, the situation is slightly more complicated.

Indeed, unless the excitons are "at rest", i.e., have total momenta zero, they cannot be classified by their angular momentum. We shall restrict ourselves to these excitons. As in Section II there have been four successively improved versions of the exciton theory with a simple potential: (1) a weak-coupling theory with no broadening effects (*22,23*), (2) a weak-coupling theory (*24*) taking broadening into account (here due to impurities rather than thermal effects), (3) a strong-coupling theory (*25*) with no broadening, and (4) a strong-coupling theory (*26*) taking broadening into account.

In the simplest version, the analogue of the Schrödinger equation for the bound state at rest is

$$\phi_{l\Omega}(\mathbf{p}) = -\int \frac{d\omega}{2\pi i}[G(\mathbf{p}, \omega_+)\, \bar{G}(\mathbf{p}, \omega_-) - F(\mathbf{p}, \omega_+)\, F(\mathbf{p}, \omega_-)] \times \int \frac{d\mathbf{p}'}{(2\pi)^3} V_l(\mathbf{p} - \mathbf{p}')\, \phi_{l\Omega}(\mathbf{p}')$$

The propagators reflect the fact that the quasi-particles have a gap in their spectrum and are linear combinations of particles and holes. If we again neglect the momentum dependence of the V_l we obtain the equation

$$0 = 1 + V_l \int \frac{d\omega}{2\pi i} \int \frac{d\mathbf{p}}{(2\pi)^3} [G(\mathbf{p}, \omega_+)\, \bar{G}(\mathbf{p}, \omega_-) - F(\mathbf{p}, \omega_+)\, F(\rho, \omega_-)]$$

for the bound-state energies. This equation for excited bound pairs is quite similar to the gap equation—the equation for ground-state bound pairs which we may write as

$$\begin{aligned} 0 &= 1 + V_0 \int \frac{d\omega}{2\pi i} \int \frac{d\mathbf{p}}{(2\pi)^3} \frac{1}{\omega^2 - (\epsilon^2 + \Delta^2)} \\ &= 1 + V_0 \int \frac{d\omega}{2\pi i} \int \frac{d\mathbf{p}}{(2\pi)^3} \frac{(\epsilon + \omega)(-\epsilon + \omega) - \Delta^2}{[\omega^2 - (\epsilon^2 + \Delta^2)][\omega^2 - (\epsilon^2 + \Delta^2)]} \\ &= 1 + V_0 \int \frac{d\omega}{2\pi i} \int \frac{d\mathbf{p}}{(2\pi)^3} [G(\mathbf{p}, \omega)\, \bar{G}(\mathbf{p}, \omega) - F(\mathbf{p}, \omega)\, F(\mathbf{p}, \omega)] \end{aligned}$$

Indeed, to avoid discussing properties far from the Fermi surface it is convenient to subtract the two, and write an equation involving an effective coupling constant and integrals whose domain lies close to the Fermi surface:

$$\frac{1}{V_l}-\frac{1}{V_0}=-\int\frac{d\omega}{2\pi i}\int\frac{d\mathbf{p}}{(2\pi)^3}\{G(\mathbf{p},\omega_+)\bar{G}(\mathbf{p},\omega_-)-F(\mathbf{p},\omega_+)F(\mathbf{p},\omega_-)$$
$$-[G(\mathbf{p},\omega)\bar{G}(\mathbf{p},\omega)-F(\mathbf{p},\omega)F(\mathbf{p},\omega)]\}$$

When we evaluate the integral we obtain

$$\frac{1}{V_l}-\frac{1}{V_0}=-N(0)\int\frac{d\omega}{2\pi i}\int d\epsilon\{\ \}=N(0)\frac{\alpha\arcsin\alpha}{\sqrt{1-\alpha^2}}$$

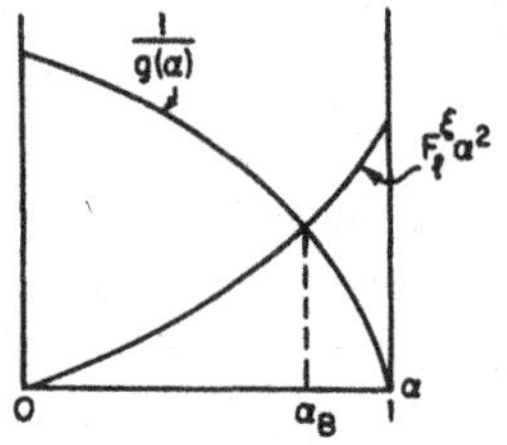

Fig. 1.

where $\alpha=\omega/2\Delta$ is the dimensionless variable which equals one at the BCS absorption edge. An excition with angular momentum l occurs when there is a solution (see Fig. 1) with $0\leqslant\alpha\leqslant 1$ to the equation

$$\left(\frac{1}{V_l}-\frac{1}{V_0}\right)\frac{1}{N(0)}\equiv\frac{1}{F_l^\xi}=\alpha^2 g(\alpha)\equiv I_1$$
$$g(\alpha)\equiv\frac{\sin^{-1}\alpha}{\alpha\sqrt{1-\alpha^2}}$$

This is essentially the theory when we have a weakly coupled system with $V_l\ll V_0$. A more accurate treatment indicates that one really should allow, even in the weak-coupling limit, for coupling of the quasi-particle pairs to quasi-particle hole pairs, which further complicates the equations to

$$\phi=-F_l^\xi[G_+\bar{G}_- - F_+F_- - (G\bar{G}-FF)]\phi - F_l^\xi[G_+F_- + F_+\bar{G}_-]\bar{\phi}$$
$$\bar{\phi}=-F_l^k[G_+F_- - G_-F_+]\phi - F_l^k[G_+G_- + F_+F_- - (GG+FF)]\bar{\phi}$$

where the usual integrals over the energy and frequency are understood. The terms in parentheses have again been subtracted to obtain equations restricted to a region near the Fermi surface. The parameters F_l^k and F_l^ξ which occur in these equations reflect both the direct quasi-particle interactions and the effective interactions that result from rapid virtual transitions to states far from the Fermi surface. The first term in the first equation gives the result we indicated above; the additional coupling alters the equation to

$$\left(\frac{1}{F_l^\xi} - I_1\right)\left(\frac{1}{F_l^k} - I_3\right) - I_2 I_4 = 0$$

where I_2, I_3, and I_4 are the integrals corresponding to I_1 in the above matrix, and where, in the weak-coupling theory,

$$[F_l^k]^{-1} = 1 + [V_l N(0)]^{-1} \qquad I_3 = 1 - g(\alpha) \qquad I_2 = -I_4 = \alpha g(\alpha)$$

This is the result (*23*) of Schrieffer and Bardasis, and it may be written more simply as

$$1 = (F_l^\xi \alpha^2 - F_l^\omega)\, g(\alpha)$$
$$(F_l^\omega)^{-1} \equiv (F_e^k)^{-1} - 1$$

Because the attraction is short range, it seems most probable that the excitons are singlet and most reasonable that the lowest exciton occurs for $l = 2$. However, reasonable estimates of F_l^ξ for $l = 2$, and of the absorption such excitons would produce suggest that α_B would have to be almost unity, and the absorption quite small.

The effect of impurities on these excitons was examined by Maki and Tsuneto (*24*). They found, as one might expect, that it does not take a very great concentration of impurities to substantially shift the exciton energy up toward the continuum. Moreover, whenever $\tau^{-1} > 2\Delta$, the binding energy of the exciton is extremely small. Thus it seems improbable that these weak coupling excitons can explain the position, strength, or impurity independence of the precursor absorption.

The effects of interactions between the quasi-particles was first examined by Larkin (*25*). He observed, in particular, that there was good reason to believe that the quantity $(F_l^\omega)^{-1} \equiv (F_l^k)^{-1} - 1$ was quite different from its weak-coupling value, $[N(0)V_l]^{-1}$. In particular, it might be negative and only slightly larger than -1, so that the exciton has large binding energy despite the small value of F_l^ξ, i.e., the slight asphericity of the electron–electron attraction. [Clearly, for there to be physical exciton (see Fig. 2) we must have

$$F_l^\xi > F_l^\omega > -1$$

the condition $F_l^\omega > -1$ being necessary for the stability of the superconducting ground state.] Larkin also showed that the absorption this exciton gives rise to increases with its binding energy and is further increased by the effective mass when F_1 is sufficiently large. It can therefore be substantially greater than in

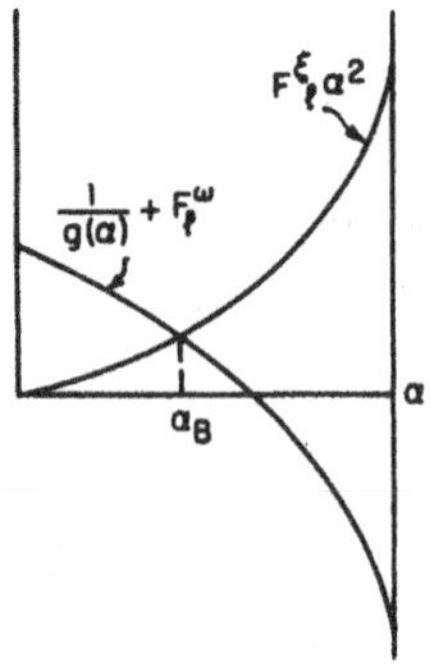

Fig. 2.

the weak-coupling theory.‡ By appropriately choosing the parameter F_2^{ω}, which is difficult to evaluate (we stress that it has been chosen and not evaluated), one can obtain an exciton at the observed position with a reasonably small aspherical interaction. In the resulting gross one-parameter theory only the mean-free-path discrepancy remains.

To understand the mean-free-path variation it is necessary to calculate the above relation taking into account the damping of the excitations. This problem was examined by Maki and Tsuneto and it has been reexamined by Fulde and Strassler (*26*), who allow for the Fermi liquid effects embodied in F_l^{ω}. These authors argue that provided $\omega\tau \gg 1$, the Fermi liquid parameters are not mean free path-dependent. Consequently, apart from choosing the parameters F_2^{ξ} and F_2^{ω} so that the exciton has the proper binding energy in a pure sample, they can use the results of Maki and Tsuneto for the integrals $I_{1,2,3,4}$. They also briefly examine the very dirty limit $\omega\tau \ll 1$.

They find that although in the strong-coupling theory the impurity dependence is somewhat less pronounced than in the weak-coupling limit studied by Tsuneto and Maki, an exciton which lies at $\omega = \Delta$ for a pure sample will be shifted upward at least to $\omega = \frac{3}{2}\Delta$ when $\tau\Delta \sim \frac{1}{13}$. A graph depicting their results is shown in Fig. 3.

All these results, it should be recalled, are based on a model in which the potential was drastically simplified. Were the interaction potential to have some structure, there might be more than one bound state with each value of *l*. This possibility has been briefly examined by McMillan (private communication), but he has not obtained any definite results.

The experimental evidence for precursors is difficult to follow and has

‡ It has been pointed out by Tinkham (private communication) that the deviations from the extreme anomalous limit of the conductivity which give rise to an increase in the absorption, which is steeper than that predicted by Bardeen and Mattis, might also increase the oscillator strength from Tsuneto's estimate.

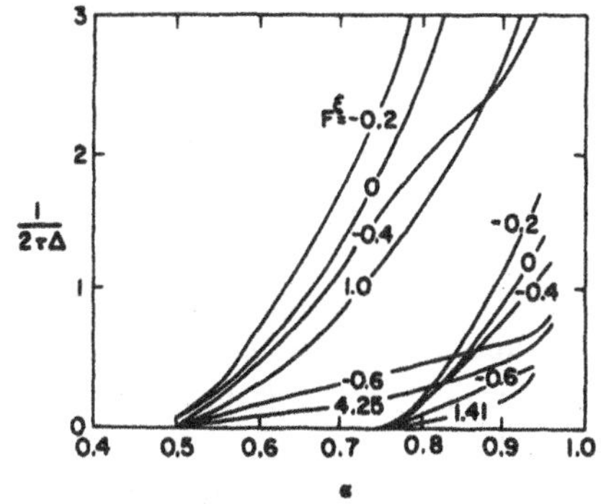

Fig. 3. Change of the energy of the exciton with $q=0$ as a function of mean free time $\tau_\xi\alpha=\Omega/2\Delta$. Different curves correspond to different values of F_2, all of which lead to an exciton energy $\alpha_\xi=0.5$ or 0.75 in the pure case. F^ω ranges from 0.23 ($F_2^\xi=4.25$) to -0.98 ($F_2^\xi=-0.6$) and from 0.22 ($F_2^\xi=1.41$) to -0.92 ($F_2^\xi=-0.6$).

changed several times during the past year. That the remarks which follow will still represent the state of affairs when this review is published is therefore quite uncertain. Experiments which demonstrate the precursor include the following: reflectivity measurements (*27,28*) of Richards and Tinkham on nominally pure bulk samples, transmission measurements (*27,29*) on very thin annealed films by Ginsberg and Tinkham, and reflection experiments (*30*) on bulk samples of alloys with mean free paths of different lengths by Leslie and Ginsberg. Taken at face value these experiments suggest that the precursor is present and is relatively independent of impurity content. A suggestion of the precursor (an unexplained small bump at the corresponding energy above the gap) has also been noted by Rowell et al. (*31*) in tunneling experiments. It would be difficult to explain all these data by an exciton, since, as we have indicated, it appears that no matter how strategically one chooses the unknown Fermi liquid parameters, there seems to be significant impurity dependence.

If we discard the data on thin samples and samples with short mean free paths, we might be able to understand the precursor absorption in terms of an exciton. This would be permissible if we discard the Ginsberg and Tinkham film measurements (*29*) and the experiments of Leslie and Ginsberg (*30*) in view of the more recent experiments of Palmer, Tinkham (*32*), and Dick and Reif,‡ which show no precursor absorption in very dirty film samples. These more accurate experiments were performed on samples similar to those of Ginsberg and Tinkham (perhaps somewhat dirtier since no attempt was made to reduce the disorder by annealing) and with mean free paths comparable to or shorter than those in the alloys of Leslie and Ginsberg. Palmer and Tinkham determine the absorption from measurements of the transmission and reflection and need no Kramers–Kronig analysis; Reif and Dick confirm these results by measuring the absorption directly in terms of the heat it produces. To explain the absence of an exciton in these dirty samples together with its appearance in samples with long mean free paths it is necessary to take particular values for the uncalculated Fermi liquid parameters, which, while not extremely unlikely, are somewhat fortuitous.

If we also discard the data on the nominally pure samples of Richards and Tinkham and of Norman and Douglass (*33*) on samples of comparable purity in favor of the recent erratum (*34*) of the latter authors, we have no precursor absorption and may forget the whole question. Theoretically, this would present no problem, since a great preponderance of plausible values for the uncalculated Fermi liquid parameters give no precursor significantly below the absorption edge.

Should the middle possibility be borne out by experiment, a detailed calculation of the Fermi liquid parameters will prove necessary before one truly understands the precursor. However, if the latest experiments are borne out, the exciton will join the longitudinal collective mode and the others discussed in Section II—for the present, a mere academic exercise with little relation to the real world.‡ All that would remain would be a few conjectured unobserved possibilities: the spin analogue of zero sound, weaker excitons lying near the gap edge, but perhaps slightly below it, and also possibly an exciton-like mode of relative phase oscillation in two-band superconductors (*35*).

ACKNOWLEDGMENTS

The author would like to thank Anthony Leggett and Michael Tinkham for extensive discussions, suggestions, and improvements. They are not to be blamed for any injustices that remain nor for the cynicism expressed.

Support from the National Science Foundation and the U.S. Air Force Office of Scientific Research during the preparation of this article is gratefully acknowledged.

REFERENCES

1. M. J. Buckingham, *Nuovo Cimento* **5**, 1763 (1957).
2. M. R. Schafroth, *Helv. Phys. Acta* **24**, 645 (1951).
3. L. P. Kadanoff and P. C. Martin, *Ann. Phys. (N.Y.)* **24**, 419 (1963).
4. P. W. Anderson, *Phys. Rev.* **110**, 827 (1958).
5. P. W. Anderson, *Phys. Rev.* **112**, 1900 (1958).
6. N. N. Bogoliubov, V. V. Tolmachev, and D. V. Shirkov, *A New Method in the Theory of Superconductivity*, Academy of Science, Moscow, 1958, Consultants Bureau, New York, 1959.
7. Y. Nambu, *Phys. Rev.* **117**, 648 (1960).

‡ Note added in proof. The experiments indicate that this is the case.

8. R. V. Lange, *Phys. Rev.* **146**, 301 (1965).

9. A. Katz and Y. Frishman, *Nuovo Cimento* **42A**, 1009 (1966), and private communication, 1966.

10. V. Ambegaokar and L. P. Kadanoff, *Nuovo Cimento* **22**, 914 (1961).

11. L. P. Kadanoff and P. C. Martin, *Phys. Rev.* **124**, 670 (1961).

12. G. A. Baym and L. P. Kadanoff, *Phys. Rev.* **124**, 287 (1961).

13. G. A. Baym, *Phys. Rev.* **127**, 1391 (1962).

14. G. Rickayzen, *Phys. Rev.* **115**, 795 (1959); **111**, 817 (1958).

15. J. M. Blatt, *Progr. Theoret. Phys (Kyoto)* **24**, 831 (1960).

16. D. J. Thouless, *Ann. Phys. (N.Y.)* **10**, 553 (1960).

17. R. E. Prange, *Phys. Rev.* **129**, 2495 (1963).

18. L. D. Landau, *Zh. Eksperim. i Teor. Fiz.* **30**, 1058 (1956); **32**, ɔ> (ı957); *Soviet Phys. JETP*

18. L. D. Landau, *Zh. Eksperim. i Teor. Fiz.* **30**, 1058 (1956); **32**, 59 (1957); **35**, 95 (1958); *Soviet Phys. JETP* **3**, 920 (1957); **5**, 101 (1957); **8**, 70 (1959).

19. A. J. Leggett, *Phys. Rev.* **140**, A1867 (1965); **147**, 119 (1966).

20. S. Schultz and G. Dunifer, *Phys. Rev. Letters* **18**, 283 (1967).

21. A. I. Larkin and A. B. Migdal, *Zh. Eksperim. i Teor. Fiz.* **49**, 1703 (1963); *Soviet Phys. JETP* **17**, 1146 (1963).

22. T. Tsuneto, *Phys. Rev.* **118**, 1029 (1960).

23. A. Bardasis and J. R. Schrieffer, *Phys. Rev.* **121**, 1050 (1961); See also V. G. Vaks, V. M. Galitskii, and A. I. Larkin, *Zh. Eksperim. i Teor. Fiz.* **41**, 1655 (1961); *Soviet Phys. JETP* **14**, 1177 (1962).

24. K. Maki and T. Tsuneto, *Progr. Theoret. Phys. (Kyoto)* **28**, 163 (1962).

25. A. I. Larkin, *Zh. Eksperim. i Teor. Fiz.* **46**, 2188 (1964); *Soviet Phys. JETP* **19**, 1478 (1964).

26. P. Fulde and S. Strassler, *Phys. Rev.* **140**, A519 (1965).

27. D. M. Ginsberg, P. L. Richards, and M. Tinkham, *Phys. Rev. Letters* **3**, 337 (1959).

28. P. L. Richards and M. Tinkham, *Phys. Rev.* **119**, 575 (1960); P. L. Richards, *Phys. Rev.* **126**, 912 (1962).

29. D. M. Ginsberg and M. Tinkham, *Phys. Rev.* **118**, 990 (1960).

30. J. D. Leslie and D. M. Ginsberg, *Phys. Rev.* **133**, A362 (1964).

31. J. M. Rowell, P. W. Anderson, and D. E. Thomas, *Phys. Rev. Letters* **10**, 334 (1963).

32. M. Tinkham, *Proc. of the Tenth Intern. Conf. on Low Temp. Physics, Moscow, 1966*, Vol. II.B (M. P. Malkov, ed. in chief), Viniti, Moscow, 1967, p. 238.

33. S. L. Norman and D. H. Douglass, *Phys. Rev. Letters* **17**, 875 (1966).

34. S. L. Norman and D. H. Douglass, *Phys. Rev. Letters* **18**, 339 (1967).

35. A. Leggett, *Progr. Theoret. Phys. (Kyoto)* **36**, 901 (1966).

8

MACROSCOPIC QUANTUM PHENOMENA

James E. Mercereau

FORD SCIENTIFIC LABORATORY
NEWPORT BEACH, CALIFORNIA
AND
CALIFORNIA INSTITUTE OF TECHNOLOGY
PASADENA, CALIFORNIA

I. INTRODUCTION

A. Macroscopic Wave Function

Superconductivity is a macroscopic quantum phenomena. This concept was first suggested by Fritz London many years ago (*1*). However, only recently have experiments revealed, in a detailed way, the unique implications of his idea. His suggestion has had such outstanding success that it is now accepted virtually without question. The theme of this chapter will be to examine his idea and describe some of the experiments establishing its validity. No attempt will be made to seek a microscopic justification for the model, since many diverse treatments of this point are available (*2*); to include them here would detract from the apparent simplicity. We will be concerned about the general properties of the macroscopic state and not the details determining how the state came about. In fact, the simplicity of the model and its complete confirmation by experiment are compelling reasons to believe that the macrostate itself may be considered fundamental—and to a great measure independent of the complex contortions of any single electron. Once a material becomes superconducting, the phenomena described here are largely independent of the type, form, or structure of matter. The results will be found to depend only on ratios of physical constants.

The central idea of the macroscopic quantum state is represented by assigning a macroscopic number of electrons to a single wave function (Ψ). These electrons are assumed to somehow have condensed into a single state. This condensation results in a macroscopic density of particles (ρ) sharing the same quantum phase (γ). The resulting wave function is then $\Psi = \sqrt{\rho}\exp(i\gamma)$. In this form ($\Psi^*\Psi$) is not the usual probability of finding a particle but due to the macroscopic number of particles involved is actually the effective particle density (ρ). Both ρ and γ may be functions of space and time and their variation will therefore determine the motion of the quantum fluid. Since, by definition, the particles are in precisely the same state and must therefore behave in an identical fashion, the equations of motion for the macrostate must also be identical to the equations of motion for any single particle in this state.

It is at this point that the importance of the phase (γ) arises. Because the phase is common to so many particles, its effects do not average out on a macroscopic scale but remain to fundamentally determine the behavior of this system and provide the experimenter the opportunity of grappling with quantum mechanics first hand. Consequently, we will adopt the point of view here that the quantum phase in these systems be considered as a real physical variable.

B. Electrodynamic Consequences—London Equations

Changes in the wave function are of course determined by the Schrödinger equation. In particular, the center of mass motion (V) can be calculated for this wave function from the velocity operator (v) common to all the particles (and thus to the macrostate),

$$v = -(1/m^*)(i\hbar\nabla + e^*A)$$

where e^* and m^* are, respectively, the effective charge and mass of the particles and A is the vector potential. This center of mass velocity is just

$$V = \tfrac{1}{2}\{\Psi v^\dagger \Psi^\dagger + \Psi^\dagger v \Psi\}/\Psi^\dagger\Psi$$

giving a current

$$J = \rho V = (\rho/m^*)(\hbar \nabla\gamma - e^*A) \tag{1}$$

Because of the interpretation of ρ as particle density for this wave function, J also determines an actual particle current. The electric current density associated with this particle flow is then

$$j = e^*J = (\rho e^*/m^*)(\hbar \nabla\gamma - e^*A)$$

Such a wave function therefore has a peculiar electrodynamic behavior that relates current and field by

$$\nabla x\,(\lambda^2 j) = -H \qquad \text{where } \lambda^2 = m^*/\rho e^{*2}\mu_0$$

This is of course one of London's celebrated equations representing the electrical behavior of a superconductor and was in fact first deduced by him from experimental behavior. It can easily be shown that this London equation satisfies experimental facts by forbidding currents or magnetic fields internal to the superconductor except within a layer roughly λ thick at the surface (the Meissner effect).

The effect of more sophisticated treatments is only to enable the calculation of λ or ρ from single-particle parameters. It was London's realization that such electrodynamic behavior was a natural consequence of quantum condensation that led him to propose superconductivity as being fundamentally a macroscopic quantum phenomena. However this could only be a speculation at the time, since the fundamental quantum behavior (the phase information) was not uniquely determinable from the available experimental data. It will be shown that new experimental techniques have recently exposed the detailed validity of his unique suggestion, which now in fact provides a mechanism for the development of new and novel quantum devices.

II. QUANTIZED BEHAVIOR OF SUPERCONDUCTORS

A. Magnetic Flux

Unusual effects can arise naturally from the topology of the macroscopic quantum state. Since the wave function can now extend to macroscopic distances, its complex nature can lead to a variety of physical consequences. From Eq. (1) the phase gradient is related to electrical quantities by

$$\nabla\gamma = (e^*/\hbar)(\mu_0\lambda^2 j + A) \tag{2}$$

Integrating any gradient completely around a closed contour must give zero for a singly connected path. However, in a multiply connected region this requirement must be relaxed to allow shifts of modulus (2π). Several physical consequences of this mathematical feature of the macroscopic wave function have been predicted and have been detected in superconductors. These experiments now serve as the first unique justification of the macroscopic quantum concept of superconductivity.

The simplest multiply connected topology to consider is the hollow superconducting cylinder. The first experiments to show the consequences of a macroscopic quantum state were done with such a topology. If Eq. (2) is summed around the cylinder, then

$$N(h/e^*) = \oint \mu_0 \lambda^2 j \, dx + \phi \tag{3}$$

where ϕ is the magnetic flux, $\oint A dx$, enclosed by the integration path and N is an integer. The current integral may be neglected for a path deep within a cylinder wall since currents and fields are confined to a thin (λ) surface layer. A thick cylinder, relative to λ, would then tend to quantize the magnetic flux enclosed by it, in units of $\phi_0 = h/e^*$. This remarkable consequence of the macroscopic quantum state was first predicted by London (*1,3*), who spoke of Eq. (3) as the quantization of the fluxoid. This equation represents the quantization of the total angular momentum for the macroscopic system. On this macroscopic scale, however, the dominant contribution is the flux term, contrary to the quantization of mechanical momentum in the microstate.

The first experimental realization of this phenomena was reported (*4*) simultaneously by B. S. Deaver and W. M. Fairbank, and R. Doll and M. Näbauer. These experiments both measured the magnetic flux enclosed by a small superconducting cylinder with sufficient precision to detect the discrete behavior impressed on it by the macroscopic quantum state (see Fig. 1). When a cylinder

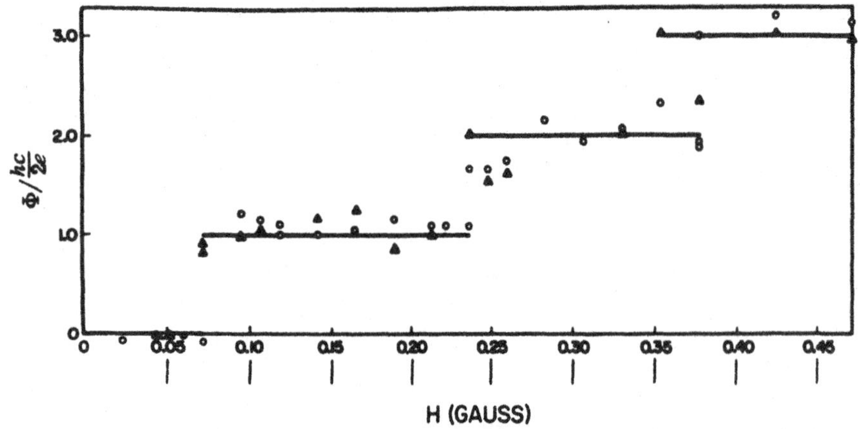

Fig. 1. Data of Deaver and Fairbank (*4*) showing trapped magnetic flux as a function of magnetic field in which the cylinder became superconducting. Such discrete behavior of trapped flux was the first experimental observation of macroscopic quantum effects.

becomes superconducting in a magnetic field it traps the flux enclosed by it. The remarkable fact exposed by these experiments was that the flux trapped exists only in discrete units of h/e^* and is not continuously variable, as would be expected classically. When the ambient field is insufficient to satisfy the quantization condition the superconducting transition is made to a current-carrying state which generates the flux required to bring the cylinder to the nearest quantum state. Both research groups found that e^* should be twice the free electron charge—probably a consequence of electron pairing. The magnitude of the flux quantum is therefore $\phi_0 = h/2e = 2.07 \times 10^{-7}$ G–cm^2. For a sufficiently thin wall cylinder the mechanical angular momentum of the electrons can no longer be avoided. The presence of this term (the current integral), has been detected (*5*) in experiments on the thermal invariance of the fluxoid.

B. Thermodynamic Effects

Another consequence of the macroscopic quantum state is observable in the thermodynamic behavior of a multiply connected system. Since Eq. (3) quantizes the angular momentum, the free energy of the system is also influenced by which quantized momentum state it is in. Thermal behavior may therefore also be expected to show a quantized periodicity. In fact, variations in the superconducting transition temperature of a thin superconducting cylinder have been observed, by Little and Parks (*6*), to be periodic in magnetic flux with a period of $h/2e$. For a sufficiently small thin-wall cylinder the magnetic flux is determined entirely by the applied external field (ϕ_{ext}). From Eq. (1) the kinetic energy density associated with the quantized angular momentum in this situation is

$$\text{KE} = \tfrac{1}{2}\rho \langle V \rangle^2 = \frac{1}{8\pi} \frac{1}{\lambda^2 r^2 \mu_0} [N(h/2e) - \phi_{\text{ext}}]^2$$

This kinetic energy can be large for a sufficiently small radius (r) cylinder and, unless the ambient flux is precisely the proper amount, a transition to multiply connected superconductivity implies the increase of kinetic energy given above. A superconducting transition must then be delayed to a lower temperature to gain this additional kinetic energy from the condensation process. From the measured variation in transition temperature it is apparent that N always takes a value to minimize this kinetic energy leading to the observed periodicity in flux of $h/2e$.

Experiments such as those described here have shown the existence of the macroscopic quantum state through the quantization of some parameter of the system. This experimental justification of some of the consequences of the macroscopic quantum state then suggests experiments designed to examine the nature of the quantum state itself.

III. JOSEPHSON QUANTUM-PHASE DETECTOR

Further examination of the quantum state itself had to await some form of quantum-phase-sensitive detector to probe the wave nature of the system. This has turned out to be the celebrated Josephson junction (*7*). Although the development of the Josephson junction is expertly treated in a separate chapter, to be consistent, a description (*8*) based on macroscopic quantization will be given here.

The equation of motion for a wave function is Schrödinger's equation relating the time rate of change to the energy operator. However, since the macroscopic wave function describes a large number of particles (density ρ), all doing precisely the same thing, the time rate of change of this macroscopic wave function must be exactly the same as the rate for any single particle. Thus the Schrödinger equation for this macroscopic state is.

$$\partial\Psi/\partial t = (-\,i/\hbar)\,\mu\Psi \tag{4}$$

where μ is the energy of the single particle. Josephson's equation results when two such macrostates are weakly coupled together; for example, two pieces of superconductor separate but close enough that particles can leak from one to the other by some tunneling process. (Tunneling is not essential but weak coupling is.) Then the wave function can change in time not only in proportion to the energy of the system itself but also by leaking particles to the other system.

The standard way of indicating such a weak-coupling situation is to write

$$\partial\Psi_1/\partial t = -\,(i/\hbar)(\mu_1\Psi_1 + \epsilon\Psi_2) \tag{4a}$$

where ϵ is a small coupling between the systems. A similar equation exists for the rate of change of the wave function of system 2. The two systems are of course connected in that the particles that disappear from one appear in the other. Performing the indicated operations [Eq. (4a)] on the functions

$$\Psi_1 = \sqrt{\rho_1}\exp(i\gamma_1) \qquad \text{and} \qquad \Psi_2 = \sqrt{\rho_2}\exp(i\gamma_2)$$

and simply collecting real and imaginary parts for systems of equal density results in two equations:

$$\dot{\rho}_1 = (2/\hbar)\,\epsilon\rho_1\sin(\gamma_2 - \gamma_1) = -\,\dot{\rho}_2 \tag{5a}$$

$$\dot{\gamma}_2 - \dot{\gamma}_1 = -\,(1/\hbar)(\mu_2 - \mu_1) \tag{5b}$$

The apparent density change involves the transfer of particles from one system to the other. If these particles are supplied from a current source, density can be maintained and $\rho_1 e^*$ represents a current from 2 to 1.

The current density (j) represented by Eq. (5a) in these circumstances is

$$j = (\epsilon\rho e^* V/\hbar A) \sin(\gamma_2 - \gamma_1)$$

where V is the volume of superconductor and A the contact area. Defining ε as coupling energy per unit area of contact, this becomes

$$j = [\varepsilon/(h/2e)] \sin(\gamma_2 - \gamma_1) = j_0 \sin(\gamma_2 - \gamma_1) \tag{5a'}$$

Equation (5a') determines the current in terms of phase and represents our quantum-phase-sensitive element. However, as it stands Eq. (5a') is not gauge-invariant and therefore cannot accurately represent the physical situation. That it is not gauge-invariant is not surprising, since the original wave function was not gauge-invariant either: phase, of course, is not an absolute quantity.

Gauge invariance is usually assured if changing the vector potential A to $A' + \nabla\chi$ and scalar potential ϕ to $\phi' - \partial\chi/\delta t$ leaves the observables unaltered. However, in this transformation process it can be easily shown [using Eq. (1)] that the phase must also transform as $\gamma = \gamma' + (e^*/\hbar)\,\chi$. Thus Eq. (5a') can easily be modified to a gauge-invariant form simply by replacing $\gamma_2 - \gamma_1$ by its gauge-invariant counterpart, $\gamma_2 - \gamma_1 - (e^*/\hbar) \int_1^2 A\, dx$. Equation (5b) represents the relative rate of phase change and is determined by the energy difference between system 1 and 2. The particles are electron pairs at the Fermi surface and therefore if only electromagnetic energies are considered here, $\mu_1 - \mu_2 = 2eV/\hbar$, where V is the potential difference between Fermi levels. Finally, the relevant equations are

$$j = j_0 \sin\left(\gamma_2 - \gamma_1 - \frac{2e}{\hbar}\int_1^2 A\, dx\right) \tag{6a}$$

$$\frac{\partial}{\partial t}\left(\gamma_2 - \gamma_1 - \frac{2e}{\hbar}\int_1^2 A\, dx\right) = \frac{-1}{\hbar}(\mu_2 - \mu_1) = \frac{2eV}{\hbar} \tag{6b}$$

These are Josephson's equations and are considered in detail in Chapter 9 from a more conventional point of view. Josephson's equations explicitly display the fundamental connection between phase and current for the situation of two weakly coupled superconductors. This is in contrast to the current in a single superconductor, where the current is related only to a phase gradient and the importance of the phase becomes obscured. It is this explicit connection between current and phase that will be useful in designing an instrument to examine the quantum-wave properties of the superconducting state.

Such structures of two superconductors nearly in contact (weakly connected) are now called Josephson junctions. Many experiments have been done with these devices, only a very few of which can be included here. The device provides a unique opportunity to study the coupling of electromagnetic and quantum waves. In this chapter, however, we will discuss only those experiments immediately relevant to the concept of the Josephson junction as a quantum-phase detector for the purpose of experimentally examining the macroscopic quantum state.

The very simplest application of Eq. (6) is adequate for our purposes. To get measurable quantities, the current density, Eq. (6a), must be integrated over some contact area. We assume the contact area is small enough that the self-field effects (*9*) may be neglected and that any voltages that appear are at such low frequency that the voltage may be considered constant over the contact. Equation (6a) relates current density to phase difference across a boundary. However, the phase changes within a single superconductor in a way given by Eq. (2), $\nabla\gamma = (2e/\hbar)(\mu_0\lambda^2 j + A)$. Consequently, in any integration over a finite area, phase variations due to surface current and fields within the contact must be properly accounted for. It is easily shown (*10*) that the result of the integration under these circumstances gives for the total current (I_J),

$$I_J = I_0 \frac{\sin(e/\hbar)\,\phi_J}{(e/\hbar)\,\phi_J} \sin\alpha = I(\phi_J)\sin\alpha \tag{7}$$

Since phase is a relative quantity, the integration must be done against some reference point in each superconductor. The angle α represents the phase difference between these symmetric reference points and henceforth will be called *relative phase*. It is no longer an arbitrary phase variable but must adjust to the experimental conditions.

The amplitude $I(\phi_J)$ shows diffraction-like behavior with flux due to phase variation within the junction enforced by the potential A. The flux (ϕ_J) is the total magnetic flux enclosed in the junction including the appropriate magnetic field penetration effects. It should again be emphasized that the only contribution a microscopic treatment can provide is a mechanism to calculate ρ or j_0 from single-particle parameters. The functional dependence on phase is independent of the details of any microscopic theory.

If a current souerce is attached to a Josephson junction, the relative phase (α) adjusts to accommodate both the magnitude and sign of the current, and, in the steady state, a zero-voltage current flows. As the current is changed, an instantaneous voltage V appears to readjust the phase, $\delta(\alpha) = (2e/\hbar)\,V\delta(t)$ to a new equilibrium value. Once equilibrium is reached, the current continues to flow and the voltage vanishes. The largest zero-voltage current which can flow, $I_{\max}$, is reached when $\alpha = \pm\pi/2$ or $I_{\max} = |I(\phi_J)|$. Currents exceeding this magnitude must flow by some other process, involving a finite voltage (for example, Giaever tunneling).

Figure 2 shows a current–voltage plot for a typical junction. When the zero voltage current is exceeded a voltage appears across the junction corresponding to that appropriate for Giaever tunneling in the junction. For this particular

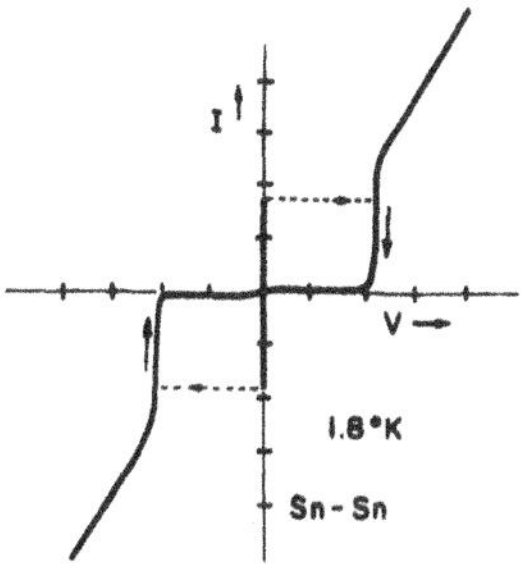

Fig. 2. Oscilloscope trace of a single Sn–Sn0_x–Sn junction showing the single-particle (Giaever) tunneling characteristic ($V \neq 0$) and the dc Josephson supercurrent at $V=0$. Current and voltage scales are 0.5 mA/div and 1 mV/div, respectively. The arrows indicate the switching path along the circuit load line taken when the applied current exceeds $I_{\max}$.

circuit, when the current is then reduced, the voltage-induced Giaever tunneling persists all the way down to zero current. Josephson has shown the maximum amplitude of zero-voltage current to be expected from a junction to be $I_0 = \pi\Delta/2R$, where Δ is the energy gap and R is the resistance of the junction to normal tunneling. Measuring these parameters from Fig. 2 it is evident that the maximum supercurrent is about 70% of the theoretical maximum. In general, Josephson's currents tend to follow the behavior expected from energy gap and resistance, with values as large as several hundred milliamperes having been observed (*11*).

The Josephson tunnel junction is actually a rather fragile thing. From Eq. (5a′) the coupling energy between superconductors is $I_0(h/2e)$, and for the junction in Fig. 2 this turns out to be about 1 eV. This is not 1 eV per electron but 1 eV of coupling for the entire macroscopic circuit. Consequently, noise and thermal fluctuations can have a profound effect on high-resistance junctions and probably explains why the maximum observed current is usually less than the theoretical maximum. For well-shielded, low-resistance junctions it is common to find the experimental maximum current within a few per cent of the theoretical maximum, giving great confidence to the adequacy of the microscopic theory.

The expected diffraction like behavior of $I_{\max}$ with flux was first shown by Anderson and Rowell (*12*). Data similar to theirs from R. C. Jaklevic are shown in Fig. 3, which displays diffraction effects out to the fifth side lobe. This junction is about $\frac{1}{4}$ mm wide, and from the assumed penetration depth and observed field periodicity a value of 2×10^{-7} G–cm^2 is obtained for $h/2e$. Considering the experimental uncertainties, this is good confirmation of the expected theoretical results. The amplitude dependence is not precisely as expected for a parallel junction and probably reflects nonuniformities in the barrier thickness. Nevertheless, data of this type are taken as strong experimental confirmation for the theoretical description of I_J.

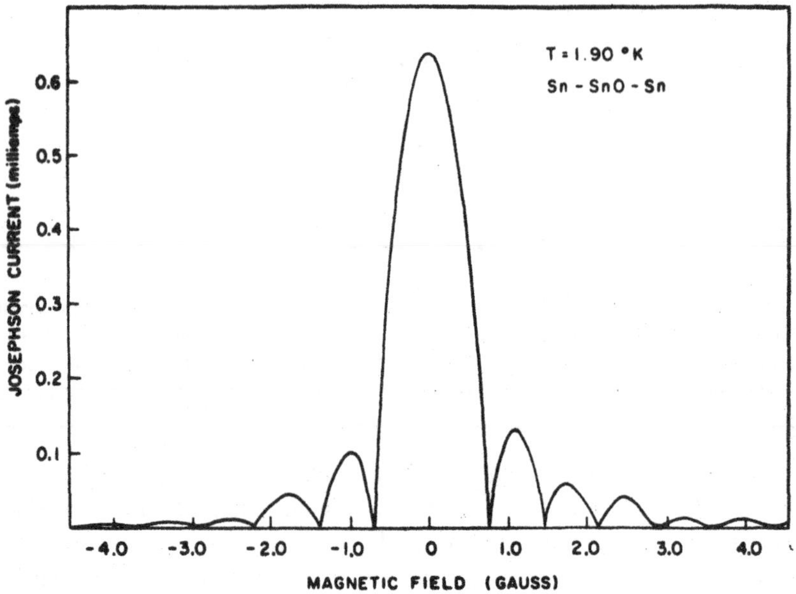

Fig. 3. Josephson current as a function of magnetic field showing the diffraction effects anticipated in Eq. (7).

The variation of current with phase has also been examined utilizing the time dependence implicit in Eq. (5b). When voltages are applied to the junction, the phase difference precesses in the manner described by Eq. (6b). In fact, the application of a dc voltage is expected to give rise to an alternating current of frequency of about 484 MHz/μV. If in addition to the dc voltage (V_0) and ac voltage ($v \cos \omega t$) is simultaneously applied, the form of the equation leads us to expect frequency-modulation effects with the attendant side bands. The mathematical formulas are

$$I = I(\phi_J) \sin\left(\frac{2e}{\hbar}\int V\,dt + \alpha\right)$$

$$= I(\phi_J) \sin\frac{2e}{\hbar}\left(V_0 t + \frac{v}{\omega}\sin\omega t + \alpha\right)$$

$$= I(\phi_J) \sum_{-\infty}^{\infty} \left\{J_n\left(\frac{2ev}{\hbar\omega}\right) \sin\left[\left(n\omega + \frac{2eV_0}{\hbar}\right)t + \alpha\right]\right\}$$

In particular, there may be side bands at zero frequencies. These will occur whenever $V_0 = n\hbar\omega/2e$. These zero-frequency currents correspond to the appearance of the Josephson dc current at a finite voltage. The result of impressing an alternating voltage on a Josephson junction would then be to alter the structure of the current–voltage characteristic from that given in Fig. 2. Current steps would appear at voltages related precisely to the impressed frequency as above. Experimental confirmation of this behavior was first obtained by Shapiro (*13*), who found the expected structure and confirmed the expected voltage–frequency relationship. His experiments were done at a high frequency, where our simple model does not directly apply. At the higher frequencies the spatial variation of both the quantum phase and the electric field must be taken into account to determine the proper coupling. Except for this point the physical consequences are the same, especially the expectation of current steps at voltages of exactly $n(\hbar\omega/2e)$.

The most precise test to which Josephson's equations have been submitted involve the verification of Eq. (6b). The oscillating currents induced at a finite voltage have a quite definite frequency. Radiation induced by these oscillating currents has been detected (*14*). Precise measurements of the applied voltage and the frequency of the induced Josephson radiation have been made. These measurements have been made with such precision by Parker et al. (*15*) that it is probable that macroscopic quantum devices utilizing Josephson's junctions will be used in the future as standards and to establish the value of fundamental physical constants.

In general, the experimental confirmation of Josephson's ideas is nearly complete. There have been many experiments. Some of them probing at the innermost detail of the theory and the theory seems to successfully withstand these examinations. Consequently we can accept the physical correctness of Josephson's prediction, in particular Eq. (6a), and utilize the Josephson junction as a device to examine the quantum phase in superconductors.

IV. SUPERCONDUCTING INTERFEROMETERS

The interferometer is an instrument for comparing phase at two points in a wave train. Implicit to the operation of the interferometer is the assumption that the wave is coherent over the interval of comparison. We have been assuming that superconductivity is represented by such a coherent quantum wave. With the Josephson quantum-phase detectors, therefore, it should be possible to examine the correctness of this assumption and perhaps fabricate superconducting de Broglie wave interferometers. This in fact has been done (*16*). The superconducting interferometer to be described here is basically two Josephson junctions connected in parallel by superconducting links. Many modifications of this arrangement are, of course, possible. In fact, in our laboratories, interferometers have been constructed containing from one to six junctions (*17*). The behavior of these interferometers is adequately explained by

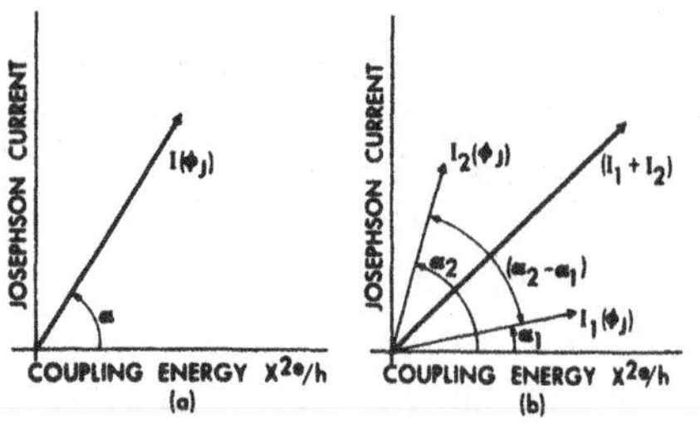

Fig. 4. Vector model of Josephson junction. (a) Single junction: vector length $I(\phi_{IJ})$ determined by applied field and reference angle (α) determined by applied current. (b) Double junction: If junctions are connected by superconducting links, the relative angle ($\alpha_2 - \alpha_1$) is determined by the enclosed fluxoid.

an obvious extension of the arguments to be given here. The superconducting links act as de Broglie wave guides which constrain the phase difference between the junctions (or across a single junction) to some unique value depending upon the de Broglie wave path difference between the junctions.

A vector model of the Josephson junction may be useful to understand the operation of the interferometer (see Fig. 4a). The length of vector $I(\phi_J)$ is determined by the coupling energy between superconductors and can be modulated by the flux in the junction, as given by Eq. (7). The angle α is the relative phase between symmetric reference points in the separate superconductors. Projecting $I(\phi_J)$ on the ordinate represents the Josephson current and the projection on the abcissa is the normalized phase-dependent binding energy. The relative phase angle is normally zero in the absence of other constraints to maximize the binding energy. However, impressing a current on the junctions specifies an additional condition on the relative phase angle (α) necessary to pass this current. Changing the current, of course, changes the relative phase (α) required. To readjust the relative phase, an instantaneous voltage V appears at the junction to give an "impulse" to the relative phase to adjust it to the proper value. At equilibrium α is a constant and a steady current flows at zero voltage. Of course, the maximum zero-voltage current that can flow is represented by the maximum projection of $I(\phi_J)$ on the ordinate which occurs when $\alpha = \pm \pi/2$. When this maximum Josephson current is exceeded, a relatively large voltage must appear across the junction to accommodate the current flow by other processes (see Fig. 2).

Current through two junctions connected in parallel can be determined easily from the same model. The total current, of course, is the vector sum of the current through the individual junctions, $I_1(\phi_J)$ and $I_2(\phi_J)$ (see Fig. 4b). Angles α_1 and α_2 are, respectively, the phase angle between the reference points in the two junctions and are in general uncorrelated. However, if the junctions are connected by superconducting links the phase angles α_1 and α_2 are no longer independent quantities. In fact α_2 is uniquely related to α_1 by the quantum-phase shift along the superconducting links. The total current, being a vector sum, depends in a periodic fashion on the relative phase angle $\alpha_2 - \alpha_1$. The vector sum of these two currents is just

$$I_T = [I_1^2 + I_2^2 + 2I_1I_2 \cos(\alpha_2 - \alpha_1)]^{1/2}$$

which explicitly displays this periodicity of the total current in the phase difference $(\alpha_2 - \alpha_1)$.

If the two currents are equal, this periodicity becomes particularly pronounced and the total current reduces to

$$I_T = 2I_1(\phi_J) \cos\left(\frac{\alpha_2 - \alpha_1}{2}\right)$$

The Josephson current is $I_T \sin[(\alpha_2 + \alpha_1)/2]$ and, as in the case of the single junction, I_T represents the maximum zero-voltage current through the device. A circulating current can also flow in the interferometer. The circulating current is represented by half the vector difference of the two currents or

$$I_c = I_1 \sin\left(\frac{\alpha_2 - \alpha_1}{2}\right) \cos\left(\frac{\alpha_1 + \alpha_2}{2}\right)$$

Circulating current and total current are out of phase. For equal junctions, at the point where the total current is maximum, the circulating current has become zero.

From Eq. (1) we know how the quantum phase varies inside the superconductor. With this expression then we can relate the phase α_1 to the phase α_2 as shown below (see Fig. 5):

$$\alpha_2 = \alpha_1 + \int_a \nabla\gamma\, dx - \int_b \nabla\gamma\, dx = \alpha_1 + \oint \nabla\gamma\, dx$$

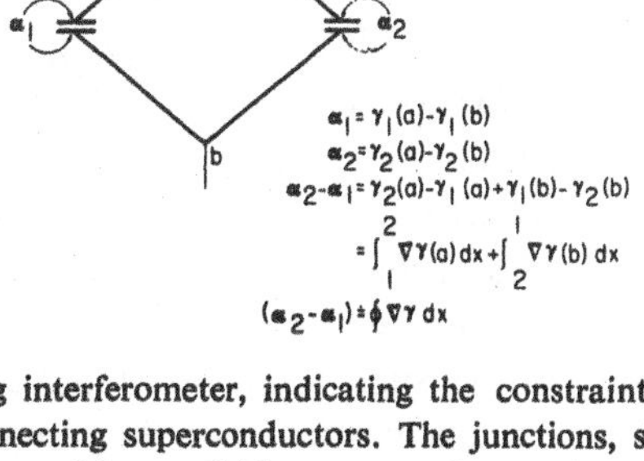

Fig. 5. Superconducting interferometer, indicating the constraint placed on the relative angles α_1, α_2 by the connecting superconductors. The junctions, shown as capacitors, are connected in parallel by superconductors (a and b).

or

$$\alpha_2 - \alpha_1 = (1/\hbar)\oint (m^*V + e^*A)\,dx = (1/\hbar)\left(\oint m^*V\,dx + e^*\phi_T\right) \tag{8}$$

The line integration actually should avoid going across the junctions themselves. However as long as this distance (~ 10 Å) is small relative to the other dimensions (about 1 cm), this contribution to the phase may be neglected, as it is already accounted for in the diffraction. In this approximation Eq. (8) is formally similar to Eq. (3). However, contrary to Eq. (3), in the present situation the phase difference $(\alpha_2 - \alpha_1)$ is not quantized. The integration path covers two separate superconductors, thereby avoiding the requirements of quantization. The phase difference $(\alpha_2 - \alpha_1)$ is thus a smoothly continuous variable except for possible "feedback" effects due to self-induced flux. This phase difference contains two terms; first, the mechanical angular momentum of the electron, and, second, the electromagnetic angular momentum or the total magnetic flux (ϕ_T) enclosed in the interferometer. It is this total canonical angular momentum that modulates the supercurrent through the interferometer. Since the mechanical angular momentum term is always small and can in principle be made zero in a perfectly symmetric interferometer, we will neglect its influence in this part of the discussion. In this symmetric limit the expression for maximum zero voltage, electron pair current thru the interferometer becomes

$$I_T = 2I_0 \left| \frac{\sin(e/\hbar)\phi_J}{(e/\hbar)\phi_j} \cos\frac{e}{\hbar}\phi_T \right| \tag{9}$$

The response of the interferometer to a magnetic flux then is a diffraction-modulated interference, an effect common to all types of wave phenomena. Here the de Broglie waves in the individual arms of the interferometer interfere with each other, while the diffraction occurs in the currents crossing the separate junctions. The flux periodicity of this modulation is seen to be $h/2e$. The implication is that it is possible to fabricate devices of macroscopic size whose operation is fundamentally determined by the ratio of physical constants rather than some material characteristic or process. Devices of this type have been fabricated in many diverse ways utilizing weak coupling of several forms; oxide barriers (*16*), point contacts (*18*), and film bridges (*19*). All the techniques yield similar results. The first experiments (*16*) utilized thin films and oxide barriers. Since this type device is still best understood, this discussion will be limited to a description of these circuits. The fact that all types of device behave in a similar way, however, emphasizes the underlying unity of principle determining their behavior. As long as the superconductor represents the macroscopic quantum state as defined previously, such periodic response is forced on it by the requirements of phase coherence.

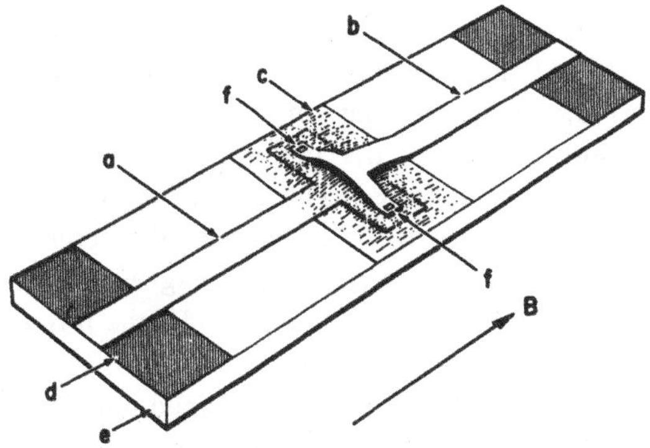

Fig. 6. Schematic of a coupled junction pair. A uniform magnetic field is applied parallel to the long dimension of the substrate e. The Formvar insulator c is applied over the base tin film a to mask out the junctions f and separate a from the second tin film b.

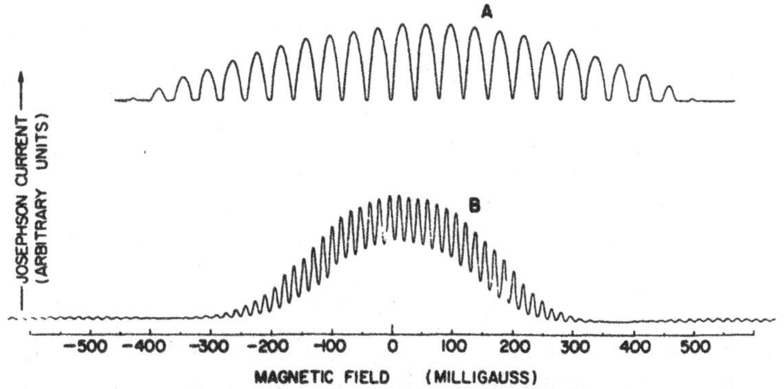

Fig. 7. Experimental trace of I_T vs. magnetic field showing interference and diffraction effects. The field periodicity is 39.5 and 16 mG for A and B, respectively. Approximate maximum currents are 1 mA (A) and 0.5 mA (B). The junction separation is $W = 3$ mm and junction width $w = 0.5$ mm for both cases. The magnetic offset is due to an incompletely compensated background field.

A typical thin-film device is shown in Fig. 6. Two T-sections of thin superconducting film (a and b) are separated from each other by an organic layer c except at the two small regions (f). In these two regions contact between the films is made through a thin oxide layer. The two oxide contacts form the junctions connected around the organic layer by the superconducting films. Current is introduced at a and b and the maximum zero-voltage current through the device is measured as a function of the applied magnetic field (B).

Figure 7 shows experimental results (*10*) from two tin thin-film interferometers. The maximum current for these interferometers was of the order of 1 mA; field periodicity in curve A is 39.5 mG. This curve clearly shows the expected diffraction modulated interference. Curve A shows only the central diffraction peak, while in curve B indications of the first side lobes are also evident. Interferometer B had a somewhat larger area, thereby giving it the smaller 16 mG periodicity shown. The amplitude of the Josephson current is temperature-dependent. However, the periodicity illustrated here is largely temperature-independent, as expected. The devices should be accurately periodic in the magnetic flux. And to the extent that the geometry is accurately defined, they should therefore also be periodic in magnetic field. To the accuracy of our area determination (at best one-tenth of 1%) the flux periodicity of all the devices so far tested has been $h/2e$. Interferometers have been fabricated from tin, indium, lead, aluminum, niobium, vanadium, and tantalium, and several alloys—some in the form of thin films, others as bulk material. The results from all these interferometers has been a periodic behavior somewhat similar to that shown in Fig. 7. The largest area so far examined has been about 3 cm^2 and the longest path examined for coherence has been $1\frac{1}{3}$ m *(23)*. In all these circumstances the simple concepts of a macroscopic quantum state as previously described seem entirely adequate to explain the periodicity observed in the experimental results.

The detailed shape of interference patterns, such as curves A or B in Fig. 7, depends so intimately on unknown geometric parameters that no attempt has been made to make a detailed analysis of the curve shape. However, some general conclusions are possible. As the interferometers get larger, the self-inductance (L) increases and shielding effects due to induced circulating currents (I_c) become important. This will be particularly troublesome when LI_c gets to be larger than $h/2e$. Curve A represents nearly the ideal effect expected from Eq. (9). In this case LI_c is less than about $\frac{1}{6}(h/2e)$. In curve B, however, shielding effects are already becoming evident in the decreased depth of modulation produced by an external field. In this case LI_c is approximately equal to $h/2e$. In fact, owing to these shielding and flux feedback effects, it is possible to show that the maximum current modulation possible, produced by an external field, is always less than or equal to $(h/2e)/L$.

This clear experimental confirmation of the flux periodicity naturally leads to a closer examination of quantum-phase effects. The full expression for the behavior of the interferometer includes effects of the electron center of mass motion. Consequently, experiments *(20)* were designed to examine this possibility. Electron center-of-mass motion implies a current. Rewriting Eq. (8) replacing center-of-mass velocity in favor of penetration depth (λ) and current gives

$$I_T = |I_0(\phi_J)\cos(e/\hbar)[(\oint \mu_0\lambda^2 j\, dx) + \phi_T]|$$

This more complete expression representing the interference modulation was examined using an experimental device illustrated in Fig. 8. Here two junctions 1

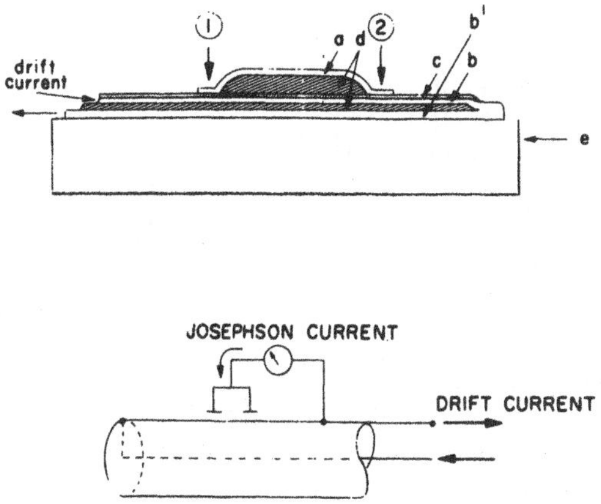

Fig. 8. Upper drawing is a cross section of a Josephson junction pair vacuum-deposited on a quartz substrate (e). A thin oxide layer (c) separates thin (~ 1100 Å) tin films (a and b). The junctions (1 and 2) are connected in parallel by superconducting thin films (a and b). A drift current return path (b), insulated from (b) by a Formvar layer (d), minimizes the magnetic field at the junctions. The lower drawing shows the idealized field-free situation. The field-free nature of the experiment was established by an actual measurement of the magnetic field in the junction region.

and 2 are connected by superconducting links in such a way that a "drift" current may be used to modulate the phase. The lower part of the drawing illustrates an idealization in which the field produced by the drift current is entirely confined to the coaxial cable. The junctions, lying outside the current path, see only the electron motion and none of the magnetic field. To the extent that the drift current flows only in the coaxial cable and the phase shifts due to the measuring currents may be neglected, the drift current period to be expected from this experiment would be $\Delta j = h/2e(\lambda^2 D)^{-1}$, where D is the distance between the junctions.

The actual experiment was done with a device such as that illustrated in the top of Fig. 8. This is a cross section of a thin-film structure. In this case magnetic field from the drift current is confined entirely to the space between films b and b′. And again the junctions 1 and 2 see only phase modulation due to electron velocity. The interferometer enclosed no magnetic flux due to the drift current. Experiments were done by measuring the maximum zero voltage current that could flow through the interferometer (for example, between points a and b) as a function of the drift current amplitude. Experimental results from such a current-modulated interferometer are shown in Fig. 9. Again the expected interference effect is clearly evident. Of course, a diffraction modulation also exists, arising from the velocity-induced phase shifts within the individual junctions. Secondary diffraction peaks could also be seen but are not illustrated here.

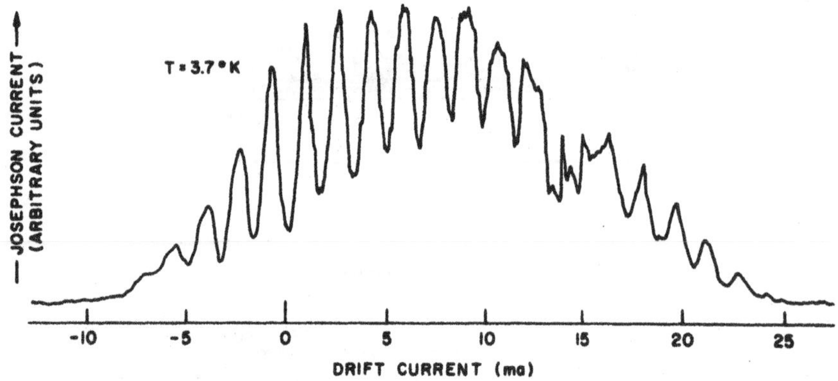

Fig. 9. Experimental trace of I_T vs. the drift current showing interference and diffraction effects. The zero offset is due to a static applied field. Maximum current is 1.5 mA.

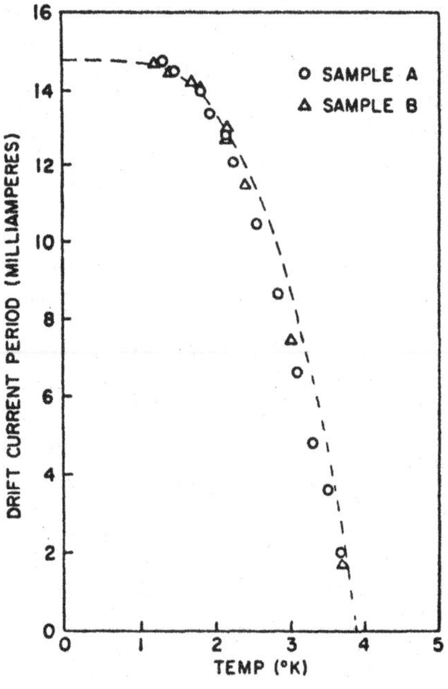

Fig. 10. Variation of observed drift-current period ΔI_d with temperature for two junction pairs of identical dimensions ($w = 0.5$ mm and $W = 8$ mm). The curve is the theoretical prediction from the text. The cross-sectional dimensions of the base film are 3 mm by 1100 ± 50 Å.

Contrary to the relative temperature independence of the flux modulation, current modulation will be strongly temperature-dependent through the temperature variation of the penetration depth. Figure 10 shows the drift current period as a function of temperature for two interferometers. To compare this with the theoretical expectation requires some assumption as to the distribution of the current in the superconducting film. The curve drawn on the figure comes from assuming a surface density (*21*)

$$j_0 = (I_0/l\lambda)\,[\sinh(\delta/\lambda)]^{-1}$$

where δ is the thickness and l the width of the base film. In this assumption the drift current periodicity will be

$$\Delta I_d = \frac{h}{2e}\frac{l}{D\delta}\frac{\delta}{\lambda}\sinh\left(\frac{\delta}{\lambda}\right)$$

The curve plotted on Fig. 10 is this expression with a value $\lambda(0) = 720$ Å and a temperature dependence of $\lambda = \lambda(0)[1-(T/T_c)^4]^{-1/2}$. The experimental results seem to follow reasonably close to this assumption and are self-consistent among themselves.

This temperature dependence of the periodicity is strong evidence for the proposed velocity modulation. However, to check that this modulation was not due to a magnetic field induced by the drift current, the fringing magnetic field was measured by another interferometer used as a magnetometer and placed 5 μ above the base film. The measured leakage field would limit any possible leakage field effect to less than 1/10 of a single period shown on Fig. 9. Consequently, these measurements are taken as separate experimental evidence of the macroscopic quantum state in which a phase shift is produced by center-of-mass motion. In this analysis no estimate of the superconducting electron mass needed to be made since a zero-temperature, thickness-corrected penetration depth of 720 Å was assumed for the tin films. If the mass were the free electron mass, the change in drift velocity required to produce a shift of one period for these devices is about 4½ cm/sec, corresponding to a de Broglie wavelength of about 1 cm.

Experiments such as those described here strongly confirm the concept of superconductivity as a macroscopic quantum state and probe in detail the spatial variation of the quantum phase. In fact, persistence of these types of interference effects in the face of all manner of experimental manipulations is itself probably the strongest possible support for the concept of a macroscopic quantum state. Interference effects such as those described here always occur between properly weakly connected superconductors independent of material parameters and depending only on the ratio of physical constants. This seems to be one of the few macroscopic examples where nature has provided such a clean system, unencumbered by statistical effects, for the experimenter to investigate.

Thermal fluctuations, however, can set a limit on the observability of these macroscopic quantum effects. Circulating currents in the interferometer represent different energy levels because of the induced flux (LI_c). This self-induced flux (LI_c) "feeds back" into the expression for circulating current as

$$I_c = I_1(\phi_J)\sin(e/\hbar)(LI_c + \phi_0)$$

The circulating current (I_c) thus becomes multiply valued, giving a spacing between allowed flux levels of $h/2e$. This spacing, however, is *not* the result of quantizing $\oint p\,dx$ around the circuit. The various allowed values of I_c determine a set of allowed values for the phase angle $(\alpha_2 - \alpha_1)$, spaced $h/2e$ in flux, but whose magnitude depends on the applied flux ϕ_0. The total current will therefore be influenced by which of the many possible circulating current states the interferometer choses. Normally the interferometer will attempt to minimize the total energy. However, energy spacing between these flux levels is roughly $(h/2e)^2/L$. When this spacing gets to be small compared to the thermal energy (kT), the interferometer can no longer uniquely determine a particular current state

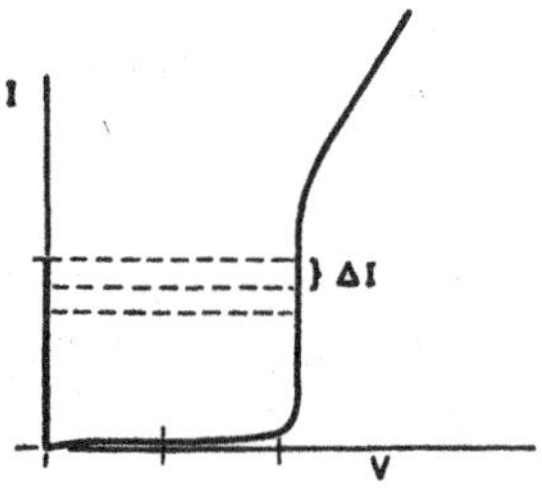

Fig. 11. Representation of experimentally observed multiple-valued maximum Josephson current indicating the onset of thermal fluctuations. The current at $V=0$ can attain any of three possible maximum values before switching along the circuit load line (dashed lines) to the normal tunneling characteristics.

against fluctuations. Figure 11 shows the current–voltage characteristic for an interferometer large enough that these thermal effects begin to appear. The curve is a composite of many tracings of the characteristic and shows the maximum total current to be a multiply valued function, as anticipated above. This whole pattern is a function of flux, being periodic in $h/2e$. From the best estimates of self-inductance for this device, $L\Delta I \sim h/2e$, as expected. As the inductance gets larger, further thermal smearing occurs among the available states and the quantum effects disappear. At 4°K this critical inductance, $L_c \sim (h/2e)^2/kT$, is $\lesssim 10^{-7}$ H. In practice severe thermal smearing begins to occur for inductances somewhat greater than 10^{-8} H. Until this limit is reached, however, it is possible to construct devices whose operation is fundamentally determined by quantum effects on a macroscopic scale and whose behavior is determined solely by the ratio of physical constants.

So far this discussion has emphasized the zero voltage current. However, many experiments have been done examining the interferometer at finite voltages (*19,22,23*). The application of a voltage implies currents at frequencies given by Eq. (6b). In these circuits the phase and amplitude of the currents will be determined not only by the electromagnetic field but also by quantum interference effects. The response of the interferometer to these alternating currents can be entirely described utilizing Maxwell theory and the simple concept of a macroscopic quantum state. Because of the unusual voltage-dependent frequency the analysis of these effects is somewhat more complex than for the zero voltage current. However, at low frequencies it is usually adequate to describe the response simply as a complicated time modulation of the zero-voltage current we have already described.

V. EXPERIMENTS UTILIZING SUPERCONDUCTING INTERFEROMETERS

These superconducting devices, utilizing spatial coherence of the de Broglie wave in interferometry, have been used as scientific instruments in several experiments. In Section IV it has been shown that such devices respond to magnetic flux and electric current in a periodic fashion. Thus they can be used as a sensitive sensor of these parameters. Their first such use as a magnetic sensor was probably in the experiment (*20*) already described on interference modulation by a drift current. In this section several additional experiments will be reviewed in which the superconductive interferometer was fundamentally used as the essential detecting element.

A. Influence of the Vector Potential

The first of these experiments to be discussed involves the physical reality of the vector potential. In the fundamental description of the quantum-phase variation [Eq. (1)], $\Delta\gamma = (1/\hbar)(m^*V + e^*A)$, it is the vector potential, A, and not the magnetic field that enters in an elemental way. All the experiments described so far however have involved the magnetic field. In 1949 Ehrenberg and Siday (*24*) pointed out that interference patterns observed in an electron microscope should depend on the magnetic flux enclosed by the accessible paths even when there is no magnetic field at the electron. This concept was later discussed extensively by Aharonov and Bohm (*25*), who showed in detailed arguments how this behavior should occur and pointed out that the effect is nonclassical in origin and arises naturally from a single-valued wave function. Their conclusions have opened discussions concerning the physical significance of the vector potential as being more than a mathematical convenience. Experiments designed to show these flux effects have been carried out by a number of experimenters utilizing electron beam techniques (*26*). Onsager (*27*) pointed out that an analogous situation occurs in multiply connected superconductors and that macroscopic quantization in superconductors is another example whereby the behavior of a charged particle depends on the flux and not on the field directly applied to the particle.

Consequently, the superconducting interferometer has been used to explore the properties of the vector potential (*28*). In these experiments a comparison was made between the response of the superconducting interferometer to a uniform magnetic field and the response of the same interferometer to a magnetic flux confined in such a way that *only* a vector potential touches the device. This flux was generated by a long, closely wound solenoid placed within the

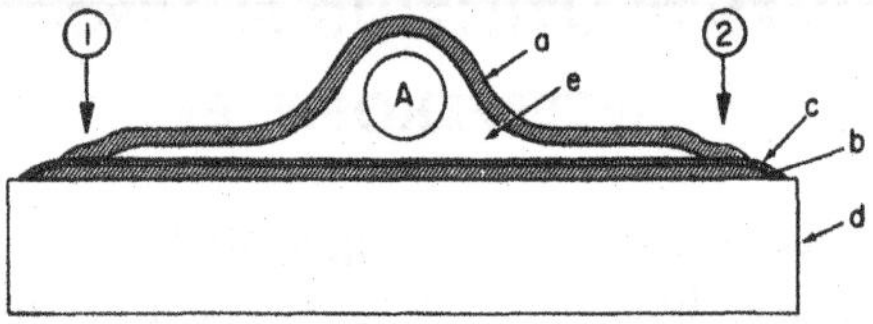

Fig. 12. Cross section of a Josephson junction pair vacuum-deposited on a quartz substrate (d). A thin oxide layer (c) separates thin (~ 1000 Å) tin films (a and b). The junctions (1 and 2) are connected in parallel by superconducting thin-film links enclosing the solenoid (A) embedded in Formvar (e). Current flow is measured between films a and b.

interferometer (see Fig. 12). No magnetic field exists external to such a solenoid, only an irrotational vector potential. The response of the interferometer to a uniform field will be the usual diffraction-modulated interference such as we have seen in Fig. 7. However, if the solenoid is perfect, there will be no leakage flux and no flux from the solenoid will link the junctions themselves. Therefore, the response of the interferometer to the solenoid will be only an interference modulation, and in fact any measured diffraction effect can be used to estimate the fringing fields.

The fundamental description of the interferometer implies response to the instantaneous value of magnetic flux. It is this characteristic we would like to test in the absence of any electric or magnetic field effects. However, since changing a magnetic flux generates an electric field, great care must be taken to eliminate these electric-field effects. Data were taken at fixed values of flux, the flux being changed to a new value only while the interferometer was warmed into the normal state. In this technique the superconducting interferometer was used to measure a static flux in the absence of any electric or magnetic fields at the interferometer. Periodic modulation of the maximum supercurrent, with flux period $h/2e$, was observed under these circumstances, evidently caused *directly* by the integrated static vector potential field.

Quasi-continuous data were also taken by switching the junctions normal with a rapidly oscillating current (much larger than the Josephson current) while measuring the zero-voltage current as a function of the slowly varying flux. These data are shown in Fig. 13. The diffraction-modulated interference produced by a uniform field is shown on the curve labeled $\phi(B)$. As expected, the interference produced by the solenoid labeled $\phi(A)$ has no associated diffraction. These data also show the integrated effects of a static vector potential and further demonstrate the absence of any magnetic fringing fields. The flux period determined from all these measurements, within the experimental accuracy of 5%, is found to be $h/2e$, as expected. This technique has been successfully utilized (*29*) at phase angles, $(e/\hbar)\oint A\,dx$), up to about 10^9.

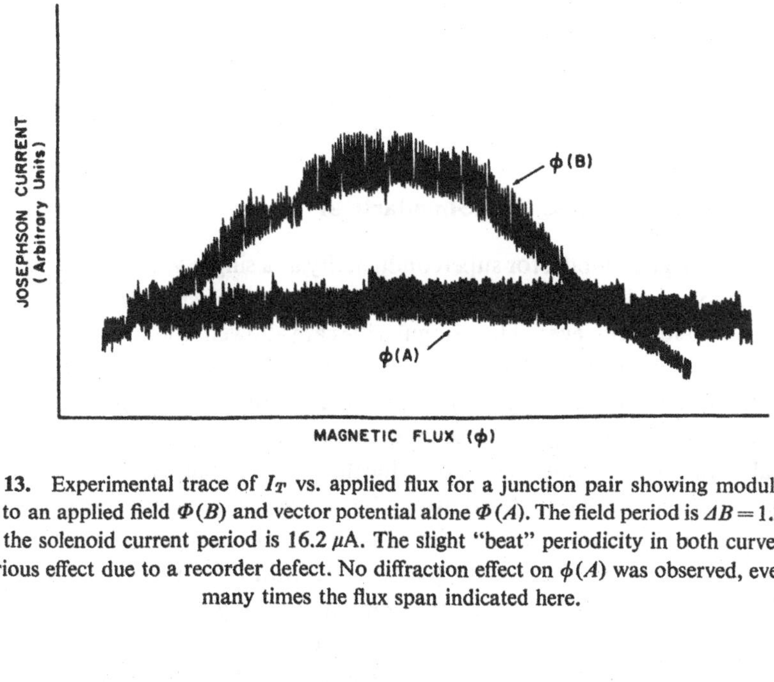

Fig. 13. Experimental trace of I_T vs. applied flux for a junction pair showing modulation due to an applied field $\Phi(B)$ and vector potential alone $\Phi(A)$. The field period is $\Delta B = 1.2$ mG and the solenoid current period is 16.2 μA. The slight "beat" periodicity in both curves is a spurious effect due to a recorder defect. No diffraction effect on $\phi(A)$ was observed, even for many times the flux span indicated here.

B. Mass of the Superconducting Electron

The response of the interferometer to electron velocity has been used as the basis for an experiment (*30*) to measure the mass of the superconducting electron. If the interferometer is put into uniform rotation, the velocity of the electrons is uniquely defined. From Eq. (8) the phase angle due to electron velocity is $(m^*/\hbar)\oint v\,dx$. Relating this to a circular interferometer, of radius r, in uniform rotation (Ω), the periodicity in rotation rate is

$$\Delta\Omega = \hbar/m^* r^2$$

An experiment to measure this rotation-induced modulation has been done with a vanadium interferometer at rotation rates up to about 10 rad/sec. In these measurements the maximum zero voltage current was measured as a function of rotation rate while the magnetic field was held fixed. From the measured periodicity and geometry a value can be determined for h/m^*. Additionally the magnetic field necessary to offset the rotation effect was determined. From these measurements a value for m^* of $1.8 \pm 0.1 \times 10^{-30}$ kg was obtained. This means that in these experiments the effective mass m^* is that of a pair of free electrons. Thus the quantum of circulation (h/m^*) in the superconductor involves electron pairs in the same manner (and for the same reason) as the flux quantum. In experiments such as these, it is the inertial mass of the electron which is important, since the electron is effectively at rest relative to the lattice. Josephson has predicted that the actual mass in such a measurement should be relativisticly corrected by the ratio of the Fermi energy to the rest mass energy of the free electron. This prediction still lies somewhat beyond our present experimental sensitivity.

C. Magnitude of Pinned Flux Singularities

London's original model for superconductivity as a single-valued macroscopic wave function produced a natural explanation for the Meissner effect (expulsion of magnetic field). However, if line singularities are allowed, the superconducting wave function can become multiply valued, but the phase can change only by modulus 2π around the singularity to preserve continuity. The physical result is the same as for a multiply connected superconductor. That is, flux is allowed internal to the superconductor but only localized at the singularity and must be integral multiples of $h/2e$. Abrikosov (*31*) showed that this type of superconductivity is possible when the decrease in the magnetic energy more than offsets the increase in kinetic energy associated with the current vortex at the singularity. He also showed that the minimum-energy state occurs for flux bundles of only one ($h/2e$). This is the so-called type II superconductivity (see Chapter 14). The superconducting interferometers have been used to measure (*32*) the flux pinned in a type II superconductor and have confirmed this aspect of Abrikosov's theory.

In these experiments the flux was pinned longitudinally in a niobium wire passing through the interferometer. Flux was pinned in a short section of the wire by cooling the wire in an applied longitudinal magnetic field. The flux thus frozen into the wire was measured by passing the wire slowly through the interferometer. Flux is constant along the length of the wire, changing only when a flux line emerges from the surface. As the wire is drawn through, the interferometer responds to such flux leakage, yielding both the sign and magni-

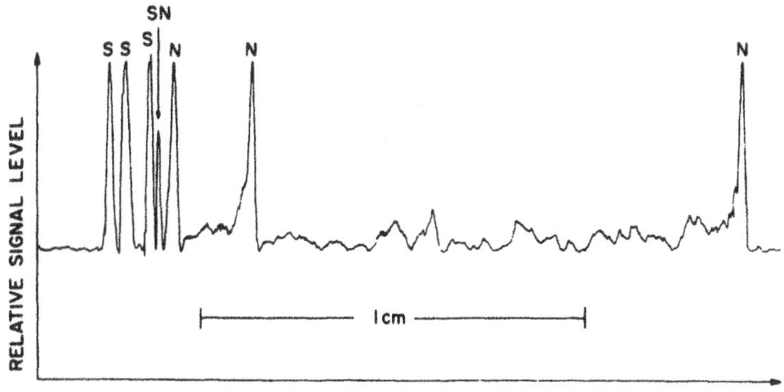

Fig. 14. Interferometer signal level as a function of wire position. Signal level is periodic in the enclosed flux with a period $h/2e$. Constant null position and the observed 2π phase shift associated with the signal peaks can only be explained by flux pinning in quantum units $h/2e$. S and N indicate the direction of the emerging field.

tude of the flux change. The essential property of the interferometer is that the maximum zero voltage current is a periodic function of the flux through the interferometer with a measured periodicity $h/2e$. Typical results are shown in Fig. 14. The output signal was proportioned to the maximum zero-voltage current, and as the wire was scanned this signal either remained nearly constant or else shifted quickly by *one* complete cycle, indicating an increase or decrease of flux in the wire of $h/2e$. By treating the wire in sufficiently low applied fields the pinned flux was entirely eliminated, as expected. As the applied field was increased the superconductor successively accepted one, then two, then three, etc., quanta depending on the field amplitude. In these experiments the magnetic flux pinned in superconducting niobium wire was always found to be localized in units of $h/2e$ and the number of such flux bundles was always determined by the original magnetic field, the number being roughly (applied flux) $\times$ $2e/h$.

VI. DYNAMICS OF THE MACROSCOPIC QUANTUM SYSTEM

For the macroscopic wave function that we have considered, $\Psi = \sqrt{\rho}\exp(i\gamma)$, we have found certain kinematic relationships between the phase γ and physical properties of the system. In Eq. (1) we have seen that the phase gradient is related to the momentum, p, and from Eq. (4) we can easily see that the variation of phase with time is related to the chemical potential μ. These equations have been experimentally verified, utilizing the Josephson junctions as described in Sections III, IV, and V and are reproduced below:

$$\nabla\gamma = (1/\hbar)(m^*V + e^*A) = p/\hbar = 1/\lambda$$

$$\partial\gamma/\partial t = -(1/\hbar)\,\mu$$

From these equations it is mathematically evident that the gradient of the chemical potential is equal to the time derivative of the gradient of the phase:

$$\nabla\mu = -\hbar\frac{\partial}{\partial t}(\nabla\gamma) = \frac{\partial}{\partial t}\frac{\hbar}{\lambda} \tag{10}$$

This relationship is nothing more than Newton's law that says a force will cause an acceleration. For a system with a constant chemical potential (force equal zero), Eq. (10) gives nothing new and says only that the phase gradient (momentum) is constant in time, leading to the London equation we have seen previously (Section I). However a nonhomogeneous chemical potential causes the phase gradient to change in time. As the system accelerates the velocity increases, the phase changes, and the de Broglie wavelength (λ) gets shorter and shorter. However, because of Abrikosov singularities, the phase may change in yet another way without necessarily changing the center-of-mass velocity.

A. Motion of Flux Singularities

As a singularity crosses between two points within the system, by definition, the relative phase between the two points changes by 2π. Integrating Eq. (10) between two points says that the difference in chemical potential between these two points is equal to $\hbar$ times the time rate of phase change between the two points. This expression can be restated, emphasizing the flow of Abrikosov singularities, by relating the difference in chemical potential $\Delta\mu$ between the two points to the average frequency $\langle\Omega\rangle$ at which Abrikosov singularities pass between the two points:

$$\langle\Omega\rangle = \Delta\mu/\hbar \tag{11}$$

These equations, relating to the motion of flux inside a superconductor, have been experimentally confirmed in various ways. One implication of this dynamic behavior is that a potential in a superconductor can be induced in an entirely nonclassical manner. Flux singularities in a current-carrying superconductor are subject to a Lorenz force. When this force exceeds the pinning forces on the flux lines, the flux will move through the superconductor. As we have seen, associated with this flux motion will be a phase change and a relative chemical potential. Therefore, the appearance of a voltage across a current-carrying superconductor does not necessarily signify the destruction of superconductivity but may imply flux motion in the superconducting state (*33*); the potential in the superconductor being exactly equal to the inductive voltage generated in the measuring circuit (*34*).

Giaever (*35*) has constructed a dc superconducting transformer based on these concepts. He arranged two superconducting films such that the flux threading one also penetrated the other. Flux motion induced by a current in one film is therefore magnetically coupled to the second. The potential difference induced by this flux motion should therefore occur simultaneously in both films even though there is no direct electrical connection between them. This behavior has in fact been observed by Giaever and can be explained only by some form of induced flux motion across the films. His ingenious experiments vividly illustrate this unique form of voltage generation but neither the frequency nor the magnitude of the flux in motion could be determined.

Additional confirmation of the details of this flux flow process in a superconductor has been obtained in other experiments. Anderson and Dayem (*36*) synchronized the motion of the vortices with an applied electromagnetic field, thereby constraining the frequency $\langle \Omega \rangle$ to be a particular value determined by the frequency of the applied radiation. By Eq. (11) this constraint on $\langle \Omega \rangle$ also implies a constraint on the voltage across the superconductor. Certain values of voltage are thereby preferred, giving rise to a step structure in the I–V characteristic reminiscent of that first observed by Shapiro in the Josephson junction. Here, however, voltage stability occurs both at harmonics and subharmonics of the applied frequency. Synchronizing $\langle \Omega \rangle$ to the radiation frequency implies the passage of one quanta per cycle, but harmonic synchronization can also occur, since it is possible to synchronize to one quanta every two cycles, two quanta per cycle, etc.

Nisenoff (*37*) has measured the magnitude of the flux bundle in motion. He has done experiments pumping flux through a thin-film superconducting ring at high frequencies. The result of this behavior is a quantized Faraday law whose analysis directly yields the magnitude of flux in motion. In his experiments the magnitude of the dynamic flux quantum is commonly $h/2e$, although often he sees multiples of this fundamental value.

B. Fluctuations and Noise

So far we have managed to avoid the problem of how the macroscopic quantum state is formed. If there are fluctuations in this formation process, of course noise will arise. However, even within the macroscopic quantum state itself fluctuations are possible due to the irregular motion of singularities. The position of a singularity is usually largely undetermined except for some inhomogeneity or pinning center. Fluctuations from equilibrium are related to the force equation [Eq. (10)] we have just introduced. This equation specifies a time dependence for the system. Without some time variation the concept of noise itself becomes obscure. In fact, a time-independent state of superconductivity is most common. The persistence of circulating currents in a superconducting ring, even though highly metastable, has been observed for years without noticeable decay. This has been one of the classic experiments to demonstrate the dramatic stability of the superconducting state. A more recent experiment emphasizing the unusual quiescence of superconductivity has been measurement (*38*) of magnetic Johnson noise in superconductors. In these experiments a superconducting interferometer was used as a magnetometer to measure magnetic flux fluctuations $\langle \phi \rangle$ produced by the thermal agitation of the conduction electrons in a metal (Johnson noise). In the normal state these thermally activated flux fluctuations, summed over all frequencies, can easily be shown to be

$$\langle \phi \rangle^2 = LkT$$

where L is the inductance of the circuit carrying the fluctuation noise currents. For a specimen of average dimension 1 cm, these fluctuations at four degrees amount to about a flux quantum. Vant-Hull et al. (*38*) measured these noise fluctuations and showed experimentally that they vanished as the material became superconducting.

The existence of singularities within the macroscopic quantum state, however, can lead to noise. Unless these singularities move with precision, a noise will be developed associated with the random phase motion. All that Eq. (11) can tell us is the average rate of motion for a singularity; unless the conditions are precisely homogeneous, the actual rate will be slightly different at each location, giving rise to a randomness and noise. As yet only relatively few experiments have been done to examine this phenomenon. Van Ooijen (*39*) has done noise measurements on current-carrying thin tin films and on vanadium foils from which he can extract the magnitude of flux in motion and its frequency distribution. These quantities seem to depend in great detail on the material properties of the superconductor. However, his data tend to confirm the concept of flux in motion and gives an independent determination for the value of the dynamic flux quantum. As we have seen before (*23,36,37*), under some circumstances it is nevertheless possible to stabilize this erratic motion by an applied electromagnetic field, and in these circumstances the random behavior can be synchronized into coherence.

REFERENCES

1. F. London, *Proc. Roy. Soc. (London)* **A152**, 24 (1935); *Phys. Rev.* **74**, 562 (1948).

2. O. Penrose and L. Onsager, *Phys. Rev.* **104**, 576 (1956); L. P. Gor'kov, *Zh. Eksperim. i Teor. Fiz.* **34**, 735 (1958); *Soviet Phys. JETP* **7**, 505 (1958); P. W. Anderson, *Phys. Rev.* **112**, 1900 (1958); *Lectures on the Many-Body Problem, Ravello, 1963,* Vol. II (E. R. Caianello, ed.), Academic Press, 1969, p. 113.

3. F. London, *Superfluids*, Vol. I, Wiley, New York, 1950.

4. B. S. Deaver and W. M. Fairbank, *Phys. Rev. Letters* **7**, 43 (1961); R. Doll and M. Näbauer, *Phys. Rev. Letters* **7**, 51 (1961).

5. J. E. Mercereau and L. T. Crane, *Phys. Rev. Letters* **12**, 191 (1964).

6. W. A. Little and R. D. Parks, *Phys. Rev. Letters* **9**, 9 (1962); *Phys. Rev.* **133**, A97 (1964).
7. B. D. Josephson, *Phys. Rev. Letters* **1**, 251 (1962).
8. R. P. Feynman, *Lectures on Physics*, Vol. III, Addison-Wesley, Reading, Mass., 1965.
9. R. A. Ferrell and R. E. Prange, *Phys. Rev. Letters* **10**, 479 (1963).
10. R. C. Jaklevic, J. Lambe, J. E. Mercereau, and A. H. Silver, *Phys. Rev.* **140**, A1628 (1965).
11. R. E. Eck, private communication, 1967.
12. P. W. Anderson and J. M. Rowell, *Phys. Rev. Letters* **10**, 230 (1963).
13. S. Shapiro, *Phys. Rev. Letters* **11**, 80 (1963).
14. I. K. Yanson, V. M. Svistunov, and I. M. Dmitrenko, *Zh. Eksperim. i Teor. Fiz.* **48**, 976 (1965); *Soviet Phys. JETP* **21**, 650 (1965); D. N. Langenberg, D. J. Scalapino, B. N. Taylor, and R. E. Eck, *Phys. Rev. Letters* **15**, 294 (1965).
15. W. H. Parker, B. N. Taylor, and D. N. Langenberg, *Phys. Rev. Letters* **18**, 287 (1967).
16. R. C. Jaklevic, J. Lambe, A. H. Silver, and J. E. Mercereau, *Phys. Rev. Letters* **12**, 159 (1964).
17. J. E. Zimmerman and A. H. Silver, *Phys. Rev.* **141**, 367 (1966).
18. J. E. Zimmerman and A. H. Silver, *Phys. Letters* **10**, 47 (1964).
19. J. Lambe, A. H. Silver, J. E. Mercereau and R. C. Jaklevic *Phys. Letters* **11**, 16 (1964); P. W. Anderson and A. H. Dayem, *Phys. Rev. Letters* **13**, 195 (1964).
20. R. C. Jaklevic, J. J. Lambe, A. H. Silver, and J. E. Mercereau, *Low Temperature Physics, LT9* (J. G. Daunt et al., eds.), Plenum Press, New York, 1965, PART A, p. 446.
21. V. L. Ginzburg, *Soviet Phys. "Doklady" (English Transl.)* **3**, 102 (1958).
22. A. H. Silver, J. E. Mercereau, and J. E. Zimmerman, *Bull. Am. Phys. Soc.* **10**, 318 (1965).
23. L. L. Vant-Hull and J. E. Mercereau, *Phys. Rev. Letters* **17**, 629 (1966); L. L. Vant-Hull, thesis, California Institute of Technology, Pasadena, 1966.
24. W. Ehrenberg and R. E. Siday, *Proc. Roy. Soc. (London)* **B62**, 8 (1949).
25. Y. Aharonov and D. Bohm, *Phys. Rev.* **115**, 485 (1959); **123**, 1511 (1961).
26. R. G. Chambers, *Phys. Rev. Letters* **5**, 3 (1960); G. Mollenstedt and W. Bayh, *Naturwiss.* **49**, 81 (1962); H. Boersch, H. Hamsich, and K. Grohmann, *Z. Physik* **169**, 263 (1962).
27. L. Onsager, *Proceedings of the International Conference of Theoretical Physics, Kyoto and Tokyo, Sept. 1953*, Science Council of Japan, Tokyo, 1954, p. 935; L. Onsager, *Phys. Rev. Letters* **7**, 56 (1961).
28. R. C. Jaklevic, J. J. Lambe, A. H. Silver, and J. E. Mercereau, *Phys. Rev. Letters* **12**, 274 (1964).
29. L. L. Vant-Hull, private communication, 1967.
30. J. E. Zimmerman and J. E. Mercereau, *Phys. Rev. Letters* **14**, 887 (1965).
31. A. A. Abrikosov, *Zh. Eksperim. i Teor. Fiz.* **32**, 1442 (1957); *Soviet Phys. JETP* **5**, 1174 (1957).
32. J. E. Zimmerman and J. E. Mercereau, *Phys. Rev. Letters* **13**, 125 (1964).
33. Y. B. Kim, C. F. Hempstead, and A. R. Strnad, *Phys. Rev. Letters* **9**, 306 (1962); *Phys. Rev.* **129**, 528 (1963); **131**, 2486 (1963); P. W. Anderson, *Phys. Rev. Letters* **9**, 309 (1962).
34. B. D. Josephson, *Phys. Letters* **16**, 242 (1965).
35. I. Giaever, *Phys. Rev. Letters* **15**, 825 (1965).
36. P. W. Anderson and A. H. Dayem, *Phys. Rev. Letters* **13**, 195 (1964).
37. M. Nisenoff, *Bull. Am. Phys. Soc.* **12**, 890 (1966).
38. L. L. Vant-Hull, R. A. Simpkins, and J. T. Harding, *Phys. Letters* **24A**, 736 (1967).
39. D. J. Van Ooijen, *Phys. Letters* **14**, 95 (1965); D. J. Van Ooijen and G. J. Van Gurp, *Phys. Letters* **17**, 230 (1965); *Philips Res. Rept.* **21**, 343 (1966).

9

WEAKLY COUPLED SUPERCONDUCTORS

B. D. Josephson

ROYAL SOCIETY MOND LABORATORY
UNIVERSITY OF CAMBRIDGE
CAMBRIDGE, ENGLAND

I. INTRODUCTION

A. Long-Range Order and Weak Coupling in Superconducting Systems

The phenomena associated with weakly coupled superconductors include some striking illustrations of the quantum nature of the ordering process in superconductors. It is convenient to begin our discussion by considering this ordering process in some detail.

Two alternative descriptions are available for the long-range order in a superconductor. The first is essentially a restatement of the property of infinite conductivity, and may be put in the form: The chemical potential or Fermi level μ at all points of a superconductor in a steady state is the same, *even if the state is a current-carrying one.* The second, more fundamental, description may be stated in a number of different ways, of which we shall choose one based on the Ginzburg–Landau theory described in Chapter 6. It will be recalled that the Ginzburg–Landau theory involves an order parameter ψ which is complex, and that the equations of the theory have symmetry properties which ensure that if $\psi(r)$ is a solution, then $\exp(i\alpha)\psi(r)$ is also a solution. Thus the Ginzburg–Landau equations do not determine the absolute phase of ψ; on the other hand, phase *differences* are determined since they are independent of α. Consequently, the phase of ψ (to be denoted by ϕ) is a quantity possessing long-range order; fixing its value at one point determines its value at any other.

The two forms of long-range order described, although apparently quite different, are in fact closely related. The connection lies in the general relation

$$\partial\phi/\partial t = -2\mu/\hbar \tag{1}$$

Equation (1) was shown to be exact for systems in equilibrium by Anderson et al. (*1*) (see also Chapter 6) and correction terms can for most purposes be ignored. If long-range order in ϕ is assumed to be present, it is easily deduced from Eq. (1) that in a steady-state situation (i.e., one in which phase differences are independent of time), μ must be independent of position.

The long-range order of the phase of the order parameter is a cooperative effect brought about by the motion of the conduction electrons from one part of the system to another. In a system consisting of two isolated pieces of superconductor, electrons cannot flow from one piece to the other, and there is no correlation between the phases of the order parameter in the two pieces. Thus two parameters appear in the solution of the Ginzburg–Landau equations for such a system, corresponding to the independent phases of the order parameter in the two regions.

Consider now a situation where two pieces of superconductor are coupled by some means which permits electrons to flow from one piece to the other. Under certain circumstances this coupling results in the appearance of correlations between the phases of the order parameter in the two pieces. For example, if the pieces are joined by a superconducting wire, the system becomes effectively a single superconducting region, and phase correlations extend over the whole of it as discussed above. The connecting wire can carry a certain critical current before the phase coherence breaks down and a finite voltage appears across it, and the magnitude of this critical current may be regarded as a measure of the strength of coupling produced by the wire. This chapter is concerned more generally with a description of the properties of any system where superconducting regions are coupled together by links which can transmit phase coherence as the superconducting wire did in the example just described. We are mainly interested in the case where the coupling is much weaker than that provided by a bulk superconducting wire, there being two ways in which the coupling strength can be reduced. In the first way a superconducting link is used, but its cross section is reduced. This may be done by using a thin-film geometry (*2*) or by using the small-area contact formed between two pieces of bulk superconductor when they are pressed into contact with each other (*3*). Alternatively, the link may be made of a substance which in bulk is not superconducting. Because of the existence of proximity effects, phase coherence can be transmitted through such a link provided it is sufficiently thin. The tunneling technique described in Chapter 11, in which the link consists of an insulating barrier approximately 20 Å thick, is the method most commonly used. However, it is also possible to use a barrier of normal metal, of thickness about 5000 Å (*4*).

B. Similarities and Differences between Bulk Superconductors and Weakly Coupled Superconductors

In general, any property of a bulk superconductor can also be observed in a link coupling two superconductors. However, if the coupling is weak, the parameters of the system will be greatly different from those of bulk superconductors. In the following, we list some of the more important properties of weak links, together with typical parameters for tunneling specimens.

1. A weak link can pass small resistanceless supercurrents, the critical current density being typically 1 A/cm^2.

2. The magnetic properties of a weak link are similar to those of a type II superconductor. There is a Meissner effect in small magnetic fields, with a penetration depth which is large ($\sim \frac{1}{2}$ mm). The critical field is small (~ 1 G), and in higher fields a mixed state, with a periodic field structure, occurs.

3. Plasma oscillations can occur in a weak link, at a frequency (~ 10 GHz) considerably below the bulk plasma frequency.

The "classical" properties of weak links described above will be discussed in detail in Section IV.A. Besides these there exist quantum-mechanical properties, some of which have been described in Chapter 8. They depend explicitly on the fact that the quantity involved in the ordering process has the nature of a phase. One group of properties, of which the alternating supercurrent (Chapter 8) produced by a potential difference is an example, are consequences of the relationship between phase and potential [Eq. (1)]. The other group is closely related to flux quantization, and the effects involve interference properties in the presence of a magnetic field. The quantum-mechanical properties of weakly coupled superconductors will be discussed in Section IV.B.

The basic theory which allows the properties of weakly coupled superconductors to be predicted falls naturally into two parts, which are described in the following two sections. First, a formula must be obtained to give the current flowing through the weak link (Section II). The current density at a point in the link turns out to be determined mainly by the local value of the difference between the phases of the order parameter on opposite sides of the link. This result must then be supplemented by equations giving the variations in space and time of the phase difference (Section III).

II. MICROSCOPIC THEORY OF THE CURRENT THROUGH A WEAK LINK

Microscopic theories for the current flowing through a weak link coupling together two superconductors may be divided into two types, based on either perturbation theory or the Green's function method. The idea behind the perturbation theory method is that if the coupling is weak, it may be treated as a small perturbation on an unperturbed system consisting of two isolated superconductors. In the Green's function method, an expression for the current through the link is obtained in terms of the Green's function describing propagation of electrons through the link. The Green's function method has so far been applied only to equilibrium situations, and in this case it has the advantage over the perturbation theory method in that the latter is valid only when the weak link is a tunneling barrier.

A. Perturbation Theory Method

1. Cooper Pair Model

We begin our discussion with a simple model due to Ferrell and Prange (*5*). Since supercurrents involve the flow of Cooper pairs rather than quasi-particles, it is reasonable to take as a model weak link one whose only effect is to transfer Cooper pairs from one superconducting region to the other. The eigenstates of such a perturbed system are of the form $\sum_{-\infty}^{\infty} a_n \psi_n$, where ψ_0 is an eigenstate of the unperturbed system and ψ_n is the state obtained from it by transferring n Cooper pairs from one region to the other. On the assumption that both regions are at the same potential, no energy is required to transfer a Cooper pair from one region to the other, so that the states ψ_n are all degenerate. The perturbation is assumed to transfer Cooper pairs one at a time, so that it connects only states whose values of n differ by unity. As Ferrell and Prange pointed out, the Hamiltonian for this problem is equivalent to that for a one-dimensional chain of atoms in the tight-binding approximation, ψ_n in that case being the wave function of an electron localized on the nth atom. In both cases the eigenfunctions are of the form

$$\Phi_\alpha = \sum_{n=-\infty}^{\infty} \exp(in\alpha)\,\psi_n \tag{2}$$

$\hbar\alpha$ being a crystal momentum canonically conjugate to n, and the corresponding eigenvalues are

$$E_\alpha = E - 2M\cos\alpha \tag{3}$$

where E is the energy of each state in the unperturbed system and $-M$ is the matrix element for transfer of Cooper pairs, assumed to be real.

a. Interpretation of the parameter α

Before proceeding further, let us consider the significance of the parameter α which labels the eigenstates of the perturbed system. For a single superconducting region, the phase of the order parameter and the number of Cooper pairs are conjugate variables, and so an equation very similar to Eq. (2) defines a state in which the phase has a definite value ϕ:

$$\chi_\phi = \sum_{\nu=-\infty}^{\infty} \exp(i\nu\phi)\,\psi_\nu \tag{4}$$

where ψ_0 is now an eigenstate in which the number of electrons is definite and ψ_ν a state obtained from it by adding ν Cooper pairs.

Now let us consider a system consisting of two isolated superconductors, with definite phases ϕ_1, ϕ_2 in each region. The wave function of the whole system is just the product of the wave functions of the two parts (we ignore the antisymmetrization, which is irrelevant here):

$$\chi = \chi_{\phi_1}\chi_{\phi_2} = \sum_{\nu_1}\sum_{\nu_2} \{\exp[i(\nu_1\phi_1 + \nu_2\phi_2)]\}\,\psi_{\nu_1}\psi_{\nu_2} \tag{5}$$

$\psi_{\nu_1}\psi_{\nu_2}$ being the wave function of a state in which ν_1 Cooper pairs have been added to one superconductor and ν_2 to the other.

χ is the wave function of a system where the order parameter has a definite phase, and so the number of electrons is correspondingly uncertain. Let us now decompose χ into states which are eigenfunctions of the number operator:

$$\chi = \sum_N [\exp(iN\phi_1)]\,\chi_N \tag{6}$$

where

$$\chi_N = \sum_n \exp[in(\phi_1 - \phi_2)]\,\psi_{N+n}\psi_{-n} \tag{7}$$

Since $\psi_{N+n}\psi_{-n}$ is obtained from $\psi_N\psi_0$ by transferring n Cooper pairs from one region to the other, χ_N is a state of the same type as Φ_α [Eq. (2)]. It is seen that α can be identified with the phase difference $\Delta\phi = \phi_1 - \phi_2$. The result of the simple model for the weak link is therefore that the weak link introduces a coupling energy $-2M\cos(\Delta\phi)$, which depends on the phase difference. This coupling energy is the means by which the weak link establishes long-range order in ϕ: If no current source is connected to the system, $\Delta\phi$ adjusts itself so as to minimize the coupling energy.

The states of the system for which $\Delta\phi$ differs from the value which makes the coupling energy a minimum are in general current-carrying states. To investigate these we use the fact that n and $\hbar(\Delta\phi)$ are conjugate variables. Then according to Hamilton's equations,

$$\dot{n} = \frac{1}{\hbar}\frac{\partial E}{\partial(\Delta\phi)} \tag{8}$$

and

$$\hbar\frac{\partial}{\partial t}(\Delta\phi) = -\frac{\partial E}{\partial n} \tag{9}$$

Since the charge of a Cooper pair is $2e$, the current I is $2e\dot{n}$. $\partial E/\partial n$ is the energy required to transfer one Cooper pair across the link, which we have previously taken to be zero; if we now generalize to the case of a finite potential difference V, this energy is $-2eV$. Equations (8) and (9) thus become

$$I = (2e/\hbar)[\partial E/\partial(\Delta\phi)] = (2e/\hbar)[2M\sin(\Delta\phi)] \tag{10}$$

$$(\partial/\partial t)(\Delta\phi) = (2e/\hbar)\,V \tag{11}$$

Equation (10) shows explicitly the dependence of the current through the link on the phase difference. From it follows the existence of a critical current

$$I_c = (2e/\hbar)\,2M \tag{12}$$

It is interesting to note that the current through the weak link involves the matrix element linearly and not quadratically; this unusual result is characteristic of processes where phase coherence is involved (*6*).

In spite of the crudeness of the model described above, it does predict many essential features correctly. In particular it illustrates the importance of the parameter $\Delta\phi$. It should be noted also that Eq. (8) is valid generally if E is taken to be the free energy of the system (*6*), and Eq. (11) is always valid as a simple consequence of Eq. (1).

2. Method of the Tunneling Hamiltonian

a. *Derivation of formula for the total current*

We shall now turn to a more realistic model, in which the primary process occurring at the weak link is taken to be the transfer of single electrons rather than pairs. The transfer of pairs, which gives rise to supercurrents, appears as a second-order process in this model, and the supercurrents can be related to properties of the weak link in the normal state. The calculation was first carried out by Josephson (*7*), and we shall describe here a simplified version due to Ambegaokar and Baratoff (*8*).

The starting point of the theory is the assumption that the Hamiltonian of the system can be written in the form

$$\mathscr{H} = \mathscr{H}_0 + \mathscr{H}_T \tag{13}$$

where

$$\mathscr{H}_T = \sum_{k,q,\sigma} (T_{kq} c^{\dagger}_{k\sigma} c_{q\sigma} + T_{qk} c^{\dagger}_{q\sigma} c_{k\sigma}) \tag{14}$$

and $\mathscr{H}_0$ is the Hamiltonian of the unperturbed system, consisting of two isolated superconductors. In Eq. (14) the indices k and q are used to distinguish electron operators in the two superconductors and σ is a spin index. Equations (13) and (14) constitute the tunneling Hamiltonian approximation introduced by Cohen et al. (*9*). Physically, it is equivalent to assuming that the weak link transfers electrons instantaneously from one superconductor to the other, and that interaction between electrons can be ignored inside the link. These assumptions are probably good in the case of a tunneling barrier, a case for which partial justification has been given by Prange (*10*).

The current through the weak link is found by relating it to the rate of change of the number of electrons on one side of the link:

$$\begin{aligned} I &= e\frac{d}{dt}\langle(\sum_{k\sigma} c^{\dagger}_{k\sigma} c_{k\sigma})\rangle = ie\langle[\mathscr{H}_0 + \mathscr{H}_T, \sum_{k\sigma} c^{\dagger}_{k\sigma} c_{k\sigma}]\rangle \\ &= -ie\sum_{kq\sigma}\langle T_{kq} c^{\dagger}_{k\sigma} c_{q\sigma} - T_{qk} c^{\dagger}_{q\sigma} c_{k\sigma}\rangle \end{aligned} \tag{15}$$

the commutator with $\mathscr{H}_0$ vanishing because in the unperturbed system (with no link present) the number of particles in each of the regions is conserved (we are now using units in which $\hbar = k_B = 1$).

The expectation value on the right side of Eq. (15) vanishes in the unperturbed system, since k and q are states in different regions. To obtain the lowest-order nonvanishing contribution it is necessary to expand the expectation value to first order in the perturbation, using the result

$$\langle A(t)\rangle = \langle A(t)\rangle_0 - i\int_{-\infty}^{t} \exp(\eta t')\langle[A(t), \mathscr{H}_T(t')]\rangle_0 \, dt' \tag{16}$$

where $\langle\ \rangle_0$ denotes the expectation value in the unperturbed system and $\eta = 0^+$. The current through the link is therefore expressible in terms of the expectation values in the unperturbed system of products of four electron operators. These products may be divided into three types, according to the distribution of creation and annihilation operators between the two regions, typical examples being $\langle c_k^\dagger c_{k'} c_q^\dagger c_{q'} \rangle$, $\langle c_k^\dagger c_{k'}^\dagger c_q c_{q'} \rangle$, and $\langle c_k c_{k'} c_q^\dagger c_{q'}^\dagger \rangle$. Since in the unperturbed system the two regions are isolated from each other, the first expectation value is equal to the product $\langle c_k^\dagger c_{k'} \rangle \langle c_q^\dagger c_{q'} \rangle$, and is therefore the product of two single-particle Green's functions. The other expectation values require a more careful discussion since they involve the question of phase coherence. If the unperturbed system consisted of an ensemble in which there was a definite number of electrons in each region, the last two expectation values would be identically zero. However, both on general grounds and from the Cooper pair model, we expect the perturbation to have the effect of producing phase coherence over the whole system. To obtain good results from lowest-order perturbation theory, it is necessary to take for the unperturbed system an ensemble in which the phase coherence is already present. For such systems operators such as $c_k^\dagger c_{k'}^\dagger c_q c_{q'}$ have a nonzero expectation value equal to the product of an F function and an F^+ function, which involves the phase difference explicitly. As will be seen, this type of term gives a finite contribution to the current even for zero applied voltage, and the total contribution from these terms may therefore be regarded as the supercurrent.

Having indicated the general form of the expression for the current, we shall proceed to the detailed calculations. These may be omitted by those readers whose main interest is in the application of the theory. The final result of the calculations is Eq. (23), which together with the defining equations (20) and (21) gives the supercurrent through the link in terms of the phase difference, the potential difference, and a quantity $I_1(\omega)$ whose evaluation is discussed in Section II.A.2.c.

It is convenient at this point to specialize somewhat to simplify the calculation. Let us assume that in the unperturbed state currents and magnetic fields are absent and the potential difference between the two regions is independent of time. We shall later be able to remove these restrictions and deal with situations where slow time variations and small currents and magnetic fields are present. We make also one further assumption—that the influence of the boundaries on the relevant Green's functions can be ignored. It is then possible to choose the states k and q in such a way that the only products of pairs of operators with nonvanishing expectation values are of the types $c_k^\dagger(t) c_k(t')$, $c_k^\dagger(t) c_{-k}^\dagger(t')$, and $c_k(t) c_{-k}(t')$, where c_{-k} is the operator obtained from c_k by the time-reversal transformation. The expression derived for the current from Eqs. (15) and (16) can be written in the form

$$\begin{aligned} I(t) = -2e \operatorname{Re} \sum_{kq\sigma} \int_{-\infty}^{t} & dt' \exp(\eta t') \\ & \times \{|T_{kq}|^2 [G_k^<(t', t)\, G_q^>(t, t') - G_k^>(t', t)\, G_q^<(t, t')] \\ & + T_{kq} T_{-k,-q} [F_k^{+>}(t, t')\, F_q^<(t', t) - F_k^{+<}(t, t')\, F_q^>(t', t)]\} \end{aligned} \tag{17}$$

where
$$G_k^>(t,t') = -i\langle c_{k\sigma}(t)\, c_{k\sigma}^\dagger(t')\rangle$$
$$F_k^>(t,t') = \langle c_{k\uparrow}(t)\, c_{-k\downarrow}(t')\rangle = -\langle c_{k\downarrow}(t)\, c_{-k\uparrow}(t')\rangle$$
$$F_k^{+>}(t,t') = \langle c_{k\uparrow}^\dagger(t)\, c_{-k\downarrow}^\dagger(t')\rangle = -\langle c_{k\downarrow}^\dagger(t)\, c_{-k\uparrow}^\dagger(t')\rangle$$

and $G_k^<$, $F_k^<$, and $F_k^{+<}$ are obtained from the above by interchanging the order of the operators and multiplying by -1. As is well known, in a grand canonical ensemble two corresponding $>$ and $<$ functions may be expressed in terms of a single spectral weight function, in the following way:

$$\begin{aligned} G_k^{\gtrless}(t,t') &= \exp[-i\mu_k(t-t')]\int_{-\infty}^{\infty} d\omega \\ &\quad\times \exp[-i\omega(t-t')](\pm i)\, A_k(\omega) f(\pm\omega) \\ F_k^{\gtrless}(t,t') &= \exp\{-i[\mu_k(t+t') - \phi_k(0)]\}\int_{-\infty}^{\infty} d\omega \\ &\quad\times \exp[-i\omega(t-t')](\pm i)\, B_k(\omega) f(\pm\omega) \\ F_k^{+\gtrless}(t,t') &= \exp\{i[\mu_k(t+t') - \phi_k(0)]\}\int_{-\infty}^{\infty} d\omega \\ &\quad\times \exp[-i\omega(t-t')](\pm i)\, B_k^+(\omega) f(\pm\omega) \end{aligned} \tag{18}$$

$f(\omega)$ being the Fermi factor $[\exp(\beta\omega)+1]^{-1}$. In the case of the normal Green's function, a derivation of the spectral representation can be found in Kadanoff and Baym (*11*). For the F and F^+ functions the same proof is applicable in the gauge $\mu_k = 0$, and the general result may be found by gauge invariance. Equations (18) have been written in such a way as to show explicitly the dependence of the Green's functions on the chemical potential and on $\phi(0)$, the phase at $t = 0$. The suffix k in μ_k and ϕ_k emphasizes the fact that the chemical potential and phase depend on which region the state k is in (although they do not depend on the individual state itself).

On inserting Eqs. (18) into Eq. (17), we obtain

$$\begin{aligned} I(t) &= -2e\,\mathrm{Re}\sum_{kq\sigma}\int_{-\infty}^{t} dt' \exp(\eta t')\int_{-\infty}^{\infty}\int_{-\infty}^{\infty} d\omega_1\, d\omega_2 [f(\omega_1) - f(\omega_2)] \\ &\quad\times \{\exp[i(t-t')(\mu_k - \mu_q + \omega_1 - \omega_2)]\,|T_{kq}|^2 A_k(\omega_1) A_q(\omega_2) \\ &\quad - \exp\{i[(t+t')(\mu_k - \mu_q) + [\phi_q(0) - \phi_k(0)] \\ &\quad + (t-t')(\omega_2 - \omega_1)]\}\, T_{kq} T_{-k,-q} B_k^+(\omega_1) B_q(\omega_2)\} \\ &= 2e\,\mathrm{Im}\sum_{kq\sigma}\int_{-\infty}^{\infty}\int_{-\infty}^{\infty} d\omega_1\, d\omega_2 [f(\omega_1) - f(\omega_2)] \\ &\quad\times \left\{|T_{kq}|^2 \frac{A_k(\omega_1) A_q(\omega_2)}{\omega_1 - \omega_2 - \omega + i\eta}\right. \\ &\quad \left. + \exp[-i\Delta\phi(t)]\, T_{kq} T_{-k,-q} \frac{B_k^+(\omega_1) B_q(\omega_2)}{\omega_1 - \omega_2 - \omega - i\eta}\right\} \end{aligned} \tag{19}$$

where

$$\omega = \mu_q - \mu_k \tag{20}$$

$$\Delta\phi(t) = \phi_k(0) - \phi_q(0) - 2(\mu_k - \mu_q)\, t \tag{21}$$

is the phase difference at time t, in view of Eq. (1).

b. *Separation into normal currents and supercurrents*

As will become apparent in the following discussion, the terms in Eq. (19) involving A and B spectral functions can be interpreted as normal currents and supercurrents, respectively. Let us first examine the normal currents. The spectral function for a normal Green's function is always real (*11*), so that the normal current reduces to

$$I_n = -\,2\pi e\,|T_{kq}|^2 \sum_{kq\sigma} \int_{-\infty}^{\infty} \int_{-\infty}^{\infty} d\omega_1\, d\omega_2 \,[f(\omega_1) - f(\omega_2)] \times A_k(\omega_1)\, A_q(\omega_2)\, \delta(\omega_1 - \omega_2 - \omega) \tag{22}$$

This expression is the generalization to finite temperatures of one obtained by Schrieffer et al. (*12*) using the golden-rule formula for transition probabilities. It can be shown from this formula that if only processes involving transfer of electrons between two particular states k, q occur, the resulting current is

$$I_{kq} = -\,2\pi e\,|T_{kq}|^2 \int_{-\infty}^{\infty} \int_{-\infty}^{\infty} d\omega_1\, d\omega_2 \,[f(\omega_1) - f(\omega_2)] \times A_k(\omega_1)\, A_q(\omega_2)\, \delta(\omega_1 - \omega_2 - \omega)$$

In the case of a weak link joining two normal metals, the total current is given simply by the expression obtained by summing I_{kq} over all pairs of states k, q, i.e., Eq. (22). With superconductors, however, the processes involving the two transitions $k\sigma \rightarrow q\sigma$, and $-q, -\sigma \rightarrow -k, -\sigma$ interfere. The extra term involving B spectral functions is the result of this interference.

As might be expected, the normal currents vanish when the applied voltage is zero. This can be deduced from the fact that the product $[f(\omega_1) - f(\omega_2)] \times \delta(\omega_1 - \omega_2 - \omega)$ vanishes when ω is zero.

Let us now consider the supercurrent contribution to Eq. (19), involving B spectral functions. As a consequence of time-reversal symmetry, $T_{kq}T_{-k,-q}$ is equal to $|T_{kq}|^2$ and is therefore real. Since in general $\exp[-i\Delta\phi(t)]\, B_k^+(\omega_1)\, B_q(\omega_2)$ is not real, the argument which showed that the normal currents vanished at zero voltage does not apply to the supercurrent term. Indeed, as long as both sides of the weak link are superconducting, the supercurrent term does tend to a nonzero value as $\omega \rightarrow 0$, and the use of the word supercurrent in describing it is thus justified.

The expression for the supercurrent can be written in the following form:

$$I_s(t) = \text{Im}\{I_1(\omega)\exp[i\Delta\phi(t)]\} = \text{Im}\{I_1(\omega)\exp[i\Delta\phi(0) + 2i\omega t]\} \quad (23)$$

In the case $\omega = 0$ (no applied voltage), there is a constant supercurrent, determined by the phase difference. The weak link thus behaves like an ordinary superconducting wire in this case. The critical current is given by $I_c = |I_1(0)|$. When $\omega \neq 0$ the supercurrent oscillates sinusoidally with angular frequency 2ω. This phenomenon, the alternating supercurrent, is a characteristic property of weakly coupled systems. It is a consequence of the fact that if the coupling is sufficiently weak the state of the system is a *single-valued* function of the phase difference. Because of this, after an interval of time during which the phase difference increases by 2π, the system must return to its original state, and periodic behavior must necessarily occur. On the other hand, when a potential difference is applied between two superconductors joined by a *macroscopic* superconducting wire, the current through the wire instead of oscillating continues to increase until the wire is driven into the normal state.

The current–voltage characteristic of a weak link, if measured at a low frequency in the usual way, consists of two parts, corresponding to zero and nonzero applied voltage. In the first case I_n is zero while I_s has a finite value, and in the second case I_n has a finite value while the dc component of I_s is zero according to Eq. (23). The current measured in a low-frequency experiment in the latter case is therefore I_n. This conclusion is not strictly correct, as we shall see later (Section IV.B.2), since it is based on the assumption of a constant $\Delta\mu$, whereas in reality the ac supercurrent causes the value of $\Delta\mu$ to oscillate.

c. ***Results for the BCS case***

In the BCS limit (Δ_k real and independent of k) simple formulas exist for B_k and B_k^+, and in some cases analytic expressions for $I_1(\omega)$ can be obtained. The results are conveniently expressed in terms of the quantity $V_s(\omega) = R_n I_1(\omega)$, where R_n is the resistance of the weak link when the metals on both sides are in the normal state. If both superconductors have the same energy gap parameter $\Delta(T)$, Ambegaokar and Baratoff (*8*) showed that

$$V_s(0) = (\pi/2)\,\Delta(T)\tanh[\tfrac{1}{2}\beta\Delta(T)] \quad (24)$$

In the case of unequal gaps Δ_1, Δ_2, Anderson (*13*) obtained the result for $T = 0$:

$$V_s(0) = \frac{\pi\Delta_1\Delta_2}{\Delta_1 + \Delta_2} K\left(\left|\frac{\Delta_1 - \Delta_2}{\Delta_1 + \Delta_2}\right|\right) \quad (25)$$

where K is a complete elliptic integral. In the general case $\omega \neq 0$, more complicated expressions involving elliptic functions have been obtained by

Werthamer (*14*). One interesting result is that if smearing of the energy gap is ignored, I_1 diverges logarithmically as the gap voltage is approached (*14,15*).

In cases other than those described above, analytic expressions have not been obtained for $I_1(\omega)$, although a transformation suitable for numerical computation has been described by Ambegaokar and Baratoff (*8*).

Experimentally, the observed zero-voltage critical currents are always less than those predicted, even when conditions are such that the calculations described in this section are expected to be applicable. The form of the temperature variation, however, seems to be in good agreement with that predicted (*16,17*).

d. *Local form of the formula for the current through a weak link*

In many cases we are interested not only in the total current through a weak link but also in its spatial distribution. For example, in the case of a tunneling barrier, situations can occur in which the normal component of current through the barrier has a different direction in different parts of the barrier. The component of the current density normal to the barrier is in fact given by a formula very similar to that for the total current:

$$J = J_s + J_n \tag{26}$$

where J_n depends only on ω, and J_s is given by

$$J_s(t) = \operatorname{Im}\{J_1(\omega)\exp[i\Delta\phi(t)]\} \tag{27}$$

Here ω and $\Delta\phi$ are determined by the *local* values of the chemical potential and phase of the order parameter.

To justify the above statements, we make the assumption that an expression similar to Eq. (15) exists for the current density at a point of the barrier, and that it is local; i.e., when expressed in real space the expression for the current density at a point involves field operators at nearby points only. When the commutator of $J(r)$ is taken with H_T to obtain the expression corresponding to Eq. (17), the result is again local, since G, F, and F^+ are all short-range quantities. The terms involving G functions and F functions can again be interpreted as normal currents and supercurrents, respectively, and the dependence of the F function on the local phase of the order parameter ensures that $J_s(t)$ is given by an expression of the form of Eq. (27). The expression for $J_1(\omega)$ involves matrix elements determined by the properties of the barrier in the vicinity of the point under consideration. It therefore depends on the local value of the resistance per unit area in the same way as $I_1(\omega)$ depends on the total resistance of the barrier.

The local form of the equation for the current through a weak link has the advantage of being applicable even in the presence of magnetic fields, when the momentum-space representation of the Green's functions is no longer applicable. Indeed, for fields small compared with the critical field, the dominant effect of the field is to cause the phases of F and F^+ to vary from place to place. In the formula for the current density, $J_n(\omega)$ and $J_1(\omega)$ can therefore be replaced by their zero-field values and the effect of the change in phase of F and F^+ incorporated into $\Delta\phi$, which must therefore be interpreted as the local value of the phase difference. Equations relating the spatial variation of $\Delta\phi$ to the magnetic field will be derived in Section III.A.

The local equations for the current density may be similarly applied to situations where the potential difference is slowly varying with time. The only modification is to use for $\Delta\phi(t)$ the value determined by Eq. (11). This approximation is valid as long as all Fourier components of the potential difference have frequencies small compared with the gap frequency. The general case where this condition is not satisfied has been considered by Werthamer (*14*).

e. *Gauge invariance*

One complication we have ignored so far is the fact that $\Delta\phi$, being the difference between the phases at two different points in space, is not gauge-invariant. The problems of gauge invariance in fact start with the perturbing term in the Hamiltonian H_T, whose gauge invariance requires that T_{kq} depend on the gauge. This in turn implies that $J_1(\omega)$ depends on the gauge. The gauge dependences of J_1 and $\Delta\phi$ must be such as to leave the combination $J_s = \text{Im}\,\{J_1 \exp[i(\Delta\phi)]\}$ invariant.

The complication of having a J_1 which depends on the gauge can be avoided by redefining $\Delta\phi$ so as to make it gauge-invariant. A suitable definition is

$$\Delta\phi = \phi(P_2) - \phi(P_1) - \frac{2e}{\hbar c}\int_{P_1}^{P_2} \mathbf{A}\cdot d\mathbf{s} \tag{28}$$

where P_1 and P_2 are two adjacent points on opposite sides of the link and the integral is taken along the straight line joining them. We shall always use this definition of $\Delta\phi$ in the following.

With the new definition of $\Delta\phi$, the J_1 required to make Eq. (27) apply is independent of the gauge. The calculations of J_1 described in Section II.A.2.c are valid in the particular gauge $\mathbf{A} = 0$ (since they involve the consequence of time-reversal symmetry $T_{-k,-q} = T_{kq}^*$), and since in this gauge the new and old definitions of $\Delta\phi$ coincide the values of J_1 obtained by these calculations are the correct ones to use in Eq. (27).

B. Green's Function Method

An alternative formula for the current through a weak link, in which the propagation of electrons through the link is described in terms of the electron

Green's function instead of the tunneling Hamiltonian, has been derived by Josephson (*6*). The method does not rely on treating the weak link as a perturbation, and effects due to interaction in the link and the finite time taken for an electron to traverse the link are automatically taken into account. It is therefore a method of more general applicability than that of the tunneling Hamiltonian with the restriction, however, that in its present form it is applicable only to equilibrium situations (i.e., calculation of the zero-voltage supercurrent). We shall content ourselves with a brief outline of the method and a description of a few general consequences which can be derived by its use. The results of the present section will not be required for Sections III and IV.

The derivation initially follows that used by Gor'kov to justify the Ginzburg–Landau equation for the current density in the London limit (*18*). By iterating the Gor'kov equations, the following expression can be obtained for the difference between the values of the single-particle Green's function in the normal and superconducting state (Chapter 6):

$$G_\omega(r, r') - \tilde{G}_\omega(r, r') = - \iint ds\, ds'\, \tilde{G}_\omega(r, s)\, \Delta(s)\, G_{-\omega}(s', s)\, \Delta^*(s')\, \tilde{G}_\omega(s', r') \qquad (29)$$

where G and $\tilde{G}$ are the statistical Green's functions for the superconducting and normal states, respectively.

Since the current density is itself a single-particle Green's function, an expression may be derived from Eq. (29) for the difference between the current density in the superconducting and normal states:

$$J(r) - \tilde{J}(r) = \iint ds\, ds'\, K(r, s, s')\, \Delta(s)\, \Delta^*(s') \qquad (30)$$

where

$$K(r, s, s') = T \sum_\omega G_{-\omega}(s', s) \times \{(ie/m)[\tilde{G}_\omega(r, s)\, \nabla_r \tilde{G}_\omega(s', r) - \tilde{G}_\omega(s', r)\, \nabla_r \tilde{G}_\omega(r, s)] - (2e^2/mc)\, A(r)\, \tilde{G}_\omega(r, s)\, \tilde{G}_\omega(s', r)\} \qquad (31)$$

A factor 2 has been included for summation over spins. Since we are dealing with systems in equilibrium $\tilde{J}(r)$, the current density in the normal state, is zero, so that the right side of Eq. (30) is actually an expression for the current density in the superconducting state.

It may be noted in passing that the short-ranged nature of K provides justification for the locality assumption used in Section II.A.2.d. In addition, Eq. (30) shows explicitly the dependence of $J(r)$ on the phase differences of the order parameter in the vicinity of r.

The most instructive result is obtained when Eq. (30) is integrated over a surface S to obtain the total current through S. A Green's function identity deducible from charge conservation allows the expression obtained to be reduced to the following:

$$I = 2ieT\left(\iint_{s'\epsilon V_1,\, s\epsilon V_2} - \iint_{s\epsilon V_1,\, s'\epsilon V_2}\right) ds\, ds' \times \sum_{\omega} G_{-\omega}(s, s')\, \tilde{G}_{\omega}(s', s)\, \Delta(s)\, \Delta^*(s') \tag{32}$$

where V_1, V_2 are the regions of space on opposite sides of S.

In Eq. (32) phase differences enter in the usual manner. However, in place of the matrix elements which appear in Eq. (19) we have the Green's functions G and $\tilde{G}$, which, because of the restrictions on the regions of integration, are correlation functions relating operators on opposite sides of S.

Time-reversal symmetry plays an important part in permitting supercurrents to flow through weak links, just as it does in the Anderson theory of dirty superconductors (*19*). To see this, consider the situation near the critical temperature, where G and $\tilde{G}$ coincide. If time-reversal symmetry is present, then

$$\tilde{G}_{\omega}(s', s) = [\tilde{G}_{-\omega}(s', s)]^*$$

so that near T_c, $G_{-\omega}(s', s)\, \tilde{G}_{\omega}(s', s)$ is the positive quantity $|\tilde{G}_{\omega}(s', s)|^2$. As a result, the rapid oscillations of sign of the Green's functions (with a period of the order of an inverse Fermi wave vector) do not significantly diminish the supercurrent in the presence of time-reversal symmetry.

A different way of avoiding the use of the tunneling Hamiltonian has been given by de Gennes (*20*). This approach is based on calculating the modification to the Ginzburg–Landau boundary condition produced by the presence of the weak coupling.

III. MACROSCOPIC EQUATIONS FOR TUNNELING BARRIERS

In the preceding section we considered the problem of finding an expression for the current density at a point of a weak link joining two superconductors. This in itself is not sufficient to determine completely the behavior of such a system; it is also necessary to have equations determining the space and time variation of the phase difference and the electromagnetic fields. This section is devoted to the problem of deriving these equations. We shall confine our attention to the case most commonly encountered in practice, that of a tunneling barrier between two superconductors of thickness large compared with the penetration depth.

A. Spatial Variation of the Phase Difference

We shall first derive an equation for the spatial variation of the phase difference using an argument due to Anderson (*13*). Let P_1, P_2 and Q_1, Q_2 be two pairs of points, the members of each pair being adjacent to each other but on opposite sides of the barrier (Fig. 1). The phase differences $\Delta\phi(P)$, $\Delta\phi(Q)$ are defined by Eq. (28):

$$\Delta\phi(P) = \phi(P_2) - \phi(P_1) - (2e/\hbar c)\int_{P_1}^{P_2} \mathbf{A}\cdot d\mathbf{s}$$

$$\Delta\phi(Q) = \phi(Q_2) - \phi(Q_1) - (2e/\hbar c)\int_{Q_1}^{Q_2} \mathbf{A}\cdot d\mathbf{s}$$

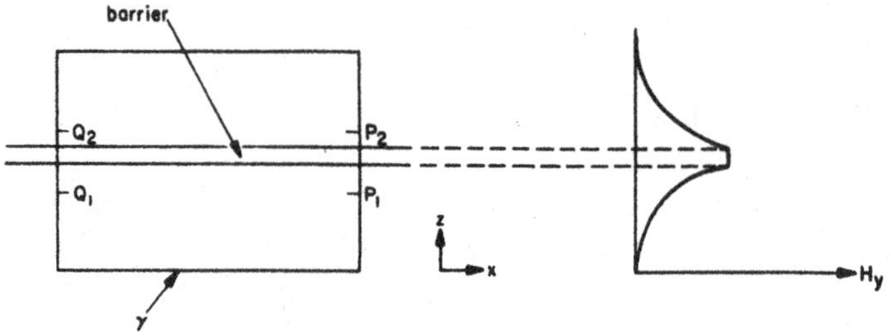

Fig. 1. Geometry involved in discussion of the spatial variation of $\Delta\phi$. The variation of the tangential component of the magnetic field with distance from the barrier is shown in the right-hand diagram.

the integrals being taken over the straight lines joining the end points. Because these definitions are gauge-invariant, any gauge may be used to calculate the quantity $\Delta\phi(P) - \Delta\phi(Q)$. Let us choose a gauge such that the order parameter is everywhere real, i.e., $\phi = 0$. In this gauge there is, in general, a large normal component of vector potential in the barrier. Indeed, since ϕ is zero, the whole of the contribution to $\Delta\phi$ comes from the term involving the integral of the vector potential, so that

$$\Delta\phi(P) - \Delta\phi(Q) = (2e/\hbar c)\left(-\int_{P_1}^{P_2} \mathbf{A}\cdot d\mathbf{s} + \int_{Q_1}^{Q_2} \mathbf{A}\cdot d\mathbf{s}\right) \tag{33}$$

To relate the sum of the integrals to a flux, let γ be a rectangle containing the line segments P_1P_2 and Q_1Q_2, of width sufficiently great that the sides parallel to the plane of the barrier avoid the region of penetration of flux near the barrier (Fig. 1). The flux through γ is given by

$$\Phi_\gamma = -\int_{P_1}^{P_2} \mathbf{A}\cdot d\mathbf{s} + \int_{Q_1}^{Q_2} \mathbf{A}\cdot d\mathbf{s} + \int_{P_1}^{Q_1} \mathbf{A}\cdot d\mathbf{s} - \int_{P_2}^{Q_2} \mathbf{A}\cdot d\mathbf{s} \tag{34}$$

all integrals being taken over portions of γ. The paths over which the last two integrals are taken lie entirely in a superconducting region, and in the particular gauge we are considering their contribution can be neglected. This is because in the gauge where the order parameter is real, the vector potential and current density are to a first approximation proportional to each other, as can be seen from the Ginzburg–Landau equation for the current density. Hence the only contributions to the last two terms of Eq. (34) come from the penetration region where **J** is nonzero, and even in these regions the contribution is small, since the component of **J** normal to the barrier is small.

If we neglect the last two terms of the right side of Eq. (34) and substitute the resulting equation into Eq. (33), we obtain finally the result

$$\Delta\phi(P) - \Delta\phi(Q) = (2e/\hbar c)\,\Phi_\gamma \tag{35}$$

In this chapter we are interested mainly in the application of Eq. (35) to a single barrier between superconductors. However, it should be noted that Eq. (35) is also applicable to the double-barrier systems described in Chapter 8. In that application P is a point of one barrier and Q a point of the other. The proof given above still applies if the long sides of γ are deformed so that they lie within the superconductors and avoid the penetration region near the surface. Φ_γ is then the flux enclosed by the two superconductors and the two barriers.

Let us now return to the case of a single barrier. Suppose that the barrier lies in a plane, and let a coordinate system be chosen such that the plane of the barrier is the xy-plane, and the rectangle γ of Fig. 1 lies in the xz-plane. We then have

$$\Phi_\gamma = \iint_\gamma B_y\,dx\,dz = \int_Q^P dx \int B_y\,dz = \int_Q^P dx\,[(\lambda_1 + \lambda_2 + t)\,H_y]$$

where λ_1, λ_2 are the penetration depths in the two superconducting regions (which will in general be frequency-dependent), t the barrier thickness, and H_y the y component of the field in the barrier. Combining this result with Eq. (35) and differentiating, we obtain

$$\frac{\partial(\Delta\phi)}{\partial x} = \frac{2ed}{\hbar c} H_y \tag{36a}$$

where $d = \lambda_1 + \lambda_2 + t$, and similarly

$$\frac{\partial(\Delta\phi)}{\partial y} = -\frac{2ed}{\hbar c} H_x \tag{36b}$$

Equations (36a) and (36b) are the required equations for the spatial variation of $\Delta\phi$ over a barrier.

B. Time Variation of the Phase Difference

The remaining equations needed for the description of a barrier may be obtained with less difficulty. First, using Eqs. (1) and (28) we have

$$\begin{aligned}\frac{\partial}{\partial t}(\Delta\phi) &= \frac{\partial}{\partial t}\phi(P_2) - \frac{\partial}{\partial t}\phi(P_1) - \frac{2e}{\hbar c}\frac{\partial}{\partial t}\int_{P_1}^{P_2}\mathbf{A}\cdot d\mathbf{s} \\ &= \frac{2e}{\hbar}\left\{\frac{1}{e}[\mu(P_1) - \mu(P_2)] - \int_{P_1}^{P_2}\frac{1}{c}\frac{\partial\mathbf{A}}{\partial t}\cdot d\mathbf{s}\right\}\end{aligned} \tag{37}$$

The quantity in braces, which we shall denote by V, is a gauge-invariant definition of potential difference. It differs by a constant from the quantity

$$V' \equiv \int_{P_1}^{P_2}\mathbf{E}\cdot d\mathbf{s} = U(P_1) - U(P_2) - \int_{P_1}^{P_2}\frac{1}{c}\frac{\partial\mathbf{A}}{\partial t}\cdot d\mathbf{s}$$

where U is the electrostatic potential. It is V and not V' which is zero for a system in equilibrium. With this definition, we may write

$$\frac{\partial}{\partial t}(\Delta\phi) = \frac{2e}{\hbar}V \tag{38}$$

Finally, we require an equation giving the effect of the current through the barrier on the magnetic field. Maxwell's equations tell us:

$$\frac{\partial H_y}{\partial x} - \frac{\partial H_x}{\partial y} = \frac{4\pi}{c}J_z + \frac{1}{c}\frac{\partial D_z}{\partial t} \tag{39}$$

A convenient expression for the displacement current may be found by regarding the barrier as the dielectric of a capacitor. If the effective capacitance of the barrier per unit area is C, then the displacement current density is $C(\partial V/\partial t)$. We therefore obtain

$$\frac{\partial H_y}{\partial x} - \frac{\partial H_x}{\partial y} - \frac{4\pi C}{c}\frac{\partial V}{\partial t} = \frac{4\pi}{c}J_z \tag{40}$$

C. Elimination of Electromagnetic Quantities

From Eqs. (36), (38), and (40) we may eliminate the electromagnetic fields to give the basic barrier equation

$$\nabla^2(\Delta\phi) - \frac{1}{v^2}\frac{\partial^2(\Delta\phi)}{\partial t^2} = \frac{J_z}{I_b} \tag{41}$$

where

$$v = c/(4\pi dC)^{1/2}$$
$$I_b = \hbar c^2/8\pi ed$$

and ∇^2 denotes the operator $\partial^2/\partial x^2 + \partial^2/\partial y^2$. Typical values of v are 2×10^9 cm/sec and of I_b 2.5×10^{-4} emu. The J_z occurring on the right-hand side of Eq. (41) is the current density through the barrier, which is itself a function of $\Delta\phi$ and its time variation, as discussed in Section II. An approximation which is useful for many applications is the zero-frequency limit,

$$J_z = \mathrm{Im}\,[J_1(0)\exp i(\Delta\phi)] = J_1(0)\sin(\Delta\phi) \tag{42}$$

[since $J_1(0)$ is real], for which the barrier equation becomes

$$\nabla^2(\Delta\phi) - \frac{1}{v^2}\frac{\partial^2}{\partial t^2}(\Delta\phi) = \lambda^{-2}\sin(\Delta\phi) \tag{43}$$

where

$$\lambda = [\hbar c^2/8\pi J_1(0)\,ed]^{1/2} \tag{44}$$

As will be seen in Section IV.A.1, λ plays the role of a penetration depth for the barrier. Typical values for tunneling specimens are of the order of $\frac{1}{2}$ mm.

In all processes occurring at a nonzero frequency, dissipative effects occur. It is clear that Eq. (43) does not take these into account, since it is unchanged on replacing t by $-t$. A number of different terms have to be added to Eq. (43) to take into account all the dissipative effects even in lowest order, and we shall not reproduce the resulting equation here.

IV. PROPERTIES OF WEAKLY COUPLED SUPERCONDUCTORS

A. Properties Shared with Bulk Superconductors

As mentioned in Section I, a weak link joining two superconductors behaves in a number of ways like a bulk superconductor. Besides the basic property of being able to pass a supercurrent, a weak link shows a Meissner effect, and a transition to a state analogous to the mixed state of a type II superconductor. As regards the nonstatic properties, in the case of tunneling barriers there is a cutoff frequency below which electromagnetic waves cannot be propagated in the plane of the barrier (analogous to the plasma cutoff frequency) and an electromagnetic disturbance resembling a plasma oscillation. It should be noted that with the exception of the zero-voltage supercurrents, none of these effects has yet been observed experimentally; however, the success of the theory in dealing with more exotic phenomena leaves little doubt that the "classical" phenomena also occur.

1. Meissner Effect

If we consider static situations and linearize Eq. (43) by replacing $\sin(\Delta\phi)$ by $\Delta\phi$, we obtain

$$\nabla^2(\Delta\phi) = \lambda^{-2}(\Delta\phi) \tag{45}$$

This is the Londons' equation in two dimensions. It follows that in small applied fields, where the linear approximation is valid, currents and magnetic fields are confined to a region near the edge of the barrier and fall off with distance from the edge as $\exp(-r/\lambda)$. In other words, there is a Meissner effect. The effect is clearly significant only for barriers whose dimensions are large compared with the barrier penetration depth λ.

2. Mixed State of a Barrier

A particular solution of the full nonlinear equation (43) is

$$\Delta\phi = 2\sin^{-1}\operatorname{sech}[(x-x_0)/\lambda] \tag{46}$$

For $x - x_0 \gg \lambda$, $\sin(\Delta\phi)$ tends to zero, so that Eq. (46) represents a disturbance localized near the line $x = x_0$. Furthermore, the total change in $\Delta\phi$ as x goes from $-\infty$ to ∞ is 2π, so that according to Eq. (35) the flux associated with the disturbance is exactly one flux quantum $hc/2e$. It is clear, then, that Eq. (46) represents a situation where a single quantized flux line is present. The free energy per unit length of such a flux line has been calculated by Josephson (*6*) to be

$$F = [2\hbar^3 J_1(0)\, c^2/\pi e^3 d]^{1/2}$$

By analogy with the treatment of type II superconductors (Chapter 14), it follows that the applied field at which it becomes energetically favorable for a single flux line to enter the barrier is

$$H_{c1} = 4\pi F/(hc/2e) = [32\hbar J_1(0)/\pi e d]^{1/2} \tag{47}$$

This is typically of the order of 1 G.

3. Plasma Oscillations and the Plasma Cutoff Frequency‡

Let us consider solutions of Eq. (43) for which $\Delta\phi$ is independent of position. These are governed by the ordinary differential equation (*13*)

$$(d^2/dt^2)(\Delta\phi) + \omega_0^2 \sin(\Delta\phi) = 0 \tag{48}$$

where $\omega_0 = v/\lambda$. This is the same as the equation of the pendulum. The frequency for small-amplitude oscillations, $\omega_0/2\pi$, is typically a few GHz. From Eq. (36) together with Maxwell's equations it is seen that the magnetic field is zero, while the current and electric field are normal to the barrier. The disturbance

‡ Note added in proof. The plasma resonance discussed in this Section has now been observed by Dahm et al. (*28*).

can therefore be thought of as a plasma oscillation, the low frequency arising from the small density of electrons in the barrier.

Let us now consider a more general type of disturbance, which propagates along the barrier with a finite velocity. If we linearize Eq. (43) by replacing $\sin(\Delta\phi)$ by $\Delta\phi$ and substitute $\Delta\phi = \exp[i(\omega t - \mathbf{k}\cdot\mathbf{r})]$ we obtain the dispersion equation

$$(\omega/\omega_0)^2 = \lambda^2 k^2 + 1 \tag{49}$$

$\omega_0/2\pi$, the frequency of the plasma oscillation, is therefore also the lowest frequency at which small-amplitude oscillations about the $\Delta\phi = 0$ state can be propagated in the plane of the barrier.

One solution of Eq. (49) which is of interest is the case $\omega = 0$, $k = i\lambda^{-1}$, which describes the static Meissner effect.

B. Properties Characteristic of Weakly Coupled Systems

1. Alternating Supercurrent and the Interference of Supercurrents

Weakly coupled superconductors possess two properties related in a very direct way to the basic equations derived in Sections II and III. These are the alternating supercurrent and the interference effect in a magnetic field. The alternating supercurrent arises because a potential difference makes the phase difference time-dependent (Eq. 38), and the interference effect because a magnetic field makes the phase difference vary in space (Eq. 36). Both effects involve periodic behavior (as a function of time in the case of the alternating supercurrent, and as a function of magnetic field in the case of the interference effect) as a direct consequence of the fact that the important physical quantity is a phase difference.

We shall not deal further with these effects here, since they have been described in some detail in Chapter 8. Instead we shall deal with a different topic—the way in which the *alternating* supercurrent may manifest itself in the *static* current–voltage characteristic.

2. Modifications to the Static Current–Voltage Characteristic

In Section II.A.2. we derived the following expression for the current density through a barrier:

$$J = \operatorname{Im}\{J_1(V)\exp[i(\Delta\phi)]\} + J_n(V) \tag{50}$$

(the variable ω has been replaced here by V, which is proportional to it). It was pointed out that if V is a nonvanishing constant, the time average of the first term on the right side of Eq. (50) is zero, since in this case $\Delta\phi$ is a linear function of time. Under this assumption it is only the normal current which is observed in a measurement of the static characteristic. In reality V is time-dependent, because of the presence of the alternating supercurrent. This has an effect even on the contribution of the normal current—rectification occurs because J_n is a nonlinear function of V. However, of greater importance is the effect which the oscillations have in causing the supercurrent to have a nonzero average.

Let us make the approximation of replacing $J_1(V)$ by its zero-voltage value $J_1(0)$. In that case the important result can be proved that the time average at a given point of the barrier of the product VJ_s is zero:

$$\begin{aligned} VJ_s &= \frac{\hbar}{2e}\frac{\partial(\Delta\phi)}{\partial t}\operatorname{Im}[J_1(0)\exp(i\Delta\phi)] \\ &= -\frac{\partial}{\partial t}\operatorname{Re}\left[\frac{\hbar}{2e}J_1(0)\exp(i\Delta\phi)\right] \end{aligned}$$

Since the quantity in brackets is bounded, the result quoted above follows. It may be regarded as an expression of the fact that no energy is dissipated in the supercurrent, since VJ_s is the rate at which energy is supplied to the supercurrent. Now let us divide up both V and J_s into two components, their average value and oscillations about that average, so that $V = V^{\text{dc}} + V^{\text{ac}}$ and $J_s = J_s^{\text{dc}} + J_s^{\text{dc}}$. The result proved above shows that

$$V^{\text{dc}}J_s^{\text{dc}} + \overline{V^{\text{ac}}J_s^{\text{ac}}} = 0$$

so that

$$J_s^{\text{dc}} = -(\overline{V^{\text{ac}}J_s^{\text{ac}}})/V^{\text{dc}} \tag{51}$$

a. ***Effects of microwaves***

There are two situations to which Eq. (51) may be applied. The first is when V^{ac} is an oscillating voltage produced by an externally applied microwave field. If the frequency of the microwave field is equal to that of the alternating supercurrent (or that of one of its harmonics, since nonlinearity causes harmonics to be produced), then $V^{\text{ac}}J_s^{\text{ac}}$ has a nonzero average and a finite J_s^{ac} results. The associated step structure in the current-voltage characteristic was first observed by Shapiro (*21*).

b. ***Effects due to the alternating supercurrent***

The second case is when no microwave field is present and V^{ac} is an oscillating voltage produced by the alternating supercurrent itself. This problem has been studied in some detail in the literature (*6,14,22–26*). Here we shall deal with only the simplest features.

The situation which is most easily analyzed is the limit of small $J_1(0)$. In this case V^{ac} is small compared with V^{dc}, so that we may obtain a first approximation to J_s by regarding V as a constant. This leads to the result

$$J_s(\mathbf{r}, t) = \mathrm{Im}\,[J_1(0)\exp\{i[\Delta\phi(\mathbf{r}, 0) + 2eV^{dc}t/\hbar]\}] \tag{52}$$

$\Delta\phi(\mathbf{r}, 0)$ being given in terms of the static magnetic field by Eq. (36). In this approximation J_s contains Fourier components of one frequency only. The only effect of changing V^{dc} is to alter the oscillation frequency of J_s^{ac}, without altering its magnitude or spatial distribution. Now according to Eq. (51), J_s^{dc} (and hence also the dc component of the total supercurrent) is determined by the in-phase component of V^{ac}, so that we may write

$$I_s^{dc} \propto [J_1(0)]^2 R(2eV^{dc}/\hbar)/V^{dc} \tag{53}$$

where $R(\omega)$ is the real (dissipative) part of the electromagnetic response function relating current and voltage [for currents with the particular spatial distribution implied by Eq. (52)]. I_s^{dc} is proportional to $[J_1(0)]^2$ because both J_s^{ac} and V^{ac} are proportional to $J_1(0)$.

The excess current I_s^{dc} is seen to be strongly dependent on the electromagnetic properties of the barrier. The first observations of excess currents of this type were made by Fiske (*16*), who found step-like characteristics such as that shown in Fig. 2. In such cases the electromagnetic response has sharp resonances at frequencies determined by the dimensions of the barrier. The resonant response is reflected as a peak in current at particular values of the applied voltage. Generally only the positive resistance portion of the peak (which is the low-voltage side) is observed, and as soon as the top of one peak is passed the specimen jumps on to the next resonant peak. This accounts for the observation of step structure rather than a series of peaks.

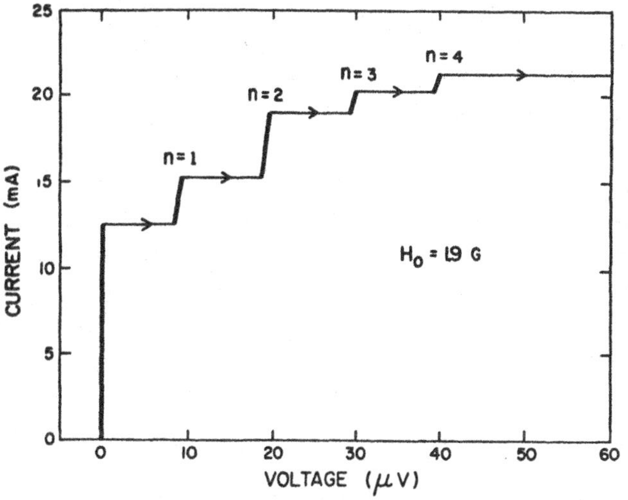

Fig. 2. Step-like structure in the I–V characteristic of a barrier. [After Langenberg et al. (*27*).]

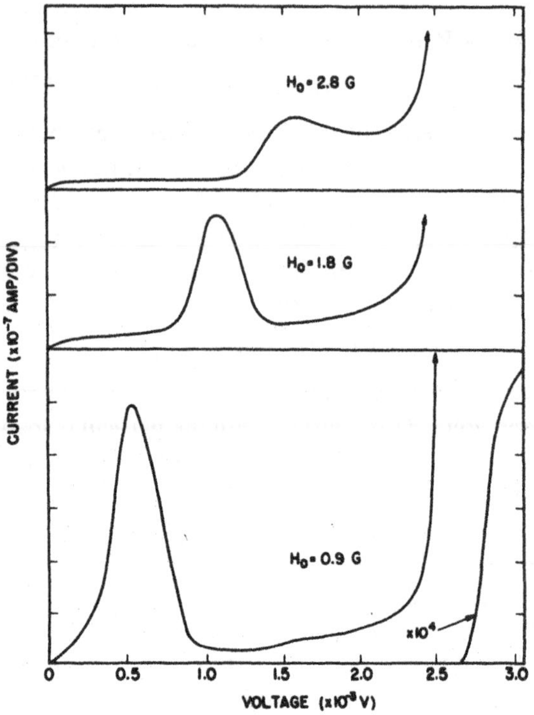

Fig. 3. Field-dependent resonant peak in the *I–V* characteristic of a barrier. [After Eck et al. (*22*).]

The above interpretation of structure such as that in Fig. 2 is confirmed in many ways. For example, when a magnetic field is applied the size of the steps changes (because of the change in the phase relationships which determine the coupling to any given resonance), but the voltages at which the steps occur remain unaltered (being determined only by the resonant frequencies). On the other hand, when the temperature is varied, the altered penetration depth affects the resonant frequencies, and hence also the step voltages.

A different effect is sometimes observed in moderate magnetic fields. This results from an enhancement of the electromagnetic response due to a velocity-matching effect. To see how this arises, consider a situation where a steady voltage V and magnetic field $(H_x, H_y, 0)$ are present simultaneously in the barrier. The phase difference is given by

$$\Delta\phi = \frac{2ed}{\hbar c}(H_y x - H_x y) + \frac{2e}{\hbar} Vt + \text{const.} \tag{54}$$

The lines of constant phase difference therefore move with velocity $Vc/|H|d$. When this velocity is close to the velocity of propagation of electromagnetic waves, which in the small J_1 limit is just v, the voltages produced by the alternating supercurrent are enhanced, with a corresponding increase in the steady current I_s^{dc}. An example of this type of behavior, observed by Eck et al. (*22*), is seen in Fig. 3. As the magnetic field is varied, the voltage of the peak changes in such a way that the phase velocity $Vc/|H|d$ remains constant. In this example the individual electromagnetic resonances are so much broadened that they overlap and only the velocity-matching resonance can be seen. It is possible to observe both multistep characteristics and characteristics with a single resonant peak in the same specimen, by varying the applied field (*23*).

ACKNOWLEDGMENT

The author wishes to thank Mr. P. C. Wraight for reading and suggesting improvements in the manuscript.

REFERENCES

Besides the detailed references given throughout the chapter, a general discussion of weakly coupled superconductors will be found in (*6*) and (*13*).

1. P. W. Anderson, N. R. Werthamer, and J. M. Luttinger, *Phys. Rev.* **138**, A1157 (1966).
2. P. W. Anderson and A. H. Dayem, *Phys. Rev. Letters* **13**, 195 (1964).
3. J. E. Zimmerman and A. H. Silver, *Phys. Letters* **10**, 47 (1964); *Phys. Rev.* **141**, 367 (1966).
4. P. H. Smith, S. Shapiro, J. L. Miles, and J. Nicol, *Phys. Rev. Letters* **6**, 686 (1961).
5. R. A. Ferrell and R. E. Prange, *Phys. Rev. Letters* **10**, 479 (1963).
6. B. D. Josephson, *Advan. Phys.* **14**, 419 (1965).
7. B. D. Josephson, *Phys. Letters* **1**, 251 (1962).
8. V. Ambegaokar and A. Baratoff, *Phys. Rev. Letters* **10**, 486 (1963); erratum **11**, 104 (1963).
9. M. H. Cohen, L. M. Falicov, and J. C. Phillips, *Phys. Rev. Letters* **8**, 316 (1962).
10. R. E. Prange, *Phys. Rev.* **131**, 1083 (1963); *Lectures on the Many-Body Problem*, Vol. 2, Academic Press, New York, 1964, p. 137.
11. L. P. Kadanoff and G. A. Baym, *Quantum Statistical Mechanics*, Benjamin, New York, 1962.
12. J. R. Schrieffer, D. J. Scalapino, and J. W. Wilkins, *Phys. Rev. Letters* **10**, 336 (1963).
13. P. W. Anderson, *Lectures on the Many-Body Problem*, Vol. 2, Academic Press, New York, 1964, p. 113.
14. N. R. Werthamer, *Phys. Rev.* **147**, 255 (1966).
15. E. Riedel, *Z. Naturforsch.* **19a**, 1634 (1964).
16. M. D. Fiske, *Rev. Mod. Phys.* **36**, 221 (1964).
17. C. B. Satterthwaite, M. G. Craford, R. N. Peacock, and R. P. Ries, *Low Temperature Physics, LT9* (J. G. Daunt et al., eds.), Plenum Press, New York, 1965, p. 443; I. K. Yanson, V. M. Svistunov, and I. M. Dmitrenko, *Zh. Eksperim. i Teor. Fiz.* **47**, 2091 (1964); *Soviet Phys. JETP* **20**, 1404 (1965).
18. L. P. Gor'kov, *Zh. Eksperim. i Teor. Fiz.* **36**, 1918 (1959); *Soviet Phys. JETP* **9**, 1364 (1959).

19. P. W. Anderson, *J. Phys. Chem. Solids* **11**, 26 (1959).
20. P. G. de Gennes, *Phys. Letters* **5**, 22 (1963); *Rev. Mod. Phys.* **36**, 225 (1964).
21. S. Shapiro, *Phys. Rev. Letters* **11**, 80 (1963).
22. R. E. Eck, D. J. Scalapino, and B. N. Taylor, *Phys. Rev. Letters* **13**, 15 (1964).
23. R. E. Eck, D. J. Scalapino, and B. N. Taylor, *Low Temperature Physics, LT9* (J. G. Daunt et al., eds.), Plenum Press, New York, 1965, p. 415.
24. D. D. Coon and M. D. Fiske, *Phys. Rev.* **138**, A744 (1965).
25. I. M. Dmitrenko and I. K. Yanson, *Soviet Phys. JETP* **22**, 1190 (1965).
26. I. O. Kulik, *Soviet Phys. JETP Letters* **2**, 84 (1965).
27. D. N. Langenberg, D. J. Scalapino, B. N. Taylor, and R. E. Eck, *Phys. Rev. Letters* **15**, 294 (1965).
28. A. J. Dahm, A. Denenstein, T. F. Finnegan, D. N. Langenberg, and D. J. Scalapino, *Phys Rev. Letters* **20**, 859 (1968).

10

THE ELECTRON–PHONON INTERACTION AND STRONG-COUPLING SUPERCONDUCTORS

Douglas J. Scalapino

DEPARTMENT OF PHYSICS
UNIVERSITY OF PENNSYLVANIA
PHILADELPHIA, PENNSYLVANIA

I. INTRODUCTION

In their original work, Bardeen, Cooper, and Schrieffer (*1*) developed the consequences of their pairing hypothesis for a simple model of an electron gas with attractive interactions near the Fermi surface. The results which they obtained depended upon one parameter, the transition temperature, and led to a law of corresponding states. The remarkable manner in which these results explained the physical properties of superconductors has been reviewed by

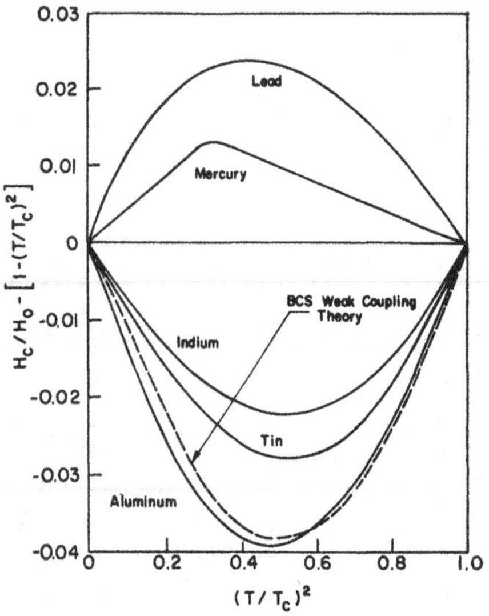

Fig. 1. Deviation of the reduced critical field from a parabolic curve as a functon of the square of the reduced temperature. The experimental curves are taken from (*5*) and (*6*).

Bardeen and Schrieffer (*2*) and is discussed in Chapter 3 of this treatise. This success implies that details of the interactions of electrons and phonons in a given metal must be sought in deviations of observed properties from the original BCS predictions. This chapter is devoted to an analysis of these deviations and their relationship to the electron–phonon interaction in metals.

One measure of the strength of the electron–phonon interaction is the ratio of the superconducting transition temperature to the Debye temperature. Another, more precise criterion is the size of the phonon enhancement of the effective electron mass at the Fermi surface. Using these criteria, one expects that Pb and Hg have large electron–phonon interactions compared with Al, Sn, or In, and should therefore show the clearest deviations from the BCS results. This is the case, and for this reason much of our discussion will be concerned with the strong-coupling superconductors Pb and Hg. As examples of the deviations which will be investigated, let us consider some thermodynamic properties. In the weak-coupling limit,‡ which is valid for most superconductors, BCS predicts that the ratio $2\Delta(0)/kT_c$ should be 3.53. The observed values of this ratio are 4.3 for Pb (*3*) and 4.6 for Hg (*4*). The observed superconducting condensation energy of Pb is about 25% lower than the predicted BCS value.

‡ Here weak coupling means that the strength of the BCS pairing interaction $N(0)V$ is less than or of order $\frac{1}{4}$.

The temperature dependence of the critical magnetic fields for a number of superconductors (*5,6*) is shown in Fig. 1. Pb and Hg have a positive deviation from the parabola predicted from the two-fluid model while BCS gives a negative deviation in the weak-coupling limit. The size of the specific-heat jump at the transition temperature is also significantly larger for Pb and Hg than the BCS prediction (*7*).

Turning to the transport properties, the ratio of the superconducting to normal thermal conductivities (*8*) of Pb and Hg has a limiting slope at T_c of order 10, while calculations (*9,10*) based upon the simple BCS model give a value of order 1.5. This latter value is in good agreement with the observed behavior of Al, Sn, and In. The tunneling characteristics of Pb (*11*) and Hg (*4*) plotted in Fig. 2 show structure near the characteristic phonon energies of the metals which deviates from the dashed curve, giving the BCS dependence. It was, in fact, these tunneling deviations in Pb (*3,12*) which provided the strongest motivation for developing a theory which could account for the details of the electron-phonon interaction in metals. In Chapter 11 a number of superconductors besides Pb and Hg are discussed which exhibit clearly resolved structure in their tunneling characteristics.However, since the size of the deviations from the BCS prediction is of order $(\Delta/\omega_D)^2$, it requires great care to observe such structure for the weak-coupling superconductors.

There are, of course, deviations in other materials and other properties. The isotopic dependence of T_c gave the first clear indication that the electron–phonon interaction was responsible for superconductivity (*13,14*). Deviations of the isotope factor from $\frac{1}{2}$ give information on the relative strengths of the phonon exchange and the direct Coulomb interactions in various metals. In addition, measurements of the specific heat when combined with measurements of the transition temperature and the isotope factor give useful information on the effective mass corrections associated with the electron–phonon coupling.

Another example of the type of deviation which bears on the problem of the underlying interactions is the nuclear spin relaxation rate. Hebel and Slichter (*15,16*) found that the ratio of the nuclear spin-lattice relaxation rate in the superconducting phase of Al to that in the normal phase initially increased below T_c, going through a peak near a reduced temperature T/T_c of order 0.8. The simple BCS model predicted that an increase of this ratio should in fact be observed initially below T_c. However, the prediction gives a logarithmic divergence of this ratio as T approaches T_c from below. This can, of course, be washed out by gap anisotropy and spatial inhomogeneities in the metal. However, it can also be washed out by damping associated with the electron–phonon interaction, as Fibich first pointed out (*17*). This latter case along with related effects which give rise to a temperature-dependent slope of the superconducting I–V characteristics at a voltage $2\Delta/e$ and to quasi-particle recombi-

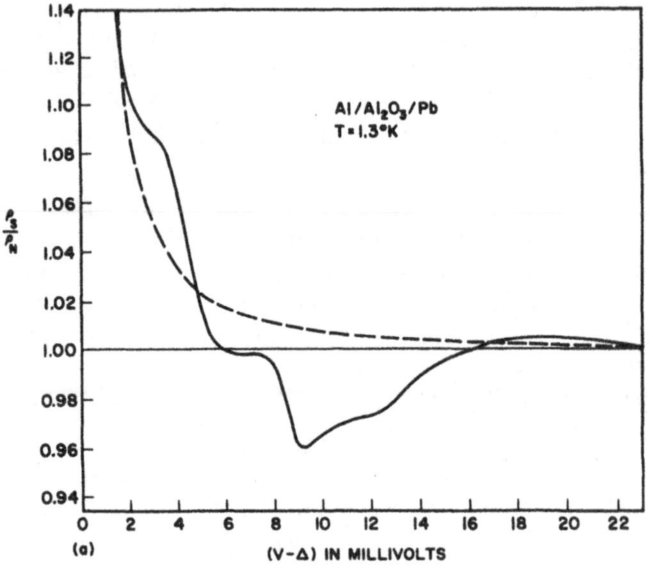

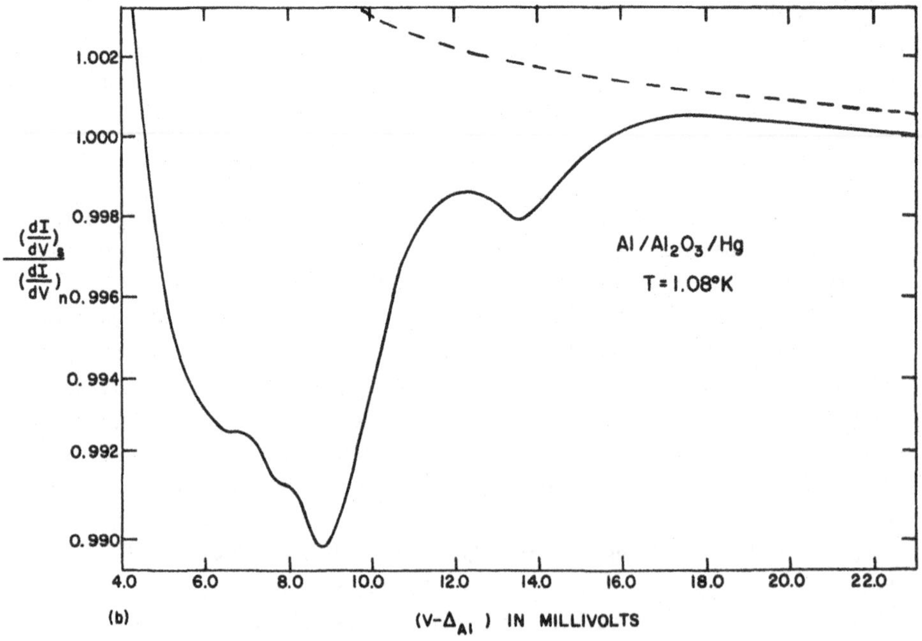

Fig. 2. (a) Normalized conductance $(dI/dV)_s/(dI/dV)_N$ vs. voltage for a Pb–I–Al junction, after Rowell and Kopf (*11*). (b) A similar plot for a Hg–I–Al junction, after Bermon and Ginsberg (*4*). In both figures the dashed line represents the BCS prediction.

nation lifetimes (*18*) depends upon the low-energy details of the electron–phonon interaction.

To understand the significance of these deviations it is important to recognize the physics contained in the BCS reduced Hamiltonian. This is a quasi-particle Hamiltonian in which, to a large degree, the normal-state correlations have been taken into account by using a modified dispersion relation and an effective electron–electron interaction. Furthermore, the pairing correlations are severely restricted by the requirement of correlated occupancy of time reversed quasi-particle states. In this way, the correlations specific to the superconducting phase are isolated and taken into account in a nonperturbative manner. Deviations from the simple BCS theory can arise in at least three ways: (1) the quasi-particle description can become inadequate if, for instance, the damping rate should become comparable with the excitation energy; (2) the assumption of an effective two-body instantaneous interaction between the quasi-particles may not provide an adequate representation of the retarded nature of the phonon-induced interaction; or (3) the pairing hypothesis may break down.

Since the attractive interaction arises from phonon exchange, it is resonant for energy transfers of order ω_D. This means that an important part of the pairing energy arises from virtually excited quasi-particles of energy ω_D relative to the Fermi energy. For the strong-coupling superconductors, the lifetime of a quasi-particle with excitation energy ω_D is so short that its level width is of the order of the excitation energy and the quasi-particle picture fails. In addition, the detailed space–time dependence of the effective electron–electron interaction is significant in these materials. Therefore, the first two possibilities for deviations clearly exist for Pb and Hg. The third possibility will not be considered in the following analysis except to the extent that the normal and pairing correlations will be treated on the same footing. The excellent agreement of the results obtained in this chapter and Chapter 11 with detailed experimental observations provides a posteriori justification for neglecting the third possibility. The fact that the deviations from the original BCS predictions can be accounted for in this manner gives further testimony, if indeed any is needed, to the sound physical basis of the BCS pairing theory of superconductivity.

The theory of the strong-coupling superconductors involves the solutions of a complicated, strongly interacting, many-body problem. In Section II, a theory of metals is reviewed with emphasis on the physical bases of the approximations which allow solutions to be constructed. These approximations rest upon the characteristic space–time properties of the system and the Pauli principle. The distinguishing feature of this theory is its almost continuous contact with experimentally measurable quantities. Besides providing a natural framework for correlating results from a number of different experiments, this feature allows the complex many-body problem to be separated into smaller, more tractable pieces. Then, experimental observations and relative comparisons rather than first-principle calculations can be used in all the pieces except the particular one of interest. In this manner, attention can be focused on a limited aspect of the overall problem and experimental data used as a starting point and a check point for the theory of that part. When this part of the problem is adequately understood, it can be used in treating further parts, until, hopefully, an accurate basis for calculating the properties of the strong-coupling superconductors is obtained. This approach is analogous to the manner in which one trouble shoots a complicated piece of electronics. A known signal is applied to a part of the circuit and its output is checked. If it is satisfactory, further parts are checked and finally the overall response is determined. The necessity of having a theoretical model which allows this type of separation, and checking can be appreciated by remembering that the energy difference between the normal and superconducting phases corresponds to roughly 10^{-7} eV per atom while kinetic energy and Coulomb interactions are of order 1 eV per atom. Added to this is the desirability of calculating superconducting properties to accuracies of a few per cent.

In Section III this theory is applied to construct equations for the electron self-energy. The self-energy forms the basic object from which a number of the physical properties of the strong-coupling superconductors can be determined. The relationship between the electron self-energy and the electron–phonon coupling is discussed and some properties of the electron–phonon interaction are noted. Then using a model of the electron–phonon interaction obtained from neutron scattering and electron tunneling data for Pb, explicit solutions for the self-energy of Pb are given. The structure of the electron–phonon and Coulomb interactions is clearly visible in the form of the self-energy. To complete this discussion, the results for the self-energy are checked against the electron tunneling data. The main features of the phonon density of states, the effective electron–phonon coupling functions as well as such details as Van Hove singularities, turn out to be clearly identifiable and good agreement is found between the theoretical predictions and the tunneling experiments. In Chapter 11, McMillan and Rowell discuss how they have developed this approach into a very sensitive probe of the electrons, phonons, and their interactions in metals.

In the final part of this chapter, Section IV, theoretical results for a number of the thermodynamic and the dynamic properties of the strong-coupling superconductors are shown to be in good agreement with the observed values. This agreement gives further support to the underlying theory and allows the deviations from the original BCS predictions to be used in obtaining information on the electron–phonon interaction.

II. THEORY OF NEARLY FREE ELECTRON METALS

During the past 10 years a physically realistic and mathematically tractable theory of the "nearly free electron" metals has evolved. To discuss the electron–phonon coupling and the detailed properties of strong-coupling superconductors it is necessary to review this theory. The underlying physical ideas are simple but some of the details are algebraically tedious. For this reason we will try to discuss the important physical aspects of the theory in this introduction, leaving a somewhat deeper review for the following subsections. Let us begin with some observations and then go on to the theoretical ideas which provide the present framework for our understanding.

That some of the alkali metals should have nearly spherical Fermi surfaces is perhaps not too surprising. However, Gold's suggestion (*19*) that the de Haas–van Alphen data on Pb could be understood on the basis of a nearly free electron picture was remarkable. Additional work by Anderson and Gold (*20*) confirmed in detail the correctness of a nearly free electron Fermi surface based on four conduction electrons per Pb atom. Further experimental Fermi surface measurements on polyvalent metals indicated that the underlying periodic ion potential had only a small effect on the conduction electron band structure. In both the alkali and polyvalent metals the band structure appears to be principally parabolic (*21*). Besides the Fermi surface, a number of thermodynamic and transport properties of these metals can be explained on the basis of an independent-electron picture. Yet one knows that the Coulomb interaction between the electrons must produce strong electron–electron correlations. It is perhaps remarkable that a Fermi surface can exist at all in a strongly interacting many-electron system.

As we noted in Section I, there is a great disparity in the energies associated with various correlations in metals. The Coulomb correlation energy in a typical metal is of order 1 eV per atom while the electron self-energy correlations associated with the electron–phonon interaction amount to about 10^{-4} eV per atom. The correlations specific to the superconducting phase are of order 10^{-7} eV per atom. In this chapter, calculations of some of these correlation effects in Pb using the phonon spectrum determined by neutron scattering and the low-temperature value of the superconducting gap as input data will be discussed. They give results for the specific heat, the superconducting transition temperature, the critical field, the tunneling I–V characteristic, etc., in good agreement with the experiment. Results from electron tunneling experiments can be used in the theory to obtain even better agreement. In addition, these same tunneling measurements can be used to determine an average electron–ion interaction for Pb which agrees with the interaction Gold and Anderson used to obtain the deviations from sphericity of their Fermi surface data.

It is clear from these remarks that the theory must explain why an independent, nearly free electron picture provides such a good description of these metals. In addition, it must clarify how some of the remnant correlations can be so accurately taken into account. The bones of the physics underlying the theory are the characteristic lengths and times appropriate to the simple metals, and the Pauli principle. These are dressed with the Green's function formalism which provides a natural description of the important ideas of elementary excitations, screening, and correlations. To begin with, the electrons and nuclei which make up the metal can be reduced to a more tractable conduction electron–ion core system by considering the length and time (energy) scales associated with the core states in the simple metals. The excitation energy of the core states is large compared with the other energies of the system, so that it is reasonable to treat the core electrons as adiabatically following the nuclei. Any influence of core polarization on the ion–ion interaction can be taken into account by modifying the effective ion–ion potential. In the alkali and polyvalent metals the radius of the ion core is small compared to the lattice dimension. Outside the core region the screened ion potential is weak, so that over the bulk of the metal the conduction electrons are plane-wave-like. In the core region where the potential is strongly attractive, the Pauli principle requires the conduction electron wave function to be orthogonal to the core states. Phillips and Kleinman (*22*) showed that the net effect of the kinetic energy arising from this orthogonalization and the attractive ion potential on the conduction electron wave function outside the core region could be simulated by a weak pseudo-potential. The fact that the pseudo-potential generates a fictitious wave function inside the ion core is of no consequence because of the great disparity of the ion-core radius to the lattice dimension. Further discussion of this is given in Section II.A and the Hamiltonian describing the simplified model of conduction electrons interacting with pseudo-potential ions is discussed in Section II.B.

The pseudo-potential ideas provide part of the answer to the origin of the nearly free electron behavior of the simple metals. However, the independent-particle picture requires justification because of the existence of strong Coulomb correlations. Here, Landau (*24*) suggested that the low-lying excited states of a strongly interacting system could be constructed from independent elementary excitations. For the electron system these excitations are quasi-particles, characterized by momentum and spin. Once again the Pauli principle enters as a restriction on the phase space available for decay, and stabilizes the low-energy quasi-particle structure associated with Coulomb correlations. The many-body correlations inherent in Landau's description modify the energy dispersion relation and the interactions between the quasi-particles. An outline of Landau's theory is given in Section II.C.

The detailed analysis of the Coulomb and electron–phonon many-body correlations is based upon the Green's function formalism discussed in Chapter 5.

The quantities of direct experimental interest are naturally expressed in terms of Green's functions and formal expressions relating a variety of properties can be constructed. The physical basis for the approximations used in obtaining many of the Green's function relations as well as the explicit solutions are once again related to the space–time characteristics of the interactions and to the Pauli principle. The characteristic distances associated with the Coulomb and the electron–phonon interactions are both of the order of the Thomas–Fermi screening length (i.e., angstroms). However, the time scales of the two interactions are grossly different. The time scale‡ of the Coulomb interactions is of the order of the inverse Fermi energy E_F^{-1}, while that of the electron–phonon interaction is of the order of the inverse Debye energy ω_D^{-1}. This means that the modifications produced by the Coulomb interaction will be spread out over a momentum range determined by the inverse of the Thomas–Fermi screening length, and an energy range (relative to the Fermi energy) determined by E_F. This is in contrast to the electron–phonon correlation effects, which have the same momentum range but are restricted to an energy range of order ω_D. For excitation energies larger than ω_D the electron–phonon renormalization effects become negligible simply because the lattice response time is too slow to follow them. These two effects can therefore be separated to order $\omega_D/E_F \sim \sqrt{m/M}$, where m is the electron mass and M is the ion mass. This is just the Born–Oppenheimer expansion parameter, and the separation reflects the ability of the electron to follow the ion motion adiabatically. Migdal (*25*) was the first to show how this parameter could be used to simplify the Green's function analysis of the electron–phonon system. This is essential, since electron excitations of the electron–phonon system with energy of order ω_D decay too rapidly for the Landau quasi-particle picture to be applicable. Section II.E is devoted to a discussion of Migdal's theorem.

Migdal's results can be used to show that the effects of the electron–phonon interaction on the band structure and the electron–phonon coupling are of order $\sqrt{m/M}$, because the important momentum transfers involve electron excitation energies much greater than ω_D. Furthermore, the frequency derivative of the electron self-energy associated with the electron–phonon interaction is large at the Fermi surface while the momentum dependence, which is spread over a scale of the Thomas–Fermi screening wave vector k_s, is negligible. This has a number of experimental consequences which have been discussed by Prange and Kadanoff (*26*). In particular, the fact that the phonon part of the electron self-energy is weakly momentum-dependent but strongly frequency-dependent near the Fermi surface implies that experiments which do not depend on the electron mass but depend only upon a mobility τ/m are unaffected by renormalization effects. This is because the mean free time is renormalized by the

‡ Throughout this chapter we use units in which $\hbar = 1$.

same factor as the mass when the self-energy is only dependent upon the frequency. Some results obtained by Englesberg and Schrieffer (*27*) for the normal electron self-energy due to the electron–phonon interaction are discussed in Section II.F.

Probably the most difficult aspect of the nearly free electron theory is the evaluation of the Coulomb correlation effects. Here there are at present two approaches: (1) the use of experimental data to replace direct calculations of these effects,‡ and (2) the high-density expansion of the electron gas properties developed from the work of Bohm and Pines (*28*) and Gell-Mann and Brueckner (*29*). This approximation also depends upon a characteristic length, the interelectron spacing. The expansion parameter r_s in this theory is the ratio of the interelectron spacing to the Bohr radius. When this is small, corresponding to a high density of electrons, the average electron kinetic energy is large compared to the screened Coulomb interaction between the electrons. Unfortunately the densities appropriate to real metals are such that r_s is greater than unity. In this case, further corrections describing the effective electron–electron interaction at close distances, large momentum transfers, become important. The restrictions due to the Pauli principle appear to be important and Hubbard (*30,31*) has suggested an approximate procedure for taking them into account. In Section II.D, some results obtained by Rice (*32*) using Hubbard's approximation are summarized.

In the analysis of the superconducting state discussed in Section III, the effective electron–electron interaction plays a central role. This can be separated into a screened Coulomb and phonon exchange part. The large momentum range $\sim k_s$ of both the Coulomb and electron–phonon interaction implies that the electron–electron interaction is short range in space. Since the important electrons are moving at the Fermi velocity, the screened Coulomb interaction is essentially instantaneous while the phonon exchange is retarded. In the superconducting state a new length, the coherence length ξ determining the range of the pairing correlations, becomes important. In addition, the energy gap Δ, related to ξ by the Fermi velocity, provides a new time dimension. The effect of the coherence length is to make the electron self-energy strongly momentum-dependent. It now has a momentum variation within a small region ξ^{-1} about the Fermi surface as well as a frequency variation over a region Δ about the Fermi energy. This structure is a direct reflection of the pairing correlations. Moreover, because the effective electron–electron interaction reflects the phonon frequency distribution, the pairing correlations are frequency-dependent. Because of the momentum variation, the effects of the frequency structure can be observed in a variety of thermodynamic and dynamic processes.

‡ Recent observations of electromagnetic waves in metals show promise of giving precise information on some of the Landau Fermi liquid parameters (*103–105*).

A. Pseudo-Potential

In the alkali and polyvalent metals the ion-core states are localized about the nuclei over distances small compared to the interatomic separation and are relatively tightly bound on the scale of conduction band energies. This means that to a good approximation the internal dynamics of the core electrons can be neglected. They can be treated as generating an ionic potential rigidly attached to the nuclei. At distances larger than the core radius r_c, the bare ionic potential has the simple Coulomb form Ze/r, where Z is the effective ionic charge. This is screened out by the conduction electrons at distances larger than the interatomic spacing. At distances less than r_c, the shielding due to the core electrons decreases until near the nucleus, the full positive potential of the nucleus is obtained.

At normal metallic densities the ion cores are sufficiently well separated that they interact through the conduction electron screened $(Ze)^2/r$ part of the potential. The conduction electrons, on the other hand, can penetrate the core and sample the strong attractive Coulomb potential inside the core radius. This interaction is so strong that any simple perturbation theory starting from plane-wave conduction electron states should be useless. However, it is known that even in the polyvalent metals such as Pb the de Haas–van Alphen data can be explained on the basis of a nearly free electron picture (*20*). In the extended zone scheme the Fermi surfaces of the alkali and polyvalent metals are remarkably spherical.

The resolution of this paradox is due to the recognition of the important role played by the Pauli principle in the core region. It requires that the conduction electron states be orthogonal to the core states. This produces strong oscillations of the conduction band states in the core region and the kinetic energy associated with these oscillations effectively cancels the strong Coulomb attraction. Phillips and Kleinman (*22*) showed that the effect of this orthogonalization on the conduction electron wave function *outside* the core region could be simulated by adding a repulsive nonlocal interaction $(E_K - H)P$ to the ion potential. Here E_k is the conduction electron energy eigenvalue and P is the core state projection operator

$$P = \sum_c |c\rangle\langle c|$$

where c is a core state. To the extent that these states form a complete set over the region of the core, the large ion potential is canceled by $(E_k - H)P$ in the core region. Figure 3 shows some results of a pseudo-potential calculation for Si by Cohen and Heine (*23*). The solid line is the effective pseudo-potential charge and the dashed line is the effective ion-potential charge. Note the cancellation for r less than r_c. Here the pseudo-potential has been reduced to a local

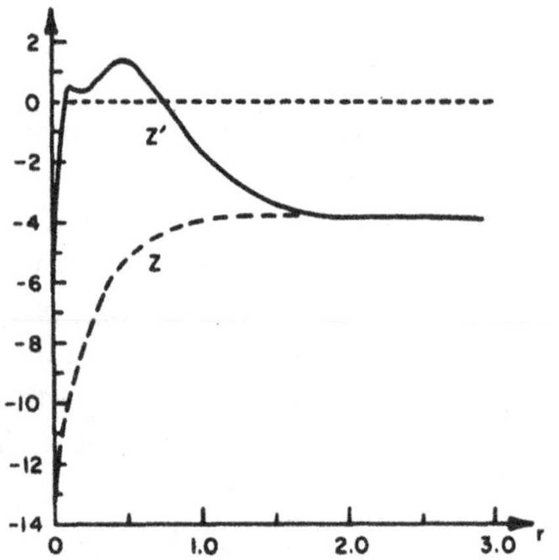

Fig. 3. Effective pseudo-potential charge Z' (solid) and the effective ion potential charge Z (dashed) vs. position for the valance electron in Si^{3+}. [After Cohen and Heine (*23*).]

potential appropriate to *s*-wave scattering. In actual practice a pseudo-potential can be constructed so that this cancellation is quite efficient for r less than r_c, and negligible for r greater than r_c, leaving a weak screened Coulomb potential outside r_c.‡ This means that the behavior of the pseudo-wave function which simulates the actual conduction band state outside the core region can be obtained by using perturbation theory. For phase-space reasons it is the behavior of the conduction electron wave function outside the ion cores which determines the properties of the system. This provides the first step in the justification for the observed nearly free electron behavior of the simple metals.

There exist a number of detailed pseudo-potential calculations. Here we want to mention two simple analytic forms which have been suggested for the bare pseudo-potential. Harrison (*33,34*) has shown that the results of orthogonalized plane-wave calculations for a number of simple metals can be represented by a potential of the form

$$-(Ze^2/r) + \beta\delta(r) \tag{1}$$

The first term represents the attractive contribution of the Coulomb field of the net ion charge Ze. The δ-function term has a positive adjustable parameter β which simulates the repulsive effects of core orthogonalization. Ashcroft (*35*) has suggested a form which takes the cancellation of the core potential into

‡ A more appropriate description of the "weakness" of the pseudo-potential is that its Fourier transform converges much more rapidly at large q than the bare interaction, because the strong attractive core part has been removed. Therefore, fewer plane waves are required to construct a reasonable representation of the electron wave function.

account in a direct manner:

$$V(r) = \begin{cases} -Ze^2/r & r > r_c \\ 0 & r < r_c \end{cases} \tag{2}$$

Here r_c is treated as an adjustable parameter which in fact turns out to have values in reasonable agreement with the ionic radii. We will return to these simple pseudo-potentials when we discuss the electron–phonon interaction. It is perhaps worth emphasizing that Eqs. (1) and (2) are approximate forms for the *bare* pseudo-potentials. In calculating the band structure of the pseudo-conduction electrons these interactions must be appropriately screened and renormalized as discussed in Section II.D.

B. Electron–Ion Hamiltonian

Using the idea of a (local) bare ion pseudo-potential $V(x)$, a simple Hamiltonian which describes the dynamics of the pseudo-conduction electrons and the ion cores is

$$\mathscr{H} = \int d^3x\, \psi^\dagger(x)\left[-\frac{\nabla^2}{2m} + \sum_\nu V(x - R_\nu)\right]\psi(x) + \sum_\nu \frac{P_\nu^2}{2M}$$
$$+ \tfrac{1}{2}\sum_{\nu \neq \mu} \tilde{V}(R_\nu - R_\mu) + \tfrac{1}{2}\int d^3x\, d^3x'\, \psi^\dagger(x)\,\psi^\dagger(x')\,\frac{e^2}{|\mathbf{x} - \mathbf{x}'|}\,\psi(x')\,\psi(x) \tag{3}$$

Here $\psi(x)$ is the pseudo-conduction electron field and R_ν and P_ν are the conjugate position and momentum variables of the ion cores. The first term in $\mathscr{H}$ contains the electron kinetic energy and the ion-core interaction. The second and third terms represent the ion kinetic and the ion–ion interaction energies, respectively. In the framework of the previous section,

$$\tilde{V}(R_\nu - R_\mu) = -ZV(R_\nu - R_\mu) \sim (Ze)^2/|R_\nu - R_\mu| \tag{4}$$

As we discussed, this neglects any interaction due to the polarization of the ion cores. In Pb such ion-core polarization effects may in fact contribute to the ion–ion interaction. The Hamiltonian [Eq. (3)] can formally include these if we allow $\tilde{V}$ to deviate from the simple Coulomb form of Eq. (4). The final term in $\mathscr{H}$ is the Coulomb interaction between the conduction electrons. Various additional interactions such as hyperfine and spin-orbit couplings have been omitted. Pb actually has a significant spin-orbit interaction, but here we have neglected it.

As outlined in the previous section, the screened effective ion potential represents a small perturbation on the pseudo-conduction electrons so that it is useful to expand $\psi(x)$ in plane waves.

$$\psi(x) = \sum_{ps} \exp(ip\cdot x)\, \chi_s c_{ps} \tag{5}$$

Here χ_s is a two-component spinor and c_{ps} destroys a plane-wave electron of momentum p and spin s. It satisfies the fermion anticommutation relations

$$\{c_{ps}, c^{\dagger}_{p's'}\} = \delta_{pp'}\delta_{ss'}$$

In the low-temperature region of interest the position variables of the ions can be expanded about their equilibrium lattice positions R^0_ν:

$$R_\nu = R^0_\nu + \delta R_\nu$$

and the motion of the bare ions treated in the harmonic approximation. Introducing in the usual way the bare phonon creation and destruction operators $b^{\dagger}_{q\lambda}$ and $b_{q\lambda}$, we have

$$\begin{aligned} \delta R_\nu &= \left(\frac{1}{NM}\right)^{1/2} \sum_{q\lambda} \boldsymbol{\epsilon}_{q\lambda} \exp(iq\cdot R^0_\nu)\left(\frac{1}{2\Omega_{q\lambda}}\right)^{1/2} (b_{q\lambda} + b^{\dagger}_{-q\lambda}) \\ P_\nu &= i\left(\frac{M}{N}\right)^{1/2} \sum_{q\lambda} \boldsymbol{\epsilon}_{-q\lambda} \exp(-\, iq\cdot R^0_\nu)\left(\frac{\Omega_{q\lambda}}{2}\right)^{1/2} (b^{\dagger}_{q\lambda} - b_{-q\lambda}) \end{aligned} \tag{6}$$

Here $b^{\dagger}_{q\lambda}$ creates a phonon of wave vector q, polarization $\boldsymbol{\epsilon}_{q\lambda}$, and energy $\Omega_{q\lambda}$. In these equations M is the ion mass and N is the number of ions per unit volume. The phonon wave vectors are restricted to the first Brillouin zone and the phonon operators obey the boson commutation relations

$$[b_{q\lambda}, b^{\dagger}_{q'\lambda'}] = \delta_{qq'}\delta_{\lambda\lambda'}$$

For q along certain symmetry directions in the crystal, the polarization vectors $\boldsymbol{\epsilon}_{q\lambda}$ are perpendicular to $\mathbf{q}$ for the transverse modes and parallel to $\mathbf{q}$ for the longitudinal modes. As $\mathbf{q}$ moves away from these special directions we will continue to label a given mode transverse or longitudinal even though the polarization vectors are no longer simply parallel or perpendicular to $\mathbf{q}$. If only the bare Coulomb interactions of the form of Eq. (4) are allowed, the frequencies $\Omega_{p'-p\lambda}$ can be calculated. In this case the sum of the squares of the transverse and longitudinal frequencies equals the ion plasma frequency squared.

$$\sum_{\lambda} \Omega^2_{q\lambda} = \Omega^2_p = 4\pi N Z^2 e^2/M$$

Since the transverse restoring force vanishes in the long-wavelength limit, the longitudinal mode must reduce to the ion plasma frequency in this limit.

Clark (*36*) has calculated $\Omega_{q\lambda}$ for a body-centered-cubic lattice and his results are shown in Fig. 4.

Using the expansions of the electron and ion operators [Eqs. (5) and (6)], the Hamiltonian [Eq. (3)] can be rewritten in the form

$$\mathscr{H} = \sum_{ps} \frac{p^2}{2m} c^{\dagger}_{ps} c_{ps} + \sum_{q\lambda} \Omega_{q\lambda} (b^{\dagger}_{q\lambda} b_{q\lambda} + \tfrac{1}{2}) + \sum_{qps} V(q)\, S(q)\, c^{\dagger}_{p+qs} c_{ps} + \sum_{\substack{pkq \\ ss'}} \frac{4\pi e^2}{q^2} c^{\dagger}_{p+qs} c^{\dagger}_{k-qs'} c_{ks'} c_{ps} + \sum_{pp'\lambda s} g_{pp'\lambda} \phi_{p'-p} c^{\dagger}_{p's} c_{ps} \tag{7}$$

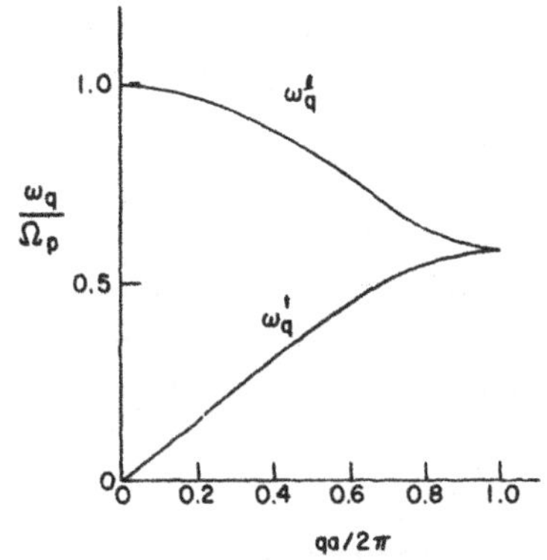

Fig. 4. Dispersion relations for transverse and longitudinal phonon branches for a body-centered-cubic lattice of positive point charges embedded in a uniform background of negative charge. Here a is the lattice spacing and Ω_p is the ion plasma frequency.

Here the first two terms are the bare electron and bare phonon energies, respectively. The third term is the electron–rigid ion interaction with $V(q)$ the Fourier transform of the bare pseudo-potential and $S(q)$ the structure factor:

$$S(q) = \sum_{\nu} \exp(iq \cdot R_\nu) = N\delta_{q,K}$$

Here N is the number of ions per unit volume and K is a reciprocal lattice vector. The fourth term is the Coulomb interaction between the electrons and the final term is the electron–phonon interaction. In the electron–phonon interaction

$$\phi_{q\lambda} = b_{q\lambda} + b^{\dagger}_{-q\lambda}$$

is the phonon field operator for wave vector q and polarization $\boldsymbol{\epsilon}_{q\lambda}$, and the bare electron–phonon coupling is given by $g_{pp'\lambda}$:

$$g_{pp'\lambda} = -\, i(N/2M\Omega_{p'-p\lambda})^{1/2} (\mathbf{p}' - \mathbf{p}) \cdot \boldsymbol{\epsilon}_{p'-p\lambda} V(p' - p) \tag{8}$$

Since the phonon wave vector is restricted to the first Brillouin zone, Eq. (8) for

$g_{pp'\lambda}$ assumes the convention that if the momentum transfer $p' - p$ falls outside the first zone, the corresponding reduced wave vector lying in the first zone is used in calculating $\Omega_{p'-p\lambda}$ and $\epsilon_{p'-p\lambda}$.

For the moment, let us consider just the dynamics of the conduction electrons in a rigid ion lattice. This corresponds to dropping the phonon terms in Eq. (7). As we have emphasized, we are actually looking at a pseudo-system which describes the conduction electrons outside the core. The pseudo-potential idea showed how the deep potential wells at the ion sites could actually produce only a feeble perturbation on the conduction electron behavior outside the core region. Therefore, as far as the effects due to ion-core scattering are concerned, a nearly free electron picture is possible. However, the Coulomb interaction between the electrons gives rise to strong two-particle correlations. The introduction of the pseudo-potential is therefore not sufficient to account for the remarkable accuracy of the nearly free electron picture in accounting for the Fermi surface and transport properties of the simple metals. In the next section we discuss some aspects of Landau's Fermi liquid theory which provides a point of view which will allow us to understand this problem. For the moment we will neglect the electron–ion interaction and concentrate on the problem of the interacting electron gas.

C. Landau's Fermi Liquid Theory

In 1956 Landau (*24*) suggested a semiphenomenological theory of a Fermi liquid which provides a useful conceptual framework for describing the interacting electron gas. Landau argued that the appropriate normal modes for describing a system of interacting Fermions were quasi-particle excitations. These can be envisioned as reducing in the noninteracting limit to the plane-wave particle states of a free Fermi gas. This implies that the quasi-particles are fermions, and the assumed translational and rotational invariance of the system means that the quasi-particle excitations can be labeled by momentum p and spin s quantum numbers. Then the low-lying excited states of the interacting system are described in terms of the quasi-particle occupation numbers in the same way the noninteracting system is described by the free-particle-state occupation numbers. In particular, the ground state of the interacting system consists of a filled Fermi sea of quasi-particles. The low-lying excited states contain a small number of excited quasi-particles (and quasi-holes). In these states the number of excited quasi-particles is sufficiently low that interactions between them are negligible and an independent-particle picture is obtained.

In such a description it is important that the quasi-particle excitations have sufficiently long lifetimes so that their level width is small compared to their excitation energy. At low excitation densities the collision of two excitations becomes unimportant and only spontaneous decay needs to be considered. For the Coulomb interaction, the simplest quasi-particle decay process leads to two quasi-particles and a quasi-hole. The Pauli principle restricts the allowed phase space for the final particles and hole so that the quasi-particle states excited at temperatures of interest have very narrow widths compared to their excitation energies.

The quasi-particle dispersion relation must of course reflect the effect of the particle interactions. Landau assumed that these could be taken into account by means of a self-consistent field. In this way, the quasi-particle excitation energy E_{ps} depends upon the occupation of all the other quasi-particle states through an interaction energy $f_{ss'}(p, p')$:

$$E_{ps} = p^2/2m + \sum_{p's'} f_{ss'}(p, p')\, n_{p's'} \tag{9}$$

In his original work Landau noted that $f_{ss'}(p, p')$ was related to the forward scattering amplitude of two quasi-particles. However, the spirit of the Landau theory is to treat $f_{ss'}(p, p')$ phenomenologically and relate various moments of it to measurable quantities, i.e., specific heat, magnetic susceptibility, compressibility, etc. It then provides a framework for relating experimental observations on an interacting fermion system. If Eq. (9) is expanded in the vicinity of the Fermi surface for $T = 0$, the quasi-particle energy, E^0_{ps}, referenced to its value at the Fermi surface, can be written as

$$\epsilon_{ps} = E^0_{ps} - E^0_{p_F s} = (p_F/m^*)(p - p_F) \qquad |p - p_F| \ll p_F \tag{10}$$

Using Galilean invariance, Landau showed that m^* and the interaction energy were related by

$$\frac{m}{m^*} = 1 - \frac{m p_F}{2(2\pi)^3} \sum_{ss'} \int d\Omega_{pp'} f_{ss'}(p, p') \cos\theta_{pp'} \tag{11}$$

Here the angular integration is taken with p and p' on the Fermi surface.

Besides a modified dispersion relation, the structure of the Landau quasi-particle states differs substantially from the excited states of the free Fermi gas. In the ground state of the free Fermi gas, the Pauli principle ensures that electrons of similar spin remain a distance of order the inverse Fermi momentum p_F^{-1} apart. In the interacting system the Coulomb interaction produces a similar dynamic correlation which keeps electrons of either spin a distance of order p_F^{-1} apart. In the excited quasi-particle states this correlation persists and the quasi-particle core is surrounded by an exchange and correlation hole which separates it from the unexcited quasi-particles. This hole moves with the quasi-particle and a back flow of the surrounding medium is produced. Exactly one unit of charge is displaced to the surface and the field of the quasi-particle is screened in a distance of order p_F^{-1}. These correlations are important in determining the effective quasi-particle interactions with each other and with the ion cores. Physically they give rise to screening and renormalization of the bare interactions. In the long-wavelength $q \to 0$ limit, Heine et al. (*37*) have used the Landau theory to show that the dressed ion potential seen by a quasi-particle is $-Z/[2N(0)]$. Here Z is the ionic charge and $N(0)$ is the single spin quasi-particle density of states at the Fermi surface. The long-range Coulomb part of the interaction has been screened out.

In its original form the Landau theory is basically a phenomenological approach, useful for describing macroscopic properties of the system. For example, the long-wavelength behavior of the quasi-particle–ion interaction was obtained by using the Boltzmann and Poisson equations in conjunction with the Landau quasi-particle picture. This is adequate to describe the long-wavelength slowly varying part of the potential; however, the shorter-wavelength parts of the potential are the most important from the point of superconductivity. In addition to this difficulty, when the interaction of electrons with the lattice vibrations is considered, one finds that the level width of quasi-particle excitations near the Debye energy ω_D is comparable with their excitation energies. Therefore, the original Landau quasi-particle picture fails for the electron–phonon system at temperatures of order ω_D. There is also a more subtle manner in which the quasi-particle picture can fail in the electron–phonon system. Even when the system is at zero temperature, if the important virtual processes involve quasi-particles with ϵ_p of order ω_D, the quasi-particle approximation fails. This is what happens in the case of the strong-coupling superconductors.

The Green's function techniques discussed in Chapter 5 will be used to treat these situations. These methods allow us to go beyond the quasi-particle approximation. Dealing as they do with the one- and two-particle correlation functions of the system, the Green's functions, like Landau's theory, are closely related to the experimental observables. This relationship suggests physical approximations, and, in turn, the mathematical structure of the Green's function formalism suggests physical relationships. For example, it is possible to construct relationships which connect band structure with the electron–phonon interaction, the electron–phonon interaction with the superconducting transition temperature, and the tunneling I–V characteristics, etc.

D. Green's Function Theory of the Interacting Electron Gas

Historically, the pioneering work of Bohm and Pines (*28*) on the interacting electron gas marked the beginning of the current many-body approach to metals. Following this, Gell-Mann and Brueckner (*29*) showed that a partial summation of infinite-order perturbation theory, keeping the most divergent terms in each order, generated a high-density theory of the electron gas. Here the expansion parameter r_s is the mean interelectron spacing in units of the Bohr radius. The high-density theory should be valid for $r_s < 1$, which unfortunately means that it cannot be relied upon to give quantitative results at metallic densities where $1.8 \leqslant r_s \leqslant 5.6$. There have been several suggested schemes for extrapolating the high-density theory into the region of metallic densities. Here we outline the Green's function approach to the electron gas, and briefly review some results obtained by Rice (*32*) using one such approximation, suggested by Hubbard (*30,31*). The Green's function techniques we will use have been discussed in Chapter 5. Besides providing a useful calculation procedure, they extend Landau's ideas and provide the best available framework for relating experimental observations.

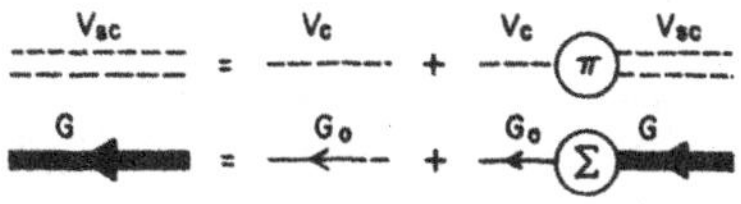

Fig. 5. Dyson's equations for the screened Coulomb Green's function V_{sc} and the one-electron Green's function G.

In Fig. 5 the diagrammatic representation of the Dyson equations for the screened Coulomb (doubled dashed line) and electron (thick solid line) propagators are shown. The single dashed line represents the fourier transform of the bare Coulomb interaction $-e^2/r$,

$$V_c(q) = 4\pi e^2/q^2$$

and a thin solid line the bare electron propagator,

$$G_0 = (\omega - \epsilon_p + i\delta_\omega)^{-1}$$

Here ω is measured relative to the Fermi energy, $\epsilon_p = p^2/2m - p_F^2/2m$, and the pole is defined for the time-ordered propagator so that $\delta_\omega = \delta\,\text{sign}(\omega)$. Using the Dyson equation, the screened Coulomb propagator can be expressed in terms of the irreducible electron polarization $\pi_c(q, \omega)$:

$$V_{sc}(q, \omega) = \frac{V_c(q)}{1 + V_c(q)\,\pi_c(q, \omega)} = \frac{V_c(q)}{\epsilon(q, \omega)} \tag{12}$$

The final form defines the electron gas dielectric constant $\epsilon(q, \omega)$.

The Dyson equation for the electron propagator is

$$G_c(\omega) = [\omega - \epsilon_p - \Sigma_c(p, \omega)]^{-1}$$

where $\Sigma_c(p, \omega)$ is the irreducible Coulomb self-energy. Near the Fermi surface the imaginary part of the self-energy is much less than the excitation energy because of phase-space restrictions imposed by the Pauli principle. Therefore, the quasi-particle approximation is appropriate and Σ_c can be expanded:

$$\Sigma_c(p, \omega) = \left(\Sigma_c + \epsilon_p \frac{1}{v_F} \frac{\partial}{\partial p} \Sigma_c + \omega \frac{\partial}{\partial \omega} \Sigma_c\right)_{\substack{p=p_F \\ \omega=0}} \tag{13}$$

The first term represents a shift of the chemical potential and the second and third terms renormalize the mass and the quasi-particle part of the propagator. Substituting Eq. (13) into the Dyson equation, the electron propagator near the Fermi surface can be written in the form

$$G_c(p, \omega) = (1/Z_c)(\omega - \tilde{\epsilon}_p + i\delta_\omega)^{-1} \tag{14}$$

with the quasi-particle energy

$$\tilde{\epsilon}_p = Z_c^{-1}\left(1 + \frac{1}{v_F} \frac{\partial \Sigma_c}{\partial p}\right)_{\substack{p=p_F \\ \omega=0}} \epsilon_p = \frac{m}{m^*} \epsilon_p \tag{15}$$

and the renormalization factor

$$Z_c = \left(1 - \frac{\partial \Sigma_c}{\partial \omega}\right)_{\substack{p=p_F \\ \omega=0}} \tag{16}$$

The inverse of the renormalization parameter Z_c^{-1} represents the overlap of the state obtained by adding a plane wave of momentum $P \sim P_F$ to the exact ground state with the normalized Landau state containing one quasi-particle of momentum p. It is less than unity, reflecting the fact that only a fraction of the plane-wave-generated state is quasi-particle-like. The remainder consists of contributions from configurations involving two or more quasi-particles. This choice of notation for the Coulomb renormalization factor Z_c is just the inverse of the standard notation. We use it because of its similarity to the way in which the phonon renormalization parameter enters the theory.

The effective electron–ion interaction can be decomposed as shown in Fig. 6. Here the shaded circle is the Coulomb vertex correction which can be separated into a screening part represented by the irreducible electron polarization π, the screened Coulomb interaction V_{sc}, and a proper Coulomb vertex part Λ.

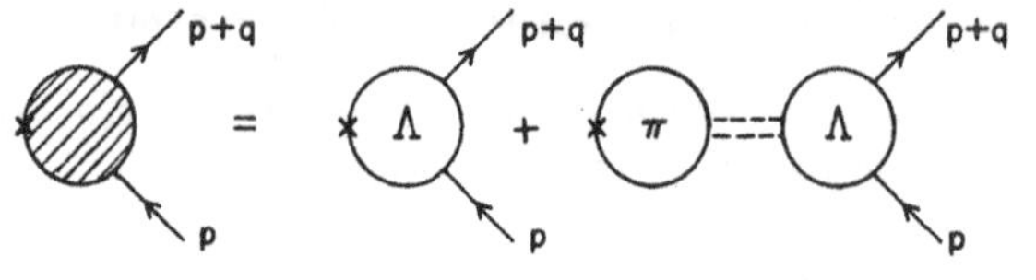

Fig. 6. Graphical separation of the screened and vertex corrected pseudo-potential $V_p(q)$. The external electron lines are included for clarity.

The proper vertex part Λ is defined to include all diagrams which cannot be separated into two parts by cutting a simple dashed V_c line. The equation for the effective electron–ion interaction shown in Fig. 6 is

$$V_p(q) = V(q)\,\Lambda_p(q)\,[1 - \pi(q)\,V_{sc}(q)] = V(q)\,\Lambda_p(q)/\epsilon(q) \tag{17}$$

Here $V(q)$ is the Fourier transform of the bare pseudo-potential and $\epsilon(q)$ is the dielectric constant of the electron gas. For an isotropic system with initial and final states on the Fermi surface, the interaction $V_{p_F}(q)$ is a function only of $|\mathbf{q}|$ which varies from zero to $2P_F$.

Now consider the potential which enters in calculating the band structure. For nearly free electron metals the band structure is determined by the lowest-order self-energy corrections shown in Fig. 7. For a reciprocal lattice reflection Q,

$$\Sigma = |V_{p_F}(Q)|^2\, G_c(p + Q, \omega)$$

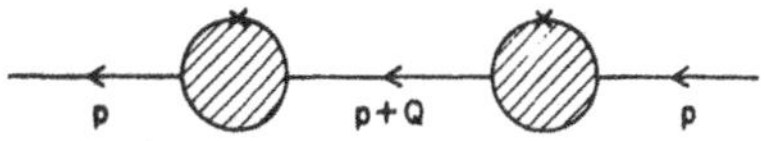

Fig. 7. Lowest order, nondiagonal, pseudo-potential correction to the electron propagator.

where G_c is the dressed Green's function of the interacting electron gas. The band structure is determined by the poles of the full Green's function, which to second order in the electron–ion interaction are given by

$$0 = G_c^{-1}(p, \omega) - |V_{p_F}(Q)|^2\, G_c(p + Q, \omega)$$

Here the diagonal part of the scattering which shifts all the levels is neglected and only one reciprocal lattice vector Q has been considered. Expanding G_c as in Eq. (14) this becomes

$$0 = (\omega - \tilde{\epsilon}_p)(\omega - \tilde{\epsilon}_{p+Q}) - |V_{p_F}(Q)|^2/Z_c^2$$

and the dressed electron–ion potential which determines the band structure is therefore

$$\Lambda_{p_F}(q)\, V(q)/\epsilon(q)\, Z_c \tag{18}$$

Luttinger and Nozières (*38*) have derived Ward identities which can be used to show that

$$\lim_{q \to 0} \frac{\Lambda_{p_F}(q)\, V(q)}{\epsilon(q)\, Z_c} \sim -\frac{Z}{2N(0)} \tag{19}$$

Here $N(0)$ is the single-spin Coulomb dressed density of states at the Fermi energy. This limiting value is the same as that obtained from the Landau theory (*37*). Migdal (*39*) and Luttinger (*40*) have shown that within the framework of perturbation theory the interacting system has a discontinuity in the occupation number $\langle c_p^\dagger c_p \rangle$ of magnitude Z_c^{-1} at a momentum value determined by

$$0 = \tilde{\epsilon}_{p_F} \tilde{\epsilon}_{p_F+Q} - |V_{p_F}(Q)|^2/Z_c^2$$

In this way, a Fermi surface of the interacting normal system can be defined in terms of the effective potential, Eq. (18). In the next section we will see that this same potential enters the electron–phonon interaction.

So far our discussion has been purely formal. To obtain explicit results it is necessary to make an approximation. From the work of Bohm and Pines and Gell-Mann and Brueckner it is known that results applicable to a high-density electron gas can be obtained by keeping only the lowest-order contribution to the irreducible polarization. This approximation is known as the random phase approximation (RPA). We will discuss some of the known results for the dielectric constant and the effective mass obtained within the RPA. Some low-order diagrams for the irreducible electron polarizability are shown in Fig. 8.

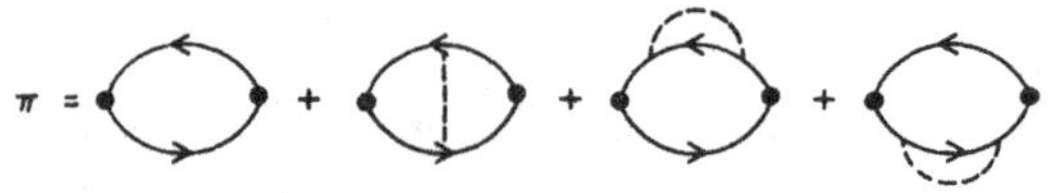

Fig. 8. Low-order graphical perturbation terms for the irreducible electron polarization.

Keeping only the contribution from the first bubble diagram generates the random phase approximation with the Lindhard form *(41)* for the frequency- and wave vector–dependent dielectric constant. In the static limit the Lindhard dielectric function is

$$\underset{\text{RPA}}{\epsilon(q)} = 1 + (k_s/q)^2\, u(q/2p_F) \tag{20}$$

with

$$u(x) = \frac{1}{2} + \frac{1 - x^2}{4x} \ln\left|\frac{1 + x}{1 - x}\right| \tag{21}$$

The long-wavelength limit of this reduces to the Thomas–Fermi result

$$\underset{\text{TF}}{\epsilon(q)} = 1 + (k_s/q)^2 \tag{22}$$

where k_s is the Thomas–Fermi screening wave vector

$$k_s^2 = 4\pi e^2 N(0)$$

Using the Lindhard dielectric constant, the most important low-momentum contributions to the self-energy are determined by the screened exchange approximation shown in Fig. 9. The effective mass and quasi-particle lifetime were calculated within this RPA approximation by Quinn and Ferrell *(42)*. They found

$$\frac{m}{m_c^*} = 1 - \frac{\alpha r_s}{2\pi}\left[2 + \ln\left(\frac{\alpha r_s}{\pi}\right)\right]$$

$$\frac{1}{\tau} = 2\,|\mathrm{Im}\,\Sigma(p, \epsilon_p)| = \alpha E_F\,|(p/p_F)^2 - 1|$$

where $\alpha = (4/9\pi)^{1/3}$. It follows that the quasi-particle states are well defined for excitation energies less than about 20% of E_F.

The ratio m_c^*/m vs. r_s is plotted in Fig. 10. The rapid increase for values of r_s larger than unity is symptomatic of the fact that RPA approximation is a high-density expansion. Unfortunately the region of metallic densities lies outside the region of validity $r_s < 1$ of the RPA expansion. Several interpolation schemes have been suggested for extrapolating the high-density theory into the

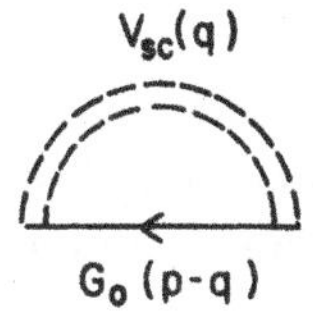

Fig. 9. High-density RPA approximation to the electron self-energy using the screened Coulomb interaction.

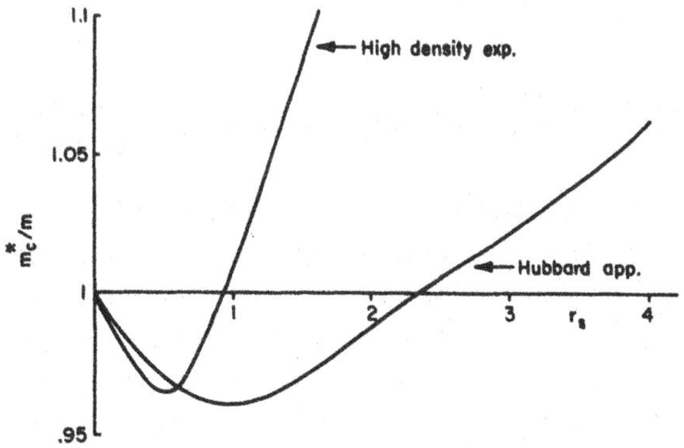

Fig. 10. Ratio of the Coulomb dressed mass to bare electron mass as a function of the electron density parameter r_s. The screened exchange approximation is labeled as the high-density result and Rice's calculation is labeled as the Hubbard approximation.

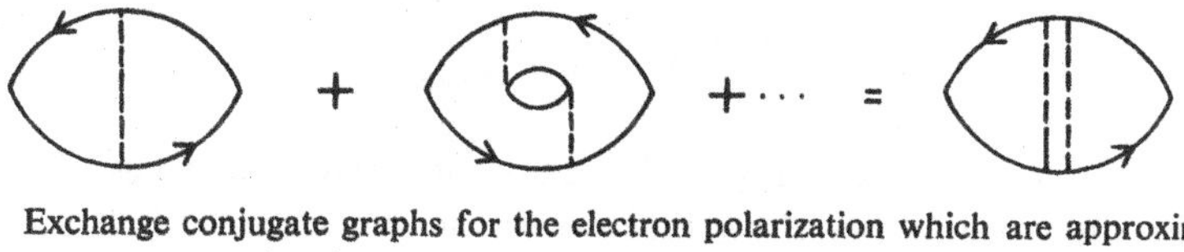

Fig. 11. Exchange conjugate graphs for the electron polarization which are approximately taken into account by Hubbard's factor $g(q)$.

region of metallic densities. Physically, one might hope that the RPA provides a useful representation at long wavelengths. However, at short wavelengths it overestimates the polarization. At short wavelengths the Pauli principle reduces the interaction of parallel spin electrons and higher-order polarization graphs such as Fig. 11 become important. Hubbard (*30*) suggested that the effects of these exchange conjugate graphs could be simulated by modifying the RPA contribution in such a way as to reduce it at large momentum transfer to one-half its value. The static screening in Hubbard's approximation is given by

$$\epsilon_s(q) = 1 + (k_s/q)^2 \, u(q/2p_F) \, [1 - g(q)]$$

where

$$g(q) = \tfrac{1}{2} q^2/(q^2 + p_F^2 + k_s^2)$$

For small momentum transfer this approaches the RPA result, while at momentum transfers of order $2p_F$ the factor $g(q)$ reduces the RPA result by about $\frac{1}{4}$.

Using this type of approximation Rice has evaluated the electron self-energy in the screened exchange approximation. This includes the RPA pieces, and, via Hubbard's g-factor, approximates additional contributions to Σ of the type

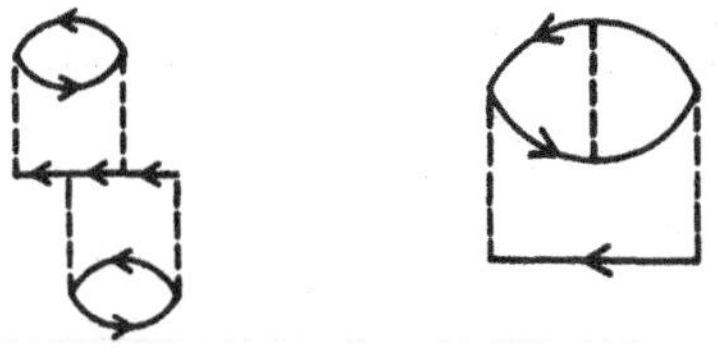

Fig. 12. Self-energy corrections to the RPA included approximately in Rice's calculation.

shown in Fig. 12. Rice's calculation of m_c^*/m is plotted as "Hubbard's" approximation in Fig. 10. For most polyvalent metals $1.8 \leqslant r_s \leqslant 2.5$ and m_c^*/m in Hubbard's approximation is nearly unity. Here it should be remembered that m is the free electron mass and m_c^* is the effective mass associated with the Coulomb interactions between the electrons. In comparing with specific-heat data, band effects and the electron–phonon interaction must be taken into account. In the next section we will see that the electron–phonon effects can give rise to large effective mass corrections.

Rice also calculated the renormalization parameter

$$Z_c = \left(1 - \frac{\partial \Sigma_c}{\partial \omega}\right)_{\substack{p=p_F \\ \omega=0}}$$

and obtained values of order $\frac{3}{2}$ in the region of metallic densities. The Ward identities discussed earlier show that in the long-wavelength limit the renormalization parameter is identically canceled by the proper vertex correction.

In view of the significant deviation of Z_c from unity, it is important to understand whether such a cancelation occurs at momentum transfers of order P_F. Rice has calculated the lowest-order contribution (Fig. 13) to the proper vertex and finds that within this approximation Λ is essentially independent of momentum transfer. This suggests the possibility that Λ and Z_c appearing in the interaction may cancel each other (ignoring factors of m_c^*/m which appear to be close

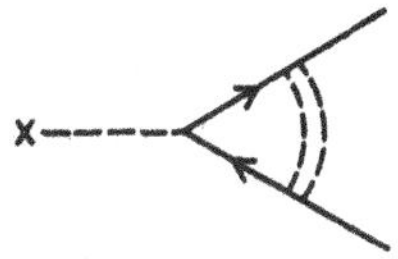

Fig. 13. Lowest-order screened Coulomb correction to the proper vertex part.

to unity for the polyvalent metals), leaving only the screened form of the ion pseudo-potential. The exact form of the static screening is of course not known.

Having reviewed so far how it is that an independent, nearly free electron picture can be used to describe the conduction electron–rigid ion system, we turn now to the effects of the electron–phonon interaction.

E. Migdal's Adiabatic Theorem

The main import of the Born–Oppenheimer theorem (*43*) is that the conduction electron wave function can be accurately described as following the motion of the ion cores adiabatically to terms of order the square root of the electron to ion mass ratio $\sqrt{m/M}$. Equivalent expansion parameters are the ratio of the velocity of sound to the Fermi velocity c/v_F or the Debye energy to the Fermi energy ω_D/E_F. The magnitude of these parameters is typically less than 1%. The Born–Oppenheimer theorem implies that any corrections to matrix elements arising from nonadiabatic electron excitations should be of order $\sqrt{m/M}$. Specifically, the Coulomb renormalized and screened matrix elements of the ion-core potential should accurately describe the electron–ion scattering to terms of this order.

The point of this discussion, and the analysis which follows, is to reexpress the Born–Oppenheimer theorem in the Green's function framework of graphical perturbation theory. This was first investigated in a classic paper by Migdal (*25*) and is sometimes referred to as "Migdal's theorem." He showed how the adiabatic expansion parameter $\sqrt{m/M}$ could be used to classify the contributions of the perturbation theory graphs which arise in analyzing the electron–phonon system. Some low-order perturbation graphs for the electron–phonon vertex are shown in Fig. 14. The first term is the electron–phonon matrix element and the

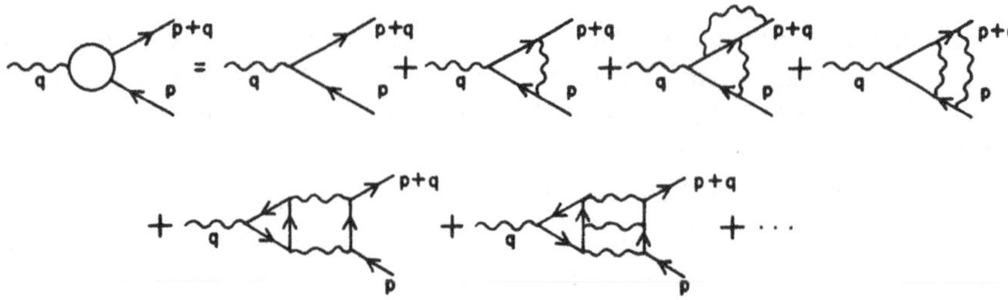

Fig. 14. Graphical perturbation expansion of the electron–phonon vertex $\Gamma(p,q)$. The external electron and phonon lines are added for clarity.

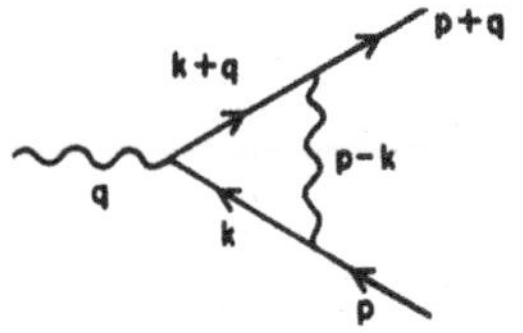

Fig. 15. Lowest-order correction to the electron–phonon vertex. Again, the external lines are for clarity.

additional terms represent corrections due to the electron–phonon interaction. Migdal argued that the individual contributions of each of the additional graphs were smaller by at least a factor of $\sqrt{m/M}$.

Consider the lowest-order vertex correction shown in Fig. 15. The intermediate quasi-particles labeled by $k+q$ and k maintain a spatial-phase coherence for distance of order $1/q$. Since they are traveling at approximately the Fermi velocity, the time over which the lattice can change to produce the final-state scattering is $1/qv_F$. Now, since the lattice responds at the Debye frequency, the change in the lattice configuration is of order ω_D/qv_F. The vertex correction, Fig. 15, is therefore of order ω_D/qv_F times the bare scattering. For the typical momentum transfers of interest $q \sim p_F$ and we have the adiabatic result. A special case occurs when q goes to zero but the energy transfer q_0 remains finite such that q_0/q becomes larger than the Fermi velocity. Under these conditions, the spatial coherence in the intermediate state is maintained so that the lattice has time to fully respond. In this case the vertex correction is of the order of the electron–phonon coupling $N(0)\,|g|^2/\omega_D$, which can be of order unity for the strong-coupling metals. Thus the importance of the vertex correction depends upon the ratio of the phase velocity of the external phonon line q_0/q to the Fermi velocity v_F. We will see that this is reflected in the fact that the electron self-energy associated with phonon exchange processes is strongly frequency-dependent and weakly momentum-dependent. In analyzing the electron–ion scattering, the important regions of phase space are where $q_0 \sim \omega_D$ and $q \sim p_F$, so that $q_0/qv_F \sim \omega_D/E_F \sim \sqrt{m/M}$ is a useful expansion parameter.

Because of the central role played by "Migdal's theorem" in the theory of strong-coupling superconductivity, we now sketch some of the analysis which goes with the description of the previous paragraph. For simplicity we treat the case in which the electron–phonon matrix element is a function of momentum transfer only. The expression for the lowest-order vertex correction shown in Fig. 15 is

$$\Gamma^{(1)}(p, q) = i\int \frac{d^4k}{(2\pi)^4}\, |g(p-k)|^2\, D(p-k)\, G(k)\, G(k+q)$$

The important range of energy and momentum transfers is set by the Debye frequency ω_D and the diameter of the Fermi sea $2P_F$, respectively. Keeping this in mind, we will be interested in $\Gamma^{(1)}$ for values of p_0 and q_0 less than or of order ω_D, and values of $\mathbf{p}$ and $\mathbf{q}$ of order P_F. Therefore, the phonon propagator vanishes like k_0^{-2} for k_0 greater than ω_D and provides a natural cutoff on the frequency integration. The electron kinetic energy denominators as well as the weakness of the effective ion potential inside the core provide a natural cutoff at large momentum transfers.

Over the important part of phase space the phonon propagator and the electron–phonon coupling are slowly varying functions of momentum transfer. Physically this expresses the fact that the electron–phonon interaction is short range, extending typically over distances of the order of the interatomic spacing. This is a consequence of (1) the flattening out of the phonon dispersion relation at large wave vectors, and (2) the Coulomb screening of the ion potential. Replacing the phonon propagator and the electron–phonon coupling by an effective average $|g|^2/\omega_D$, the vertex correction reduces to

$$\Gamma^{(1)} \sim i\,\frac{|g|^2}{\omega_D}\int_{-\omega_D}^{\omega_D} \frac{dk_0}{2\pi}\int \frac{d^3k}{(2\pi)^3}\, \frac{1}{k_0 - \epsilon_k + i\delta_{k_0}}\, \frac{1}{k_0 + q_0 - \epsilon_{k+q} + i\delta_{k_0+q_0}} \tag{23}$$

Here the cutoff at large momentum is obtained from the electron kinetic energies alone. Dividing the momentum integration into two regions determined, for example, by $|k| \gtrless 2p_F$, it is easy to see that the integration for $|k| > 2k_F$ gives a contribution of order

$$\frac{N(0)\,|g|^2}{\omega_D}\,\frac{\omega_D}{E_F} \tag{24}$$

which contains the adiabatic parameter ω_D/E_F.

Finally, consider the part of Eq. (23) with $|k| < 2k_F$. As long as $q_0 \ll kq/m$, the kinetic energies of the electrons may be treated as independent variables and the overlap of the singularities in the electron propagators is unimportant. To see this in detail, we make a change of variables from k and $\mu = \mathbf{k}\cdot\mathbf{q}/|k|\,|q|$ to

ϵ_k and ϵ_{k+q}, obtaining

$$\Gamma^{(1)} \sim i \frac{|g|^2}{\omega_D} N(0) \frac{m^*}{qp_F} \frac{1}{4\pi} \int_{-\omega_D}^{\omega_D} dk_0 \int d\epsilon_k \int d\epsilon_{k+q}$$
$$\times \frac{1}{k_0 - \epsilon_k + i\delta_{k0}} \frac{1}{k_0 + q_0 - \epsilon_{k+q} + i\delta_{k_0+q_0}}$$

Here $N(0)$ is the quasi-particle density of states at the Fermi surface:

$$N(0) = m^* p_F / 2\pi^2$$

where m^* contains the band and Coulomb effects but does *not* include any electron–phonon renormalization. The $(\epsilon_k, \epsilon_{k+q})$ integration region is shown in

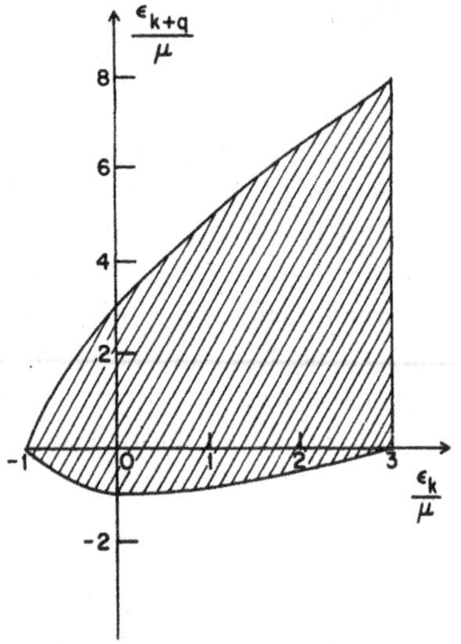

Fig. 16. $(\epsilon_k, \epsilon_{k+q})$ integration region for evaluating the lowest-order electron–phonon vertex.

Fig. 16 for $q = P_F$. The structure of the energy denominators which occurs for ϵ_k and ϵ_{k+q} of order ω_D would be about the thickness of the x-axis in this figure. It is clear that in this region ϵ_k and ϵ_{k+q} are indeed independent, except for negligible regions within ω_D/E_F of -1 and 3. Extending the end points of the integrations from $-\infty$ to ∞ introduces errors which are the same order as Eq. (24) and can therefore be neglected. Then using

$$\int_{-\infty}^{\infty} d\epsilon_k \frac{1}{k_0 - \epsilon_k + i\delta_{k_0}} = -i\pi\delta \operatorname{sgn}(k_0) \tag{25}$$

the $|k| < 2k_F$ contribution to $\Gamma^{(1)}$ is found to be of order

$$\frac{N(0)\,|g|^2}{\omega_D}\,\frac{\omega_D}{E_F}$$

This is Migdal's result for the condition corresponding to $q_0/q \ll v_F$.

In the opposite limit, where q_0/q is large compared to v_F, Migdal and Engelsberg and Schrieffer (*27*) have noted that this vertex correction is of order $N(0)|g|^2/\omega_D$. This result can be simply obtained by observing that the vertex correction in this limit can be written as

$$\Gamma^{(1)}_{q_0/q\to\infty}(p, q) \sim i\,\frac{|g|^2\,N(0)}{\omega_D}\int_{-\omega_D}^{\omega_D}\frac{dk_0}{2\pi}\int d\epsilon_k \times \frac{1}{k_0 - \epsilon_k + i\delta_{k_0}}\,\frac{1}{k_0 + q_0 - \epsilon_k + i\delta_{k_0+q_0}}$$

If the integrand is expressed in partial fractions, and integrated as shown in Eq. (25), one obtains

$$\Gamma^{(1)}_{q_0/q\to\infty}(p, q) \sim i\,\frac{|g|^2\,N(0)}{\omega_D}\int_{-\omega_D}^{\omega_D} dk_0\,\frac{\operatorname{sgn}(k_0) - \operatorname{sgn}(k_0 + q_0)}{q_0} \sim \frac{|g|^2\,N(0)}{\omega_D}$$

F. Green's Function Theory of the Normal Electron–Phonon System

Further analysis of the type just outlined shows that order by order in perturbation theory the phonon vertex corrections for $v_F \gg q_0/q$ are individually small by factors of $\sqrt{m/M}$. There is, of course, no guarantee that an infinite summation of such corrections is small. In fact, the onset of the superconducting transition is signaled by a singularity in the contribution of one such infinite subset of vertex corrections to Γ. This subset involves particle–particle scattering graphs, two of which are shown as the last two graphs in Fig. 14. In the next section, this singularity is removed and the remaining vertex corrections are once again formally of order $\sqrt{m/M}$. Whether there exist further instabilities is not known; however, the present close agreement between the theoretical and experimental results suggests that the theory is adequate for the simple metals. However, as discussed in Chapter 13, a class of Coulomb vertex corrections associated with critical spin fluctuations is found to play an important role in the superconductivity of the transition metals.

Based upon the perturbation theory result that $\Gamma \sim 1 + 0(\sqrt{m/M})$, Migdal (*25*) showed how the phonon contribution to the electron self-energy of a *normal* metal could be calculated to order $\sqrt{m/M}$. This calculation will be discussed later in this section. The resulting self-energy is of order ω_D for frequencies within ω_D of the Fermi energy and is essentially momentum-independent. For frequencies larger than ω_D it rapidly goes to zero. Numerous consequences of Migdal's theorem have been discussed in the literature (*26*). Here we review several results pertinent to the theory of the electron–phonon interaction. First, the phonon corrections to the electron polarization π are negligible. Migdal's theorem implies that the phonon vertex corrections to π are of order $\sqrt{m/M}$. Furthermore, the typical electron-hole pairs which contribute to the polarization have energies of order E_F, so that the phonon dressing of the electron Green's functions gives rise to ω_D/E_F corrections. Second, the effective electron–ion interaction which determines the band structure is unaffected by phonon corrections. Most of the band gaps occur at energies much greater than ω_D from the Fermi energy, so that the phonon renormalization effects on the electron are unimportant; and, once again, the phonon corrections to the vertex are negligible. Furthermore, the phonon part of the self-energy is tied to the Fermi energy, so that any changes in the Fermi surface produced by the electron–phonon interaction are of order ω_D/E_F. Finally, the dressed phonon frequencies are determined by the effective electron–ion interaction, the band structure, and the electron polarizability. Each of these is unaffected by phonon dressing to order $\sqrt{m/M}$. This is again nothing more than an acknowledgment that the phonon frequencies could be determined by making an adiabatic distortion of the lattice.

It is simplest to begin an analysis of the phonon and normal electron Green's functions by considering a jellium-like model in which band effects are neglected. Modifications associated with the ion lattice are discussed later. The Dyson

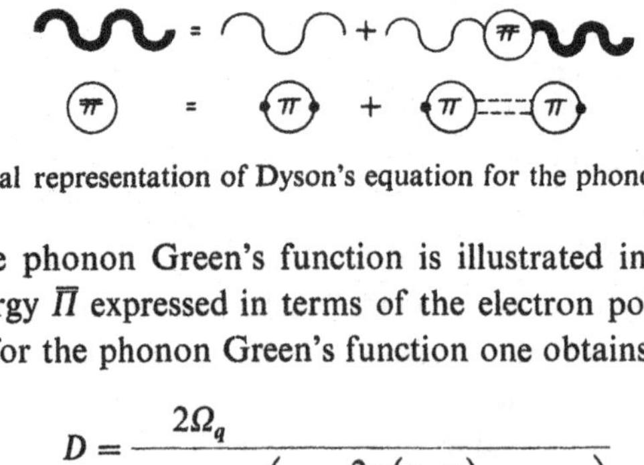

Fig. 17. Graphical representation of Dyson's equation for the phonon Green's function.

equation for the phonon Green's function is illustrated in Fig. 17 with the phonon self-energy Π expressed in terms of the electron polarization as illustrated. Solving for the phonon Green's function one obtains

$$D = \frac{2\Omega_q}{\omega^2 - \Omega_q^2\left(1 - \dfrac{2\pi(q, \omega)}{\Omega_q \epsilon(q, \omega)} |g(q)|^2\right)} \tag{26}$$

Here $\epsilon = 1 + \pi V_c$ is the Coulomb dielectric constant and π is the irreducible electron polarizability. For a pure jellium model $2|g(q)|^2/\Omega_q = V_c$ and Eq. (26) reduces to

$$D = 2\Omega_q/[\omega^2 - \Omega_q^2/\epsilon(q, \omega)]$$

Introducing the dressed phonon frequency

$$\omega_q = \Omega_q/[\epsilon_1(q, 0)]^{1/2}$$

and the damping width

$$\Gamma_q(\omega) = \tfrac{1}{2}\omega_q[\epsilon_2(q, \omega)/\epsilon_1(q, 0)]$$

the phonon propagator takes the standard form

$$D = \frac{\Omega_q}{\omega_q} \frac{2\omega_q}{\omega^2 - \omega^2 + i2\omega_q\Gamma_q(\omega)} \tag{27}$$

Here ω_q/Ω_q is the renormalization factor and ϵ_1 and ϵ_2 are the real and imaginary parts of the electron dielectric function. The static approximation for ϵ_1 is accurate to order ω_D/E_F. Using the Lindhard expression for the dielectric constant one finds $\Gamma_q(\omega_q)/\omega_q \sim \Omega_q/E_F \ll 1$, so that the phonons are well-defined excitations of the system.

The phonon contribution to the electron self-energy was originally discussed by Migdal on the basis of Fröhlich's model (*44*). In this model the effects of Coulomb interactions are approximately taken into account by using dressed electron and phonon excitations and a screened electron–ion interaction from the start. However, in this approximation, the direct Coulomb interaction between the electrons is neglected. Recently Batyev and Pokrovskii (*45*) have analyzed this same problem when the Coulomb interaction is present. In this case, it is convenient to use skeleton diagrams in which dressed electron, phonon, and screened Coulomb propagators form the basic elements. To avoid further complicating the figures, the single-line representation previously used for the

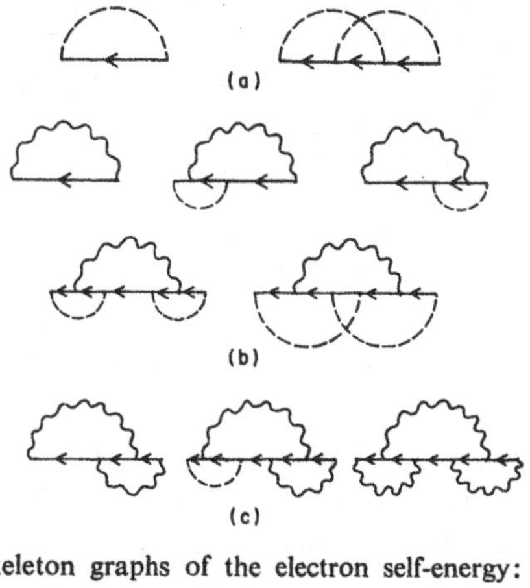

Fig. 18. Representative skeleton graphs of the electron self-energy: (a) shows Coulomb contribution, (b) one-phonon contribution with Coulomb corrections, and (c) multiphonon contributions.

bare Green's functions will be used from now on to denote the dressed Green's functions.

Batyev and Pokrovskii divide the electron self-energy into three groups of which representative types are shown in Fig. 18. The first class, shown in Fig. 18a, has only electron and Coulomb lines; the second class, Fig. 18b, contains only one phonon line; and the third class, Fig. 18c, contains graphs with more than one phonon line. The first class gives rise to the Coulomb part of the electron self-energy Σ_c discussed in the previous section. The second class can be re-

Fig. 19. Separation of the one-phonon self-energy graphs into a type containing an intermediate electron and phonon propagator and a type containing an intermediate electron, phonon, and screened Coulomb propagator.

grouped into two types as shown in Fig. 19. The first of these corresponds to using a Coulomb screened and vertex corrected electron–phonon interaction

$$\Lambda_{p_F}(q)\, g(q)/\epsilon(q, 0)$$

in place of $g(q)$, and gives a contribution

$$\Sigma_{\text{ph}}(p) = i \int \frac{d^4 q}{(2\pi)^4} \left| \frac{\Lambda_{p_F}(q)\, g(q)}{\epsilon(q, 0)} \right|^2 D(p-q)\, G(q) \tag{28}$$

The static form of the proper Coulomb vertex and dielectric constant are appropriate since the important energy transfers are less than or of order ω_D. This also means that the Coulomb part of the electron self-energy can be expanded, giving a shift of the chemical potential, a renormalization of the electron mass, and a renormalization of the propagator [see Eq. (14)], so that

$$G = \frac{1}{Z_c} \frac{1}{\omega - \tilde{\epsilon}_p - \tilde{\Sigma}_{\text{ph}}} \tag{29}$$

Here $\tilde{\Sigma}_{\text{ph}}$ is defined by

$$\tilde{\Sigma}_{\text{ph}} = \Sigma_{\text{ph}}/Z_c \tag{30}$$

and Z_c and $\tilde{\epsilon}_p$ are the previously discussed Coulomb renormalization parameter and quasi-particle energy given by Eqs. (15) and (16), respectively. It is also useful to exhibit the phonon renormalization explicitly, defining a reduced phonon propagator $\hat{D}$ by

$$D(q, \omega) = (\Omega_q/\omega_q)\, \hat{D}(q, \omega) \tag{31}$$

Putting all this algebra together, the Coulomb renormalized phonon part of the self-energy satisfies the integral equation

$$\tilde{\Sigma}_{\text{ph}}(p) = i \int \frac{d^4 q}{(2\pi)^4} |\bar{g}(q)|^2 \,\hat{D}(p-q) \frac{1}{q_0 - \tilde{\varepsilon}_q - \tilde{\Sigma}_{\text{ph}}(q)} \tag{32}$$

with the dressed electron–phonon coupling

$$\bar{g}(q) = -i\left(\frac{N}{2M\omega q}\right)^{1/2} \mathbf{q}\cdot\boldsymbol{\epsilon}_q \frac{\Lambda(q)\,V(q)}{\epsilon(q)\,Z_c} \tag{33}$$

Comparing this with the bare electron–phonon coupling

$$g(q) = -i\left(\frac{N}{2M\Omega q}\right)^{1/2} \mathbf{q}\cdot\boldsymbol{\epsilon}_q V(q)$$

we note that the Coulomb renormalization of the phonon dispersion relation has changed Ω_q to ω_q and the bare ion pseudo-potential to the Coulomb screened, vertex-corrected, and renormalized ion pseudo-potential. Furthermore, this is the *same*‡ dressed potential that occurs in the band calculations [see Eq. (18)].

In addition to the Σ_{ph} one-phonon graph, there is another type of one-phonon graph, shown as the second graph in Fig. 19, which cannot be separated into two parts connected only by the electron and phonon propagators. Examination of the intermediate states which contribute to this type of graph shows that their contribution to the self-energy is a slowly varying function of energy and momentum. It can be absorbed into the chemical potential and small shifts of Z_c and $\tilde{\varepsilon}_p$. Finally, according to Migdal's theorem, contributions from the third class of self-energy graphs containing more than one phonon line are of order $\sqrt{m/M}$.

In extending these ideas to the nearly free electron metals, effects of the underlying ion lattice must be taken into account. Through the structure factor $S(q)$ and the dressed pseudo-potential, the ion lattice determines the band structure. The structure factor, pseudo-potential, and band structure, along with the Coulomb correlations, determine the phonon dispersion relations. For the nearly free electron metals the band effects can be taken into account by perturbation theory. In the electron–ion interactions, crystal momentum is conserved modulo a reciprocal lattice vector K. For the nearly free electron it is convenient to treat the electrons in an extended zone scheme and the phonons in a reduced zone scheme which is formally extended periodically throughout

‡ The same dressed pseudo-potential enters here because the energy transfer is small compared to the band widths and the Fermi energy.

momentum space to allow for Umklapp processes. The dressed electron–phonon matrix element has the form

$$\bar{g}_{pp'\lambda} = \left(\frac{N}{2M\omega_{p-p'\lambda}}\right)^{1/2} \boldsymbol{\epsilon}_{p-p'\lambda}\cdot\langle p|\,\nabla v\,|p'\rangle \tag{34}$$

where ∇v is the gradient of the dressed pseudo-potential whose plane-wave matrix elements are given by‡

$$-i(\mathbf{p}-\mathbf{p}')\frac{\Lambda_{p_F}(p-p')\,V(p-p')}{Z_c\epsilon(p-p',0)} \tag{35}$$

The periodic extension of the phonon system is such that if $\mathbf{p}-\mathbf{p}'$ falls outside the first Brillouin zone, the corresponding reduced wave vector in the first zone is used in determining $\omega_{p-p'\lambda}$ and $\boldsymbol{\epsilon}_{p-p'\lambda}$. The full momentum transfer is used in the proper vertex part, the static dielectric constant, and the Fourier transform of the bare pseudo-potential. The states $|p\rangle$ and $|p'\rangle$ are the Bloch pseudo-conduction electron states determined by the dressed pseudo-potential v. Over most of the Fermi surface these Bloch states are well represented by a single plane wave and $\bar{g}_{pp'\lambda}$ is given by combining Eqs. (34) and (35). However, where the Fermi surface intersects a zone boundary, it is essential to use the correct linear combinations for $|p\rangle$ and $|p'\rangle$ in evaluating the matrix element which enters $\bar{g}_{pp'\lambda}$. In this region p' and p can differ by a reciprocal lattice vector K so that $\omega_{p'-p}$ can vanish. Sham and Ziman (*46*) have shown that when the Bloch states appropriate to the mixing produced by v are used, the matrix element $\langle p|\,\bar{\nabla} v\,|p'\rangle$ vanishes in the vicinity of a zone boundary in such a way that $|\bar{g}_{pp'\lambda}|^2$ is proportional to the reduced-phonon wave vector. This is, of course, just the conclusion one would reach using a deformation potential approach which relates the long-wavelength coupling to the observable effects of homogeneous strains.

Also, in the nearly free electron metals, additional electron self-energy graphs shown in Fig. (20) arise from intermediate electron states involving a reciprocal lattice momentum transfer. These Umklapp graphs contribute to the irreducible self-energy because the momenta of the intermediate electron lines are not equal to the momentum of the external electron. Their contribution relative to the single electron–phonon skeleton $\tilde{\Sigma}_{\text{ph}}$ is smaller by a fraction of order $\omega_D/\tilde{\epsilon}_{p+Q}$ for each intermediate electron state. In discussing correlation effects associated with phonons we will be interested in the electron self-energy for p near the Fermi surface. Therefore, the intermediate states of momentum $p+K$, where K

‡ It is worth pointing out that in actual practice the dressed pseudo-potential

$$\Lambda_{p_F}(p')\,V(p-p')/\epsilon(p-p',0)$$

may be simply a convenient analytic form fit to the Fermi surface data.

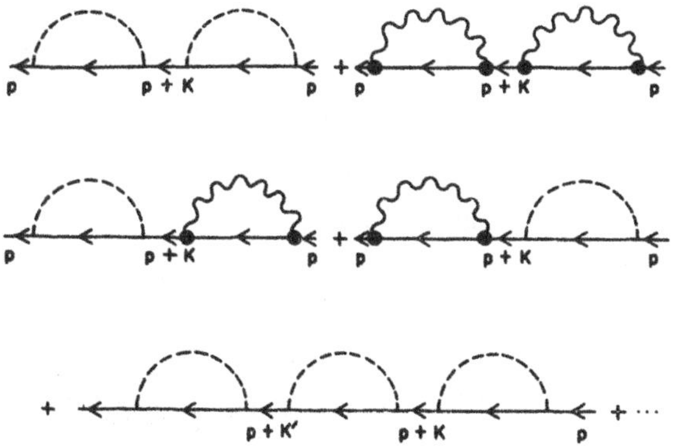

Fig. 20. Umklapp corrections to the irreducible self-energy. Here Q and Q' are reciprocal lattice vectors.

is a reciprocal lattice vector, are in general far from the Fermi surface (unless p happens to be very near a zone boundary). Therefore, except for the small amount of phase space near the zone boundaries, these particular Umklapp corrections are negligible.

Engelsberg and Schrieffer (*27*) have evaluated the phonon contribution to the normal electron-self energy $\tilde{\Sigma}_{\text{ph}}$ for both Debye and Einstein models of the phonon spectra. Here we review some of their results for the simple case of the Einstein model. The phonon dispersion relation in this case is a constant and the effective electron–phonon coupling $\bar{g}(q)$ is taken as a constant $\bar{g}$. Then the integral equation (32) for $\tilde{\Sigma}_{\text{ph}}$ becomes

$$\tilde{\Sigma}_{\text{ph}}(p, \omega) = i\bar{g}^2 \int \frac{d^4q}{(2\pi)^4} \frac{2\omega_0}{(\omega - q_0)^2 - \omega_0^2 + i\delta} \frac{1}{q_0 - \tilde{\epsilon}_q - \tilde{\Sigma}_{\text{ph}}(q, q_0)}$$

As first pointed out by Migdal, this nonlinear integral equation can be solved to order $\sqrt{m/M}$. First we note that $\tilde{\Sigma}_{\text{ph}}$ is independent of momentum p. This is a consequence of the Einstein model which corresponds to a zero-range interaction. In a more realistic model, $\tilde{\Sigma}_{\text{ph}}$ is a slowly varying function of momentum because the screening reduces the range of the interaction to the Thomas–Fermi screening length and the analysis is similar to the present case. Writing $d^3q = N(\tilde{\epsilon}_q)\, d\tilde{\epsilon}_q$ and noting that the phonon propagator restricts the important contributions to near the Fermi surface so $N(\tilde{\epsilon}_q)$ can be replaced by its value $N(0)$ at the Fermi surface, we have

$$\tilde{\Sigma}_{\text{ph}}(\omega) = i\bar{g}^2 N(0) \int dq_0 \int d\tilde{\epsilon}_q \frac{2\omega_0}{(\omega - q_0)^2 - \omega_0^2 + i\delta} \frac{1}{q_0 - \tilde{\epsilon}_q - \tilde{\Sigma}_{\text{ph}}(q_0)}$$

Extending the $d\tilde{\epsilon}_q$ integral so that it runs from $-\infty$ to ∞ introduces a correction of order ω_D/E_F. Then using the analytic properties of $\tilde{\Sigma}_{\text{ph}}(q_0)$ which imply that the sign of $\operatorname{Im} \tilde{\Sigma}_{\text{ph}}(q_0)$ is negative for $q_0 > 0$ and positive for $q_0 < 0$, the $d\tilde{\epsilon}_q$ integration gives

$$\int d\tilde{\epsilon}_q \frac{1}{q_0 - \tilde{\epsilon}_q - \tilde{\Sigma}_{\text{ph}}(q_0)} = -i\pi \operatorname{sgn}(q_0)$$

Carrying out the final q_0 integration one obtains

$$\operatorname{Re} \tilde{\Sigma}_{\text{ph}}(\omega) = -\bar{g}^2 N(0) \ln \left| \frac{\omega_0 + \omega}{\omega_0 + \omega} \right|$$

$$\operatorname{Im} \tilde{\Sigma}_{\text{ph}}(\omega) = \begin{cases} 0 & |\omega| < \omega_0 \\ -\pi \bar{g}^2 N(0) \operatorname{sgn} \omega & |\omega| > \omega_0 \end{cases} \tag{36}$$

The onset of the imaginary part for $|\omega| > \omega_0$ corresponds to the fact that phonon emission processes are not possible until the energy is equal to or greater than the phonon frequency ω_0.

For $|\tilde{\epsilon}_p| \ll \omega_0$, the spectral weight function

$$A(p, \omega) = \frac{1}{\pi} |\operatorname{Im} G(p, \omega)|$$

has a delta-function spike at

$$\omega = \frac{\tilde{\epsilon}_p}{1 + [2\bar{g}^2 N(0)/\omega_0]} \tag{37}$$

The phonon renormalization of the mass $2\bar{g}^2N(0)/\omega_0$ is a measure of the strength of the electron–phonon coupling. For the weak-coupling superconductors it may be of order $\frac{1}{4}$ to $\frac{1}{2}$, while for Pb and Hg it is greater than 1. The ratio of the width $2|\operatorname{Im} \tilde{\Sigma}_{\text{ph}}|$ to the peak position for $\omega \sim \omega_0$ is proportional

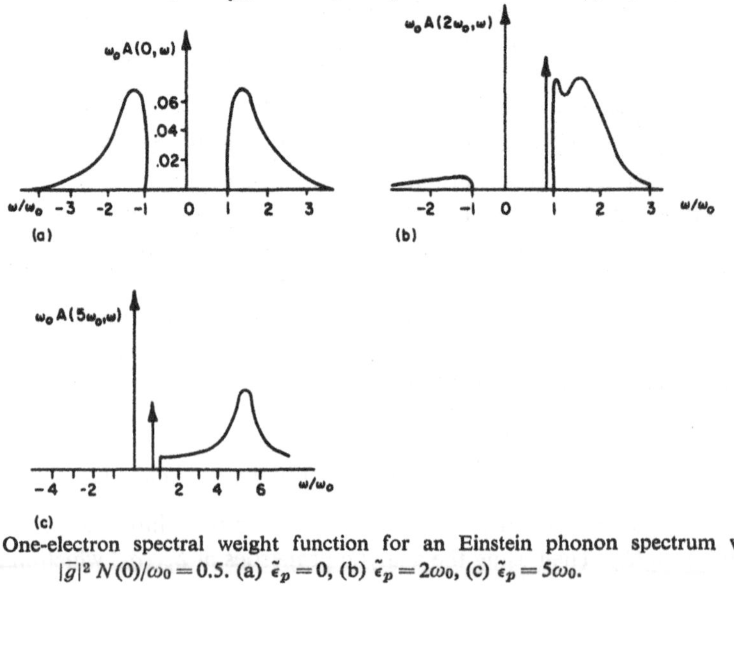

Fig. 21. One-electron spectral weight function for an Einstein phonon spectrum with $|\bar{g}|^2 N(0)/\omega_0 = 0.5$. (a) $\tilde{\epsilon}_p = 0$, (b) $\tilde{\epsilon}_p = 2\omega_0$, (c) $\tilde{\epsilon}_p = 5\omega_0$.

to this same factor, indicating that quasi-particles are not well defined in this frequency region. In Fig. 21, some spectral weights for the Einstein model obtained by Schrieffer and Englesberg are plotted. For $\tilde{\varepsilon}_p \gg \omega_0$, the broad peak near $\omega \sim \tilde{\varepsilon}_p$ is associated with the quasi-particle. The center of this peak approaches $\tilde{\varepsilon}_p$ for ω greater than ω_0, showing that the phonon renormalization effects become negligible for ω several times larger than ω_0. The weight of the δ-function just below ω_0 goes exponentially to zero as ω increases beyond several times ω_0. Therefore, for $\omega \ll \omega_0$ or $\omega \gg \omega_0$, the quasi-particle picture is useful for the electron–phonon system. However, as Fig. 21b shows, when $\omega \sim \omega_0$ the spectral weight is a complicated multipeaked function and the simple quasi-particle picture is misleading. If a more realistic phonon relation is used, the continuum extends down to the origin and the quasi-particle peak is slightly broadened even for $\omega \ll \omega_0$. For this case, the structure of the spectral weight when ω is near the characteristic peak frequencies of the phonon density of states is similar to that of the Einstein model for $\omega \sim \omega_0$, and the quasi-particle approximation is not appropriate.

III. ELECTRON SELF-ENERGY OF A STRONG-COUPLING SUPERCONDUCTOR

Since the determination of the physical properties of the strong-coupling superconductors and the relation of the electron–phonon interaction to these properties is based upon the self-energy equations, it is essential to develop some physical insight into their structure. For this reason, before developing the theory in detail, it seems worthwhile to begin with a simple perturbation calculation of the energy change which occurs when an electron is added to the ground state of a normal metal. Using this, the physical significance of the terms which appear in the normal electron self-energy can be interpreted as discussed by Schrieffer (*47*). Furthermore, it turns out that the electron–phonon matrix elements and the phonon density of states enter this expression in the same form that they appear in the self-energy of the superconducting phase. With this background, the zero-temperature form of the self-energy equations for the superconductor are introduced. Proceeding by analogy, the various terms are related to the underlying physical processes and some feeling for the physics represented in the algebraic structure of the equations can be obtained. Following this introduction, a detailed derivation of the finite temperature equations for the superconducting electron self-energy is presented. Some properties of the electron–phonon interaction are discussed with some detailed results for Pb given. Then solutions for the electron self-energy of Pb are compared with the results of electron tunneling experiments.

A. Eliashberg Equations

When an electron of momentum p is added to the ground state of a normal system, the zeroth-order change in energy relative to the chemical potential is simply

$$\epsilon_p = (p^2/2m) - \mu$$

The lowest-order energy shift due to the vitual emission and reabsorption of a phonon by the added electron is given by the golden rule:

$$\sum_{p'} |\bar{g}_{pp'}|^2 \frac{1 - f_{p'}}{\epsilon_p - \epsilon_{p'} - \omega_{p-p'}} \tag{38}$$

Here $\bar{g}_{pp'}$ is the dressed electron–phonon coupling and $\omega_{p-p'}$ is the energy of the virtual dressed phonon. A typical process which contributes to this sum is illustrated in Fig. 22a. The Fermi factor $(1 - f_{p'})$ ensures that the state to which

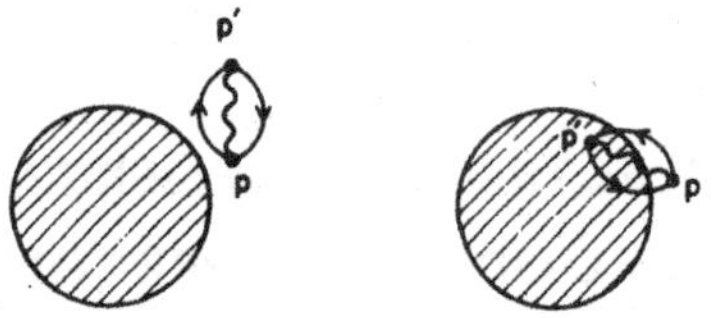

Fig. 22. (a) Fluctuation in which the electron in state p virtually emits a phonon, goes to an unoccupied state p', absorbs the phonon, and returns to p. (b) Similar virtual transition of an electron from the Fermi sea to the state p which is blocked by the addition of an electron of momentum p.

the electron fluctuates is not occupied. A further consequence of the Pauli principle is that the added electron in state **p** blocks transitions of the type illustrated in Fig. 22b, in which an electron in the Fermi sea emits a phonon and goes virtually to **p**. These transitions previously contributed to lowering the ground-state energy by an amount

$$\sum_{p'} \frac{|g_{pp'}|^2 f_{p'}}{\epsilon_{p'} - \epsilon_p - \omega_{p-p'}} \tag{39}$$

Because they no longer can occur, the energy ω required to add a particle of momentum p is given by taking ϵ_p minus the blocking energy, Eq. (39), plus the transition energy, Eq. (38):

$$\omega = \epsilon_p + \sum_{p'} |\bar{g}_{pp'}|^2 \left(\frac{f_{p'}}{\epsilon_p + \omega_{p-p'} - \epsilon_{p'}} - \frac{1 - f_{p'}}{\epsilon_{p'} - \epsilon_p + \omega_{p-p'}} \right) \tag{40}$$

The correction to ϵ_p represents an approximation to the electron self-energy Σ. From the discussion in Section II we know that an expression for the normal self-energy which is accurate to order $\sqrt{m/M}$ can be obtained by making the Brillouin–Wigner-like substitution $\epsilon_p \to \omega$ in the golden rule approximation. For this reason, we replace ϵ_p by ω in Eq. (40) and consider the self-energy expression

$$\Sigma(p, \omega) = \sum_{p'} |\bar{g}_{pp'}|^2 \left(\frac{f_{p'}}{\omega + \omega_{p-p'} - \epsilon_{p'} - i\delta} - \frac{1 - f_{p'}}{\epsilon_{p'} - \omega + \omega_{p-p'} - i\delta} \right) \tag{41}$$

Here a small imaginary part $i\delta$ has been subtracted from each denominator. The real part of Σ is the energy shift and the imaginary part of Σ is given by

$$\operatorname{Im} \Sigma = -\pi \sum_{p'} |\bar{g}_{pp'}|^2 (1 - f_{p'})\, \delta(\epsilon_{p'} + \omega_{p-p'} - \omega) \qquad \omega > 0$$

The golden rule expression for the lifetime of the added electron is

$$1/\tau = 2\,|\operatorname{Im} \Sigma|$$

so that $|\operatorname{Im} \Sigma|$ is clearly just the level-width parameter.

The electron excitations of interest have ω less than or of order ω_D and p of order p_F. Writing the sum as an integration

$$\sum_{p'} = \int d\epsilon_{p'} \int \frac{d\Omega_{p'}}{(2\pi)^3}\, mp'$$

and changing $\epsilon_{p'}$ to $-\epsilon_{p'}$ in the first term one obtains

$$\Sigma(p, \omega) = \int_0^\infty d\epsilon_{p'} \int \frac{d\Omega_{p'}}{(2\pi)^3}\, mp' |\bar{g}_{pp'}|^2 \times \left(\frac{1}{\omega + \omega_{p-p'} + \epsilon_{p'} - i\delta} - \frac{1}{\epsilon_{p'} - \omega + \omega_{p-p'} - i\delta} \right) \tag{42}$$

Since ω and $\omega_{p-p'}$ are less than or of order ω_D, the dominant contribution to the ϵ_p integration comes for $\epsilon_{p'} \gtrsim \omega_D$ and therefore p' is essentially restricted to the Fermi surface. This rapid convergence of the $\epsilon_{p'}$ integration has already been used in a trivial way to extend the limit of integration of the first term from μ to ∞. More importantly, the restriction of p' to the region of the Fermi surface means that the angular integration in Eq. (42) can be written as an average over the Fermi surface S_F:

$$\int d\Omega_{p'}\, mp' = \int_{S_F} \frac{dp'^2}{v'_F}$$

Here v'_F is the Fermi velocity at p'_F on the Fermi surface S_F.

Two more steps are necessary to cast this into the standard form used in the gap equations. First, a δ-function is inserted to separate the phonon contribution in a convenient manner:

$$\Sigma(p,\omega)=\int_0^\infty d\epsilon_{p'}\int_0^\infty d\omega_0\int\frac{dp'^2}{v_F'(2\pi)^3}|g_{pp'}|^2\,\delta(\omega_0-\omega_{p-p'})\times\left(\frac{1}{\omega+\omega_0+\epsilon_{p'}-i\delta}-\frac{1}{\epsilon_{p'}-\omega+\omega_0-i\delta}\right)$$

and finally $\Sigma(p,\omega)$ is averaged over the Fermi surface, giving a self-energy $\Sigma(\omega)$ appropriate to an isotropic or dirty material:

$$\Sigma(\omega)=\int_0^\infty d\epsilon_{p'}\int_0^\infty d\omega_0\,\alpha^2(\omega_0)\,F(\omega_0)\times\left(\frac{1}{\epsilon_{p'}+\omega+\omega_0-i\delta}-\frac{1}{\epsilon_{p'}-\omega+\omega_0-i\delta}\right)\tag{43}$$

Here $F(\omega_0)$ is the phonon density of states

$$F(\omega_0)=\sum_\lambda\int\frac{d^3p}{(2\pi)^3}\,\delta(\omega_0-\omega_{p\lambda})\tag{44}$$

and $\alpha^2(\omega_0)$ is an effective electron phonon coupling defined by

$$\alpha^2(\omega_0)\,F(\omega_0)=\int_{S_F}d^2p\int_{S_F}\frac{dp'^2}{(2\pi)^3\,v_F'}\times\sum_\lambda|g_{pp'\lambda}|^2\,\delta(\omega-\omega_{p-p'\lambda})\Big/\int_{S_F}d^2p\tag{45}$$

In this final form the phonon polarization index λ has been added.

With this notation established, we turn to an interpretation of the zero-temperature Eliashberg equations for the renormalization parameter Z and the gap Δ which will be derived in Section III.B.

$$[1-Z(\omega)]\,\omega=\int_{\Delta_0}^\infty d\omega'\,\mathrm{Re}\left[\frac{\omega'}{(\omega'^2-\Delta'^2)^{1/2}}\right]\times\int_0^\infty d\omega_0\,\alpha^2(\omega_0)\,F(\omega_0)\left(\frac{1}{\omega'+\omega+\omega_0-i\delta}-\frac{1}{\omega'-\omega+\omega_0-i\delta}\right)\tag{46}$$

$$\Delta(\omega)=\frac{1}{Z(\omega)}\int_{\Delta_0}^\infty d\omega'\,\mathrm{Re}\left[\frac{\Delta'}{(\omega'^2-\Delta'^2)^{1/2}}\right]\times\int_{\Delta_0}^\infty d\omega_0\,\alpha^2(\omega_0)\,F(\omega_0)\left(\frac{1}{\omega'+\omega+\omega_0-i\delta}+\frac{1}{\omega'-\omega+\omega_0-i\delta}\right)-\frac{\mu^*}{Z(\omega)}\int_{\Delta_0}^{\omega_c}d\omega'\,\mathrm{Re}\left[\frac{\Delta'}{(\omega'^2-\Delta'^2)^{1/2}}\right]\tag{47}$$

Here $\Delta_0 = \Delta(\Delta_0)$ is the gap at the gap edge which is real and μ^* is a Coulomb pseudo-potential which takes account of the screened Coulomb interaction in a convenient manner for numerical calculations. The factors $\alpha^2(\omega)\,F(\omega)$ are just the same as those defined by Eqs. (44) and (45) for the normal self-energy.

The "normal" part of the self-energy $(1 - Z)\,\omega$ can be easily understood within the framework of our discussion of the normal state. Comparing Eq. (46) with the previous expression Eq. (43) for the normal-state electron self-energy, one sees $\epsilon_{p'}$ has been written as ω', and a superconducting density of states factor $\mathrm{Re}\,[\omega'/(\omega'^2 - \Delta'^2)^{1/2}]$ has been introduced to take into account the modifications of the ground state. The underlying physical significance of the contributions remains unchanged: The first represents the blocking of virtual transitions to p and the second represents the contribution of the virtual transitions from p to unoccupied states. Note that just as before, Z will have an imaginary part associated with real energy-conserving processes.

The terms appearing in the equation for Δ can be given a similar interpretation. The replacement of $\mathrm{Re}\,[\omega'/(\omega'^2 - \Delta'^2)^{1/2}]$ by $\mathrm{Re}\,[\Delta'/(\omega'^2 - \Delta'^2)^{1/2}]$ and the relative sign change between the energy denominators have their origins in the coherence factor Δ/ω associated with pair-scattering processes. The interpretation of the two terms is simplified by considering the special case of a particle added at the Fermi surface (momentum $\mathbf{p}_F'$) with energy $\omega = \Delta_0$. Part of this energy arises from blocking of virtual pair transitions to the state $(\mathbf{p}_F', -\mathbf{p}_F')$ and has the energy denominator $\omega' + \Delta_0 + \omega_0$. The remainder of the energy corresponds to virtual processes in which a pair breaks up with one member going to form a pair with the added electron and the other member forming an excited quasi-particle. These processes give the second energy denominator $\omega' - \Delta_0 + \omega_0$ appearing in Eq. (47).

B. Self-Energy Equations

At the superconducting transition temperature the theory for the normal metals, Section II, breaks down. The particle–particle scattering vertex corrections shown in Fig. 23, while individually proportional to powers of the adiabatic expansion parameter $\sqrt{m/M}$, sum to a contribution which diverges

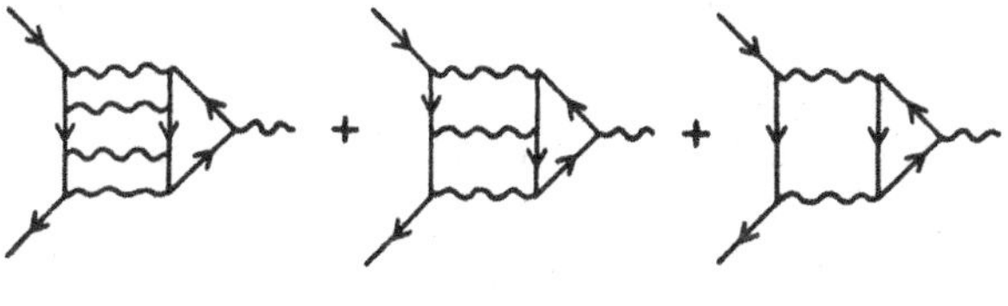

Fig. 23. Particle–particle scattering vertex corrections which sum to a divergent expression at the superconducting transition temperature.

at the transition temperature. Here, an expression for the electron self-energy will be derived using the Nambu (*48*)–Gor'kov (*49*) formalism, which removes this divergence while preserving the simplicity of the usual graphical method. Within this framework, Eliashberg (*50*) extended Migdal's techniques to provide a formal treatment of the electron–phonon interaction to order $\sqrt{m/M}$. Morel and Anderson (*51*) considered the additional effects of the repulsive screened Coulomb interaction on the "pairing" self-energy, but treated the "normal" self-energy in the effective mass approximation. To construct a theory of the strong-coupling superconductors it is necessary to treat the "normal" and "pairing" parts of the self-energy on the same footing. Nambu's two-component notation provides a natural way of consistently calculating both the normal and pairing parts of the electron self-energy. Using this formalism, Scalapino, Schrieffer, and Wilkins (*52*) have constructed equations for the electron self-energy which take account of the Coulomb as well as the phonon interaction between electrons. McMillan and Rowell (*53*) have shown how the Coulomb renormalization parameter enters the theory in a consistent manner. For the reader who is not interested in the details of the following analysis, the main results have been summarized on page 500.

In Nambu's scheme, a two-component electron field operator

$$\Psi_p = \begin{pmatrix} c_{p\uparrow} \\ c^\dagger_{-p\downarrow} \end{pmatrix}$$

and its adjoint

$$\Psi_p^\dagger = (c^\dagger_{p\uparrow}, c_{-p\downarrow})$$

are introduced. Here $c^\dagger_{p\uparrow}$ creates an electron in a state of momentum p and spin up and $c_{-p\downarrow}$ destroys an electron in the time-reversed state. Introducing the Pauli spin matrices τ_i, and subtracting μN, the Hamiltonian discussed in Section II.B for the electron ion system can be written as

$$\mathscr{H} = \sum_p \epsilon_p \Psi_p^\dagger \tau_3 \Psi_p + \sum_{q\lambda} \Omega_{q\lambda} b^\dagger_{q\lambda} b_{q\lambda} + \sum_{qp} V(q)\, S(q)\, \Psi^\dagger_{p+q} \tau_3 \Psi_p$$
$$+ \sum_{pp'\lambda} g_{pp'\lambda} \phi_{p-p'\lambda} \Psi^\dagger_{p'} \tau_3 \Psi_p + \frac{1}{2} \sum_{pkq} \frac{4\pi e^2}{q^2} (\Psi^\dagger_{p-q} \tau_3 \Psi_p)(\Psi^\dagger_{k+q} \tau_3 \Psi_k) \tag{48}$$

Using Nambu's notation (*48*), a thermodynamic electron Green's function matrix is defined by

$$G(p, \tau) = -\langle UT\Psi_p(\tau)\, \Psi_p^\dagger(0)\rangle \tag{49}$$

The average represents a trace over the grand canonical ensemble generated from $\mathscr{H}$, and the fields evolve in imaginary time $i\tau$ according to

$$\Psi_p(\tau) = \exp(\mathscr{H}\tau)\, \Psi_p(0) \exp(-\mathscr{H}\tau)$$

The symbol T specifies a τ-ordered product and the operator U is given by

$$U = 1 + P^+ + P$$

where P^+ acting on a state of the N-particle system generates the corresponding state for the $(N+2)$-particle system. Writing G out in component form,

$$G = -\begin{pmatrix} \langle Tc_{p\uparrow}(\tau)\, c^\dagger_{p\uparrow}(0)\rangle & \langle P^\dagger Tc_{p\uparrow}(\tau)\, c_{-p\downarrow}(0)\rangle \\ \langle PT^\dagger_{c-p\downarrow}(\tau)\, c^\dagger_{p\uparrow}(0)\rangle & \langle Tc^\dagger_{-p\downarrow}(\tau)\, c_{-p\downarrow}(0)\rangle \end{pmatrix}$$

one sees that the diagonal components G_{11} and G_{22} are the normal Green's functions for up-spin electrons and down-spin holes, respectively. The off-diagonal components G_{12} and G_{21} are just the anomalous Green's functions F and $\bar{F}$ introduced by Gor'kov. Physically they describe the pairing correlations as discussed in Chapter 5, where Gor'kov's treatment of the superconductor was developed. The importance of Nambu's formalism is that in this matrix form, the familiar Feynman–Dyson rules for perturbation theory hold. Schrieffer (*47*) has given a detailed discussion of this and here we will simply proceed directly to the calculation of the electron self-energy.

The Fourier series representation of the electron and phonon propagators are

$$G(p, \tau) = 1/\beta \sum_{n=-\infty}^{\infty} \exp(-i\omega_n\tau)\, G(p, i\omega_n)$$

$$D_\lambda(q, \tau) = 1/\beta \sum_{\nu=-\infty}^{\infty} \exp(-i\omega_\nu\tau)\, D_\lambda(q, i\omega_\nu)$$

where a convention will be used in which the Roman-indexed subscripts are fermion-like, $\omega_n = (2n+1)\pi/\beta$, and the Greek-indexed subscripts are bosons, $\omega_\nu = 2\nu\pi/\beta$, with n and ν integers. The Dyson equation for the electron Green's function is now a 2×2 matrix equation

$$G^{-1}(p, i\omega_n) = i\omega_n\mathbf{1} - \epsilon_p\tau_3 - \Sigma(p, i\omega_n) \tag{50}$$

where Σ has the canonical form

$$\Sigma(p, i\omega_n) = (1 - Z(p, i\omega_n))\, i\omega_n\mathbf{1} + \phi(p, i\omega_n)\,\tau_1 + \phi'(p, i\omega_n)\,\tau_2 + \chi(p, i\omega_n)\,\tau_3 \tag{51}$$

For our purposes, the pair phase can be chosen so that ϕ' vanishes. It should be kept clearly in mind that the "phase" of the gap, which is so important in Josephson tunneling, is associated with the relative τ_1, τ_2 weights. It is not associated with the ratio of the real and imaginary parts of ϕ.

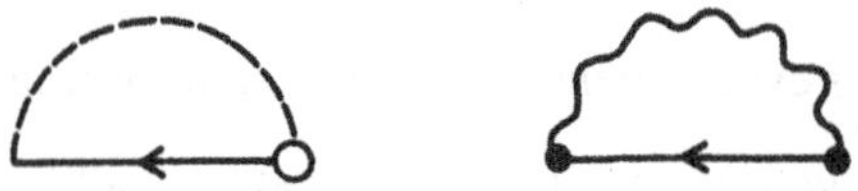

Fig. 24. Electron self-energy diagrams for the screened Coulomb (dashed line) and dressed phonon (wavy line) exchange by the self-consistently dressed electron propagator (solid line). The Coulomb diagram contains the proper Coulomb vertex Λ and the electron–phonon vertices are understood to be dressed by Λ divided by the Coulomb dielectric constant.

The skeleton diagrams for the electron self-energy are formally identical to those for the normal metal. With the particle–particle scattering divergence eliminated by the pairing self-energy ϕ which appears in the Nambu–Gor'kov formalism, our approximations are once again motivated by the same characteristic space–time properties of the interactions as in the normal metal case. The two basic self-energy diagrams discussed in Section II.F are repeated in Fig. 24. The only modifications in the rules for evaluating the contribution of a diagram are the insertion of τ_3 Pauli matrices for each electron–Coulomb and electron–phonon vertex. A closed electron loop contributes a factor (-1) and is traced over. With this in mind, the contributions to the Coulomb part of the self-energy Σ_c and the phonon part Σ_{ph} are constructed in the standard way from the diagrams shown in Fig. 24.

$$\Sigma_c(p, i\omega_n) = -1/\beta \sum_{mp'} V_{sc}(p-p', i\omega_n - i\omega_m)\, \Lambda(p, p'; i\omega_n, i\omega_m)\, \tau_3 G(p', i\omega_m)\, \tau_3 \tag{52}$$

$$\Sigma_{\text{ph}}(p, i\omega_n) = -1/\beta \sum_{mp'\lambda} \left| \frac{\Lambda(p, p')\, g_\lambda(p-p')}{\epsilon(p-p')} \right|^2 \times D_\lambda(p-p', i\omega_n - i\omega_m)\, \tau_3 G(p', i\omega_m)\, \tau_3 \tag{53}$$

Here Λ is the proper Coulomb vertex, and the static limits of Λ and the Coulomb dielectric constant $\epsilon(p-p')$ have been used in constructing the electron–phonon vertices. These vertices, the screened Coulomb interaction V_{sc}, and the phonon propagators D_λ are essentially unchanged from their normal-state values. Separating the integral equation for the Coulomb parts into its various components one obtains

$$\begin{aligned}
(1 - Z_c(p, i\omega_n))\, i\omega_n &= -\frac{1}{\beta} \sum_{mp'} \frac{V_{sc}(p-p', i\omega_n - i\omega_m)\, \Lambda Z(p', i\omega_m)\, i\omega_m}{[Z(p', i\omega_m)\, i\omega_m]^2 - \epsilon_{p'}^2 - \phi^2(p', i\omega_m)} \\
\chi_c(p, i\omega_n) &= -\frac{1}{\beta} \sum_{mp'} \frac{V_{sc}(p-p', i\omega_n - i\omega_m)\, \Lambda \epsilon_{p'}}{[Z(p', i\omega_m)\, i\omega_m]^2 - \epsilon_{p'}^2 - \phi^2(p', i\omega_m)} \\
\phi_c(p, i\omega_n) &= \frac{1}{\beta} \sum_{mp'} \frac{V_{sc}(p-p', i\omega_n - i\omega_m)\, \Lambda \phi(p', i\omega_m)}{[Z(p', i\omega_m)\, i\omega_m]^2 - \epsilon_{p'}^2 - \phi^2(p', i\omega_m)}
\end{aligned} \tag{54}$$

In the first two equations, the dominant contributions come from energies of the order of the Fermi energy, so that Z_c and χ_c can be replaced by their normal-state values to order Δ_0/E_F. Before dealing with the Coulomb contribution to the gap parameter ϕ_c, the phonon contribution to Σ will be simplified.

To convert the frequency sum to an integral it is convenient to introduce spectral representations for G and D:

$$G(p, i\omega_n) = -\int_{-\infty}^{\infty} \frac{d\omega'}{\pi} \frac{\operatorname{Im} G(p, \omega' + i\delta)}{i\omega_n - \omega'} \tag{55}$$

$$D_\lambda(q, i\omega_\nu) = \int_0^{\infty} d\Omega B_\lambda(q, \Omega) \frac{2\Omega}{(i\omega_\nu)^2 - \Omega^2} \tag{56}$$

Here $G(p, \omega' + i\delta)$ is the bounded analytic continuation of $G(p, i\omega_n)$ onto the real axis from the upper half ω'-plane, and the phonon spectral weight $B_\lambda(q, \Omega)$ is given by

$$B_\lambda(q, \omega) = [1 - \exp(\beta\omega)] \times \sum_{nm} \exp(-\beta E_n) |\langle m| \phi_{q\lambda}^{+} |n\rangle|^2 \, \delta(\omega - E_{mn}) / \sum_n \exp(-\beta E_n) \tag{57}$$

where $\mathscr{H}|n\rangle = E_n|n\rangle$. We could, of course, have expressed $B_\lambda(q, \omega)$ in terms of the analytic continuation of D making the two spectral representations appear more similar. However, while G will be self-consistently calculated, the expression for D_λ eventually enters as a phonon density of states which will be estimated from neutron scattering and electron tunneling experiments. The explicit form for the phonon spectral weight given by Eq. (57) lends itself more naturally to this approach.

The summation which appears in Eq. (53) for Σ_{ph} is now evaluated as follows:

$$-\frac{1}{\beta}\sum_m G(i\omega_m) D(i\omega_n - i\omega_m) = \int_{-\infty}^{\infty} \frac{d\omega'}{\pi} \int_0^{\infty} d\Omega \, B(\Omega) \operatorname{Im} G(\omega' + i\delta) \times -\frac{1}{\beta}\sum_m \frac{1}{i\omega_m - \omega'} \frac{2\Omega}{(i\omega_n - i\omega_m)^2 - \Omega^2}$$

Using the Poisson summation procedure

$$-\frac{1}{\beta}\sum_m \frac{1}{i\omega - \omega'} \frac{2\Omega}{(i\omega - i\omega)^2 - \Omega^2} = \int_c \frac{dz}{2\pi i} f(z) \frac{1}{z - \omega'} \frac{2\Omega}{(i\omega_n - z)^2 - \Omega^2}$$

Here the contour C encircles all the points $i\omega_m$ in a counterclockwise manner. Folding out the contour and evaluating the residues, this reduces to

$$[f(-\omega') + n(\Omega)] \frac{1}{\omega' + \Omega - i\omega_n} + [f(\omega') + n(\Omega)] \frac{1}{\omega' - \Omega - i\omega_n}$$

with $f(\omega')$ and $n(\Omega)$ the Fermi $[\exp(\beta\omega')+1]^{-1}$ and Bose $[\exp(\beta\Omega)-1]^{-1}$ factors, respectively. In obtaining this result we have set $f(i\omega_n-\Omega)=-n(-\Omega)$, which is necessary to obtain the correction analytic continuation. Substituting this into the original equation for Σ_{ph} and analytically continuing $i\omega_n$ to $\omega(\text{Im}\,\omega=0^+)$ one finds

$$\Sigma_{\text{ph}}(p,\omega)=\int_{-\infty}^{\infty}\frac{d\omega'}{\pi}\int_0^{\infty}d\Omega\int\frac{d^3p'}{(2\pi)^3}\tau_3\,\text{Im}\,G(p',\omega'+i\delta)\,\tau_3\sum_\lambda B_\lambda(p-p',\Omega)$$
$$\times\left|\frac{\Lambda(p,p')}{\epsilon(p,p')}g_\lambda(p,p')\right|^2\left[\frac{f(-\omega')+n(\Omega)}{\omega'+\Omega-\omega}+\frac{f(\omega')+n(\Omega)}{\omega'-\Omega-\omega}\right] \tag{58}$$

The excitation frequencies ω of interest are small compared to the Fermi energy, so that the phonon energy denominators provide a natural cutoff on the ω'-integration at energies of order ω_D. This implies through the variation of $\text{Im}\,G(p',\omega'+i\delta)$ that the major contribution comes from momentum states p' near the Fermi surface. In this region the Coulomb contribution $Z_c(p,\omega)$ can be replaced by its value at the Fermi surface Z_c and $\chi_c(p,\omega)$ can be expanded to give a shift of the chemical potential and a Coulomb mass renormalization. The resulting Green's function has the form

$$G^{-1}(p,\omega)\simeq Z_c(p_F)\,[\tilde{Z}_{\text{ph}}(p,\omega)\,\omega\mathbf{1}-(\tilde{\epsilon}_p+\tilde{\chi}_{\text{ph}}(p,\omega))\,\tau_3-\tilde{\phi}(p,\omega)\,\tau_1] \tag{59}$$

where

$$\begin{aligned}1-\tilde{Z}_{\text{ph}}&=\frac{1-Z_{\text{ph}}}{Z_c}\\ \tilde{\epsilon}_p&=\frac{\epsilon_p}{Z_c}\left(1+\frac{\partial\chi_c}{v_F\partial p}\right)\\ \tilde{\chi}_{\text{ph}}&=\chi_{\text{ph}}/Z_c\\ \tilde{\phi}&=\phi/Z_c\end{aligned} \tag{60}$$

Now, although the notation may appear to be getting out of hand, if Eq. (58) is divided by Z_c we obtain the basic integral equation which will be used to determine the Coulomb-renormalized, phonon part of the self-energy $\tilde{\Sigma}_{\text{ph}}=\Sigma_{\text{ph}}/Z_c$.

$$\tilde{\Sigma}_{\text{ph}}(p,\omega)=\int_{-\infty}^{\infty}\frac{d\omega'}{\pi}\int_0^{\infty}d\Omega\int\frac{d^3p'}{(2\pi)^3}\tau_3\,\text{Im}\,\tilde{G}(p',\omega'+\delta)\,\tau_3$$
$$\times\sum_\lambda B_\lambda(p-p',\Omega)\,|\bar{g}_\lambda(p,p')|^2\left[\frac{f(-\omega')+n(\Omega)}{\omega'+\Omega-\omega}+\frac{f(\omega')+n(\Omega)}{\omega'-\Omega-\omega}\right] \tag{61}$$

Here

$$\tilde{G}^{-1}=\tilde{Z}_{\text{ph}}(p,p_0)\,\mathbf{1}-(\tilde{\epsilon}_p+\tilde{\chi}_{\text{ph}})\,\tau_3-\tilde{\phi}\tau_1 \tag{62}$$

and $\bar{g}_\lambda(p, p')$ is the effective electron–phonon matrix element introduced in Section II:

$$\bar{g}_\lambda(p, p') = \frac{\Lambda(p, p')}{Z_c} \frac{g_\lambda(p, p')}{\epsilon(p, p')} \tag{63}$$

The form of Eq. (61) shows explicitly how the Coulomb renormalization effects enter the phonon part of the self-energy. It is identical to that used by Scalapino et al. (*52*), who argued that the effect of the Coulomb corrections would lead to essentially constant scale factors multiplying the electron–phonon matrix elements and an effective mass and chemical potential shift. McMillan was the first to show how this happens in detail, as we have just discussed.

The dependence of $\tilde{\Sigma}_{\text{ph}}$ on momentum p arises from the variation of the electron–phonon interaction and the phonon spectral weight with the momentum transfer $\mathbf{p} - \mathbf{p}'$. Since the variation of these is on a scale of p_F we can set $|p| = p_F$ and consider only the angular variation as p moves on the Fermi surface. For the nearly free electron metals, band structure effects near zone boundaries are important only over a small part of the Fermi surface. It is therefore possible to neglect the angular dependence of the self-energy parameters which appear in the integral for $\tilde{\Sigma}_{\text{ph}}$. Then, after solving for $\tilde{\Sigma}_{\text{ph}}$ and obtaining the anisotropy due to the electron–phonon and phonon spectral weight, the mixing effects of the electronic band structure are used to give the appropriate weighted average of $\tilde{\Sigma}_{\text{ph}}(p, \omega)$.‡ With this in mind, the momentum integration can be carried out by changing variables from p' to $\epsilon_{p'}$, $q = |\mathbf{p} - \mathbf{p}'|$ and the azimuthal angle ϕ as shown in Fig. 25. The azimuthal integration gives 2π and

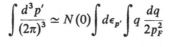

$$\int \frac{d^3p'}{(2\pi)^3} \simeq N(0) \int d\epsilon_{p'} \int q \frac{dq}{2p_F^2}$$

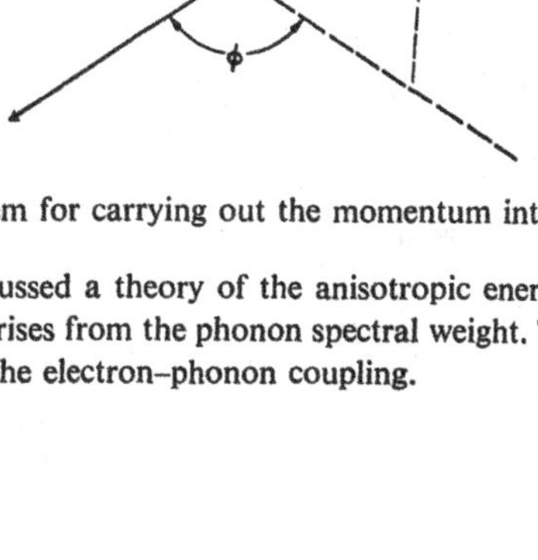

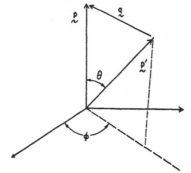

Fig. 25. Coordinate system for carrying out the momentum integral which occurs in $\tilde{\Sigma}_{\text{ph}}$.

‡ Bennett (*54*) has discussed a theory of the anisotropic energy gap for Pb in which he finds that the anisotropy arises from the phonon spectral weight. This calculation ignores the effect of the variation of the electron–phonon coupling.

where $N(0)$ is the single-particle Bloch density of states at the Fermi surface dressed by the Coulomb interactions. For $|\mathbf{p}|$ equal to p_F, the $(\epsilon_{p'}, q)$ region of integration is shown in Fig. 26. The dominant contribution to the integrand occurs over a region of the $\epsilon_{p'}$ variable $\pm\omega_D$ about the Fermi surface. This is of order the thickness of the x-axis in the sketch. The q integration therefore runs from 0 to $2p_F$ and corresponds to averaging p' over the Fermi surface, and the $\epsilon_{p'}$ integration can be formally extended from $-\infty$ to ∞ so that

$$\int_{-\infty}^{\infty} d\epsilon_{p'}\,\tilde{G}(p', \omega) = \frac{-i\pi\,[\omega\tilde{Z}(\omega) + \tilde{\phi}(\omega')\,\tau_1]}{[\omega'^2\tilde{Z}^2(\omega') - \tilde{\phi}^2(\omega')]^{1/2}} \tag{64}$$

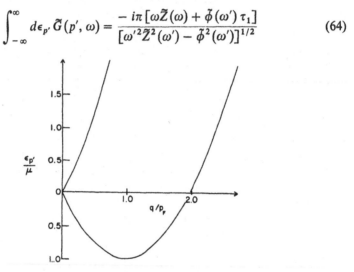

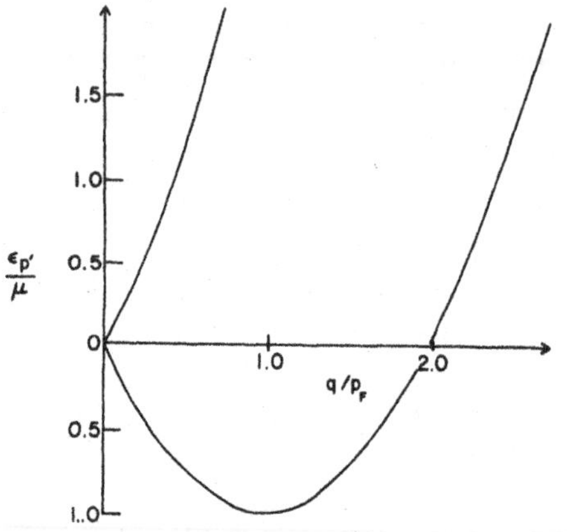

Fig. 26. $(\epsilon_{p'}, q)$ region of integration for $|p| = p_F$.

with the square root defined to have a positive imaginary part for Im $\omega' > 0$. Here the τ_3 part has been eliminated, since it was odd in $\epsilon_{p'}$. This implies $\tilde{\chi}_{\text{ph}}$ contributes an extremely small shift to the chemical potential in the superconducting state, which is given by

$$\mu_N - \mu_s = \mu_N\left(\frac{1}{4}\right)\left(\frac{\Delta}{\mu_N}\right)^2\left[\ln\left(\frac{2\omega_c}{\Delta}\right) - 1\right]$$

Here μ_N and μ_s are the chemical potentials in the normal and superconducting states, respectively, Δ is the gap, and ω_c is an energy of order several times the Debye energy.

Writing the dq integration as an explicit average of p' over the Fermi surface,

$$\int \frac{dq\,q}{2p_F^2} = \int_{FS} \frac{dp'^2}{4\pi p_F^2}$$

the final part of the momentum integration gives

$$\sum_{\lambda}\int_{S_F}\frac{dp'^2}{4\pi p_F^2}\,|\bar{g}_{pp'\lambda}|^2\,B_\lambda(p,p',\Omega)=\sum_{\lambda}\alpha_{p\lambda}^2(\Omega)\,F_{p\lambda}(\Omega)\equiv\alpha_p^2(\Omega)\,F_p(\Omega) \tag{65}$$

where, with the spectral representation, Eq. (57), for B_λ in mind, an effective phonon density of states $F_{p\lambda}(\Omega)$ has been introduced. It gives the density of phonon states of energy Ω and polarization λ whose wave vectors connect the point $\mathbf{p}$ of the Fermi surface with another point $\mathbf{p}'$ of the Fermi surface. With this in mind, Eq. (65) really serves to define an effective electron–phonon coupling $\alpha_{p\lambda}(\omega)$. Collecting all of this

$$\Sigma_{\text{ph}}(p,\omega)=\int_{-\infty}^{\infty}d\omega'\int_0^{\infty}d\Omega\,\alpha_p^2(\Omega)\,F_p(\Omega)\,\text{Re}\left\{\frac{\tau_3\left[\omega'\tilde{Z}_{\text{ph}}(\omega')-\tilde{\phi}(\omega')\,\tau_1\right]\tau_3}{\sqrt{\omega'^2\tilde{Z}^2(\omega')-\tilde{\phi}^2(\omega')}}\right\}\times\left[\frac{f(-\omega')+n(\Omega)}{\omega'+\Omega-\omega}+\frac{f(\omega')+n(\Omega)}{\omega'-\Omega-\omega}\right] \tag{66}$$

For an isotropic or dirty superconductor the self-energy is determined by averaging the momentum p over the Fermi surface. In this case $\alpha_p^2(\Omega)$ goes over to the effective coupling function $\alpha^2(\Omega)$ times the phonon density of states $F(\omega)$ as discussed in the introduction to this section [see Eqs. (44) and (45)]. Before separating Σ_{ph} into its components, we turn to an analysis of the Coulomb contribution to ϕ.

In simplifying the form of Coulomb contribution to the gap parameter, the frequency dependence of V_{sc} and Λ will be neglected. It could, in fact, be handled in a similar way to the frequency dependence of the phonon contribution. For the simple metals this procedure leads at most to a slight modification of the Coulomb pseudo-potential. However, near the filled end of the transition metals, where the susceptibility is strongly exchanged enhanced, critical spin fluctuations can produce important frequency variations in these factors. The treatment of the frequency dependence of V_{sc} and Λ_c for this case is important to the understanding of the superconductivity properties in transition metals (see Chapter 13). Transforming the frequency summation to an integration as in the phonon case, the equation for the Coulomb part of the gap parameter [Eq. (54)] takes the form

$$\phi_c(p)=\frac{1}{\pi}\int_0^{\infty}d\omega'\int\frac{d^3p'}{(2\pi)^3}\,\Lambda(p,p')\,V_{sc}(p,p')\times\text{Im}\left[\frac{\phi(p',\omega')\tanh(\beta\omega'/2)}{Z^2(p,\omega')\,\omega'^2-\epsilon_{p'}^2-\tilde{\phi}^2(p',\omega')}\right] \tag{67}$$

This will be reduced to a one-dimensional integral, by introducing a Coulomb pseudo-potential as done in (*52*). Such an approach was first proposed by Bogoliubov et al. (*55*) and was used by Morel and Anderson (*51*) in their treatment of the superconductor.

Introducing a cutoff frequency $\omega_c \sim 10\omega_D$, the frequency integration can be separated into two parts. Over the (ω_c, ∞) interval, the phonon contribution to the self-energy parameters Z and ϕ is negligible and $\tanh(\beta\omega'/2) \sim 1$ for the temperatures at which superconductivity occurs. Carrying out the ω'-integration over the (ω_c, ∞) region, the equation for ϕ_c becomes

$$\phi_c(p) + \int \frac{d^3p'}{(2\pi)^3} \frac{\Lambda(p, p')\, V_{sc}(p, p')}{Z_c^2} \frac{\theta(|\tilde{\epsilon}_{p'}| - \omega_c)}{2|\tilde{\epsilon}_{p'}|} \phi_c(p')$$
$$= \int \frac{d^3p'}{(2\pi)^3} \frac{\Lambda(p, p')\, V_{sc}(p, p')}{Z_c^2} \int_0^{\omega_c} \frac{d\omega'}{\pi}$$
$$\times \operatorname{Im}\left[\frac{\phi(p', \omega')}{\tilde{Z}^2(\omega')\,\omega'^2 - \tilde{\epsilon}_{p'}^2 - \tilde{\phi}^2(p', \omega')}\right] \tanh\left(\frac{\beta\omega'}{2}\right) \quad (68)$$

Here the integrated term has been put on the left side. The θ-function restricts the scattering to states outside a shell of width $2\omega_c$ centered at the Fermi surface. Scattering processes associated with the states inside this shell are taken into account by the remaining term on the right side. The Coulomb renormalization parameter Z_c is explicitly exhibited in both terms.

Equation (68) for ϕ_c can be expressed symbolically in the matrix form

$$(1 + \Omega)\,\phi_c = VF$$

if the momentum indices p and p' are taken as indices and $\int d^3p'/(2\pi)^3$ is replaced by $\Sigma_{p'}$. The matrix elements of Ω and V are

$$\Omega_{pp'} = \frac{\theta(|\tilde{\epsilon}_{p'}| - \omega_c)\,\Lambda(p, p')\, V_{sc}(p - p')}{2|\tilde{\epsilon}_{p'}|\, Z_c^2}$$
$$V_{pp'} = \Lambda(p, p')\, V_{sc}(p - p')/Z_c^2$$

respectively. The vector components of ϕ_c are $\phi_c(p')$ and those of F are given by

$$F_{p'} = \int_0^{\omega_c} \frac{d\omega}{\pi} \operatorname{Im}\left[\frac{\phi(p', \omega')}{Z^2(p', \omega')\,\omega'^2 - \tilde{\epsilon}_{p'}^2 - \tilde{\phi}^2(p', \omega')}\right] \tanh\left(\beta\frac{\omega'}{2}\right)$$

The effective Coulomb potential V is repulsive, so that the inverse of $1 + \Omega$ exists. A formal solution for ϕ_c is therefore

$$\phi_c = (1 + \Omega)^{-1}\, VF \equiv UF$$

with the Coulomb pseudo-potential U determined by

$$(1 + \Omega)\, U = V$$

In component form the equation for ϕ_c is

$$\phi_c(p) = \int \frac{d^3p'}{(2\pi)^3} U(p, p') F(p')$$

and the Coulomb pseudo-potential is determined by the integral equation

$$U(p, p') = V(p, p') - \int \frac{d^3p''}{(2\pi)^3} V(p, p'') \frac{\theta(|\tilde{\epsilon}_{p''}| - \omega_c)}{2|\tilde{\epsilon}_{p''}|} U(p'', p') \qquad (69)$$

The frequency cutoff ω_c appearing in the expressions for $F_{p'}$ means that it decreases rapidly for $\epsilon_{p''}$ greater than ω_c. Therefore the major contribution to the pseudo-potential expression for ϕ_c comes from momentum states p' near the Fermi surface so that the same integration procedure used for the phonon contribution to Σ can be applied. In this way, a one-dimensional integral for $\tilde{\phi}_c = \phi_c/Z_c$ is obtained:

$$\tilde{\phi}_c(p, \omega) = - N(0)\, U(p) \int_0^{\omega_c} d\omega' \, \mathrm{Re} \left\{ \frac{\tilde{\phi}(p_F, \omega') \tanh[\beta(\omega'/2)]}{\sqrt{Z_{ph}^2(p_F, \omega')\,\omega'^2 - \tilde{\phi}^2(p_F, \omega')}} \right\} \qquad (70)$$

Here $|p| = p_F$ and $U(p)$ is the average of the Coulomb pseudo-potential over the Fermi surface.

$$U(p) = \int (dp'^2/4\pi p_F^2)\, U(p, p') \qquad (71)$$

The integral equation for U is of the t-matrix form, taking into account the multiple Coulomb scattering of two electrons outside the energy region $\pm\,\omega_c$. It is just these scattering processes which have been excluded from Eq. (70) by the cutoff at ω_c. Because the Coulomb interaction is repulsive, the correlations induced by these multiple scatterings reduce the probability that the two electrons are within the range of the screened Coulomb interaction, making U smaller than the average of the renormalization, screened Coulomb interaction over the Fermi surface.

For the dirty limit, the s-wave average of the pseudo-potential

$$U = \int_{SF} (d^2p/4\pi p_F^2)\, U(p) \qquad (72)$$

determines the Coulomb contribution to the gap. Taking the s-wave average of Eq. (67), the integral equation for $\langle U(p, p')\rangle_s$ is

$$\langle U(p, p')\rangle_s = \langle U(p, p')\rangle_s - N(0) \int d\tilde{\epsilon}_{p''} \theta(|\tilde{\epsilon}_{p''}| - \omega_c) \langle U(p, p'')\rangle_s \langle U(p'', p')\rangle_s$$

If $A(p, p'')/Z_c^2$ is set equal to unity and the screened Coulomb interaction is treated in the Thomas–Fermi approximation, the value of $\langle V\rangle_s$ on the Fermi surface is

$$N(0)\langle V\rangle_s = a^2 \ln[(1+a^2)/a^2]$$

where $a^2 = k_s^2/4k_F^2$. Finally, if $\langle V(p,p')\rangle_s$ is replaced in the integral equation by the separable form

$$V = \begin{cases} \langle V\rangle_s & |\epsilon_{p'}|, |\epsilon_p| < E_F \\ 0 & |\epsilon_{p'}|, |\epsilon_p| > E_F \end{cases}$$

the integral equation can be solved and

$$U = \langle V\rangle_s / [1 + N(0)\langle V\rangle_s \ln(E_F/\omega_c)] \tag{73}$$

This completes the construction of the self-energy equations. Here we collect and summarize the results for the dirty limit in which the appropriate self-energy is determined by an average over the Fermi surface. Combining the τ_2 components of Eq. (66) for Σ_{ph} with the Coulomb contribution given by Eq. (70), an integral equation for the gap $\Delta(\omega) = [\bar{\phi}_{\text{ph}}(\omega) + \bar{\phi}_c(\omega)]/\tilde{Z}_{\text{ph}}(\omega)$ is obtained:

$$\begin{aligned}\Delta(\omega) = {} & \frac{1}{Z(\omega)} \int_0^\infty d\omega' \,\mathrm{Re}\left\{\frac{\Delta(\omega')}{[\omega'^2 - \Delta^2(\omega')]^{1/2}}\right\} \times \int_0^\infty d\Omega\, \alpha^2(\Omega) F(\Omega) \\ & \times \left[\frac{f(-\omega') + n(\Omega)}{\omega' + \omega + \Omega} + \frac{f(-\omega') + n(\Omega)}{\omega' - \omega + \Omega} - \frac{f(\omega') + n(\Omega)}{-\omega' + \omega + \Omega} - \frac{f(\omega') + n(\Omega)}{-\omega' - \omega + \Omega}\right] \\ & - \frac{\mu^*}{Z(\omega)} \int_0^{\omega_c} d\omega' \,\mathrm{Re}\left\{\frac{\Delta(\omega')}{[\omega'^2 - \Delta^2(\omega')]^{1/2}}\right\} \tanh\left(\frac{\beta\omega'}{2}\right)\end{aligned} \tag{74}$$

The **1** component of Eq. (66) for Σ_{ph} gives an equation for the phonon renormalization parameter:

$$\begin{aligned}[1 - Z(\omega)]\,\omega = {} & \int_0^\infty d\omega' \,\mathrm{Re}\left\{\frac{\omega'}{[\omega'^2 - \Delta^2(\omega')]^{1/2}}\right\} \\ & \times \int_0^\infty d\Omega\, \alpha^2(\Omega) F(\Omega) \left[\frac{f(-\omega') + n(\Omega)}{\omega' + \omega + \Omega} - \frac{f(-\omega') + n(\Omega)}{\omega' - \omega + \Omega}\right. \\ & \left. + \frac{f(\omega') + n(\Omega)}{-\omega' + \omega + \Omega} - \frac{f(\omega') + n(\Omega)}{-\omega' - \omega + \Omega}\right]\end{aligned} \tag{75}$$

Here the notation has been simplified by writing $\tilde{Z}_{\text{ph}}(\omega)$ as $Z(\omega)$. This should not cause confusion, since the only other renormalization parameter which enters in the remainder of this chapter is the Coulomb renormalization at the Fermi surface, which is denoted by Z_c. We have continued to suppress the temperature variable in Δ and Z. The electron–phonon interaction enters in the form

$$\alpha^2(\omega) F(\omega) = \int_{SF} d^2p \int_{SF} \frac{d^2p'}{(2\pi)^3 v_F'} \sum_\lambda |\bar{g}_{pp'\lambda}|^2 \,\delta(\omega - \omega_{p-p'\lambda}) \Big/ \int_{SF} d^2p \tag{76}$$

Here $\bar{g}_{pp'\lambda}$ is the dressed electron–phonon matrix element previously discussed in Eq. (63), v_F' is the Fermi velocity, $\omega_{p-p'\lambda}$ is the phonon energy for reduced momentum $p - p'$ and polarization λ, and the integrations are over the Fermi surface. The phonon density of states $F(\omega)$ is given by

$$F(\omega) = \sum_{\lambda} \int \frac{d^3q}{(2\pi)^3} \delta(\omega - \omega_{q\lambda})$$

so that Eq. (76) defines the effective electron–phonon coupling function $\alpha^2(\omega)$. The Coulomb pseudo-potential has been written as

$$\mu^* = N(0)\, U \tag{77}$$

where U is approximately given by Eq. (73).

C. Electron–Phonon Interaction

We have seen that the electron–phonon matrix element $\bar{g}_{pp'\lambda}$ enters the self-energy equations through the effective coupling function $\alpha^2(\omega)$. In the plane-wave approximation, which is useful for the nearly free electron metals, the square of the effective electron–phonon matrix element can be written as

$$|\bar{g}_{pp'\lambda}|^2 = \frac{N}{2M\omega_{p-p'\lambda}} |(\mathbf{p} - \mathbf{p}')\cdot \boldsymbol{\epsilon}_{p-p'\lambda} v(p - p')|^2$$

where M is the ion mass, N is the number of ions per unit volume, and $v(q)$ is the dressed pseudo-potential

$$v(q) = \Lambda(q)\, V(q)/Z_c \epsilon(q)$$

Here Z_c is the Coulomb renormalization, $\Lambda(q)$ is the Coulomb renormalization, $\Lambda(q)$ is the proper Coulomb vertex, $\epsilon(q)$ the dielectric constant, and $V(q)$ the bare ion pseudo-potential.

Here we summarize some of the things that are known about the dressed pseudo-potential $v(q)$ which enters the electron–phonon interaction.

1. For a uniform electron gas, the long-wavelength limit of $v(q)$ is given by

$$\lim_{q\to 0} v(q) = -Z/2N(0) \tag{78}$$

Here Z is the valency of the atom and $N(0)$ is the single spin band and Coulomb dressed density of states at the Fermi surface.

2. The band structure and the Fermi surface are determined by $v(q)$. This means that Fermi surface measurements provide information on $v(q)$ at certain reciprocal lattice vectors. Using a parametrized form for $v(q)$ adjusted to fit the

known data, a useful semiempirical form for $v(q)$ can be obtained. There are now a variety of such forms and we will consider two of these.

Harrison (*33,34*) has suggested that $v(q)$ can be approximated by a Coulomb attraction and a repulsive δ-function (representing the effects of orthogonalization to the core electrons) all screened by a Hartree dielectric constant. Here we will approximate the dielectric constant by the Fermi–Thomas form so that

$$v(q) = \frac{-(4\pi e^2 Z/q^2) + \beta}{1 + (k_s/q)^2} \tag{79}$$

with k_s the Fermi–Thomas screening factor, Z the valence ion charge and β the strength of the repulsive core. Harrison estimated that β times the volume of the unit cell is 60 Ry–au^3.‡ Ashcroft (*35*) has suggested another parametrization of the pseudo-potential which was mentioned in Section II, Eq. (2). Here the pseudo-potential incorporates the cancelation effects by vanishing inside some core radius R_c and becoming Coulomb-like outside. Screening this with the Lindhard dielectric function, Ashcroft's pseudo-potential is

$$V(q) = \frac{-(4\pi e^2/q^2)\cos R_c q}{1 + (k_s/q)^2\, u(q/2p_F)} \tag{80}$$

Here $u(x)$ is given by Eq. (21) and R_c for Pb is taken as 0.57 Å. Setting x equal to $q/2p_F$ and measuring $V(q)$ in units of $V(0)$, Ashcroft's pseudo-potential is plotted in Fig. 27. The circled points give the principal (111) and (200) Fourier coefficients determined from Fermi surface data of Anderson and Gold (*20*).

McMillan and Rowell (see Chapter 11) have used the results of electron tunneling to obtain information on the electron–phonon coupling. They point out that since $|\bar{g}_{pp'\lambda}|^2$ varies as $(\omega_{p-p'})^{-1}$, the first moment of $\alpha^2(\omega)F(\omega)$ is independent of the phonon frequencies:

$$\int_0^\infty d\omega\, \omega\alpha^2(\omega)F(\omega) = \frac{N}{2Mv_F}\frac{1}{(2\pi)^2}\int_0^{2p_F} dq\, q^3 v^2(q)$$

Using the long-wavelength limit [Eq. (78)] and the relation $N(0) = \frac{3}{4} NZ/E_F$ this moment can be written as

$$\int_0^\infty d\omega\, \omega\alpha^2(\omega)F(\omega) = \frac{2}{3}\frac{mZ}{M} E_F^2 \overline{v^2} \tag{81}$$

where the average value of the squared pseudo-potential is defined by

$$\overline{v^2(q)} = \int_0^{2p_F} v^2(q)\, q^3\, dq \Big/ \int_0^{2p_F} v^2(0)\, q^3\, dq \tag{82}$$

‡ Using the measured value of $\overline{\alpha^2}$ from electron tunneling in Pb, McMillan estimates that $\beta \sim 40$ Ry–au^3.

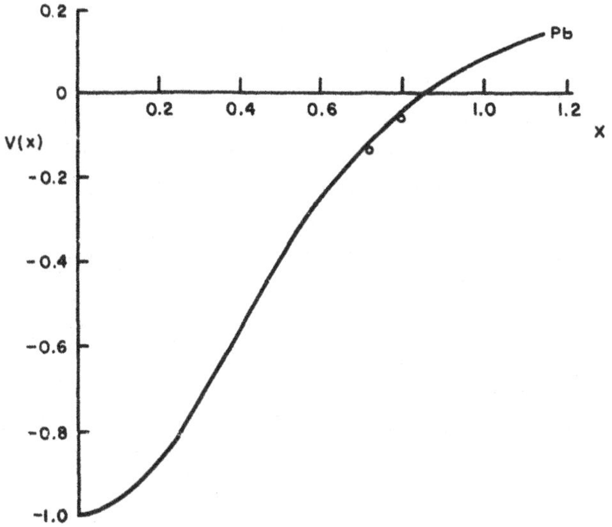

Fig. 27. Ashcroft's pseudo-potential for Pb. The circled points give the principal (111) and (200) Fourier coefficients determined from the Anderson and Gold (*20*) de Haas–van Alphen measurements.

From the tunneling data for Pb, McMillan and Rowell find $\overline{v^2(q)} = 0.037$. Using Ashcroft's pseudo-potential, Eq. (80), we find that $\overline{v^2(q)} \sim 0.04$, in surprisingly good agreement with the experimental value.‡ Zone boundary as well as spin-orbit effects could easily lead to errors of 25%. Nevertheless, this argument gives strong support to the notion that the pseudo-potential which determines the Fermi surface also determines the electron–phonon interaction to the extent that the system is a nearly free electron metal.

The effective mass enhancement due to the electron–phonon interaction is given by

$$\frac{m^*}{m} = 1 + \lambda = 1 + 2\int_0^\infty d\omega \, \frac{\alpha^2(\omega)\, F(\omega)}{\omega} \tag{83}$$

The parameter λ provides a direct measure of the electron–phonon coupling strength. McMillan (*57*) has recently used the strong-coupling predictions and the measured values of the transition temperature, the electronic specific heat

‡ Using the Heine–Abarenkov model potential (*56*) fit to the Fermi surface of Pb, McMillan and Rowell obtain a value of 0.038.

(m^*/m), the Debye temperature, and the isotope shift to empirically study the behavior of λ for certain classes of metals. He found that

$$\lambda = 2\int_0^\infty d\omega \frac{\alpha^2(\omega)\, F(\omega)}{\omega} = \frac{C}{M\overline{\omega_{\rm av}^2}} \tag{84}$$

where C is a constant for a given class of materials, M is the ionic mass, and $\omega_{\rm av}^2$ is an average phonon frequency squared. For Pb, In, and Al, $C \simeq 2\ {\rm eV/Å}^2$.

For the nearly free electron polyvalent metals McMillan pointed out that this remarkable result was a consequence of the extreme sensitivity of the phonon frequencies to small changes in the pseudo-potential. Here we review his analysis. Introducing an average phonon frequency squared $\omega_{\rm av}^2$ by

$$\omega_{\rm av}^2 = \int_0^\infty d\omega\, \omega\alpha^2(\omega)\, F(\omega) \Big/ \int_0^\infty d\omega \frac{\alpha^2(\omega)\, F(\omega)}{\omega} \tag{85}$$

the expression for λ can be written as

$$\lambda = \frac{\pi}{2} \frac{E_F}{e^2 k_F} \frac{\overline{v^2}}{\omega_{\rm av}^2/\Omega_p^2} \tag{86}$$

Here Ω_p is the ion-plasma frequency $(4\pi NZ^2e^2/M)^{1/2}$ and v^2 is the average of the pseudo-potential, Eq. (82), previously discussed. The factor $E_F/k_F e^2$ is equal to $0.96/r_s$, where r_s is the ratio of the radius of a sphere containing one electron to the Bohr radius, so that

$$\lambda = \frac{1.51}{r_s} \frac{\overline{v^2}}{(\omega_{\rm av}^2/\Omega_p^2)} \tag{87}$$

The phonon frequencies ω_q are determined by the poles of the phonon Green's function

$$\omega_q^2 = \Omega_q^2 - 2\Omega_q \Pi \tag{88}$$

Here Ω_q is the bare ion frequency and Π is the phonon self-energy. Neglecting band structure, the phonon self-energy is given by

$$\Pi = \frac{|g_{q\lambda}|^2}{V_c(q)} \left[1 - \frac{1}{\epsilon(q)}\right] \tag{89}$$

This is proportional to the square of the pseudo-potential which would also be the case if band structure effects were taken into account in lowest order. In Pb, the ratio of ω_q^2/Ω_q^2 at the band is about $\frac{1}{10}$, so there is a large cancelation between the bare ion frequencies and the self-energy term which represents the electronic contribution. This means that the resulting phonon frequencies are very sensitive to small changes in $v(q)$, a fact empirically established by a number of authors (*58*) in attempts to calculate ω_q. Thus the dominant dependence of λ on the pseudo-potential arises from ω_{av}^2 rather than the $\overline{v^2}$ which appears in the numerator of Eq. (87).

Given the crystal structure and the pseudo-potential, the effective electron–phonon coupling function $\alpha^2(\omega)$ can, in principle, be determined. However, because of the extreme sensitivity of the phonon frequencies to the form of the pseudo-potential, this is a difficult procedure at present. Another, more empirical approach has been used by Scalapino, Wada, and Swihart (*59*) to determine $\alpha^2(\omega)$ for Pb. In this work, in addition to the crystal structure and the pseudo-potential, a simple analytic form for $F(\omega)$ is chosen to fit the gross features of the phonon spectrum determined by neutron scattering and electron tunneling data. Then, working back from this and the pseudo-potential, $\alpha^2(\omega)$ can be approximately evaluated. The specification of the crystal structure is necessary to determine the reciprocal lattice vectors which contribute to the Umklapp processes. Here we review some recent results obtained for Pb by Wada using this approach.

The main features of the phonon density of states for Pb are two peaks at approximately 4.4 meV and 8.5 meV. We will refer to these as the transverse and longitudinal peaks, respectively, and use cutoff Lorentzians to approximate them. The low-frequency region will be approximated by a Debye part varying as ω^2. Throughout this analysis a normalized density of states for each polarization λ will be used:

$$\int_0^\infty F_\lambda(\omega)\, d\omega = 1 \tag{90}$$

This choice is for convenience and does nothing to alter the product $\alpha^2(\omega)\,F(\omega)$ which is given by

$$\alpha^2(\omega)\,F(\omega) = \sum_\lambda \alpha_\lambda^2(\omega)\,F_\lambda(\omega) \tag{91}$$

Approximating the first Brillouin zone by a sphere of radius equal to the Debye wave vector q_D,

$$F_\lambda(\omega) = \int \frac{d^3q}{(2\pi)^3}\,\delta(\omega - \omega_{q\lambda}) \Big/ \frac{4\pi}{3}\, q_D^3/(2\pi)^3 \tag{92}$$

and we have

$$F_\lambda(\omega) = \frac{3}{q_D^3}\,\frac{q^2}{d\omega_{q\lambda}/dq}\bigg|_{\omega_{q\lambda}=\omega} \tag{93}$$

If the phonon spectrum were a monotonic function of q, the phonon dispersion relation associated with a given $F_\lambda(\omega)$ could be obtained by integrating Eq. (93) to give ω_q vs. q. Although this is not the case, the important ingredient entering the determination of $\alpha^2(\omega)$ is $F(\omega)$, and the results are relatively insensitive to the phonon dispersion relation as long as it faithfully reproduces the actual phonon density of states $F(\omega)$.

Treating the Fermi surface as spherical and using the plane-wave form of the electron–phonon matrix elements, the effective coupling function for polarization λ is

$$\alpha_\lambda^2(\omega)\,F_\lambda(\omega) = \int d^2p \int \frac{d^2p'}{v_F(2\pi)^3}\, |g_{p-p'\lambda}|^2\, \delta(\omega - \omega_{p-p'\lambda}) \Big/ \int d^2p \tag{94}$$

Substituting the expression given in Eq. (92) for $F_\lambda(\omega)$, and using the δ-function to reduce the expression, one finds

$$\alpha_\lambda^2(\omega) = \frac{q_D^3}{6\pi^2} \sum_K \int \frac{d\Omega_q}{2\pi} \frac{|g_{q+K\lambda}|^2}{|q+K|}\, \theta(2p_F - |q+K|)\Big|_{\omega_{q\lambda}=\omega} \tag{95}$$

Here K is a reciprocal lattice vector and the condition $\omega_{q\lambda} = \omega$. In general, this would not be a unique prescription, but, as we have discussed, the approximate dispersion relation obtained by integrating Eq. (93) leads to a monatonic phonon dispersion relation.

Once the phonon density of states of polarization λ, the electron–phonon matrix element $g_{pp'\lambda}$, and the reciprocal lattice vectors K are specified, $\alpha_\lambda^2(\omega)$ can be determined from Eqs. (93) and (95). For Pb the assumed form for $F_\lambda(\omega)$ is

$$F_\lambda(\omega) = \begin{cases} 3\omega^2/S_\lambda^3 q_D^3 & \omega \leqslant \omega_0^\lambda \\ A_\lambda\left[\dfrac{1}{(\omega-\omega_1^\lambda)^2 + \omega_2^{\lambda 2}} - \dfrac{1}{\omega_3^{\lambda 2} + \omega_2^{\lambda 2}}\right] & \omega_3^\lambda \leqslant \omega \leqslant \omega_1^\lambda + \omega_3^\lambda \\ 0 & \omega_1^\lambda + \omega_3^\lambda \leqslant \omega \end{cases} \tag{96}$$

Here S_λ is the velocity of sound for polarization λ, and the values of the parameters ω_1^λ, ω_2^λ, and ω_3^λ are selected to approximate the gross structure of the phonon peaks. The parameters ω_0^λ and A_λ are determined by requiring that $F_\lambda(\omega)$ be continuous and satisfy the normalization condition Eq. (90). In Table I the values used for Pb are given. The values of ω_1^λ and ω_2^λ are the same as those used in the work of Scalapino et al. (*52*). This choice was originally made on the basis of the neutron scattering data of Brockhouse et al. (*60*) and upon the observed structure in the *I–V* characteristic of Pb tunnel junctions. At the time

TABLE I

Values of the Parameters Used in Eq. (96) for the Phonon Density of States in Pb.[a]

	ω_1^λ, meV	ω_2^λ, meV	ω_0^λ, meV	S_λ, cm/sec	$A_\lambda/\omega_2^\lambda$
t	4.4	0.75	2.5	1.07×10^5	0.52
l	8.5	0.50	7.1	2.42×10^5	0.50

[a] ω_3^λ was set equal to $3\omega_2^\lambda$ and q_D was taken as 1.25×10^8 cm^{-1}.

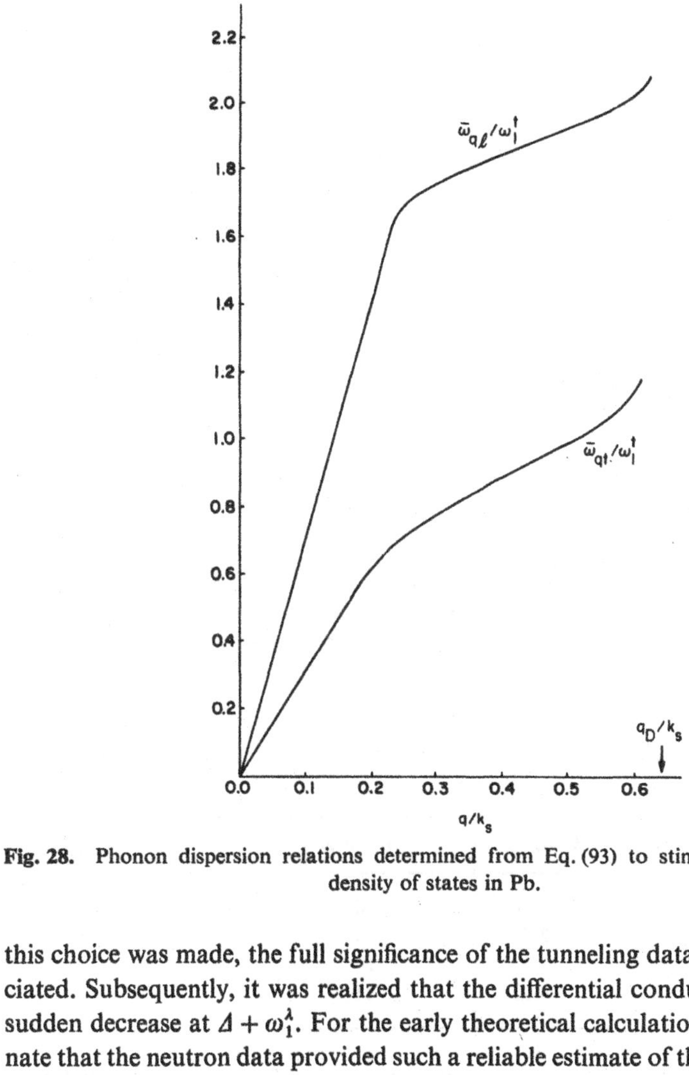

Fig. 28. Phonon dispersion relations determined from Eq. (93) to stimulate the phonon density of states in Pb.

this choice was made, the full significance of the tunneling data was not appreciated. Subsequently, it was realized that the differential conductance shows a sudden decrease at $\Delta + \omega_1^\lambda$. For the early theoretical calculations, it was fortunate that the neutron data provided such a reliable estimate of these parameters. The phonon dispersion relations ω_{qt} and ω_{ql} which are determined from Eq. (93) are plotted in Fig. 28. The unphysical increase of $\omega_{q\lambda}$ in the vicinity of the Debye momentum q_D is due to the sharp cutoff in the density of states $F_\lambda(\omega)$ at the high frequencies.

Harrison's form for the pseudo-potential was used with a charge Z of 4 and a β value of 60 in atomic units. The dielectric constant was approximated by the Thomas–Fermi form, giving the electron–phonon matrix element

$$g_{q+K} = -i\sqrt{\frac{N}{2M}}(\mathbf{q}+\mathbf{K})\cdot\boldsymbol{\epsilon}_{q\lambda}\left(\frac{-4\pi e^2 Z}{q^2}+\beta\right)\Big/1+(q/k_s)^2$$

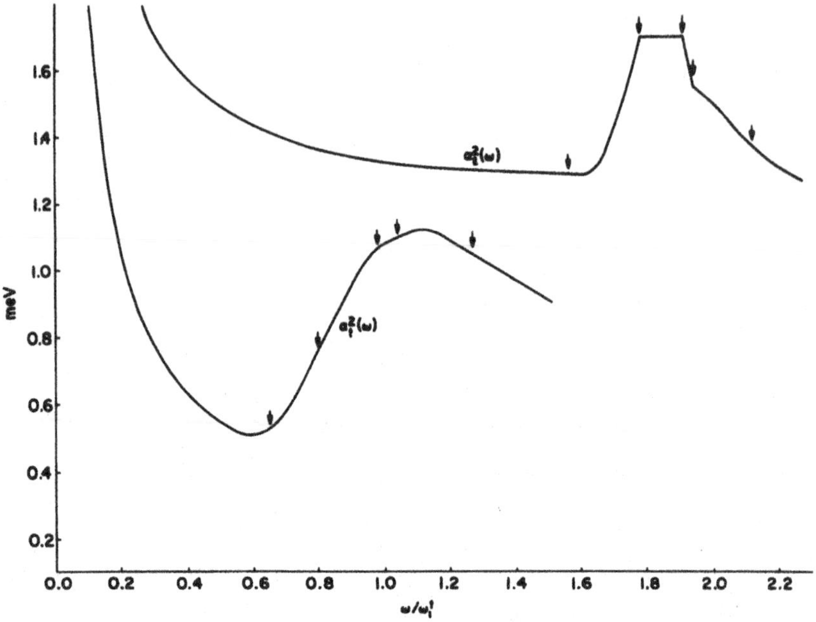

Fig. 29. Longitudinal and transverse effective coupling functions calculated for Pb vs. ω.

TABLE II

Fifty Reciprocal Lattice Vectors Used in Calculating the Umklapp Contribution to $\alpha^2(\omega)$

K, 10^8 cm^{-1}	No. of K
2.20	8
2.54	6
3.59	12
4.22	24

For Pb, the free electron value of the Fermi momentum is $k_F = (3\pi^2 NZ)^{1/3} = 1.57\ \text{Å}^{-1}$, the Debye momentum is $q_D = (2/Z)^{1/3} k_F = 1.25\ \text{Å}^{-1}$, and the Thomas–Fermi screening constant is $k_s = 1.95\ \text{Å}^{-1}$. The crystal structure of Pb is face-centered-cubic with a unit cell edge of 4.94 Å, and a nearest-neighbor distance of 3.49 Å. The 50 reciprocal lattice vectors which must be taken into account in determining the Umklapp contributions are listed in Table II. All other reciprocal lattice vectors are such that $K - q_D$ is greater than $2p_F$ and so they do not contribute.

Performing the integrals and carrying out the sum over the reciprocal lattice vectors leads to the results for $\alpha_t^2(\omega)$ and $\alpha_l^2(\omega)$ shown in Fig. 29. At low energies, the increase in $\alpha^2(\omega)$ reflects the presence of Umklapp processes. In the absence of Umklapp processes $|g_q^2|_{\omega_q=\omega} \sim \omega$ in this low-energy region and $\alpha^2(\omega)\,F(\omega)$ varies as ω^2. However, to the extent that $|g_{q+k}|^2_{\omega_q=\omega}$ is independent of ω, $\alpha^2(\omega)\,F(\omega) \sim \omega$ and $\alpha^2(\omega) \sim \omega^{-1}$, as shown in Fig. 29. As previously discussed, one expects that the electron–phonon matrix will have zeros for momentum transfers, corresponding to reciprocal lattice vectors, and that these will be approached continuously. This means that below some frequency, $\alpha^2(\omega)$ should approach a constant value. This frequency can be estimated by remembering that the matrix elements of the pseudo-potential which enter the determination of $\alpha^2(\omega)$ should be taken between the Bloch states appropriate to the pseudo-potential. Sham and Ziman have shown that the Umklapp matrix elements between Bloch states vanish like the reduced momentum q for small q. The criterion for small q is illustrated in Fig. 30. Here K is a reciprocal lattice vector and q^* is the separation between the two pieces of Fermi surface at the zone boundary. When the momentum transfer $K+q$ is such that q is less than q^*, the mixing of the plane-wave states separated by K become important and the Umklapp process begins to vary as q. The frequencies at which $\alpha^2(\omega)$ reflects the loss of Umklapp processes are determined by $\omega_{q^*}^{\lambda}$ for the various q^* values associated with the Fermi surface. McMillan and Rowell (*61*) have pointed out that Van Hove–like singularities in $\alpha^2(\omega)$ arise at these points. Referring again to Fig. 30, we note that when q becomes greater than q^* an Umklapp from p to p' can occur. The relative flatness of the Fermi surface at p and p' imply that a large number of Umklapp processes with $\omega = \omega_{q^*}$ are possible. Furthermore, the transverse phonon density of states is about 30 times that of the longitudinal density of states in this low-frequency region. Because the Fermi surface is nearly spherical, the transverse phonons are only weakly coupled to the electrons until Umklapp processes can occur.

Although the effective coupling functions $\alpha_\lambda^2(\omega)$ may appear to be rather

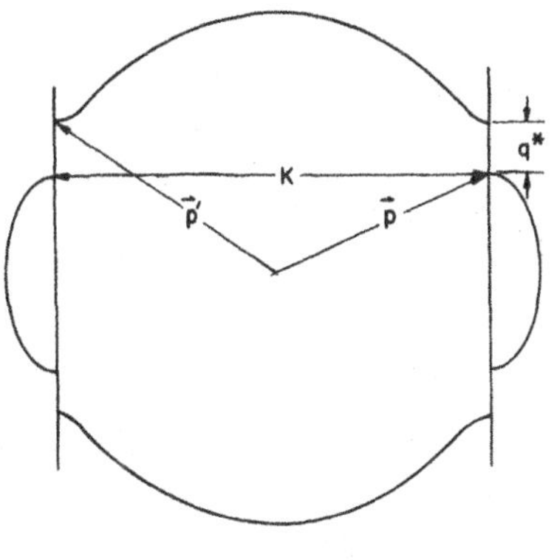

Fig. 30. Umklapp processes for small reduced momentum transfer.

rapidly varying with ω, this variation is slow compared to that of $F(\omega)$. In Fig. 31 we have plotted

$$\alpha^2(\omega)\, F(\omega) = \sum_\lambda \alpha_\lambda^2(\omega)\, F_\lambda(\omega)$$

as the solid curve and

$$\sum_\lambda \alpha_\lambda^2(\omega_1^\lambda)\, F_\lambda(\omega)$$

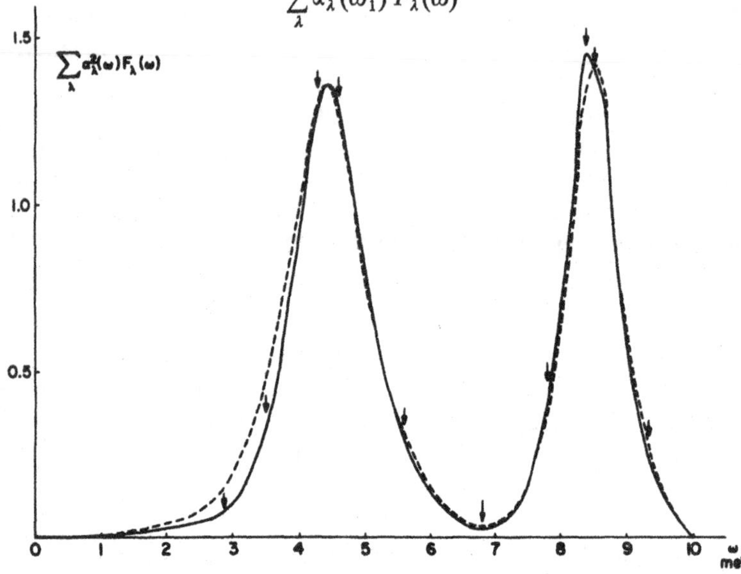

Fig. 31. $\alpha^2(\omega)\, F(\omega)$ calculated for Pb vs. ω. The solid curve is $\sum_\lambda \alpha^2(\omega)\, F_\lambda(\omega)$ and the dashed curve is $\sum_\lambda \alpha^2(\omega_1^\lambda)\, F_\chi(\omega)$.

as the dashed curve. In this latter case, the effective coupling functions have been replaced by their values at the peak phonon energies ω_1^λ, $\alpha_z^2(\omega_1^t) = 1.08$ meV and $\alpha_l^2(\omega_1^l) = 1.57$ meV. It is obvious that the differences between these results are negligible compared to the errors introduced in the approximate determination of $\alpha_\lambda^2(\omega)$. The arrows on these curves indicate points at which changes in slope of $\alpha_\lambda^2(\omega)$ are possible due to the onset of contributions from new reciprocal lattice vectors. Two of these occur at 2.8 and 3.5 meV, corresponding to wave numbers such that $q = 2k_F - k_2$ and $2k_F - k_1$. Within the present treatment, the phase space is such that the kinks produced at these points are quite small. Van Hove singularities of $F(\omega)$ associated with the phonon critical points as well as the singularities in $\alpha^2(\omega)$ associated with the Fermi surface q^* values of Pb do not appear because of our approximation.

Finally, it is interesting to compare these results with the measured values of $F(\omega)$ and $\alpha^2(\omega)\, F(\omega)$. Figure 32 is a plot of $F(\omega)$ for Pb obtained from an analysis of neutron scattering data (*62*). It clearly shows the two dominant phonon peaks. However, the phonon density between the two peaks is con-

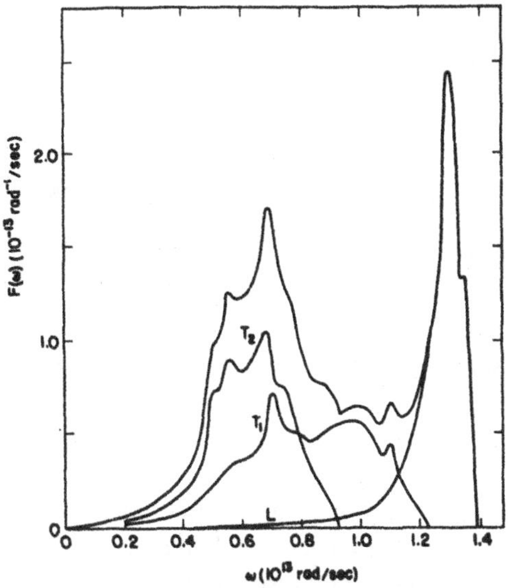

Fig. 32. Phonon density of states in Pb determined from an analysis of neutron scattering data by Stedman et al. (*62*).

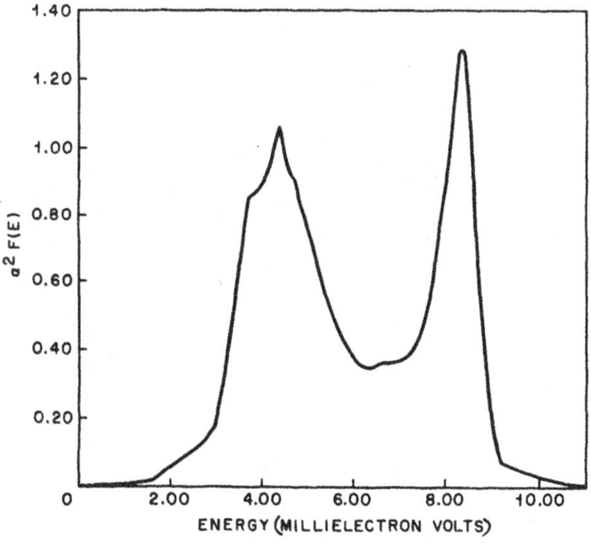

Fig. 33. $\alpha^2(\omega)\,F(\omega)$ vs. ω, obtained from electron tunneling in Pb by McMillan and Rowell (*61*).

siderably greater than that of the cutoff Lorentzian model, and the upper peak is slightly lower than the value of 8.5 meV used in our model and the calculations discussed below. In Chapter 11 McMillan and Rowell describe how electron tunneling data can be analyzed to obtain $\alpha^2(\omega)\,F(\omega)$, which is plotted in Fig. 33.

The structure at 1.6 meV and 3.0 meV is associated with the Umklapp singularities previously discussed. By combining the results shown in Figs. 32 and 33, we estimate that $\alpha_t^2(\omega_1^t) = 1.3$ meV and $\alpha_l^2(\omega_1^l) = 1.2$ meV.

D. Solutions for Pb and Electron Tunneling Results

Before discussing the computer solutions of the self-energy equations for Pb, it is useful to have a rough physical picture of what to expect. To simplify the discussion, consider the case in which the phonon density of states has only a single peak at ω_0 (see Fig. 34a). At excitation energies ω less than ω_0, the bulk of the phonons which can be exchanged have frequencies larger than ω. In this situation the positive ion cores are being driven below their natural frequencies and overrespond, creating a net attractive electron–electron interaction. For ω much larger than ω_0, the majority of the phonons involved in the electron–electron interaction have frequencies less than ω. Now the ion cores are driven above their natural frequencies and respond out of phase, producing a repulsive

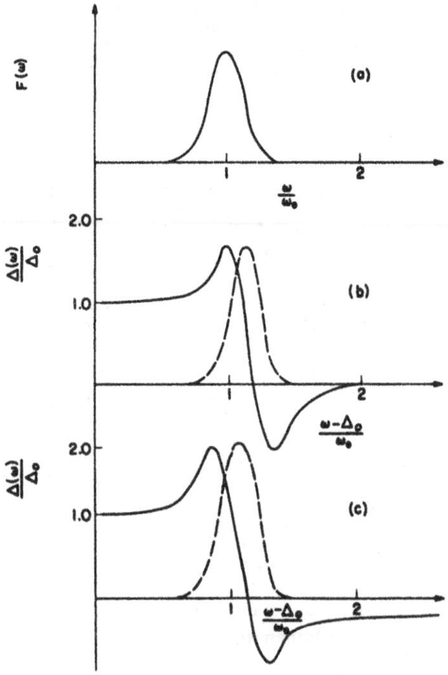

Fig. 34. Single-phonon peak model as an illustration of the manner in which structure in the phonon density of states is reflected in the gap. The phonon density of states $F(\omega)$ is plotted in (a); the real (solid) and imaginary (dashed) parts of the gap $\Delta(\omega)$ for the case of zero Coulomb pseudo-potential in (b); and $\Delta(\omega)$ in the presence of a Coulomb pseudo-potential μ^* in (c).

effective interaction between electrons. Since the real part of the gap is a measure of the strength of this effective electron–electron interaction, we should expect that Re $\Delta(\omega)$ will be positive below ω_0 and negative above ω_0. In the vicinity of ω_0 the resonant nature of the phonon exchange should enhance the positive and negative swing of the interaction and hence of Re $\Delta(\omega)$. Furthermore, for ω in the vicinity of the peak at ω_0, the probability for real phonon emission increases giving rise to a peak in the complex part of the gap, Im $\Delta(\omega)$, which reflects the damping processes. This peak in the damping should actually occur nearer $\omega_0 + \Delta_0$, since the lowest energy state available to the electron in a superconductor is Δ_0. Here $\Delta_0 = \text{Re}\, \Delta(\Delta_0)$ is the real part of the gap at the gap edge. This contortion of words and algebra is necessary when one deals with a complex frequency- and temperature-dependent energy gap. In the zero-temperature limit the imaginary part of the gap vanishes at a frequency equal to the gap edge, since there is no lower state into which a single electron can decay. At finite temperatures, as we will discuss, scattering by thermal phonons and recombination with a thermally excited electron to form a pair give rise to a complex gap at the gap edge.

Figure 34b shows the zero-temperature solution for Δ appropriate to the $F(\omega)$ plotted in Fig. 34a for the case in which there is no Coulomb pseudo-potential. Here weak structure at the harmonics $n\omega_0 + \Delta_0$, associated with multiphonon processes, has been neglected for clarity. The main features which were anticipated on physical grounds are clearly evident. The positive peak in the real part of Δ actually occurs at $\omega_0 + \Delta_0$ and the peak in the imaginary part of Δ occurs at a slightly higher frequency. If the Coulomb pseudo-potential is present, the real part of the gap reflects this and remains negative beyond ω_0, as shown in Fig. 34c.

Turning now to Pb, the computer solutions which will be discussed here are based upon a cutoff Lorentzian density of states:

$$F_\lambda(\omega) = \begin{cases} A_\lambda \left[\dfrac{1}{(\omega - \omega_1^\lambda)^2 + (\omega_2^\lambda)^2} - \dfrac{1}{(\omega_3^\lambda)^2 + (\omega_2^\lambda)^2} \right] & |\omega - \omega_1^\lambda| < \omega_3^\lambda \\ 0 & |\omega - \omega_1^\lambda| > \omega_3^\lambda \end{cases} \tag{97}$$

which differs from the more general form, Eq. (96), discussed in Section III.C by the neglect of the low-frequency Debye part of the spectrum. Here A_λ normalizes $F_\lambda(\omega)$ to unity and ω_3^λ is chosen to be $3\omega_2^\lambda$. An approximate fit to the neutron scattering data was obtained by taking the lower "transverse" peak, designated by $\lambda = t$ to have $\omega_1^t = 4.4$ meV and $\omega_2^t = 0.75$ meV and the upper "longitudinal" peak designated by $\lambda = l$ to have $\omega_1^l = 8.5$ meV and $\omega_l^2 = 0.5$ meV. A plot of the resulting phonon density of states,

$$F(\omega) = 2F_t(\omega) + F_l(\omega) \tag{98}$$

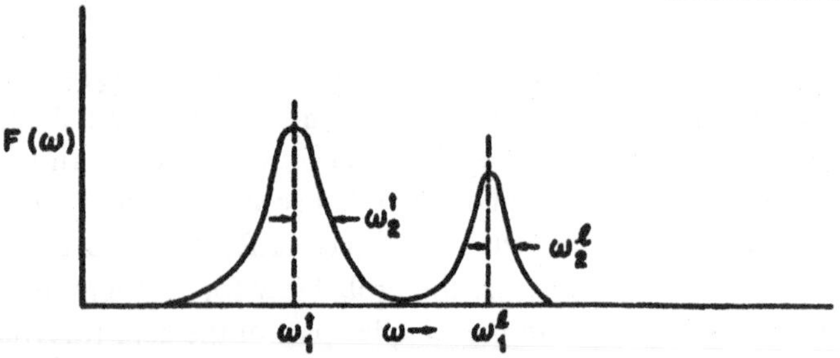

Fig. 35. Model for the lead phonon density of states $F(\omega)$. Here $\omega_1^t = 4.4$ meV, $\omega_2^t = 0.75$ meV, $\omega_1^l = 8.5$ meV, and $\omega_2^l = 0.5$ meV.

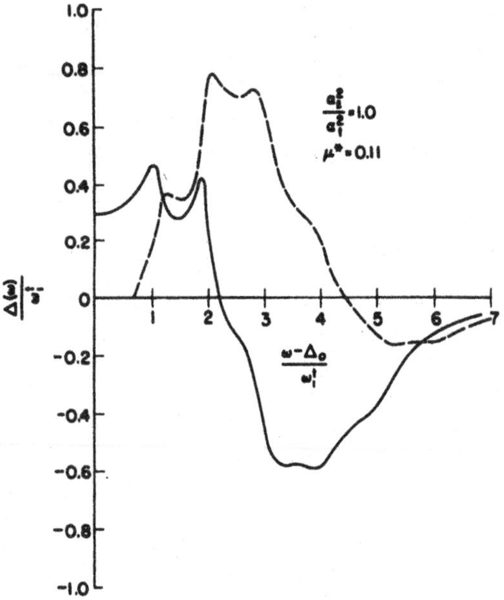

Fig. 36. Plot of the real (solid) and imaginary (dashed) parts of $\Delta(\omega)/\omega_1^t$ vs. $(\omega - \Delta_0)/\omega_1^t$ for the model of the Pb phonon density of states shown in Fig. 35. Here $\omega_1^t = 4.4$ meV, $\Delta_0 = 1.34$ meV, $\alpha_t^2 = 1.2$ meV, $\alpha_l^2/\alpha_t^2 = 1.0$, and $\mu^* = 0.11$.

in which two "transverse" modes were taken and each $F_\lambda(\omega)$ is normalized to unity is shown in Fig. 35. The effective electron–phonon coupling function $\alpha^2(\omega)$ was approximated by its values at the peaks $\alpha_\lambda^2 = \alpha^2(\omega_1^\lambda)$ and two ratios $\alpha_l^2 = \alpha_t^2$ and $\alpha_l^2 = 2\alpha_t^2$ were tried. In each case the value of α_t^2 was fit by requiring that the gap at the gap edge Δ_0 agree with the experimental value of 1.34 meV. The Coulomb pseudo-potential μ^* was estimated to be 0.11 and computations were carried out for values of μ^* equal to 0.11 and 0.

In Figs. 36 and 37 results from a numerical solution of the self-energy equations for Δ and Z at zero temperature are plotted. These solutions are for equal

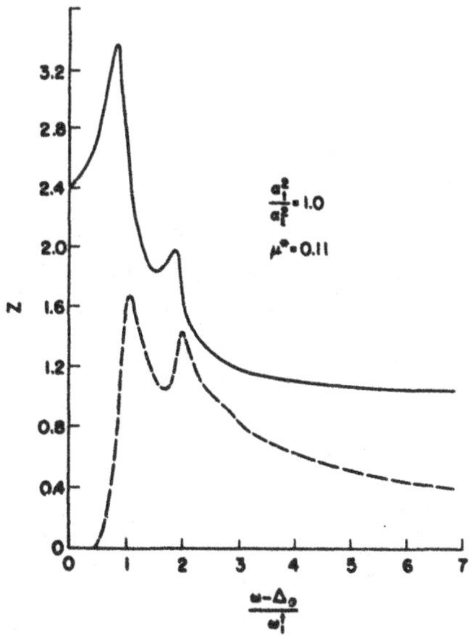

Fig. 37. Plot of the real (solid) and imaginary (dashed) parts of the renormalization parameter $Z(\omega)$ vs. $(\omega - \Delta_0)/\omega_1^t$ for the model parameters given in Fig. 36.

coupling constants $\alpha_l^2 = \alpha_t^2 = 1.2$ and $\mu^* = 0.11$. First consider the real part of the gap. The two peaks which occur for values of $(\omega - \Delta_0)/\omega_1^t$ near 1 and 2 For $(\omega - \Delta_0)/\omega_1^t$ greater than 2, the bulk of the phonons which are exchanged lay below ω and the effective electron–electron interaction induced by the lattice is repulsive. This produces, as we previously discussed, a large negative value of Re Δ in this region. Additional structure occurs at the values $n\omega_1^t + n\omega_1^l + \Delta_0$ (n and m integers) and reflects the multiphonon processes. Because of the repulsive Coulomb pseudo-potential, the real part of the gap remains negative, asymptotically approaching a value proportional to μ^* at high frequency. The imaginary part of the gap exhibits resonances near $\omega_1^\lambda + \Delta_0$ as well as structure at the higher harmonics.

The real and imaginary parts of the renormalization parameter Z, shown in Fig. 37, also provide a reflection of the underlying electron–electron interaction. The peaks in the imaginary part of Z occur at ω equal to $\omega_1^\lambda + \Delta_0$, where the damping is greatest. From the plots of Δ_2/ω_1^t and Z_2 it is clear that the quasi-particle approximation fails badly in the vicinity of the phonon peaks where the widths of the levels are comparable with their excitation energies. This breakdown of the quasi-particle approximation does not affect any of our results, since the present Green's function formalism does not make use of this concept or approximation. One final point worth noting is that Z goes to unity rapidly beyond the region of the phonon excitations. At these high energies, the phonon-induced renormalization effects have been stripped away. At a frequency near the gap edge the phonon-induced renormalization is of order 2.4. We will see in Section IV that this implies an effective mass enhancement at the Fermi surface of 2.4, owing to the electron–phonon interaction.

Before looking at further results for different choices of α_l^2/α_t^2 and μ^*, it seems appropriate to pause for a moment to see how the structure in the electron self-energy can be observed. In Section IV, a number of thermodynamic and dynamic properties showing deviations from the BCS predictions which can be explained on the basis of these results for $Z(\omega)$ and $\Delta(\omega)$ will be discussed. Many of these properties either depend upon averages of the self-energy over ω or are experimentally difficult to probe over a frequency region extending from Δ_0 to ω_D. There is, however, one measurement, electron tunneling, which provides a very direct probe of the self-energy at a given frequency. Because the energy of the injected electrons can be varied by simply changing the applied bias voltage, tunneling measurements give information over the entire frequency region of interest. Schrieffer et al. (*63*) showed that the ratio of the differential conductance in the superstate to the normal state at a voltage $eV = \omega$ was equal to the normalized single-particle tunneling density of states $N_T(\omega)/N(0)$.

$$\frac{(dI/dV)_s}{(dI/dV)_N} = \frac{N_T(\omega)}{N(0)} = \mathrm{Re}\left[\frac{\omega}{\sqrt{\omega^2 - \Delta^2(\omega)}}\right] \tag{99}$$

Here $N(0)$ is the Bloch single spin density of states at the Fermi surface and $\Delta(\omega)$ is the complex, frequency-dependent gap. A complete discussion of electron tunneling will be found in Chapter 11. Here we will only give a brief outline showing how it reflects the structure in the electron self-energy.

Once again, let's start with the simple example in which the phonon density of states has only a single peak at ω_0. Since $\omega_0 > \Delta(\omega_0)$, it is possible to obtain a qualitative understanding of the normalized conductance by expanding the expression for $N_T(\omega)$ in powers of $(\Delta/\omega)^2$:

$$\frac{N_T(\omega)}{N(0)} \sim 1 + \frac{\Delta_1^2}{2\omega^2} - \frac{\Delta_2^2}{2\omega^2} \tag{100}$$

where $\Delta = \Delta_1 + i\Delta_2$. As ω increases toward ω_0 (see Fig. 34), the real part of the gap increases and $N_T(\omega)$ initially rises above the BCS prediction based upon a constant real gap Δ_0. In the vicinity of $\omega_0 + \Delta_0$, the phonons can be resonantly transferred and the imaginary part of the gap increases. Now, just above $\omega_0 + \Delta_0$, Δ_2 is increasing to its peak value while Δ_1 is rapidly going to zero. The combination of these effects produces a sharp decrease in $N_T(\omega)$, causing it to drop below the BCS value. It can, in fact, drop below the normal-state value $N(0)$. This characteristic behavior is shown in Fig. 38 for the model

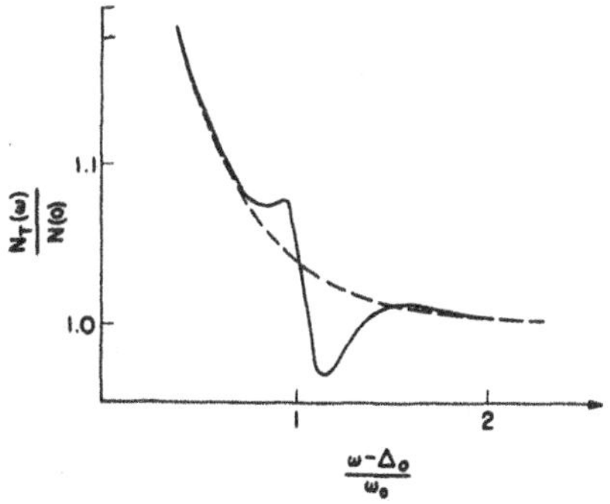

Fig. 38. Normalized tunneling density of states $N_T(\omega)/N(0)$ (solid) compared with the BCS form (dashed) for the case of a phonon density of states with peak at ω_0 (see Fig. 34).

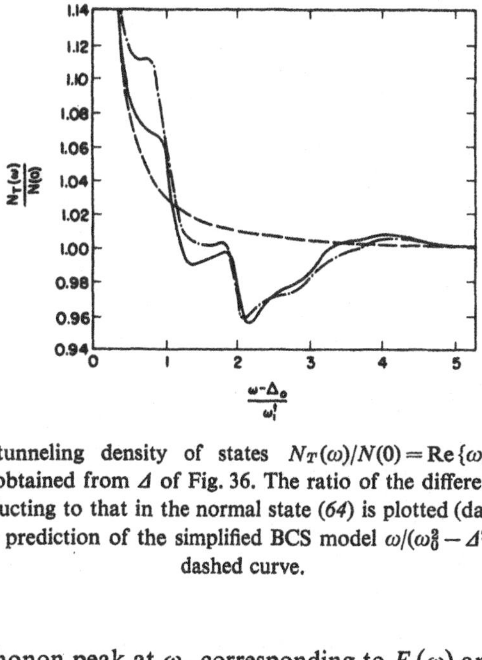

Fig. 39. Effective tunneling density of states $N_T(\omega)/N(0) = \text{Re}\{\omega/[\omega^2 - \Delta^2(\omega)]^{1/2}\}$ vs. $(\omega - \Delta_0)/\omega_1^t$ (solid) obtained from Δ of Fig. 36. The ratio of the differential conductance of Pb in the superconducting to that in the normal state (*64*) is plotted (dash-dot) as a function of $(\omega - \Delta_0)/\omega_1^t$. The prediction of the simplified BCS model $\omega/(\omega_0^2 - \Delta^2)^{1/2}$ is shown as the dashed curve.

case of a single phonon peak at ω_0 corresponding to $F(\omega)$ and $\Delta(\omega)$ shown in Figs. 34a and b.

In Fig. 39, the normalized single-particle density of states for Pb calculated from $\Delta(\omega)$ shown in Fig. 36 is plotted as a solid line. The experimental tunneling conductance data observed by Rowell et al. (*64*) is shown as the dash-dot curve. This was obtained by plotting the ratio of the differential conductance dI/dV in the superstate to that in the normal state as a function of bias voltage ($eV = \omega$). The dashed curve in Fig. 39 is the BCS constant-gap prediction. As discussed for the simple model, these results clearly show the reflection of the peaks of

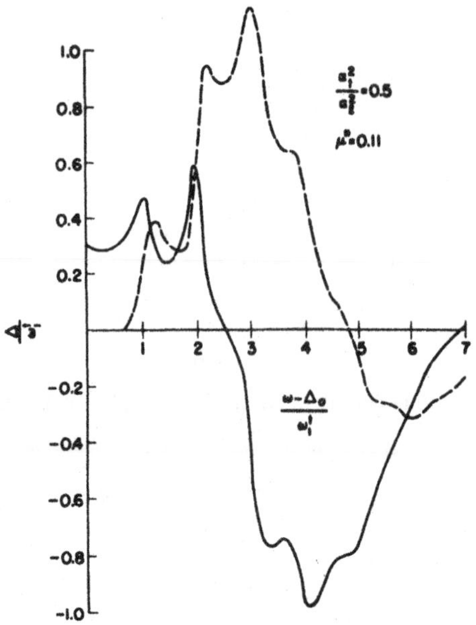

Fig. 40. Plot of the real (solid) and imaginary (dashed) parts of $\Delta(\omega)/\omega_1^t$ vs. $(\omega-\Delta_0)/\omega_1^t$ for the model of the Pb phonon density of states shown in Fig. 35. Here $\omega_1^t=4.4$ meV, $\Delta_0=1.34$ meV, $\alpha_l^2=1.6$ meV, $\alpha_t^2/\alpha_l^2=0.5$, and $\mu^*=0.11$.

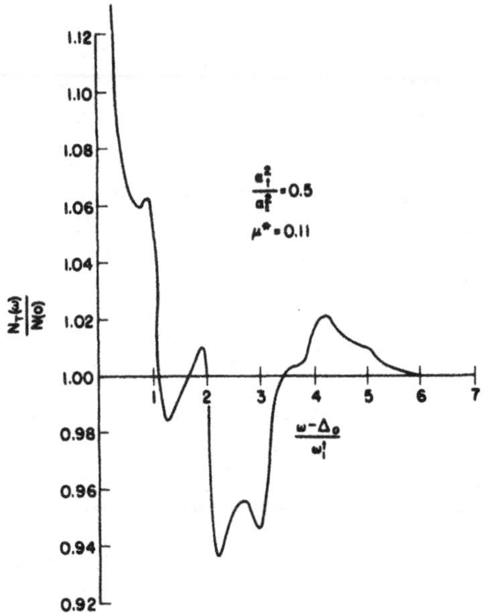

Fig. 41. Effective tunneling density of states $N_T(\omega)/N(0)=\text{Re}\{\omega/[\omega^2-\Delta^2(\omega)]^{1/2}\}$ vs. $(\omega-\Delta_0)/\omega_1^t$ for Δ obtained from Fig. 40. Comparison with Fig. 39 shows the increase in the size of the structure near $\omega_1^t+\Delta_0$ which arises from the choice of $\alpha_t^2/\alpha_l^2=0.5$ instead of 1.0.

the phonon density of states at ω_1^λ as sharp decreases in $N_T(\omega)/N(0)$ at $(\omega - \Delta_0)/\omega_t$ values near 1 and 2.

One way to see how sensitive such measurements are to the details of the electron–phonon interaction and the Coulomb pseudo-potential is to calculate $\Delta(\omega)$ for different α_l^2/α_t^2 and μ^*-values. Figures 40 and 41 show the results for $\Delta(\omega)$ and $N_T(\omega)/N(0)$ for a case in which $\alpha_l^2 = 2\alpha_t^2 = 1.6$ and $\mu^* = 0.11$. Since the size of α_t^2 is determined by demanding that Δ_0 equal the experimentally observed value of 1.34 meV, the effect of increasing the α_l^2/α_t^2 ratio is to put more weight in the longitudinal peak at ω_1^l. This is clearly visable in the increase of Δ_1 near ω_1^l in this case compared to the previous case shown in Fig. 36. The large knee near $\omega_1^l + \Delta_0$ in the resulting density of states is in disagreement with the experimental data and shows that the previous $\alpha^2(\omega)\,F(\omega)$ form more nearly represented the electron–phonon interaction. The effect of Coulomb pseudo-potential μ^* can be understood by considering the results for $\alpha_t^2/\alpha_l^2 = 0.5$ and $\mu^* = 0$ shown in Figs. 42 and 43. In this case, the size of the structure which appears in $\Delta(\omega)$ and $N_T(\omega)/N(0)$ is reduced. This simply reflects the fact that the size of the electron–phonon coupling α_l^2 can be reduced if $\mu^* = 0$. This leads to a reduction in the size of the resonant phonon effects. As noted in the figure captions, the strength of α_l^2 is 1.3 for $\mu^* = 0$ compared to the previous value of 1.6 for $\mu^* = 0.11$.

It is clear from these comparisons that the gross features of the phonon

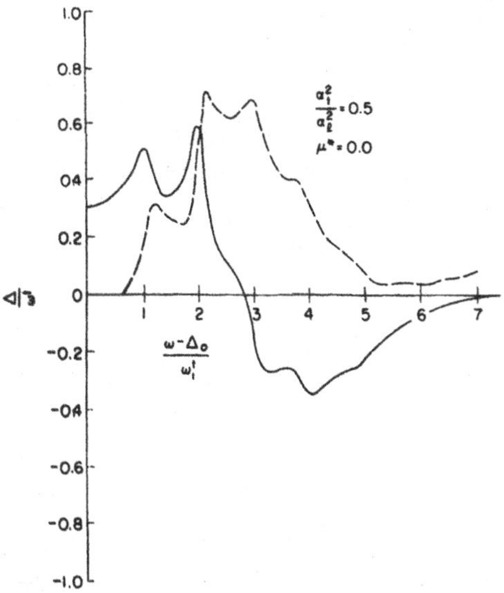

Fig. 42. Plot of the real (solid) and imaginary (dashed) parts of $\Delta(\omega)/\omega_1^t$ vs. $(\omega - \Delta_0)/\omega_1^t$ for the model of the Pb phonon density of states shown in Fig. 35. Here $\omega_1^t = 4.4$ meV, $\Delta_0 = 1.34$ meV, $\alpha_1^t = 1.3$ meV, $\alpha_t^2/\alpha_l^2 = 0.5$, and $\mu^* = 0$.

density of states and the strength of the electron–phonon coupling are reflected in the tunneling characteristics. In addition, an analysis (*65,66*) of the gap equations shows that Van Hove critical points produce log and square-root singularities in d^2I/dV^2. In Chapter 11, McMillan and Rowell discuss techniques which allow the complete $\alpha^2(\omega)\,F(\omega)$ curve to be unfolded from tunneling data.

Finally, to conclude this discussion of the electron self-energy, the behavior of Δ and Z at finite temperature should be considered. Scalapino, Wada, and

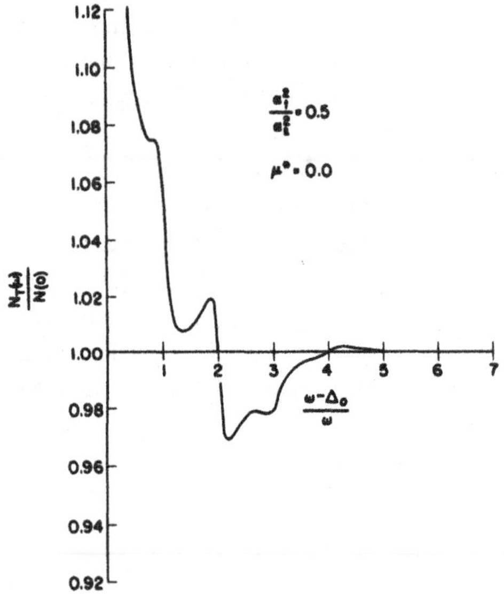

Fig. 43. Effective tunneling density of states $N_T(\omega)/N(0) = \mathrm{Re}\{\omega/[\omega^2 - \Delta^2(\omega)]^{1/2}\}$ vs. $(\omega - \Delta_0)/\omega$ for $\Delta(\omega)$ obtained from Fig. 42. Comparison with Fig. 41 shows that μ^* enhances the structure in the effective tunneling density of states.

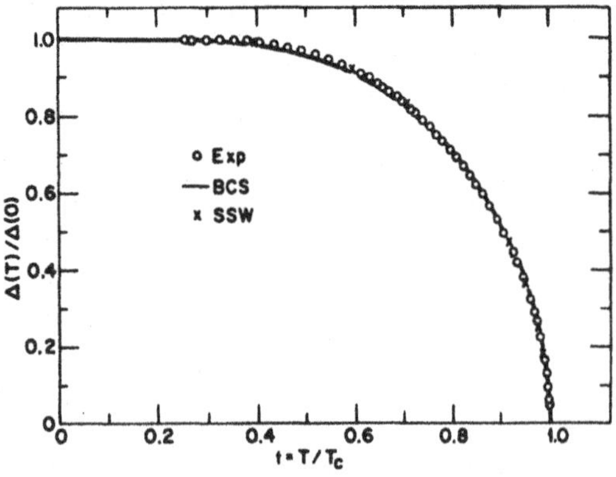

Fig. 44. Reduced energy gap $\Delta(T)/\Delta(0)$ vs. the reduced temperature T/T_c for Pb. [After Gasparovic et al. (*67*).]

Swihart (*59*) have studied this problem using the finite-temperature self-energy equations derived in Section III. For a weak-coupling superconductor such as Al, it was found that $\Delta(\omega, T)$ was very closely approximated by

$$\Delta(\omega, T) \sim \Delta(\omega, 0)\,[\Delta_0(T)/\Delta_0(0)]$$

That is, the zero temperature gap $\Delta(\omega, 0)$ simply scaled with the ratio $\Delta_0(T)/\Delta_0(0)$ as T increased toward T_c. The temperature dependence of the gap at the gap edge,

$$\Delta_0(T) = \mathrm{Re}\,\Delta\,[\Delta_0(T), T]$$

followed the BCS curve. For the strong-coupling superconductors the behavior of $\Delta_0(T)/\Delta_0(0)$ was also amazingly close to the BCS curve. In Fig. 44, some results for Pb are compared to experimental measurements by Gasparovic et al. (*67*). The fact that the strong-coupling theoretical results (SWS, Fig. 44) seem to agree with the small deviations from the BCS behavior should not be taken very seriously, but the fact that the strong-coupling theory predicts that $\Delta_0(T)/\Delta_0(0)$ vs. T/T_c follows very closely the BCS curve is significant.

At frequencies greater than Δ_0, the strong-coupling solution for $\Delta(\omega, T)$ shows important deviations from a simple $\Delta_0(T)/\Delta_0(0)$ scaling of the zero-temperature gap. Figure 45 shows the real and imaginary parts of Δ for T/T_c equal to 0.98. New resonance structures occur near the frequencies $\omega_1^\lambda - \Delta_0(T)$. This can be clearly seen at $\omega_1^t - \Delta_0(T)$ (3.9 meV), where $\Delta_2(\omega, T)$ has a negative peak. Physically, these new resonances correspond to additional decay channels which open up at finite temperatures. At nonzero temperatures, the injected electron can combine with an excited quasi-particle, emit a phonon, and form a ground-state pair. Since there is a high density of states for quasi-particles at

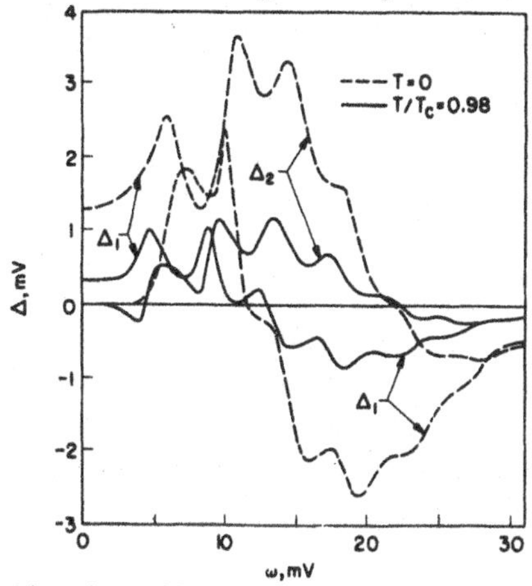

Fig. 45. Real (solid) and imaginary (dashed) parts of $\Delta(\omega, T)$ for Pb as a function of ω. The dashed curves are for $T=0$ and the solid curves are for a reduced temperature $T/T_c = 0.98$.

the gap edge $\Delta_0(T)$, the rate of recombination is a maximum when the energy of the injected electron is equal to $\omega_1^\lambda - \Delta_0(T)$. In this case the phonon is emitted into a peak in the phonon density of states. This additional structure in Δ produces a small modification in the tunneling density of states. Unfortunately, the thermal smearing makes it difficult to detect (*68,69*).

IV. PROPERTIES OF STRONG-COUPLING SUPERCONDUCTORS

Although electron tunneling experiments provide the most direct probe of the internal dynamics of the strong-coupling superconductors, there are a number of other measurements which yield information on the electron–phonon interaction. With the "vanishing" of the infamous precursor (*70,71*) it seems reasonable to suggest that the theory of the strong-coupling superconductors which was developed in the previous section has proved sufficient to account for all of the presently excepted experimental measurements on the strong-coupling superconductors. In discussing these measurements we have separated them into thermodynamic and dynamic properties.

A. Thermodynamic Properties

The basic thermodynamic properties follow from the free-energy which can be evaluated in terms of the electron and phonon Green's functions. The difference in free energy between the normal and superconducting phases is directly measured by the critical magnetic field. In Section IV.A.2 two approaches which allow $H_c(T)$ to be calculated are given and some numerical results based upon these relations are compared with experiment. In Section IV.A.3 the transition temperature and its dependence on the electron–phonon coupling is reviewed. All these results involve averages of the single-particle properties over virtual or thermal excited states. Therefore, before smearing these single-particle properties by averaging, it is appropriate to look at them in a little more detail.

1. Single-Particle Properties‡

From an "$F = ma$" point of view, probably the most physical single-particle question one can ask is: What is the probability of transferring energy ω and momentum p to the system by injecting or extracting an electron? To add a single particle to the ground state and transfer momentum p, we apply $c_p^\dagger$ to the N-particle ground state

$$c_p^\dagger \, |\psi_0\rangle_N$$

‡ The numerical results given in this section were obtained by Swihart (*72*).

Since this state is not necessarily normalized to unity, the *relative* probability that the system is in an energy eigenstate $|\psi_n\rangle_{N+1}$ of the $(N+1)$-particle system is

$$|_{N+1}\langle\psi_n| c_p^\dagger |\psi_0\rangle_N|^2$$

In this state the energy transfer would be

$$\begin{aligned} E_n^{N+1} - E_0^N &= (E_n^{N+1} - E_0^{N+1}) + (E_0^{N+1} - E_0^N) \\ &= \omega_{n0} + \mu \end{aligned}$$

Measuring the energy transfer relative to the chemical potential μ, the relative probability density for momentum transfer p and energy transfer ω is given by

$$A^+(p, \omega) = \sum_n |_{N+1}\langle\psi_n| c_p^\dagger |\psi_0\rangle_N|^2 \, \delta(\omega - \omega_{n0}) \tag{101}$$

In the same way, it follows that the relative probability density for momentum transfer $-p$ and energy transfer ω, when an electron is extracted, is given by

$$\sum_n |_{N-1}\langle\psi_n| c_p |\psi_0\rangle_N|^2 \, \delta(\omega - \omega_{n0})$$

Here the energy transfer ω is measured relative to $-\mu$ and is positive. For a system with reflection invariance the matrix elements are unchanged if c_p is replaced by c_{-p}, so that the momentum transfer is p. To map both the electron injection and extraction as a single function along the ω-axis, it is conventional to measure the energy transfer along the negative ω-axis and define a relative probability density for momentum and energy transfer associated with electron extraction by

$$A^-(p, \omega) = \sum_n |_{N-1}\langle\psi_n| c_p |\psi_0\rangle_N|^2 \, \delta(\omega + \omega_{n0}) \tag{102}$$

In this way, the single-particle spectral weight,

$$A(p, \omega) = A^+(p, \omega) + A^-(p, \omega) \tag{103}$$

characterizes the single-particle energy-momentum transfer properties for electron injection when $\omega > 0$ and electron extraction when $\omega < 0$.

The spectral weight satisfies a useful sum rule

$$\int_{-\infty}^{\infty} d\omega \, A_p(\omega) = \langle\psi_0| \{c_p, c_p^\dagger\}_+ |\psi_0\rangle = 1 \tag{104}$$

which follows directly from the normalization of the ground state and the anticommutation relations of the c_p operators. It has the simple physical interpretation that the single-particle state p must be either occupied or unoccupied. The zero moments are also directly related to the single-particle properties of the ground state

$$\int_{-\infty}^{0} d\omega\, A(p, \omega) = \langle \psi_0 | c_p^\dagger c_p | \psi_0 \rangle = n_p \tag{105}$$

$$\int \frac{d^3 p}{(2\pi)^3} A(p, \omega) = N(\omega) \tag{106}$$

Here n_p is the momentum occupation of the single-particle states in the ground state and $N(\omega)$ is the single-particle density of states at energy ω.

For the case of a spherical band model of the bare single-particle energies ϵ_p, the variable p can be replaced by ϵ_p, so that the spectral weight is a function of the two-energy variables ϵ_p and ω. Plotting A as a surface over the $\epsilon_p - \omega$ plane gives a useful way of visualizing the single-particle properties and, through them, the underlying correlations present in the ground state. For example, the spectral weight of a noninteracting system consists of a ridge of δ-functions running along the line $\omega = \epsilon_p$ as shown in Fig. 46. Here, when a given momentum p is transferred (labeled by $\epsilon_p = p^2/2m - \mu$), a definite energy $\omega = \epsilon_p$ is also transferred. Note that the sum rule is trivially satisfied and that

$$n_p = \int_{-\infty}^{0} d\omega\, \delta(\omega - \epsilon_p) = \begin{cases} 1 & p < p_F \\ 0 & p > p_F \end{cases}$$

$$N(\omega) = \int \frac{d^3 p}{(2\pi)^3} \delta(\omega - \epsilon_p) = \frac{1}{\pi^2} \left(\frac{m}{2}\right)^{1/2} (\epsilon_p + \mu)^{1/2}$$

as one would expect. The ground state is just the filled Fermi sea with only the Pauli principle correlations.

Interactions modify the structure of the $A(\epsilon_p, \omega)$ surface in a number of ways. The sharp ridge is broadened and shifted by processes which couple an excited particle to the other particles of the system. If the breadth $\Delta\omega_p$ in the ω-direction for fixed ϵ_p is small compared with the peak position ω_p, it is meaningful to describe the single-particle properties in terms of quasi-particles having a dispersion relation ω_p and a lifetime $\tau = \Delta\omega_p^{-1}$. It can happen that the interactions produce correlations in the ground state which give rise to qualitative changes in the single-particle dispersion relation. For example, one can think of He^4, in which the low momentum dispersion relation becomes linear or the superconductor in which the ground-state pairing correlations are reflected by a gap in the single-particle dispersion. Besides such modifications, the interactions couple other modes into A. For example, adding an electron and transferring momentum p can occur by the electron going into the state ϵ_{p-q} and a phonon ω_q being created. For a given energy ω and momentum p a number of states may satisfy $\omega = \epsilon_{p-q} + \omega_q$ so that the phonon structure in A is not as sharply defined as the quasi-particle ridge. Once again it is worth emphasizing that these

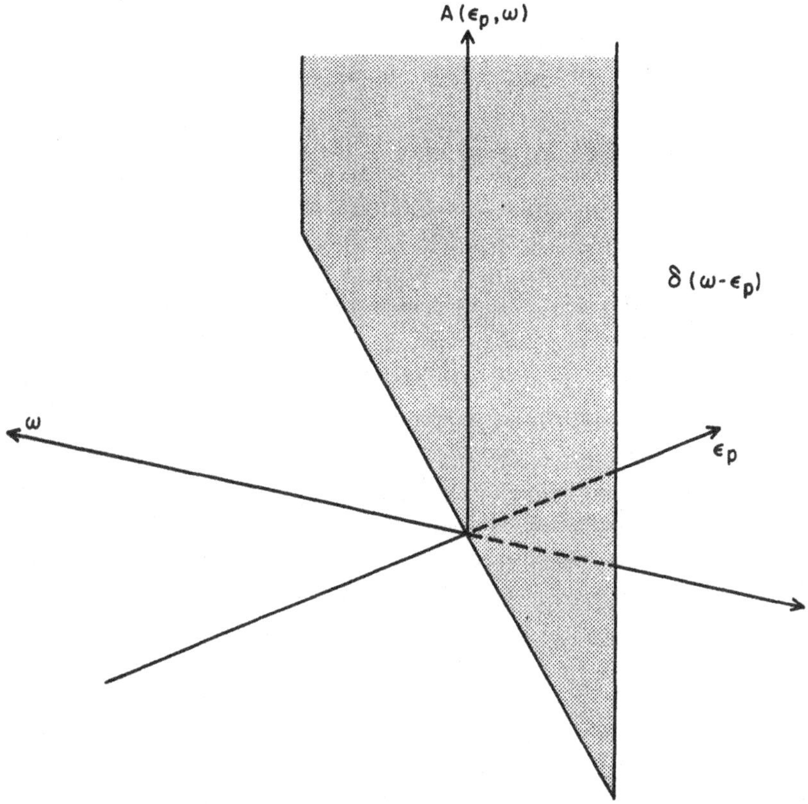

Fig. 46. Spectral weight of a noninteracting electron gas.

processes also reflect the correlations in the ground state. In the language of perturbation theory, the ground state contains virtual fluctuations arising from the electron–phonon interaction in which particle-hole pairs are created and a phonon excited. The phonon is then absorbed and the particle and hole recombine. If an electron is injected into the hole "during" this virtual fluctuation, with energy equal to the remaining electron and phonon, the process can contribute to A. Thus the structure of A measures the correlations in $|\psi_0\rangle$ which can be probed by single-particle injection or extraction.

A direct calculation of A from the definitions of A^+ and A^- is a formidable task for a strongly interacting system. It would require that the exact energy eigenstates coupled to the ground state through $c_p^\dagger$ and c_p be obtained, the matrix elements evaluated and the sum over states determined. Fortunately, as discussed in Chapter 5, the spectral weight A is simply related to the single-particle Green's function $G_p(\omega)$,

$$A(\epsilon_p, \omega) = \frac{1}{\pi} |\text{Im}\, G_p(\omega)| \tag{107}$$

Here $G_p(\omega)$ is the one-to-one component of the Nambu Green's function:

$$G_p(\omega) = \frac{Z(\omega)\,\omega + \epsilon_p}{Z^2(\omega)\,\omega^2 - \epsilon_p^2 - [Z(\omega)\,\Delta(\omega)]^2} \tag{108}$$

where $Z(\omega)$ and $\Delta(\omega)$ were evaluated in Section III. For the case of a normal metal Δ is zero and

$$G_p(\omega) = (Z(\omega)\cdot\omega - \epsilon_p)^{-1} \tag{109}$$

Here $Z(\omega)$ is determined from Eq. (75) with Δ set equal to zero.

Just as in the discussion of the self-energy, it is simplest to begin by looking at A for a system having a single peak in the phonon density of states centered at ω_0. In Fig. 47, photographs of a model for A in the *normal* state are shown. Because of the assumed particle-hole symmetry, the negative part of the spectral weight is obtained by a 180° rotation of the figure,

$$A^-(\epsilon_p, \omega) = A^+(-\epsilon_p, -\omega) \tag{110}$$

The lines inked on the photograph indicate the part of the spectral weight which can be identified as quasi-particle-like. At large ϵ_p/ω_0 values, the quasi-particle

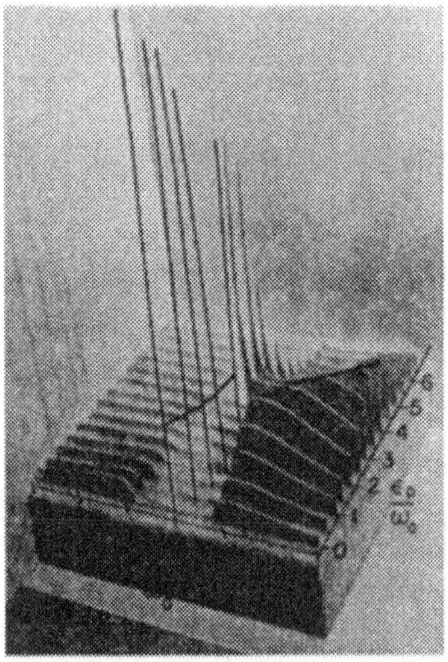

Fig. 47. Spectral weight $A^+(\epsilon_p, \omega)$ for an electron–phonon system in the normal state. The phonon density of states is assumed to have a peak at ω_0. The solid line indicates the part of the spectral weight which can be identified as a quasi-particle.

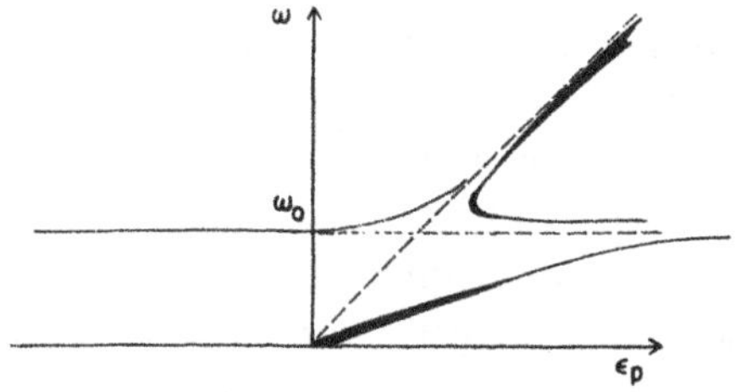

Fig. 48. Normal-state ridges of $A^+(\epsilon_p, \omega)$ are shown as solid lines. The height of the ridge is proportional to the thickness of the solid line. The dashed lines $\omega = \epsilon_p$ and $\omega = \omega_0$ correspond to the dispersion relations of the noninteracting electron and phonon.

peak has moved beyond the region plotted in this model. Looking down on the $A^+(\epsilon_p, \omega)$ surface, the peaks form several well-defined ridges which are sketched as solid lines in Fig. 48. As is evident from the spectral weight model, the height of A^+ along these ridges varies over a wide range. In Fig. 48 the thickness of the branch lines is used as an indication of this relative weight. The dashed lines $\omega = \epsilon_p$ and $\omega = \omega_0$ represent the noninteracting electron and phonon dispersion relations.

At low energy ($\epsilon_p/\omega_0 \ll 1$) and at high energy ($\epsilon_p/\omega_0 \gg 1$) the spectral weight is dominated by a single peak. In the low-energy region, the slope of this ridge in the $\epsilon_p\omega$-plane is less than the free electron dispersion relation, indicating an enhanced effective mass due to the electron–phonon interaction. In the high-energy region $\epsilon_p/\omega_0 \gg 1$, the ridge approaches the noninteracting dispersion, showing that the electron–phonon mass renormalization vanishes in this region. The width of the ridge is due to the electron–phonon coupling. At still higher energies the effects of Coulomb induced particle-hole production dominate this phonon emission damping. In the vicinity where the noninteracting electron and phonon dispersion relations cross, the coupling lifts the degeneracy giving rise to considerable structure in A. In this region, which is important for the pairing interaction in the strong-coupling superconductors, it is easy to see that the spectral weight is not well described by a quasi-particle peak. Weak structure persists for $\omega \sim \omega_0$, even when ϵ_p is negative. This reflects the phonon modes as discussed before.

At low-energy transfer the quasi-particle peak narrows and for a normal system approaches a δ-function as ω goes to zero. This narrowing arises because the Pauli principle restricts the available phase space for creating more complicated excitations. In the model shown, a δ-function appears in the spectral weight at low frequencies where the single peak used to represent the phonon density of states vanishes. In a more realistic model the phonon density of states would not vanish until ω equals zero. For example, in a jellium model of the electron–ion system, the width of the peak is found to vary as ω^3. For p just below p_F, the negative-energy part of the spectral weight contains this spike, while for p just greater than p_F this spike has moved to the positive-energy part. Since the momentum occupation n_p is given by the area under the negative part of $A(\epsilon_p, \omega)$ for fixed p, there is a jump discontinuity in n_p at the Fermi surface. In the interacting system this discontinuity is less than unity. A simple calculation shows that it is given by $Z(0)^{-1}$ for the case of the electron–phonon interaction. An example computed for Pb will be given below. It must be kept

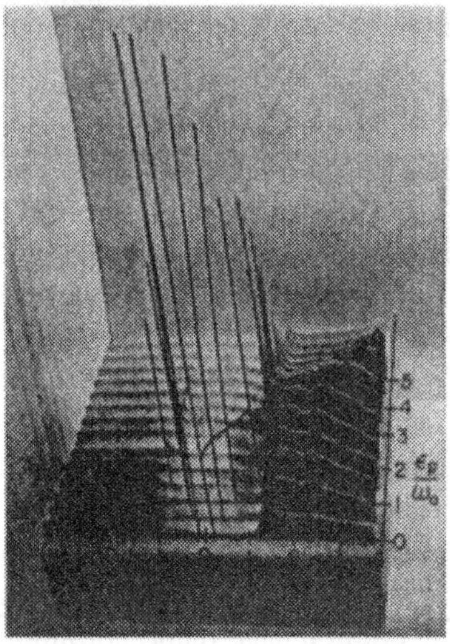

Fig. 49. Photograph of the spectral weight $A^+(\epsilon_p, \omega)$ for an electron–phonon system in the superconducting state. The phonon density of states is assumed yo have a peak at ω_0. The solid line indicates the part of the spectral weight which can be identified as quasi-particle-like.

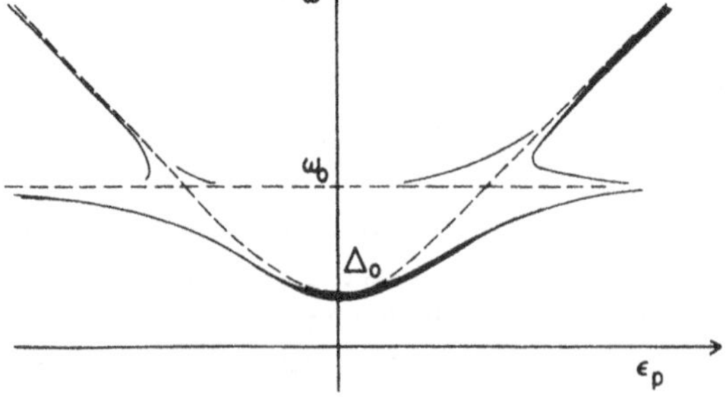

Fig. 50. Superconducting-state ridges of $A^+(\epsilon_p, \omega)$ are shown as solid lines. The height of the ridge is proportional to the thickness of the solid line. The dashed curves $\omega = \sqrt{\epsilon_p^2 + \Delta_0^2}$ and $\omega = \omega_0$ correspond to the BCS quasi-particle dispersion relation and the noninteracting phonon dispersion relation.

in mind that these results are based upon Migdal's approximation of the vertex and it is possible that further corrections could eliminate this discontinuity.

Using this same, single-peak model for the phonon density of states, the spectral weight for the superconducting state has been computed. Photographs of a model for the superconducting case are shown in Fig. 49. The ridge topology is illustrated in Fig. 50 along with dashed lines representing the BCS quasi-particle dispersion relation $\omega^2 = \epsilon_p^2 + \Delta^2$ and the unperturbed phonon dispersion $\omega = \omega_0$. Just as before, the spectral weight has particle-hole symmetry

$$A^-(\epsilon_p, \omega) = A^+(-\epsilon_p, \omega)$$

so that the complete A surface can be generated by rotating the figures through 180°.

The appearance of a gap and the associated double set of quasi-particle spikes characteristic of the correlations in the superconducting state are clearly evident. The amplitude of the spike in the positive part of ω-plane drops off as ϵ_p/Δ_0 becomes negative because of the decreasing probability of adding an electron below the Fermi surface. However, unlike the model of the normal metal, the quasi-particle ridge associated with adding (extracting) an electron remains in the positive (negative) ω-part of the plane. Therefore, since the amplitude of A along this ridge varies continuously, $\langle n_p \rangle$ for a superconductor is a smooth function at the Fermi surface. Just as in the normal metal, the quasi-particle structure is complicated for $\omega \sim \omega_0 + \Delta_0$ and exhibits multiple peaks associated with phonon excitations. For ω large compared to ω_0, the topology is again

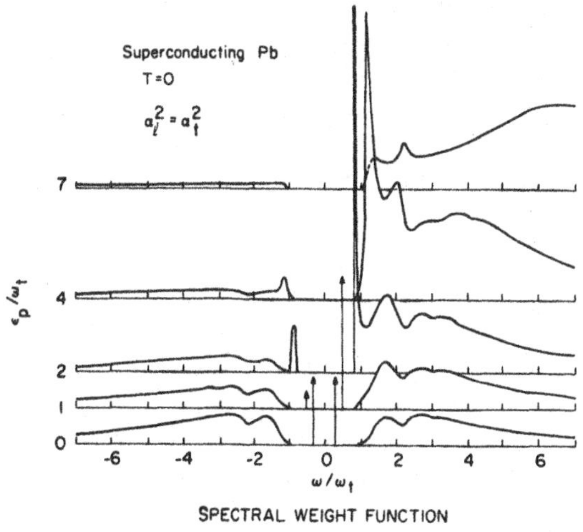

Fig. 51. Typical crossections of $A(\epsilon_p, \omega)$ for the equal coupling constant model of Pb discussion in Section III.

dominated by a broad quasi-particle peak following the free-particle dispersion relation and broadened by phonon emission processes.

Turning now to the calculation of $A(\epsilon_p, \omega)$ for the Pb model, our discussion will be brief. Figure 51 shows some typical cross sections of $A(\epsilon_p, \omega)$ in the superconducting state for fixed ϵ_p, and in Fig. 52 the spectral weight surface is sketched. The multiple bumps in the continuum part are due to the two peaks in the phonon density of states. The arrows in Fig. 52 indicate δ-function quasi-particle contributions which arise because the cutoff Lorentzian model of the phonon density of states vanishes below $\omega_1^t - 3\omega_2^t$. The general features

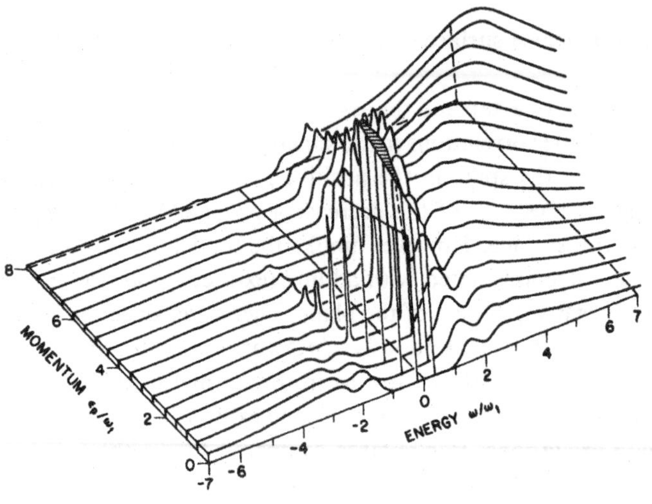

Fig. 52. Spectral weight surface $A^+(\epsilon_p, \omega)$ for the equal-coupling-constant model of superconducting Pb. The arrows indicate δ-function pieces.

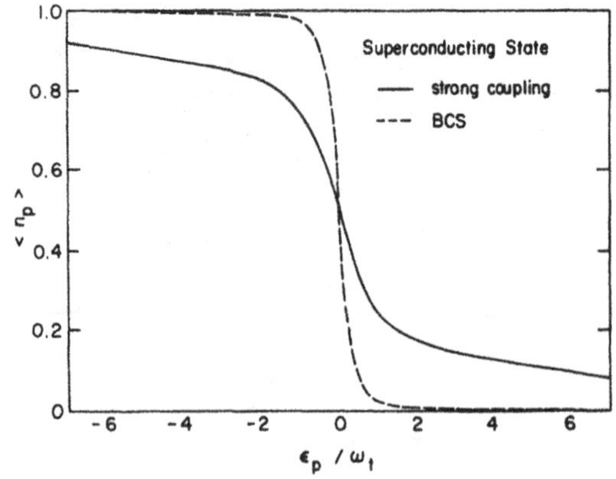

Fig. 53. Single-particle momentum occupation n_p (solid line) in the superconducting ground state. The BCS result $\frac{1}{2}(1 - \epsilon_p/E_p)$ is shown as the dashed line for comparison.

previously discussed are clearly illustrated. The zeroth ϵ_p moment,

$$N(0)\int d\epsilon_p\, A(\epsilon_p, \omega) = N_T(\omega)$$

gives the single-particle density of states and has been discussed in Section III in connection with the tunneling results. The zeroth, negative ω moment,

$$\int_{-\infty}^{0} d\omega\, A(\epsilon_p, \omega) = n_p$$

has been plotted as the solid line in Fig. 53. The BCS result,

$$n_p = \tfrac{1}{2}(1 - \epsilon_p/E_p)$$

with $E_p = \sqrt{\epsilon_p^2 + \Delta^2}$ is shown as the dashed line for comparison. Both curves show the continuity of n_p across the Fermi surface. The deviation of the strong-coupling result from the BCS curve arises from the coupling to the phonon modes.

Finally, using this same model, results for Pb in the normal state were computed. The momentum occupation in the normal ground state, $\langle n_p \rangle$, plotted

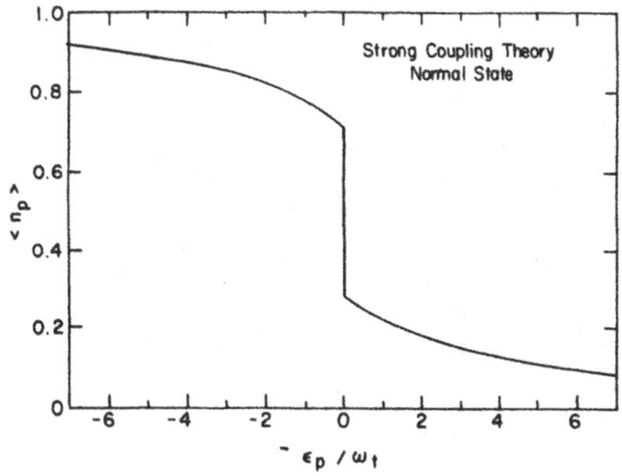

Fig. 54. Momentum occupation n_p in the normal ground state. The discontinuity at p_F is equal to $Z(0)^{-1}$.

in Fig. 54, shows a discontinuity of about 0.4 at the Fermi surface with a tail extending above the Fermi surface. Effects of the Coulomb interaction which were not treated in detail are expected to modify this result slightly over the range of energies illustrated. At energies large compared to ω_1^l the phonon effects become negligible and the Coulomb corrections to the results shown in Fig. 54 must be taken into account.

TABLE III

Coupling Constants Used in the Calculations and Comparison of the Results with Experiment

Superconductor	Phonon peak ω_1, mV	Peak width ω_2, mV	Electron–phonon coupling α^2, mV	Coulomb interaction μ^*	Calcd. $Z_{n1}(\omega=0, T=0)$	Exptl. m^*/m	Calcd. $2\Delta_p/kT_c$	Exptl. $2\Delta_0/kT_c$
Pb						2.1 (*74, 75*)		4.3 (*3*)
Case 1				0.11	2.4		4.40	
Transverse	4.4	0.75	1.198					
Longitudinal	8.5	0.50	1.198					
Case 2				0.11	2.2		4.33	
Transverse	4.4	0.75	0.807					
Longitudinal	8.5	0.50	1.614					
Hg				0.14	2.5	1.9 (*76*)	4.8	4.6 (*77*)
Transverse	1.80	0.5	0.71					
Longitudinal	7.60	0.5	0.71					
Al				0.11	1.4	1.5 (*6*)	3.5	3.5 (*78*)[a]
Longitudinal	34.0	1.0	5.60					

[a] The experiments that are pertinent to this calculation are those with the silver impurity for which the aluminum is a "dirty" superconductor with an isotropic gap.

Another quantity of direct experimental interest which can be determined from the normal metal calculations is the phonon effective mass enhancement at the Fermi surface. Near the Fermi surface, the quasi-particle ridge is determined by the poles of G:

$$G_p^{-1}(\omega) = Z(\omega)\cdot\omega - \epsilon_p = 0$$

At the Fermi surface this becomes

$$\omega = \epsilon_p/Z(0)$$

so that the effective mass is simply determined by‡

$$\frac{m^*}{m_c} = Z(0) = 1 + \int_0^{\infty} d\omega \, \frac{2\alpha^2(\omega)\,F(\omega)}{\omega} \tag{111}$$

Here m_c is the Coulomb and band dressed effective mass. Using $\alpha^2 F$ determined for the superconducting problem, Swihart et al. (*73*) compute m^*/m_c for Pb, Hg, and Al, obtaining the results shown in Table III. In comparing these results to the experimental specific-heat measurements, it is important to remember that m_c is the Coulomb and band dressed effective mass. From Rice's work, the Coulomb dressing is expected to be small. Using the Gold–Anderson Fermi surface data for Pb one can estimate that the band mass is about 10% less than the free electron mass, bringing the phonon-induced effective mass results in good agreement with the specific-heat measurements. Ashcroft and Wilkins (*79*) have used the pseudo-potential and the known phonon frequencies to compute m^*/m_c and have discussed the band and Coulomb corrections in detail.

2. Critical Magnetic Field

The critical magnetic field $H_c(T)$ of a type I superconductor is a direct measure of the condensation energy (i.e., the free energy difference between the normal and superconducting phases). Since the free energy can be determined in terms of the electron and phonon Green's functions, the problem of calculating $H_c(T)$ reduces to determining the changes in these Green's functions when the system goes from the normal to the superconducting state. As we have discussed, detailed calculations of the electron self-energy in both states have been carried out. To proceed it is necessary to take into account the modification of the phonon propagator. One expects that the changes are small, but the sum of their contributions can be important. In fact, a theorem proved by Chester (*80*) states that for a superconductor having a normal isotope factor of 0.5, the change in the ion kinetic energy between the normal and superconducting states is equal to the condensation energy!

‡ The parameter $Z(0)$ determines the mass renormalization, the Fermi velocity renormalization $v_F = p_F/Z(0)\,m$, and the discontinuity in the momentum occupation $\Delta\langle n_{p_F}\rangle = Z(0)^{-1}$.

Two procedures for dealing with this problem have been proposed. The first involves using Chester's theorem to eliminate the phonon propagator from the condensation energy. This was applied by Scalapino and Schrieffer (*81*) to determine the condensation energy at zero temperature and by Wada (*82*) to obtain an expression for the temperature dependence of the critical field and specific heat for strong-coupling superconductors. A second, more elegant procedure, was applied by Bardeen and Stephen (*83*), who used an expression for the free energy derived by Eliashberg (*84*) which is stationary with respect to variations in the electron and phonon self-energies. Using this stationary property, the phonon self-energy in the superconducting state can be replaced by that in the normal state correct to second order in the deviation, which is small. Eliashberg's expression for the free energy includes the electron–phonon interaction but omits effects of Coulomb interactions except as they enter in renormalizing the electron mass, so this theory is applicable to systems with an isotope coefficient of 0.5. Bardeen and Stephen showed how the expression they obtained could be approximately related to Wada's when the isotope coefficient appearing in Wada's expression was set equal to 0.5, and numerical calculations for Pb using both expressions are in close agreement. However, the form of the Bardeen–Stephen expression is more rapidly convergent and hence more useful in numerical calculations.

To begin with, consider the condensation energy U_0 at zero temperature:

$$U_0 = \langle \mathscr{H} \rangle_n - \langle \mathscr{H} \rangle_s \equiv \Delta \langle \mathscr{H} \rangle \tag{112}$$

where the subscripts n and s refer to the normal and superconducting ground states. In this section the symbol Δ means the difference between the normal and superconducting state. Experimentally, the system is forced into the normal ground state by applying a magnetic field, and the condensation energy per unit volume is determined by the low-temperature limit of the critical magnetic field:

$$U_0 = \lim_{T \to 0} [H_c^2(T)/8\pi]$$

Theoretically, the ground-state energy of the normal system can be obtained by using "normal-state" propagators in computing $\langle \mathscr{H} \rangle_n$. The small effect of the external magnetic field on the normal propagators is negligible.

The method for calculating the condensation energy based on Chester's theorem proceeds as follows. Using the equations of motion, the expectation value of the electron–ion Hamiltonian [Eq. (7)] can be expressed in terms of the electron and phonon Green's functions:

$$\langle \mathscr{H} \rangle = \int \frac{d^3p}{(2\pi)^3} \oint \frac{dp_0}{2\pi i} G(p)(p_0 + \epsilon_p) - \sum_\lambda \int \frac{d^3q}{(2\pi)^3} \oint \frac{dq_0}{2\pi i} D_\lambda(q) \frac{q_0^2}{2\Omega q_\lambda} \tag{113}$$

Here the frequency integrals are closed in the upper-half plane and G and D are the dressed electron and phonon Green's functions. The last term in the expression for $\langle \mathscr{H} \rangle$ is just twice the expectation value of the ion kinetic energy:

$$\langle \sum_\nu P_\nu^2/2M \rangle = \tfrac{1}{2} \sum_\lambda \int \frac{d^3q}{(2\pi)^3} \oint \frac{dq_0}{2\pi i} D_\lambda(q) \frac{q_0^2}{2\Omega q_\lambda} \tag{114}$$

The change in this between the normal and superconducting states can be related to the condensation energy by Chester's theorem. In this way an expression for U_0 depending only upon the changes in the electron Green's functions is obtained. Chester argued that if the critical field H_c has an isotopic dependence $M^{-\alpha}$, then

$$\frac{1}{M} \frac{\partial}{\partial(1/M)} \left(\frac{H_c^2}{8\pi} \right) = 2\alpha U_c = \frac{1}{M} \frac{\partial}{\partial(1/M)} \Delta \langle \mathscr{H} \rangle \tag{115}$$

Using the stationary property of an energy eigenstate, the last equality can be written in the form

$$2\alpha U_0 = \frac{1}{M} \Delta \left\langle \frac{\partial}{\partial(1/M)} \mathscr{H} \right\rangle$$

It is clear in the configuration space representation of $\mathscr{H}$ that the M dependence is contained in the ion kinetic energy. This means that

$$\frac{1}{M} \left\langle \frac{\partial}{\partial(1/M)} \mathscr{H} \right\rangle = \langle \sum_\nu P_\nu^2/2M \rangle$$

and Eq. (115) reduces to Chester's theorem:

$$2\alpha U_0 = \Delta \langle \sum_\nu P_\nu^2/2M \rangle \tag{116}$$

Here we will treat the simple isotope effect case in which $\alpha = 0.5$. Substituting this expression for the term in $\langle \mathscr{H} \rangle$ containing D, the condensation energy can be expressed in terms of the electron Green's functions only:

$$\Delta \langle \mathscr{H} \rangle = \int \frac{d^3p}{(2\pi)^3} \oint \frac{dp_0}{2\pi i} [G_s(p) - G_n(p)] (\epsilon_p + p_0) \tag{117}$$

The normal Green's function has a branch cut along the real axis. The branch cut associated with G_s runs from $-\infty$ to $-\Delta_0$ and Δ_0 to ∞, since there are no single-particle states in the gap. Since the p_0 integration is closed in the upper half plane, one finds

$$U_0 = 2 \,\mathrm{Im} \int \frac{d^3p}{(2\pi)^3} \int_{-\infty}^{0} dp_0 (p_0 + \epsilon_0)(G_s - G_n) \tag{118}$$

Because of the natural cutoff $\omega_c(\Delta_0 \ll \omega_c \ll \mu)$ for the p_0 integration, the electron states of energy $\epsilon_p < \omega_c$ provide the dominant contribution. The 3-momentum integration in Eq. (118) may therefore be transformed by replacing d^3p by $N(0)\, d\epsilon_p$, and evaluated by residues to give

$$U_0 = N(0)\,\mathrm{Re}\int_0^\infty dp_0 \times\left\{[Z_n(p_0)+1]\,p_0 - Z_s(p_0)\sqrt{p_0^2-\Delta^2(p_0)} - \frac{p_0^2}{\sqrt{p_0^2-\Delta^2(p_0)}}\right\} \tag{119}$$

Here Δ is the gap parameter and Z_n and Z_s are the normal and superconducting renormalization parameters.

As an example, consider a system with

$$\Delta(p_0) = \begin{cases} \Delta_0 & |p_0| < \omega_0 \\ 0 & |p_0| > \omega_0 \end{cases}$$

and $Z_n = Z_s = 1$. The expression Eq. (119) for the condensation energy reduces to

$$U_0 = N(0)\left[\int_0^{\Delta_0} dp_0\, 2p_0 + \int_{\Delta_0}^{\omega_0} dp_0\left(2p_0 - \frac{2p_0^2-\Delta_0^2}{\sqrt{p_0^2-\Delta_0^2}}\right)\right]$$

and integrating, we obtain

$$U_0 = N(0)\,\Delta_0^2 + N(0)\,\omega_0^2\{1-[1+(\Delta_0/\omega_0)^2]\}^{1/2}$$

The first term, Δ_0^2, arises from the fact that the cut for G_n extends along the entire real axis while that for G_s stops at $\pm\Delta_0$. In the weak-coupling limit $\Delta_0/\omega_0 \ll 1$, this expression reduces to $N(0)\,\Delta_0^2/2$. Formally, this agrees with the well-known BCS result. However, in making comparisons between the BCS prediction and the strong-coupling results one must remember that $N(0)$ in BCS denoted the fully dressed single spin density of states at the Fermi surface obtained from a specific-heat measurement. Therefore, in the notation used in this chapter

$$U_{0(\mathrm{BCS})} = Z_n(0)\,N(0)\,\Delta_0^2/2 \tag{120}$$

where $Z_n(0)$ is the normal-state renormalization parameter evaluated at the Fermi surface.

In the case of a strong-coupling retarded interaction, the correction to the weak-coupling BCS result arises primarily from the difference between Z_n and Z_s. From Eq. (119) it follows that this correction has the approximate form

$$N(0)\,\mathrm{Re}\int dp_0\,(Z_n - Z_s)\,p_0 \tag{121}$$

with the dominant contribution arising from energies in the neighborhoods of the peaks ω_1^λ in the phonon spectrum. Since the peak in the real part of Z_s is shifted to higher energies by the gap parameter Δ_0 relative to that in Z_n, this integral produces a negative contribution to the condensation energy. The net effects of retardation is to reduce the condensation energy below the BCS prediction of $N(0)\,\Delta_0^2/2$. The fact that the condensation energy of the strong couplers is *reduced* may at first seem contrary to one's intuition. The point is that the BCS estimate depends upon the experimental value of Δ_0 and gives full recognition to the increased strength of the effective two-particle binding in the strong-coupling superconductors. However, the expression $N(0)\,\Delta_0^2/2$ does not account for the change in the energy due to the modification of the "normal" correlations. This is what the correction indicated in Eq. (121) is bringing in. Roughly one can say that virtual particle-hole fluctuations with phonon emission and reabsorption, Fig. 22b, lower the ground-state energy *less* in the superconducting state than in the normal state. Because of the energy gap, these fluctuations are further off the energy shell and therefore persist over shorter times in the superconducting phase. The pairing correlations disrupt these correlations which reduce the ground-state energy. It is natural that in the strong-coupling materials, where this reduction is substantial, the effect should be most easily observed as a decrease in the condensation energy below the BCS prediction.

Using the equal-coupling constant results for Pb discussed in Section III, the strong-coupling expression Eq. (119) gives $U_0 = 2.49 \times 10^4$ ergs/cm^3. The experimental critical field measurement gives a value of 2.56×10^4 ergs/cm^3. Evaluating the weak-coupling BCS result by taking experimental values for $Z_n(0)\,N(0)$ from specific heat measurements and Δ_0 from tunneling one finds

$$Z_n(0)\,N(0)\,\Delta_0^2/2 = 3.10 \times 10^4 \text{ ergs/cm}^3 .$$

In this case the strong-coupling effects have reduced the condensation energy by about 25% from the BCS predictions and give a result in good agreement with experiment.

Wada (*82*) extended this method to finite temperature and obtained an expression for the critical magnetic field of the form

$$\frac{H_c^2(T)}{8\pi} = N(0)\,I(\beta) = N(0)\,\mathrm{Re}\int_0^\infty dp_0 \tanh\left(\frac{\beta p_0}{2}\right) \times \left\{[Z_n(p_0)+1]\,p_0 - Z_s(p_0)\sqrt{p_0^2 - \Delta^2(p_0)} - \frac{p_0^2}{\sqrt{p_0^2 - \Delta^2(p_0)}}\right\} \tag{122}$$

Wada also showed that the jump in the specific heat of a strong coupling superconductor with $\alpha = 0.5$ was given by

$$\Delta C = k_B N(0)\,\beta_c^3\,(d^2 I/d\beta_c^2) \tag{123}$$

Although the specific-heat jump has not been evaluated to date, arguments can be given which show that the present theory should give a larger jump in the specific heat for the strong-coupling superconductors than obtained from the simple BCS theory. One such argument will be reviewed after the discussion of temperature dependence of $H_c(T)$.

Rather than derive the expression for the critical field obtained by Bardeen and Stephen, the transformation between Wada's form and theirs will be given. This depends upon the validity of an approximate relation derived by Bardeen and Stephen (*83*):

$$N(0)\,\mathrm{Re}\int_0^\infty dp_0 \tanh\left(\frac{\beta p_0}{2}\right)\left\{p_0[Z_s(p_0)-1] - \frac{p_0^2[Z_n(p_0)-1]}{\sqrt{p_0^2-\Delta^2(p_0)}}\right\} = 0$$

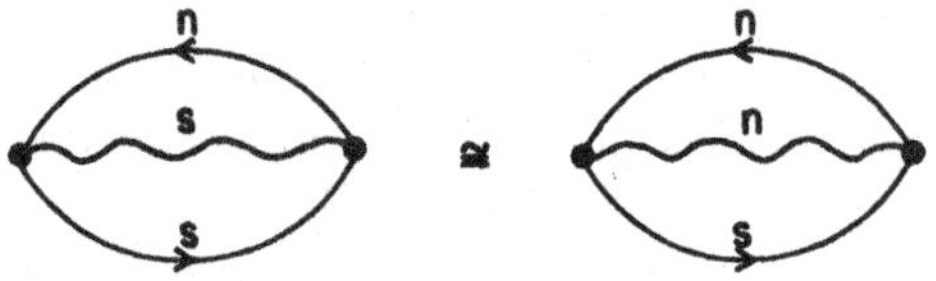

Fig. 55. Approximate identity relating the Wada result for $H_c^2(T)$ to that of Bardeen and Stephen. Here s and n signify that the corresponding propagator is that of the superconducting or normal state, respectively.

which follows, if the contributions of the two diagrams shown in Fig. 55 are equated. Adding this expression to Wada's form for the free energy, the Bardeen–Stephen expression is obtained:

$$\frac{H_c^2(T)}{8\pi} = N(0)\,\mathrm{Re}\int_0^\infty dp_0 \tanh\left(\frac{\beta p_0}{2}\right)\left\{[Z_s(p_0)+Z_n(p_0)] \times\left(p_0 - \sqrt{p_0^2-\Delta^2(p_0)} - \frac{\Delta^2(p_0)}{2\sqrt{p_0^2-\Delta^2(p_0)}}\right) + \frac{[Z_s(p_0)-Z_n(p_0)]\,\Delta^2(p_0)}{2\sqrt{p_0^2-\Delta^2(p_0)}}\right\} \tag{124}$$

Using the finite-temperature solution for Z and Δ, the temperature dependence of the critical magnetic field for Pb has been studied. Wada's expression gives a result which approaches the Bardeen–Stephen result as the cutoff frequency ω_c is extended to about $10\omega_D$ to obtain convergence. The Bardeen–Stephen form converges more rapidly and provides the most useful form for evaluating $H_c(T)$. A plot of some points determined by these numerical calculations for Pb is compared with the experimental results in Fig. 56. Whereas weak-coupling superconductors have a temperature dependence of $H_c(T)$, which falls below a parabolic curve by an amount in rough agreement with the BCS prediction, Pb and Hg show a positive deviation from a parabola. The present numerical

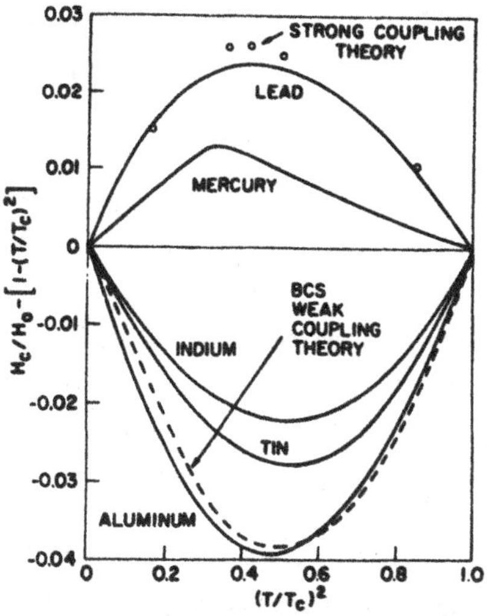

Fig. 56. Deviation of the reduced critical field from a parabolic curve as a function of the square of the reduced temperature. The experimental curves are taken from (*5*) and the circles are obtained from the strong-coupling calculation using the equal-coupling-constant results for Δ and Z.

results for Pb are in good agreement with experiment. However, it should be emphasized that the magnitudes of the deviation from the parabolic behavior are small, so that very precise calculations of $H_c(T)$ and T_c are necessary to make a meaningful comparison. Further work on Pb, Hg, and In is in progress at the present time (*85*).

The agreement of the critical field calculation with the experimental results for Pb implies (*74*) that the observed specific heat should also be obtained from the strong-coupling theory. From thermodynamics it follows that

$$C_s - C_n = \frac{T}{4\pi}\left[H_c \frac{d^2H_c}{dT^2} + \left(\frac{dH_c}{dT}\right)^2\right] \tag{125}$$

In the low-temperature limit this reduces to

$$C_n = \frac{1}{2\pi}\left(\frac{H_0}{T_c}\right)^2 (1-\alpha)\, T$$

where α is the low-temperature slope of the critical field deviation shown in Fig. 57. Setting $C_n = \gamma T$ this gives the relation

$$\gamma = \frac{1}{2\pi}\left(\frac{H_0}{T_c}\right)^2 (1-\alpha) \tag{126}$$

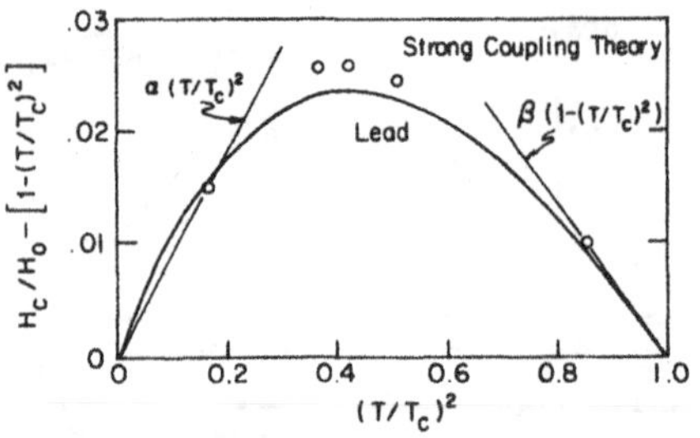

Fig. 57. Critical-field deviation showing the angles used in discussing the specific heat of Pb.

In the limit where T approaches T_c, the specific-heat jump is

$$\Delta C = C_s - C_n \bigg|_{T=T_c} = \frac{1}{\pi} \frac{H_0^2}{T_c} (1+\beta)^2 \tag{127}$$

where $-\beta$ is the high-temperature slope of the critical field deviation indicated in Fig. 57. From these relations, Eqs. (126) and (127), we find

$$\frac{\Delta C}{\gamma T_c} = \frac{2(1+\beta)^2}{1-\alpha} \tag{128}$$

$$\frac{C_s}{\gamma T_c}\bigg|_{T=T_c} = 1 + \frac{2(1+\beta)^2}{1-\alpha} \tag{129}$$

Both Pb and Hg have positive α and β coefficients, giving larger values of these ratios than the weak-coupling limit of the BCS theory, which leads to negative α- and β-values. The BCS prediction for $C_s/\gamma T_c$ is 2.5, while the value of this ratio obtained by using the critical field *computed* for Pb is 3.6, in good agreement with recent experimental measurements (7) which give 3.7.

3. Transition Temperature

According to the BCS theory, the transition temperature T_c depends upon the product

$$\lambda = N(0)\,V \tag{130}$$

of the single spin density of states at the Fermi surface $N(0)$ with the pairing potential V, and a cutoff frequency ω_c of order the Debye frequency. When λ is small, in practice less than $\frac{1}{4}$, the BCS theory predicts that

$$kT_c = 1.14\omega_c \exp(-1/\lambda) \tag{131}$$

In this same weak-coupling limit, the gap at zero temperature is given by

$$\Delta(0) = 2\omega_c \exp(-1/\lambda) \tag{132}$$

so that

$$2\Delta(0)/kT_c = 3.53 \tag{133}$$

For a number of superconductors this relation between $\Delta(0)$ and T_c is observed to hold. However, for the strong-coupling superconductors Pb and Hg this ratio turns out to be 4.3 and 4.6, respectively. Even in the extreme strong-coupling limit $\lambda \to \infty$, the BCS ratio saturates at 4. Swihart solved the BCS gap equation using an interaction similar to the Bardeen–Pines form for $V_{kk'}$ but found that it was difficult to fit the experimental results with reasonable values of the parameters which enter the interaction. It appeared that these large $2\Delta_0/kT_c$ ratios could not be satisfactorily obtained in models with a real gap function.

Wada and Schrieffer (*86*) pointed out that it was important to take account of the inelastic phonon processes which give rise to quasi-particle damping. They argued that damping would reduce both $\Delta(0)$ and T_c from their undamped values. However the reduction of T_c would be relatively greater because of the thermal phonons present at temperature T_c. Carrying out an approximate calculation of the effects of damping, Wada found that this could give rise to large values of the ratio $2\Delta(0)/kT_c$. In fact, for Pb, a value of 5 was obtained. Although this gave strong support to the underlying importance of damping, it also suggested that a more exact treatment of the gap equation was necessary.

About this same time, Schrieffer, Scalapino, and Wilkins had shown that solutions of the zero-temperature Eliashberg equations were in good agreement with the tunneling data for Pb. It was therefore of interest to see whether the extension of these equations to finite temperature would provide an explanation of the anomalous values of $2\Delta(0)/kT_c$ for Pb and Hg. Swihart, Wada, and the author investigated this finite-temperature problem for Pb, Hg, and Al. Since Al is a weak-coupling superconductor with an experimental $2\Delta(0)/kT_c$ ratio of 3.5, it provided a useful boundary condition for which the results should reduce to those of BCS. In this calculation the phonon density of states was estimated from the neutron scattering and known tunneling data. The coupling was adjusted to give the experimental value of the zero-temperature gap at the gap edge $\Delta(0) = \Delta_1[\Delta(0), 0]$ as discussed in Section III. The parameters entering

$$\alpha^2(\omega)\, F(\omega) = \sum_{\lambda} \alpha_\lambda^2(\omega_1 \lambda)\, F_\lambda(\omega)$$

are given in Table III. The notation t for transverse and l for longitudinal are somewhat arbitrary and simply represented a convenient parametrization procedure. The values of the Coulomb pseudo-potential shown in this table were taken from the work of Morel and Anderson (*51*).

The transition temperature T_c for each model was determined by numerically solving the finite-temperature version of the Eliashberg equations derived in Section III. A plot of the square of the real part of the gap at the gap edge $\Delta(T) = \Delta_1[\Delta(T), T]$ vs. the temperature extrapolated as a straight line to zero gap at $T = T_c$. The slope of this line was found to be essentially the same as that predicted by the BCS theory for both the weak and the strong superconductors. This is consistent with the ideas of Schrieffer and Wada and has been experimentally verified for Pb (*67,68*). The results obtained for $2\Delta(0)/kT_c$ are shown in Table III. The agreement with the experimentally observed ratios gave further proof that the Eliashberg gap equations provided an excellent quantitative description of superconductivity. In this formulation the damping and retardation effects so important to the properties of the strong-coupling superconductors are accurately taken into account.

Neglecting the Coulomb interaction for the moment, the main modification in the expression for T_c introduced by the strong-coupling theory is to renormalize λ by a factor of $(1+\lambda)^{-1}$:

$$T_c \propto \omega_c \exp\left[-\left(\frac{1+\lambda}{\lambda}\right)\right] \tag{134}$$

Mathematically this arises from the factor Z^{-1} appearing in the Eliashberg gap equation. Physically this reflects the fact that the electron–phonon matrix element was computed between normalized plane-wave states, while in the interacting system only a fraction Z^{-1} of the spectral weight of $c_p^\dagger|0\rangle$ is plane-wave-like.‡ In the strong-coupling theory the value λ is directly related to the phonon enhancement of the effective mass:

$$\frac{m^*}{m} = 1 + \lambda = 1 + 2\int_0^\infty \frac{\alpha^2(\omega_0)\,F(\omega_0)}{\omega_0}\,d\omega_0$$

If the Coulomb interaction is included, Morel and Anderson showed that in the weak-coupling limit $\lambda \ll 1$, the transition temperature is given by

$$T_c = \omega_0 \exp\left[-\left(\frac{1}{\lambda-\mu^*}\right)\right] \tag{135}$$

where μ^* is $N(0)$ times the Coulomb pseudo-potential [Eq. (73)]. Recently McMillan (*87*) has extended these results to the case of strong-coupling superconductors. He finds that the numerical results for T_c obtained by solving the Eliashberg equations§ can be fit by the form

$$T_c = \frac{\Theta}{1.45}\exp\left[-\frac{1.04(1+\lambda)}{\lambda-\mu^*(1+0.62\lambda)}\right] \tag{136}$$

Here the Debye temperature Θ is used as the typical phonon frequency. In the limit where $\mu^* \ll 1$ this essentially reduces to Eq. (134) and when $\lambda \ll 1$ it reduces to the Morel–Anderson result. The most interesting feature of McMillan's

‡ Since the pairing interaction V involves the square of a matrix element, one might expect that $V \to V/Z^2$. Using this type of argument one must also note that $N(0) \to ZN(0)$, so that finally $N(0)\,V \to N(0)\,V/Z$.

§ This solution assumed a phonon density of states similar to niobium.

expression is the fact that the phonon contribution is effectively reduced by $0.62\mu^*\lambda$. This arises because the Coulomb interaction changes the structure of the gap equation. Physically, the time correlations between paired electrons are distorted by the repulsive Coulomb interaction so that in the presence of the Coulomb interaction a member of a pair cannot take full advantage of the attractive lattice polarization produced by its partner.

Since λ is independent of the isotopic mass, the simple BCS form for T_c gives

$$T_c \sim M^{-1/2}$$

and one finds an isotope effect of $\frac{1}{2}$. When the Coulomb interaction is taken into account, the M dependence of the cutoff frequency appearing in μ^* modifies the isotope effect. Setting $T_c \propto M^{-\alpha}$ and using Eq. (136) one obtains for the strong-coupling case

$$\alpha = \frac{1}{2}\left\{1 - \frac{(1+\lambda)(1+0.62\lambda)\,\mu^{*2}}{[\lambda - \mu^*(1+0.62\lambda)]^2}\right\} \tag{137}$$

To conclude this section, one might ask whether the theory of strong-coupling superconductivity can be used to determine a maximum superconducting transition temperature for metals. The fact that T_c can be calculated for Pb, Hg, and Al suggests that the Eliashberg equations provide an adequate formulation of the theory of strong-coupling superconductors. The problem is to determine the interrelationship of the parameters λ, μ^*, and ω_0. In principle λ and ω_0 can be determined once the pseudo-potential is known and μ^* can be satisfactorily estimated from the screened Coulomb interaction. As we have mentioned, the difficulty of such an approach in practice is that the typical phonon frequency ω_0 is very sensitive to small changes in the pseudo-potential because of the large cancellation between the bare ion and the electronic parts of the phonon energy. As discussed in Section III, McMillan noted that this apparent handicap could in fact be exploited to suggest that

$$\lambda = C/M\omega_{\text{av}}^2 \tag{138}$$

where C is a constant for a given class of materials, M is the ion mass, and ω_{av} is a typical phonon frequency which we will take as ω_0. Neglecting μ^* and combining Eq. (138) with the strong-coupling result, McMillan found that

$$T_c = \omega_0 \exp[-(M\omega_0^2/C) - 1] \tag{139}$$

This has a maximum value as a function of ω_0 for $\omega_0 = \sqrt{C/2M}$, and

$$T_c^{\max} = \sqrt{C/2M}\, e^{-3/2} \tag{140}$$

As ω_0 is initially increased T_c increases, but at the same time the coupling λ decreases and for $\lambda < 2$ the exponential dependence causes T_c to decrease. An account of the results using the full expression for T_c is given in (*87*). For Pb-like materials, McMillan's analysis suggests a maximum T_c of 9.2°K, while for Nb-like materials the maximum T_c is about twice the observed value. Whether or not these maximum values will be observed depends upon the stability of the lattice. As the coupling is varied toward a higher T_c, the lattice may become unstable with respect to a different crystal structure. This remains to be investigated along with a further study of the relationship between λ, ω_0, and μ^*.

B. Dynamic Properties

The agreement of the BCS predictions with the observed transport properties of superconducting systems provided a striking confirmation of the pairing hypothesis. Here we will review some of the deviations from the BCS predictions observed in the dynamics of the strong-coupling superconductors.

1. Electromagnetic Properties

In the original BCS paper (*1*), the electrodynamic properties were calculated in the transverse gauge. The Meissner effect was obtained, and in the long-wavelength limit the response reduced to the London form. Mattis and Bardeen (*88*) determined the transverse response for general values of q, ω, and T. To obtain a gauge-invariant theory the longitudinal plasmon mode must be treated consistently within the pairing scheme. This takes into account the back-flow cloud, which dresses the quasi-particle cores and ensures charge conservation. To do this, Anderson (*89*) and Rickayzen (*90*) extended the random phase approximation and obtained a gauge-invariant electrodynamics. Following this, Nambu (*48*) showed that a generalized Ward identity could be used to ensure local charge conservation and hence gauge invariance. Recently, Nam (*91*) has applied Nambu's methods to treat the electrodynamics of strong-coupling superconductors.

Here we will outline a calculation of the electromagnetic absorption in the Pippard limit for a strong-coupling superconductor at zero temperature. The real part of the conductivity $\sigma_1(\omega)$ determined by this analysis shows structure at $2\Delta + \omega_1^\lambda$ associated with the electron–phonon interaction. This structure is not sharp and is difficult to distinguish from the Mattis–Bardeen expression with present experimental techniques. However, Tinkham observed that Nam's numerical result for σ_1 enclosed more area than the Mattis–Bardeen result. Using the Kramers–Kronig relations and a sum rule (*92,93*), he estimated that the strong-coupling σ_1 results implied a reduction in the imaginary part of the conductivity $\sigma_2(\omega)$ at low frequencies. Experimentally, Palmer and Tinkham (*71*) observe a reduction of about 25% and unpublished calculations‡ of σ_2 by Nam

‡ This reduction is not shown in the graphs published by Nam because of an error in his numerical evaluation.

appear to give a theoretical value for Pb in good agreement with these results. It should be noted that Nam's numerical work is based upon the early data for Δ obtained in (*63*). It will be interesting to see whether the present agreement remains when the $\Delta(\omega)$ data obtained by McMillan and Rowell is used in evaluating σ_1.‡

Just as in the tunneling calculation, the expression for the electromagnetic absorption rate of a superconductor in a weak transverse field follows directly from perturbation theory. To first order in the vector potential A, the coupling of electrons to a transverse electromagnetic field of frequency q_0 and wave vector $\mathbf{q}$ is

$$\mathscr{H}_1 = -\frac{1}{c}\,\mathbf{j}(q)\cdot\mathbf{A}(q)\exp(iq_0 t) + \text{h.c.}$$

with

$$\mathbf{j}(q) = -\frac{e}{m}\sum_{ps}(\mathbf{p} + \mathbf{q}/2)\, c^{\dagger}_{ps} c_{p+qs}$$

The transition rate for photon absorption is

$$w = 2\pi \sum_n \left| \langle \psi_n | \frac{1}{c}\,\mathbf{j}\cdot\mathbf{A}\, |\psi_0\rangle \right|^2 \delta(q_0 - E_n + E_0) \tag{141}$$

where $|\psi_n\rangle$ is an exact energy eigenstate of the many-body system with energy E_n and $|\psi_0\rangle$ is the ground state of energy E_0. This transition rate times q_0 is just the power absorption at zero temperature, which is phenomenologically described in terms of a frequency-dependent conductivity tensor. For a normally incident field polarized in the x-direction the power absorption is

$$P = (q_0/c)^2\, \sigma_1(q)\, |A_x(q)|^2$$

Here $\sigma_1(q) = \text{Re}\,\sigma_{xx}(q)$ is the real part of the xx-component of the conductivity tensor. From Eq. (141) we see that

$$\sigma_1(q) = \frac{2\pi}{q_0}\sum_n |\langle \psi_n | j_x(q)\, |\psi_0\rangle|^2\, \delta(q_0 - E_n + E_0)$$

Invoking closure this can be written as

$$\sigma_1(q) = \frac{2}{q_0}\,\text{Im}\,\langle \psi_0 | j_x(q)\, \frac{1}{H - E_0 - q_0 - i\delta}\, j_x(q)\, |\psi_0\rangle \tag{142}$$

and the diagrammatic techniques discussed in Chapter 5 can be used to evaluate it. For $q_0 > 0$ the amplitude appearing in Eq. (142) is graphically represented in

‡ Shaw and Swihart (*106*) have obtained excellent agreement between the strong-coupling σ_2 calculation and the experimental results (*71*) for Pb.

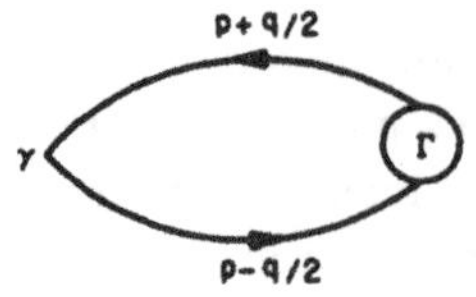

Fig. 58. Graphical representation of the amplitude which determines the conductivity.

Fig. 58. The solid lines are the dressed Nambu propagators, $\Gamma_x(p+q/2, p-q/2)$ the dressed vertex, and γ_x the bare vertex

$$\gamma_x(p-q/2, p+q/2) = -\frac{e}{m}\, p_x \mathbf{1}$$

with **1** the unit matrix. Using the usual rules, the real part of the conductivity is

$$\sigma_1(q) = -\frac{2}{q_0}\,\mathrm{Im}\,\mathrm{Tr}\int \frac{d^4p}{(2\pi)^4 i}\,\Gamma_x(p+q/2, p-q/2) \times G(p+q/2)\,\gamma_x(p-q/2, p+q/2)\,G(p-q/2) \qquad (143)$$

Here Tr stands for a trace over the Nambu G-matrices.

Since the self-energy in G has been calculated by summing all noncrossed line graphs, a gauge-invariant result for σ_1 requires that Γ be determined within the ladder approximation. However, in the Pippard limit where the penetration depth or film thickness δ is much smaller than the coherence length ξ_0, the nonresonant vertex correction in the transverse gauge can be neglected. The possibility of resonant vertex corrections associated with a collective mode having energy less than $2\Delta_0$ have been discounted in Chapter 7 and are not presently observed experimentally. We will therefore replace Γ by γ in Eq. (143):

$$\sigma_1(q) = -\frac{2e^2}{m^2 q_0}\,\mathrm{Im}\,\mathrm{Tr}\int \frac{d^3p}{(2\pi)^3}\, p_x^2 \int \frac{dp_0}{2\pi i}\, G(p+q/2)\,G(p-q/2)$$

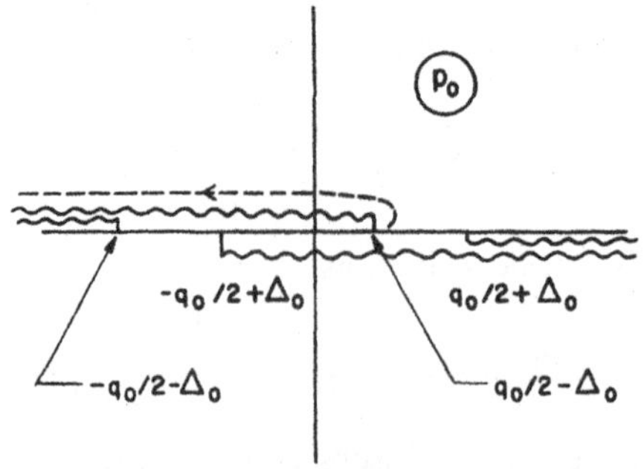

Fig. 59. Branch cuts and contour in the p_0 plane for the σ_1-evaluation. The p_0-integration for σ_1 goes along the real axis. The branch cuts $G(p+q/2)$ and $G(p-q/2)$ have been displaced from the real axis for clarity and the deformation of the p_0-contour from the real axis is shown as a dashed line.

The branch cuts of $G(p \pm q/2) \equiv G_\pm$ are shown in Fig. 59. Closing the contour in the upper half-plane we find that the p_0-integral becomes

$$\int_{-\infty}^{-q_0/2+\Delta_0} \frac{dp_0}{\pi} \operatorname{Im} G_+ G_- + \int_{-q_0/2+\Delta_0}^{q_0/2-\Delta_0} \frac{dp_0}{\pi} G_+ \operatorname{Im} G_- \, \theta(q_0/2 - \Delta_0)$$

where $\theta(x)$ is a step function which is unity for $x > 0$ and zero for $x < 0$. The first integral is pure real and does not contribute (at zero temperature) to the absorptive part of the conductivity. For $q_0 < 2\Delta_0$, the second integral does not enter, so that no absorption occurs. For $q_0 > 2\Delta_0$, quasi-particle pair production is possible and the second integral contributes. The real part of the conductivity at zero temperature is therefore given by

$$\sigma_1(q) = -\frac{2e^2}{\pi m^2 q_0} \operatorname{Tr} \int_{-q_0/2+\Delta_0}^{q_0/2-\Delta_0} dp_0 \int \frac{d^3p}{(2\pi)^3} p_x^2 \operatorname{Im} G(p + q/2) \operatorname{Im} G(p - q/2) \tag{144}$$

Next the momentum integration is carried out, leaving only the p_0 integration, which was done numerically using the results for the Green's function renormalization parameter $Z(p_0)$ and the gap parameter $\Delta(p_0)$ determined in Section III. In the Pippard limit, appropriate to dirty materials (or thin films) the coherence length ξ_0 is much greater than the penetration depth (or film thickness), so that the important q-values are much greater than Δ_0/v_F. This means that for frequencies $q_0 \ll qv_F$, the dominant contributions from the Green's functions in Eq. (144) occur for $\mu = \mathbf{p}\cdot\mathbf{q}/pq \approx 0$ and $p \approx p_F$. We can take advantage of this by changing variables to $\epsilon_\pm = \{[p \pm (q/2)]^2/2m\} - \mu_F$ and replacing μ by 0 and p by p_F in the slowly varying parts of the integral. Carrying out this change of variables, one finds

$$\sigma_1 = \frac{-e^2 p_F^2}{\pi^3 q q_0} \operatorname{Tr} \int_{-q_0/2+\Delta_0}^{q_0/2-\Delta_0} dp_0 \int d\epsilon_+ \int d\epsilon_- \times \operatorname{Im} G(\epsilon_+, p_0 + q_0/2) \operatorname{Im} G(\epsilon_-, p_0 - q_0/2)$$

Now since $qv_F \gg \Delta_0$, the region of integration can be extended so that ϵ_+ and ϵ_- both run from $-\infty$ to ∞. These integrals can then be evaluated by residues, giving

$$\int_{-\infty}^{\infty} d\epsilon \operatorname{Im} G(\epsilon, p_0) = -\pi \operatorname{Re}\left\{\frac{p_0 + \tau_1 \Delta(p_0)}{[p_0^2 - \Delta^2(p_0)]^{1/2}}\right\}$$

Shifting the p_0 by $q_0/2$ and taking the trace, we obtain as our final expression for the real part of the conductivity at zero temperature

$$\frac{\sigma_1}{\sigma_N} = \frac{1}{q_0} \int_{\Delta_0}^{q_0-\Delta_0} dp_0 \left[\operatorname{Re}\left\{\frac{p_0}{[p_0^2 - \Delta^2(p_0)]^{1/2}}\right\} \operatorname{Re}\left\{\frac{q_0 - p_0}{[(q_0 - p_0)^2 - \Delta^2(q_0 - p_0)]^{1/2}}\right\} - \operatorname{Im}\left\{\frac{\Delta(p_0)}{[p_0^2 - \Delta^2(p_0)]^{1/2}}\right\} \operatorname{Im}\left\{\frac{\Delta(q_0 - p_0)}{[(q_0 - p_0)^2 - \Delta^2(q_0 - p_0)]^{1/2}}\right\}\right] \tag{145}$$

with σ_N the normal-state conductivity in the Pippard limit:

$$\sigma_N = p_F^2 e^2 / 2\pi q$$

The expression‡ given by Eq. (145) reduces directly to the BCS limit when the complex, frequency-dependent gap parameter $\Delta(\omega)$ is replaced by Δ_0.

In Fig. 60 the frequency dependence of σ_1/σ_N obtained by Nam using the equal-coupling-constant solution for $\Delta(\omega)$ appropriate to Pb is compared with the BCS prediction obtained by Mattis and Bardeen. The structure at $2\Delta + \omega_1^\lambda$ arises from the electron–phonon coupling. Some recent experimental results for

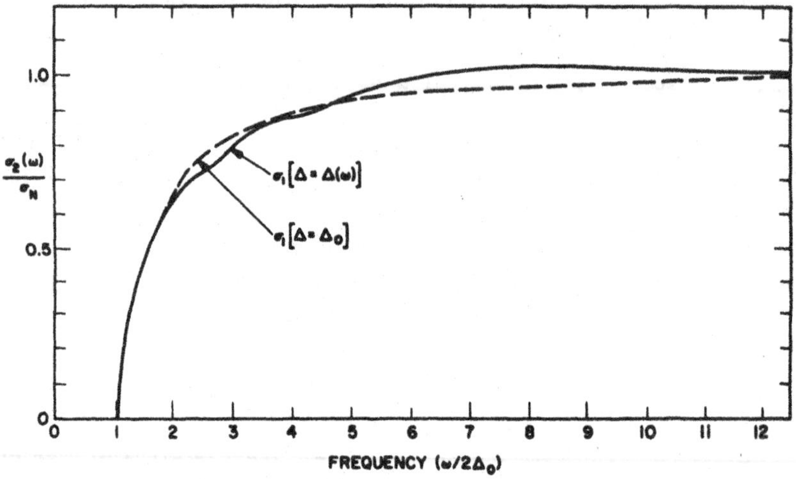

Fig. 60. Real part of the conductivity using the strong-coupling $\Delta(\omega)$ for Pb (solid) (*91*) compared with the Mattis–Bardeen result (*88*) based on a constant gap.

σ_1 measured in thin Pb films at 2°K by Palmer and Tinkham are plotted in Fig. 61. The scatter of the data makes it difficult to say with any certainty that the deviations in σ_1 predicted by the strong-coupling theory are observed. Actually, the major deviation from the Mattis–Bardeen behavior occurs for values of ω beyond the region in which σ_1 was experimentally observed. However, as we have noted, Tinkham (*71*) has pointed out that the strong-coupling result of Nam encloses more area than the Mattis–Bardeen expression. This means that the strength of the zero-frequency δ-function associated with σ_1 is reduced and hence, via the Kramers–Kronig relation, the low-frequency value of σ_2 is reduced. In Fig. 62 experimental results for σ_2 are compared with the BCS prediction of Mattis and Bardeen and a recent calculation of Nam. Nam's result is based upon the strong-coupling expression for σ_2 given in (*91*).

‡ This expression was obtained independently by Fibich and Scalapino, and unpublished results of Swihart using Eq. (145) are similar to Nam's results shown in Fig. 60.

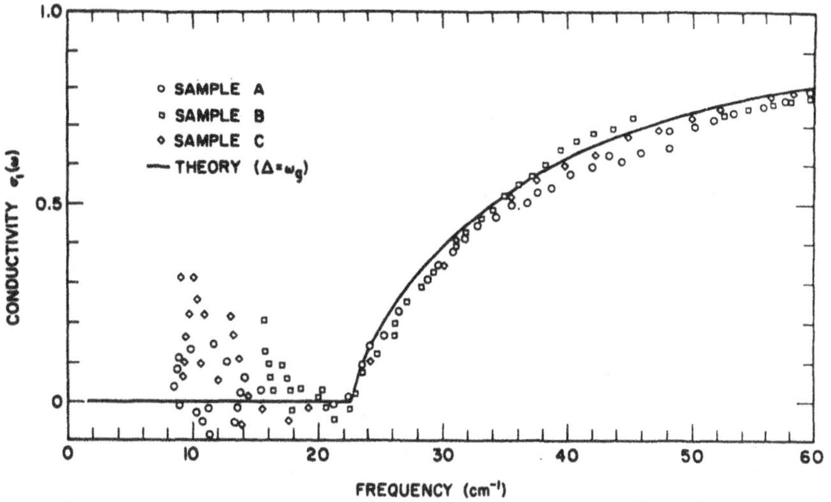

Fig. 61. Measured behavior of σ_1 in three thin Pb films at 2°K taken from (*91*). The solid curve is the Mattis–Bardeen result fit to a gap of $2\Delta_0 = 22.5\ \text{cm}^{-1}$.

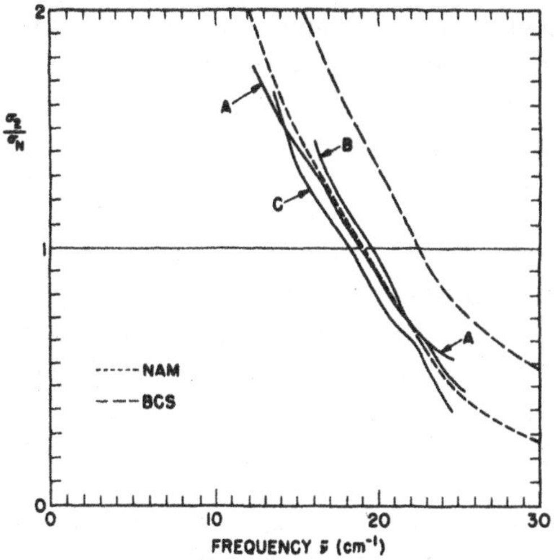

Fig. 62. Imaginary part of the conductivity σ_2 compared with the Mattis–Bardeen and the strong-coupling results of Nam (*71*).

2. Thermal Conductivity

Except in extremely impure metals, the electrons are responsible for most of the heat transport near T_c. As the temperature is lowered below T_c, the gap opens up decreasing the number of electron excitations until, below about $0.25T_c$, the heat is carried primarily by the phonons. The temperature region of most interest for superconductivity is just below T_c, and a convenient quantity to consider is the reduced thermal conductivity K_s/K_n as a function of the reduced temperature T/T_c. Here K_s and K_n are the thermal conductivities of the super and normal phases, respectively. When the electron mean free path is determined by impurity scattering, the slope of K_s/K_n plotted vs. T/T_c is zero at the transition temperature. Bardeen et al. (*94*) analyzed this case using the simple BCS theory and obtained results in good agreement with experiment. They showed that in the superconducting phase, the quasi-particle relaxation lifetime due to impurity scattering was given by $|E_p/\epsilon_p|\tau_n$, where τ_n is the relaxation time in the normal state and $E_p = \sqrt{\epsilon_p^2 + \Delta_p^2}$. Since the group velocity $\nabla_p E_p$ of the quasi-particles in the superconducting state varies as $\epsilon_p \mathbf{p}/mE_p$, the mean free path for impurity scattering is the same in the normal and superconducting states.

For the case in which the heat flux is limited by phonon scattering of the quasi-particles, the mean free path is different in the superconducting state and K_s/K_n has a finite limiting slope at the transition temperature. For the weak-coupling superconductors such as tin and indium, this limiting slope is of order 1.5, while for Pb and Hg the initial decrease is much steeper. Experiments by Watson and Graham (*8*) give a limiting slope of about 9 for Pb, while calculations based upon the simple BCS theory (*9*) give a slope of about 1.5 for this ratio. Work by Ambegaokar and Woo (*95*) based upon an expression for K_s/K_n derived by Ambegaokar and Tewordt (*96*) shows that the anomalously steep slope for Pb can be understood in the framework of the strong-coupling theory. In particular, these calculations show that the quasi-particle kinetic theory form of K_s gives meaningful results if the strong-coupling expressions for the temperature and frequency dependence of the gap and renormalization parameter are used to determine the quasi-particle group velocity and lifetime. For this reason, we will begin with the simpler kinetic theory calculation and then discuss the Green's function approach.

The results of this treatment show that the steep slope of the K_s/K_n ratio below T_c for Pb and Hg is due to the strong electron–phonon interaction in these materials. It gives rise to a large $2\Delta(0)/kT_c$ ratio, which means that the number of quasi-particles excited at a given reduced temperature is less than in the weak-coupling superconductors. In addition, recombination of two quasi-particles with the emission of a phonon to form a pair provides an important decay channel for the electron excitations in strong-coupling superconductors. This uniquely superconducting relaxation process makes the quasi-particle lifetime in the superconductor less than in the normal metal. Furthermore, since the recombination process varies as the phase space available for a phonon of energy $2\Delta(T)$, the quasi-particle lifetime decreases with frequency as well as, initially, with temperature, as the gap opens up. The combined effect of the more rapid initial decrease of excitations plus the frequency and temperature variation of the recombination decay are responsible for the anomalous thermal conductivity of superconducting Pb and Hg.

The thermal conductivity K relates the heat flux $\mathbf{Q}$ to the gradient of the temperature. For an isotopic system

$$\mathbf{Q} = -K\nabla T \tag{146}$$

Suppose that T varies along the z-axis and consider the electronic part of the heat flow associated with the transfer of quasi-particles across an xy-plane where the temperature is T. The energy flow per unit area in the positive z-direction is given by

$$Q_+ = 2 \sum_{p(p_z>0)} E_p v_{p_z} f_T(E_p) \tag{147}$$

Here E_p is the quasi-particle energy $\sqrt{\epsilon_p^2 + \Delta^2}$, v_{p_z} is the group velocity in the z-direction

$$\partial E_p/\partial p_z = p_z \epsilon_p / m E_p$$

and $f_T(E_p) = [\exp(\beta E_p) + 1]^{-1}$ is the Fermi distribution for the quasi-particles at temperature T.

In the same way the energy flow per unit area along the negative z-axis is given by

$$Q_- = 2 \sum_{p(p_z<0)} E_p v_{p_z} f_{T+\Delta T}(E_p) \tag{148}$$

Since the particles come from a region which is a mean free path farther along the z-axis, the Fermi factor contains a temperature $T + \Delta T$ with

$$\Delta T = v_{p_z} \tau (\partial T/\partial z) \tag{149}$$

Here τ is the quasi-particle lifetime due to phonon scattering and pair recombination processes. Expanding the Fermi factor

$$f_{T+\Delta T}(E_p) = f_T(E_p) - \frac{E_p}{T} \frac{\partial f_T(E_p)}{\partial E_p} \Delta T$$

and using Eq. (148), the net heat flow per unit cross-sectional area along the z-axis is

$$-2\frac{v_F^2}{T} \sum_{p(p_z<0)} \epsilon_p^2 \tau \cos^2\theta_p \frac{\partial f_T}{\partial E_p} \frac{\partial T}{\partial z}$$

Comparing this with Eq. (146), the electronic part of the thermal conductivity is given by

$$K = \frac{2v_F^2}{T} \sum_{p(p_z<0)} \epsilon_p^2 \tau \cos^2\theta_p \frac{\partial f_T}{\partial E_p}$$

Carrying out the angular integration and changing variables to the quasi-particle energy $E \equiv E_p$ this reduces to

$$K = \frac{2N(0)\, v_F^2}{3T} \int_{\Delta}^{\infty} dE\, E \sqrt{E^2 - \Delta^2}\, \tau \frac{\partial f_+}{\partial E} \tag{150}$$

Because of the derivatives of the Fermi factor, the important excitation energies contributing to K are of order several $k_B T$. For $T \leqslant T_c$, this is much less than ω_D. The Green's function analysis shows that this is simply handled by using the poles of the single-particle Green's function to define the quasi-particle energy and lifetime. The poles are determined by

$$\omega^2 = [\epsilon_p / Z(\omega)]^2 + \Delta^2(\omega)$$

Setting

$$Z = Z_1 + iZ_2 \qquad \Delta = \Delta_1 + i\Delta_2 \qquad \omega = E - i\Gamma/2$$

and solving to lowest order in the imaginary parts one finds the quasi-particle energy

$$E = \sqrt{[\epsilon_p / Z_1(E)]^2 + \Delta_1^2(E)} \tag{151}$$

and a level width

$$\Gamma = \frac{2Z_2(E)\,[E^2 - \Delta_1^2(E)] - 2\Delta_1(E)\,\Delta_2(E)\,Z_1(E)}{EZ_1(E)} \tag{152}$$

A plot of the ratio of the level width to excitation energy $\Gamma(E)/E$ for the range of quasi-particles energy important in evaluating K is shown in Fig. 63. This clearly shows that the quasi-particle expansion should be adequate. This, of course, does not mean that the strong electron–phonon interaction has been neglected. For example, the renormalization parameter Z_1 is of order 2.35 in the region of interest.

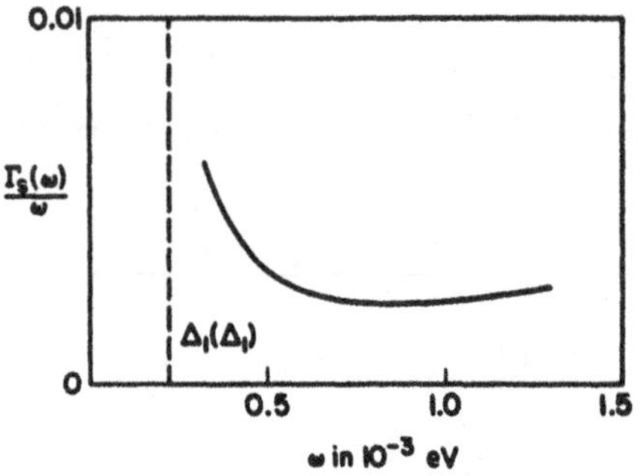

Fig. 63. Ratio of the level width to quasi-particle energy $\Gamma(\omega)/\omega$ plotted vs. ω for T/T_c of 0.978.

The fact that Z and Δ depend upon frequency but not momentum means that the previous analysis goes through simply and the electronic part of the thermal conductivity in the superconducting state is

$$K_s = \frac{2N(0)\,v_F^2}{3T}\int_{\Delta_1(\Delta_1,\,T)}^{\infty} dE\,\frac{E\sqrt{E^2-\Delta_1^2(E,T)}}{Z_1(E,T)\,\Gamma(E,T)}\frac{\partial f_T}{\partial E} \tag{153}$$

Here we have explicitly exhibited the fact that Δ, Z, and Γ are functions of both frequency E and temperature T. Their dependence is determined by the Eliashberg equations, so that the strong-coupling effects are implicitly included. The corresponding thermal conductivity K_n in the normal state is obtained by setting Δ equal to zero and using the renormalization parameter $Z_n(\omega, T)$ appropriate to the normal state. Ambegaokar and Woo have used the numerical results for Pb discussed in Section III to obtain the results for K_s/K_n shown in Fig. 64. Near the critical temperature they find a limiting slope of 11 for K_s/K_n vs. T/T_c. The absolute value of the calculated conductivity is also in good agreement with experiment. This slope should be compared with an early result of 1.6 obtained by Tewordt (*9*) using the simple BCS theory to account for the virtual processes and a Debye spectrum to treat the real processes.

The anomalously steep decrease of K_s/K_n for the strong-coupling superconductors arises from several sources. The large value of $2\Delta(0)/kT_c$ (4.3 for Pb as opposed to 3.5 for weak-coupling superconductors) implies a more rapid freezing out of the quasi-particle excitations as T/T_c decreases. In this connection it is important to remember that the reduced Pb gap $\Delta(T)/\Delta(0)$ vs. T/T_c follows the BCS curve, so that $2\Delta(0)/kT_c$ is the significant parameter. Besides this, the ratio of the superconducting quasi-particle lifetime to the normal quasi-particle lifetime decreases for Pb. This is associated with the pair recombination processes

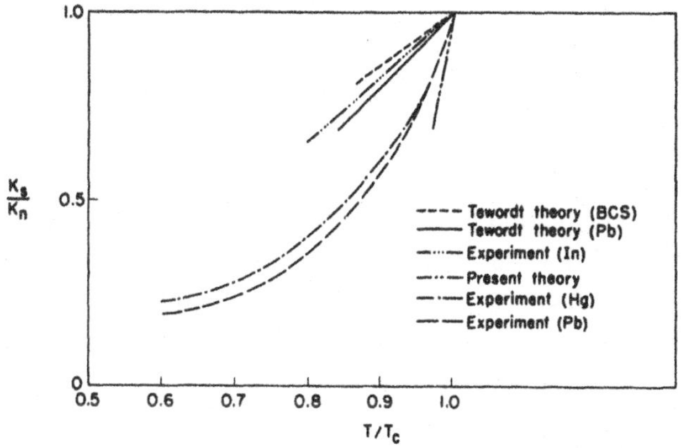

Fig. 64. Reduced conductivity vs. reduced temperature. Experimental points for In from Teword (*9*), that for Pb from Watson and Graham (*8*). [After Ambegaokar and Woo (*95*).]

which can occur in the superconducting state. As discussed in Section III, the gap parameter Δ essentially scales with $\Delta(T)/\Delta(0)$ for a weak-coupling superconductor such as Al. However, for Pb, new structure due to pair recombination in Δ_2 is clearly evident. Finally, the quasi-particle lifetime decreases with frequency more rapidly in lead than in the weak-coupling superconductors because of the dependence of the lifetime on the pair recombination process. This process varies as the phase space for phonons of frequencies $2\Delta(T)$. Further discussion of these effects and details of the numerical results for Pb are given in the paper of Ambegaokar and Woo. We now turn to a discussion of the Green's function approach.

The Green's function calculation proceeds from the Kubo expression for the electronic part of the thermal conductivity

$$K = \frac{1}{3T}\frac{\partial}{\partial\omega}\operatorname{Im} P(q, i\omega_\nu = \omega + i\delta)_{\substack{\omega=0\\ q=0}} \tag{154}$$

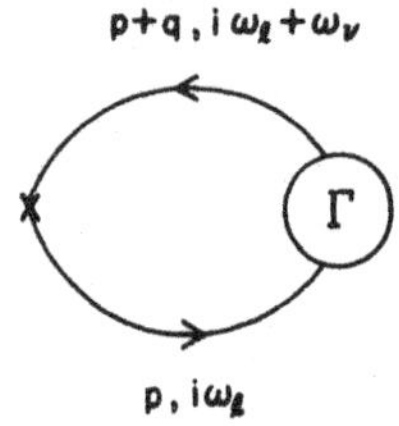

Fig. 65. Diagramatic representation of the amplitude for the thermal conductivity.

Here $P(q, i\omega)$ is shown diagrammatically in Fig. 65. As before, Roman-indexed frequency variables are fermion-like, $\omega_l = (2l+1)\pi/\beta$, and Greek-indexed variables are boson-like, $\omega_\nu = 2\pi\nu/\beta$. The bare vertex denoted by an x contributes a factor

$$\tau_3\left[i\omega_l\left(\frac{\mathbf{p}+\mathbf{q}}{2m}\right)_z + i(\omega_l+\omega_\nu)\frac{\mathbf{p}_z}{2m}\right]$$

and the two-component Nambu notation is used. Within the ladder approximation an integral equation for the vertex can be constructed. The inhomogeneous term is the bare vertex and the remaining part represents the "scattering in" corrections which change the scattering cross section to the transport cross section. This correction can be shown to be unimportant, so that

$$P(q, i\omega_\nu) = -\frac{1}{4m^2}\int\frac{d^3p}{(2\pi)^3}\frac{1}{\beta}\sum_l\left[i\omega_l\frac{(p+q)_z}{2m} + i(\omega_l+\omega_\nu)\frac{p_z}{2m}\right]^2 \times \operatorname{Tr}\left[\tau_3 G(p+q, i\omega_l + i\omega_\nu)\,\tau_3 G(p, i\omega_l)\right] \tag{155}$$

Changing the l-sum to an integration in the frequency plane in the standard way and carrying out the limiting process given in Eq. (154), one finds

$$K = -\frac{1}{m^2 T}\int \frac{d^3p}{(2\pi)^3}\int \frac{d\omega}{2\pi} p_z^2 \omega^2 \frac{\partial f}{\partial \omega} \operatorname{Tr}\left[\tau_3 \operatorname{Im} G(p,\omega)\, \tau_3 \operatorname{Im} G(p,\omega)\right]$$

Because Z and Δ are independent of momentum to order $\sqrt{m/M}$, the momentum integration can be evaluated with the result

$$K = \frac{N(0)\, v_F^2}{3T}\int_{\Delta_1(\Delta_1, T)}^{\infty} d\omega \frac{\partial f}{\partial \omega} \frac{\omega^2\left[1 + \dfrac{\omega^2 - |\Delta(\omega, T)|^2}{|\omega^2 - \Delta^2(\omega, T)|}\right]}{\operatorname{Im}\{Z(\omega, T)\,[\omega^2 - \Delta^2(\omega, T)]^{1/2}\}} \tag{156}$$

This is the strong-coupling expression for K first obtained by Ambegaokar and Tewordt (*96*). To lowest order in Z_2 and Δ_2 this reduces to the quasi-particle form previously discussed. Explicit numerical calculations comparing Eqs. (56) and (53) show that K is accurately represented by the quasi-particle approximation as one would expect from the small value of the level width over the excitation energy shown in Fig. 63.

3. Nuclear Spin Relaxation

One of the first measurements to substantiate the microscopic aspects of the BCS pairing correlations was the observation that the nuclear spin relaxation rate increased as T dropped below T_c. This is in contrast to the rapid decrease in ultrasonic attenuation in the same region. In the BCS theory these processes have different coherence factors and the large density of quasi-particle states near the Fermi surface is exactly canceled out of the ultrasonic attenuation relation, while for the nuclear spin relaxation rate it remains and within the simple BCS theory gives rise to a logarithmic singularity at T_c. Hebel and Slichter (*15,16*) observed an increase in Al of about 2, and fit their data by smoothing the BCS density of states. It can be argued that this smoothing is physically associated with effects of gap anisotropy and spacial inhomogeneity. Following the work on tunneling which showed that the gap had an imaginary part, Fibich (*17*) suggested that this could provide a dynamic explanation of the Hebel and Slichter data. Although this effect is not specifically associated with the strong-coupling superconductors, it does arise from the dynamics of the electron–phonon interaction and we have therefore included it in this chapter.

In Fig. 66 the ratio R_s/R_n of nuclear spin relaxation rates for superconducting and normal aluminum are plotted vs. the reduced temperature T/T_c. The circles represent experimental data by Hebel and Slichter, the crosses data by Redfield and Anderson (*97,98*) and the solid curve is Fibich's calculation. Near the transition temperature, the thermal phonon broadening of the quasi-particle states has washed out the logarithmic divergence.

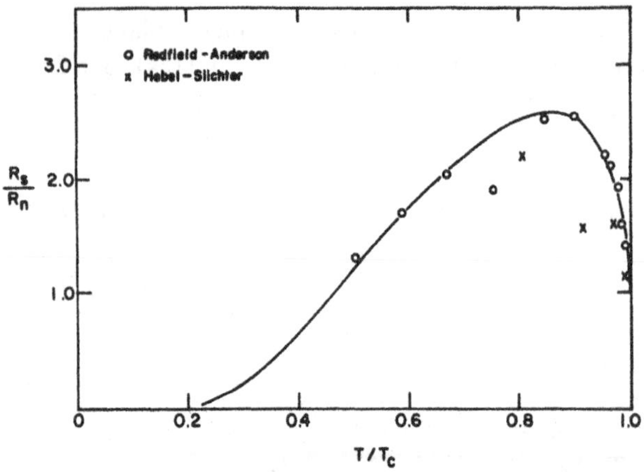

Fig. 66. Ratio of nuclear spin relaxation R_s/R_n between superconducting and normal aluminum as a function of the reduced temperature T/T_c. The circles represent experimental data by Hebel and Slichter (*15*), the crosses data by Redfield and Anderson (*97, 98*), and the solid curve was calculated by Fibich (*17*) using Eq. (157).

The analysis of this is particularly simple when the coupling between the nuclear spins and the conduction electrons is via a contact hyperfine interaction. Then the ratio R_s/R_n is given by

$$\frac{R_s}{R_n} = 2\int_0^\infty d\omega\left(-\frac{\partial f}{\partial\omega}\right)\left\{\left[\mathrm{Re}\left\{\frac{\omega}{[\omega^2 - \Delta^2(\omega, T)]^{1/2}}\right\}\right]^2 + \left[\mathrm{Re}\left\{\frac{\Delta(\omega, T)}{[\omega^2 - \Delta^2(\omega, T)]^{1/2}}\right\}\right]^2\right\} \tag{157}$$

Here the momentum independence of Z, Δ, and the interaction have allowed the momentum integrations to be carried out in the standard manner. In this expression, the small Zeeman energy associated with the spin flip has been set equal to zero. If it were retained and used as a cutoff energy for the BCS result, the ratio R_s/R_n would increase by a factor of 10 in the superconducting state. As we will see, a larger cutoff can be provided by phonon broadening.

If $\Delta(\omega, T)$ is replaced by the BCS gap $\Delta(T)$, Eq. (157) becomes

$$\frac{R_s}{R_n} = 2\int_{\Delta(T)}^\infty d\omega\left(-\frac{\partial f}{\partial\omega}\right)\frac{\omega^2 + \Delta^2(T)}{\omega^2 - \Delta^2(T)}$$

This has a logarithmic singularity at the lower end of the integration. When the complex gap parameter is introduced, Fibich showed that

$$\frac{R_s}{R_n} \approx 2f(\Delta_1(T))\left\{1 + \frac{\Delta_1(T)}{2kT}[1 - f(\Delta_1(T))]\ln\left|\frac{4\Delta_1(T)}{\Delta_2(T)}\right|\right\}$$

Here Δ_1 and Δ_2 are, as usual, the real and imaginary part of the gap evaluated at the gap edge. Using a ratio of Δ_2/Δ_1 obtained from the Eliashberg equations, Fibich calculated the ratio R_s/R_n shown as the solid curve in Fig. 66. Although the agreement is good, it should be remembered that a small amount of gap anisotropy and spatial inhomogeneity can give a similar kind of curve. Furthermore, because Δ_2/Δ_1 enters in a logarithm, the results are relatively insensitive to the detailed behavior of the quasi-particle damping.‡

4. Ultrasonic Attenuation

It seems appropriate to conclude this chapter by noting one case in which strong-coupling electron–phonon effects are not responsible for deviations from the BCS prediction. Deaton (*99*) observed anomalously low attenuation of longitudinal sound in pure crystals of Pb when $ql \gg 1$. This was originally attributed to strong-coupling effects, in analogy with the anomalous decrease in the thermal conductivity of Pb below T_c.

When the phonon wavelength is short compared to the electron mean free path ($ql \gg 1$), the electrons feel only the average ion motion. In this case, collision drag effects can be neglected and the ultrasonic attenuation is determined by the density–density correlations. Therefore, the attenuation is determined by the phase space available for the phonon to decay into a pair of quasi-particles. Using the momentum independence of the gap and renormalization parameter, Ambegaokar (*100*) showed that the ratio of attenuation in the superconducting to the normal state for $ql \gg 1$ reduced to the BCS prediction

$$\alpha_s/\alpha_n = 2f\,(\Delta\,(T))$$

It therefore appears that the strong-coupling effects are not responsible for the deviations observed by Deaton in the $ql \gg 1$ region. In fact, recent measurements (*101*) of the longitudinal sound attenuation in Hg for $ql \gg 1$ fit the BCS prediction. It has been suggested that the anomalous behavior of Pb may be associated with the scattering of phonons from dislocations (*102*).

ACKNOWLEDGMENTS

The author gratefully acknowledges support from the National Science Foundation during the preparation of this article.

‡ Recent calculations by Scalapino and Taylor (*107*) show that Fibich (*17*) overestimated the effect of Δ_2 so that gap anisotropy and spatial inhomogeneity provide the dominant observed effects as originally suggested by Hebel and Slichter (*15*).

REFERENCES

1. J. Bardeen, L. N. Cooper, and J. R. Schrieffer, *Phys. Rev.* **108**, 1175 (1957).
2. J. Bardeen and J. R. Schrieffer, *Progress in Low Temperature Physics*, Vol. II (C. J. Gorter, ed.), North-Holland, Amsterdam, 1961.
3. I. Giaever, H. R. Hart, Jr., and K. Megerle, *Phys. Rev.* **126**, 941 (1962).
4. S. Bermon and D. M. Ginsberg, *Phys. Rev.* **135**, A306 (1964).
5. D. K. Finnemore, D. E. Mapother, and R. W. Shaw, *Phys. Rev.* **118**, 127 (1960).
6. N. E. Phillips, *Phys. Rev.* **114**, 676 (1959).
7. J. E. Neighbor, J. F. Cochran, and C. A. Shiffman, *Phys. Rev.* **155**, 384 (1967).
8. J. H. P. Watson and G. M. Graham, *Can. J. Phys.* **41**, 1738 (1963).
9. L. Tewordt, *Phys. Rev.* **129**, 657 (1963).
10. L. Tewordt, *Phys. Rev.* **128**, 12 (1962).
11. J. M. Rowell and L. Kopf, *Phys. Rev.* **137**, 907 (1965).
12. J. M. Rowell, A. G. Chynoweth, and J. C. Phillips, *Phys. Rev. Letters* **9**, 59 (1962).
13. E. Maxwell, *Phys. Rev.* **78**, 477 (1950).
14. C. A. Reynolds, B. Serin, W. H. Wright, and L. B. Nesbitt, *Phys. Rev.* **78**, 487 (1950).
15. L. C. Hebel and C. P. Slichter, *Phys. Rev.* **113**, 1504 (1959).
16. L. C. Hebel, *Phys. Rev.* **116**, 79 (1959).
17. M. Fibich, *Phys. Rev. Letters* **14**, 561 (1965).
18. B. I. Miller and A. H. Dayem, *Phys. Rev Letters* **18**, 1000 (1967).
19. A. V. Gold, *Phil. Trans. Roy. Soc. London* **A251**, 85 (1958).
20. J. R. Anderson and A. V. Gold, *Phys. Rev.* **139**, A1459 (1965).
21. W. A. Harrison, *Pseudopotentials in the Theory of Metals*, Benjamin, New York, 1966, Chap. 3.
22. J. C. Phillips and L. Kleinman, *Phys. Rev.* **116**, 287 (1959).
23. M. H. Cohen and V. Heine, *Phys. Rev.* **122**, 1821 (1961).
24. L. D. Landau, *Zh. Eksperim. i Teor. Fiz.* **30**, 1058 (1956); *Soviet Phys. JETP* **7**, 996 (1958).
25. A. B. Migdal, *Zh. Eksperim. i Teor. Fiz.* **34**, 1438 (1958); *Soviet Phys. JETP* **7**, 996 (1958).
26. R. E. Prange and L. P. Kadanoff, *Phys. Rev.* **134**, A566 (1964).
27. S. Engelsberg and J. R. Schrieffer, *Phys. Rev.* **131**, 993 (1963).
28. D. Bohm and D. Pines, *Phys. Rev.* **92**, 609 (1953).
29. M. Gell-Mann and K. A. Brueckner, *Phys. Rev.* **106**, 364 (1957).
30. J. Hubbard, *Proc. Roy. Soc. (London)* **A240**, 539 (1957).
31. J. Hubbard, *Proc. Roy. Soc. (London)* **A243**, 336 (1957).
32. T. M. Rice, *Ann. Phys. (N.Y.)* **31**, 100 (1965).
33. W. A. Harrison, *Phys. Rev.* **131**, 2433 (1963).
34. W. A. Harrison, *Rev. Mod. Phys.* **36**, 256 (1964).
35. N. W. Ashcroft, *Phys. Letters* **23**, 48 (1966).
36. C. B. Clark, *Phys. Rev.* **109**, 1133 (1958).
37. V. Heine, P. Nozières, and J. W. Wilkins, *Phil. Mag.* **13**, 741 (1966).
38. J. M. Luttinger and P. Nozières, *Phys. Rev.* **127**, 1423, 1431 (1962).
39. A. B. Migdal, *Zh. Eksperim. i Teor. Fiz.* **32**, 399 (1957); *Soviet Phys. JETP* **5**, 333 (1957).
40. J. M. Luttinger, *Phys. Rev.* **119**, 1153 (1960).
41. J. Lindhard, *Kgl. Danske Videnskab. Selskab, Mat. Fys. Medd.* **28**, 8 (1954).
42. J. J. Quinn and R. A. Ferrell, *Phys. Rev.* **112**, 812 (1958).
43. M. Born and J. R. Oppenheimer, *Ann. Physik* **84**, 457 (1927).
44. H. Fröhlich, *Phys. Rev.* **79**, 845 (1950).

45. E. G. Batyev and V. L. Pokrovskii, *Zh. Eksperim. i Teor. Fiz.* **46**, 262 (1964); *Soviet Phys. JETP* **19**, 181 (1964).
46. L. J. Sham and J. M. Ziman, *Solid State Phys.* **15**, 221 (1963).
47. J. R. Schrieffer, *Theory of Superconductivity*, Benjamin, New York, 1964.
48. Y. Nambu, *Phys. Rev.* **117**, 648 (1960).
49. L. P. Gor'kov, *Zh. Eksperim. i Teor. Fiz.* **34**, 735 (1958); *Soviet Phys. JETP* **7**, 505 (1958).
50. G. M. Eliashberg, *Zh. Eksperim. i Teor. Fiz.* **38**, 966 (1960); *Soviet Phys. JETP* **11**, 696 (1960).
51. P. Morel and P. W. Anderson, *Phys. Rev.* **125**, 1263 (1962).
52. D. J. Scalapino, J. R. Schrieffer, and J. W. Wilkins, *Phys. Rev.* **148**, 263 (1966)
53. Chapter 11 of this treatise.
54. A. J. Bennett, *Phys, Rev.* **140**, A1902 (1965).
55. N. N. Bogoliubov, V. V. Tolmachev, and D. V. Shirkov, *A New Method in the Theory of Superconductivity*, Academy of Science, Moscow, 1958, Consultants Bureau, New York, 1959.
56. V. Heine and I. Abarenkov, *Phil. Mag.* **9**, 451 (1964).
57. W.L. McMillan, *Phys. Rev.*, **167**, 331 (1968).
58. S. H. Vosko, R. Taylor, and G. H. Keech, *Can. J. Phys.* **43**, 1187 (1965).
59. D. J. Scalapino, Y. Wada, and J. C. Swihart, *Phys. Rev. Letters* **14**, 102 (1965).
60. B. N. Brockhouse, T. Arase, G. Caglioti, K. R. Rao, and A. D. B. Woods, *Phys. Rev.* **128**, 1099 (1962).
61. W. L. McMillan and J. M. Rowell, *Phys. Rev. Letters* **14**, 108 (1965).
62. R. Stedman, L. Almquist, and G. Milsson, to be published.
63. J. R. Schrieffer, D. J. Scalapino, and J. W. Wilkins, *Phys. Rev. Letters* **10**, 336 (1963).
64. J. M. Rowell, P. W. Anderson, and D. E. Thomas, *Phys. Rev. Letters* **10**, 334 (1963).
65. D. J. Scalapino and P. W. Anderson, *Phys. Rev.* **133**, A921 (1964).
66. D. J. Scalapino, *Rev. Mod. Phys.* **36**, 205 (1964).
67. R. F. Gasparovic, B. N. Taylor, and R. E. Eck, *Solid State Commun.* **4**, 59 (1966).
68. J. P. Franck and W. J. Keeler, to be published.
69. B. N. Taylor and R. E. Eck, private communication.
70. S. L. Norman and D. H. Douglass, Jr., *Phys. Rev. Letters* **18**, 339 (1967).
71. L. H. Palmer and M. Tinkham, to be published.
72. D. J. Scalapino, J. C. Swihart, and Y. Wada, to be published.
73. J. C. Swihart, D. J. Scalapino, and Y. Wada, *Phys. Rev. Letters* **14**, 106 (1965).
74. D. L. Decker, D. E. Mapother, and R. W. Shaw, *Phys. Rev.* **112**, 1888 (1958).
75. P. H. Keesom and B. J. C. Van der Hoeven, Jr., *Phys. Letters* **3**, 360 (1963).
76. N. E. Phillips, M. H. Lambert, and W. R. Gardner, *Rev. Mod. Phys.* **36**, 131 (1964).
77. S. Bermon and D. M. Ginsberg, *Phys. Rev.* **135**, A306 (1964).
78. M. A. Biondi, M. P. Garfunkel, and W. A. Thompson, *Low Temperature Physics, LT9* (J. C. Daunt et al., eds.), Plenum Press, New York, 1965.
79. N. W. Ashcroft and J. W. Wilkins, *Phys. Letters* **14**, 285 (1965).
80. G. V. Chester, *Phys. Rev.* **103**, 1693 (1956).
81. D. J. Scalapino and J. R. Schrieffer, *Proceedings of the 1963 Eastern Theoretical Conference*, p. II-2.
82. Y. Wada, *Phys. Rev.* **135**, A1481 (1964).
83. J. Bardeen and M. Stephen, *Phys. Rev.* **136**, A1485 (1964).
84. G. M. Eliashberg, *Zh. Eksperim. i Teor. Fiz.* **43**, 1005 (1962); *Soviet Phys. JETP* **16**, 780 (1963).
85. J. C. Swihart, to be published.

86. Y. Wada, *Rev. Mod. Phys.* **36**, 253 (1964).
87. W. L. McMillan, *Phys. Rev.*, to be published.
88. D. C. Mattis and J. Bardeen, *Phys. Rev.* **111**, 412 (1958).
89. P. W. Anderson, *Phys. Rev.* **112**, 1900 (1958).
90. G. Rickayzen, *Phys. Rev.* **115**, 795 (1959).
91. S. B. Nam, *Phys. Rev.* **156**, 470, 487 (1967).
92. R. A. Ferrell and R. E. Glover, III, *Phys. Rev.* **109**, 1398 (1958).
93. M. Tinkham and R. A. Ferrell. *Phys. Rev. Letters* **2**, 331 (1959).
94. J. Bardeen, G. Rickayzen, and L. Tewordt, *Phys. Rev.* **113**, 982 (1959).
95. V. Ambegaokar and J. Woo, *Phys. Rev.* **139**, A1818 (1965).
96. V. Ambegaokar and L. Tewordt, *Phys. Rev.* **134**, A805 (1964).
97. A. G. Redfield, *Physica* **24**, 5150 (1957).
98. A. G. Anderson and A. G. Redfield, *Bull. Am. Phys. Soc.* **2**, 388 (1957).
99. B. C. Deaton, *Phys. Rev. Letters* **16**, 577 (1966).
100. V. Ambegaokar, *Phys. Rev. Letters* **16**, 1047 (1966).
101. R. B. Ferguson and J. H. Burgess, *Phys. Rev. Letters* **19**, 494 (1967).
102. W. P. Mason, *Phys. Rev.* **143**, 229 (1966).
103. P.M. Platzman and P.A. Wolff, *Phys. Rev. Letters* **18**, 280 (1967).
104. S. Schultz and G. Dunifer, *Phys. Rev. Letters* **18**, 283 (1967).
105. P.M. Platzman and W.M. Walsh Jr., *Phys. Rev. Letters* **19**, 514 (1967); Erattum, *Phys. Rev. Letters* **20**, 89 (1968).
106. W. Shaw and J.C. Swihart, *Phys. Rev. Letters* **20**, 1000 (1968).
107. D.J. Scalapino and B.N. Taylor, *Phys. Rev.*, to be published.

11

TUNNELING AND STRONG-COUPLING SUPERCONDUCTIVITY

W. L. McMillan and J. M. Rowell

BELL TELEPHONE LABORATORIES, INCORPORATED
MURRAY HILL, NEW JERSEY

I. HISTORICAL INTRODUCTION

In this chapter we will outline how the tunneling technique has developed from the exciting discovery of obvious potential into the most sensitive probe of the superconducting state. The measurements have shown not only that we have a theory of superconductivity accurate to a few per cent but also that the technique is a valuable tool in the study of normal metal properties.

The quantum-mechanical tunneling current flowing through a thin insulating layer between metal electrodes was first investigated in detail by Fisher and Giaever (*1*). They measured the current (I) vs. voltage (V) characteristics for junctions made with evaporated electrodes of aluminum and a thermally grown aluminum oxide insulator. Proof that the current was actually tunneling through the insulator was obtained (*2*) on replacing the second aluminum electrode by lead. Below the superconducting transition temperature of the lead film, the current–voltage plot changed progressively with decreasing temperature and at the lowest temperature showed conclusively that the density of electron states in the superconductor was reflected in the tunneling characteristics. The BCS predictions (*3*) of both an energy gap and an inverse square-root singularity in the one-electron excitation spectrum were confirmed. The gap measurement was made more definite by Nicol et al. (*4*) and Giaever (*5*) when they reduced the temperature below the transition temperature of the aluminum film and observed the negative resistance region that results. Further work on the measurement of the energy gap and the effect of magnetic field, temperature, and magnetic impurities will not be discussed here as they are covered in Chapter 3 and in the review of Douglass and Falicov (*6*).

On extending the investigation to lower temperatures and to other superconductors Giaever et al. (*7*) found that the "weak-coupling" superconductors Sn, In, and Al have a density of states vs. energy dependence very close to the BCS prediction of $N_S(E)/N(0) = E/\sqrt{E^2 - \Delta^2}$. However, in the case of the "strong-coupling" or "bad-actor" superconductor lead, they observed small but distinct deviations from the BCS dependence at energies comparable to the Debye energy ($\Theta_D \sim 9$ meV). Although the full significance of their result was perhaps not appreciated by the authors, it was apparent to Anderson, who in 1960 had pointed out (*8*) that in the bad-actor heavy metals there might be effects both of phonon scattering and of variation of energy gap with energy. Improvement of the derivative techniques being used by Rowell et al. (*9*) was encouraged by Anderson and the development of the technique since that time is due to the stimulus provided by interested theorists, particularly Anderson, Schrieffer, Scalapino, and Swihart. The "phonon-induced structure" has now been studied in several of the weak-coupling (*10*) and the two strong-coupling superconductors [lead (*11*) and mercury (*12*)] as well as in a number of alloys (*13*). Similar investigations of the transition metals tantalum and niobium (*14*) indicate that evidence for a non-phonon-mediated electron–electron attraction in superconductivity is slight indeed.

The fundamental importance of these experiments can be appreciated by examining the progress made in the theory of superconductivity in the past 10 years. Since the rather remarkable success of Bardeen, Cooper, and Schrieffer in explaining nearly all the properties of nearly all superconductors using the model of an electron gas with attractive interactions near the Fermi surface, one has been left with the task of explaining the deviations of a few superconductors from the predictions of the BCS model. For example, two obvious deviations are the critical field vs. temperature dependences for lead and mercury and the ratio $2\Delta_0/kT_c$, which is as large as 4.6 and 4.3 for mercury and lead compared to the BCS value of 3.52. It was evident that experimentally the superconducting metals with low Debye energies were the "bad actors" from the point of view of the BCS predictions. Thus it appeared that the electron–phonon interaction must be dealt with more realistically than in the BCS model.

The role of this electron–phonon interaction in metals has become clear from the work of Migdal (*15*) on the normal metal and Eliashberg and others (*16*–20) on the superconductor. In particular, one can sum the perturbation series for the electron–phonon interaction and obtain self-energy equations which are accurate to lowest order in the square root of the electron to ion mass ratio (*15,16*), an expansion parameter of order 10^{-2}. One finds an effective interaction between electrons which is retarded in time and local in space, in contrast to the BCS model interaction, which is instantaneous and nonlocal. The formalism is sufficiently powerful to treat accurately the case of strong interactions where the width of an electronic state due to phonon emission is comparable to its energy and even the quasi-particle concept breaks down.

At this stage in the development of the theory a number of solutions of the gap equation were made which attempted to allow for differences in the normal metal properties of superconductors. Morel and Anderson (*18*) pointed out that the electron–electron coupling was largely mediated by short-wavelength phonons and therefore assumed that the effective phonon density could be approximated by an Einstein peak at the longitudinal phonon frequency ω_L. They obtained an energy-dependent gap parameter $\Delta(\omega)$ which exhibits structure at ω_L, $2\omega_L, \ldots$, and this harmonic structure should be reflected in the density of states $N(\omega)$.

Further energy-dependent gap parameters were obtained by Swihart (*21*), who assumed an Einstein peak or a Debye spectrum as approximations to the effective phonon density. These solutions also exhibit structure at the energy of the Einstein peak or Debye cutoff and their harmonics.

An attempt to explain the tunneling results of Giaever et al. (*7*) was made by Culler et al. (*22*) using a Debye spectrum approximation; the result bears little overall resemblance to the experimental curve but is significant in that the calculated density of states shows a sharp drop in the vicinity of Θ_D and a broader peak at $2\Theta_D$, features reproduced in the tunneling density of states. In addition, the computer techniques to be used later were established in this work.

The measurements of the density of states in lead were improved by Rowell et al. (*11*), but a more striking result was obtained by taking the second derivative d^2I/dV^2 of the tunneling characteristic of a Pb–I–Pb junction. Not only were the two large deviations from the BCS density of states resolved as negative peaks in this derivative but additional fine structure was observed within these peaks, which left no doubt that both transverse and longitudinal phonon branches must be accounted for in detail in any solution of the gap equation. The fine structure was explained by Scalapino and Anderson (*23*) as a reflection of Van Hove (*24*) critical points in the tunneling characteristic.

After consideration of the above results Schrieffer et al. (*19*) undertook a solution of the gap equation which is notable in four respects:

1. The effective phonon density was a realistic approximation to that in the normal metal and was estimated not only from the tunneling result but also from the dispersion curves measured by neutron scattering.

2. An expression was derived for the tunneling density of states when the gap parameter is not only energy-dependent but also complex,

$$\frac{N(\omega)}{N(0)} = \mathrm{Re}\left[\frac{|\omega|}{\sqrt{\omega^2 - \Delta^2(\omega)}}\right] \tag{2}$$

3. The Coulomb repulsion term was varied to study its effect on the tunneling characteristic.

4. The calculated density of states was in excellent general agreement with the experimental result. [It is important to note here that the tunneling experiment measures properties evaluated at the surface of the superconductor. This result is implicit in (2) but is of little importance when the superconductors are uniform across their thickness. In films with contaminated surfaces or in proximity sandwiches (*25*) this is not the case, and some discretion in the application of (2) is required.]

The calculation of Schrieffer et al., while very successful in reproducing the main features of the tunneling density of states, was not able to account for the fine structure mentioned above with the model phonon density of two Lorentzian peaks. The influence of critical points in the phonon spectrum on the derivatives of the tunneling characteristic was studied in detail by Scalapino and Anderson (*23*). Using their results one can infer the position of many features of the phonon spectrum from second derivative curves but it was obvious that much more information was contained in the tunneling characteristics than could be obtained by such analysis and guess work. The experimental technique had by this time been extended to mercury by Bermon and Ginsberg (*12*); to tin, indium, and thallium (*10*); and to lead–bismuth alloys (*13*). Calculations of the thermodynamic properties of the strong-coupling superconductors (*26*) had also been made using the model phonon spectrum mentioned above and results were in good agreement with experiment. At this point the strong-coupling aspect of the theory appeared accurate and well developed.

The obvious and logical step in this development of the theory of superconductivity, which by now appeared accurate to $\sim 1\%$, would have been to calculate the properties of superconductors to this accuracy using the newly developed theory and knowledge of normal metals. It is at this point, the information about normal metal properties, that difficulties are encountered. One needs to know the Fermi surface, the matrix elements of the electron–phonon interaction, and the phonon dispersion curves of a particular metal to find the parameters to be used in the Eliashberg gap equations. With the development of the pseudo-potential approach (*27–30*) to the band structure of polyvalent metals and the availability of a variety of experimental data on the Fermi surface (*31–33*), a quantitative picture of the pseudo-potential in metals is emerging. A knowledge of the pseudo-potential is sufficient to determine both the Fermi surface and the electron–phonon interaction. For a number of metals the phonon dispersion curves have been measured by inelastic neutron scattering (*34*). The information necessary to compute the superconducting properties of the polyvalent metals is therefore becoming available, but only slowly.

On considering the lack of knowledge of the normal metal properties of lead and the possible accuracy of the superconducting tunneling data the authors decided it would be of interest to examine the data critically with two aims. First, it seemed probable that the parameters entering the Eliashberg gap equation could be extracted from the data. These parameters are, in fact, the phonon density of states times an average of the square of the electron–phonon matrix elements (α^2F) and the Coulomb pseudo-potential μ^*. Second, the accuracy of the theory of superconductivity could be tested, particularly by examination of the tunneling results for $E > \Theta_D$. A procedure was devised to accomplish these ends and has been applied to lead in great detail and to mercury, tin, indium, thallium, lead–indium, and indium–thallium alloys somewhat less critically.

In brief the method is as follows. With the aid of a computer the parameters mentioned above are adjusted until the computed electronic density of states accurately fits the density of states measured in the tunneling experiment for $E < \Theta_D$. By examination of the fit at $E > \Theta_D$ the accuracy of the gap equation can be confirmed; the final input parameters give α^2F, μ^*, and in addition the electronic self-energies in the normal and superconducting states and the thermodynamic properties of the superconductor can be computed.

In Section II we will cover the necessary theory of normal metals and superconductivity, and in Section III the theory of the tunneling process will be given; some repetition of material in previous chapters is unavoidable. The description of experimental techniques in Section IV may seem out of place in this book, but as the tunneling technique is relatively new and no recent review exists, some detail is necessary. In addition, the claims made for the method depend solely on the accuracy of the measurements. The procedures used in the solution and inversion of the gap equation will be described in Section V and the significance of the results discussed in Section VI.

II. THEORY OF SUPERCONDUCTIVITY

In this section we present a limited discussion of the theory of superconductivity. Since we will apply the theory to a quantitative study of real metals, our primary concern is whether or not the underlying physical model contains all the physics necessary to describe superconductivity in metals. We review here the theoretical developments leading to the accurate self-energy equations for the strong-coupled superconductor. The reader is referred to Chapter 10 by Scalapino and to Schrieffer's excellent book (*20*) for a more detailed discussion of the calculations.

The physical model of a metal considered by Bardeen, Cooper, and Schrieffer (*3*) in their original paper is the picture that was formalized by Landau (*35*) and known as the Fermi liquid theory. It is assumed that the low-lying electronic excitations are long-lived quasi-particles, in one-to-one correspondence with the excitations of the free Fermi gas, and that these fully dressed quasi-particles have some residual interaction. In a metal there are two contributions to this interaction: (1) the exchange of virtual phonons which produces an attractive interaction between quasi-particles provided their energy is less than the phonon energy, and (2) the screened Coulomb interaction, which is repulsive. BCS showed by a variational calculation that, when this residual interaction is attractive at the Fermi surface, a new superconducting ground state, with pairs of electrons bound into Cooper pairs, is energetically more favorable than the normal state. For this purpose BCS chose a model interaction equal to $-V$ when the kinetic energy of both quasi-particles is less than a typical phonon energy and zero otherwise. With the interaction parameter chosen phenomenologically the BCS model is in remarkably good agreement with a wide variety of experiments. The underlying physical model, the Landau Fermi liquid theory, is valid only for temperatures and excitation energies much less than a typical phonon energy. The lifetime of an electron with enough energy to emit real phonons is very short so that the quasi-particle is not well defined. One finds deviations from the BCS model for the "strong-coupling" superconductors, Pb and Hg, where the transition temperature is high enough and Debye temperature low enough for the Fermi liquid theory to fail.

Since the BCS paper appeared in 1957 there has been a great deal of progress in understanding the role of the electron–phonon interaction in metals.

Migdal (*15*) was able to show that, within the Fröhlich (*36*) Hamiltonian, lowest-order perturbation theory in the electron–phonon interaction, performed self-consistently, is quite accurate for metals. Migdal used the fact that the ions move very slowly compared to the electrons so that one has an expansion parameter (a typical phonon energy ω_0 divided by the Fermi energy E_F) which can be used to simplify the theory even though the dimensionless electron–phonon coupling constant is of order unity. Migdal's result for the electron

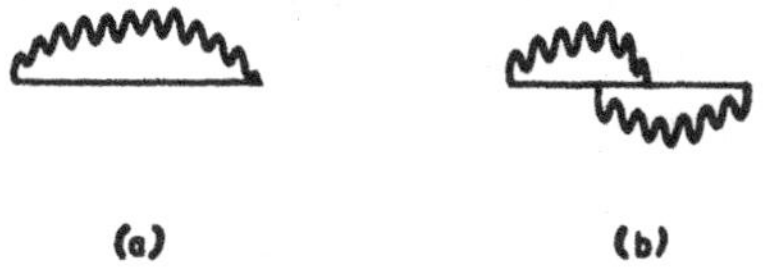

Fig. 1. (a) Self-energy diagram included in the theory; (b) self-energy diagram which is negligible. The solid line represents a fully dressed electron propagator and the wavy line a phonon propagator.

self-energy Σ is shown in Fig. 1a, where the wavy line is the dressed phonon propagator and the solid line is the dressed electron propagator

$$G^{-1} = G_0^{-1} - \Sigma \tag{1}$$

Higher-order graphs such as Fig. 1b are of order $\hbar\omega_0/E_F$ compared with Fig. 1a and are neglected. Thus for the model of free fermions interacting with phonons one can write down accurate self-energy equations for the normal state even though the coupling is strong.

Migdal's work has been extended in several ways; we discuss first the extension to the superconducting state. Following the BCS variational calculation it was realized by a number of people that the superconducting state could be found by a generalization of the Hartree self-consistent-field method. In the Hartree method one takes into account the scattering of an electron by a potential V,

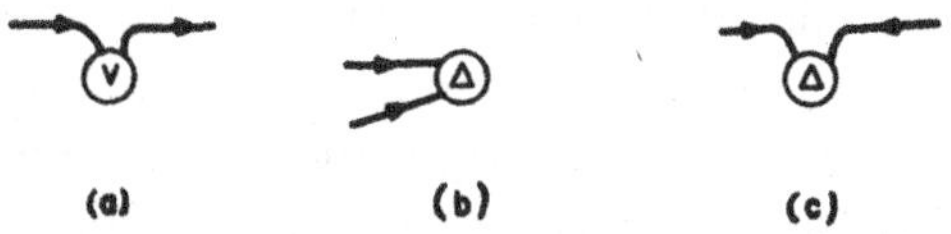

Fig. 2. (a) Electron scattering from the Hartree potential; (b) two excited electrons forming a Cooper pair; (c) an excited electron pairing with a ground-state electron leaving a hole excitation, i.e., a scattering event by the BCS potential.

represented pictorially in Fig. 2a, and this potential is determined self-consistently from the interaction with the other electrons. In the superconductor one must take into account processes (Fig. 2b) in which two electrons interact, bind themselves into a Cooper pair, and disappear into the condensate. One can twist this diagram around (Fig. 2c) into a single-particle scattering process in which an excited electron pairs itself with a ground-state electron leaving a hole excitation. One has then an "anomalous" self-consistent potential (which we denote the BCS potential) which scatters electrons into holes. It is then natural to introduce a two-component "spinor" Green's function to describe the propagation of single-particle excitations where the diagonal part of the propagator describes the scattering of electrons into electrons (Fig. 2a) and holes into holes and the off-diagonal part describes the scattering of electrons into holes (Fig. 2c) and holes into electrons. With this generalization of the Green's function technique one has a diagrammatic expansion for the Green's functions of the superconductor in one-to-one correspondence with the usual diagrammatic technique for the normal state. In this way Eliashberg (*16*) and Nambu (*17*) obtained the self-energy equations for the superconductor within the Fröhlich Hamiltonian using second-order self-consistent perturbation theory (Fig. 1a) and Eliashberg showed that the higher-order corrections were negligible. Migdal's theorem was thus generalized to the superconductor.

These results were obtained for the Fröhlich Hamiltonian, which includes the electron–phonon interaction, assumed to be screened, but does not include the Coulomb interaction between electrons. Batyev and Pokrovskii (*37*) have extended the Migdal calculation for the normal state by including the Coulomb interaction between electrons and treating the phonons within essentially the jellium model. This model calculation shows explicitly the way in which Coulomb interactions renormalize the quasi-particles and the interactions. It is readily generalized to the superconducting state and we discuss it in some detail here. The calculation splits naturally into two parts. First, one solves the electron gas problem with the lattice held rigid—that is, with the electron–phonon interaction turned off. One calculates formally the electron self-energy, the polarization propagator, etc., and observes that for $T \ll E_F$ the interacting electron gas behaves as a Landau Fermi liquid. The phonon frequencies can be calculated formally, making use of the adiabatic approximation, from the bare electron–ion interaction and the exact dielectric function. Second, one calculates the modification of the Landau quasi-particle propagator due to the phonon interaction and one finds for the quasi-particle self-energy just Migdal's equation. The parameters entering Migdal's equation—the phonon frequencies, the fermion energies, and the electron–phonon interaction—are fully dressed by the Coulomb interaction, and we will discuss that dressing below. Finally, since the phonon modification of the electron propagator is only important very near the Fermi surface, this propagator modification has a negligible effect on the phonon frequencies, the polarization propagator, and the proper vertex function, which all depend on the electron propagator far from the Fermi surface. Thus an elegant picture of interacting electrons and phonons in the normal state has emerged which depends only on the existence of the expansion parameter $\hbar\omega_0/E_F$ and which is therefore applicable to real metals. With the lattice held rigid, one has a Landau Fermi liquid theory for the Coulomb dressed electrons for $T \ll E_F$ in which the low-energy excitations are single-particle-like quasi-particles with residual interactions. The Coulomb dressed electrons behave as single particles in low-energy encounters simply because there is not enough energy available to excite the internal structure—the dressing—of the particle. This is the same reason that, for example, He^4 atoms behave as single particles—bosons—in liquid helium. When the lattice is allowed to vibrate the quasi-particle propagator near the Fermi surface is modified according to Migdal's equations, which are valid even when the electron–phonon coupling is strong.

We turn now to the diagrammatic derivation of the self-energy equations for the normal state, working, for simplicity, within the jellium model. Our discussion will be brief; the reader is referred to the papers by Batyev and Pokrovskii and by Prange and Sachs (*38*) for a more detailed discussion. The Hamiltonian for the model is

$$\mathcal{H} = \sum_k \epsilon_k^0 c_k^\dagger c_k + \sum_{\substack{k_1 k_2 \\ q}} V_q c_{k_1-q}^\dagger c_{k_2+q}^\dagger c_{k_2} c_{k_1} + \sum_q \omega_q^0 a_q^\dagger a_q + \sum_{k_1 k_2} g_{k_1-k_2} (a_{k_1-k_2} + a_{k_2-k_1}^\dagger)\, c_{k_1}^\dagger c_{k_2} \qquad (2)$$

where the $c^\dagger(a^\dagger)$ create bare electrons (phonons) with energy $\epsilon^0(\omega^0)$ and $V^0(g^0)$ is the bare Coulomb (electron–phonon) interaction. Spin and polarization indices are suppressed. We first dress the Coulomb interaction and phonon propagator by replacing all graphs beginning and ending with a bare Coulomb or phonon line by the full polarization propagator. The polarization propagator is the sum of Coulomb propagator, containing no bare phonon lines, and the phonon propagator, containing at least one bare phonon line. The total interaction is the sum of the screened Coulomb interaction $V^1(q, \omega)$ and the dressed phonon propagator $D^1(q, \omega)$, where within the jellium model

$$V^1(q, \omega) = \frac{4\pi e^2}{q^2 \epsilon(q, \omega)} \qquad (3)$$

$$D^1(q, \omega) = \frac{2\omega_q^1}{\omega^2 - \omega_q^{1^2}}\, g_q^{1^2} \qquad (4)$$

Here $\epsilon(q, \omega)$ is the dielectric function,

$$\omega_q^{1^2} = \omega_q^{0^2}/\epsilon(q, \omega = 0)$$

$$g_q^{1^2} = \frac{g_q^{0^2}}{\epsilon(q, \omega_q^1)^2}\, \frac{\omega_q^0}{\omega_q^1}$$

The electron self-energy Σ is given as a perturbation series in the screened Coulomb interaction, the screened phonon propagator, and the dressed Green's function, which satisfies Dyson's equation $G^{-1} = G^{\circ^{-1}} + \Sigma$. The self-energy diagrams are classified into three groups: (1) the Coulomb contributions Σ_c (Fig. 3), containing no phonon lines; (2) the lowest-order phonon contributions

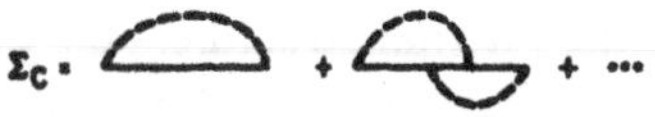

Fig. 3. Diagrams summed in the Coulomb self-energy.

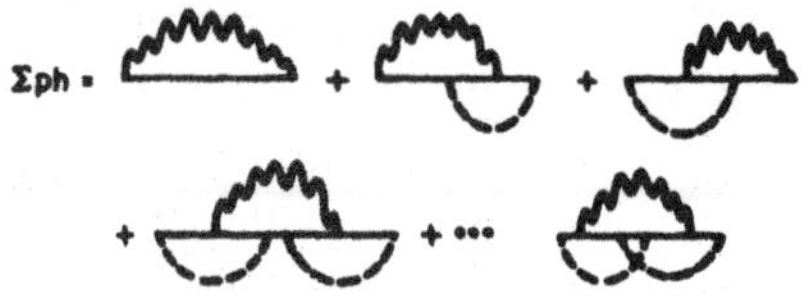

Fig. 4. Diagrams summed in the self-energy due to phonon interactions.

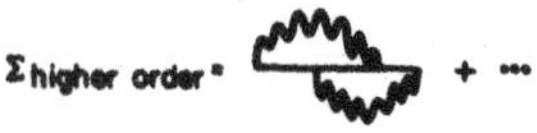

Fig. 5. Higher-order diagrams which are neglected.

Σ_{ph} (Fig. 4), containing one phonon line; and (3) the higher-order phonon contributions (Fig. 5), which are negligible. The Coulomb self-energy is

$$\Sigma_c(p) = \int \frac{d^4p'}{(2\pi)^4} G(p')\, V_c^1(p-p')\, \Gamma^1(p, p') \tag{5}$$

where $\Gamma^1(p, p')$ is the proper vertex correction for the Coulomb problem. In evaluating Σ_c one can in addition neglect the phonon modification to G so that Σ_c is just the electron self-energy in the absence of phonons. The phonon contribution is

$$\Sigma_{\text{ph}}(p) = \int \frac{d^4p'}{(2\pi)^4} G(p')\, g_{p-p'}^{12} D^1(p-p')\, \Gamma^1(pp')\, \Gamma^1(p'p) + \mathcal{O}\left(\frac{\hbar\omega_{\text{ph}}}{E_F}\right) \tag{6}$$

which is just the lowest-order graph with both vertices corrected by the Coulomb interaction. Graphs, such as the last one of Fig. 4, which are not summed in this way are constant near the Fermi surface and are absorbed into the chemical potential. The vertex correction in Eq. (6) is taken for zero frequency and for p and p' on the Fermi surface and is a function of momentum transfer $q = p - p'$. The sum of Eqs. (5) and (6) gives the electron self-energy accurate to lowest order in $\hbar\omega_{\text{ph}}/E_F \sim 10^{-3}$. It is convenient at this point to go over to the quasi-particle representation which is valid for $\omega \ll E_F$. We expand Σ_c about the

Fermi surface

$$\Sigma_c(p, \omega) = \Sigma_c(p_F, 0) + \left.\frac{\partial \Sigma_c}{\partial \omega}\right|_{\substack{p=p_F \\ \omega=0}} \omega + \left.\frac{\partial \Sigma_c}{v_F\, \partial p}\right|_{\substack{p=p_F \\ \omega=0}} \epsilon_p \tag{7}$$

and absorb the constant term into the chemical potential. Writing

$$z_{p_F} = \left(1 - \frac{\partial \Sigma_c}{\partial \omega}\right)^{-1} \qquad \epsilon_p^1 = z_{p_F}\left(1 + \frac{\partial \Sigma_c}{v_F\, \partial p}\right)\epsilon_p^0 \tag{8}$$

and $\Sigma_{\text{ph}}^{qp}(\omega) = z_{p_F}\Sigma_{\text{ph}}(\omega)$ we have

$$G(p\omega) = \frac{z_{p_F}}{\omega - \epsilon_p^1 - \Sigma_{\text{ph}}^{qp}(\omega)} = z_{p_F} G^{qp}(p, \omega) \tag{9}$$

where G^{qp} is the quasi-particle propagator and

$$\Sigma_{\text{ph}}^{qp}(p) = \int \frac{d^4 p'}{(2\pi)^4} G^{qp}(p')\, D^1(p - p')\, (z_{p_F}\Gamma^1(p - p')\, g_{p-p'}^1)^2 \tag{10}$$

This is just Migdal's equation for the self-energy with a fully dressed quasi-particle–phonon interaction

$$g_p^{qp} = \frac{g_q^0}{\epsilon(q, 0)} \sqrt{\frac{\omega_q^1}{\omega_q^0}}\, z_{p_F}\Gamma^1(q) \tag{11}$$

For simple (nearly free electron) metals we can write

$$g_{q\lambda}^{qp} = -\, i v_q^{qp}(q \cdot \epsilon_{q\lambda}) \sqrt{\hbar/2NM\omega_{q\lambda}^1} \tag{12}$$

(where $\epsilon_{q\lambda}$ is the polarization vector for phonons of polarization λ and M and N are the mass and number of the ions), where

$$v_q^{qp} = [v_q^0/\epsilon(q, 0)]\, z_{p_F}\Gamma(q) \tag{13}$$

is the fully dressed quasi-particle–ion interaction which determines the Fermi surface. Thus one can find reasonably accurate quasi-particle–phonon matrix elements for those simple metals for which the Fermi surface is known.

This calculation justifies for jellium the physical picture of an electron Fermi liquid interacting with the phonons according to Migdal's equations. This result did not depend on any special property of the theoretical model but only on the existence of the expansion parameter $\hbar\omega_{\text{ph}}/E_F \ll 1$.

The above discussion for the normal state can be extended to the super state in two sentences. We first note that Nambu and Eliashberg have shown that the diagrammatic perturbation theory for the superstate is identical to that for the normal state, provided one uses the spinor propagators for the electrons. The second statement is that the electron propagator is modified in the superconducting state only very near the Fermi surface for energies of order Δ and the effects of this propagator modification on the phonon frequencies, the quasi-particle renormalization factor, the vertex function, etc., can be safely neglected. The quasi-particle (matrix) self-energy in the superstate is

$$\Sigma^{qp}_{\text{ph}}(p) = i\int \frac{d^4p'}{(2\pi)^4}\,\bar{\tau}_3 \bar{G}^{qp}(p')\,\bar{\tau}_3 D(p-p')\,(g^{qp}_{p-p'})^2 \tag{14}$$

which was derived by Eliashberg and Nambu for the Fröhlich model. There is, in addition, a small off-diagonal Coulomb contribution (from the residual interactions in the Fermi liquid) which cannot be neglected.

$$\Sigma^{qp}_{c}(p) = \text{O.D.}\left[\int \frac{d^4p'}{(2\pi)^4}\,\tau_3 G^{qp}(p')\,\tau_3 V^{\text{eff}}_c(p-p')\right] \tag{15}$$

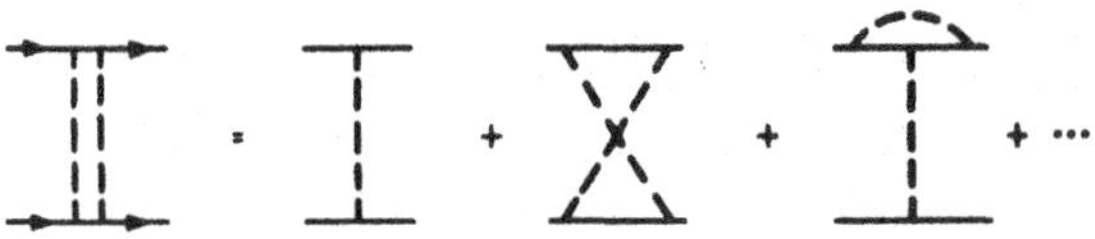

(a)

(b)

Fig. 6. (a) Coulomb scattering processes summed in defining the effective Coulomb scattering potential; (b) scattering processes not summed to avoid double counting.

Here V^{eff}_c is an effective scattering potential which is $z^2_{p_F}$ times an irreducible four-vertex which sums the scattering processes of Fig. 6. The integral equations automatically sum the electron–electron ladder diagrams for V^{eff}_c and, to avoid the double counting of diagrams, one must not include diagrams such as Fig. 6b. One can estimate V^{eff}_c from the lowest-order diagram of Fig. 6a and find $\mu \equiv N(0)\,V^{\text{eff}}_c \approx 0.3$. The self-energy diverges logarithmically for large energy and we apply an energy cutoff at $\omega = \pm E_F$. It is convenient numerically to use an energy cutoff $\omega_c \sim 5\omega_0$ and replace μ by a Coulomb "pseudo-potential" (*18*)

$$\mu^* = \mu/(1 + \mu \ln E_F/\omega_0) \tag{16}$$

This replacement introduces no error in the solution of the integral equations provided $\Delta(\omega)$ has approached its asymptotic value at $\omega = \omega_c$. For lead this is the case for $\omega_c = 50$ meV and the theoretical estimate for μ^* is 0.11. If we believe the theoretical estimate for μ within $\pm 50\%$, then $\mu^* = 0.11 \pm 0.02$. To a very good approximation the self-energies are independent of the momentum variable **p** and the momentum integration in Eqs. (10) and (14) can be performed first, resulting in one-dimensional nonlinear integral equations for the diagonal (normal) and off-diagonal (pairing) self-energies as functions of energy. The solution of these equations is discussed in Section V.

III. THEORY OF TUNNELING

In this section we discuss the theory of electron tunneling using the concept of the effective tunneling Hamiltonian (*39–42*). The physical system that we wish to describe is fabricated by evaporating a metal film (~ 2000 Å), allowing it to oxidize to form an insulating layer some 20 Å thick, and then evaporating a second metal film onto the oxide to form a metal–insulator–metal junction. The 20 Å has become a conventional figure which no one appears to have measured; in practice, one makes a junction of convenient resistance rather than known thickness.

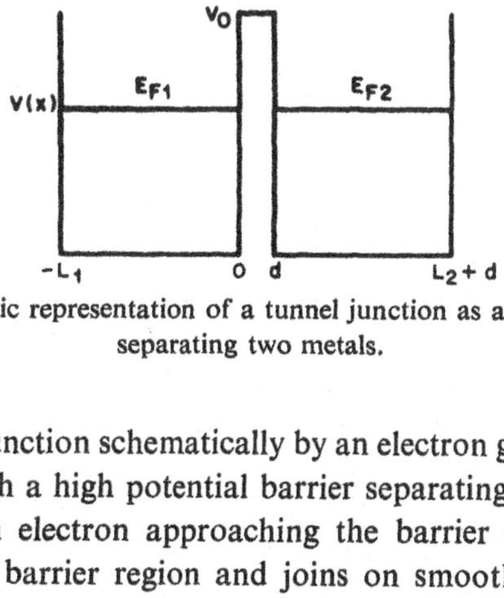

Fig. 7. Schematic representation of a tunnel junction as a potential barrier separating two metals.

We represent the junction schematically by an electron gas in a single-particle potential (Fig. 7) with a high potential barrier separating the two metals. The wave function of an electron approaching the barrier from the left decays exponentially in the barrier region and joins on smoothly to a propagating wave function on the right. This is the textbook (*43*) example of tunneling through a one-dimensional square barrier and the transmission probability, assumed small, is given by

$$\tau = 16\,\frac{E_x(V_0 - E_x)}{V_0^2}\exp(-2Bd) \qquad B = \sqrt{\frac{2m(V_0 - E_x)}{\hbar^2}} \tag{17}$$

where E_x is the kinetic energy of the electron in the x-direction $\hbar^2 k_x^2/2m$, and V_0 and d are the height and width of the barrier. This small transmission probability of the barrier can be simulated by an effective tunneling Hamiltonian which plucks an electron from the left metal near the surface of the barrier and injects it into the right metal near the barrier surface:

$$H_T = \sum_{kp_x} T_{kp} c_k^\dagger c_p + \text{h.c.} \tag{18}$$

Here $c_k^\dagger$ and c_k are the creation and annihilation operators for electrons in single-particle eigenstates Φ_k of the left metal and $c_p^\dagger$ and c_p are the same for the right metal. Note that the potential is independent of y and z so that the y- and z-components of momentum are conserved but not the x-component. To define the eigenstates Φ_k for the left metal we imagine that the right metal is removed and that $V(x) = V_0$ for $x > 0$. The eigenstates defined in this way for the left metal are not orthogonal to those of the right metal and one must use some care in calculating matrix elements between them. We want to choose the matrix elements T_{kp} to reproduce the transmission probability of the barrier. Suppose that we place an electron in eigenstate Φ_k of the left metal and ask how fast it leaks out into the right metal. The transition probability per unit time is

$$W = (2\pi/\hbar)\, \rho(p)\, |T_{kp}|^2 \tag{19}$$

where $\rho(p) = mL_2/v_{2x}\pi$ is the density of states per unit energy in the right metal and v_{2x} is the electron velocity. Now the transition probability per unit time is the number of times the electron strikes the barrier per unit time $v_{1x}/2L_1$ times the transmission probability τ on each approach:

$$W = v_{1x}\tau/2L_1 \tag{20}$$

comparing Eqs. (19) and (20) we find for the tunneling matrix elements

$$|T_{kp}|^2 = \frac{\hbar}{4m} \frac{v_{1x}v_{2x}}{L_1 L_2} \tau \tag{21}$$

We are interested in electrons with kinetic energies near the Fermi energy $E_F = \hbar^2 k_F{}^2/2m$ for which the tunneling matrix elements are peaked sharply for electrons moving in the x-direction:

$$|T_{kp}|^2 \simeq \frac{\hbar}{4m} \frac{v_F^2}{L_1 L_2} 16 \frac{E_F(V_0 - E_F)}{V_0^2} \times \exp\left[-2d\sqrt{\frac{2m(V_0 - E_F)}{\hbar^2}}\left(1 - \frac{E_F}{V_0 - E_F}\frac{\theta^2}{2}\right)\right] \tag{22}$$

where θ is the angle between the electron momentum and the normal to the barrier. For a typical junction with an area of 10^{-5} cm^2 and a resistance of 30 Ω the exponential factor is roughly $\exp(-20 - 20\theta^2)$.

We now want to calculate the tunneling current with a given voltage across the barrier including the effects of electron–electron and electron–phonon interactions in each metal. We write for the total Hamiltonian

$$\mathscr{H} = \mathscr{H}_1 + \mathscr{H}_2 + \mathscr{H}_T = \mathscr{H}^0 + \mathscr{H}_T \tag{23}$$

where $\mathscr{H}_1$ is the total Hamiltonian for metal 1 which is the sum of the kinetic and single-particle potential energies of the ions, the Coulomb electron–electron interaction and the electron–ion interaction. We are not including interactions with phonons in the barrier or the Coulomb interaction between an electron in metal 1 and electron in metal 2. Treating $\mathscr{H}_T$ as a perturbation we calculate the transition probability per unit time for transferring an electron from the metal 1 to metal 2.

$$W_{1-2} = \frac{2\pi}{\hbar} \sum_l |\langle 0| \mathscr{H}_T |l\rangle|^2 \, \delta(E_l - V) \tag{24}$$

The eigenfunctions of the total Hamiltonian $\mathscr{H}_1 + \mathscr{H}_2$ are products of eigenfunctions of $\mathscr{H}_1$ and $\mathscr{H}_2$ $[|l\rangle = |m\rangle_1 \, |n\rangle_2]$ and we have

$$W_{1-2} = \frac{2\pi}{\hbar} \sum_{mn} \left| \sum_{kp_x} T_{kp} \langle 0| c_k |m\rangle \langle 0| c_p^\dagger |n\rangle \right|^2 \delta(E_m + E_n - V) \tag{25}$$

Consider the sum $\sum_{p_x} T_{kp} \langle 0| c_p^\dagger |n\rangle$. The tunneling matrix elements are approximately independent of p_x over a wide range of momentum and the phase of the matrix elements is independent of p_x provided we choose the phase of the wave functions so that the exponential tail of Φ_p in the barrier is positive. We can bring T_{kp} outside the summation and sum over some momentum range $p_F - \Delta p < p_x < p_F + \Delta p$, where $\Delta p \ll p_F$. Then we have $T_{k\perp} \langle 0| \sum_{p_x} c_p^\dagger |n\rangle$; but $\sum_{p_x} c_p^\dagger = \tilde{c}_{p\perp}^\dagger$ creates electrons within a few wavelengths of the barrier with transverse momentum $p\perp$. Writing $\delta(E_m + E_n - V) = \int d\omega \, \delta(E_m - \omega) \, \delta(E_n + \omega - V)$ and noting that $|n\rangle$ and $|m\rangle$ are eigenfunctions of $p\perp$, we have

$$\begin{aligned} W_{1-2} &= \frac{2\pi}{\hbar} \int d\omega \sum_{p\perp} |T_{kp}|^2 \\ &\quad \times \sum_m \langle 0| \tilde{c}_{k\perp} |m\rangle \langle m| \tilde{c}_{k\perp}^\dagger |0\rangle \, \delta(E_m - \omega) \\ &\quad \times \sum_n \langle 0| \tilde{c}_{p\perp}^\dagger |n\rangle \langle n| \tilde{c}_{p\perp} |0\rangle \, \delta(E_n + \omega - V) \\ &= \frac{2\pi}{\hbar} \int_{-\infty}^{\infty} d\omega \sum_{p\perp} T_{p\perp}^2 A_1^-(p_\perp, \omega) \, A_2^+(p\perp, V - \omega) \end{aligned} \tag{26}$$

where

$$A_2^+(p\perp, \omega) = \sum_m \langle 0| \tilde{c}_{p\perp} |m\rangle \langle m| \tilde{c}^\dagger_{p\perp} |0\rangle \, \delta(E_m - \omega)$$
$$A_2^-(p\perp, \omega) = \sum_m \langle 0| \tilde{c}^\dagger_{p\perp} |m\rangle \langle m| \tilde{c}_{p\perp} |0\rangle \, \delta(E_m - \omega) \tag{27}$$

Here $A_2^+(p\perp, \omega)$ is the density of available electron states of energy ω and transverse momentum $p\perp$ near the barrier surface and A_1^- is similarly the density of occupied states of metal 1. This is a very intuitive result. We extract an electron from the surface of metal 1 with energy ω and transverse momentum $p\perp$ and inject it into the surface of metal 2 with energy $V - \omega$ and the same transverse momentum. The transition probability is proportional to the initial density of states A_1^-, the matrix elements T^2, and the final density of states A_2^+ summed over energy and $p\perp$. We turn now to the calculation of A_2^+ for several interesting cases: the uniform free electron gas (normal metal), the BCS superconductor, the strong-coupling superconductor, and the nonuniform superconductor.

A. Free Electron Gas

The one-electron energy levels for fixed $p\perp$ shown in Fig. 8a are filled up to the Fermi energy in the ground state. The one-particle excitation spectrum above that ground state is shown in Fig. 8b, where the solid line indicates electron excitations (available states) and the dashed line hole excitations (filled states). The excited state is just $|p\rangle = c_p^\dagger |0\rangle$, so that the matrix element $\langle m| c_p^\dagger |0\rangle = \delta_{mp}$ and we have

$$A_2^+(p\perp, \omega) = \sum_{p_x > p_F} \delta(E_p - \omega) = \begin{cases} \dfrac{L_2}{\pi\hbar v_{2x}} & \omega > 0 \\ 0 & \omega < 0 \end{cases} \tag{28}$$

In calculating the tunneling current the factors L_2 and v_{2x} in A_2^+ will cancel the same quantities in T^2. Thus the tunneling current is not simply proportional to the density of states in this case.

B. BCS Superconductor

We next consider the uniform superconductor with the BCS model interaction. The single-particle excitations (quasi-particles) are linear combinations of electrons and holes:

$$|p\rangle = u_p c_p^\dagger |0\rangle - v_p c_{-p} |0\rangle \tag{29}$$

where $|0\rangle$ is the BCS ground state and the coherence factors are given by

$$u_p^2 = \tfrac{1}{2}\left(1 + \frac{\epsilon_p}{E_p}\right) \qquad v_p^2 = \tfrac{1}{2}\left(1 - \frac{\epsilon_p}{E_p}\right) \tag{30}$$

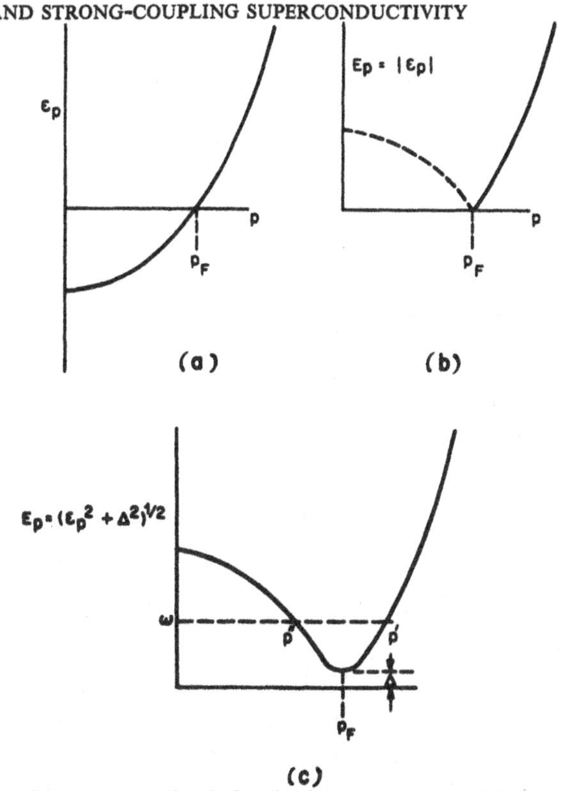

Fig. 8. (a) One-electron energy levels for the free electron gas; (b) one-particle excitation spectrum for the normal state; (c) one-particle excitation spectrum for the superconducting state.

where $E_p = \sqrt{\epsilon_p^2 + \Delta^2}$ is the quasi-particle energy and $\epsilon_p = (p^2/2m) - \mu$. The quasi-particle of momentum p is an electron u_p^2 of the time and a hole $v_p^2 = 1 - u_p^2$ of the time. Referring to the excitation spectrum (Fig. 8c) we see that an electron injected with energy $\omega > \Delta$ creates a linear combination of a quasi-particle of momentum $p' = p_{F_x} + \sqrt{\omega^2 - \Delta^2}/v_{x2}$ and a quasi-particle of momentum $p'' = p_{F_x} - \sqrt{\omega^2 - \Delta^2}/v_{x2}$. The matrix elements are $\langle m|\, c_p^\dagger \,|0\rangle = u_{p'}\delta_{p'1m} + u_{p''}\delta_{p''1m}$, where $u_{p''}^2 = v_{p'}^2$

$$
\begin{aligned}
A_2^+(p\perp, \omega) &= \sum_{p_x > p_F} (u_p^2 + v_p^2)\, \delta(E_p - \omega) \\
&= \sum_{p_x > p_F} \delta(E_p - \omega) \\
&= \frac{L_2}{\pi\hbar v_{2x}} \frac{\omega}{\sqrt{\omega^2 - \Delta^2}} \qquad \omega > \Delta \\
&= 0 \qquad \omega < \Delta
\end{aligned}
\tag{31}
$$

For an energy $\omega = 2\Delta$ the quasi-particle at p' is 0.93 electron and 0.07 hole and the quasi-particle at p'' is 0.07 electron and 0.93 hole. An electron injected at this energy has an overlap $u_{p'}^2 = 0.93$ with the p' quasi-particle and an overlap $u_{p''}^2 = 0.07$ with the p'' quasi-particle. The coherence factors sum to unity and the density of available states is proportional to the quasi-particle density of states $\omega/\sqrt{\omega^2 - \Delta^2}$. Thus the picture of tunneling into the uniform superconductor is that an electron is taken from the left metal and injected into the superconductor (right metal) as a mixture (of energy ω) of p' quasi-particle and p'' quasi-particle. For $\omega \gg \Delta$ the electron tunnels predominantly into the p' branch.

C. Strong-Coupling Superconductor

For the case of the strong-coupling superconductor with a realistic electron–phonon interaction the single-particle states are lifetime-broadened by phonon emission and are no longer eigenstates. The problem can be handled using Green's function techniques and the tunneling current calculated from the Green's functions. The electron Green's function is defined as

$$G(\mathbf{r}t, \mathbf{r}'t') = -i\langle 0|\, T[c(\mathbf{r}t), c^{\dagger}(\mathbf{r}'t')]\, |0\rangle \tag{32}$$

where $|0\rangle$ is the exact ground state in the Heisenberg representation and T is the Wick time-ordering operator. When the metal is translationally invariant in the y- and z-directions we can Fourier transform G:

$$\begin{aligned} G(p\perp, x, x', \omega) &= \int \exp[-ip_{\perp}\cdot(\mathbf{r} - \mathbf{r}') + i\omega(t - t')]\, G(\mathbf{r}t, \mathbf{r}'t') \times dy\, dz\, dt \\ &= -\sum_m \frac{\langle 0|\, \tilde{c}_{p\perp}(x)\, |m\rangle\, \langle m|\, \tilde{c}_{p\perp}^{\dagger}(x')\, |0\rangle}{\omega - E_m + i\alpha} \\ &\quad + \sum_m \frac{\langle 0|\, \tilde{c}_{p\perp}^{\dagger}(x')\, |m\rangle\, \langle m|\, \tilde{c}_{p\perp}(x)\, |0\rangle}{\omega - E_m - i\alpha} \end{aligned} \tag{33}$$

where we have defined

$$\tilde{c}_{p\perp}^{\dagger}(x') = \int dy\, dz\, \exp(ip_{\perp}\cdot\mathbf{r}')\, c^{\dagger}(\mathbf{r}') \tag{34}$$

which creates electrons with transverse momentum $p\perp$ a distance x' from the tunneling surface and $|m\rangle$ and E_m are the exact eigenstates and eigenvalues for the metal. Then we find

$$A_1^{+(-)}(p_{\perp}, \omega) = \frac{1}{\pi} \operatorname{Im} G_1(p_{\perp}, x, x', \pm\omega)|_{x=x'=0} \tag{35}$$

where the Green's function is evaluated with $x = x'$ at the tunneling surface.

This is the usual electron Green's function which is the 11-component of the Nambu spinor Green's function. For the uniform superconductor the spinor Green's function in momentum space is

$$G(p, \omega) = \frac{Z(\omega)\,\omega\tau_0 + \epsilon_p\tau_3 + \Phi(\omega)\,\tau_1}{[Z(\omega)\,\omega]^2 - \epsilon_p^2 - \Phi^2(\omega) + i\delta} \tag{36}$$

Fourier transforming on the x-component of momentum we find

$$\begin{aligned} G(p\perp, x, x', \omega) = {} & \frac{i\exp[i(\Omega|x - x'|/\hbar v_{F_x})]}{2\hbar v_{F_x}\Omega} \\ & \times [\exp(ip_{F_x}|x - x'|)(Z\omega\tau_0 + \Phi\tau_1 + \Omega\tau_3) \\ & + \exp(-ip_{F_x}|x - x')|(Z\omega\tau_0 + \Phi\tau_1 - \Omega\tau_3)] \end{aligned} \tag{37}$$

where

$$\Omega = \sqrt{Z^2\omega^2 - \Phi^2}$$

Finally, using Eq. (35),

$$A^{\pm}(p_{\perp}, \omega) = \frac{1}{\pi\hbar v_{F_x}} \operatorname{Re}\left[\frac{\omega}{\sqrt{\omega^2 - \Delta^2(\omega)}}\right] \tag{38}$$

which is the central result.

D. Nonuniform Superconductor

Finally, we consider the case of a superconductor with spatial variations of the BCS potential $\Phi(x, \omega)$ (*44*). Tomasch and Wolfram (*45,46*) and McMillan and Rowell (*47*) have performed tunneling experiments on nonuniform superconductors—a film of a superconductor in good metallic contact with a normal metal film—and have observed in the tunneling density of states a term which oscillates with both energy and film thickness. These oscillations are the analogue in the superconductor of the interference effect which gives rise to the Friedel (*48*) oscillations in the normal metal. If one has in the normal metal a sharp variation of the Hartree potential $V(x)$ an incident electron in the state $\exp(ip_F \cdot x)$ is partially reflected into the state $\exp(-ip_F \cdot x)$ and the interference between the incident and reflected waves gives rise to an oscillatory component $\sim \cos(2p_F x)$ in the electron density. The characteristic length of the Friedel oscillations is $2\pi/2p_F \approx 3$ Å. If in the superconducting state one has a sharp variation of the BCS potential, an incident quasi-particle in the state p' (Fig. 8c) is partially reflected into the state p'' and the interference effect gives rise to an oscillatory component $\sim \cos(p' - p'')\,x = \cos(2x\sqrt{\omega^2 - \Delta^2}/\hbar v_F)$ in, for example, the tunneling density of states $N(\omega)$. Here ω is the energy of the quasi-particle and x is the distance of the tunnel junction from the perturbation. Note that the characteristic length $2\pi\hbar v_F/2\sqrt{\omega^2 - \Delta^2}$ is a function of energy ω and varies from

~ 1000 Å for $\omega = 10$ meV to several microns for energies near the gap edge. To show this we write the BCS potential as a constant term plus a small spatially varying term

$$\Phi(x, \omega) = \Phi(\omega) + \delta\Phi(x, \omega) \tag{39}$$

and treat the spatially varying term in first order. The Green's function to first order is

$$\begin{aligned} G(p_\perp, x, x', \omega) &= G^0(p_\perp, x, x', \omega) \\ &+ \int dx''\, G^0(p_\perp, x, x'', \omega)\, \delta\Phi(x'', \omega)\, \tau_1 G^0(p_\perp, x'', x', \omega) \end{aligned} \tag{40}$$

where G^0 is the Green's function for the uniform superconductor [Eq. (37)]. Then using Eq. (35) we find

$$\delta A^+(p_\perp, \omega) = \frac{-1}{\pi} \operatorname{Im} \int dx'' \frac{\exp(2i\Omega x''/\hbar v_{F_x})}{(\hbar v_{F_x})^2} \frac{Z\omega}{\Omega} \frac{\Phi}{\Omega} \delta\Phi(x'', \omega) \tag{41}$$

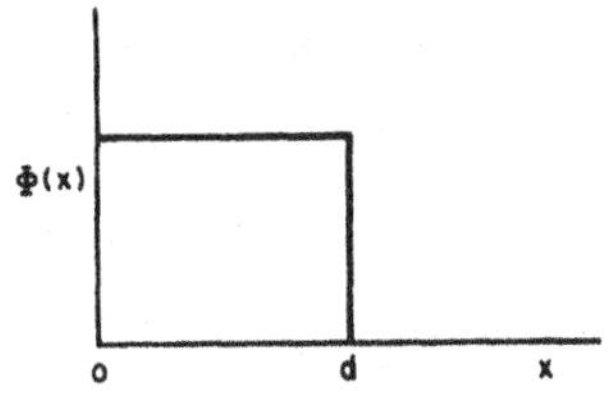

Fig. 9. BCS potential for the Tomasch experiment; the tunnel junction is at $x = 0$.

For the Tomasch experiment, tunneling into a superconductor of thickness d backed by a normal metal, we take for the BCS potential (Fig. 9)

$$\delta\Phi(x, \omega) = \begin{cases} 0 & 0 < x < d \\ -\Phi(\omega) & d < x \end{cases} \tag{42}$$

and find

$$\delta A^+(p_\perp, \omega) \simeq -\frac{1}{\pi\hbar v_{F_x}} \frac{\Delta^2}{2\omega^2} \cos\left(\frac{2x\sqrt{\omega^2 - \Delta^2}}{\hbar v_{F_x}}\right) \tag{43}$$

We have treated the reflection from the interface in the Born approximation which is valid at high energy $\omega \gg \Delta$. We note that the oscillations are not present for a uniform BCS potential and that the observation (*46*) of the oscillations in a tunnel experiment implies a spatially varying BCS potential.

When the energy of the tunneling electron is comparable to a typical phonon energy the mean free path due to phonon emission $l_{ph} = \hbar v_F/2 \operatorname{Im}(Z\omega)$ becomes short and according to Eq. (41) the tunneling experiment examines the BCS potential only within a distance l_{ph} from the tunneling surface. For Pb with

$\omega > 9$ meV, $l_{ph} \approx 500$ Å. For the transition metals (with a smaller Fermi velocity) this distance is ~ 100 Å and tunneling measurements on Nb and Ta have not succeeded in measuring bulk properties, although phonon-induced structure has been observed. Tunneling experiments into more exotic materials (such as Nb_3Sn) are even more difficult as $l_{ph} \sim 30$ Å. At low energies and in high-purity samples the electron mean free path is very long and the tunneling experiment is sensitive to variations of the BCS potential over very large distances. The Tomasch oscillations have been observed (*46*) in indium films with the tunneling surface 33 μ from the perturbation.

For the Rowell experiment, tunneling into a normal metal of thickness d backed by a superconductor, we take for the BCS potential (Fig. 10)

$$\Phi(x, \omega) = \begin{cases} 0 & 0 < x < d \\ \Phi(\omega) & d < x \end{cases}$$

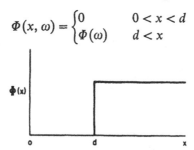

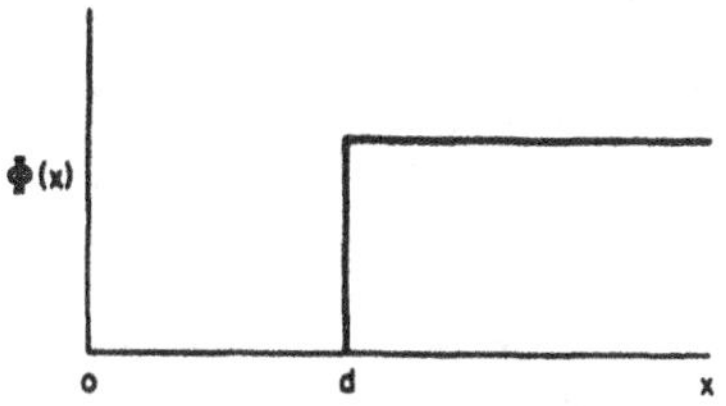

Fig. 10. BCS potential for the Rowell experiment with the tunnel junction at $x = 0$.

and treat $\Phi(x, \omega)$ as a perturbation. The zeroth-order Green's functions are those of the normal metal and the first-order term [Eq. (41)] vanishes. The second-order term which requires two reflections from the N–S interface gives

$$\delta A^{+}(p_{\perp}, \omega) = +\frac{1}{2\pi\hbar v_{F_x}} \operatorname{Re}\left[\frac{\Delta^2}{2\omega^2} \exp(4i\omega d/\hbar v_{F_x})\right] \tag{44}$$

In (*47*) this expression was averaged over $p_{\perp}$ neglecting the dependence of the tunneling matrix elements on $p_{\perp}$ and gave

$$N(\omega) = 1 + \operatorname{Re}\left[\frac{\Delta^2(\omega)}{2\omega^2} F\left(\frac{4\omega d}{\hbar v_F}\right)\right] \tag{45}$$

where

$$F(y) \equiv \int_1^{\infty} \exp(ixy)\,(dx/x^2) \tag{46}$$

The oscillatory term is proportional to the real part of the product of the complex interference factor $F(y)$ and the gap function squared, which is complex and strongly energy dependent. Figure 11 shows the measured and calculated densities of states for tunneling into a 1400-Å silver film backed by a

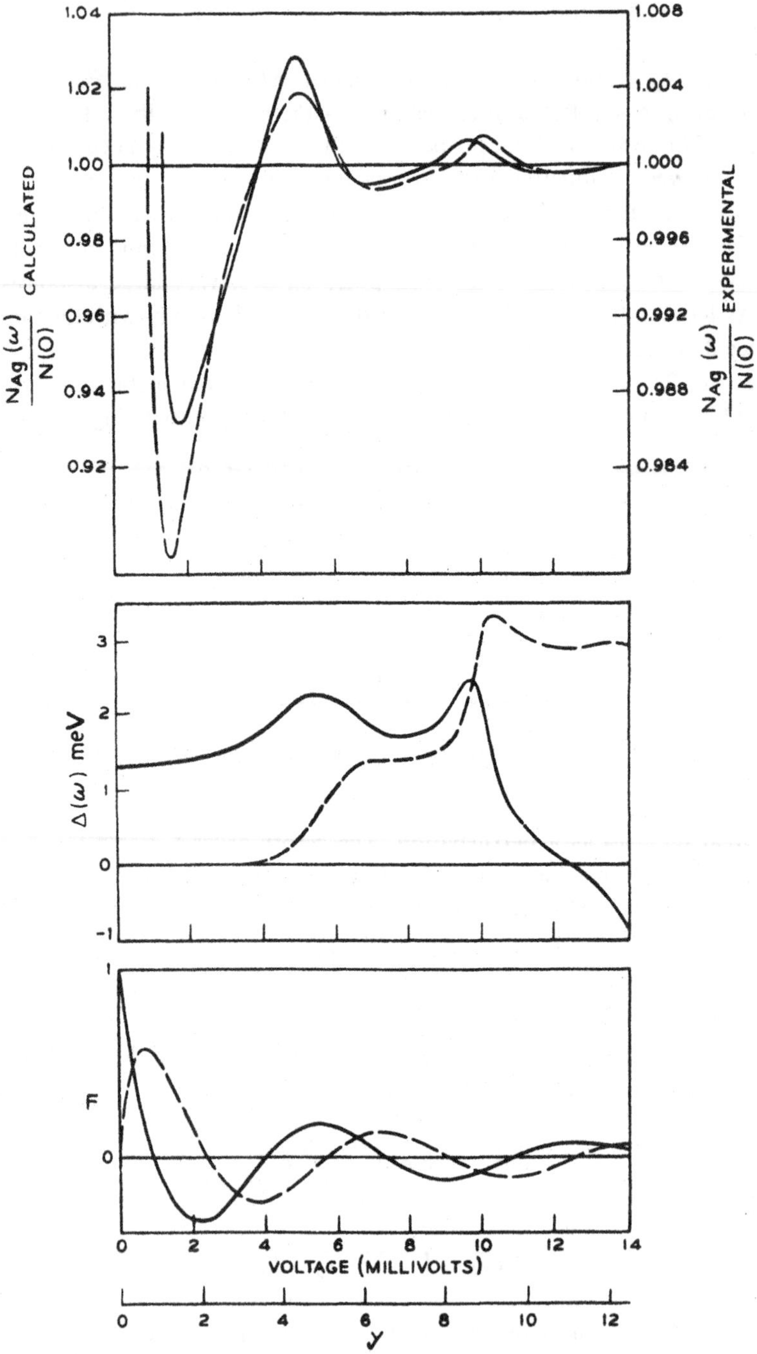

Fig. 11. Lower curves show the real (solid line) and imaginary (dashed line) parts of the interference term $F(y)$ plotted vs. $y(=4\omega d/\hbar v_F)$ and vs. voltage for $d=1460$ Å. The middle curves are the real (solid line) and imaginary (dashed line) parts of the gap function $\Delta(\omega)$ for lead. The upper solid curve is a measured tunneling density of states for a 1460-Å Ag film backed by 1010 Å of Pb. The upper dashed curve is a calculated density of states obtained from Eq. (45) using $F(y)$ and $\Delta(\omega)$ shown below.

1000-Å lead film. The lower graph shows the interference function, the middle graph the gap function for lead, and the upper graph the densities of states. The measured magnitude of the structure is about five times smaller than the calculated magnitude, owing, probably, to the neglect of impurity scattering in the calculation. The shapes of the curves are in good agreement for a number of film thicknesses verifying the form of Eq. (45) and exhibiting the interplay between the energy-dependent strong-coupled [$\Delta^2(\omega)$] and space-dependent [$F(y)$] aspects of the theory.

IV. EXPERIMENT

A. Sample Preparation

Considering the basic simplicity of the tunneling experiment, one might have expected that the technique would have been exploited more extensively than has been the case. Difficulties in the preparation of junctions seem to be common enough to explain this lack of progress.

The tunnel junction may be represented as in the upper part of Fig. 12, where two metals A and B are separated by insulator I to form an A–I–B junction.

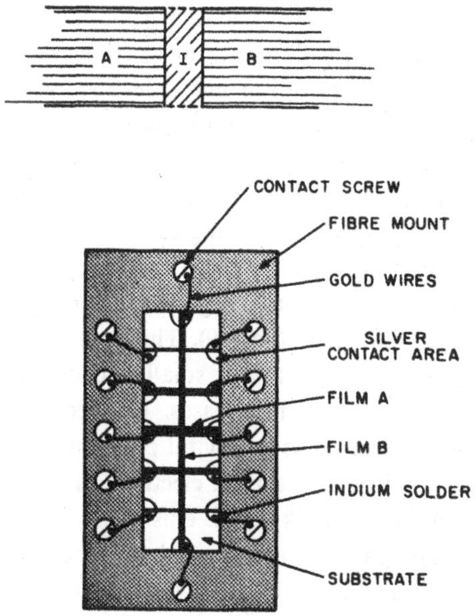

Fig. 12. Upper part, an abstraction of a tunnel junction; lower half, the actual appearance of the junctions as fabricated.

More realistically, the junctions, when comprised of thin films, appear as the lower part of Fig. 12. Throughout this chapter we adopt the convention that an A–I–B junction indicates that the insulator I is an oxide (generally of indeterminate composition) of the metal A. This all-important insulating layer I is generally prepared by the thermal oxidation of the first metal electrode A. If this electrode is a freshly evaporated film, the surface is clean and active and the chances of producing a thin uniform oxide layer are maximized. Aluminum is not only the easiest metal to oxidize but is also very convenient from many points of view for use in the tunneling measurements, as will become clear later in this chapter. Aluminum can generally be oxidized by exposure of the evaporated film to air from the laboratory; tin and lead films are oxidized by heating in air or oxygen. Variations in technique are considerable but some obvious general rules may be worth stating.

1. It is desirable to keep one evaporation station solely for the preparation of junctions. If the evaporator is contaminated with a low-vapor-pressure material, it is possible for this to partially cover the film before oxidation commences.

2. If a high-resistance (> 10 kΩ) "capacitor junction" can be prepared on the metal of interest, it is usually not difficult to vary time or temperature of oxidation to produce a tunnel junction of convenient resistance (1Ω–1 kΩ). Hence one's first aim with a new material is to see if a thick insulating layer can be produced by any variation of known oxidation procedures.

3. If junctions are consistently shorted, the junction area should be reduced to the minimum possible. Evaporation marks $\sim 25\ \mu$ wide can be made by etching if saw cuts are too wide ($\sim 150\ \mu$). Stainless steel is preferred as a mask material.

4. Junctions keep indefinitely when immersed in liquid nitrogen but only aluminum junctions survive frequent cycling to room temperature.

5. Junctions are easily destroyed by large currents or transients.

It has been reported that oxidation in a glow discharge is more reproducible and controlled than thermal oxidation (*49*). The number of people who have successfully used this technique is rather small and one doubts whether the extra equipment is a necessity in most cases. For recent studies (*50*) of the junction oxide, however, it seems to be a vital procedure.

The preparation of junctions on a bulk electrode is further complicated by the necessity of first cleaning the bulk metal. This was avoided by Zavaritskii (*51*) in his study of the energy gap in single-crystal tin by casting the tin plates in vacuum. The plate took a surface finish from the glass mold. The method could probably also be applied to lead. In the case of tantalum and niobium the bulk sample has been cleaned by heating in high vacuum (*52*) or by ion bombardment (*53*).

The completed sample, usually consisting of 3 to 5 junctions, is mounted on a holder and electrical contacts are made to both ends of each film. Pure indium will solder to many films; often it is smeared on the substrate before evaporation. Figure 12 shows indium solder connections of fine gold wires to silver contact areas which were evaporated onto the substrate before junction preparation. Silver paste or pressure contacts can also be used but the solder contacts seem to be best for low-level measurements.

After mounting, the sample is cooled to helium temperatures and for measurements below 4.2°K can be immersed directly into liquid helium.

B. Junction Evaluation

After producing junctions with a resistance suitable for measurement it is important to examine them closely before time and effort is expended in taking and interpreting detailed derivative plots. Junctions should therefore satisfy the following criteria.

1. In Fig. 12 the junctions made at one time have different areas. The junction resistance should be inversely proportional to area to $\pm 20\%$.

2. When a normal metal–insulator–superconductor junction is cooled well below the transition temperature, the conductance at zero bias in the superconducting state $(dI/dV)_S$ should be very small compared to that at zero bias in the normal state $(dI/dV)_N$. The ratio of these conductances (R) has been tabulated by Bermon (*54*). If the ratio is large it is questionable whether even a valid gap measurement can be made on such a junction. In fact, a large ratio is evidence for "gaplessness" or for the presence of a metallic short. We generally require that Al–I–Pb, Al–I–Sn, or Al–I–In junctions at 1°K have this ratio $R < 10^{-3}$; of course, the aluminum is also superconducting at this temperature. This implies that any current flowing through the junction by a "nontunneling" path is less than 10^{-3} of the tunneling current.

3. If the junction comprises two dissimilar superconductors 1 and 2, the cusp in the I–V characteristic at $\Delta_1 - \Delta_2$ should be reasonably sharp and the negative resistance region from $\Delta_1 - \Delta_2$ to $\Delta_1 + \Delta_2$ well defined. Smearing of this region usually occurs in junctions which do not satisfy requirement (2) above.

C. Measurement of the Transition Temperature and Energy Gap

After deciding that the junctions are sufficiently well behaved, the first measurements to be made are those of transition temperature and energy gap. The transition temperature should be measured on the film used in the junction. The best way to do this is to measure the ratio R as a function of temperature just below T_c and extrapolate to $R = 1$, which may be used to define T_c for tunneling purposes. This method gives a T_c which is very close ($< 10^{-2}$°K) to that found resistively for the same film (*55*).

Determination of the energy gap presents more difficulty than might be expected. The tunneling technique is the most direct method of measuring an energy gap but rather indiscriminate interpretation of tunneling characteristics has led to a spread of gap values for any one material which is beyond the limits of acceptable experimental error. A summary of some methods of determining the gap seems to be in order.

1. Figure 13 shows a simple construction which has been used to estimate the gap in a M–I–S junction, by determination of the voltage at which the conductance equals that of the normal state. This is a useful rough check at low temperatures but is in serious error at high temperatures. A study of Bermon's tabulation (*54*) of this conductance vs. Δ/kT and eV/Δ, and of Mühlschlegel's (*56*) of $\Delta(T)/\Delta(0)$ vs. T/T_c shows that for an Al–I–Pb junction at 6°K

$$\left(\frac{dI}{dV}\right)_N = \left(\frac{dI}{dV}\right)_S$$

at $V = 1.4\Delta(6°\mathrm{K})$, whereas at 1°K this occurs at $V = 0.92\Delta(1°\mathrm{K})$.

2. A useful and accurate method, the only one for a M–I–S junction at high temperatures near T_c, is to measure the conductance at zero bias $(dI/dV)_S$ as a

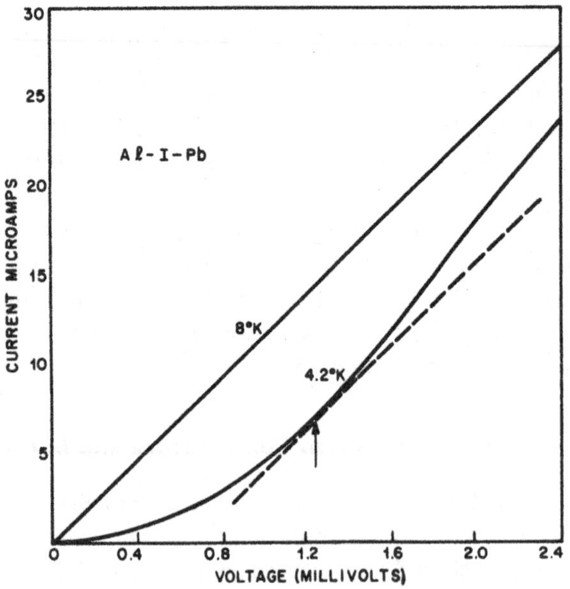

Fig. 13. Simple construction used to measure the energy gap of the superconductor in a M–I–S junction.

fraction of that in the normal state $(dI/dV)_N$. Then

$$R = \left(\frac{dI}{dV}\right)_S \Big/ \left(\frac{dI}{dV}\right)_N$$

at $V = 0$ depends on Δ/kT as in Eq. (17) of Chapter 3. This dependence of R on Δ/kT is available from the tables of Bermon, a plot has been made by Douglass and Falicov (*6*) (Chapter 3, Fig. 10). The expression [Eq. (17) of Chapter 3] has been derived assuming an exact BCS density of states in the superconductor. This method cannot, therefore, be used accurately for tunneling into superconductors which do not have this density of states dependence, such as superimposed layers used for proximity effect measurements or films with magnetic impurities. At low temperatures a background conductance due to metallic shorts or trapped magnetic flux (*57*) leads to difficulties but other methods are available in this range.

3. Junctions comprising two identical superconductors should have a discontinuous jump in current at a voltage 2Δ. In practice this is not observed and the value of 2Δ has to be taken from a current jump which is smeared by about $\frac{1}{15}(2\Delta)$ in the case of Pb–I–Pb. Gasparovic et al. (*58*) have used the construction shown in Fig. 11 of Chapter 3 for finding the minimum excitation energy which is in the strict sense the energy gap. For our purposes an "average gap" gives a better fit to the density of states just above the gap; further details will be given in Section V.

4. If the junction is comprised of two different superconductors 1 and 2, then the I–V characteristic should exhibit a cusp at $\Delta_1 - \Delta_2$ and discontinuous current jump at $\Delta_1 + \Delta_2$. Again these features are not perfectly sharp in real junctions. Douglass and Meservey (*59*) measured the smaller gap Δ_2 by drawing parallel lines tangent to the I–V trace near these voltages, but this is not accurate for the measurement of Δ_1, which is the gap of interest in our case.

5. As our junctions are often Al–I–S, where S is a superconductor of higher transition temperature and larger gap than Al, the following technique is useful for finding Δ_S. The current–voltage characteristic of the junction is monitored as the helium is slowly pumped down. At the aluminum transition temperature a very small sharp break appears in the curve exactly at Δ_S as the aluminum gap opens. By adjusting the bath temperature the location of this break can be made with considerable accuracy if the junction characteristic is sharp. By measuring both directions of bias any thermal emfs are eliminated. The gap Δ_S is now known at the transition temperature of the aluminum film which should be measured at the same time. Correction to zero temperature is made using the BCS dependence of Δ on T (*56*); for Pb, Sn, and In this correction is very small. This method appears to be useful for a study of small changes in the gap, for example, changes due to film thickness, strain, substrate material, alloying, and

grain size. However if the I–V characteristic is somewhat smeared, the aluminum gap appears only as a kink in the trace and choosing the point of minimum slope seems to give a gap value that is too small.

D. Electrical Measurements

1. Current–Voltage Characteristic

The simple circuit of Fig. 14 is used to display I–V characteristics on an XY-recorder, amplifiers with high imput impedance should be inserted on the X- and Y-axis if the junction resistance approaches the input impedance of the recorder.

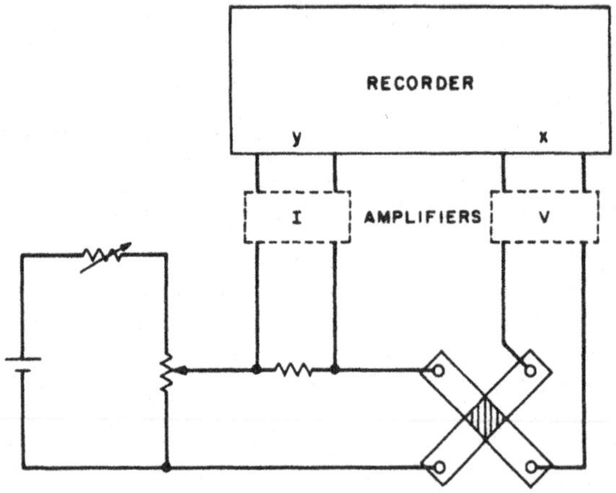

Fig. 14. Circuit to display the I–V characteristic of a junction on an XY-recorder.

2. Derivative Measurements

The first extensive application of derivative techniques in the study of current–voltage characteristics was made by Hall et al. (*60*) and by Chynoweth et al. (*61*) in their studies of semiconductor tunnel junctions. The power of the second derivative technique was realized by Chynoweth et al. (*62*) and exploited very successfully using equipment developed by Thomas. In the derivative circuits described by Hall‡ and Thomas and Klein (*63*), the modulation or sensing signal generally was too large for accuracy in superconductor tunneling, where the energy scale is smaller and finer structure is observed. However, more recent circuits used for semiconductor work (*64*) could easily be applied to this problem.

‡ The method used by Hall was suggested by Tiemann but apparently is unpublished [see (*60*)].

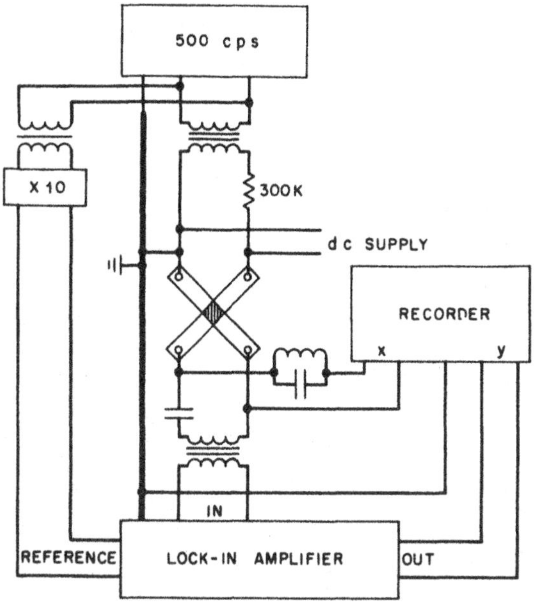

Fig. 15. Circuit to measure directly *dV/dI* of a junction by the ac modulation technique.

a. *First derivative measurement*

A dc bridge network has been used by Adler and Rogers (*65*) and described by Rogers et al. (*66*). The system is accurate and sensitive, but because of the use of a galvanometer amplifier it is slow and reduction of the raw data into normalized derivative curves extremely tedious.

The simplest way of obtaining a first derivative is to modulate the applied dc bias with a small constant current ac signal δI and to pick up the generated voltage δV at the junction with a lock-in amplifier. The signal δV should be $\sim 50\ \mu V$ rms for junctions at 1°K. This measures dV/dI rather than the conductance dI/dV but as all measurements have to be normalized the extra effort of taking the reciprocal is small. A suitable circuit is shown in Fig. 15, the only precautions necessary are to avoid ground loops and to make sure that all connections to the junction are high impedance to both ac and dc. By using a stable amplitude oscillator and a good lock-in amplifier, measurements of dV/dI can be made with an accuracy ~ 1 in 10^3. Sensitivity is not the main difficulty but rather drift in the oscillator or amplifier. This is important because the derivative trace is first taken with the film in the superconducting state and then in the normal state. As will be seen later, these two measurements at any particular voltage may be taken as much as 1 hr apart, and it is the *normalized* derivative which must be known accurately. To make full use of the gap inversion techniques to be described later the accuracy of the first derivative measurement should be approximately 1% of the density of states deviations from the BCS dependence. An approximate estimate of whether an experiment is possible can be made by noting that the strength of the strong coupling effects—the deviations from BCS—are proportional to $(T_c/\Theta_D)^2$. Table I gives the transition temperature, Debye temperature, and strength of the deviation for a number of superconductors. The deviations have been normalized to the 5% value measured for Pb.

TABLE I

Expected Deviation of the Tunneling Density of States from the BCS Density of States

	T_c, °K	Θ_D, °K	Deviation, %
Pb	7.2	96	5
Hg	4.15	69	3
La	6.0	132	2
In	3.4	109	1
Tl	2.4	88	0.75
Sn	3.7	195	0.5
Zn	0.88	235	0.01
Ga	1.1	317	0.01
Al	1.2	420	0.008
Cd	0.54	300	0.003
Nb	9.5	320	0.8
Ta	4.5	255	0.25
V	5.3	338	0.25

It can be seen that for elements other than Pb and Hg direct measurement of dV/dI will not be accurate enough to utilize gap inversion because of drift, and for the weaker-coupled superconductors a bridge circuit must be used. If this is well designed a large voltage is subtracted from the signal from the junction and the oscillator and amplifier drift affects only the difference rather than the full signal. A suitable circuit has been described by Adler and Jackson (*67*) which conveniently measures both first and second derivatives with a minimum of alteration. The circuit used by the authors is a bridge circuit designed by D. E. Thomas; it is shown schematically in Fig. 16. The performance of this network seems to be at least as good as that used by Adler and conductance measurements approaching a stability of 1 part in 10^5 over a period of 1 hr are possible.

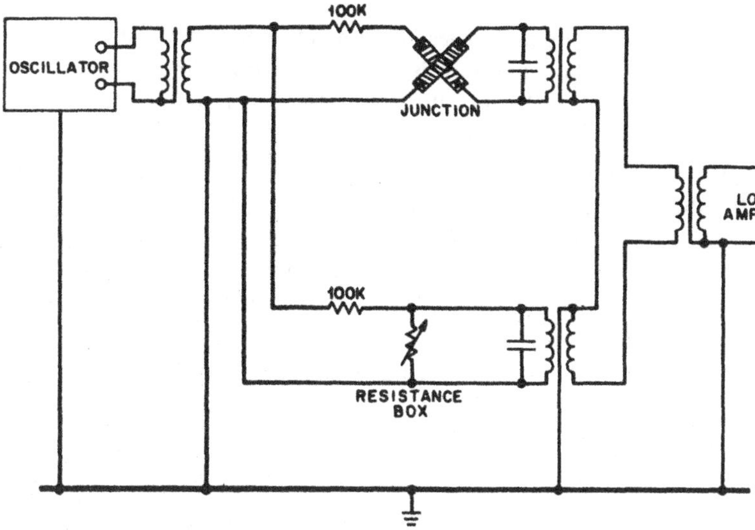

Fig. 16. Ac bridge circuit used to measure dV/dI of a junction.

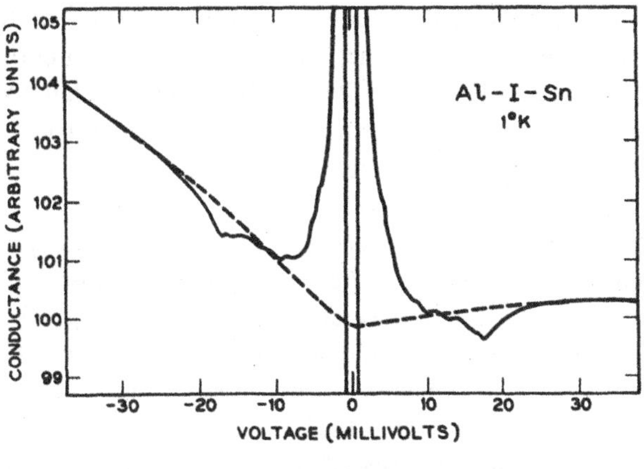

Fig. 17. Conductance of an Al–I–Sn junction in the normal (----) and superconducting (——) states.

The normalization of the first derivative measurement is made by taking the ratio of the conductance in the superconducting state to that in the normal state at each voltage. An example of the raw data for the case of an Al–I–Sn junction is shown in Fig. 17. The figure vividly illustrates (1) the necessity of measuring the normal state with an accuracy equal to that of the superconducting state, (2) the interesting fact that aluminum junctions are quite asymmetrical at low biases, (3) that the Al negative direction is the most convenient for normalization purposes, and (4) that the normal conductance cannot be assumed linear in voltage. These normal-state properties of the junction have been studied by the authors (*68*); the important question is whether the tunneling matrix element $|T|^2$ remains unchanged in the transition from the superconducting to normal states. This point will be discussed in more detail later.

The normal state is obtained by the application of either a magnetic field or increase in temperature above T_c. Because of changes in the normal state with temperature, magnetic field is preferable. It is important to watch the junction conductance at H_c or T_c; occasionally an abrupt change in conductance occurs as the films go normal. This is probably due to the lack of a perfect four-terminal connection to the junction (*7,69*); that is, the resistances of the films of the junction in the normal state is not negligible compared to the insulator resistance. In practice, the voltage on the junction is increased to $\sim 6\Theta_D$ (60 meV for Pb) and any conductance change is measured as the films go normal. This small change is applied as a correction to make the conductances equal in the superconducting and normal states at this high bias.

b. *Second derivative measurement*

The first measurements of the second derivative d^2I/dV^2 of the tunneling characteristics of superconductor junctions were obtained using the equipment described by Thomas and Klein (*63*). Although the sensing signal was too large by today's standards the observation of the structure arising from critical points in the lead transverse phonon peak created great interest and encouraged development of the technique. The method now in use is to detect the harmonic of the applied ac modulation using a lock-in amplifier. Circuits to accomplish this have been published by Adler (*67*) and by Thomas and Rowell (*70*). Data on a Pb–I–Pb junction are shown in the results of Section VI (Fig. 20). The applied 500-Hz signal must be kept small and was 20 μV rms. The criticism could be made that the circuit of Thomas and Rowell uses a two-terminal connection to the junction but it is unlikely that the leads and connections to the junction are nonlinear in the second derivative of current with respect to voltage. In any case, the second derivative plot is only used to locate fine structure and not to determine magnitudes. The authors' greatest difficulty in taking derivative traces was the problem of external interference, probably airborne, although the 110-V supply and dc ground connection can also carry rf disturbance. The interference usually appears as sudden spikes in the derivative trace, especially the second derivative. These spikes cannot be eliminated conveniently by using long time constants in the circuitry. The problem became impossible and a shielded room was installed to enclose the cryostats and electronics; now no external interference is ever observed. In large institutions, where no thought is given to the effect of rf furnaces in one room on low-level experiments next door, such an installation seems to be a necessity.

E. Data Handling

To summarize the experimental technique, the measurements to be made are of the energy gap and transition temperature of the superconducting film and the first and second derivatives of the tunneling characteristic of the junction in both superconducting and normal states as a function of voltage. This information is generally on *XY*-recorder sheets but for ease of handling on the computer the authors record the derivative data on punched paper tape. This is done following a method suggested to us by D. E. Thomas.

The dc bias on the junction is swept at a rate which is almost independent of the junction resistance. As an example, the first derivative signal from the lock-in amplifier (dc output) is changed to a frequency which is proportional to the dc level. This frequency is counted for 10 sec and the total count punched on tape. Thus the first derivative signal is averaged over small voltage intervals corresponding to 10 sec of the bias sweep. The noise is integrated with a 10-sec time constant, and as a short time constant can now be used on the lock-in amplifier the effect of transient signals can be minimized. Generally 300 points are taken across the derivative trace which corresponds to a voltage grid of $\sim 50\ \mu V$ for an Al–I–Pb junction. The second derivative is taken with exactly the same number of points at the same voltages. Thus the value of both derivatives at all the grid points is known. The normal state first and second derivative characteristics are taken with a coarser grid. It is essential to note that if a magnetic field is applied to produce the normal state, then the sample must be briefly warmed above T_c in zero field to eliminate trapped flux before any further superconducting results are taken. The paper tape is read onto cards which can be used for the computer analysis to be discussed below.

V. INVERTING THE GAP EQUATION

In this section we describe the numerical procedures used to extract from the tunneling measurement the parameters, $\alpha^2F(\omega)$ and μ_c^*, which enter the Eliashberg equations. We first describe the method for solving the integral equations for a given set of parameters and then discuss the technique of adjusting the parameters to fit the measured electronic density of states.

The integral equations for the normal and pairing self-energies of a dirty superconductor are

$$\xi(\omega) = [1 - Z(\omega)]\,\omega = \int_{\Delta_0}^{\infty} d\omega' \,\mathrm{Re}\left[\frac{\omega'}{(\omega'^2 - \Delta'^2)^{1/2}}\right] \times \int d\omega_q\, \alpha^2(\omega_q)\, F(\omega_q)\, [D_q(\omega' + \omega) - D_q(\omega' - \omega)] \qquad (47)$$

$$\phi(\omega) = \int_{\Delta_0}^{\omega_c} d\omega' \, \mathrm{Re}\left[\frac{\Delta'}{(\omega'^2 - \Delta'^2)^{1/2}}\right]$$
$$\times \left\{ \int d\omega_q \, \alpha^2(\omega_q) \, F(\omega_q) \, [D_q(\omega' + \omega) + D_q(\omega' - \omega)] - \mu^* \right\} \tag{48}$$

where $D_q(\omega) = (\omega + \omega_q - i0^+)^{-1}$, $\Delta(\omega) = \phi(\omega)/Z(\omega)$, and $\Delta_0 = \Delta(\Delta_0)$. $F(\omega)$ is the phonon density of states

$$F(\omega) = \sum_{\lambda} \int \frac{d^3q}{(2\pi)^3} \, \delta(\omega - \omega_{q\lambda}) \tag{49}$$

and $\alpha^2(\omega)$ is an effective electron–phonon coupling function for phonons of energy ω:

$$\alpha^2(\omega) \, F(\omega) = \int_S \frac{d^2p}{v_F} \int_{S'} \frac{d^2p'}{(2\pi)^3 \, v'_F} \sum_{\lambda} g^2_{pp'\lambda} \delta(\omega - \omega_{p-p'\lambda}) \Big/ \int_S \frac{d^2p}{v_F} \tag{50}$$

where $g^2_{pp'\lambda}$ is the dressed electron–phonon matrix element, $\omega_{q\lambda}$ is the phonon energy for polarization λ and wave number q (reduced to the first zone), and v_F is the Fermi velocity. The two surface integrations are performed over the Fermi surface.

For the uniform isotropic superconductor the electronic density of states (which is measured directly in the tunneling experiment) is

$$N(\omega) = \mathrm{Re}\left[\omega/\sqrt{\omega^2 - \Delta^2(\omega)}\right] \tag{51}$$

The density of states is normalized to unity in the normal state.

The integral equations can be solved conveniently by the straightforward procedure of simultaneously iterating both equations. To start the iteration we make a zeroth-order guess for the gap function:

$$\Delta^{(0)}(\omega) = \begin{cases} \Delta_0 & \omega < \omega_0 \\ 0 & \omega > \omega_0 \end{cases} \tag{52}$$

where Δ_0 is the measured energy gap and ω_0 is the maximum phonon frequency. Using this gap function in the right side of Eqs. (47) and (48) we perform the indicated integrations to find $\xi^{(1)}(\omega)$ and $\phi^{(1)}(\omega)$ and gap function $\Delta^{(1)} = \phi^{(1)}/Z^{(1)}$. This procedure is repeated until $\Delta^{(n)}(\omega)$ converges and we find that after six to eight iterations $\Delta(\omega)$ has converged to three decimal places. We compute $N(\omega)$ from $\Delta(\omega)$ using Eq. (51).

We now have a straightforward and accurate method for calculating $N(\omega)$ from an arbitrary $\alpha^2F(\omega)$ and μ^*. Our next task is to devise a method for adjusting $\alpha^2F(\omega)$ and μ^* so that the calculated density of states $N_c(\omega)$ will fit the experimentally measured quantity $N_e(\omega)$. To accomplish this we begin by choosing a zeroth-order $\alpha^2F^{(0)}$ and $\mu^{*(0)}$ and calculate the density of states $N_c^{(0)}(\omega)$. In addition, we compute the linear response of $N_c(\omega')$ to a small change in $\alpha^2F(\omega)$, that is, the functional derivative $\delta N_c(\omega')/\delta\alpha^2 F(\omega)$. We can now calculate the change in α^2F necessary to fit the experimental density of states:

$$\delta\alpha^2F(\omega) = \int d\omega' \left[\frac{\delta N(\omega')}{\delta\alpha^2F(\omega)}\right]^{-1} [N_e(\omega') - N_c^{(0)}(\omega')] \tag{53}$$

The first-order approximation to $\alpha^2F(\omega)$ is

$$\alpha^2F^{(1)}(\omega) = \alpha^2F^{(0)}(\omega) + \delta\alpha^2F(\omega) \tag{54}$$

Since the equations are not linear we must iterate this procedure until $\alpha^2F^{(n)}(\omega)$ converges, that is, until we find an $\alpha^2F(\omega)$ which exactly reproduces the measured density of states when substituted in the gap equation. The Coulomb term μ^* is adjusted at each stage of the iteration so that the calculated energy gap $\Delta_0 = \Delta(\Delta_0)$ agrees with the measured energy gap. In addition to finding $\alpha^2F(\omega)$ and μ^* we find from the solution of the integral equation the self-energies, or $\Delta(\omega)$ and $Z(\omega)$, for the superconductor at zero temperature. We easily compute $Z(\omega)$ in the normal state from Eq. (47) with $\Delta(\omega) = 0$.

VI. RESULTS

Having discussed the theory, the experiment, and the numerical fitting procedure, we are now ready to put the three together and present some results. Our aims are fourfold.

1. We first want to fit the experimental results for Pb and obtain accurate parameters $\alpha^2F(\omega)$ and μ^*; as a by-product we will find the electron self-energies in the normal and super states.
2. Second, we will discuss the reasonableness of the parameters which we have obtained and, further, we will calculate several properties which have not been used in the fit and compare these properties with experiment.
3. Third, we want to put the Eliashberg equations through a stringent experimental test to determine whether the theory is as accurate as we believe it to be.
4. Finally, we will briefly present results on a number of other materials.

A. Information Obtained Directly from the Gap Equation Inversion—Lead

Our most extensive and accurate work has been on lead and we will present a detailed account of the experiments and calculations for that metal. We first measure the current I vs. voltage V for a Pb–I–Pb tunnel junction, which according to Eq. (38) is given by

$$I(V) = \frac{1}{R}\int_{\Delta_0}^{V-\Delta_0} N(\omega)\,N(V-\omega)\,d\omega \tag{55}$$

where $N(\omega)$ is the normalized electronic density of states in the superconducting state,

$$N(\omega) = \mathrm{Re}\,[\omega/\sqrt{\omega^2 - \Delta^2(\omega)}] \tag{56}$$

R is junction resistance in the normal state, and $\Delta_0 = \Delta(\Delta_0)$ is the energy gap. The density of states vanishes for $\omega < \Delta_0$ and for ω somewhat greater than Δ_0 is given by

$$N(\omega) \approx \frac{\omega}{\sqrt{\omega^2 - \Delta_0^2}}\left(1 + \frac{1}{2}\frac{\partial \Delta}{\partial \omega}\bigg|_{\omega=\Delta_0}\right) \tag{57}$$

which is just the BCS density of states enhanced by the strong-coupling factor in parentheses, which is ~ 1.025 for lead. According to Eq. (55), the current should vanish for $V < 2\Delta_0$, jump to the value

$$I(2\Delta_0) = \frac{2\Delta_0}{R}\frac{\pi}{4}\left(1 + \frac{1}{2}\frac{\partial \Delta}{\partial \omega}\right)^2 \tag{58}$$

at $V = 2\Delta_0$, and slowly approach the normal-state current at higher voltage. The experimental current jump is somewhat rounded but the midpoint of current jump, $V = 2.79 \pm 0.01$ meV, provides an accurate measurement of the average energy gap and the magnitude of the current jump,

$$I_{\text{jump}} = \frac{2\Delta_0}{R}\frac{\pi}{4}(1.05)$$

measures the density of states at the gap edge. Figure 18 shows the construction used to find this midpoint, we extrapolate the I–V characteristic from voltages somewhat greater than Δ, where there is an extended region of almost constant slope and find the voltage (2.793 ± 0.005) meV, where the junction current is half the extrapolated current. The construction used by Gasparovic et al. (*58*) is also shown in Fig. 18 and gives a minimum excitation energy of (2.707 ± 0.005) meV. This measurement is made for each junction for which derivative data is taken.

The measurement of dI/dV (Fig. 19) by the ac modulation technique exhibits the structure in the density of states at typical phonon energies, the sharp drop at $V \sim 7$ meV (11 meV) being due to transverse (longitudinal) phonons. This voltage is measured from zero and the phonon energy corresponds to this value minus 2Δ. We require for the analysis an accurate measurement of the first

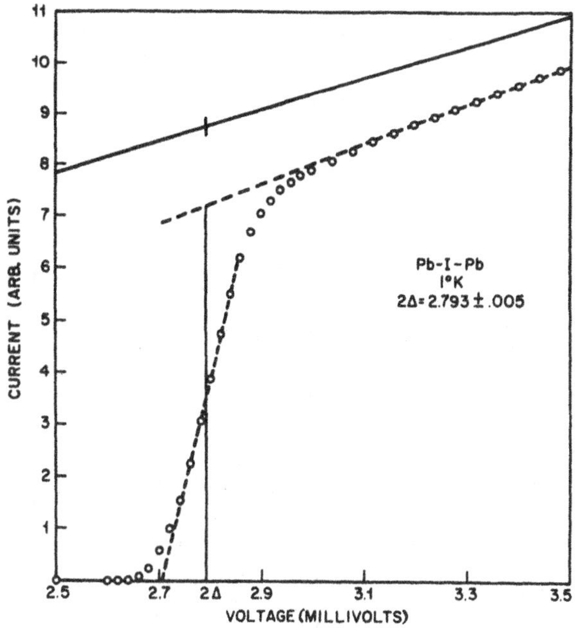

Fig. 18. *I–V* characteristic of a Pb–I–Pb junction showing the construction used to find the energy gap. The solid line and open circles are the current in the normal and superconducting states, respectively.

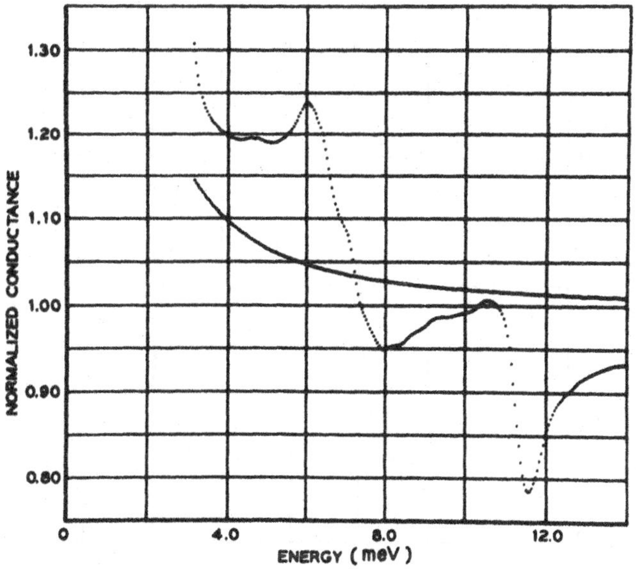

Fig. 19. Conductance dI/dV of a Pb–I–Pb junction in the superconducting state normalized by the conductance in the normal state vs. voltage. Also shown is the two-superconductor conductance calculated from the BCS density of states which contains no phonon structure.

derivative in the super state relative to the first derivative in the normal state. This ratio deviates from unity by 20% for lead in the fundamental phonon region and a direct four-terminal measurement with a precision of ± 0.001 is sufficient. The first derivation is reproducible from sample to sample within ± 0.001. The second derivative (Fig. 20) shows considerable structure which will be reflected in the $\alpha^2 F(\omega)$.

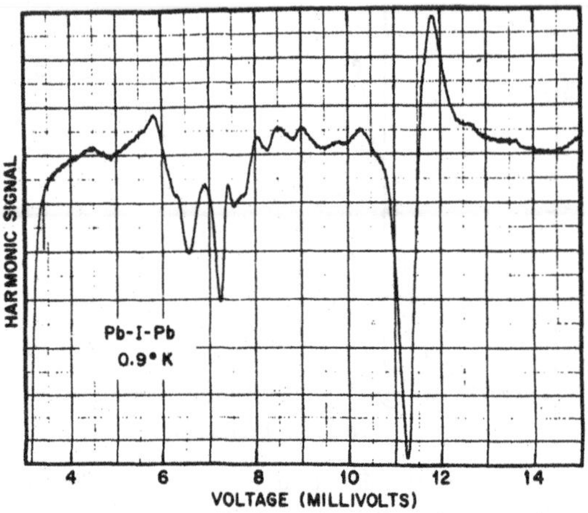

Fig. 20. Second derivative d^2I/dV^2 (actually the second harmonic signal) for a Pb–I–Pb junction in the superconducting state vs. voltage.

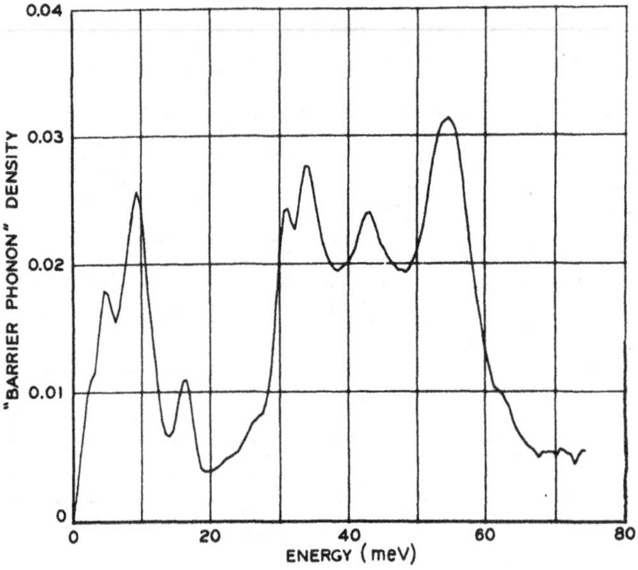

Fig. 21. Second derivative d^2I/dV^2 for a Pb–I–Pb junction in the normal state which is proportional to the phonon density of states in the barrier region.

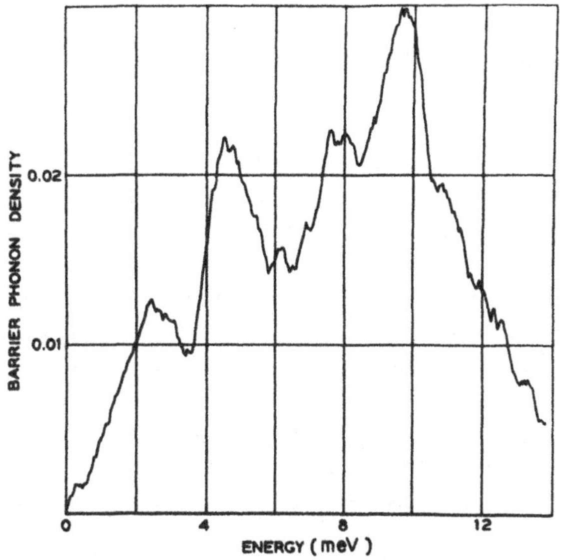

Fig. 22. Barrier phonon density of states at low energy for a Pb–I–Pb junction exhibiting the peaks at the longitudinal and transverse phonon energies of lead.

The simple tunneling theory of Section III predicts that dI/dV in the normal state should be constant within a few parts in 10^3. In fact we observe that, for Pb–I–Pb junctions, the first derivative increases by 5% up to 60 meV. The increase is not a smooth ($\propto V^2$) term which could arise from change in shape of the barrier but has well-defined kinks indicating that a new tunneling process, the emission of phonons in the barrier region, is taking place (*50,71*). If there is a group of phonons of energy ω_0 in the barrier, tunneling with the emission of these phonons can take place for $|V| > \omega_0$, and we expect a step function increase in dI/dV at $V = \omega_0$. In fact, d^2I/dV^2 in the normal state is proportional to the phonon density of states in the barrier times some electron–phonon interaction matrix elements squared. For a Pb–Pb junction, this second derivative (Fig. 21) shows a complex series of well-defined peaks characteristic of the lattice vibrations of the insulating layer, which is some oxide of lead. This well-defined, complex spectrum suggests that the oxide is formed from complex lead oxide molecules, possibly Pb_3O_4, and that we are observing the internal vibrations of these molecules. We observe, as well (Fig. 22), peaks very close to the longitudinal and transverse phonon energies of the metal, which could result from the coupling of the surface monolayer of metal atoms into the tunneling process. It is a simple matter to correct the superconducting tunneling results for this additional process and we do this. Imagine an electron tunneling between two normal metals with emission of a phonon of energy ω_b, obviously this process can only occur for $eV \geqslant \omega_b$. If one metal now becomes superconducting with energy gap Δ the process occurs for $eV \geqslant \omega_b + \Delta$ and is strongly peaked for electrons tunneling into the peak in density of states at Δ. Thus the correction of the normalization takes account of these two factors. The approximation of using the ratio of the first derivatives at a given voltage is in error by only 1 part in 10^3, which is negligible for Pb but not for the weaker-coupling materials, Sn and In, which will be mentioned at the end of the chapter.

To obtain $N(\omega)$ we use the I vs. V and dI/dV measurements to obtain a precise $I(V)$ containing all of the phonon structure and unfold the convolution integral [Eq. (55)] numerically. [In the case of the weaker-coupling superconductors, where the measurements of dI/dV are not so precise, the d^2I/dV^2 result is combined with dI/dV to obtain the precise $I(V)$ which contains all the phonon fine structure. In the case of Pb we discovered the dI/dV measurement was accurate enough to generate all the details of the phonon density without using the measured d^2I/dV^2.] The density of states obtained in this way is shown in Fig. 23 and is in good agreement with the density of states obtained from Al–I–Pb junctions (with the aluminum superconducting) and from normal metal–I–Pb junctions. In the case of the Al–I–Pb junction we assume a BCS density of states in the aluminum and allow for thermal excitation by putting a delta function of quasi-particles above the gap. The normal metal–I–Pb junctions measure the density of states except for thermal smearing and the plots only need to be normalized. There are several advantages to the use of Pb–I–Pb junctions:

1. The measurements are made at 1°K, so that $T/T_c \approx 0.15$, $\Delta \sim 14\, kT$; hence

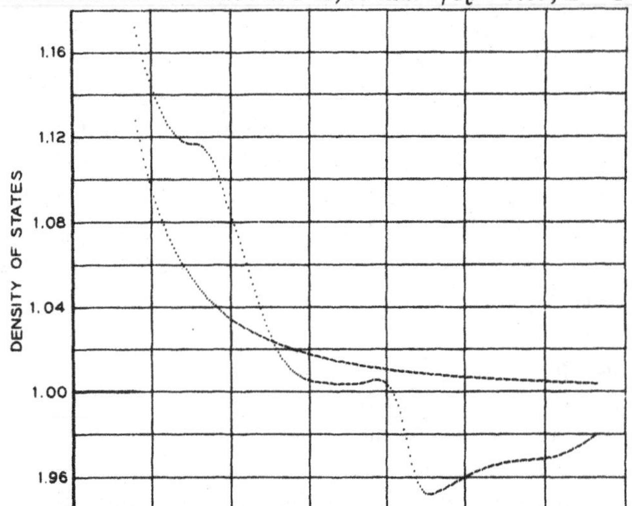

Fig. 23. Electronic density of states $N(E)$ for lead vs. $E - \Delta_0$ obtained from the data of Fig. 19. The smooth curve is the BCS density of states.

there is no smearing of the results due to thermal excitation.

2. The structure is more than a factor of 2 larger than in the Al–I–Pb case and is more easily measured.

3. dI/dV is less singular for energies just above the gap edge and the density of states at low energy can be measured accurately.

4. The energy gap can easily be measured accurately.

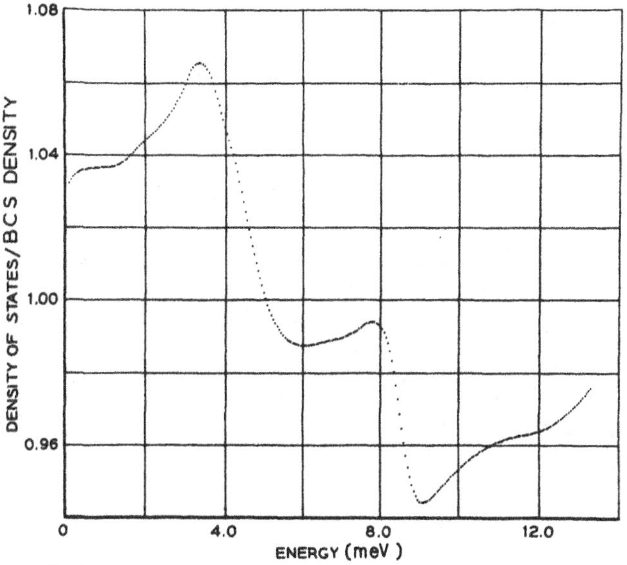

Fig. 24. Electronic density of states of Pb divided by the BCS density of states vs. $E-\Delta_0$.

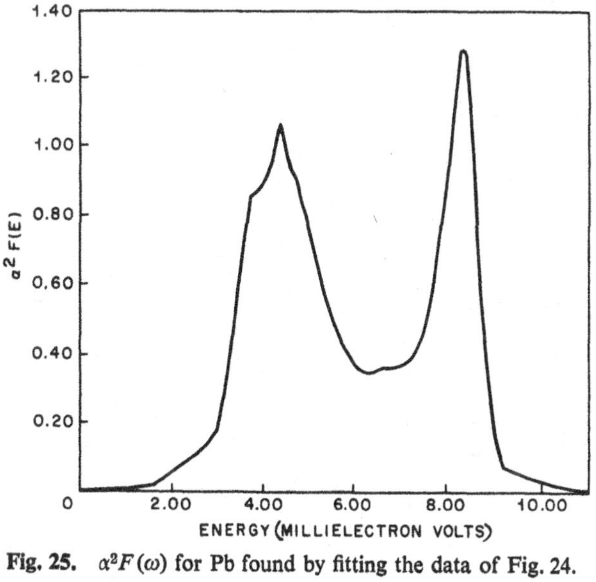

Fig. 25. $\alpha^2F(\omega)$ for Pb found by fitting the data of Fig. 24.

In Fig. 24 we plot the measured density of states divided by the BCS density of states

$$N_{BCS}(\omega) = \omega/\sqrt{\omega^2 - \Delta_0^2} \tag{59}$$

which shows that the deviations from BCS are of order 0.05. This is the information that is actually fed into the computer. The gap value Δ_0 which is used in Eq. (59) is the "average gap" as found above.

The first result of inverting this $N(\omega)/N_{BCS}(\omega)$ information is shown in Fig. 25. This is $\alpha^2F(\omega)$, which is determined by fitting to the measured $N(\omega)/N_{BCS}(\omega)$ over the energy range 1.6–11 meV. $\alpha^2F(\omega)$ is assumed to be proportional to ω^2 for $\omega < 1.6$ meV and to vanish for $\omega > 11$ meV. The first encouraging result is that $\alpha^2F(\omega)$ exhibits the two peaks in the phonon density of states for longitudinal and transverse phonons expected for the fcc lattice. Note, in addition, that the Van Hove critical points in the transverse region of the spectrum, which are observed very easily in the d^2I/dV^2 plot (Fig. 20), are faithfully reproduced in $\alpha^2F(\omega)$ as discontinuities in slope. In the longitudinal region it appears that the critical points are too close in energy to be resolved. From the same fitting procedure the Coulomb term μ^* is found to be 0.12, which is in good agreement with the theoretical estimate of 0.11. We have found that the value obtained for μ^* is very sensitive to the accuracy of the experimental density of states determination.

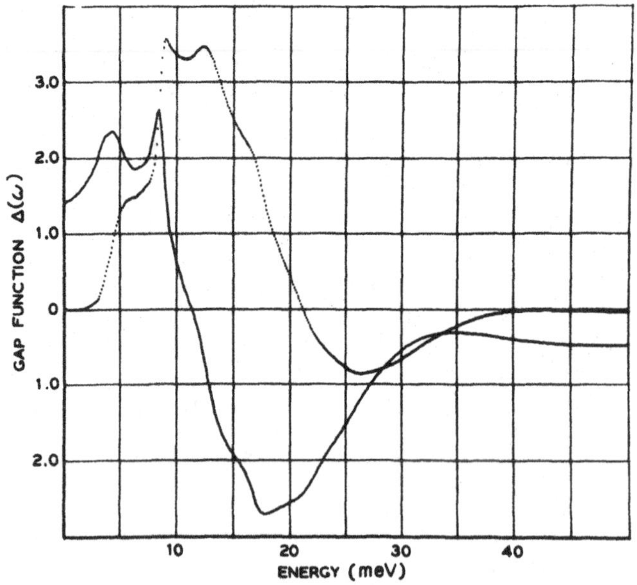

Fig. 26. Real (——) and imaginary (···) parts of the computed energy gap function $\Delta(\omega)$ for Pb vs. $\omega - \Delta_0$.

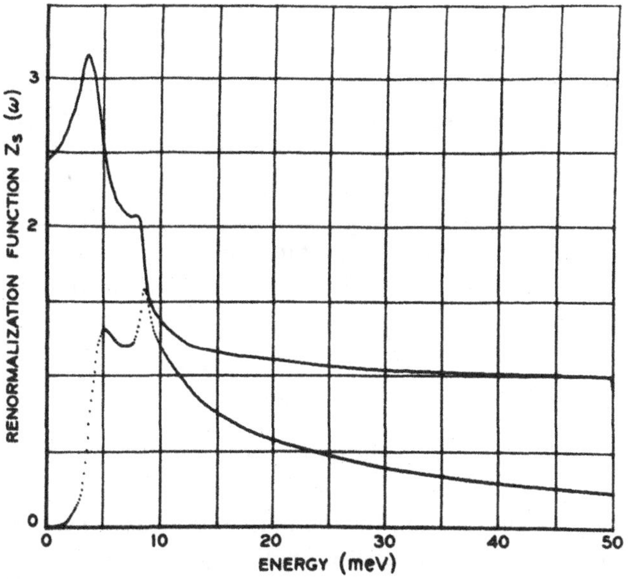

Fig. 27. Real (——) and imaginary (···) parts of the computed renormalization function $Z_s(\omega)$ for superconducting Pb vs. $\omega - \Delta_0$.

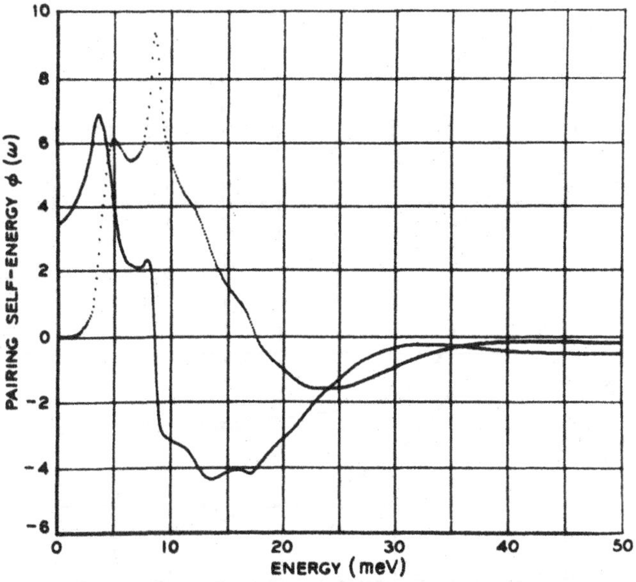

Fig. 28. Real (——) and imaginary (···) parts of the computed pairing self-energy $\Phi(\omega)$ for Pb vs. $\omega - \Delta_0$.

It is apparent from Sections II and IV that in the course of obtaining an $\alpha^2F(\omega)$ and μ^* which exactly reproduce the measured density of states we must also obtain a solution for the renormalized complex gap function $\Delta(\omega)$, the renormalization function $Z_s(\omega)$ and also the pairing self-energy $\Phi(\omega)$. The energy gap function $\Delta(\omega)$ obtained in this way is shown in Fig. 26. Of particular interest is the behavior of the real and imaginary parts of $\Delta(\omega)$ near the fundamental phonon frequencies. The real part has peaks at these energies implying strong coupling while the imaginary part increases at these energies as damping due to phonon emission increases. The renormalization function $Z_s(\omega)$ and the pairing self-energy $\Phi(\omega)$ shown in Figs. 27 and 28 also show strong structure at the fundamental phonon energies. Note also that by using experimental results only in the region of the phonon energies we have uniquely determined these parameters up to high energies where multiphonon processes give rise to rather smeared structures. We will return to this important region of higher energies below.

The renormalization function for the normal state $Z_n(\omega)$ can be calculated from the information discussed above simply by putting $\Delta = 0$ in Eq. (47). This result is shown in Fig. 29. It may be worth mentioning that Figs. 23 through 28 (excepting Fig. 25) are exact reproductions of the computer output and that the results of the program are conveniently presented to us in this form.

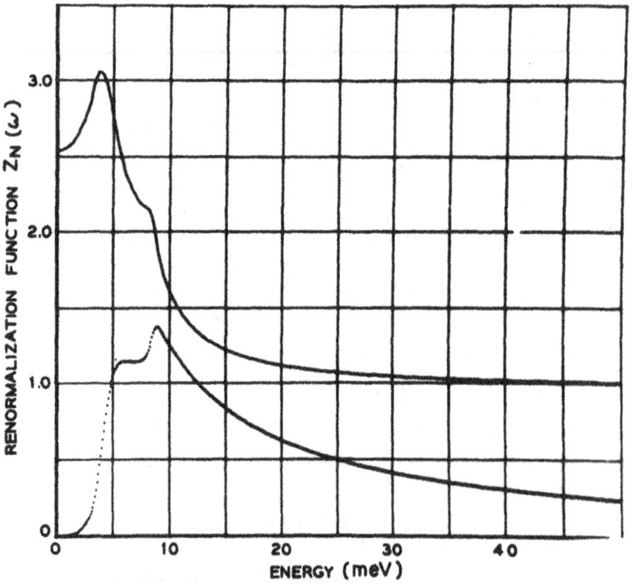

Fig. 29. Real (——) and imaginary (···) parts of the computed renormalization function $Z_n(\omega)$ for Pb in the normal state vs. ω.

B. Discussion of the Parameters and Comparison with Other Experiments

We now wish to examine the results obtained in Section VI.A to check whether there is any disagreement with relevant information that can be deduced from other experiments.

We have found a good fit to the measured electronic density of states by a judicious choice of the parameters entering the Eliashberg equations. The density of states calculated from the $\alpha^2F(\omega)$ of Fig. 25 with $\mu_c^* = 0.12$ agrees with the measured density of states within ± 0.0005 over the energy range $1.6\ \text{meV} < \omega - \Delta_0 < 11\ \text{meV}$. The energies of the transverse and longitudinal peaks in $\alpha^2F(\omega)$ are in reasonable agreement with the energies one expects from the dispersion curves obtained by inelastic neutron scattering. Where the singularities in $\alpha^2 F(\omega)$ can be identified with a critical point in a particular phonon branch the agreement with the neutron measurements is excellent (*34*). According to the dispersion curves measured along symmetry directions the longitudinal branch is very nearly spherically symmetric near its maximum at $\omega \approx 9$ meV. We are a little puzzled by the small tail on $\alpha^2F(\omega)$ extending to 11 meV, well above what we believe to be the maximum phonon energy for bulk lead. We note that the tunneling experiment at this energy probes metal properties within ~ 500 Å of the tunneling surface. The frequencies of the atoms within a few atomic layers of the tunneling surface are disturbed by the presence of the surface and we would expect to see a small contribution to $\alpha^2F(\omega)$ which is characteristic of the surface rather than the bulk phonons. As can be seen from Fig. 22, which we believe represents the surface and oxide phonon density, there is a strong peak at ~ 10 meV which might account for this tail beyond 9 meV.

The final reasonable direct result is that the Coulomb term is in good agreement with the theoretical estimate.

Electrons in the normal metal near the Fermi surface are dressed with a cloud of "virtual phonons" and this dressed electron has a Fermi velocity $v_F = v_F^0/Z_n(0)$ which is reduced from the band structure velocity v_F^0 by the factor $Z_n(0) = 1 + \lambda$, where λ is the dimensionless electron–phonon interaction strength:

$$\lambda = \int_0^\infty 2\frac{\alpha^2F(\omega)\,d\omega}{\omega} \tag{60}$$

For lead we calculate $\lambda = 1.5$. The electronic heat capacity and cyclotron mass are enhanced by this same factor, $Z_n(0) = 2.5$.

We can get independent estimates of the coupling constant λ from the enhancement of the electronic heat capacity and cyclotron masses over the calculated band structure values. Anderson and Gold (*31*) fit their de Haas–van Alphen measurements of the Fermi surface of lead with the pseudo-potential model and calculated the electronic density of states at the Fermi surface and the cyclotron masses of several orbits. The experimental electronic heat capacity is a factor of 2.3 larger than the band structure value and the experimental cyclotron masses are 2.0 to 2.4 times larger than the calculated masses. We find 2.5 for this enhancement factor. We should note that the de Haas–van Alphen measurements determine only the dimensions of the Fermi surface and not the energy scale. In this pseudo-potential calculation the energy scale was set by choosing the mass in the kinetic energy term to be the free electron mass. In actual fact, the effective mass may differ somewhat from the free mass and this will affect the above estimate of the enhancement factor.

We can calculate the condensation energy at zero temperature using the expression derived by Bardeen and Stephen (*72*):

$$E_{\text{cond}} = \operatorname{Re} N(0)\left[\int_0^\infty (Z_S + Z_n)\left(\omega - \sqrt{\omega^2 - \Delta^2} - \frac{\Delta^2}{2\sqrt{\omega^2 - \Delta^2}}\right) d\omega - \int_0^\infty (Z_n - Z_S)\frac{\Delta^2}{2\sqrt{\omega^2 - \Delta^2}}\, d\omega\right] \tag{61}$$

If we neglect the difference in Z between the normal and super states and, further, neglect the energy dependence of Z and Δ, we find the BCS expression for the condensation energy,

$$E_{\text{cond}}^{\text{BCS}} = \tfrac{1}{2} N(0)\, Z_n(0)\, \Delta_0^2 \tag{62}$$

where $N(0)\, Z_n(0)$ is the density of states for fully dressed electrons which can be determined from the electronic heat capacity. Using Eq. (61) and the Z_S, Z_n, and Δ in Figs. 26–29, we find

$$E_{\text{cond}} = 0.78\, E_{\text{cond}}^{\text{BCS}}$$

Taking the experimental values (*73*) $\gamma = 3.03 \pm 0.04$ mJ/mole–deg^2, $H_c = 802.5$ G, and $2\Delta_0 = 2.79 \pm 0.01$ meV, we find an experimental value for the condensation energy:

$$H_c^2/8\pi = (0.76 \pm 0.02)\, E_{\text{cond}}^{\text{BCS}}$$

in good agreement with the calculated number.

The most critical test of our results would be to perform a detailed calculation of $\alpha^2 F(\omega)$. This would require a knowledge of the electron band structure, the matrix elements of the electron–phonon interaction, and the phonon dispersion curves, information which is not yet fully available. However, we can extract from $\alpha^2 F(\omega)$ one average of the matrix elements which can be compared with that average calculated from the pseudo-potential model for lead. The matrix elements of electron–phonon interaction between states $|P\rangle$ and $|P'\rangle$ are given by (*74*)

$$g^2_{PP'\lambda} = \frac{\hbar N}{2M\omega_{P-P'\lambda}} \left| \epsilon_{P-P'\lambda} \langle P' | \nabla_i U_i | P \rangle \right|^2 \tag{63}$$

where U_i is the fully dressed potential of the ith ion, M and N are the mass and number density of the ions, and $\epsilon_{P-P'\lambda}$ and $\omega_{P-P'\lambda}$ are the polarization vector and energy of the phonon. If we take plane waves for the wave functions and the pseudo-potential for U we find

$$g^2_{q\lambda} = \frac{\hbar}{2M\omega_{q\lambda}N} v^2_q (q \cdot \epsilon_{q\lambda})^2 \tag{64}$$

Now from the definition of $\alpha^2 F(\omega)$ [Eq. (50)] the first moment is independent of the phonon frequencies. If we perform the Fermi surface integrals over the free electron Fermi surface we find

$$\int_0^\infty \omega\alpha^2 F(\omega)\, d\omega = \frac{2}{3}\frac{mZ}{M} E_F^2 \langle v_q^2 \rangle \tag{65}$$

where Z is the valence of the ion and $\langle v_q^2 \rangle$ is a normalized average of the dressed electron–ion pseudo-potential.

$$\langle v_q^2 \rangle = \int_0^{2K_F} q^3 v_q^2 \, dq \Big/ \int_0^{2K_F} q^3 v_0^2 \, dq \tag{66}$$

We find for lead that $\int \omega\alpha^2 F/\omega = 24$ meV^2 and using Eq. (65) our "experimental" value is $\langle v_q^2 \rangle = 0.037$. We have calculated $\langle v_q^2 \rangle$ from the Heine–Abarenkov (*30*) model potential which fits the Anderson and Gold Fermi surface data and find $\langle v_q^2 \rangle_{HA} = 0.038$. The close numerical agreement between $\langle v_q^2 \rangle$ deduced from the tunneling experiment and that calculated from the model pseudopotential is accidental—the errors inherent in the use of the free electron model here are $\sim 20\%$. However, it does indicate that the pseudo-potential which fits the Fermi surface also provides accurate electron–phonon matrix elements.

In view of the comparisons outlined above we conclude that the parameters which we find for lead are quite reasonable.

C. Test of the Eliashberg Gap Equation

Although in fact the reasonableness of the parameters deduced above is good evidence for the accuracy of the Eliashberg gap equation there is one more stringent test of the equations that we can perform. In the fitting procedure we did not make use of the measured electron density of states at high energy, $V - \Delta_0 > 11$ meV, where the sum and harmonic structure is observed as shown in Fig. 30. The comparison of the calculated and measured $N(\omega)$ in this multiple-phonon-emission region provides the most detailed test of the structure of the Eliashberg theory, which retains certain multiphonon processes (Fig. 31a) and neglects others (Fig. 31b). To obtain an accurate measurement of this small structure at high bias we measure on an ac bridge the first derivative dI/dV of a normal metal–Pb junction [the normalized first derivative is equal to $N(V)$ smeared a little by the thermal distribution of electrons in the normal metal]. The measured and calculated densities of states are shown in Fig. 32 and agree within ± 0.001, providing our strongest confirmation of the theory.

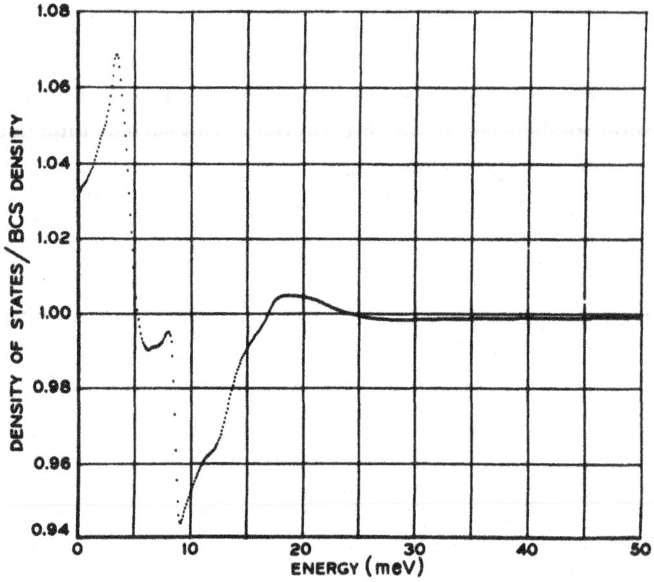

Fig. 30. Computed electronic density of states $N(\omega)$ divided by the BCS density of states for Pb vs. $\omega - \Delta_0$.

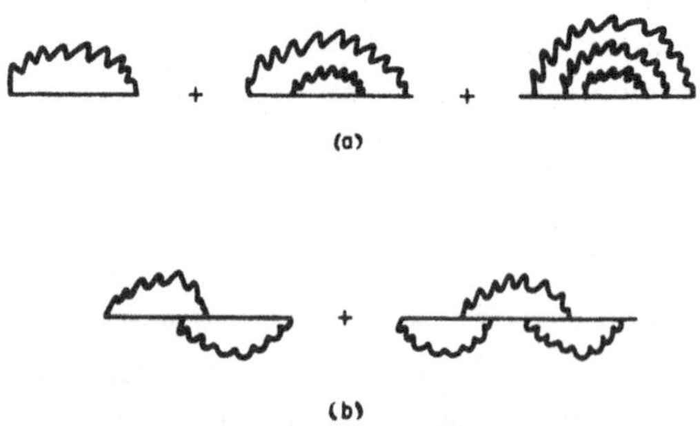

Fig. 31. (a) Self-energy diagrams included in the theory of superconductivity corresponding to single and multiple phonon emission. Here the straight line is the bare electron propagator —this diagram summation collapses to the first diagram when one uses dressed electron propagators. (b) Self-energy diagrams involving multiple phonon emission which are omitted.

Considering these results as a whole, we must conclude that the analysis of the tunneling experiments provides strong experimental support for the strong-coupled theory of superconductivity as embodied in the Eliashberg equations.

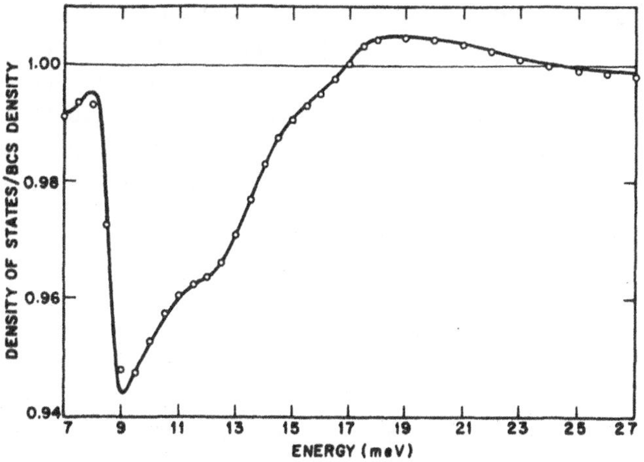

Fig. 32. Calculated (——) and measured (○○○) electronic density of states $N(E)$ for Pb normalized by the BCS density of states vs. $E-\Delta_0$. The measured density of states for $E-\Delta_0 > 11$ meV was not used in the fitting procedure and a comparison of theory and experiment in this "multiple-phonon-emission" region is a valied tst of the theory. In the experiment the sharp drop near 9 meV is affected by thermal smearing.

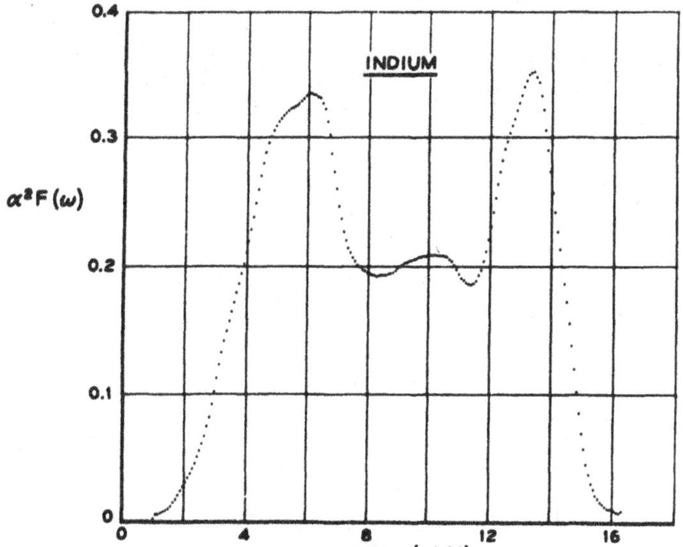

Fig. 33. $\alpha^2 F(\omega)$ for indium.

D. Weaker-Coupling Materials and Alloys

To conclude this section we present rather preliminary results for several other metals which we are investigating. Indium (Fig. 33) has a face-centered-tetragonal crystal structure which is a small distortion of fcc and the phonon density of states is similar to lead. Metallic tin (Fig. 34) has a complex tetragonal lattice with two atoms per unit cell. There are both acoustic and optic phonon modes and we find a phonon density of states rich in structure. Our results for mercury (Fig. 35) are based on the tunneling data of Bermon and Ginsberg (*12*). This rhombihedral structure has a soft shear direction and we find a very low

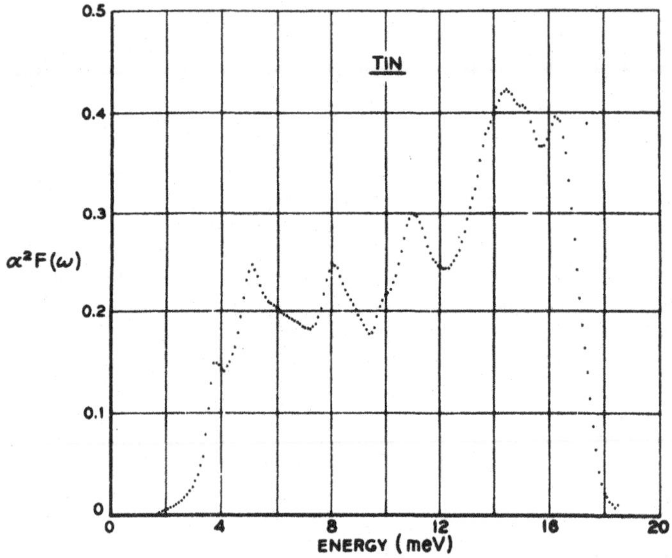

Fig. 34. $\alpha^2 F(\omega)$ for tin.

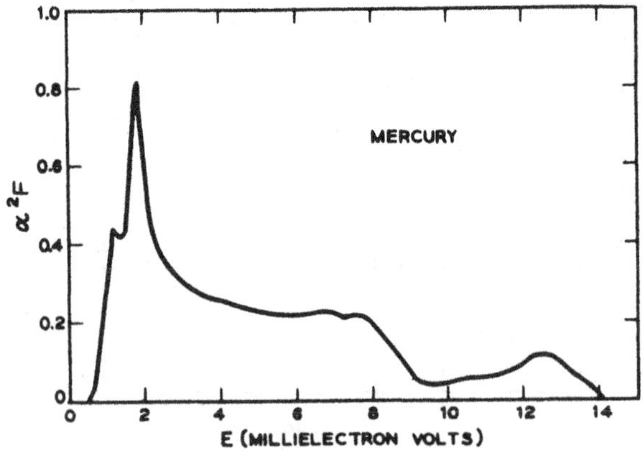

Fig. 35. $\alpha^2 F(\omega)$ for mercury determined from the experimental data of (*12*).

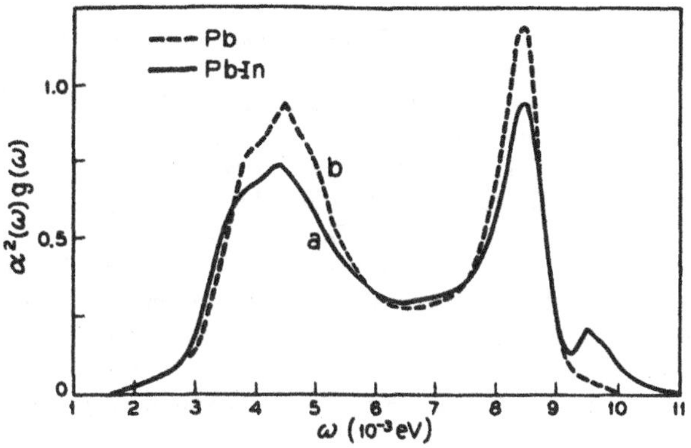

Fig. 36. $\alpha^2 F(\omega)$ for $Pb_{0.97}In_{0.03}$ from (*13*) compared with $\alpha^2 F(\omega)$ for Pb.

energy transverse phonon strongly coupled to the electrons. Finally, we show results for an alloy, $Pb_{0.97}$ $In_{0.03}$ (*13*) (Fig. 36), showing the localized phonon mode of the light indium impurity embedded in the heavy lead host lattice.

VII. CONCLUSIONS

The analysis of the tunneling experiments on strong-coupling superconductors has provided a new source of information about the electrons, the phonons, and the electron–phonon interaction in metals. The first result of this analysis is the determination of the parameters which enter the theory of superconductivity, $\alpha^2 F(\omega)$ and μ^*. From these parameters one can calculate the equilibrium properties of the metal in the normal and super states and find the electron self-energies, the critical field, etc. Also, from $\alpha^2 F(\omega)$ one finds (1) the peaks in the phonon density of states and, possibly, the energies of a few critical points; (2) the magnitude of the electron–phonon matrix elements; and (3) the electron–phonon coupling constant λ which determines the enhancement of the specific heat and cyclotron mass. The most important result, however, is that we have, for the first time, a critical experimental test of the structure of the self-energy equations of the strong-coupling theory of the electron–phonon interaction in metals. The theory is believed to be correct to lowest order in the expansion parameter $\hbar\omega_0/E_F \sim 10^{-2}$–$10^{-3}$ and we are able to show experimentally that the errors are not greater than a few per cent.

REFERENCES

1. J. C. Fisher and I. Giaever, *J. Appl. Phys.* **32**, 172 (1961).

2. I. Giaever, *Phys. Rev. Letters* **5**, 147 (1966).

3. J. Bardeen, L. N. Cooper, and J. R. Schrieffer, *Phys. Rev.* **106**, 162 (1957), **108**, 1175 (1957).

4. J. Nicol, S. Shapiro, and P. H. Smith, *Phys. Rev. Letters* **5**, 461 (1960).

5. I. Giaever, *Phys. Rev. Letters* **5**, 464 (1960).

6. D. H. Douglass, Jr., and L. N. Falicov, *Progress in Low Temperature Physics*, Vol. IV (C. J. Gorter, ed.), North-Holland, Amsterdam, 1964.

7. I. Giaever, H. R. Hart, and K. Megerle, *Phys. Rev.* **126**, 941 (1962).

8. P. W. Anderson, *Proc. of the Seventh Intern. Conf. on Low Temp. Physics* (G. M. Graham and A. C. Hollis Hallett, eds.), Univ. Toronto Press, Toronto, 1961, p. 304.

9. J. M. Rowell, A. G. Chynoweth, and D. E. Thomas, *Phys. Rev. Letters* **9**, 59 (1962).

10. J. M. Rowell and L. Kopf, *Phys. Rev.* **137**, 907 (1965); J. M. Rowell and W. L. McMillan, to be published; J. G. Adler, J. S. Rogers, and S. B. Woods, *Can. J. Phys.* **43**, 557 (1965).

11. J. M. Rowell, P. W. Anderson, and D. E. Thomas, *Phys. Rev. Letters* **10**, 334 (1963); W. L. McMillan and J. M. Rowell, *Phys. Rev. Letters* **14**, 108 (1965).

12. S. Bermon and D. M. Ginsberg, *Phys. Rev.* **135**, A306 (1964).

13. J. G. Adler and S. C. Ng, *Can. J. Phys.* **43**, 594 (1965); J. M. Rowell, W. L. McMillan, and P. W. Anderson, *Phys. Rev. Letters* **14**, 633 (1965); J. G. Adler, J. E. Jackson, and B. S. Chandrasekhar, *Phys. Rev. Letters* **16**, 53 (1966); T. Claeson, *Solid State Commun.* **5**, 119 (1967).

14. A.F.G. Wyatt, *Phys. Rev. Letters* **13**, 160 (1964); L.Y.L. Shen, private communication, 1966.

15. A. B. Migdal, *Zh. Eksperim. i Teor. Fiz.* **34**, 1438 (1958); *Soviet Phys. JETP* **7**, 996 (1958).

16. G. M. Eliashberg, *Zh. Eksperim. i Teor. Fiz.* **38**, 966 (1960); *Soviet Phys. JETP* **11**, 696 (1960).

17. Y. Nambu, *Phys. Rev.* **117**, 648 (1960).

18. P. Morel and P. W. Anderson, *Phys. Rev.* **125**, 1263 (1962).

19. J. R. Schrieffer, D. J. Scalapino, and J. W. Wilkins, *Phys. Rev. Letters* **10**, 336 (1963); D. J. Scalapino, J. R. Schrieffer, and J. W. Wilkins, *Phys. Rev.* **148**, 263 (1966).

20. J. R. Schrieffer, *Theory of Superconductivity*, Benjamin, New York, 1964.

21. J. C. Swihart, *IBM J. Res. Develop.* **6**, 14 (1962).

22. G. J. Culler, B. D. Fried, R. W. Huff, and J. R. Schrieffer, *Phys. Rev. Letters* **8**, 399 (1962).

23. D. J. Scalapino and P. W. Anderson, *Phys. Rev.* **133**, A291 (1964).

24. L. Van Hove, *Phys. Rev.* **89**, 1189 (1953).

25. J. M. Rowell and W. L. McMillan, to be published.

26. J. C. Swihart, D. J. Scalapino, and Y. Wada, *Phys. Rev. Letters* **14**, 106 (1965).

27. J. C. Phillips and L. Kleinman, *Phys. Rev.* **116**, 287 (1959).

28. B. J. Austin, V. Heine, and L. J. Sham, *Phys. Rev.* **127**, 276 (1962).

29. W. A. Harrison, *Phys. Rev.* **126**, 497 (1962); *Pseudopotentials in the Theory of Metals*, Benjamin, New York, 1966.

30. V. Heine and I. Abarenkov, *Phil. Mag.* **9**, 451 (1964).

31. J. R. Anderson and A. V. Gold, *Phys. Rev.* **139**, A1459 (1965).

32. G. B. Brandt and J. A. Rayne, *Phys. Rev.* **148**, 644 (1966).

33. N. W. Ashcroft, *Phil. Mag.* **8**, 2055 (1963).

34. B. N. Brockhouse, T. Arase, G. Caglioti, K. R. Rao, and A. D. B. Woods, *Phys. Rev.* **128**, 1099 (1962); J. M. Rowe, B. N. Brockhouse, and E. C. Swensson, *Phys. Rev. Letters* **14**, 554 (1965).

35. L.D. Landau, *Zh. Eksperim. i Teor. Fiz.* **30**, 1058 (1956); *Soviet Phys. JETP* **3**, 920 (1957); A. A. Abrikosov, L. P. Gor'kov, and I. E. Dzyaloshinskii, *Methods of Quantum Field Theory in Statistical Physics*, Prentice-Hall, Englewood Cliffs, N.J., 1963.

36. H. Fröhlich, *Phys. Rev.* **79**, 845 (1950).
37. E. G. Batyev and V. L. Pokrovskii, *Zh. Eksperim. i Teor. Fiz.* **46**, 262 (1963); *Soviet Phys. JETP* **19**, 181 (1964).
38. R.E. Prange and A. Sachs, *Phys. Rev.* **158**, 672 (1967).
39. J. Bardeen, *Phys. Rev. Letters* **6**, 57 (1961); **9**, 147 (1962).
40. W. A. Harrison, *Phys. Rev.* **123**, 85 (1961).
41. M. H. Cohen, L. M. Falicov, and J. C. Phillips, *Phys. Rev. Letters* **8**, 316 (1962).
42. R. E. Prange, *Phys. Rev.* **131**, 1083 (1963).
43. L. I. Schiff, *Quantum Mechanics*, McGraw-Hill, New York, 1955.
44. W. L. McMillan and P. W. Anderson, *Phys. Rev. Letters* **16**, 85 (1966).
45. W. J. Tomasch, *Phys. Rev. Letters* **15**, 672 (1965); **16**, 16 (1966).
46. W. J. Tomasch and T. Wolfram, *Phys. Rev. Letters* **16**, 352 (1966).
47. J. M. Rowell and W. L. McMillan, *Phys. Rev. Letters* **16**, 453 (1966).
48. J. Friedel, *Phil. Mag.* **43**, 153 (1952); *Nuovo Cimento Suppl.* **2**, 287 (1958).
49. J. L. Miles and P. Smith, *J. Electrochem. Soc.* **110**, 1240 (1963).
50. R. C. Jaklevic and J. Lambe, *Phys. Rev. Letters* **17**, 1139 (1966).
51. N. V. Zavaritskii, *Zh. Eksperim. i Teor. Fiz.* **45**, 1839 (1963); **48**, 837 (1965); *Soviet Phys. JETP* **18**, 1839 (1963); **21**, 557 (1965).
52. P. Townsend and J. Sutton, *Phys. Rev.* **128**, 591 (1962).
53. J. M. Rowell and L. Y. L. Shen, *Phys. Rev. Letters* **17**, 15 (1966).
54. S. Bermon, *Tech. Rept. 1*, University of Illinois, Urbana, National Science Foundation Grant NSF GP1100, 1964.
55. W. L. Feldmann and J. M. Rowell, unpublished data, 1966.
56. B. Mühlschlegel, *Z. Physik* **155**, 313 (1959).
57. G. B. Donaldson, *Proc. of the Tenth Intern. Conf. on Low Temp. Physics, Moscow, 1966*, Vol. IIB. (M.P. Malkov, ed. in chief), Viniti, Moscow, 1967, p. 291.
58. R. F. Gasparovic, B. N. Taylor, and R. E. Eck, *Solid State Commun.* **4**, 59 (1966).
59. D. H. Douglass, Jr., and R. Meservey, *Phys. Rev.* **135**, A19 (1964).
60. R. N. Hall, J. H. Racette, and H. Ehrenreich, *Phys. Rev. Letters* **4**, 456 (1960).
61. A. G. Chynoweth, G. H. Wannier, R. A. Logan, and D. E. Thomas, *Phys. Rev. Letters* **5**, 57 (1960).
62. A.G. Chynoweth, R.A. Logan, and D.E. Thomas, *Phys. Rev.* **125**, 877 (1962).
63. D. E. Thomas and J. M. Klein, *Rev. Sci. Instr.* **34**, 920 (1963).
64. R. T. Payne, *Phys. Rev.* **139**, A570 (1965).
65. J. G. Adler and J. S. Rogers, *Phys. Rev. Letters* **10**, 217 (1963).
66. J. S. Rogers, J. G. Adler, and S. B Woods, *Rev. Sci. Instr.* **35**, 208 (1964).
67. J. G. Adler and J. E. Jackson, *Rev. Sci. Instr.* **37**, 1049 (1966).
68. J. M. Rowell and W. L. McMillan, *Bull. Am. Phys. Soc.* **12**, 77 (1967); to be published.
69. R. J. Pederson and F. L. Vernon, Jr., *Appl. Phys. Letters* **10**, 29 (1967); see also (7).
70. D. E. Thomas and J. M. Rowell, *Rev. Sci. Instr.* **36**, 1301 (1965).
71. J. M. Rowell and W. L. McMillan, *Bull. Am. Phys. Soc.* **12**, *77* (1967).
72. J. Bardeen and M Stephen, *Phys Rev.* **136**, A1485 (1964).
73. D. L. Decker, D. E. Mapother, and R. W. Shaw, *Phys. Rev.* **112**, 1888 (1958); B. J. C. van der Hoeven, Jr., and P. H. Keesom, *Phys. Rev.* **137**, A103 (1965).
74. J.M. Ziman, *Electrons and Phonons*, Clarendon, Oxford, 1960.

12

SUPERCONDUCTIVITY IN LOW-CARRIER-DENSITY SYSTEMS: DEGENERATE SEMICONDUCTORS

Marvin L. Cohen

DEPARTMENT OF PHYSICS
UNIVERSITY OF CALIFORNIA
BERKELEY, CALIFORNIA

I. INTRODUCTION

In the early days of research on superconductivity the question of whether materials other than metals were superconducting was raised. Although it appeared that superconductivity was limited to metals, it was puzzling that some metals which were bad conductors were superconducting at relatively high temperatures while other very good conductors were not superconducting at all. So it seemed natural to look for superconductivity in very poor conductors such as insulators or semiconductors. These early investigations did not prove fruitful, but the search for a semiconductor or insulator with superconducting properties was not abandoned. As an indication of the extent to which this work was pursued, in 1935, pure germanium was tested (*1*) for superconductivity down to 50 mdeg with negative results. These results were not surprising to most people, since at such low temperatures, there are very few free electrons in this system, and it could be argued that this material was not even a conductor let alone a superconductor.

This point of view was consistent with the BCS theory (*2*), which was published in 1957. The BCS theory with a simplified model (BCS model) gives an expression for the superconducting transition temperature, T_c:

$$T_c \sim T_D \exp[-1/N(0)V] \quad (1)$$

where $N(0)$ is the electronic density of states at the Fermi energy E_F, T_D is the Debye temperature, and V is the effective, attractive, electron–electron interaction. The density of states has the form

$$N(0) \sim m^* n^{1/3} \quad (2)$$

where m^* is the effective mass and n is the carrier concentration. The results for pure germanium are therefore very reasonable, since in a pure semiconductor $n \to 0$ as $T \to 0$.

The case for a degenerate semiconductor is not as obvious. In this case $n \nrightarrow 0$

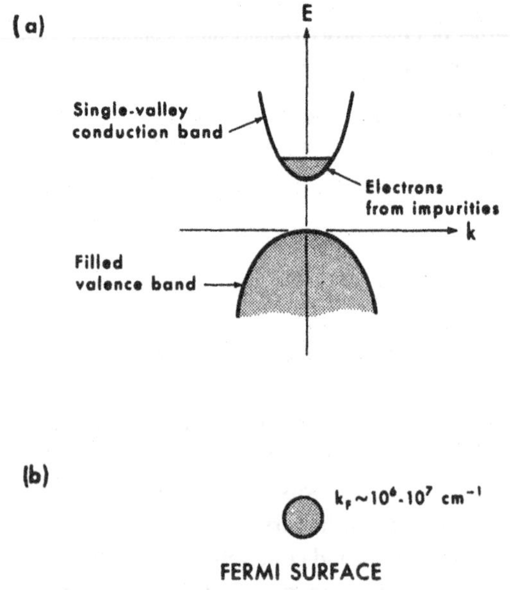

Fig. 1. (a) Two-band single-valley model for a semiconductor; (b) Fermi surface for a doped semiconductor having the single-valley band structure.

as $T \to 0$; impurity banding is present and hence free electrons or holes are available to carry currents. Such a system is still not as promising as a metal, since the carrier concentration and hence the density of states are usually much smaller in most semiconductors than in metals. The latter is evident if one takes the classic two-band single-valley model for a semiconductor (Fig. 1a). For typical densities the Fermi wave vector $k_F \sim 10^6$–10^7 cm^{-1}, giving a small Fermi sphere as shown in Fig. 1b and the density of states is only about one-tenth that of a metal. The situation for a degenerate semiconductor is therefore still not promising if one uses these simple models. The fact that crude estimates of T_c based on simple models were not encouraging was probably the reason for the paucity of theoretical publications on this problem. Some post-BCS theoretical work was done, however, and some statements were made in the literature. For example, Pines (*3*) in 1958 mentioned in a paper dealing with the occurrence of superconductivity in the periodic system that degenerate semiconductors should not be ruled out. It appears that the general trend of thought was that extremely strong interactions would be necessary to compensate for the small density of states. One can see an example of this approach in the paper by Gurevich et al. (*4*) in which these authors considered the coupling of electrons to optical phonons and piezoacoustic phonons.

There were no full-scale experimental programs in this field before 1963. In fact, in the first few post-BCS years experimental activity probably decreased. Outside of some tests made by people primarily interested in other properties of these materials, the search for superconductivity in semiconductors was not very active.

We reopened (*5,6*) this problem motivated by the idea that semiconductors and semimetals could be useful systems for studying superconductivity. This is primarily because these materials are better understood than metals and it is possible to make large changes in the carrier densities and band structures (through alloying and pressure) to study the dependence of the superconducting properties on the normal-state properties.

This work was based on the BCS theory and we considered a many-valley semiconductor band structure model. [See Fig. 2, in which the Fermi surface for a sample band structure with minima in the cubic (100) direction is given.] A many-valley band structure has several advantages over the single-valley case. This type of band structure allows both intravalley and intervalley electron–phonon scattering processes. The latter are important for the following reasons:

1. Intervalley scattering processes involve large momentum transfers and it is therefore difficult to weaken this interaction through screening.

2. An electron has a larger density of states to scatter to in an intervalley process because of the many Fermi pockets in k-space.

3. Many-valley semiconductors are also favorable, because the masses for

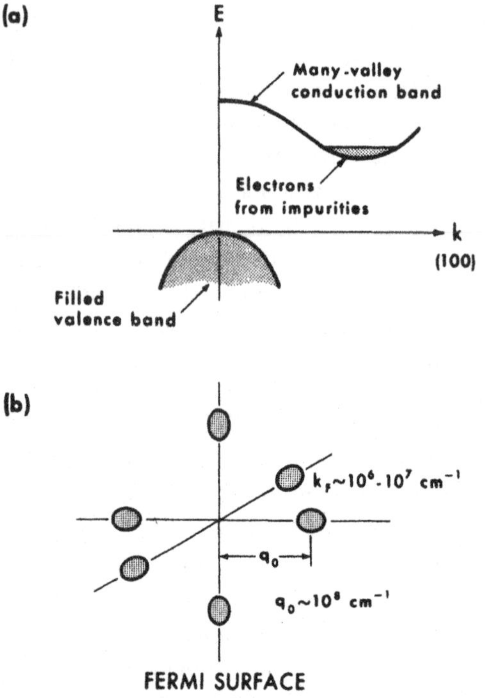

Fig. 2. (a) Many-valley band structure model for a semiconductor with valleys along (100); (b) Fermi surface for a doped many-valley semiconductor with valleys along (100).

valleys away from $k = 0$ are usually larger than $k = 0$ masses. This again gives a larger density of states for electron scattering.

To achieve a quantitative understanding of the role and importance of the various interactions, calculations were done for germanium, silicon, and an alloy of these elements. These crystals were chosen as prototypes not because they were the most promising materials for observing superconductivity, but primarily because they have been studied extensively and are well understood. The calculations indicated that superconductivity was possible. They also yielded information about the nature of the interactions involved and how these should be treated in the superconducting gap equation. It was hoped that these calculations would stimulate experimental work and help determine the direction of this work. An attempt to achieve this was made in the form of a list of normal-state properties which a semiconductor should have if it is to be a promising choice for superconductivity. This list did, in fact, play the central role in the choosing of the three existing superconducting semiconductors.

It was also possible to predict on the basis of the model calculations that the expected transition temperature of a degenerate semiconductor should be in the range of 0.1°K (although higher temperatures are certainly possible), that these materials should be type II superconductors, and that the transition temperatures of these materials should be sensitive to changes in carrier density and strain.

The degenerate semiconductors which exhibit superconducting properties are germanium telluride (*7*), strontium titanate (*8*), and tin telluride (*9*). All these materials are superconducting in 0.1°K range, all are type II superconductors, and the transition temperatures of these materials are sensitive to changes in carrier concentration. Strain data are not available as yet for these materials.

In Section II the theory will be described. In particular, the superconducting gap equation, kernels of this equation, and its solution will be discussed. Section III will deal with some of the properties of superconducting semiconductors, and section IV will contain discussion and speculation.

II. THEORY

A. BCS Gap Equation

The theory of superconductivity in degenerate semiconductors as presented here is based on the BCS theory. According to the BCS theory the mechanism causing superconductivity is the electron–lattice interaction. This interaction can cause two electrons to be attracted to one another to form a Cooper pair if the electron–lattice–electron interaction is larger than the repulsive Coulomb interaction between electrons. It is this phenomenon of Cooper pair formation which is believed to be the origin of the impressive properties of superconductors. The binding energy of a pair depends on the superconducting energy gap function or the order parameter. This gap function (*10*) is wave vector– and/or energy–dependent and it is given by the BCS gap equation,

$$\Delta_k = -\frac{1}{2}\sum_{k'} \frac{V_{kk'}\Delta_{k'}}{E_{k'}} \tanh\left(\frac{E_{k'}}{2k_BT}\right) \tag{3}$$

where Δ_k is the superconducting energy gap, $E_k = (\epsilon_k^2 + \Delta_k^2)^{1/2}$ is the superconducting quasi-particle energy, ϵ_k is the normal-state electron energy measured from the Fermi energy E_F, T is the temperature, k_B is Boltzmann's constant and $V_{kk'}$ is the matrix element for electron–electron scattering from the state k to the state k'.

The free carriers of a degenerate semiconductor arise from the doping with impurities. At high carrier concentrations and at low temperatures these impurities scatter electrons with typical scattering times $\tau \sim 10^{-13}$ sec. The resulting $\hbar/\tau$ is much larger than the superconducting energy gap and these materials are "dirty" in the Anderson (*11*) sense. To account for this effect, we make the changes suggested by Anderson for a dirty system with nonmagnetic impurities. To do this, we consider the problem of the wave functions of the electrons in the presence of the scatterers to be solved exactly and assume that the wave function and its time reversed state are‡

$$\psi_{n\sigma} = \sum_k (n|k)\,\phi_{k,\sigma}$$
$$(\psi_{n\sigma})^* = \sum_k (n|k)^*\,\phi_{-k,-\sigma} \tag{4}$$

where $\phi_{k,\sigma}$ are the Bloch waves and the $(n|k)$'s are the unitary transformation coefficients solving the scattering problem. These new wave functions are now used to compute the electron–electron interactions and the integral equation for the superconducting energy gap Δ_n, which is given by

$$\Delta_n = -\frac{1}{2}\sum_{n'} \frac{V_{nn'}\Delta_{n'}}{E_{n'}} \tanh\left(\frac{E_{n'}}{2k_BT}\right) \tag{5}$$

where $E_n = (\epsilon_n^2 + \Delta_n^2)^{1/2}$ is the superconducting quasi-particle energy, ϵ_n is the normal-state energy of the exact one-electron states, measured from the Fermi energy, and $V_{nn'}$ is a matrix element of the effective interaction for scattering between state n and its time-reversal conjugate, and n' and its time-reversal conjugate.

We choose the $(n|k)$'s of random phase and normalized so that

$$\sum_k |(n|k)|^2 = 1 \tag{6}$$

$$|(n|k)|^2 = \frac{d(\epsilon_n - \epsilon_k)}{N^t(\epsilon_n)} \tag{7}$$

where $d(\epsilon_n - \epsilon_k)$ is a spread-out delta function with a half-width $\approx \hbar/\tau$, ϵ_k is the normal-state energy, measured from the Fermi energy, of the initial plane-wave states, and $N^t(\epsilon_n)$ is the total density of states for electrons of energy ϵ_n. Equations (6) and (7) yield the following form for $V_{nn'}$:

$$V_{nn'} = \sum_{kk'} \frac{d(\epsilon_n - \epsilon_k)\,d(\epsilon_{n'} - \epsilon_{k'})}{N^t(\epsilon_n)\,N^t(\epsilon_{n'})} V_{kk'} \tag{8}$$

where $V_{kk'}$ is the usual BCS matrix element between plane-wave states. For the case in which the valleys are all equivalent (i.e., Ge) Eq. (8) reduces approximately to

$$V_{nn'} = \bar{V}_{kk'} d(\epsilon_n - \epsilon_k)\,d(\epsilon_{n'} - \epsilon_{k'}) \tag{9}$$

‡ Henceforth, the spin index will be suppressed.

where $\bar{V}_{kk'}$ is an average value of $V_{kk'}$, k and k' are now considered to be average values of k and k' in an energy shell of width $\hbar/\tau$ about the Fermi energy, and ϵ_k and $\epsilon_{k'}$ are the electron energies for the average wave numbers k and k' measured from the Fermi energy. In the above derivation we have implicitly assumed that the average value of the product of the superconducting energy gap and the BCS matrix element is equal to the product of the averages of these quantities taken separately.

In general, we will be considering many-valley semiconductors; when these materials are highly doped they become degenerate systems of a pseudo-metallic type. The resulting Fermi surface in such a system is composed of several pockets of electrons (n-type) or holes (p-type) in the Brillouin zone. Commonly these pockets are spheres or ellipsoids having radii only about one-tenth of a typical Fermi sphere found in a metal. The pockets are usually well separated in the zone and the electronic processes can be treated separately as intravalley and intervalley processes. This greatly simplifies the calculations compared to a metal in which such a division cannot usually be made.

One can treat the dirty many-valley semiconductors by taking $d(\epsilon_n - \epsilon_k) \approx \delta(\epsilon_n - \epsilon_k)$. A semiconductor with ν equivalent valleys can then be treated as a clean superconductor with an energy gap $\Delta(\epsilon_k)$ arising from both intravalley and intervalley processes. If k space is divided into ν regions (ν valleys), Eq. (5) takes the following form:

$$\Delta_k = -\frac{1}{2} \sum_{\substack{k' \\ k,\,k' \text{ in the} \\ \text{same region}}} \frac{V_{kk'}^{ra}}{E_{k'}} \tanh\left(\frac{E_{k'}}{2k_B T}\right) - \left(\frac{\nu - 1}{2}\right) \sum_{\substack{k' \\ k,\,k' \text{ in} \\ \text{different} \\ \text{regions}}} \Delta_{k'} \frac{V_{kk'}^{er}}{E_{k'}} \tanh\left(\frac{E_{k'}}{2k_B T}\right) \tag{10}$$

where superscripts ra and er mean intravalley and intervalley, respectively.

We now transform the sum over k' to integrals over $q = |k' - k|$ and $\epsilon_{k'}$, and after making the appropriate changes to include both intravalley and intervalley processes, Eq. (10) becomes

$$\Delta_k = -\frac{\Omega}{2(2\pi)^3} \int \frac{\Delta_{k'}}{E_{k'}} \left[\frac{1}{v\hbar} \int_{|k-k'|}^{|k+k'|} q V^{ra}(\epsilon_k, \epsilon_{k'}, q)\, dq \right.$$
$$\left. + \frac{\nu - 1}{v\hbar} \int_{|k-k'|}^{|k+k'|} q V^{er}(\epsilon_k, \epsilon_{k'}, q + q_0)\, dq \right] \tanh\left(\frac{E_{k'}}{2k_B T}\right) d\epsilon_{k'} \tag{11}$$

where the origin for k and k' are in different valleys, q_0 is the separation of the valleys in k-space, Ω is the crystal volume, and v is the velocity $\hbar k/m^*$, m^* being the density of states effective mass of these valleys which are assumed to be spherical. We follow the BCS theory and assume $V^{ra}(\epsilon_k, \epsilon_{k'}, q)$ and

$V^{er}(\epsilon_k, \epsilon_{k'}, q+q_0)$ are each composed of an attractive phonon-induced interaction and a repulsive, screened-Coulomb interaction.

Before we discuss these interactions, we will make some convenient changes in our integral equation. Letting $D_k = (k/k_F)\,\Delta_k$ and $D_{k'} = (k'/k_F)\,\Delta_{k'}$, Eq. (11) becomes

$$D_k = -\int \frac{D_{k'}}{E_{k'}} K(c, \delta) \tanh\left(\frac{E_{k'}}{2k_B T}\right) d\epsilon_{k'} \tag{12}$$

The kernel‡ $K(c, \delta)$ of the integral equation corresponds to half of the "$N(0)\,V$" parameter of the BCS theory and is defined by

$$K(c, \delta) = K^{ra}(c, \delta) + K^{er}(c, \delta) \tag{13}$$

$$K^{ra}(c, \delta) = \frac{k}{k'} \frac{\Omega}{2(2\pi)^2} \frac{1}{v\hbar} \int_{|k-k'|}^{|k+k'|} q V^{ra}(\epsilon_k, \epsilon_{k'}, q)\, dq$$

$$= \frac{\Omega k_F^3}{4\pi^2 E_F} \frac{1}{\sqrt{\delta + c^2}} \int_{\frac{1}{2}|c-\sqrt{\delta+c^2}|}^{\frac{1}{2}|c+\sqrt{\delta+c^2}|} V^{ra}(\beta, \delta)\, \beta\, d\beta \tag{14}$$

$$K^{er}(c, \delta) = \frac{(v-1)\,\Omega k_F^3}{4\pi^2 E_F} \frac{1}{\sqrt{\delta + c^2}} \int_{\frac{1}{2}|c-\sqrt{\delta+c^2}|}^{\frac{1}{2}|c+\sqrt{\delta+c^2}|} \beta V^{er}(\beta, \delta)\, d\beta \tag{15}$$

where $c = k/k_F$, $\beta = q/2k_F$, $\delta = \hbar\omega/E_F = (\epsilon_k - \epsilon_{k'})/E_F$, and E_F and k_F are the Fermi energy and wave number of a specific valley for a given carrier concentration.

The BCS integral equation (12) with the kernels given by Eqs. (14) and (15) does not include the effects of renormalization and lifetime effects. Such equations can be derived using modern techniques (*12*). These effects can be important especially for strongly coupled superconductors, but because the systems we will discuss are generally treated in the weak-coupling limit and because the simpler version of the BCS equation given in this section contains almost all the essential features of the more exact equation, we will ignore renormalization and lifetime effects in this paper.

B. Solution of the Gap Equation

We will deal with the problem of solving the gap equation before taking up the task of computing the correct kernels which go into the equation. This latter problem will be dealt with in the next sections along with results for various models. In this section we will assume that a kernel has been computed and that it is composed of an attractive phonon part and a repulsive Coulomb part.

‡ The term "kernel" will refer to the gap-independent part of the integral equation (12).

Once a form for the gap equation is taken and a kernel is assumed, all that remains to be done is to solve the gap equation with this kernel. At this point all the "physics" is in the problem, and one is faced with the job of extracting useful information from the gap equation, e.g., the transition temperature and its dependence on normal-state properties.

The gap equation can be solved using various computer techniques. Iteration is widely used with much success. Another method which we will discuss is the square-well method. Although the square-well method can give less accurate results for some cases than iteration, this method more readily illustrates how the size and shape of the kernels affect the superconducting properties. This is because this method is an extension of the BCS (one-square-well) and Tolmachev (*13*) (two-square-well) models which can be solved algebraically. We will deal with these two models first and then discuss the extension to the many-square-well case.

Basically, the problem is to solve Eq. (12) for a specific kernel which is the sum of attractive and repulsive kernels. In the BCS model it is assumed that both kernels have the same cutoff energy $\hbar\omega_c$. If we choose $K(\epsilon, \epsilon')$ instead of $K(k, k')$ or $K(c, \delta)$ as our kernel (to conform to the usual notation), at $T = T_c$ Eq. (12) has the form

$$D(\epsilon) = -\int_{-\hbar\omega_c}^{\hbar\omega_c} \frac{D(\epsilon')}{\epsilon'} K(\epsilon, \epsilon') \tanh\left(\frac{\epsilon'}{2k_B T_c}\right) d\epsilon' \qquad (16)$$

where $K(\epsilon, \epsilon')$ is assumed to be half of the BCS $N(0)V$ parameter and both $K(\epsilon, \epsilon')$ and D are constant for energies less than the phonon cutoff energy $\hbar\omega_c$. With this model Eq. (16) can be solved to give

$$\frac{1}{N(0)V} = \int_0^{\hbar\omega_c} \frac{\tanh(\epsilon'/2k_B T_c)}{\epsilon'} d\epsilon' = \ln\left(\frac{1.14\hbar\omega_c}{k_B T_c}\right) \qquad (17)$$

which is essentially Eq. (1). This model has been a very important one in superconductivity, since it roughly illustrates the dependence of the transition temperature on the interactions and the density of states. However, if this model is used naively by taking the sum of phonon and Coulomb kernels with an energy cutoff at kT_D for both kernels, a gross underestimate of the superconducting transition temperature is obtained. This results from the fact that the value of the Coulomb kernel above the phonon cutoff is very important.

A much more appropriate model to most physical systems which considers the Coulomb kernel at higher energies is the two-square-well model introduced by Tolmachev. This model still gives only a rough approximation to the kernels when one considers the large amount of structure the kernels can have (next section), but this approach does illustrate many of the important features of the many-square-well model and it is easily solved.

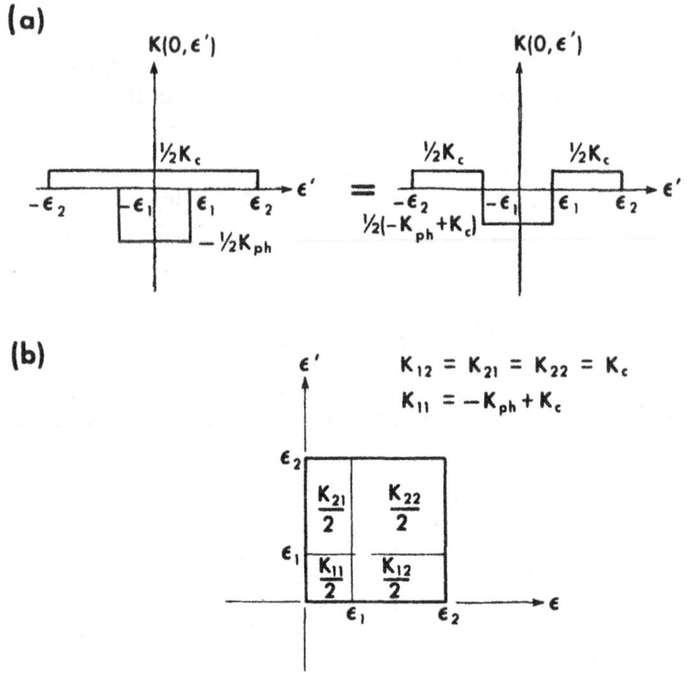

Fig. 3. Two-square-well kernel as a function of one energy variable ϵ'; (b) (ϵ, ϵ') plane of the two-square-well kernel.

The phonon and Coulomb kernels‡ K_{ph} and K_c are assumed symmetric in ϵ and ϵ' and constant in energy with cutoffs ϵ_1 and ϵ_2, respectively (Fig. 3). The phonon cutoff is assumed to be approximately equal to the Debye energy and the Coulomb cutoff is of the order of the plasma energy. This is a good approximation to the kernels computed in the next section. If we assume two gaps $D(\epsilon) = D_1$, for $\epsilon < \epsilon_1$, and $D(\epsilon) = D_2$ for $\epsilon_1 < \epsilon < \epsilon_2$, and take $\epsilon_1, \epsilon_2 < E_F$, Eq. (16) becomes

$$-D_1 = K_{11}D_1Z_1 + K_{12}D_2Z_2 \tag{18}$$

$$-D_2 = K_{21}D_1Z_1 + K_{22}D_2Z_2 \tag{19}$$

where

$$Z_1 = \int_0^{\epsilon_1} \frac{\tanh(\epsilon'/2k_BT_c)}{\epsilon'}\, d\epsilon' \tag{20}$$

$$Z_2 = \int_{\epsilon_1}^{\epsilon_2} \frac{\tanh(\epsilon'/2k_BT_c)}{\epsilon'}\, d\epsilon' \tag{21}$$

‡ To conform to the usual notation in this section K_{ph} and K_c correspond to $N(0)\,V$ without a factor of $\frac{1}{2}$.

In weak coupling

$$Z_1 = \log\left(\frac{1.14\epsilon_1}{k_B T_c}\right) \tag{21a}$$

$$Z_2 = \log(\epsilon_2/\epsilon_1) \tag{22}$$

Solving,

$$k_B T_c = 1.14\epsilon_1 \exp(-Z_1) \tag{23}$$

$$Z_1 = \left(-K_{11} + \frac{K_{12}K_{21}Z_2}{K_{22}Z_2 + 1}\right)^{-1} \tag{24}$$

For a physical system $K_{11} = -K_{\text{ph}} + K_c$ and $K_{12} = K_{21} = K_{22} = K_c$, so that Eq. (24) becomes

$$Z_1 = \frac{1}{K_{\text{ph}} - K_c^*} \tag{25}$$

where

$$K_c^* = \frac{K_c}{1 + Z_2 K_c} \tag{26}$$

Hence the effect of the higher repulsive well is to change the Coulomb kernel K_c to an effective Coulomb kernel K_c^*, which is weaker. The high-energy repulsive well therefore *increases* the transition temperature. This point will be discussed further in the last section.

As an aside, to illustrate how useful this model is, it is interesting to note that one can compute the isotope effect *(14,15)* using two square wells. If we take T_c proportional to $M^{-\lambda}$ and assume $\epsilon_1 \sim kT_D \sim M^{-1/2}$, where M is the ionic mass, then from Eqs. (23) and (24),

$$d \ln T_c / d \ln M = -\lambda = -\tfrac{1}{2}(1 - Z_1^2 K_c^{*2}) \tag{27}$$

The reduced isotope effect for transition metals has been explored by Garland *(15)* using similar models. He noted that the reduced band width in these metals reduces Z_2. This can cause K_c^* to increase and $|\lambda|$ to decrease [see Eq. (27)]. There is also the possibility of a sign change.

The extension to the many-square-well case is trivial.

$$\begin{aligned}
-D_1 &= K_{11}D_1Z_1 + K_{12}D_2Z_2 + \cdots + K_{1n}D_nZ_n \\
-D_2 &= K_{21}D_1Z_1 + K_{22}D_2Z_2 + \cdots + K_{2n}D_nZ_n \\
&\vdots \\
-D_n &= K_{n1}D_1Z_1 + K_{n2}D_2Z_2 + \cdots + K_{nn}D_nZ_n
\end{aligned} \tag{28}$$

where

$$Z_n = \int_{\epsilon_{n-1}}^{\epsilon_n} \frac{\tanh(\epsilon'/2k_BT_c)}{\epsilon'}\, d\epsilon' \tag{29}$$

The solving of the integral equation therefore reduces to the solving of determinants.

C. Kernels and Results for Nonpolar Interactions

The essential part of the gap equation is the kernel. The kernel contains the basic mechanisms causing the superconductivity. Once the kernel is known, the gap equation can be solved to give the gap function and the gap function contains the information about the superconducting state. The procedure to follow is then to determine the important interactions for the superconductivity of a specific material and then to compute the kernels corresponding to these interactions. After this is done, the integral equation for the gap is solved using the methods described in the last section or by other methods.

In this section we will discuss nonpolar interactions. The nonpolar interactions treated in this section are also present in polar semiconductors and the theory goes through without much modification; it is simpler, however, in a polar semiconductor to treat the intravalley Coulomb and polar coupling together rather than separately as in the nonpolar case. For this reason the polar interaction will be treated as a special case in the next section.

Through most of this section germanium will be used as an example of a nonpolar semiconductor. Calculations will be done for germanium and an alloy of germanium and silicon.

In a nonpolar many-valley semiconductor it is convenient to divide the electron–electron interactions and kernels into intravalley Coulomb, intravalley phonon, intervalley Coulomb, and intervalley phonon interactions.

The intravalley Coulomb kernel arises from the screened intravalley Coulomb interaction V_C^{ra}. This interaction has the form

$$V_c^{ra} = \frac{4\pi e^2}{\Omega q^2 \epsilon(q, \omega)} = \frac{4\pi e^2}{\Omega (2k_F)^2 \beta^2 \epsilon(\beta, \delta)} \tag{30}$$

where $\epsilon(\beta, \delta)$ is a dynamic dielectric function modeled after the dielectric function derived first by Lindhard (*16*). In many cases the Lindhard dielectric function is a good approximation to the actual dielectric function for a degenerate semiconductor, as can be seen by examining the parameter r_s. In Ge, for example, the static dielectric constant ϵ_∞ is 15.8 and $m^* = 0.22 m_0$, so that the value for r_s at a carrier concentration n of 10^{20} carriers/cm^3 is

$$r_s = (3/4\pi n)^{1/3} (m^* e^2/\hbar^2 \epsilon_\infty) \approx 0.35$$

We are, therefore, dealing with a high-density electron gas and expect the

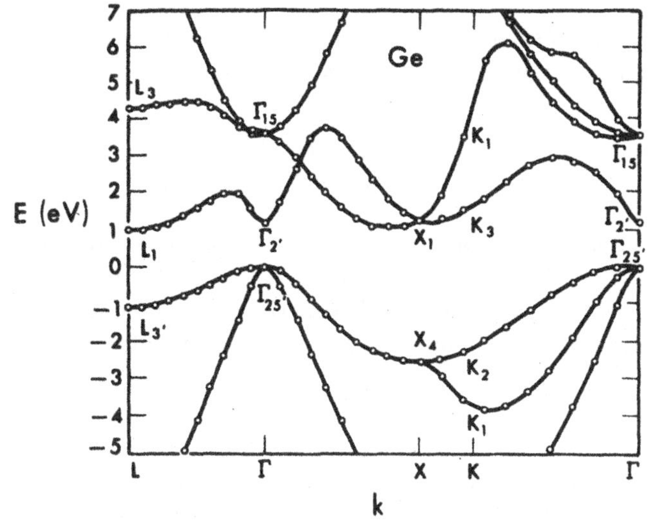

Fig. 4. Band structure of germanium.
[Reproduced by permission from Cohen and Bergstresser (*18*).]

Lindhard dielectric function to give the Gell-Mann and Brueckner (*17*) result for the correlation energy to a good approximation.

For a degenerate semiconductor with ν (i.e., Ge, $\nu = 4$, see Fig. 4) equivalent valleys, the Lindhard dielectric function takes the following form:

$$\epsilon(\beta, \delta) = \epsilon_1(\beta, \delta) + i\epsilon_2(\beta, \delta)$$

where

$$\epsilon_1(\beta, \delta) = \epsilon_\infty + \frac{\nu 3\pi e^2 n_\nu \hbar^2}{32 m^* E_F^2 \beta^3}\left\{\left[1 - \left(\beta + \frac{\delta}{4\beta}\right)^2\right] \ln\left|\frac{1 + \beta + \delta/4\beta}{1 - \beta - \delta/4\beta}\right| + [1 - (\beta - \delta/4\beta)^2] \ln\left|\frac{1 + \beta - \delta/4\beta}{1 - \beta + \delta/4\beta}\right| + 4\beta\right\} \tag{31}$$

$$\epsilon_2(\beta, \delta) = -\frac{\nu 3\pi^2 e^2 n_\nu \hbar^2}{32 m^* E_F^2 \beta^3}\begin{cases} \delta & \text{when } \beta < 1 \text{ and } \delta < |4\beta^2 - 4\beta| \\ 1 - (\beta - \delta/4\beta)^2 & \text{when } |4\beta^2 - 4\beta| < \delta < |4\beta^2 + 4\beta| \\ 0 & \text{when } \delta > |4\beta^2 + 4\beta| \\ 0 & \text{when } \beta < 1 \text{ and } \delta < |4\beta^2 - 4\beta| \end{cases}$$

n_ν is the carrier concentration in each valley and ϵ_∞ is the static background dielectric constant arising from interband transitions.

Using Eqs. (14), (30), and (31), the contribution of $V_C^{ra}(\beta, \delta)$ to the kernel becomes (taking only the real part)

$$K_C^{ra}(c, \delta) = \frac{k_F e^2}{4\pi E_F}\frac{1}{\sqrt{\delta + c^2}}\int_{\frac{1}{2}|c - \sqrt{\delta + c^2}|}^{\frac{1}{2}|c + \sqrt{\delta + c^2}|} \frac{\epsilon_1(\beta, \delta)}{\epsilon_1^2(\beta, \delta) + \epsilon_2^2(\beta, \delta)}\frac{d\beta}{\beta}$$

A Bardeen–Pines (*19*) interaction is chosen to evaluate the attractive electron–phonon–electron intravalley contribution to superconductivity. This interaction has the form

$$V_{\text{ph}}^{ra} = -\sum_i \frac{2\hbar\omega_{k-k'}|M_{k-k'}^i|^2}{(\hbar\omega_{k-k'}^i)^2 - (\epsilon_k - \epsilon_{k'})^2} \tag{32}$$

where the sum is over i phonon modes, $\omega_{k-k'}^i$ is the phonon frequency of wave number, $q = |k' - k|$, and $M_{k-k'}^i$ is the matrix element for the scattering of an electron from k to k' by a phonon characterized by the mode index i.

We consider coupling to both the acoustical and optical phonon modes. The electron–phonon matrix elements involved in calculating these contributions via Eq. (32) are evaluated by expressing the matrix elements in terms of the appropriate deformation potentials, the values of which are then taken from experiment.

For the acoustical phonon modes we take a q-dependent matrix element of deformation potential theory (*20*),

$$|M_q|_{Ac}^2 = \frac{\alpha_i \hbar^2 q^2 E_i^2}{2MN\hbar\omega_q} \tag{33}$$

where E_i is the coupling constant for the ith mode, α_i is the degeneracy of the mode, N is the total number of unit cells in the lattice, and M is the ion mass. The coupling constants E_{LA} and E_{TA} can be expressed in terms of the crystal deformation potentials Ξ_u and Ξ_d by taking the appropriate angular averages of these constants (*21*) [i.e., the values for these constants for Ge are $\Xi_u = 19$ eV (*22*) and $\Xi_d = -5.8$ eV (*23*)]. The strength of the deformation potentials and hence the E_i's are reduced because of the screening of the electron–phonon interaction by the free carriers of the crystal. The above coupling constants are considered to be the "bare" values which must be screened. Because intravalley phonon processes involve small momentum transfer, the screening of these processes is very important. The reduction of the magnitudes of the deformation potentials arising from the screening can be evaluated by treating the phonon wave as an external perturbation of the nearly free electron system and thus by multiplying the intravalley electron–phonon interaction by $|\epsilon(\beta, \delta)|^{-2}$, where $\epsilon(\beta, \delta)$ is the dynamic dielectric function described by Eq. (31).

The resulting intravalley contribution to the kernel arising from acoustic phonons is

$$K_{Ac}^{ra}(c, \delta) = \frac{k_F^3}{4\pi^2 E_F} \frac{\hbar^2\Omega}{MN} \left[\left\{ \frac{E_{LA}^2}{\hbar^2 u_{LA}^2} \frac{1}{\sqrt{\delta + c^2}} \int_{\frac{1}{2}|c - \sqrt{\delta + c^2}|}^{\frac{1}{2}|c + \sqrt{\delta + c^2}|} \frac{\beta^3 \, d\beta}{|\epsilon(\beta, \delta)|^2 (\beta^2 - \eta_{LA}^2 \delta^2)} \right. \right.$$
$$\left. \left. + \frac{2E_{TA}^2}{\hbar^2 u_{TA}^2} \frac{1}{\sqrt{\delta + c^2}} \int_{\frac{1}{2}|c - \sqrt{\delta + c^2}|}^{\frac{1}{2}|c + \sqrt{\delta + c^2}|} \frac{\beta^3 \, d\beta}{|\epsilon(\beta, \delta)|^2 (\beta^2 - \eta_{TA}^2 \delta^2)} \right\} \right] \tag{34}$$

where

$$\eta_{LA} = \frac{E_F}{2\hbar u_{LA} k_F} \qquad \eta_{TA} = \frac{E_F}{2\hbar u_{TA} k_F}$$

We have assumed a spherical Debye distribution for our model of the acoustical phonons with u_{LA} and u_{TA} as the sound velocities for the longitudinal acoustical and transverse acoustical phonon modes, respectively.

For the optical phonon modes we choose a q independent matrix element (*24,25*) similar in form to Eq. (33):

$$|M_q|^2_{\text{op}} = \alpha_i \frac{\hbar^2 R^2}{2MN} \frac{E_i^2}{\hbar\omega_{\text{op}}} \tag{35}$$

where R is a reciprocal lattice vector and ω_{op} is the frequency of the longitudinal and transverse optical modes for $q \to 0$; these are degenerate at $q = 0$. The coupling parameter $R^2 E_i^2$ can be expressed as a function of the optical deformation potential D_0. This deformation potential is difficult to measure accurately. [For Ge (*6*), we use the value $D_0 = 0.5 \times 10^9$ eV/cm, which is consistent with most experimental measurements.] This deformation potential will also be screened by the free carriers of the crystal, and again this effect is computed by using the dynamic dielectric function $\epsilon(\beta, \delta)$, yielding the following expression for the intravalley optic contribution to the kernel:

$$K^{ra}_{\text{op}}(c, \delta) = \frac{k_F^3}{4\pi^2 E_F^3} \frac{\hbar^2 \Omega}{MN} R^2 \frac{(E_{LO}^2 + 2E_{TO}^2)}{\delta_{\text{op}}^2 - \delta^2} \frac{1}{\sqrt{\delta + c_2}} \times \int_{\frac{1}{2}|c - \sqrt{\delta + c^2}|}^{\frac{1}{2}|c + \sqrt{\delta + c^2}|} \frac{\beta \, d\beta}{\epsilon_1^2(\beta, \delta) + \epsilon_2^2(\beta, \delta)} \tag{36}$$

where $\delta_{\text{op}} = \hbar\omega_{\text{op}}/E_F$. We have assumed an Einstein distribution for our model of the optical phonons with frequency ω_{op}. The above intravalley optical kernel is added to the intravalley acoustical kernel Eq. (33), and their sum K^{ra}_{ph} is plotted‡ in Fig. 5 as a function of δ for $c = 1$.

The intervalley Coulomb interaction is evaluated in the same manner as the intravalley Coulomb interaction and has the form

$$V_c^{er} = 4\pi e^2/\Omega(2k_F)^2 \beta_0^2 \epsilon^{\text{inter}} \tag{37}$$

where $\beta_0 = q_0/2k_F$ and ϵ^{inter} is the intervalley dielectric constant. The intervalley Coulomb interaction involves only short-range correlations, and both the dynamic and the static dielectric constant ϵ_∞ must be modified (*26*) to screen large

‡ Note that Figs. 3 and 5 of (*6*) have been accidently interchanged.

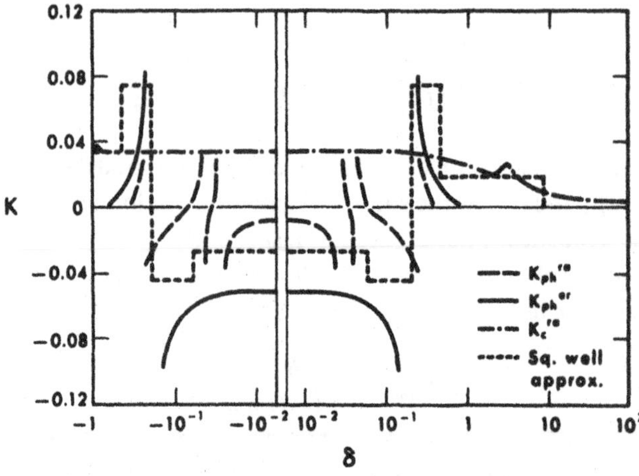

Fig. 5. Intravalley-Coulomb, intravalley-phonon, and intervalley phonon kernels for Ge. A square-well approximation to the total kernel is also given.

q processes correctly. The contribution of intervalley Coulomb interactions to the total kernel is

$$K_c^{er}(c, \delta) = \frac{(\nu - 1)\, k_F e^2}{4\pi E_F} \frac{1}{\sqrt{\delta - c^2}} \operatorname{Re}\left(\int_{\frac{1}{2}|c - \sqrt{\delta + c^2}|}^{\frac{1}{2}|c + \sqrt{\delta + c^2}|} \frac{\beta \, d\beta}{\epsilon^{\text{inter}} \beta_0^2} \right) \tag{38}$$

Despite the reduction of the dielectric function and the larger density of states for intervalley processes, K_C^{er} is much smaller than $K_C^{ra}(c, \delta)$, because $\beta_0 \approx 10k_F$. We have, therefore, included this contribution in the intravalley Coulomb kernel plotted in Fig. 5 as a small additive constant.

The intervalley phonon processes contribute a large attractive part to the total kernel, and we have evaluated this contribution by using the Bardeen–Pines interaction given by Eq. (32). Since q does not vary much in an intervalley transition and is approximately q_0, we choose a q-independent electron–phonon matrix element similar to that used to describe the intravalley optical phonon modes. This matrix element has the form *(27,28)*

$$|Mq|_{er}^2 = \frac{\alpha \hbar^2 q_0^2 \xi^2}{2MN\hbar\omega_{er}} \tag{39}$$

where α is a degeneracy factor to be discussed shortly and ω_{er} is the frequency of the intervalley deformation potential. We consider ξ to be the screened value of the intervalley deformation potential; however, the change in ξ arising from the screening of the free carriers is usually small because of the large momentum transfer involved in an intervalley process. Since the contribution of the intervalley phonons is large, the superconducting transition temperature is a strong function of the intervalley deformation potential ξ. This deformation potential must therefore be known accurately if it is to yield a reliable value for the transition temperature. To evaluate ξ we can use optical measurements of the semiconducting energy gap and a calculation of the electron–phonon self-energies in Ge (*29*). We identify the deformation potential derived in this way with the deformation potential required for a calculation of the contribution of intervalley processes to the superconductivity. Even though they may differ and this method does not give the accuracy required, we usually have no better measurements of the intervalley deformation potentials.

The degeneracy factor α of Eq. (39) can be evaluated using the group-theoretic analysis of Lax and Hopfield (*30*). For example, their work has shown that the intervalley phonons involved in transferring electrons between the L_1 valleys in Ge transform like X_1, which is composed of the *LA* and *LO* modes degenerate at the zone edge (*31*). The value of α for Ge is, therefore, 2, and the assumption is made that for high carrier concentrations the above selection rule is still essentially correct.

If the intervalley electron–phonon matrix element M_q is assumed to be independent of q, if screening is ignored, and if an Einstein phonon distribution is used for the intervalley phonons, $V_{\text{ph}}^{er}(\beta, \delta)$ is q-independent, and the integral to obtain the kernel becomes a trivial one, yielding

$$K_{\text{ph}}^{er}(c, \delta) = (\nu - 1)\frac{k_F^3 c}{4\pi^2 E_F^3}\frac{\hbar^2\Omega}{MN}\frac{\xi^2 q_0^2}{\delta_{er}^2}\frac{1}{1-(\delta/\delta_{er})^2} \tag{40}$$

where δ_{er} is the energy of an intervalley phonon in units of the Fermi energy. This kernel for $\xi = 8$ eV is plotted along with the intravalley kernel in Fig. 5 as a function of δ for $c = 1$. It represents the largest attractive contribution to the total kernel and is larger than the Coulomb kernel, implying that a superconducting state exists for the value of $\xi = 8$ eV.

The intravalley phonon kernel is smaller than the Coulomb kernel and would not, by itself, produce a superconducting transition in the case of Ge. This contribution can be larger in other semiconductors, e.g., polar semiconductors. The peaks appearing in the intravalley phonon and intervalley phonon kernels arise from the singularities in the Bardeen–Pines interaction. In describing the intravalley acoustical phonon modes a Debye distribution is used resulting in a logarithmic singularity in the kernel. For the intravalley optical phonons and the intervalley phonons the singularity is of the $1/\delta$-type, which is a consequence of the use of an Einstein distribution for these modes. The peaks in the intravalley acoustical and optical phonon kernels do contribute to the superconductivity, but because of the small size of these kernels their contribution is very small and is therefore neglected. The intervalley phonon peaks can give a large contribution to the superconductivity and this contribution is evaluated by means of some approximations which will be discussed.

We note that in the phonon kernel presented in the present work and in other forms for this kernel (*12*) the large variation of the superconducting energy-gap function in the region of the singularity in the kernel is spread out over an energy width of approximately the Debye energy. We approximate the kernel in this region by a square well of width roughly equal to the Debye energy.

Using the square-well approximation of Fig. 5, a numerical solution to the integral equation can be obtained yielding a transition temperature for Ge of about 5 mdeg. This value of the transition temperature should be considered to be only a rough estimate of the actual transition temperature because of the uncertainties in the kernels; however, this analysis indicates that degenerate Ge is not a very good choice of a semiconductor for observing a superconducting transition. The investigation of Ge as a model semiconductor does, however, serve to point out the important properties required of a good choice. These properties will be discussed later in the text.

It was shown through our investigation of degenerate Ge that intervalley phonon processes contribute a large attractive electron–electron interaction in many-valley semiconductors, and that an increase in the number of available intervalley processes will enhance the superconducting transition temperature. We therefore examine the case of a degenerate Ge–Si alloy with the six Δ_1-valleys degenerate in energy with the four L_1-valleys. These valleys become degenerate in energy in alloys composed of approximately 15% Si (*32–34*).

We assume that the only difference between Ge and this Ge–Si alloy is the above difference in band structure, and that the changes in the phonon dispersion curves of Ge arising from the addition of Si to make the 15% Ge–Si alloy are small (*35*). Consequently, we can follow the formalism described for Ge, making the appropriate changes arising from band structure differences between Ge and Ge–Si alloy. The intravalley phonon processes will not be investigated in detail in our description of the alloy as we did in the case of Ge. These processes, which were small in Ge, are even smaller in the alloy because the existence of a larger density of electron states in the alloy implies that there is more screening in this case. We can, therefore, put an upper bound on the contribution of these processes to be equal to their contribution in the Ge case, and this contribution is much smaller than that we can expect from intervalley phonon processes in the alloy. Since there are no experimental measurements of the intervalley deformation potential ξ in the alloy, ξ will be treated as a variable in this section. We can, consequently, assume that the attractive intervalley phonon contribution for a given value of ξ contains in addition the small attractive intravalley contribution.

The L- and Δ-valleys can be described by a nearly free electron model, characterized by the density of states effective masses, m_L and m_Δ. Because these valleys have different effective masses, they are not equivalent, and the Anderson theory of "dirty" superconductors yields a different kernel in the "dirty" case than in the "clean" case. Since carrier concentrations $n \geqslant 5 \times 10^{19}$ will be considered, the Ge–Si alloy falls into the class of a "dirty" superconductor and will, therefore, be considered to have only one average superconducting energy gap Δ_n.

Following Section II.A we define a new energy gap with wave vector measured relative to the L-valleys,

$$D_n \equiv (k_L/k_L^F)\,\Delta_n \qquad \text{and} \qquad D_{n'} \equiv (k_L'/k_L^F)\,\Delta_{n'} \tag{41}$$

obeying the integral equation

$$D_n = -\int D_{n'} K_{nn'}(c, \delta) \tanh\left(\frac{E_{n'}}{2k_B T}\right) d\epsilon_{n'} \tag{42}$$

$$K_{nn'}(c, \delta) = K_{nn'}^{ra}(c, \delta) + K_{nn'}^{er}(c, \delta)$$

where k_L, k_L', and k_L^F are the electron wave vectors in the valleys for energies ϵ_n, ϵ_n', and E_F, respectively,

$$c = k_L/k_L^F \qquad \delta = (\epsilon_{n'} - \epsilon_n)/E_F \qquad \beta = q/2k_F^L$$

The intravalley kernel contains the intravalley Coulomb interaction alone and can be evaluated using a dynamic dielectric function which includes the screening arising from the presence of the valleys. The resulting intravalley kernel containing both the L- and Δ-valley contributions is

$$\begin{aligned} K_c^{ra}(c, \delta) = \frac{e^2}{2\pi\hbar^2 k_F^L \epsilon_0} \Bigg\{ & \frac{4m_L^{5/2}}{4m_L^{3/2} - 6m_\Delta^{3/2}} \\ & \times \left[\frac{1}{(\delta + c^2)^{1/2}} \int_{\frac{1}{2}|c-(\delta+c^2)^{1/2}|}^{\frac{1}{2}|c+(\delta+c^2)^{1/2}|} \frac{\epsilon_1(\beta, \delta)}{\epsilon_1^2(\beta, \delta) + \epsilon_2^2(\beta, \delta)} \frac{d\beta}{\beta}\right] \\ & + \frac{6m_\Delta^2 m_L^{1/2}}{4m_L^{3/2} + 6m_\Delta^{3/2}} \left[\frac{1}{(\delta + c^2)^{1/2}} \int_{(1/2x)|c-(\delta+c^2)^{1/2}|}^{(1/2x)|c+(\delta+c^2)^{1/2}|} \frac{\epsilon_1(\beta, \delta)}{\epsilon_1^2(\beta, \delta) + \epsilon_2^2(\beta, \delta)} \frac{d\beta}{\beta}\right]\Bigg\} \end{aligned} \tag{43}$$

where $x = k_L/k_\Delta = (m_L/m_\Delta)^{1/2}$, k_Δ is the wave vector for the Δ-valley, and $\epsilon(\beta, \delta)$ is the dynamic dielectric function for the Ge–Si alloy.

The above intravalley Coulomb kernel is smaller in the Ge–Si alloy for a given carrier concentration than in Ge. This occurs because the larger density of electron states causes more screening in this situation than in Ge as discussed above.

The intervalley kernel contains contributions arising from intervalley phonon scattering between different L_1 valleys, between different Δ_1 valleys, and between the L_1 valleys and Δ_1 valleys. To calculate these contributions, the degeneracy factor α of the phonons involved must be determined.

The first of the above three types of scattering is identical to Ge, and the intervalley phonons in this example are the double degenerate X_1-phonons or $\alpha = 2$. There exist two types of intervalley scattering between two Δ_1-valleys; the scattering can be along the cubic axis and off the cubic axis. The Δ_1-valleys have their minima at approximately 85% of the way to the Brillouin zone edge going from Γ to X. The intervalley scattering between Δ_1-valleys along the cubic axis will, therefore, involve an Umklapp process. The phonon involved is a Δ_1-phonon which is singly degenerate, $\alpha = 1$. The intervalley process between Δ_1-valleys off the cubic axis uses a singly degenerate Σ_1 phonon (*30*).

Although a complete analysis of the selection rules which exist for intervalley scattering between Δ_1- and L_1-valleys has not been made, there is one phonon involved and it must be nondegenerate; consequently, the degeneracy factor is $\alpha = 1$.

Even though each of the above intervalley processes involves different coupling constants, we expect them to be not very different among themselves and approximately equal to the Ge intervalley (L_1 to L_1) coupling constant. We have, therefore, assumed an average intervalley deformation potential ξ and have expressed all the above intervalley phonon processes in terms of ξ. The constant ξ is then treated as a parameter. (We ignore screening.)

We now choose a Bardeen–Pines interaction with a q-independent electron–phonon matrix element and an Einstein distribution for the intervalley phonons of frequency $\omega = \omega_{x_1}$. The above assumptions lead to the following form for the intervalley phonon kernel:

$$K_{\text{ph}}^{er}(c, \delta) = \frac{\hbar^2 \Omega}{MN} \frac{c(k_L^F)^3}{8\pi E_F^3} \frac{f(x)\,\xi^2 q_0^2}{\delta_0^2} \frac{1}{1 - (\delta/\delta_0)^2} \tag{44}$$

where

$$f(x) = \frac{12x^6 + 24x^3 + 15}{2x^3 + 3} \qquad \delta_0 = \hbar\omega_0 / E_F$$

Although the intervalley Coulomb kernel can be neglected in the case of Ge, in the alloy, because of the larger density of states, it can contribute a larger repulsive interaction. This has a form similar to Eq. (38), with corrections arising from the presence of the extra valleys.

The intravalley Coulomb, intervalley phonon, and intervalley Coulomb, kernels were evaluated for three different values of n, the carrier concentration. These values were $n = 5 \times 10^{19}$ carriers/cm^3, $n = 10^{20}$ carriers/cm^3, and $n = 5 \times 10^{20}$ carriers/cm^3. The intervalley phonon kernel was computed for five values of ξ ($\xi = 6$, 7, 8, 9, and 10 eV) for each value of n. The resulting kernels were approximated by a series of square wells. As in the case of Ge, the contribution of the peaks in the intervalley phonon kernel were evaluated by assuming a square-well kernel in the region of the peaks. Using this square-well approximation, the contribution of the peaks to the transition temperature can be estimated with a maximum error of about 40%. This uncertainty in the contribution of the peaks is equivalent to approximately a 4% change in the coupling constant.

By means of square-well approximations for the kernels, the integral equations were solved for the superconducting transition temperature. As an example of these kernels, we show in Fig. 6 the kernel for $\xi = 8$ eV and $n = 10^{20}$ carriers/cm^3.

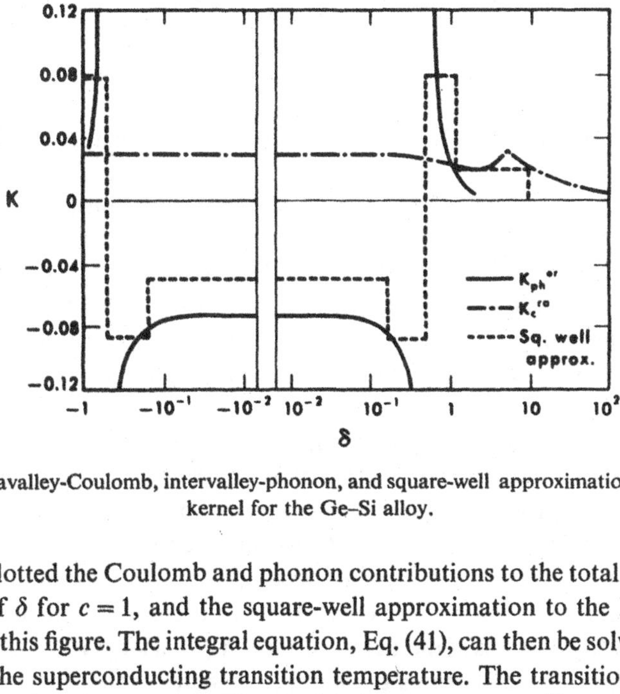

Fig. 6. Intravalley-Coulomb, intervalley-phonon, and square-well approximation to the total kernel for the Ge–Si alloy.

We have plotted the Coulomb and phonon contributions to the total kernel as a function of δ for $c = 1$, and the square-well approximation to the kernel also appears in this figure. The integral equation, Eq. (41), can then be solved numerically for the superconducting transition temperature. The transition temperature for the sample kernel in Fig. 6 with $\xi = 8$ and $n = 10^{20}$ is $T_c = 0.16$°K. Although as in the case of Ge, this transition temperature is only to be considered approximate because of the uncertainties in the kernels, it is worth noting that T_c is in a measurable range for a reasonable choice of ξ and n ($\xi = 8$ eV, $n = 10^{20}$ carriers/cm^3). We also note that for $n = 10^{20}$ carriers/cm^3 the minimum ξ for which the phonon kernel is larger than the Coulomb kernel is $\xi \approx 4.9$ eV.

We have solved in a similar way the integral equation for the other values of ξ and n mentioned above. By interpolating between these values, we have plotted isotherms for three values of the transition temperature T_c: $T_c = 0.002$°K,

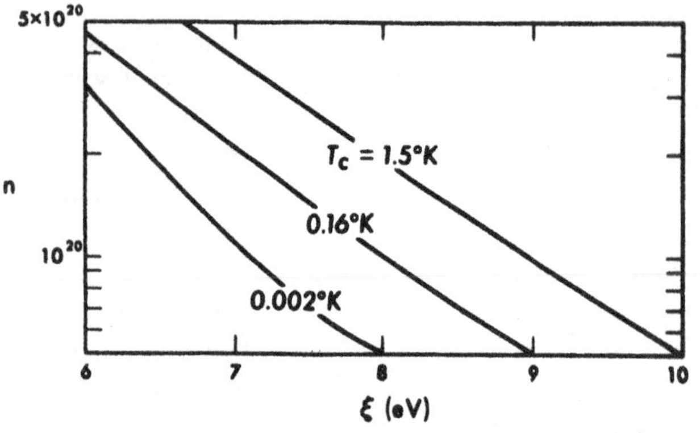

Fig. 7. Approximate isotherms for the superconducting transition temperature T_c of the Ge–Si alloy as a function of the carrier concentration n and the intervalley coupling constant ξ.

$T_c = 0.16$°K, and $T_c = 1.5$°K. These isotherms appear in Fig. 7 and indicate that T_c is a strong function of both ξ and n, as is expected, and can vary by orders of magnitude for a 20% change in either of these constants.

We also note in Fig. 7 that for very high concentrations ($n \approx 5 \times 10^{20}$) the transition temperature is still in the 0.01–0.1°K range for an average coupling constant, ξ, 25% less than that expected for Ge (*29*). This indicates that at high concentrations of impurities the Ge–Si alloy can show superconducting properties and therefore seems to be an interesting choice for an experimental investigation.

The investigations of Ge and the Ge–Si alloy indicate the important properties required of a semiconductor in which a superconducting transition is expected to occur. These properties can be best illustrated by discussing the band-structure parameters of the semiconductor and considering them adjustable variables which maximize the superconducting transition temperature.

A maximization of the carrier concentration n is essential. The larger n, the larger the density of states at the Fermi surface, $N(0)$; a large $N(0)$ is important for two reasons: (1) it increases the number of states available for scattering, and (2) it makes the dielectric screening more complete. The increase in the number of available states is equivalent to effectively increasing the "$N(0)\,V$" parameter of the BCS theory, mainly through the effect on the kernel of the attractive intervalley processes. The effect of a better screening is twofold: it reduces the repulsive intravalley-Coulomb interaction but at the same time the attractive inttravalley-phonon interaction is also reduced. The latter is not very important in our case, as the dominant phonon processes are those arising from intervalley scattering, which, because of the large momentum transfer, are essentially unscreened.

The number of degenerate valleys should also be maximized to contribute as many intervalley processes as possible. The presence of more valleys increases the density of states at the Fermi energy, which, as is stated in the above discussion, is a desirable situation.

The density of states and screening of the intravalley-Coulomb interaction are also increased by maximizing m^*, the single-valley effective mass, and ϵ_0, the static dielectric constant. Let us incidentally remark that the intervalley Coulomb interaction is also reduced significantly by a large ϵ_0.

Finally, ξ, the intervalley coupling constant, and α, the phonon degeneracy factor, should both be as large as possible. The degeneracy factor can be determined by a group-theoretical analysis of intervalley scattering in the crystal. The coupling constant ξ can be extracted from a measurement of the change of the semiconducting energy gap as a function of temperature. Despite the fact that the value of ξ is difficult to obtain accurately and has not been determined for most materials, the values of the other parameters mentioned can easily be obtained, giving an indication of how promising a choice a semiconductor will be for observing superconductivity.

D. Kernels and Results for Polar Interactions

1. Kernels

In the case of a polar semiconductor, a doped ferroelectric, or a doped pseudo-ferroelectric (i.e., $SrTiO_3$) the possibility of strongly coupling electrons to optical phonons arises (*36,37*). This problem can be formulated in the same manner as described for the nonpolar case, i.e., by computing the screened Coulomb and phonon kernels, but for the polar case it is more convenient to treat some of the intravalley interactions in a slightly different way. In particular, a dielectric function is evaluated which contains both the intravalley Coulomb and optical phonon–electron interactions. It is convenient to treat these interactions in this way because of the mixing of the optical phonon modes and the plasma modes. For example, at long wavelengths, the electrons (quasi-particles) interact via the coupled "phoplasmon" modes.

We consider here only the optical phonons. The contributions of the acoustic phonons and the intervalley phonons can be evaluated as in the last section. The intravalley Coulomb and electron–optical phonon interactions are now described in terms of a dielectric function $\epsilon(q, \omega)$, which is a sum of electronic and ionic contributions:

$$\epsilon(q, \omega) = \epsilon_{el} + \alpha_{ionic} \tag{45}$$

The ionic part is given by

$$\alpha_{ionic} = \sum_i \frac{4\pi z_i^2}{\omega_i^2 - \omega^2 + i\gamma_i\omega} \tag{46}$$

where the z_i's are the effective charges of the transverse-optic (TO) phonon modes of frequency ω_i; the γ_i's are the damping constants and i is the mode index.

For simplicity we will limit ourselves to the diatomic case and explore the "*one-phonon*" problem (i.e., a single nondegenerate LO-branch and a single doubly degenerate TO-branch). For this case it is convenient to write Eq. (46) in terms of experimentally measured dielectric constants (*38*). Equation (45) becomes (without damping)

$$\epsilon(q, \omega) = \epsilon_{\text{el}} + \frac{(\epsilon_0 - \epsilon_\infty)\, \omega_{TO}^2}{\omega_{TO}^2 - \omega^2} \tag{47}$$

where ϵ_0 and ϵ_∞ are the low- and high-frequency dielectric constants for the undoped material and ω_{TO} is the frequency of the TO-mode. This dielectric function should give the *total* electron–electron intravalley interaction when it is used in Eq. (30) (recall we are ignoring here acoustical phonon contributions).

As an aside, it is instructive (and reassuring) to show that computing the electron–electron interaction with the above dielectric function is equivalent to taking the Coulomb and Bardeen–Pines terms separately as was done in the previous section. To do this we use Eq. (45) and break up (*36*) V^{ra} in the following way:

$$V^{ra} = \frac{4\pi e^2}{q^2}\left[\frac{1}{\epsilon_{\text{el}}} - \frac{\alpha_{\text{ionic}}}{\epsilon_{\text{el}}(\epsilon_{\text{el}} + \alpha_{\text{ionic}})}\right] \tag{48}$$

which for the single-mode case, Eq. (47) gives

$$V^{ra} = \frac{4\pi e^2}{\Omega q^2}\left[\frac{1}{\epsilon_{\text{el}}} - \frac{(\epsilon_0 - \epsilon_\infty)\, \omega_{TO}^2}{\epsilon_{\text{el}}(\epsilon_{\text{el}}\omega_{TO}^2 - \epsilon_{\text{el}}\omega^2 + (\epsilon_0 - \epsilon_\infty)\, \omega_{TO}^2)}\right] \tag{49}$$

Using the Lyddane–Sachs–Teller relation $\omega_{LO}^2 = (\epsilon_0/\epsilon_\infty)\, \omega_{TO}^2$, Eq. (49) becomes

$$V^{ra} = \frac{4\pi e^2}{\Omega q^2}\left[\frac{1}{\epsilon_{\text{el}}} - \frac{[(1/\epsilon_\infty) - (1/\epsilon_0)]\, \omega_{LO}^2}{(\epsilon_{\text{el}}/\epsilon_\infty)^2\, (\tilde{\omega}^2 - \omega^2)}\right] \tag{50}$$

where ω_{LO} is the "bare" (undoped crystal) LO-frequency and $\tilde{\omega}$ is the renormalized frequency given by

$$[\tilde{\omega}(q, \omega)]^2 = \omega_{TO}^2\left[1 - \frac{\epsilon_0 - \epsilon_\infty}{\epsilon_{\text{el}}(q, \omega)}\right] = \omega_{TO}^2\left[\frac{\epsilon_\infty}{\epsilon_0} + \frac{\epsilon_\infty}{\epsilon_{\text{el}}} - \frac{\epsilon_\infty^2}{\epsilon_0 \epsilon_{\text{el}}}\right] \tag{51}$$

Inspection of Eq. (50) shows that we have proved what we set out to prove. The interaction V^{ra} separates into a repulsive electron–electron term screened by the electronic dielectric function and an electron–phonon–electron term, having the Bardeen–Pines form, screened by the square of the electronic dielectric function. This was the same form that was assumed in the last section and in (*6*).

It is interesting to note that the poles of the phonon term occur at the renormalized phonon frequencies $\tilde{\omega}$ rather than the bare *LO*-frequencies ω_{LO}. The ω_{LO} are the longitudinal resonance of the undoped crystal, i.e., at $\omega = \omega_{LO}$, the dielectric function for the undoped crystal goes to zero. The $\tilde{\omega}$ now give the zeros of the new dielectric function with free carriers. This can be seen by using Eq. (31),

$$\epsilon_{el} = \epsilon_\infty + 4\pi\alpha_L(q, \omega) \tag{52}$$

where $\alpha_L(q, \omega)$ is the Lindhard polarizability and this gives an expression for

$$\tilde{\omega}^2(q, \omega) = \omega_{LO}^2\left(\frac{1 + 4\pi\alpha_L/\epsilon_0}{1 + 4\pi\alpha_L/\epsilon_\infty}\right) = \omega_{TO}^2\left(\frac{\epsilon_0 + 4\pi\alpha_L}{\epsilon_\infty + 4\pi\alpha_L}\right) \tag{53}$$

which satisfies $\epsilon(q, \tilde{\omega}) = 0$. Note that $\tilde{\omega} \to \omega_{LO}$ as $\alpha_L \to 0$ (i.e., n small or q large) but $\tilde{\omega} \to \omega_{TO}$ for $\alpha_L \to \infty$ (i.e., n large or q small) and the Lyddane–Sachs–Teller relation is "shorted out" by the free carriers.

At this point the problem is to find the kernel for this interaction. Unfortunately, this is not straightforward. The complexities lie in evaluating the integrals for the Lindhard part of the dielectric function. The same problem existed in the last section and the approach there was to rely on electronic computers. The physics is more transparent in the case of nonpolar materials since the modes are more or less distinct and one can separate them and screen in an appropriate way. For the polar case the modes mix, and it is convenient to explore the limiting expressions for the dielectric function‡ and discuss the resonant modes, the interactions, and the kernels within these approximations. After an analysis of this type, the results of computer calculations with the full Lindhard function are more understandable physically.

a. *Long-wavelength approximation*

In the long-wavelength limit ($q \to 0$) or "classical limit" $4\pi\alpha_L = -\omega_p^2/\omega^2$, where $\omega_p^2 = 4\pi ne^2/m^*$ and Eq. (47) can easily be solved for the coupled phonon–plasmon (*4,39,42*) modes,

$$2\omega_\pm^2 = \omega_{LO}^2 + \tilde{\omega}_p^2 \pm \sqrt{(\omega_{LO}^2 + \tilde{\omega}_p^2)^2 - 4\tilde{\omega}_p^2\omega_{TO}^2} \tag{54}$$

where $\tilde{\omega}_p^2 = \omega_p^2/\epsilon_\infty$. It is interesting to examine the coupled modes for three limiting cases of Eq. (54). First, for low density,

$$\tilde{\omega}_p < \omega_{TO} < \omega_{LO}$$

$$\omega_+^2 \simeq \omega_{LO}^2 + \omega_p^2\left(\frac{1}{\epsilon_\infty} - \frac{1}{\epsilon_0}\right) = \omega_{LO}^2 + \frac{\tilde{\omega}_p^2}{\omega_{LO}^2}(\omega_{LO}^2 - \omega_{TO}^2) \tag{55}$$

$$\omega_-^2 \simeq \omega_p^2/\epsilon_0$$

‡ Throughout this section we will ignore the effects of damping.

The upper, "phonon-like," mode is raised by the coupling while the lower, "plasmon-like," mode is depressed by the phonon or ionic screening. Second, for high density, $\omega_t < \omega_{LO} < \tilde{\omega}_p$ and

$$\begin{aligned} \omega_+^2 &\simeq \tilde{\omega}_p^2 + \omega_{LO}^2 - \omega_{TO}^2 \\ \omega_-^2 &\simeq \omega_{TO}^2 - \frac{\omega_{TO}^2}{\tilde{\omega}_p^2}(\omega_{LO}^2 - \omega_{TO}^2) = \omega_{TO}^2 - \frac{\omega_{TO}^4}{\omega_p^2}(\epsilon_0 - \epsilon_\infty) \end{aligned} \tag{56}$$

For this case, the upper, "plasmon-like," mode is pushed higher than $\tilde{\omega}_p$ while the lower, "phonon-like," mode is pinned below ω_{TO}. This is the case described before where the Lyddane–Sachs–Teller relation is shorted out and the low longitudinal mode is trapped below the *TO*-mode. A sketch of the coupled $q = 0$ phoplasmon mode frequencies as a function of n is given in Fig. 8. Third, for very polar materials such as "pseudo-ferroelectrics" (i.e., $SrTiO_3$), ω_{TO} is very small at low temperatures. In fact, it is the decrease of ω_{TO} with temperature which causes the system to have ferroelectric-like properties. For this case $\omega_{TO} \ll \omega_{LO}$ and $\omega_{TO} < \omega_p$ (i.e., $SrTiO_3$ $n \sim 10^{18}$–10^{21} cm^{-2}) and $\tilde{\omega}_p^2\omega_{TO}^2 \ll (\omega_{LO}^2 + \tilde{\omega}_p^2)^2$. These conditions and Eq. (54) give

$$\begin{aligned} \omega_+^2 &\simeq \omega_{LO}^2 + \tilde{\omega}_p^2 - \frac{\tilde{\omega}_p^2\omega_{TO}^2}{\omega_{LO}^2 + \tilde{\omega}_p^2} \simeq \omega_{LO}^2 + \tilde{\omega}_p^2 \\ \omega_-^2 &\simeq \frac{\tilde{\omega}_p^2\,\omega_{TO}^2}{\omega_{LO}^2 - \tilde{\omega}_p^2} \end{aligned} \tag{57}$$

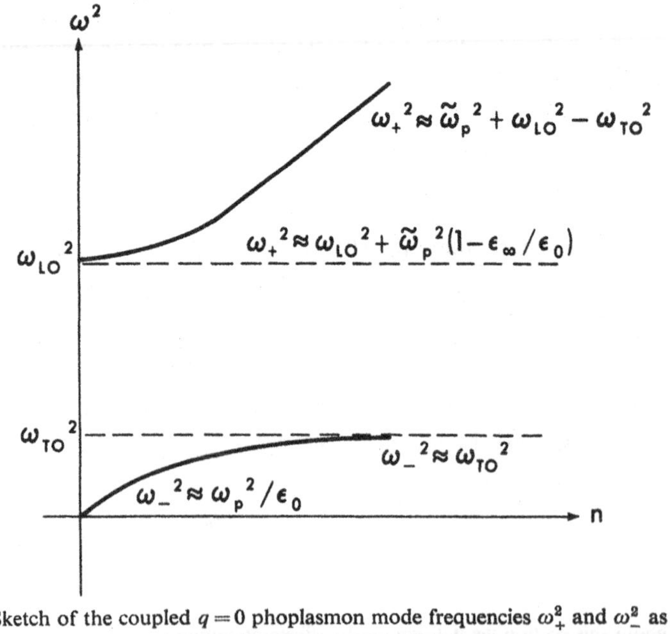

Fig. 8. Sketch of the coupled $q = 0$ phoplasmon mode frequencies ω_+^2 and ω_-^2 as a function of carrier concentration.

It is now possible to compute the interactions for electrons coupled to the mixed phonon–plasmon modes in the $q=0$ limit.

$$V(q, \omega) = \frac{4\pi e^2}{\Omega q^2}\left[\epsilon_\infty + \frac{\epsilon_0 - \epsilon_\infty}{\omega_{TO}^2 - \omega^2}\,\omega_{TO}^2 - \frac{\omega_p^2}{\omega^2}\right]^{-1} \tag{58}$$

$$V(q, \omega) = -\frac{4\pi e^2}{\Omega \epsilon_\infty q^2}\left[\frac{\omega^4 - \omega^2(\omega_{LO}^2 + \tilde{\omega}_p^2) + \tilde{\omega}_p^2\omega_{TO}^2}{\omega^2(\omega_{TO}^2 - \omega^2)}\right]^{-1} \tag{59}$$

$$V(q, \omega) = -\frac{4\pi e^2}{\Omega \epsilon_\infty q^2}\left[\frac{\omega^2(\omega_{TO}^2 - \omega^2)}{(\omega^2 - \omega_+^2)(\omega^2 - \omega_-^2)}\right] \tag{60}$$

Equation (60) demonstrates that the total $q=0$ interaction can be given in terms of simple poles at ω_+ and ω_- with $\omega_- < \omega_{TO} < \omega_{LO} < \omega_+$. The ω_- and ω_+ are given in certain limits in Eqs. (55), (56), and (57). For all three cases the interaction has a $1/X$ type singularity at ω_+ and ω_- and it goes to zero at ω_{TO}. The "strength" of the interaction is proportional to

$$\left|\frac{\omega_\pm(\omega_{TO}^2 - \omega_\pm^2)}{\omega_+^2 - \omega_-^2}\right|$$

and it is therefore stronger for ω_+ than ω_-.

Equation (60) can be combined with Eq. (14) and the kernel can be evaluated for this case. Since the interaction is of the $1/q^2$ form the limits give a singularity at $\delta = 0$. This singularity represents a failure of this approximation.

b. *Fermi–Thomas approximation*

A more appropriate (free of singularities at $\delta = 0$) but still highly approximate form of α_L is the Fermi–Thomas limit, i.e., $\hbar\omega/E_F \to 0$ and q/k_F small. In this limit

$$4\pi\alpha_L = k_s^2/q^2 = (4\pi e^2/q^2)\,N(E_F) \tag{61}$$

where $N(E_F)$ is the electronic density of states at E_F. This approximation yields only one phonon-like solution for $\tilde{\omega}(q, \omega)$, since the "static" limit omits collective electronic excitations such as plasmons. The equation for the resonant frequencies, Eq. (53), becomes

$$\tilde{\omega}^2(q) = \omega_{TO}^2\left(\frac{q^2\epsilon_0 + k_s^2}{q^2\epsilon_\infty + k_s^2}\right) = \omega_{LO}^2\left(\frac{q^2 + k_s^2/\epsilon_0}{q^2 - k_s^2/\epsilon_\infty}\right) \tag{62}$$

In this limit $\tilde{\omega} \to \omega_{TO}$ for q/k_s small and $\tilde{\omega} \to \omega_{LO}$ for large q/k_s. So ω_{LO} and ω_{TO} are the high and low limit of the resonant frequency in this limit (no plasmons). Note that Eq. (62) yields the usual jellium result if $\epsilon_\infty \to 1$ and $\epsilon_0 \to \infty$, i.e., metallic behavior.

The intravalley interaction in the Fermi–Thomas approximation has the form

$$V(q, \omega) = \frac{4\pi e^2}{\Omega q^2}\left(\epsilon_\infty + \frac{k_s^2}{q^2} + \alpha_{\text{ionic}}\right)^{-1}$$

$$= \frac{4\pi e^2}{\Omega \epsilon_{\text{ph}}}\left(q^2 + \frac{k_s^2}{\epsilon_{\text{ph}}}\right)^{-1} = \frac{4\pi e^2}{\Omega \epsilon_{\text{ph}} (2k_F)^2} (\beta^2 + k_s'^2/\epsilon_{\text{ph}})^{-1} \tag{63}$$

where $k_s'^2 = k_s^2/(2k_F)^2$ and $\epsilon_{\text{ph}} = \epsilon_\infty + \alpha_{\text{ionic}}$ is the "bare" dielectric function of the undoped crystal assumed to be independent of wave vectors. Using the above interaction, the intervalley kernel can be computed in this approximation:

$$K^{ra}(c, \delta) = \frac{\Omega k_F^3}{4\pi^2 E_F} \frac{1}{\sqrt{\delta + c^2}} \int_{\frac{1}{2}|c - \sqrt{\delta + c^2}|}^{\frac{1}{2}|c + \sqrt{\delta + c^2}|} \frac{4\pi e^2}{\Omega \epsilon_{\text{ph}} (2k_F)^2} \left(\beta^2 + \frac{k_s'^2}{\epsilon_{\text{ph}}}\right) \beta \, d\beta \tag{64}$$

It is possible to express this kernel in terms of a "pseudo-electron gas parameter,"

$$r_s = (3/4\pi n)^{1/3} (m^* e^2/\hbar^2 \epsilon_\infty)$$

but the interpretation in terms of r_s is a bit tricky. The contribution of α_{ionic} must also be considered; otherwise, $r_s \gg 1$ would mean that no free electrons were available from the impurities. A true "electron gas parameter" in the usual sense probably does not exist, since the dynamics of α_{ionic} must always be considered in a description of the electron gas. The parameter r_s is therefore used here only for convenience. Integrating Eq. (64) we get

$$K^{ra}(c, \delta) = \frac{\alpha r_s \epsilon_\infty}{4\nu \epsilon_{\text{ph}}} \frac{1}{\sqrt{\delta + c^2}} \log \left| \frac{\frac{1}{4}|c + \sqrt{\delta + c^2}|^2 + k_s'^2/\epsilon_{\text{ph}}}{\frac{1}{4}|c - \sqrt{\delta + c^2}|^2 + k_s'^2/\epsilon_{\text{ph}}} \right| \tag{65}$$

where $k_s'^2 = \epsilon_\infty \alpha r_s$, $\alpha = (4\nu^4/9\pi^4)^{1/3}$.

As an example of a limiting form of Eq. (65) for high densities, $\delta \to 0$ or $E_F \gg \hbar\omega_{LO}$, Eq. (65) becomes $(c = 1)$

$$K^{ra}(\delta) = \frac{\alpha r_s}{4\nu} \frac{\omega^2 - \omega_{TO}^2}{\omega^2 - \omega_{LO}^2} \log \left| 1 + \frac{1}{\alpha r_s} \frac{\omega^2 - \omega_{LO}^2}{\omega^2 - \omega_{TO}^2} \right| \tag{67}$$

A sketch of this kernel appears in Fig. 9a along with kernels for $E_F \sim \hbar\omega_{LO}$ and $E_F \ll \hbar\omega_{LO}$ in Figs. 9b and c. In all three approximations, the kernel is repulsive at $\omega = 0$ and goes to zero at ω_{TO}. The kernels then exhibit negative and positive peaks at frequencies which make the denominator and numerator of the argument of the log in Eq. (65) vanish. In the high-density limit, the attractive peak occurs at ω_{TO}, and since the coefficient of the log vanishes at this frequency, the pole does not appear.

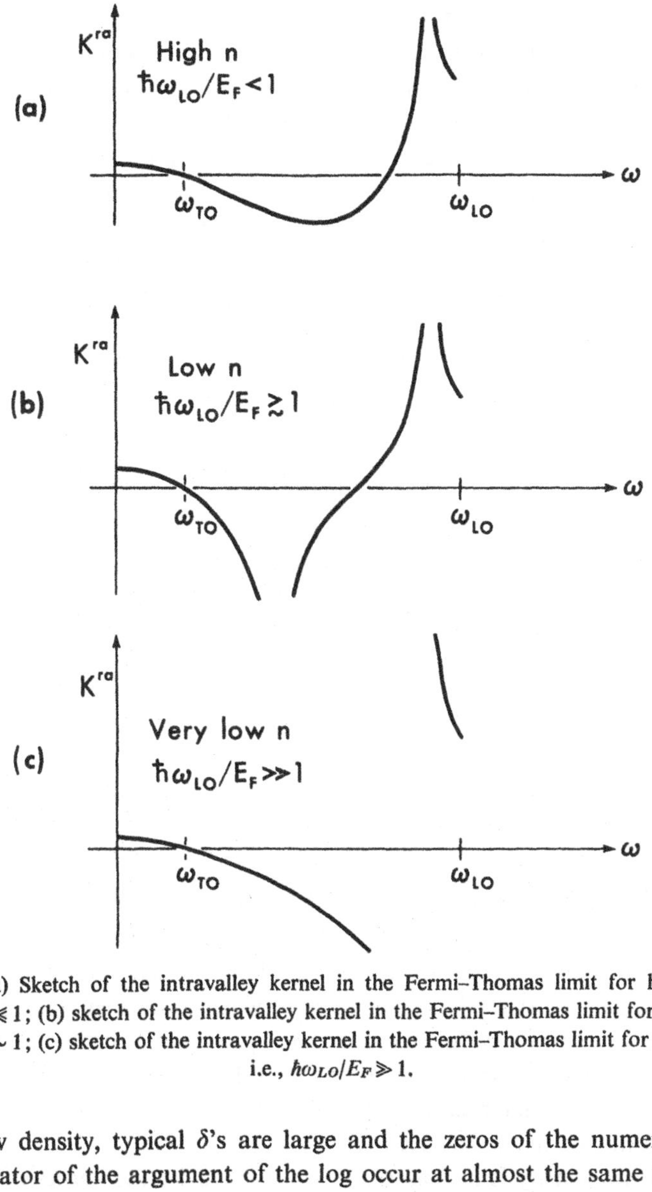

Fig. 9. (a) Sketch of the intravalley kernel in the Fermi–Thomas limit for high n, i.e., $\hbar\omega_{LO}/E_F \ll 1$; (b) sketch of the intravalley kernel in the Fermi–Thomas limit for low n, i.e., $\hbar\omega_{LO}/E_F \sim 1$; (c) sketch of the intravalley kernel in the Fermi–Thomas limit for very low n, i.e., $\hbar\omega_{LO}/E_F \gg 1$.

At low density, typical δ's are large and the zeros of the numerator and denominator of the argument of the log occur at almost the same frequency $\omega \approx \omega_{LO}$. This causes the resonance to look more and more like a $1/X$ singularity as the density is decreased. This is shown in Figs. 9b and c.

c. *Lindhard approximation and discussion*

The long-wavelength approximation and the Fermi–Thomas approximation have both been discussed to give some feeling for the kernels and the effects of mode mixing. The Fermi–Thomas approximation is the more appropriate of the two for the kernel and the results of using the full Lindhard (*16*) dielectric function of Eq. (31) are similar in many respects to the Fermi–Thomas results.

The problem is to solve Eq. (14) using the full Lindhard dielectric function [Eqs. (52) and (31) added to the ionic contribution as in Eq. (45)]. The resulting dielectric function is too complicated to allow an analytic solution of Eq. (14) and one is forced to use the computer. Sample results of computer calculations of K^{ra} are given in Figs. 10 and 11. Both kernels were computed assuming a three-valley semiconductor with $m^* = 1.6\, m_e$. For the kernel shown in Fig. 10

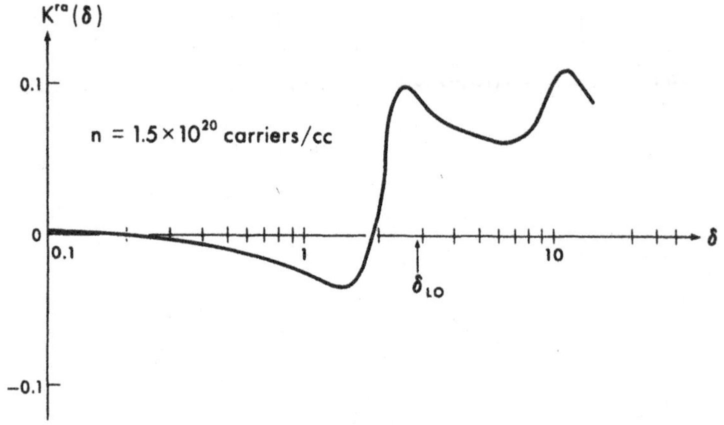

Fig. 10. Sample intravalley kernel in the Lindhard approximation for $n = 1.5 \times 10^{20}$ carriers/cm^3.

the carrier density was assumed to be $n = 1.5 \times 10^{20}$ carriers/cm^3. At this carrier concentration and for the parameters chosen we are somewhere between the limits taken in Fig. 9a and b for the Fermi–Thomas case. The attractive dip is shallow and smaller than the repulsive peak which occurs at frequencies just below ω_{LO}. The extra repulsive peak above ω_{LO} comes from the point in the integration where the phoplasmon mode enters the continuum. The peak is also present in nonpolar materials as a plasma peak (see Fig. 6 at $\delta \sim 8$). This peak is discussed further in the last section.

If the carrier density is increased above the value for the kernel of Fig. 10, the attractive dip becomes smaller and tends to disappear. The repulsive peak is still present, but it is decreased in size. The $\delta_{LO} = \hbar\omega_{LO}/E_F$ is also reduced because of the increasing Fermi energy.

At lower densities the attractive dip begins to grow. The phoplasmon peak gets closer to the repulsive phonon peak and we get kernels such as the one shown in Fig. 11. This case corresponds to a limit somewhere between that

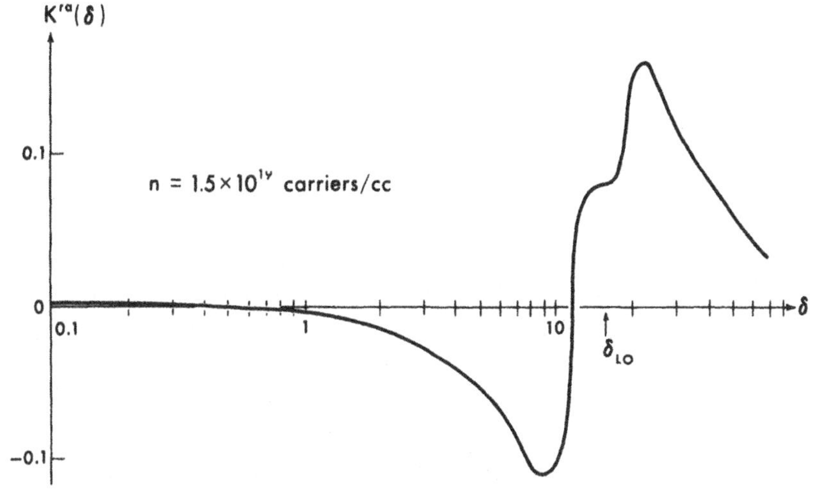

Fig. 11. Sample intravalley kernel in the Lindhard approximation for $n = 1.5 \times 10^{19}$ carriers/cm^3.

shown in Figs. 9b and c for the Fermi–Thomas case. The extra peak just above ω_{LO} is the phoplasmon peak which is absent in the Fermi–Thomas approximation of Fig. 9. This kernel, being larger and having a large attractive dip which is favorable for superconductivity, gives a higher transition temperature than the higher density of kernel Fig. 10. The transition temperature for the polar interaction therefore decreases with increasing density.

The "Lindhard" kernels of Figs. 10 and 11 resemble the Fermi–Thomas kernels of Fig. 9 in the frequency range up to ω_{LO}. One can use the analysis of the Fermi–Thomas calculation of the last section to understand the computer results given in Figs. 10 and 11.

2. Calculation of Gurevich, Larkin, and Firsov

In 1962, Gurevich, Larkin, and Firsov (*4*) (GLF) explored the possibility of superconductivity in a polar semiconductor. This work represents a pioneering attempt at a quantitative examination of the problem of superconductivity in low-density systems. There are, however, several assumptions and conclusions in this paper which are not entirely consistent with the theory given in the last section. We will discuss this paper in some detail using the theory developed in previous sections and try to point out some of the problems.

GLF begin by discussing the problem of obtaining a degenerate electron gas from the impurity electrons and conclude that r_s must be less than unity to ensure impurity banding. Actually this is a complex problem, since the parameter r_s contains the dielectric constant and some question exists as to which dielectric constant to use in r_s. The r_s defined by Eq. (66) uses ϵ_∞, but, as is stated, this parameter is only used for convenience and it is not implied that an $r_s > 1$ will give impurity electrons bound to the impurity centers. This is because $r_s' = r_s(\epsilon_\infty/\epsilon_0) \ll 1$ and the phonons can provide some screening. One can, therefore, have systems where $r_s \gg 1$ while $r_s' \ll 1$ (i.e., $SrTiO_3$) and the exact nature of the metallic transition is not understood (*43*). The problem of considering r_s is complicated by the fact that the phonon contribution is frequency-dependent and the electron gas constant is itself not constant with frequency. The problem then becomes an interesting study of the dynamics of a coupled electron–lattice system. Fortunately, from the point of view of discussing the superconducting properties of a specific material, this problem becomes an academic one. One can test the system experimentally to see if it is a degenerate electron gas. In fact, since the kernels of the integral equation represent electron–electron interactions, it may be possible in the future to use the superconducting properties to shed some light on the exact mechanisms of the transition to the metallic state.

In calculating the kernel, GLF assume that $E_F \gg \hbar\omega_{LO}$ and $q^2 \ll k_s^2$. These conditions are essentially equivalent to the Fermi–Thomas conditions on the dielectric function, and the kernel in this limit is given in Fig. 9a. To show this more clearly, we can incorporate the above conditions into Eq. (67). The second term dominates the argument of the log and the kernel has the form (for $v = 1$)

$$K_{\mathrm{GLF}} = \alpha_{\mathrm{GLF}} \frac{\omega^2 - \omega_{TO}^2}{\omega^2 - \omega_{LO}^2} \tag{68}$$

where‡

$$\alpha_{\mathrm{GLF}} = \frac{\alpha r_s}{4} \log \left| \frac{\epsilon_0}{k_s'^2} \right| \tag{69}$$

Equation (68) has limited validity; the frequency dependence of the argument of the log has been neglected, and, as we saw in the last section, this can be very important. In addition, we are still working only in the Fermi–Thomas limit. The kernel is now dominated by

$$\epsilon_{\mathrm{ph}}^{-1} = \left[\frac{\epsilon_\infty(\omega^2 - \omega_{LO}^2)}{\omega^2 - \omega_{TO}^2} \right]^{-1}$$

A pole occurs at ω_{LO} which yields a $1/X$ type of singularity in the kernel. GLF neglect the attractive part of this singularity and replace the kernel by two *repulsive* square wells. This approximation and a sketch of the kernel of Eq. (68) is shown in Fig. 12. The omission of the attractive part of the kernel is not as

‡ This is Eq. (13) of (*41*).

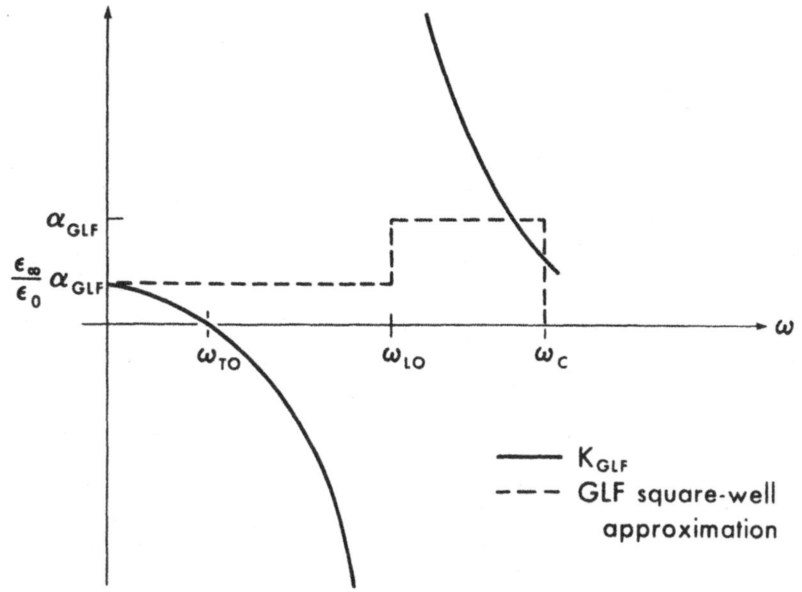

Fig. 12. GLF kernel of Eq. (68) and the GLF square-well approximation to the kernel.

bad an approximation as Fig. 12 suggests, since at high densities, i.e., at the GLF limit $E_F \gg \hbar\omega_{LO}$ the kernel of Fig. 9a is a more appropriate approximation than K_{GLF}. In the case of very high density the attractive part of the kernel of Fig. 9a gets very small.

The square-well approximation in Fig. 12 is $K = (\epsilon_\infty/\epsilon_0)\,\alpha_{GLF}$ for $\omega < \omega_{LO}$, $K = \alpha_{GLF}$ for $\omega_{LO} < \omega < \omega_c$ and $K = 0$ for $\omega > \omega_c$. The cutoff frequency ω_c, is taken to be approximately equal to $E_F/\hbar$. This estimate is somewhat low *(15,44)*. Using Eqs. (18) and (19) and assuming the gap to be $\Delta = \Delta_1 + \Delta_2$ for $\omega < \omega_{LO}$ and Δ_2 for $\omega_{LO} < \omega < \omega_c$, we get

$$\begin{aligned} \Delta_1 &= -\alpha\left[\Delta_1 \log\left(\frac{\hbar\omega_c}{\Delta}\right) + \Delta_2 \log\left(\frac{\hbar\omega_{LO}}{\Delta}\right)\right] \\ \Delta_2 &= \alpha\Delta(1 - \epsilon_\infty/\epsilon_0)\log\left(\frac{\hbar\omega_{LO}}{\Delta}\right) \end{aligned} \tag{70}$$

which are equations (15) of GLF. Solving for Δ [i.e., Eq. (24)] we obtain the GLF‡ result,

$$1 = \alpha\left[\left(1 - \frac{\epsilon_\infty}{\epsilon_0}\right) - \frac{1}{1 + \alpha\log(E_F/\hbar\omega_{LO})}\right]\log\left(\frac{\hbar\omega_{LO}}{\Delta}\right) \tag{71}$$

‡ It is unlikely that the gap anisotropy predicted by GLF will be observed because of the Anserson theory. See *(11)* and the discussion of Section II. A.

GLF then make the observation that for $E_F < \hbar\omega_{LO}$ there is no superconductivity. The reason for their conclusion can be seen in Fig. 12. For this limit their approximation for the kernel becomes a single repulsive square well. This condition is, therefore, not a general one but results from the approximations. In fact, as can be seen in Fig. 9, as $\hbar\omega_{LO}/E_F$ gets larger the attractive peak begins to appear and it gets stronger for lower densities. This peak, which is lost in the GLF model, can contribute strongly to the superconductivity.

So the GLF two-repulsive-well model is not a realistic one at lower n and the conclusion that $E_F \gg \hbar\omega_{LO}$ must be satisfied results from the model. If the attractive part of the kernel is included, an acceptable solution of the two-square-well model is obtained.

The GLF model yields a solution for the transition temperature which decreases with increasing carrier concentration. This is also found in the last section for the high-n (GLF) limit. In addition, at lower n it was shown that the transition temperature decreased when n increased, not only because the kernel size was changing but also because of the shape, e.g., an attractive well appeared.

One effect which tends to destroy superconductivity at lower densities in this model is lifetime effects. These effects have not be included here. It is expected that for $E_F < \hbar\omega_{LO}$, lifetime effects could cut off the kernels at energies less than $\hbar\omega_{LO}$, resulting in a loss of the most important contributions of the kernel to the superconductivity. (In addition, the cutoff for negative frequencies will be reduced.)

The fundamental question of the form of the integral equation for the case where the phonon frequency is greater than the Fermi energy of the impurity electron gas system has not been answered as yet. The role of Migdal's theorem for this type of system with a low carrier density gas interacting via the phonons of the host crystal should be explored.

III. SOME PROPERTIES OF LOW-CARRIER-DENSITY SUPERCONDUCTORS

Using the results of the calculations described in the previous sections it was possible to encourage experimental searches for superconductivity in degenerate semiconductors and semimetals. In particular, it was suggested that materials be chosen having favorable normal-state properties, i.e., those which maximize the attractive kernels and minimize the repulsive ones. The calculations also indicated that a superconductive transition should occur in a temperature range which can be reached using adiabatic demagnetization techniques, i.e., 0.01–0.1°K. These materials were expected to show type II or hard superconductor properties and the superconducting transition temperature was expected to be very sensitive to carrier density charges and strain.

The first degenerate semiconductor to exhibit superconducting properties were GeTe (*7*). It was chosen for investigation because the normal-state properties of this material were very favorable. In particular, crystals having very high carrier densities could be obtained. The transition temperature was in the 0.1°K range, the magnetic properties indicated that this material was a type II or hard superconductor and the transition temperature was found to depend on carrier density.

The idea was then advanced that "pseudo-ferroelectrics" (i.e., crystals in which the static dielectric constant increases with decreasing temperature, but a ferroelectric transition does not occur) would be good choices for investigation. $SrTiO_3$ was the first to be tested and it was found (*8*) superconducting in the 0.1°K range. Again, the superconducting transition temperature was a function of carrier density.

SnTe was next discovered (*9*) to be a superconductor with properties very similar to GeTe.

All three materials were carefully investigated to show that the superconductivity was a bulk property of the material.

A. Germanium Telluride and Tin Telluride

Both GeTe and SnTe have similar normal-state properties and superconducting properties. In both cases it is the *p*-type semiconductor which is superconducting and samples can be obtained with high carrier concentration by doping with excess tellurium. The fact that the holes carry the supercurrent make these materials seem more exotic than *n*-type semiconductors, but from the theoretical point of view both cases are essentially the same.

Because of the large number of carriers the fundamental gap of GeTe was difficult to measure. In fact, it was spectulated that GeTe was a semimetal (overlapping bands) rather than a semiconductor. Recent tunneling experiments have shown (*45*) that GeTe is a semiconductor with a small gap. Above 400°C this material is cubic (fcc) but at lower temperature it is rhombohedral (fcr).

SnTe is also a semiconductor. The crystal structure of SnTe has been assumed to be cubic (fcc) at all temperatures; however, recently it has been proposed (*53*) that SnTe changes from fcc to fcr at low temperatures.

The electronic band structure of SnTe has recently been computed (*46*) using the empirical pseudo-potential method (*47*). In addition to giving bands in agreement with the ultraviolet reflectivity, the electronic structure near the fundamental gap appears to be consistent with experiment. The valence band maximum is at the point L giving a many-valley band structure with four degenerate valleys. This calculation also reveals the presence of another maximum along the Σ-direction slightly lower in energy than the L-maxima. It is interesting to speculate (*36*) that once this crystal is doped, the valleys can be populated and more intervalley processes are possible, enhancing the superconductivity. Preliminary calculations show that GeTe has similar valence band structure, but that in the lead salts these maxima are lower in energy.

The above suggestion should be regarded as speculation, since experiments have not yet revealed these extra maxima, and the band structure calculations are expected to give only approximate results for small energy differences. In addition, the nonexistence of superconductivity in the lead salts in the same temperature range may not be the result of the lowering of the Σ-valleys, since there are other differences between these materials and GeTe and SnTe. The extra maxima should be explored, however, as they allow interesting possibilities.

To show that superconductivity was a bulk property, the superconducting GeTe samples were investigated spectroscopically for superconducting impurities and none were found in any substantial amount. Experiments were done to test the original Ge and Te as well as the GeTe samples down to 0.04°K and a superconducting transition was observed only for the compound. This ruled out the possible effects of any foreign impurity. To rule out the possibility that some other phase (formed in the fabrication of the compound) was responsible for the observed superconducting properties, the samples were powdered and tested magnetically. Resistive measurements would not be conclusive since in

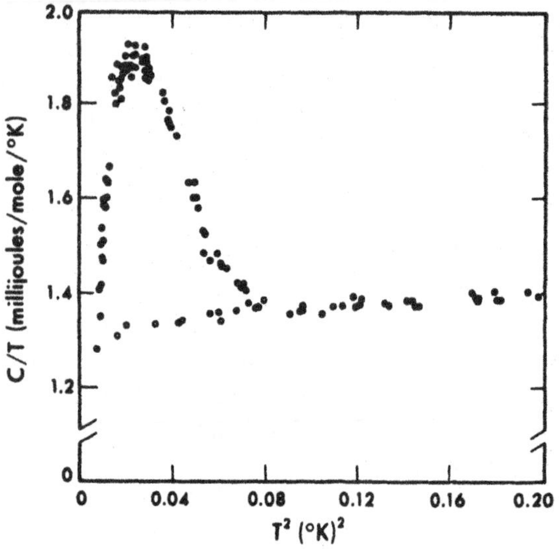

Fig. 13. Heat capacity of germanium telluride at low temperatures, with no field (circles) and with 500 G applied (squares). [Reproduced by permission from Finegold (*48*).]

this case (depending on the current) a transition could be observed when a single filament of superconductor was present across the sample. Magnetic susceptibility measurements of bulk samples could also be called into question, as it was argued that a collection of superconducting filaments shielding the interior could give the appearance of bulk superconductivity. This was the reason for measurements on powdered samples.

Although the positive results for powdered crystals should be conclusive, heat-capacity measurements were done (*48, 49*) (Fig. 13). Such measurements show whether superconductivity is a bulk property of a sample or only a property of a small part of a sample. For the case of GeTe, the measurements showed that the major part of the GeTe sample was superconducting. This experiment firmly established GeTe as a superconductor.

SnTe was subjected to similar investigations. Powdered samples were superconducting and heat-capacity measurements (*50*) (Fig. 14) showed that superconductivity was a bulk effect.

The first superconducting property of these materials which was explored was the dependence of the superconducting transition temperature on carrier density. For GeTe superconductivity was found (*7*) to exist in a range of densities from 8.5×10^{20} cm^{-3} to approximately 15×10^{20} cm^{-3}. The transition temperature increases monatonically with density, but the T_c (n)-curve rises less sharply at high densities. More recent experiments (*51*) have extended this range. These experiments also showed that not only samples with excess tellurium were

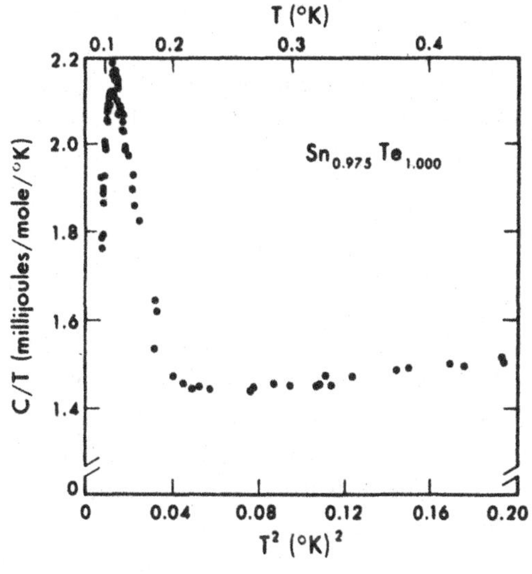

Fig. 14. Heat capacity of tin telluride at low temperatures. [Reproduced with permission from Finegold et al. (*50*).]

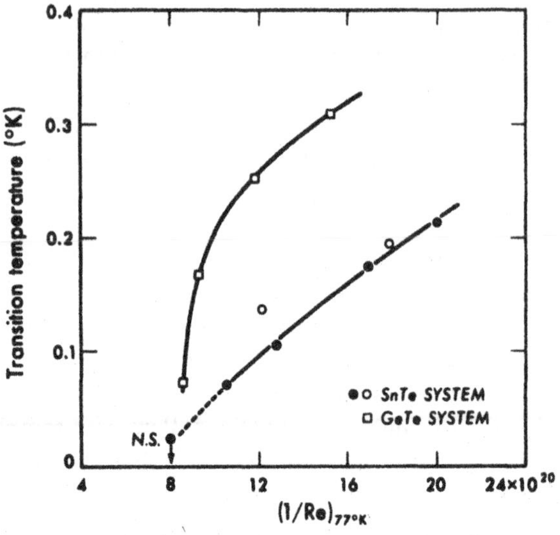

Fig. 15. Transition temperature vs. carrier concentration for germanium telluride and tin telluride. [Reproduced with permission from Hein et al. (*52*).]

superconducting, but silver-doped samples were also superconducting. The highest transition temperature reported for GeTe was 0.420°K. The dependence of the superconducting transition temperature on carrier density was also measured (*9*) for SnTe between 8×10^{20} and 20×10^{20} cm^{-3}. The T_c (n)-function is very similar to the GeTe case but the transition temperature rises less sharply with density (Fig. 15).

Theoretical calculations of the dependence of the superconducting transition temperature on carrier density, T_c (n), for GeTe and SnTe have not as yet been done. The main reason being that the normal-state electronic band structure was not well understood. Recent developments such as normal-state tunneling (*45*), Shubnikov–de Haas measurements (*53*), and band structure calculations (*46*) have given much new information about the normal state, and theoretical calculations of $T_c(n)$ may be possible in the near future.

One expects that the electrons couple to all the phonons and almost all the kernels discussed in the previous sections should be considered. In particular, since these materials are polar, electron–optical phonon couplings should be strong. The experimental $T_c(n)$-curve, however, suggests that the intervalley phonons dominate. The general features of this curve as discussed above is that T_c increases with n over the entire range measured, and the increase is less sharp at higher densities. This is the same general form one gets from the screened electron–intervalley phonon interaction. These couplings are weakened by screening only at very high densities. The screening parameters, the shape

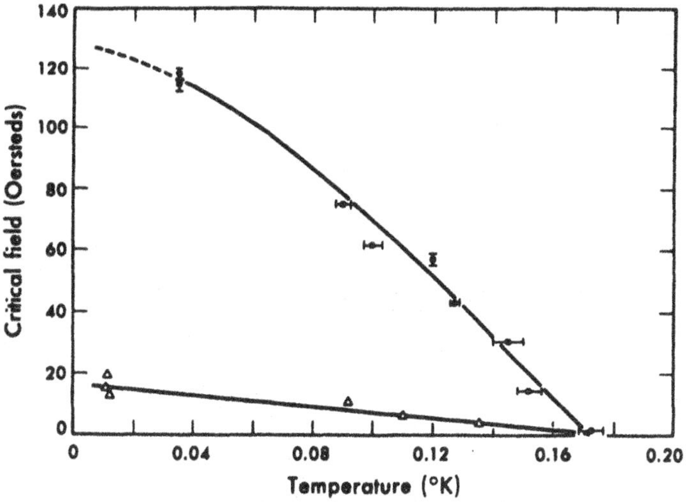

Fig. 16. Critical fields $Hc_1(T)$ and $Hc_2(T)$ for germanium telluride. [Reproduced with permission from Hein et al. (*51*).]

of the $T_c(n)$-curve, and the magnitude of T_c are all consistent with models in which intervalley phonons dominate, but a detailed calculation has not been done. It has been stated (*9*) that the rise in T_c with n in SnTe is also consistent with the theory of Gurevich, Larkin, and Firsov (*4*), but, as was shown in the last section, the GLF kernel gives a *decrease* of T_c with increasing n.

The magnetic properties of GeTe have been investigated by Hein et al. (*51*). These properties indicate that this material is a type II, or hard, superconductor. The samples display nonreversible magnetization curves and "flux jumping" has been observed. The values for the upper and lower critical fields, Hc_1 and Hc_2, have been measured and the temperature dependence of these fields have been obtained (Fig. 16). The Landau–Ginzburg κ was also estimated and it was found to be in the range 5–10, depending on sample composition. Theoretical estimates of κ are also in this range, and in general the magnetic properties which have been measured are consistent with the Landau–Ginzburg theory. It is expected that the κ's for SnTe (and other type II properties) will differ from GeTe because of mean-free-path effects, sample preparation, and the difference in electronic structure.

Recently the superconducting energy gap of GeTe has been observed (*54*) by the tunneling technique. The tunnel junction was composed of Al–A_2O_3–GeTe, and the conductance (dI/dV) was measured as a function of voltage from 2.50 to 0.085°K. The latter temperature is probably the lowest temperature at which a tunneling experiment has been done. The ratio of the superconducting energy gap to kT_c was found to be approximately 4.3.

B. Strontium Titanate

$SrTiO_3$ is a complex material. The lattice structure is perovskite, but it undergoes some transitions which destroy its cubic symmetry. As far as anyone can tell at this time, $SrTiO_3$ does not undergo a ferroelectric transition (i.e., $\omega_{TO} < 0$ for zero field). The band structure has been calculated (*55*) and these calculations conclude that this material has a many-valley conduction band with valleys in the 100-direction (similar to Si) and the maximum in the valence band is at $k = 0$. The band structure calculation is based on the tight binding method. The presence of the *d*-bands of Ti make other methods (like pseudo-potential or OPW calculations) difficult. Most of the normal-state experiments on $SrTiO_3$ are consistent with this picture (*56*), although the question of the existence of a Γ-valley has been raised (*57*). This crystal can be doped *n*-type either by partial reduction at high temperatures after growth or by adding impurities (i.e., Nb), during growth, but, as mentioned before, the actual mechanism by which the electron gas becomes degenerate is not yet fully understood. The complexities of the crystal structure, lattice waves, band structure, transport properties, impurity electrons, and superconductivity have inspired the comment (*58*): "If $SrTiO_3$ had magnetic properties, a complete study of this material would require a thorough knowledge of all of solid state physics." With this comment serving as a caveat, we will discuss some of the superconducting properties of this interesting material.

As in the cases of GeTe and SnTe, much effort was expended to show that superconductivity in $SrTiO_3$ was a bulk effect. Magnetic measurements were performed on powdered samples with positive results. Heat-capacity measurements (*59*) were also made to 0.3°K to show that the observed superconductivity was a bulk effect; the transition temperature for this sample was 0.38°K and a peak in the heat capacity was found at this temperature. This peak was destroyed by a magnetic field equal to the critical field.

Superconductivity is observed in $SrTiO_3$ for a wide range of carrier density. Transitions have been seen by Schooley et al. (*36,60*) using magnetic techniques from 10^{18} to almost 10^{21} cm^{-3}. Resistive transitions have been observed for lower densities $\sim 7 \times 10^{17}$ cm^{-3} and higher densities $\sim 10^{21}$ cm^{-3}. One expects the transition temperature measured resistively to be higher, because in this type of measurement a transition is observed as soon as a superconducting path is available in the sample. Since the samples tested can be inhomogeneous, a part of the sample could be superconducting at a temperature higher than the bulk. In the case of low-density samples, the crystals are still transparent.‡ A plot of

‡ It may be possible to use Raman scattering to explore the superconducting quasi-particle spectrum.

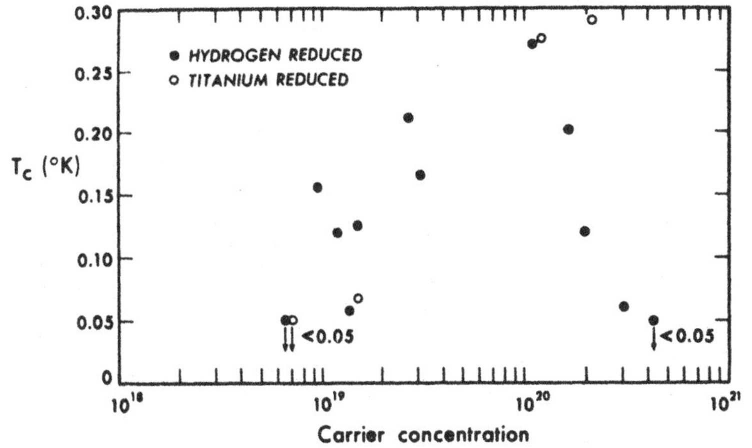

Fig. 17. Transition temperature vs. carrier concentration for $SrTiO_3$.

$T(n)$ appears in Fig. 17. The general feature of this curve is that it rises, reaches a maximum near $n = 10^{20}\ cm^{-3}$, and then falls sharply. The dependence of $T_c(n)$ is then different from GeTe and SnTe, since these materials do not exhibit a maximum. The highest transition temperature reported for $SrTiO_3$ (doped with Nb) is 0.43°K.

In a preliminary calculation the $T_c(n)$-curve for $SrTiO_3$ was fit *(60)* over the entire range assuming that intervalley phonons dominate. This fit requires only one adjustable parameter, the intervalley deformation potential. At low carrier density the rise in T_c with n is characteristic of this type of interaction, as discussed previously. At higher densities the screening becomes more critical, and because the attractive phonon contribution is screened by the square of the dielectric function, the transition temperature decreases very rapidly in this range.

The above preliminary calculation shows that the magnitude of T_c can be accounted for on the basis of an attractive interaction arising from intervalley phonons alone; however, it is expected that the optical phonons should contribute. Calculations *(61)* are now in progress to include all the phonons and adjust the intervalley deformation potential to give the observed $T_c(n)$-curve.

The magnetization and critical fields of $SrTiO_3$ have been studied by Ambler et al. *(59)*. The specimens which were investigated were those having densities yielding the highest transition temperature, i.e., $n = 1$ to $2 \times 10^{20}\ cm^{-3}$. The magnetization was measured as a function of field and of temperature. The material was shown to be a type II, or hard, superconductor. The Landau-Ginzburg κ was determined for the samples tested and found to be of the order of $\kappa \sim 10$. In particular, for the case of sample HR 24 for which the magnetization curves are given in Figs. 18 and 19, $\kappa = 8.4$. Other samples with higher mobilities have smaller kappas. The dependence of Hc_1 and Hc_2 on temperature

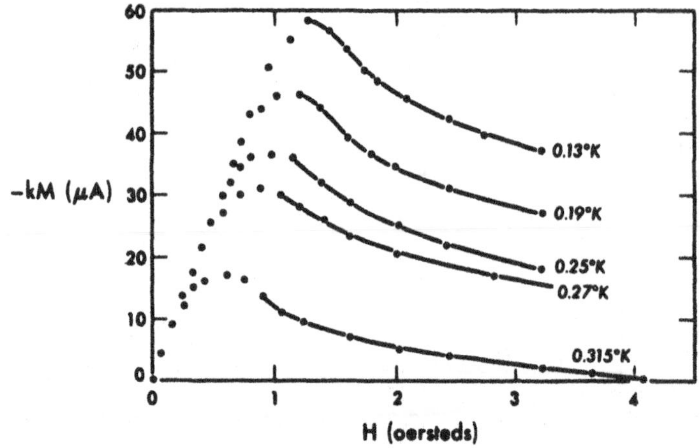

Fig. 18. Low-field magnetization of specimen HR24 of strontium titanate. [Reproduced with permission from Ambler et al. (*59*).]

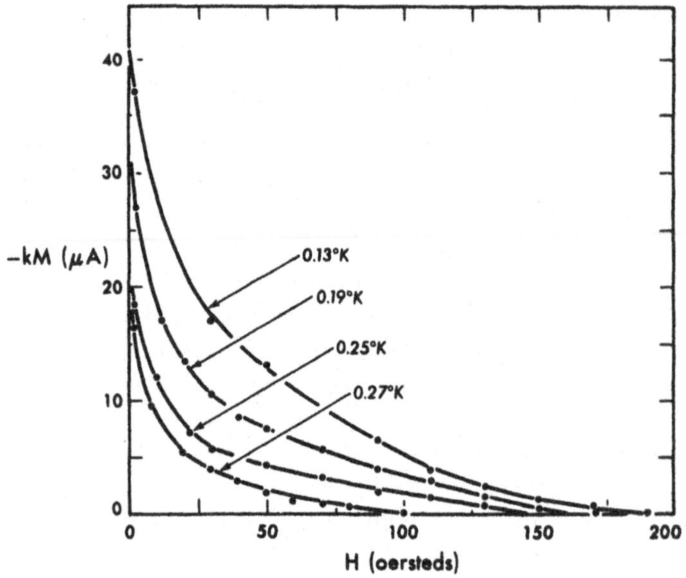

Fig. 19. Complete magnetization of specimen HR24 of strontium titanate. [Reproduced with permission from Ambler et al. (*59*).]

is given in Fig. 20, and a comparison is made with Maki's theory. The penetration depth of $SrTiO_3$ has also been measured (*62*) and shown to be anomalously large, i.e., $\lambda > 10^{-3}$ cm. One expects large values for λ for a degenerate semiconductor because n is small, but the observed value is even larger than one would estimate for these systems. One possible explanation is that samples

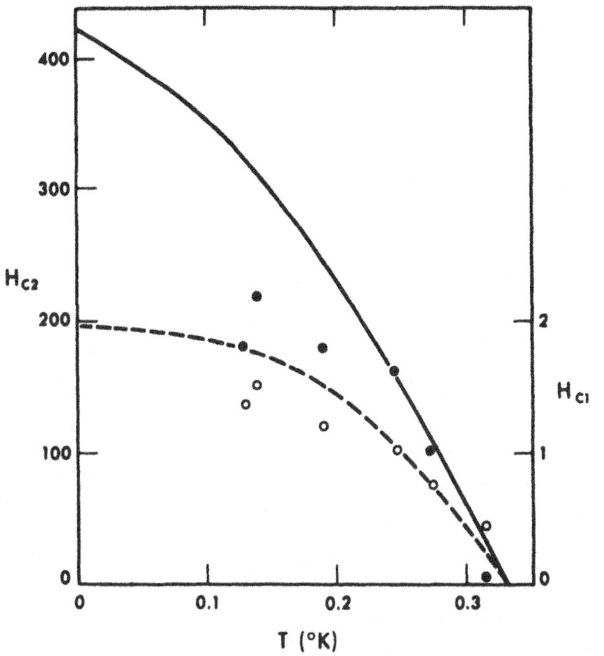

Fig. 20. $Hc_1(T)$ (crosses), $Hc_2(T)$ (solid circles), Hc_1(calcd.) (dashed curve), and Hc_2(calcd.) (solid curve) for specimen HR24 of strontium titanate. [Reproduced with permission from Ambler et al. (*59*).]

of $SrTiO_3$ tend to reoxidize at the surface, giving a lower surface carrier concentration than in the bulk and hence longer penetration depths. This is only one possibility and the anomalously large penetration depth may be a more fundamental question. Outside of this question about the penetration depth, $SrTiO_3$ appears to have much the same characteristics as other type II, or hard, superconductors. In general, agreement between experiment and theory for this type of superconductor is as good as it is for metal superconductors.

Some interesting experiments have been done (*63*) on the superconducting properties of ceramic, mixed titanates. Mixed compounds of $(Ba_xSr_{1-x})\,TiO_3$ and $(Ca_ySr_{1-y})\,TiO_3$, where $x \leqslant 0.1$ and $y \leqslant 0.3$, have been found superconducting in approximately the same transition temperature range as $SrTiO_3$. It is believed that the addition of Ba or Ca changes the normal-state properties mostly through the crystal structure and the static dielectric constant. The superconducting properties (i.e., the transition temperature) are also changed, and the effect of adding Ba or Ca depends on carrier density. The transition temperatures for the mixed crystals can be somewhat higher than for $SrTiO_3$. The highest reported is 0.55°K.

It is not clear at present how the addition of Ba and Ca affects the superconducting properties of $SrTiO_3$. The dielectric constant for the mixed crystal is expected to change in some cases and this could affect the electron–optical phonon coupling. It will be interesting to see how the $T_c(n)$-curves look for these crystals as there should be deviations especially at lower densities from the $SrTiO_3$ curve. Experiments of this type might help determine the role of the various phonon processes in the superconductivity.

IV. SPECULATIONS, SUMMARY, AND DISCUSSION

A. Speculations on Increasing the Superconducting Transition Temperature of Low-Carrier-Density Superconductors

This section contains speculation. In it we will discuss the possibility of enhancing the superconducting transition temperature of a low-carrier-density superconductor. The main point which will be exploited will be the fact that the superconducting transition temperature depends not only on the size of the attractive and repulsive kernels but also on their shape. The machinery for dealing with the question of the influence of the shape of ther kernels on the superconducting transition temperature for a two-square-well model has been developed in Section II.B. The two-square-well model suffices to illustrate the main points.

The gap equation in general relates the gap function at a given energy, say E_F, to the same function at another energy through a kernel which is a function of both energies, i.e., Eq. (12) or (16). The form of the equation is such that an integral or a sum is taken of the *product* of the kernel times the gap function to evaluate the gap at the Fermi energy. In addition, the gap equation has the property that when the kernel changes sign the gap function changes sign at approximately the same energy. This means that the product of the kernel and the gap function has the same sign over the entire energy range, even though the kernel may be changing its sign. When the product of the kernel and gap function is summed or integrated over the energy range of interest, all regions contribute to the gap at E_F even in the energy range in which the kernel is repulsive. This is not really a case of a repulsive potential causing superconductivity. The real space potential is attractive; it is the Fourier transform (of time) which we are examining here.

For example, consider the case of the sample kernel sketched in Fig. 21. This kernel is attractive below the first cutoff frequency, ϵ_1, and repulsive above up to ϵ_2. Our aim in producing a high transition temperature is to make the kernel as attractive as possible for $0 \leqslant \epsilon \leqslant \epsilon_1$ and as repulsive as possible for $\epsilon_1 < \epsilon \leqslant \epsilon_2$. The shape of the kernel is therefore very important. A repulsive peak in the

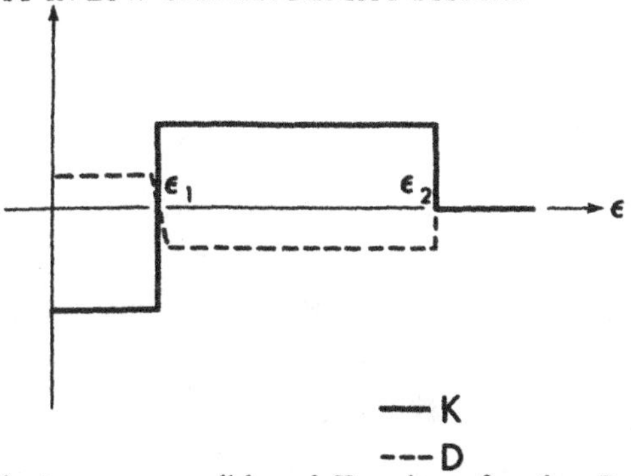

Fig. 21. Sketch of the two-square-well kernel K, and gap function D, to illustrate the role of the repulsive part of the kernel.

correct energy range enhances superconductivity and in the wrong energy range destroys it.

The above features are found in physical kernels for basically two reasons. One is the energy dependence of the interactions involved. For example, the Coulomb, phonon, or any other interactions contain resonances at characteristic frequencies. These resonances show up as attractive or repulsive peaks in the kernels for these interactions. If two of these kernels are made to "line-up" so that the resonances occur at the same frequency, the transition temperature is greatly enhanced over the case where they do not line up. Another origin for the structure desired in the kernel can come from the structure in the density of states. For example, peaks can arise from bands with large effective masses or localized states within the region of interest (i.e., *f*-bands or *d*-bands). We note that it is not a necessary condition that the peaks in the density of states occur at E_F or at energies within a Debye frequency of E_F. The presence of peaks further away in energy can also affect the superconductivity. It should be mentioned that this structure also changes the interaction. In fact, attractive interactions are possible because of antiscreening in the Coulomb interaction. This may contribute to the superconductivity in some materials (i.e., La, U) in addition to density of states considerations and to the isotope effect.

We will discuss here a few ways of getting a favorable kernel using the interaction resonances instead of density of states peaks. The methods chosen are not necessarily the best methods nor do they represent an exhaustive search, but they are illustrative of the possibilities.

1. Plasma Peak

Although the plasma resonance is damped ($\epsilon_2 \neq 0$) in the region of integration for computing the Coulomb kernel, the $\epsilon_1 = 0$ curve does affect the kernel in the frequency range $\Delta\omega$, where this curve intersects the continuum (see Fig. 22). In this region the kernel has a dip followed by a peak (see Fig. 6). Mathemati-

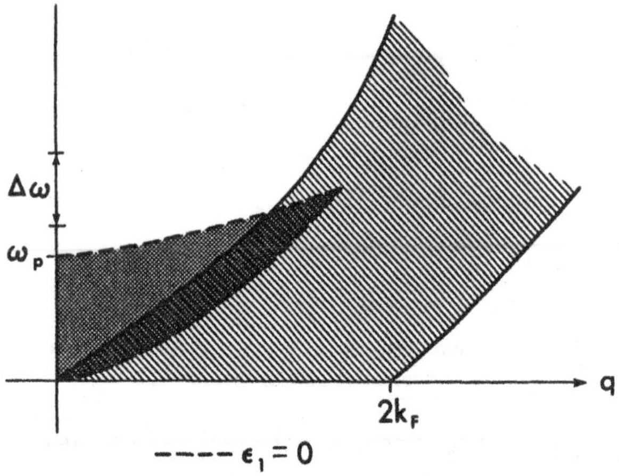

Fig. 22. *ωq*-plane for the dielectric function. In the shaded area $\epsilon_2 \neq 0$, and this is the region of q integration for a given ω in computing the kernel. The cross-hatched region gives the area over which $\epsilon_1 < 0$. The frequency range $\Delta\omega$ is the range over which the "plasma" structure appears in the kernel.

cally this behavior arises from the integration over the plasma pole at finite q. Physically this corresponds to the resonance in the electron–electron interaction arising from the exchange of plasmons.

In a degenerate semiconductor it is possible by changing the carrier density to cause this structure in the Coulomb kernel to coincide with the phonon resonance. When this is done, the transition temperature is raised even though the coupling to plasmons is not unusually strong, since the kernel becomes more attractive below the resonant frequency and more repulsive above it. The energy widths of the phonon and plasmon peaks are different and it is not possible to make the structure exactly coincide. This means that the two-square-well model only gives a rough estimate of the enhancement, but the effect should be observable.

The influence of this peak depends on r_s, the electron gas parameter, since $r_s \sim (\hbar\omega_p/E_F)^2$ and we want to move the plasma resonance to match the phonon resonance. Another dependence on r_s occurs because the Coulomb kernel is proportional to r_s. This means that a reduction in r_s by reducing the carrier density to make $\hbar\omega_{\text{plasmon}} \sim \hbar\omega_{\text{phonon}}$ also reduces the magnitude of the effect. It is better therefore to choose materials with large effective masses. In this way $\hbar\omega_p$ can be reduced while r_s is increased.

In polar materials it is possible to think of this plasma-type mechanism while including the optical phonon–plasmon coupling. As we demonstrated in Section II.D, the kernel has a repulsive resonance (particularly for high n) which arises from the coupled plasmon–phonon mode. As was discussed in

Section II.D this repulsive resonance is actually the origin of superconductivity in the Gurevich–Larkin–Firsov model.

2. Coulomb Resonances in Semimetals

One possibility which could enhance the superconducting transition temperature considerably is the Coulomb resonance in a semimetal. The basic idea would be to take advantage of the strong screening at low frequencies arising from the small band gap which occurs in materials such as Bi, Sb, As, and their alloys. At low frequencies the dielectric constant is very large (~ 100) and the Coulomb interaction is small. At frequencies above the characteristic resonance frequency the dielectric function becomes smaller and this causes the Coulomb interaction to become very large. This is exactly the behavior desired. In particular, if the resonance were made to coincide with the phonon resonance of a superconducting semimetal the enhancement would be very large. There are at present no semimetals which exhibit these properties, or superconductivity, but intervalley scattering is present in all semimetals and superconductivity is possible on the basis of these processes alone. The semimetals Bi, Sb, As, and their alloys should be investigated; especially doped crystals, p-type and n-type. These are specifically mentioned because they are the best known semimetals; there may be others which are more promising.

The case of semimetals provides a good argument for ultra-low-temperature research. In the cases discussed the lining up of resonances are critical in enhancing the superconducting transition temperature. If superconductivity were found in a doped semimetal, even if it existed at temperatures below a hundredth of a degree, the prescription for raising the transition temperature is available, and the effort in going to low temperatures to find such a superconductor would be worth expending.

B. Summary and Discussion

The usual objection against the possibility of superconductivity in a degenerate semiconductor is that the carrier concentration is too low. A low carrier concentration implies a small density of states at E_F since $N(0) \sim m^* n^{1/3}$, and the BCS model gives a very low transition temperature unless an extremely strong interaction is assumed. These points have been discussed in the text and the above objections circumvented for some cases. Some of the arguments are (1) a large density of states at lower carrier densities can occur for systems with large masses or many valleys, (2) the BCS model is not really applicable for kernels which arise in the case of a degenerate semiconductor, and (3) the interactions in a semiconductor can be strong both because of strong-coupling and screening arguments. It is therefore possible for superconductivity to exist in low-density systems, i.e., semiconductors and semimetals. (Although semiconductors were considered, the theory discussed applies equally well to semimetals.)

Semiconductors and semimetals can serve as host materials for a low density electron or hole gas and calculations can be done on such a system with less difficulty than in the case of metals because the electrons or holes usually occupy only small pockets in k-space. The interactions in these pockets and between these pockets can be determined because unlike metals the lattice waves can be treated with simple models and the electron states, electron–electron, and electron–phonon couplings are better understood in semiconductors. Such a system (host material with an electron gas) can provide a small laboratory for testing the dependence of superconductivity on the normal-state properties since these properties are variable. For example, the band structure and carrier density can be varied independently of one another to determine the role of each separately in the superconductivity.

One point which should be stressed is that in computing the properties of a superconductor from the normal-state properties one is limited by the fact that the normal-state properties must be known very well. All the physics goes into the kernels once a superconductivity theory is available. The kernels are transformed in solving the gap equation to quantities which appear in an exponential, e.g., the $N(0)\,V$ parameter of BCS. This means that the kernels must be known very accurately if one is to trust estimates of the critical temperature and other superconducting properties. This is the severest limitation on application of the theory to specific materials. In fact, the strong dependence of the superconducting properties on the normal state implies that a law of corresponding states should not be well satisfied for these materials. For example, it is expected that GeTe and SnTe will be different from $SrTiO_3$ and similar to each other. Different samples of the same material or similar materials (i.e., with different carrier densities) should vary in their superconducting properties, especially type II properties. Since the superconducting properties depend so critically on the normal state, it is hoped that in the future this problem can be turned around and the superconducting properties will help to determine normal-state properties like electronic structure, electron–electron couplings, and electron–phonon couplings.

In the previous section the possibility of increasing the transition temperature was discussed. Several speculations were presented (clearly labeled speculation, since this paper is primarily a review). It is very difficult for the theorist to pick the exact material to be tested. It is even difficult to choose the most promising "peaks" to exploit. The purpose of the brief discussion given was to show that such effects can exist, and to point out some possible directions for future research.

ACKNOWLEDGMENTS

The author would like to thank C. S. Koonce for many discussions and for his collaboration on Section II.D. He would also like to express his gratitude to T. K. Bergstresser for helpful comments on the presentation of this material and to Dr. V. Heine and the theoretical group at the Cavendish Laboratory, University of Cambridge, for the hospitality extended to him while the last part of this manuscript was being prepared. The support of the National Science Foundation and the Alfred P. Sloan Foundation is also gratefully acknowledged.

REFERENCES

1. N. Kurti and F. Simon, *Proc. Roy. Soc. (London)* **A151**, 610 (1935).

2. J. Bardeen, L. N. Cooper, and J. R. Schrieffer, *Phys. Rev.* **108**, 1175 (1957).

3. D. Pines, *Phys. Rev.* **109**, 280 (1958).

4. V. L. Gurevich, A. I. Larkin, and Y. A. Firsov, *Fiz. Tverd. Tela* **4**, 185 (1962); *Soviet Phys.-Solid State* **4**, 131 (1962).

5. M. L. Cohen, *Rev. Mod. Phys.* **36**, 240 (1964).

6. M. L. Cohen, *Phys. Rev.* **134**, A511 (1964).

7. R. A. Hein, J. W. Gibson, R. L. Mazelsky, R. C. Miller, and J. K. Hulm, *Phys. Rev. Letters* **12**, 320 (1964).

8. J. F. Schooley, W. R. Hosler, and M. L. Cohen, *Phys. Rev. Letters* **12**, 474 (1964).

9. R. A. Hein, J. W. Gibson, R. S. Allgaier, B. B. Houston, Jr., R. L. Mazelsky, and R. C. Miller, *Low Temperature Physica, LT9* (J. G. Daunt et al., eds.), Plenum Press, New York, 1965, p. 604.

10. D. H. Douglass, Jr., and L. M. Falicov, *Progress in Low Temperature Physics*, Vol. IV (C. J. Gorter, ed.), North-Holland, Amsterdam, 1964, p. 97.

11. P. W. Anderson, *Phys. Chem. Solids* **11**, 26 (1959).

12. J. R. Schrieffer, *Theory of Superconductivity*, Benjamin, New York, 1964.

13. N. N. Bogoliubov, V. V. Tolmachev, and D. V. Shirkov, *A New Method in the Theory of Superconductivity*, Academy of Science, Moscow, 1958, Consultants Bureau, New York, 1959.

14. J. C. Swihart, *IBM J. Res. Develop.* **6**, 14 (1962).

15. J. W. Garland, *Phys. Rev. Letters* **11**, 114 (1963).

16. J. Lindhard, *Kgl. Danske Videnskab. Selskab, Mat. Fys. Medd.* **28**, 8 (1954).

17. M. Gell-Mann and K. A. Brueckner, *Phys. Rev.* **106**, 181 (1957).

18. M. L. Cohen and T. K. Bergstresser, *Phys. Rev.* **141**, 789 (1966).

19. J. Bardeen and D. Pines, *Phys. Rev.* **99**, 1140 (1955).

20. W. Shockley and J. Bardeen, *Phys. Rev.* **80**, 72 (1950).

21. C. Herring and E. Vogt, *Phys. Rev.* **101**, 944 (1955).

22. H. Fritzsche, *Phys. Rev.* **115**, 336 (1959).

23. R. W. Keyes, *Solid State Phys.* **11**, 149 (1960).

24. J. M. Ziman, *Electrons and Phonons*, Oxford, New York, 1960, p. 439.

25. F. Seitz, *Phys. Rev.* **73**, 549 (1948).

26. D. R. Penn, *Phys. Rev.* **128**, 2093 (1962).

27. S. M. Ziman, *op. cit.*, p. 443.

28. C. Herring, *Bell. System Tech. J.* **34**, 237 (1955).

29. M. L. Cohen, *Phys. Rev.* **128**, 131 (1962).

30. M. Lax and J. J. Hopfield, *Phys. Rev.* **124**, 115 (1961).

31. B. N. Brockhouse and P. K. Iyengar, *Phys. Rev.* **111**, 747 (1958).

32. F. Herman, *Phys. Rev.* **95**, 847 (1954).

33. M. Glicksman, *Phys. Rev.* **100**, 1146 (1955).

34. R. Braunstein, A. R. Moore, and F. Herman, *Phys. Rev.* **109**, 695 (1958).

35. R. Braunstein, *Phys. Rev.* **130**, 879 (1963).

36. M. L. Cohen and C. S. Koonce, *Proc. Intern. Conf. Semicond. Phys., Kyoto, 1966*, 633.

37. C. S. Koonce and M. L. Cohen, to be published.

38. H. Fröhlich, *Polarons and Excitons* (C. G. Kuper and G. D. Whitfield, eds.), (Oliver & Boyd, London, 1963, p. 1.

39. B. B. Varga, *Phys. Rev.* **137**, A1896 (1965).

40. A. S. Barker, *Proc. Intern. Colloq. on Optical Properties and Electronic Structure of Metals and Alloys, 1966*, 45.

41. K. S. Singwi and M. P. Tosi, *Phys. Rev.* **147**, 658 (1966).

42. A. Mooradian and G. B. Wright, *Phys. Rev. Letters* **16**, 999 (1966).

43. N. F. Mott, private communication.

44. J. W. Garland, to be published.

45. L. Esaki *Proc. Intern. Conf. Semicond. Phys., Kyoto, 1966*, 589.

46. P.J. Lin, W. Saslow, and M.L. Cohen, *Solid State Comm.* **5**, 893 (1967).

47. M. L. Cohen and T. K. Bergstresser, *Phys. Rev.* **141**, 789 (1966).

48. L. Finegold, *Phys. Rev. Letters* **13**, 233 (1964).

49. B. B. Goodman and S. Marcucci, *Proc. of the Low Temperature Calorimetry Conf. Helsinki, 1966*, to be published.

50. L. Finegold, J. K. Hulm, R. L. Mazelsky, N. E. Phillips, and B. B. Triplett, *Proc. of the Low Temperature Calorimetry Conf., Helsinki, 1966*, to be published.

51. R. A. Hein, J. W. Gibson, R. L. Falge, Jr., R. L. Mazelsky, R. C. Miller, and J. K. Hulm, *Proc. Intern. Conf. on the Semiconduct. Physics, Kyoto, 1966*, p. 643.

52. R. A. Hein, J. W. Gibson, R. S. Allgaier, B. B. Houston, Jr., R. L. Mazelsky, and R. C. Miller, *Low Temperature Physics, LT9* (J. Daunt et al., eds.), Plenum Press, New York, 1965, p. 604.

53. J. R. Burke, Jr., B. B. Houston, Jr., H. T. Savage, J. Babiskin, and P. G. Siebenmann, *Proc. Intern. Conf. Semiconduct. Phys., Kyoto, 1966*, 384.

54. P. J. Stiles, L. Esaki, and J. F. Schooley, *Phys. Letters* **23**, 206 (1966).

55. A. H. Kahn and A. J. Leyendecker, *Phys. Rev.* **135**, A1321 (1964).

56. H. P. R. Frederikse, W. R. Hosler, and W. R. Thurber, *Proc. Intern. Conf. Semiconduct. Phys., Kyoto, 1966*, 32.

57. O. N. Tufte and E. L. Stelzer, *Phys. Rev.* **141**, 675 (1966).

58. J. F. Schooley, H. P. R. Frederikse, and M. L. Cohen, unpublished.

59. E. Ambler, J. H. Colwell, W. R. Hosler, and J. F. Schooley, *Phys. Rev.* **148**, 280 (1966).

60. J. F. Schooley, W. R. Hosler, E. Ambler, J. H. Becker, M. L. Cohen, and C. S. Koonce, *Phys. Rev. Letters* **14**, 305 (1965).

61. C. S. Koonce and M. L. Cohen, to be published.

62. J. F. Schooley and W. R. Thurber, *Proc. Intern. Conf. Semiconduct. Phys., Kyoto, 1966*, 639.

63. H. P. R. Frederikse, J. F. Schooley, W. R. Thurber, E. Pfeiffer, and W. R. Hosler, *Phys. Rev. Letters* **16**, 579 (1966).

Printed and bound by CPI Group (UK) Ltd, Croydon, CR0 4YY
23/10/2024
01778245-0004